FOUR**TH** EDITION

Basic Biomechanics of the Musculoskeletal System

Margareta Nordin, PT, Dr Med Sci

Professor (Research)
Departments of Orthopaedic Surgery and
Environmental Medicine
NYU School of Medicine
Director
Occupational and Industrial Orthopaedic Center
NYU Hospital for Joint Diseases
NYU Langone Medical Center
New York University
New York, New York

Victor H. Frankel, MD, PhD, KNO

Professor
Department of Orthopaedic Surgery
NYU School of Medicine
President Emeritus
NYU Hospital for Joint Diseases
NYU Langone Medical Center
New York University
New York, New York

Dawn Leger, PhD, *Developmental Editor*

Wolters Kluwer | Lippincott Williams & Wilkins
Health

Philadelphia · Baltimore · New York · London
Buenos Aires · Hong Kong · Sydney · Tokyo

Senior Publisher: Julie K. Stegman
Senior Acquisitions Editor: Emily Lupash
Senior Managing Editor: Heather A. Rybacki
Marketing Manager: Allison Powell
Manufacturing Coordinator: Margie Orzech-Zeranko
Designer: Holly McLaughlin
Artwork: Kim Batista
Compositor: Aptara, Inc.
Printer: C&C Offset

Printed in China
Not authorised for Sale in United States, Canada, Australia and New Zealand

Library of Congress Cataloging-in-Publication Data

Basic biomechanics of the musculoskeletal system / [edited by] Margareta Nordin, Victor H. Frankel ; Dawn Leger, developmental editor. – 4th ed.
   p. ; cm.
   Includes bibliographical references and index.
   ISBN 978-1-4511-1709-7 (alk. paper)
   I. Nordin, Margareta. II. Frankel, Victor H. (Victor Hirsch), 1925-
   [DNLM: 1. Biomechanics. 2. Musculoskeletal Physiological Phenomena. WE 103]
   612.76–dc23

                                                                            2011040629

DISCLAIMER
Care has been taken to confirm the accuracy of the information presented and to describe generally accepted practices. However, the authors, editors, and publisher are not responsible for errors or omissions or for any consequences from application of the information in this book and make no warranty, expressed or implied, with respect to the currency, completeness, or accuracy of the contents of the publication. Application of this information in a particular situation remains the professional responsibility of the practitioner; the clinical treatments described and recommended may not be considered absolute and universal recommendations.

The publishers have made every effort to trace the copyright holders for borrowed material. If they have inadvertently overlooked any, they will be pleased to make the necessary arrangements at the first opportunity.

To purchase additional copies of this book, call our customer service department at **(800) 638-3030** or fax orders to **(301) 223-2400**. International customers should call **(301) 223-2300**.

Visit Lippincott Williams & Wilkins on the Internet: http://www.LWW.com. Lippincott Williams & Wilkins customer service representatives are available from 8:30 am to 5:00 pm, EST.

2 3 4 5 6 7 8 9 10

FOURTH EDITION

Basic Biomechanics of the Musculoskeletal System

*In Memory of Lars Eric Frankel 1959–2008 and
Anna Ingrid Forssen Nordin 1916–2009*

Contributors

Gunnar B. J. Andersson, MD, PhD
Professor and Chairman Emeritus
Department of Orthopaedic Surgery
Rush-Presbyterian-St. Luke's Medical Center
Rush University Medical Center
Chicago, Illinois

Sherry I. Backus, PT, DPT, MA
Clinical Supervisor and Research Associate
Leon Root MD, Motion Analysis Laboratory
Rehabilitation Department
Hospital for Special Surgery
New York, New York

Ann E. Barr, PT, DPT, PhD
Vice Provost and Executive Dean
College of Health Professions
Pacific University
Hillsboro, Oregon

Jane Bear-Lehman, PhD, OTR, FAOTA
Associate Professor and Department Chair
Department of Occupational Therapy
Steinhardt School of Education, Culture and Human
   Development
New York University
New York, New York

Allison M. Brown, PT, PhD
Adjunct Instructor
Department of Rehabilitation & Movement Sciences
Doctoral Program in Physical Therapy
School of Health Related Professions
University of Medicine & Dentistry of New Jersey
Newark, New Jersey

Florian Brunner, MD, PhD
Consultant
Department of Physical Medicine and Rheumatology
Balgrist University Hospital
Zurich, Switzerland

Marco Campello, PT, PhD
Clinical Associate Professor
Department of Orthopaedic Surgery
NYU School of Medicine
Associate Director
Occupational and Industrial Orthopaedic Center
NYU Hospital for Joint Diseases
NYU Langone Medical Center
New York University
New York, New York

Dennis R. Carter, PhD
Professor
Departments of Mechanical Engineering and
   Bioengineering
Stanford University
Stanford, California

Michael S. Day, MD, MPhil
Department of Orthopaedic Surgery
NYU School of Medicine
NYU Hospital for Joint Diseases
NYU Langone Medical Center
New York University
New York, New York

Carlo D. de Castro, PT, MS, OCS
Senior Physical Therapist/Clinical Specialist
Occupational and Industrial Orthopaedic Center
NYU Hospital for Joint Diseases
NYU Langone Medical Center
New York University
New York, New York

Kharma C. Foucher, MD, PhD
Assistant Professor and Co-Director
Motion Analysis Laboratory
Department of Orthopaedic Surgery
Rush Medical College
Rush University Medical Center
Chicago, Illinois

Victor H. Frankel, MD, PhD, KNO
Professor
Department of Orthopaedic Surgery
NYU School of Medicine
President Emeritus
NYU Hospital for Joint Diseases
NYU Langone Medical Center
New York University
New York, New York

Marshall A. Hagins, PT, PhD, DPT
Professor
Department of Physical Therapy
School of Health Professions
Long Island University
Brooklyn, New York

Clark T. Hung, PhD
Professor
Department of Biomedical Engineering
Fu Foundation School of Engineering and Applied Science
Columbia University
New York, New York

Laith M. Jazrawi, MD
Associate Professor
Department of Orthopaedic Surgery
NYU School of Medicine
NYU Hospital for Joint Diseases
NYU Langone Medical Center
New York University
New York, New York

Charles J. Jordan, MD
Fellow, Orthopaedic Trauma Service
Florida Orthopaedic Institute
Tampa, Florida

Owen Kendall, BA, MFA
Student, Class of 2014
School of Medicine
Boston University
Boston, Massachusetts

Frederick J. Kummer, PhD
Professor
Department of Orthopedic Surgery
NYU School of Medicine
Associate Director
Musculoskeletal Research Center
NYU Hospital for Joint Diseases
NYU Langone Medical Center
New York University
New York, New York

Dawn Leger, PhD
Adjunct Assistant Professor
Department of Orthopaedic Surgery
NYU School of Medicine
New York University
New York, New York

Angela M. Lis, PT, PhD
Associate Clinical Director
Occupational and Industrial Orthopaedic Center
NYU Hospital for Joint Diseases
NYU Langone Medical Center
New York, New York

Tobias Lorenz, MD, MSc
Research Hospital Specialist
Center for Rehabilitation
Klinik Adelheid
Unteraegeri, Switzerland

Goran Lundborg, MD
Professor
Department of Hand Surgery
Skåne University Hospital
Malmö, Sweden

Ronald Moskovich, MD, FRCS
Assistant Professor and Associate Chief of Spinal
  Surgery
Department of Orthopaedic Surgery
NYU School of Medicine
NYU Hospital for Joint Diseases
NYU Langone Medical Center
New York University
New York, New York

Van C. Mow, PhD
Stanley Dicker Professor of Biomedical Engineering, and
  Orthopaedic Bioengineering
Department of Biomedical Engineering
Fu Foundation School of Engineering and
  Applied Science
Columbia University
New York, New York

Robert R. Myers, PhD
Professor
Department of Anesthesiology
Department of Pathology, Division of
  Neuropathology
University of California San Diego
La Jolla, California

Margareta Nordin, PT, Dr Med Sci
Professor (Research)
Departments of Orthopaedic Surgery and
  Environmental Medicine
NYU School of Medicine
Director
Occupational and Industrial Orthopaedic Center
NYU Hospital for Joint Diseases
NYU Langone Medical Center
New York University
New York, New York

Kjell Olmarker, MD, PhD
Professor
Musculoskeletal Research
Department of Medical Chemistry and Cell Biology
Institute of Biomedicine
Sahlgrenska Academy
University of Gothenburg
Gothenburg, Sweden

Nihat Özkaya, PhD (deceased)
Research Associate Professor
Departments of Orthopaedic Surgery and Environmental
    Medicine
NYU School of Medicine
New York University
New York, New York

Evangelos Pappas, PT, PhD, OCS
Associate Professor and Chair
Department of Physical Therapy
School of Health Professions
Long Island University-Brooklyn Campus
Brooklyn, New York

Bjorn Rydevik, MD, PhD
Professor
Department of Orthopaedics
University of Gothenburg
Sahlgrenska University Hospital
Gothenburg, Sweden

Ali Sheikhzadeh, PhD
Research Associate Professor
Departments of Orthopaedic Surgery and Environmental
    Medicine
NYU School of Medicine
Program Director
Program of Ergonomics and Biomechanics
Occupational and Industrial Orthopaedic Center
NYU Hospital for Joint Diseases
NYU Langone Medical Center
New York University
New York, New York

Peter S. Walker, PhD
Professor of Orthopaedic Surgery
NYU Hospital for Joint Diseases
NYU Langone Medical Center
New York University
Professor
Mechanical and Aeronautical Engineering
New York University—Polytechnic
New York, New York

Shira Schecter Weiner, PT, PhD
Assistant Professor (Clinical)
Department of Orthopaedic Surgery
School of Medicine
Coordinator, Master Program
Program of Ergonomics and Biomechanics
Occupational and Industrial Orthopaedic Center
NYU Hospital for Joint Diseases
NYU Langone Medical Center
New York University
New York, New York

Markus A. Wimmer, PhD
Associate Professor and Director
Department of Orthopaedic Surgery
Rush University Medical Center
Chicago, Illinois

Brett H. Young, MD
Attending Surgeon
Department of Orthopaedic Surgery
Cayuga Medical Center
Ithaca, New York

Joseph D. Zuckerman, MD
Professor and Chair
Department of Orthopaedic Surgery
NYU School of Medicine
NYU Hospital for Joint Diseases
NYU Langone Medical Center
New York University
New York, New York

Reviewers

Kevin A. Ball
University of Hartford

Sébastien Boyas
University of Ottawa

Michael Buck
Ithaca College

Christopher Hughes
Slippery Rock University

Wei Liu
Walsh University

Karen Lomond
University of Vermont

Sharon McCleave
Seneca College of Applied Arts and Technology

Patrick S. Pabian
University of Central Florida

Krystyna Gielo-Perczak
Worcester Polytechnic Institute

Daniel Poulsen
Texas Tech University Health Sciences Center

Donald Rodd
University of Evansville

Roberta L. Russell
Eastern Washington University

Jane Worley
Lake Superior College

Jim Youdas
Mayo Clinic College of Medicine

Foreword

Mechanics and biology have always fascinated humankind. The importance of understanding the biomechanics of the musculoskeletal system cannot be underestimated. Much attention has been paid in recent years to genetic, biological, and biomolecular research, but the study of the mechanics of structure and of the whole body system is still of immense importance. Musculoskeletal ailments are among the most prevalent disorders in the world and will continue to grow as the population ages.

This text seeks to integrate biomechanical knowledge into clinical training for patient care. This is not a simple task but by relating the basic concepts of biomechanics to everyday life, rehabilitation, orthopedics, traumatology, and patient care are greatly enhanced. Biomechanics is a multidisciplinary specialty, and so we have made a special effort to invite national and international contributors from many disciplines so that individuals from different fields may feel comfortable reading this book. This book is now translated into Cantonese, Dutch, Japanese, Korean, Portuguese, and Spanish.

Together with an invaluable team, we have produced this fourth edition of *Basic Biomechanics of the Musculoskeletal System*. The new edition is sharpened and improved, thanks to the input from the students, residents, instructors, and faculty in different disciplines that have used the text during the past 20 years. This book is written for students and with a major input from students and will hopefully be used to educate students and residents for many years to come. Although the basic information contained in the book remains largely unchanged, a considerable amount of updated information has been provided throughout. We have also made a special point to document with the key references any significant changes in the field of biomechanics, orthopedics, and rehabilitation.

It has always been our interest to bridge the gap between engineering knowledge and clinical care and practice. This book is written primarily for clinicians such as orthopedists, physiatrists, physical and occupational therapists, physician assistants, clinical ergonomists, chiropractors, athletic trainers, and other health professionals who are acquiring a working knowledge of biomechanical principles for use in the evaluation and treatment of musculoskeletal dysfunction. We only hope that if you find this book interesting, you will seek more in-depth study in the field of biomechanics. We have always said, "Know basic biomechanics principles and you will understand musculoskeletal ailments better."

Victor H. Frankel, MD, PhD, KNO and
Margareta Nordin, PT, Dr Sci

Preface

Biomechanics uses physics and engineering concepts to describe the motion undergone by the various body segments and forces acting on these body parts during normal activities. The interrelationship of force and motion is important and must be understood if rational treatment programs are to be applied to musculoskeletal disorders. Deleterious effects may be produced if the forces acting on the areas with disorders rise to high levels during exercise or other activities of daily living.

The purpose of this text is to acquaint the readers with the force-motion relationship within the musculoskeletal system and the various techniques used to understand these relationships. The fourth edition of *Basic Biomechanics of the Musculoskeletal System* is intended for use as a textbook either in conjunction with an introductory biomechanics course or for independent study. The fourth edition has been updated to reflect changes in knowledge, but it is still a book that is designed for use by students who are interested in and want to learn about biomechanics. It is primarily written for students who do not have an engineering background but who want to understand the most basic concepts in biomechanics and physics and how these apply to the human body.

Input from students has greatly improved this edition. We have used the book for 20 years in the Program of Ergonomics and Biomechanics at New York University, and it is the students and residents who have suggested the changes and who have continuously showed an interest in developing and improving this book. This edition has been further strengthened by the contribution of the students over the past year. We formed focus groups to understand better what the students wanted and applied their suggestions wherever possible. We retained the selected examples to illustrate the concepts needed for basic knowledge of the musculoskeletal

biomechanics; we also have kept the important engineering concepts throughout the volume. The three chapters on applied biomechanics topics have been updated. Patient case studies and calculation boxes have been added to each chapter. Flowcharts appear throughout the book as teaching tools.

The text will serve as a guide to a deeper understanding of musculoskeletal biomechanics gained through further reading and independent research. The information presented should also guide the reader in assessing the literature on biomechanics. We have attempted to provide therapeutic examples but it was not our purpose to cover this area; instead, we have described the underlying basis for rational therapeutic or exercise programs.

An introductory chapter describes the importance of the study of biomechanics, and an appendix on the international system of measurements serves as an introduction to the physical measurements used throughout the book. The reader needs no more than basic knowledge of mathematics to fully comprehend the material in the book, but it is important to review the appendix on the SI System and its application to biomechanics.

The body of the fourth edition is divided into three sections. The first section is the Biomechanics of Tissues and Structures of the Musculoskeletal System and covers the basic biomechanics of bone, ligaments, cartilage, tendons, muscles, and nerves.

The second section covers the Biomechanics of Joints, including every joint system in the human body. We have organized the chapters going from the simplest to the most complex joint in the body. While there are many ways to arrange the chapters, for example, starting from the spine and working down to the ankle, we have found that the best approach for teaching is to begin with the easiest system and progress to the most complicated. In this case, therefore, we begin the section with a chapter about the knee and end with the wrist and hand. Obviously, others may choose to teach this topic in a different order and that certainly is best left to the instructor to decide.

The third section covers some topics in Applied Biomechanics, including chapters on fracture fixation, arthroplasty, and gait. These are basic chapters that serve to introduce topics in applied biomechanics; they are not in-depth explorations of the subject.

Finally, we hope that the revision and expansion of this fourth edition of *Basic Biomechanics of the Musculoskeletal System* will bring about an increased awareness of the importance of biomechanics. It has never been our intention to completely cover the subject, but instead provide a basic introduction to the field that will lead to further study of this important topic.

Margareta Nordin, PT, Dr Sci and
Victor H. Frankel, MD, PhD, KNO

Acknowledgments

This book was made possible through the outstanding contributions of many individuals. The chapter authors' knowledge and understanding of the basic concepts of biomechanics and their wealth of experience have brought both breadth and depth to this work. Over the past 20 years, questions raised by students and residents have made this book a better teaching tool. There are too many names to list here, but we thank every student who asked a question or made a suggestion during the course of his or her studies.

We are honored and grateful for the contributions of everyone who has worked to prepare this new edition. We can honestly say that this edition is written for the student and by students, residents, instructors, and faculty who leave the classroom with the knowledge to enhance our life and existence. We listened carefully and have tried to accommodate all suggestions.

A book of this size with its large number of figures, legends, and references cannot be produced without a strong editorial team. As project editor, Dawn Leger's continuous effort and perseverance and thoughtfulness shine through the entire book. She has contributed not just to the editing but also to logistics, and as a stylist, as an innovator, and a friend.

We remain grateful to the team at Lippincott Williams & Wilkins at Wolters Kluwer Health, especially for a development grant provided by the company to finance this effort.

Our colleagues at the Occupational and Industrial Orthopaedic Center (OIOC) and the Department of Orthopaedics of the NYU Hospital for Joint Diseases Orthopaedic Institute, New York University Langone Medical Center functioned as critical reviewers and contributors to the chapters. Special thanks to the staff at OIOC who have been managing the center while we were absorbed with the book.

The fourth edition of *Basic Biomechanics of the Musculoskeletal System* was supported throughout its production by the Research and Development Foundation of the NYU Hospital for Joint Diseases Orthopaedic Institute and the hospital administration, to whom we forward our sincere gratitude.

To all who helped, we say again, thank you and TACK SA MYCKET.

Margareta Nordin, PT, Dr Sci and
Victor H. Frankel, MD, PhD, KNO

Contents

PART 1

Biomechanics of Tissues and Structures of the Musculoskeletal System

PART 2

Biomechanics of Joints

PART 3

Applied Biomechanics

FOURTH EDITION

Basic Biomechanics of the Musculoskeletal System

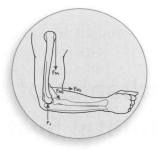

CHAPTER **1**

Introduction to Biomechanics: Basic Terminology and Concepts

Nihat Özkaya and Dawn Leger

Introduction

Biomechanics is considered a branch of bioengineering and biomedical engineering. Bioengineering is an interdisciplinary field in which the principles and methods from engineering, basic sciences, and technology are applied to design, test, and manufacture equipment for use in medicine and to understand, define, and solve problems in physiology and biology. Bioengineering is one of several specialty areas that come under the general field of biomedical engineering.

Biomechanics concerns the applications of classical mechanics to the analysis of biologic and physiologic systems. Different aspects of biomechanics utilize different parts of applied mechanics. For example, the principles of statics have been applied to analyze the magnitude and nature of forces involved in various joints and muscles of the musculoskeletal system. The principles of dynamics have been utilized for motion description, gait analysis, and segmental motion analysis and have many applications in sports mechanics. The mechanics of solids provides the necessary tools for developing the field constitutive equations for biologic systems that are used to evaluate their functional behavior under different load conditions. The principles of fluid mechanics have been used to investigate blood flow in the circulatory system, air flow in the lungs, and joint lubrication.

Research in biomechanics is aimed at improving our knowledge of a very complex structure—the human body. Research activities in biomechanics can be divided into three areas: experimental studies, model analyses, and applied research. Experimental studies in biomechanics are done to determine the mechanical properties of biologic materials, including the bone, cartilage, muscle, tendon, ligament, skin, and blood as a whole or as parts constituting them. Theoretical studies involving mathematical model analyses have also been an important component of research in biomechanics. In general, a model that is based on experimental findings can be used to predict the effect of environmental and operational factors without resorting to laboratory experiments.

Applied research in biomechanics is the application of scientific knowledge to benefit human beings. We know that musculoskeletal injury and illness is one of the primary occupational hazards in industrialized countries. By learning how the musculoskeletal system adjusts to common work conditions and by developing guidelines to ensure that manual work conforms more closely to the physical limitations of the human body and to natural body movements, these injuries may be combatted.

Basic Concepts

Biomechanics of the musculoskeletal system requires a good understanding of basic mechanics. The basic terminology and concepts from mechanics and physics are utilized to describe internal forces of the human body. The objective of studying these forces is to understand the loading condition of soft tissues and their mechanical responses. The purpose of this section is to review the basic concepts of applied mechanics that are used in biomechanics literature and throughout this book.

SCALARS, VECTORS, AND TENSORS

Most of the concepts in mechanics are either scalar or vector. A scalar quantity has a magnitude only. Concepts such as mass, energy, power, mechanical work, and temperature are scalar quantities. For example, it is sufficient to say that an object has 80 kilograms (kg) of mass. A vector quantity, conversely, has both a magnitude and a direction associated with it. Force, moment, velocity, and acceleration are examples of vector quantities. To describe a force fully, one must state how much force is applied and in which direction it is applied. The magnitude of a vector is also a scalar quantity. The magnitude of any quantity (scalar or vector) is always a positive number corresponding to the numeric measure of that quantity.

Graphically, a vector is represented by an arrow. The orientation of the arrow indicates the line of action, and the arrowhead denotes the direction and sense of the vector. If more than one vector must be shown in a single drawing, the length of each arrow must be proportional to the magnitude of the vector it represents. Both scalars and vectors are special forms of a more general category of all quantities in mechanics called tensors. Scalars are also known as "zero-order tensors," whereas vectors are "first-order tensors." Concepts such as stress and strain, conversely, are "second-order tensors."

FORCE VECTOR

Force can be defined as mechanical disturbance or load. When an object is pushed or pulled, a force is applied on it. A force is also applied when a ball is thrown or kicked. Forces acting on an object may deform the object, change its state of motion, or both. Forces may be classified in various ways according to their effects on the objects to which they are applied or according to their orientation as compared with one another. For example, a force may be internal or external, normal (perpendicular) or tangential; tensile, compressive, or shear;

gravitational (weight); or frictional. Any two or more forces acting on a single body may be coplanar (acting on a two-dimensional plane surface); collinear (having a common line of action); concurrent (having lines of action intersecting at a single point); or parallel. Note that weight is a special form of force. The weight of an object on Earth is the gravitational force exerted by Earth on the mass of that object. The magnitude of the weight of an object on Earth is equal to the mass of the object times the magnitude of the gravitational acceleration, which is approximately 9.8 meters per second squared (m/s^2). For example, a 10-kg object weighs approximately 98 newtons (N) on Earth. The direction of weight is always vertically downward.

TORQUE AND MOMENT VECTORS

The effect of a force on the object it is applied on depends on how the force is applied and how the object is supported. For example, when pulled, an open door will swing about the edge along which it is hinged to the wall. What causes the door to swing is the torque generated by the applied force about an axis that passes through the hinges of the door. If one stands on the free end of a diving board, the board will bend. What bends the board is the moment of the body weight about the fixed end of the board. In general, torque is associated with the rotational and twisting action of applied forces, whereas

moment is related to the bending action. However, the mathematical definition of moment and torque is the same.

Torque and moment are vector quantities. The magnitude of the torque or moment of a force about a point is equal to the magnitude of the force times the length of the shortest distance between the point and the line of action of the force, which is known as the lever or moment arm. Consider a person on an exercise apparatus who is holding a handle that is attached to a cable (Fig. 1-1). The cable is wrapped around a pulley and attached to a weight pan. The weight in the weight pan stretches the cable such that the magnitude F of the tensile force in the cable is equal to the weight of the weight pan. This force is transmitted to the person's hand through the handle. At this instant, if the cable attached to the handle makes an angle θ with the horizontal, then the force $\underline{F}$ exerted by the cable on the person's hand also makes an angle θ with the horizontal. Let O be a point on the axis of rotation of the elbow joint. To determine the magnitude of the moment due to force $\underline{F}$ about O, extend the line of action of force $\underline{F}$ and drop a line from O that cuts the line of action of $\underline{F}$ at right angles. If the point of intersection of the two lines is Q, then the distance d between O and Q is the lever arm, and the magnitude of the moment $\underline{M}$ of force $\underline{F}$ about the elbow joint is $M = dF$. The direction of the moment vector is perpendicular to the plane defined by the line of action

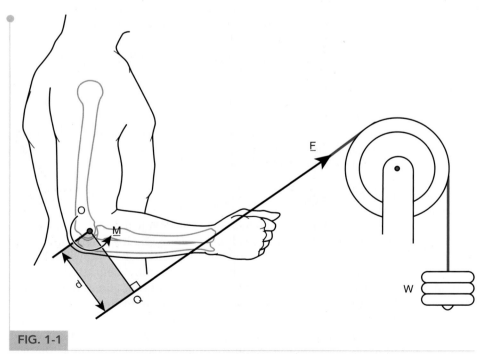

FIG. 1-1

Definition of torque. *Adapted from Özkaya, N. (1998). Biomechanics. In W.N. Rom, Environmental and Occupational Medicine (3rd ed.). New York: Lippincott-Raven, 1437–1454.*

of $\underline{F}$ and line OQ, or for this two-dimensional case, it is counterclockwise.

NEWTON'S LAWS

Relatively few basic laws govern the relationship between applied forces and corresponding motions. Among these, the laws of mechanics introduced by Sir Isaac Newton (1642–1727) are the most important. Newton's first law states that an object at rest will remain at rest or an object in motion will move in a straight line with constant velocity if the net force acting on the object is zero. Newton's second law states that an object with a nonzero net force acting on it will accelerate in the direction of the net force and that the magnitude of the acceleration will be proportional to the magnitude of the net force. Newton's second law can be formulated as $\underline{F} = m\underline{a}$. Here, $\underline{F}$ is the applied force, m is the mass of the object, and $\underline{a}$ is the linear (translational) acceleration of the object on which the force is applied. If more than one force is acting on the object, then $\underline{F}$ represents the net or the resultant force (the vector sum of all forces). Another way of stating Newton's second law of motion is $\underline{M} = I\underline{\alpha}$, where $\underline{M}$ is the net or resultant moment of all forces acting on the object, I is the mass moment of inertia of the object, and $\underline{\alpha}$ is the angular (rotational) acceleration of the object. The mass m and mass moment of inertia I in these equations of motion are measures of resistance to changes in motion.

The larger the inertia of an object, the more difficult it is to set in motion or to stop if it is already in motion.

Newton's third law states that to every action there is a reaction and that the forces of action and reaction between interacting objects are equal in magnitude, opposite in direction, and have the same line of action. This law has important applications in constructing free-body diagrams.

FREE-BODY DIAGRAMS

Free-body diagrams are constructed to help identify the forces and moments acting on individual parts of a system and to ensure the correct use of the equations of mechanics to analyze the system. For this purpose, the parts constituting a system are isolated from their surroundings and the effects of surroundings are replaced by proper forces and moments.

The human musculoskeletal system consists of many parts that are connected to one another through a complex tendon, ligament, muscle, and joint structure. In some analyses, the objective may be to investigate the forces involved at and around various joints of the human body for different postural and load conditions. Such analyses can be carried out by separating the body into two parts at the joint of interest and drawing the free-body diagram of one of the parts. For example, consider the arm illustrated in Figure 1-2.

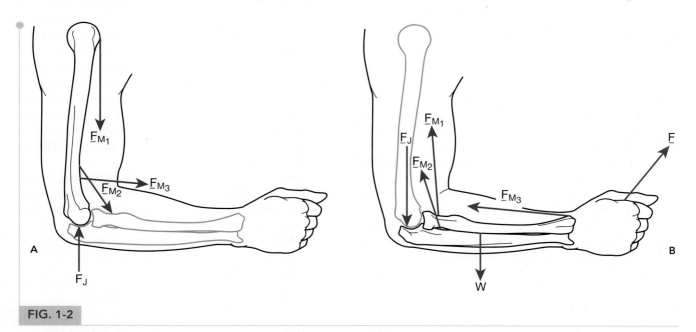

FIG. 1-2

A. Forces involved at and around the elbow joint. B. The free-body diagram of the lower arm.
Adapted from Özkaya, N. (1998). Biomechanics. In W.N. Rom, Environmental and Occupational Medicine (3rd ed.). New York: Lippincott-Raven, 1437–1454.

Assume that the forces involved at the elbow joint are to be analyzed. As illustrated in Figure 1-2, the entire body is separated into two at the elbow joint and the free-body diagram of the forearm is drawn (Fig. 1-2B). Here,

$\underline{F}$ is the force applied to the hand by the handle of the cable attached to the weight in the weight pan,

$\underline{W}$ is the total weight of the lower arm acting at the center of gravity of the lower arm,

$\underline{F}_{M_1}$ is the force exerted by the biceps on the radius,

$\underline{F}_{M_3}$ is the force exerted by the brachioradialis muscles on the radius,

$\underline{F}_{M_2}$ is the force exerted by the brachialis muscles on the ulna, and

$\underline{F}_J$ is the resultant reaction force at the humeroulnar and humeroradial joints of the elbow. Note that the muscle and joint reaction forces represent the mechanical effects of the upper arm on the lower arm. Also note that as illustrated in Figure 1-2A (which is not a complete free-body diagram), equal magnitude but opposite muscle and joint reaction forces act on the upper arm as well.

CONDITIONS FOR EQUILIBRIUM

Statics is an area within applied mechanics that is concerned with the analysis of forces on rigid bodies in equilibrium. A rigid body is one that is assumed to undergo no deformations. In reality, every object or material may undergo deformation to an extent when acted on by forces. In some cases, the amount of deformation may be so small that it may not affect the desired analysis and the object is assumed to be rigid. In mechanics, the term equilibrium implies that the body of concern is either at rest or moving with constant velocity. For a body to be in a state of equilibrium, it has to be both in translational and rotational equilibrium. A body is in translational equilibrium if the net force (vector sum of all forces) acting on it is zero. If the net force is zero, then the linear acceleration (time rate of change of linear velocity) of the body is zero, or the linear velocity of the body is either constant or zero. A body is in rotational equilibrium if the net moment (vector sum of the moments of all forces) acting on it is zero. If the net moment is zero, then the angular acceleration (time rate of change of angular velocity) of the body is zero, or the angular velocity of the body is either constant or zero. Therefore, for a body in a state of equilibrium, the equations

of motion (Newton's second law) take the following special forms:

$$\Sigma \underline{F} = 0 \text{ and } \Sigma \underline{M} = 0$$

It is important to remember that force and moment are vector quantities. For example, with respect to a rectangular (Cartesian) coordinate system, force and moment vectors may have components in the x, y, and z directions. Therefore, if the net force acting on an object is zero, then the sum of forces acting in each direction must be equal to zero ($\Sigma F_x = 0$, $\Sigma F_y = 0$, $\Sigma F_z = 0$). Similarly, if the net moment on an object is zero, then the sum of moments in each direction must also be equal to zero ($\Sigma M_x = 0$, $\Sigma M_y = 0$, $\Sigma M_z = 0$). Therefore, for three-dimension force systems there are six conditions of equilibrium. For two-dimensional force systems in the xy-plane, only three of these conditions ($\Sigma F_x = 0$, $\Sigma F_y = 0$, and $\Sigma M_z = 0$) need to be checked.

STATICS

The principles of statics (equations of equilibrium) can be applied to investigate the muscle and joint forces involved at and around the joints for various postural positions of the human body and its segments. The immediate purpose of static analysis is to provide answers to questions such as the following: What tension must the neck extensor muscles exert on the head to support the head in a specified position? When a person bends, what would be the force exerted by the erector spinae on the fifth lumbar vertebra? How does the compression at the elbow, knee, and ankle joints vary with externally applied forces and with different segmental arrangements? How does the force on the femoral head vary with loads carried in the hand? What are the forces involved in various muscle groups and joints during different exercise conditions?

In general, the unknowns in static problems involving the musculoskeletal system are the magnitudes of joint reaction forces and muscle tensions. The mechanical analysis of a skeletal joint requires that we know the vector characteristics of tensions in the muscles, the proper locations of muscle attachments, the weights of body segments, and the locations of the centers of gravity of the body segments. Mechanical models are obviously simple representations of complex systems. Many models are limited by the assumptions that must be made to reduce the system under consideration to a statically determinate one. Any model can be improved by considering the contributions of other muscles, but that will increase the number of unknowns and make the model a statically indeterminate one. To analyze the improved model, the researcher would need additional information related to the muscle forces. This information can be gathered through electromyography measurements of muscle signals or by applying certain optimization techniques. A similar analysis can be made to investigate forces involved at and around other major joints of the musculoskeletal system.

MODES OF DEFORMATION

When acted on by externally applied forces, objects may translate in the direction of the net force and rotate in the direction of the net torque acting on them. If an object is subjected to externally applied forces but is in static equilibrium, then it is most likely that there is some local shape change within the object. Local shape change under the effect of applied forces is known as deformation. The extent of deformation an object may undergo depends on many factors, including the material properties, size, and shape of the object; environmental factors such as heat and humidity; and the magnitude, direction, and duration of applied forces.

One way of distinguishing forces is by observing their tendency to deform the object they are applied on. For example, the object is said to be in tension if the body tends to elongate and in compression if it tends to shrink in the direction of the applied forces. Shear loading differs from tension and compression in that it is caused by forces acting in directions tangent to the area resisting the forces causing shear, whereas both tension and compression are caused by collinear forces applied perpendicular to the areas on which they act. It is common to call tensile and compressive forces normal or axial forces; shearing forces are tangential forces. Objects also deform when they are subjected to forces that cause bending and torsion, which are related to the moment and torque actions of applied forces.

A material may respond differently to different loading configurations. For a given material, there may be different physical properties that must be considered while analyzing the response of that material to tensile loading as compared with compressive or shear loading. The mechanical properties of materials are established through stress analysis by subjecting them to various experiments such as uniaxial tension and compression, torsion, and bending tests.

NORMAL AND SHEAR STRESSES

Consider the whole bone in Figure 1-3A that is subjected to a pair of tensile forces of magnitude F. The bone is in static equilibrium. To analyze the forces induced within the bone, the method of sections can be applied by hypothetically cutting the bone into two pieces through a plane perpendicular to the long axis of the bone. Because the bone as a whole is in equilibrium, the two pieces must individually be in equilibrium as well. This requires that at the cut section of each piece there is an internal force that is equal in magnitude but opposite in direction to the externally applied force (Fig. 1-3B). The internal force is distributed over the entire cross-sectional area of the cut section, and F represents the resultant of the distributed force (Fig. 1-3C). The intensity of this distributed force (force per unit area) is known as stress. For the case shown in Figure 1-3, because the force resultant at the cut section is perpendicular to the plane of the cut, the corresponding stress is called a normal or axial stress. It is customary

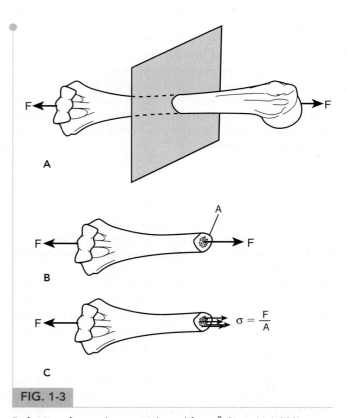

FIG. 1-3

Definition of normal stress. *Adapted from Özkaya, N. (1998). Biomechanics. In W.N. Rom, Environmental and Occupational Medicine (3rd ed.). New York: Lippincott-Raven, 1437–1454.*

to use the symbol σ (sigma) to refer to normal stresses. Assuming that the intensity of the distributed force at the cut section is uniform over the cross-sectional area A of the bone, then $\sigma = F/A$. Normal stresses that are caused by forces that tend to stretch (elongate) materials are more specifically known as tensile stresses; those that tend to shrink them are known as compressive stresses. According to the Standard International (SI) unit system (see Appendix), stresses are measured in newton per square meter (N/m^2), which is also known as pascal (Pa).

There is another form of stress, shear stress, which is a measure of the intensity of internal forces acting tangent (parallel) to a plane of cut. For example, consider the whole bone in Figure 1-4A. The bone is subject to several parallel forces that act in planes perpendicular to the long axis of the bone. Assume that the bone is cut into two parts through a plane perpendicular to the long axis of the bone (Fig. 1-4B). If the bone as a whole is in equilibrium, its individual parts must be in equilibrium as well. This requires that there must be an internal force at the cut section that acts in a direction tangent to the cut surface. If the magnitudes of the external forces are known, then the magnitude F of the internal force can be calculated by considering the translational and rotational equilibrium of one of the parts constituting the bone. The intensity of the internal force tangent to the cut section is known as the shear stress. It is customary to use the symbol τ (tau) to refer to shear stresses (Fig. 1-4C). Assuming that the intensity of the force tangent to the cut section is uniform over the cross-sectional area A of the bone, then $\tau = F/A$.

NORMAL AND SHEAR STRAINS

Strain is a measure of the degree of deformation. As in the case of stress, two types of strains can be distinguished. A normal strain is defined as the ratio of the change (increase or decrease) in length to the original (undeformed) length, and is commonly denoted with the symbol ε (epsilon). Consider the whole bone in Figure 1-5. The total length of the bone is l. If the bone is subjected to a pair of tensile forces, the length of the bone may increase to l' or by an amount $\Delta l = l' - l$. The normal strain is the ratio of the amount of elongation to the original length, or $\varepsilon = \Delta l / l$. If the length of the bone increases in the direction in which the strain is calculated, then the strain is tensile and positive. If the length of the bone decreases in the direction in which the strain is calculated, then the strain is compressive and negative.

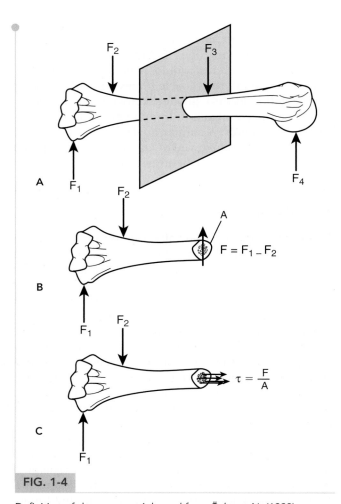

FIG. 1-4

Definition of shear stress. *Adapted from Özkaya, N. (1998). Biomechanics. In W.N. Rom, Environmental and Occupational Medicine (3rd ed.). New York: Lippincott-Raven, 1437–1454.*

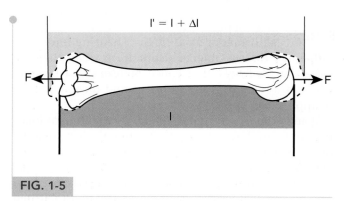

FIG. 1-5

Definition of normal strain. *Adapted from Özkaya, N. (1998). Biomechanics. In W.N. Rom, Environmental and Occupational Medicine (3rd ed.). New York: Lippincott-Raven, 1437–1454.*

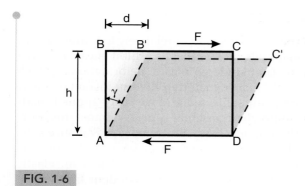

FIG. 1-6

Definition of shear strain. *Reprinted with permission from Özkaya, N. (1998). Biomechanics. In W.N. Rom, Environmental and Occupational Medicine (3rd ed.). New York: Lippincott-Raven, 1437–1454.*

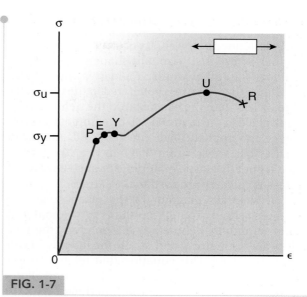

FIG. 1-7

Stress-strain diagrams. *Reprinted with permission from Özkaya, N. (1998). Biomechanics. In W.N. Rom, Environmental and Occupational Medicine (3rd ed.). New York: Lippincott-Raven, 1437–1454.*

Shear strains are related to distortions caused by shear stresses and are commonly denoted with the symbol γ (gamma). Consider the rectangle (ABCD) shown in Figure 1-6 that is acted on by a pair of tangential forces that deform the rectangle into a parallelogram (AB'C'D). If the relative horizontal displacement of the top and the bottom of the rectangle is d and the height of the rectangle is h, then the average shear strain is the ratio of d and h, which is equal to the tangent of angle γ. The angle γ is usually very small. For small angles, the tangent of the angle is approximately equal to the angle itself measured in radians. Therefore, the average shear strain is $\gamma = d/h$.

Strains are calculated by dividing two quantities measured in units of length. For most applications, the deformations and consequently the strains involved may be very small (e.g., 0.001). Strains can also be given in percentages (e.g., 0.1%).

STRESS-STRAIN DIAGRAMS

Different materials may demonstrate different stress-strain relationships. Consider the stress-strain diagram shown in Figure 1-7. There are six distinct points on the curve, which are labeled as O, P, E, Y, U, and R. Point O is the origin of the stress-strain diagram, which corresponds to the initial (no load, no deformation) state. Point P represents the proportionality limit. Between O and P, stress and strain are linearly proportional and the stress-strain diagram is a straight line. Point E represents the elastic limit. Point Y is the yield point, and the stress σ_y corresponding to the yield point is called the yield strength of the material. At this stress level, considerable elongation (yielding) can occur without a corresponding increase of load. U is the highest stress point on the stress-strain diagram. The stress σ_u is the ultimate strength of the material. The last point on the stress-strain diagram is R, which represents the rupture or failure point. The stress at which the failure occurs is called the rupture strength of the material. For some materials, it may not be easy to distinguish the elastic limit and the yield point. The yield strength of such materials is determined by the offset method, which is applied by drawing a line parallel to the linear section of the stress-strain diagram that passes through a strain level of approximately 0.2%. The intersection of this line with the stress-strain curve is taken to be the yield point, and the stress corresponding to this point is called the apparent yield strength of the material.

Note that a given material may behave differently under different load and environmental conditions. If the curve shown in Figure 1-7 represents the stress-strain relationship for a material under tensile loading, there may be a similar but different curve representing the stress-strain relationship for the same material under compressive or shear loading. Also, temperature is known to alter the relationship between stress and strain. For some materials, the stress-strain relationship may also depend on the rate at which the load is applied on the material.

ELASTIC AND PLASTIC DEFORMATIONS

Elasticity is defined as the ability of a material to resume its original (stress-free) size and shape on removal of applied loads. In other words, if a load is applied on a material such that the stress generated in the material is equal to or less than the elastic limit, the deformations that took place in the material will be completely recovered once the applied loads are removed. An elastic material whose stress-strain diagram is a straight line is called a linearly elastic material. For such a material, the stress is linearly proportional to strain. The slope of the stress-strain diagram in the elastic region is called the elastic or Young's modulus of the material, which is commonly denoted by E. Therefore, the relationship between stress and strain for linearly elastic materials is $\sigma = E\varepsilon$. This equation that relates normal stress and strain is called a material function. For a given material, different material functions may exist for different modes of deformation. For example, some materials may exhibit linearly elastic behavior under shear loading. For such materials, the shear stress τ is linearly proportional to the shear strain γ, and the constant of proportionality is called the shear modulus, or the modulus of rigidity. If G represents the modulus of rigidity, then $\tau = G\gamma$. Combinations of all possible material functions for a given material form the constitutive equations for that material.

Plasticity implies permanent deformations. Materials may undergo plastic deformations following elastic deformations when they are loaded beyond their elastic limits. Consider the stress-strain diagram of a material under tensile loading (Fig. 1-7). Assume that the stresses in the specimen are brought to a level greater than the yield strength of the material. On removal of the applied load, the material will recover the elastic deformation that had taken place by following an unloading path parallel to the initial linearly elastic region. The point where this path cuts the strain axis is called the plastic strain, which signifies the extent of permanent (unrecoverable) shape change that has taken place in the material.

Viscoelasticity is the characteristic of a material that has both fluid and solid properties. Most materials are classified as either fluid or solid. A solid material will deform to a certain extent when an external force is applied. A continuously applied force on a fluid body will cause a continuous deformation (also known as flow). Viscosity is a fluid property that is a quantitative measure of resistance to flow. Viscoelasticity is an example of how areas in applied mechanics can overlap because it utilizes the principles of both fluid and solid mechanics.

VISCOELASTICITY

When they are subjected to relatively low stress levels, many materials such as metals exhibit elastic material behavior. They undergo plastic deformations at high stress levels. Elastic materials deform instantaneously when they are subjected to externally applied loads and resume their original shapes almost instantly when the applied loads are removed. For an elastic material, stress is a function of strain only, and the stress-strain relationship is unique (Fig. 1-8). Elastic materials do not exhibit time-dependent behavior. A different group of materials, such as polymer plastics, metals at high temperatures, and almost all biologic materials, exhibit gradual deformation and recovery when subjected to loading and unloading. Such materials are called viscoelastic. The response of viscoelastic materials depends on how quickly the load is applied or removed. The extent of deformation that viscoelastic materials undergo depends on the rate at which the deformation-causing loads are applied.

The stress-strain relationship for a viscoelastic material is not unique but is a function of time or the rate at which the stresses and strains are developed in the material (Fig. 1-9). The word "viscoelastic" is made of two words. Viscosity is a fluid property and is a measure of resistance to flow. Elasticity is a solid material property. Therefore, viscoelastic materials possess both fluid- and solid-like properties.

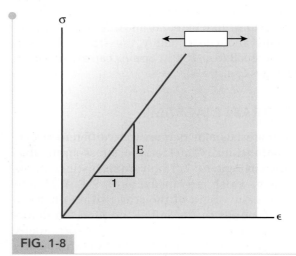

FIG. 1-8

Linearly elastic material behavior. *Reprinted with permission from Özkaya, N. (1998). Biomechanics. In W.N. Rom, Environmental and Occupational Medicine (3rd ed.). New York: Lippincott-Raven, 1437–1454.*

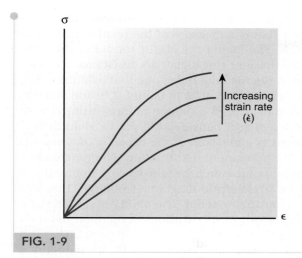

FIG. 1-9

Strain rate-dependent viscoelastic material behavior. *Reprinted with permission from Özkaya, N. (1998). Biomechanics. In W.N. Rom, Environmental and Occupational Medicine (3rd ed.). New York: Lippincott-Raven, 1437–1454.*

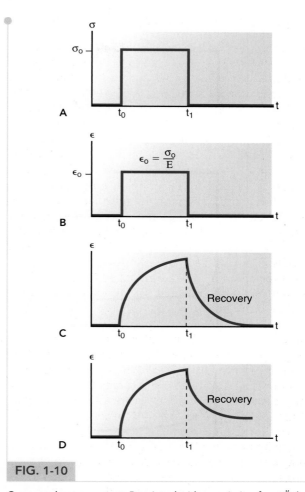

FIG. 1-10

Creep and recovery test. *Reprinted with permission from Özkaya, N. (1998). Biomechanics. In W.N. Rom, Environmental and Occupational Medicine (3rd ed.). New York: Lippincott-Raven, 1437–1454.*

For an elastic material, the energy supplied to deform the material (strain energy) is stored in the material as potential energy. This energy is available to return the material to its original (unstressed) size and shape once the applied load is removed. The loading and unloading paths for an elastic material coincide, indicating no loss of energy. Most elastic materials exhibit plastic behavior at high stress levels. For elasto-plastic materials, some of the strain energy is dissipated as heat during plastic deformations. For viscoelastic materials, some of the strain energy is stored in the material as potential energy and some of it is dissipated as heat regardless of whether the stress levels are small or large. Because viscoelastic materials exhibit time-dependent material behavior, the differences between elastic and viscoelastic material responses are most evident under time-dependent loading conditions.

Several experimental techniques have been designed to analyze the time-dependent aspects of material behavior. As illustrated in Figure 1-10A, a creep and recovery test is conducted by applying a load on the material, maintaining the load at a constant level for a while, suddenly removing the load, and observing the material response. Under a creep and recovery test, an elastic material will respond with an instantaneous strain that would remain at a constant level until the load is removed (Fig. 1-10B). At the instant when the load is removed, the deformation will instantly and completely recover. To the same constant loading condition, a viscoelastic material will respond with a strain increasing and decreasing gradually. If the material is a viscoelastic solid, the recovery will eventually be complete (Fig. 1-10C). If the material is a viscoelastic fluid, complete recovery will never be achieved and there will be a residue of deformation left in the material (Fig. 1-10D). As illustrated in Figure 1-11A, a stress-relaxation experiment is conducted by straining the material to a level and maintaining the constant strain while observing the stress response of the material. Under a stress-relaxation test, an elastic material will respond with a stress developed instantly and maintained at a constant level (Fig. 1-11B). That is, an elastic material will not exhibit a stress-relaxation behavior. A viscoelastic material, conversely, will respond with an

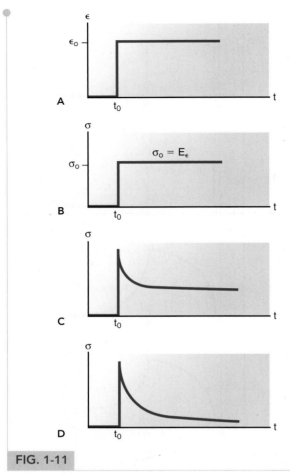

FIG. 1-11

Stress-relaxation experiment. *Reprinted with permission from Özkaya, N. (1998). Biomechanics. In W.N. Rom, Environmental and Occupational Medicine (3rd ed.). New York: Lippincott-Raven, 1437–1454.*

diagram in the elastic region represents the elastic modulus that is a measure of the relative stiffness of materials. The higher the elastic modulus, the stiffer the material and the higher its resistance to deformation. A ductile material is one that exhibits a large plastic deformation prior to failure. A brittle material, such as glass, shows a sudden failure (rupture) without undergoing a considerable plastic deformation. Toughness is a measure of the capacity of a material to sustain permanent deformation. The toughness of a material is measured by considering the total area under its stress-strain diagram. The larger this area, the tougher the material. The ability of a material to store or absorb energy without permanent deformation is called the resilience of the material. The resilience of a material is measured by its modulus of resilience, which is equal to the area under the stress-strain curve in the elastic region.

Although they are not directly related to the stress-strain diagrams, other important concepts are used to describe material properties. For example, a material is called homogeneous if its properties do not vary from location to location within the material. A material is called isotropic if its properties are independent of direction. A material is called incompressible if it has a constant density.

initial high stress level that will decrease over time. If the material is a viscoelastic solid, the stress level will never reduce to zero (Fig. 1-11C). As illustrated in Figure 1-11D, the stress will eventually reduce to zero for a viscoelastic fluid.

MATERIAL PROPERTIES BASED ON STRESS-STRAIN DIAGRAMS

The stress-strain diagrams of two or more materials can be compared to determine which material is relatively stiffer, harder, tougher, more ductile, or more brittle. For example, the slope of the stress-strain

PRINCIPAL STRESSES

There are infinitely many possibilities of constructing elements around a given point within a structure. Among these possibilities, there may be one element for which the normal stresses are maximum and minimum. These maximum and minimum normal stresses are called the principal stresses, and the planes whose normals are in the directions of the maximum and minimum stresses are called the principal planes. On a principal plane, the normal stress is either maximum or minimum, and the shear stress is zero. It is known that fracture or material failure occurs along the planes of maximum stresses, and structures must be designed by taking into consideration the maximum stresses involved. Failure by yielding (excessive deformation) may occur whenever the largest principal stress is equal to the yield strength of the material, or failure by rupture may occur whenever the largest principal stress is equal to the ultimate strength of the material. For a given structure and loading condition, the principal stresses may be within the limits of operational safety. However, the structure must also be checked for critical shearing stress, called the maximum shear stress. The maximum shear stress occurs on a material element for which the normal stresses are equal.

FATIGUE AND ENDURANCE

Principal and maximum shear stresses are useful in predicting the response of materials to static loading configurations. Loads that may not cause the failure of a structure in a single application may cause fracture when applied repeatedly. Failure may occur after a few or many cycles of loading and unloading, depending on factors such as the amplitude of the applied load, mechanical properties of the material, size of the structure, and operational conditions. Fracture resulting from repeated loading is called fatigue.

Several experimental techniques have been developed to understand the fatigue behavior of materials. Consider the bar shown in Figure 1-12A. Assume that the bar is made of a material whose ultimate strength is σ_u. This bar is first stressed to a mean stress level σ_m and then subjected to a stress fluctuating over time, sometimes tensile and other times compressive (Fig. 1-12B). The amplitude σ_a of the stress is such that the bar is subjected to a maximum tensile stress less than the ultimate strength of the material. This reversible and periodic stress is applied until the bar fractures and the number of cycles N to fracture is recorded. This experiment is repeated on specimens having the same material properties by applying stresses of varying amplitude. A typical result of a fatigue test is plotted in Figure 1-12C on a diagram showing stress amplitude versus number of cycles to failure. For a given N, the corresponding stress value is called the fatigue strength of the material at that number of cycles. For a given stress level, N represents the fatigue life of the material. For some materials, the stress amplitude versus number of cycles curve levels off. The stress σ_e at which the fatigue curve levels off is called the endurance limit of the material. Below the endurance limit, the material has a high probability of not failing in fatigue, regardless of how many cycles of stress are imposed on the material.

The fatigue behavior of a material depends on several factors. The higher the temperature in which the material is used, the lower the fatigue strength. The fatigue behavior is sensitive to surface imperfections and the presence of discontinuities within the material that can cause stress concentrations. The fatigue failure starts with the creation of a small crack on the surface of the material, which can propagate under the effect of repeated loads, resulting in the rupture of the material.

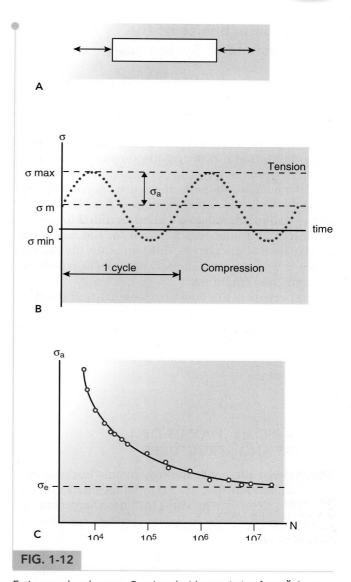

FIG. 1-12

Fatigue and endurance. *Reprinted with permission from Özkaya, N. (1998). Biomechanics. In W.N. Rom, Environmental and Occupational Medicine (3rd ed.). New York: Lippincott-Raven, 1437–1454.*

Orthopaedic devices undergo repeated loading and unloading as a result of the activities of the patients and the actions of their muscles. Over a period of years, a weight-bearing prosthetic device or a fixation device can be subjected to a considerable number of cycles of stress reversals as a result of normal daily activity. This cyclic loading and unloading can cause fatigue failure of the device.

Basic Biomechanics of the Musculoskeletal System

Understanding even a simple task executed by the musculoskeletal system requires a broad, in-depth knowledge of various fields that may include motor control, neurophysiology, physiology, physics, and biomechanics. For example, based on the purpose and intention of a task and the sensory information gathered from the physical environment and orientation of the body and joints, the central nervous system plans a strategy for a task execution. According to the strategy adopted, muscles will be recruited to provide the forces and moments required for the movement and balance of the system. Consequently, the internal forces will be changed and soft tissues will experience different load conditions.

The purpose of this book is to present a well-balanced synthesis of information gathered from various disciplines, providing a basic understanding of biomechanics of the musculoskeletal system. The material presented here is organized to cover three areas of musculoskeletal biomechanics.

PART I: BIOMECHANICS OF TISSUES AND STRUCTURES

The material presented throughout this textbook provides an introduction to basic biomechanics of the musculoskeletal system. Part I includes chapters on the biomechanics of bone, articular cartilage, tendons and ligaments, peripheral nerves, and skeletal muscle. These are augmented with case studies to illustrate the important concepts for understanding the biomechanics of biologic tissues.

PART II: BIOMECHANICS OF JOINTS

Part II of this textbook covers the major joints of the human body, from the spine to the ankle. Each chapter contains information about the structure and functioning of the joint, along with case studies illuminating the clinical diagnosis and management of joint injury and illness. The chapters are written by clinicians to provide an introductory level of knowledge about each joint system.

PART III: APPLIED BIOMECHANICS

The third section of this book introduces important topics in applied biomechanics. These include the biomechanics of fracture fixation, arthroplasty, and gait. It is important for the beginning student to understand the application of biomechanical principles in different clinical areas.

Summary

- Biomechanics is a young and dynamic field of study based on the recognition that conventional engineering theories and methods can be useful for understanding and solving problems in physiology and medicine. Biomechanics concerns the applications of classical mechanics to biologic problems. The field of biomechanics flourishes from the cooperation among life scientists, physicians, engineers, and basic scientists. Such cooperation requires a certain amount of common vocabulary: An engineer must learn some anatomy and physiology, and medical personnel need to understand some basic concepts of physics and mathematics.

- The information presented throughout this textbook is drawn from a large scholarship. The authors aim to introduce some of the basic concepts of biomechanics related to biologic tissues and joints. The book does not intend to provide a comprehensive review of the literature, and readers are encouraged to consult the list of suggested reading that follows to supplement their knowledge. Some basic textbooks are listed here, and students should consult peer-reviewed journals for in-depth presentations of the latest research in specialty areas.

SUGGESTED READING

Bartel, D.L., Davy, D.T., Keaveny, A.M. (2006). *Orthopaedic Biomechanics: Mechanics and Design in Musculoskeletal Systems.* New York: Pearson/Prentice Hall.

Chaffin, D.B., Andersson, G.B.J., Martin, B.J. (2006). *Occupational Biomechanics (3rd ed.).* New York: Wiley-Interscience.

Mow, V.C., Huiskes, R. (2004). *Basic Orthopaedic Biomechanics (3rd ed.).* New York: Raven Press.

Nordin, M., Andersson, G.B.J., Pope, M.H. (Eds.). (2007). *Musculoskeletal Disorders in the Workplace (2nd ed.).* Philadelphia: Mosby–Year Book.

Özkaya, N., Nordin, M. (1999). *Fundamentals of Biomechanics: Equilibrium, Motion, and Deformation (2nd ed.).* New York: Springer-Verlag.

Whiting, W.C., Zernicke, R.F. (2008). *Biomechanics of Musculoskeletal Injury (2nd ed.).* New York: Human Kinetics.

Williams, M., Lissner, H.R. (1992). *Biomechanics of Human Motion (3rd ed.).* Philadelphia: WB Saunders.

Winter, D.A. (2005). *Biomechanics and Motor Control of Human Movement (3rd ed.).* New York: John Wiley and Sons.

Wright, T.M., Maher, S.A. (2008). Musculoskeletal biomechanics. In J.D. Fischgrund (Ed.): *Orthopaedic Knowledge Update 9.* New York: American Academy of Orthopaedic Surgeons.

The International System of Measurement (Le Système International d'Unites)

Dennis R. Carter

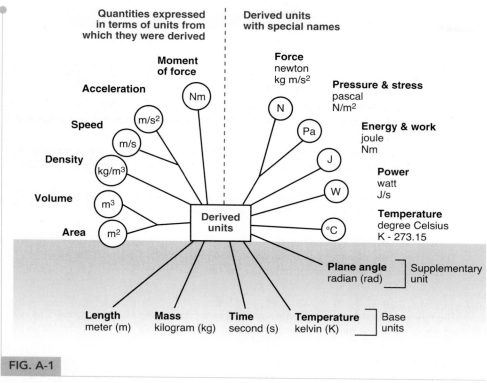

FIG. A-1

The International System of Units.

The SI Metric System

The International System of Measurement (LeSys-tème International d'Unites [SI]), the metric system, has evolved into the most exacting system of measures devised. In this section, the SI units of measurement used in the science of mechanics are described. SI units used in electrical and light sciences have been omitted for the sake of simplicity.

BASE UNITS

The SI units can be considered in three groups: (1) the base units, (2) the supplementary units, and (3) the derived units (Fig. A-1). The base units are a small group of standard measurements that have been arbitrarily defined. The base unit for length is the meter (m), and the base unit of mass is the kilogram (kg). The base units for time and temperature are the second (s) and the kelvin (K), respectively. Definitions of the base units have become increasing sophisticated in response to the expanding needs and capabilities of the scientific community (Table A-1). For example, the meter is now defined in terms of the wavelength of radiation emitted from the krypton-86 atom.

SUPPLEMENTARY UNITS

The radian (rad) is a supplementary unit to measure plane angles. This unit, like the base units, is arbitrarily defined (Table A-1). Although the radian is the SI unit for plane angle, the unit of the degree has been retained for general use, since it is firmly established and is widely used around the world. A degree is equivalent to $\pi/180$ rad.

DERIVED UNITS

Most units of the SI system are derived units, meaning that they are established from the base units in accordance with fundamental physical principles. Some of these units are expressed in terms of the base units from which they are derived. Examples are area, speed, and acceleration, which are expressed in the SI units of square meters (m^2), meters per second (m/s), and meters per second squared (m/s^2), respectively.

Specially Named Units

Other derived units are similarly established from the base units but have been given special names (see Fig. A-1 and Table A-1). These units are defined through

TABLE A-1	
Definitions of SI Units	
Base SI Units	
meter (m)	The meter is the length equal to 1,650,763.73 wavelengths in vacuum of the radiation corresponding to the transition between the levels $2p_{10}$ and $5d_5$ of the krypton-86 atom.
kilogram (kg)	The kilogram is the unit of mass and is equal to the mass of the international prototype of the kilogram.
second (s)	The second is the duration of 9,192,631,770 periods of the radiation corresponding to the transition between the two hyperfine levels of the ground state of the cesium-133 atom.
kelvin (k)	The kelvin, a unit of thermodynamic temperature, is the fraction 1/273.16 of the thermodynamic temperature of the triple point of water.
Supplementary SI Units	
radian (rad)	The radian is the plane angle between two radii of a circle that subtend on the circumference of an arc equal in length to the radius.
Derived SI Units with Special Names	
newton (N)	The newton is that force which, when applied to a mass of one kilogram, gives it an acceleration of one meter per second squared. $1\ N = 1\ kg\ m/s^2$.
pascal (Pa)	The pascal is the pressure produced by a force of one newton applied, with uniform distribution, over an area of one square meter. $1\ Pa = 1\ N/m^2$.
joule (J)	The joule is the work done when the point of application of a force of one newton is displaced through a distance of one meter in the direction of the force. $1\ J = 1\ Nm$.
watt (W)	The watt is the power that in one second gives rise to the energy of one joule. $1\ W = 1\ J/s$.
degree Celsius (°C)	The degree Celsius is a unit of thermodynamic temperature and is equivalent to $K - 273.15$.

the use of fundamental equations of physical laws in conjunction with the arbitrarily defined SI base units. For example, Newton's second law of motion states that when a body that is free to move is subjected to a force, it will experience an acceleration proportional to that force and inversely proportional to its own mass. Mathematically, this principle can be expressed as follows:

$$\text{force} = \text{mass} \times \text{acceleration}$$

The SI unit of force, the newton (N), is therefore defined in terms of the base SI units as

$$1\ N = kg \times 1\ m/s^2$$

The SI unit of pressure and stress is the pascal (Pa). Pressure is defined in hydrostatics as the force divided by the area of force application. Mathematically, this can be expressed as follows:

$$\text{pressure} = \text{force}/\text{area}$$

The SI unit of pressure, the pascal (Pa), is therefore defined in terms of the base SI units as follows:

$$1\ pa = IN/1\ m^2$$

Although the SI base unit of temperature is the kelvin, the derived unit of degree Celsius (°C or c) is much more commonly used. The degree Celsius is equivalent to the kelvin in magnitude, but the absolute value of the Celsius scale differs from that of the Kelvin scale such that $°C = K - 273.15$.

When the SI system is used in a wide variety of measurements, the quantities expressed in terms of the base, supplemental, or derived units may be either very large or very small. For example, the area on the head of a pin is an extremely small number when expressed in terms of square meters (m^2). On the other hand, the weight of a whale is an extremely large number when expressed in terms of newtons (N). To accommodate the convenient representation of small or large quantities, a system of prefixes has been incorporated into the SI system

TABLE A-2

SI Multiplication Factors and Prefixes

Multiplication Factor	SI Prefix	SI Symbol
1 000 000 000 = 10^9	giga	G
1 000 000 = 10^6	mega	M
1 000 = 10^3	kilo	k
100 = 10^2	hecto	h
10 = 10	deka	da
0.1 = 10^{-1}	deci	d
0.01 = 10^{-2}	centi	c
0.001 = 10^{-3}	milli	m
0.000 001 = 10^{-6}	micro	μ
0.000 000 001 = 10^{-9}	nano	n
0.000 000 000 001 = 10^{-12}	pico	p

From Özkaya, N., Nordin, M. (1999). Fundamentals of Biomechanics: Equilibrium, Motion, and Deformation (2nd ed.). New York: Springer-Verlag, 10.

(Table A-2). Each prefix has a fixed meaning and can be used with all SI units. When used with the name of the unit, the prefix indicates that the quantity described is being expressed in some multiple of 10 times the unit used. For example, the millimeter (mm) is used to represent one thousandth (10^{-3}) of a meter and a gigapascal (Gpa) is used to denote one billion (10^9) pascals.

Standard Units Named for Scientists

One of the more interesting aspects of the SI system is its use of the names of famous scientists as standard units. In each case, the unit was named after a scientist in recognition of his or her contribution to the field in which that unit plays a major role. Table A-3 lists various SI units and the scientist for which each was named.

For example, the unit of force, the newton, was named in honor of the English scientist Sir Isaac Newton (1642–1727). He was educated at Trinity College at Cambridge and later returned to Trinity College as a professor of mathematics. Early in his career, Newton made fundamental contributions to mathematics that formed the basis of differential and integral calculus. His other major discoveries were in the fields of optics, astronomy, gravitation, and mechanics. His work in gravitation was purportedly spurred by being hit on the head by an apple falling from a tree. It is perhaps poetic justice that the SI unit of one newton is approximately equivalent to the weight of a medium-sized apple. Newton was knighted in 1705 by Queen Anne for his monumental contributions to science.

TABLE A-3

SI Units Named after Scientists

Symbol	Unit	Quantity	Scientist	Country of Birth	Dates
A	ampere	electric current	Ampere, Andre-Marie	France	1775–1836
C	coulomb	electric charge	Coulomb, Charles-Augustin de	France	1736–1806
°C	degree celsius	temperature	Celsius, Anders	Sweden	1701–1744
F	farad	electric capacity	Faraday, Michael	England	1791–1867
H	henry	inductive resistance	Henry, Joseph	United States	1797–1878
Hz	hertz	frequency	Hertz, Heinrich Rudolph	Germany	1857–1894
J	joule	energy	Joule, James Prescott	England	1818–1889
K	kelvin	temperature	Thomson, William (Lord Kelvin)	England	1824–1907
N	newton	force	Newton, Sir Isaac	England	1642–1727
Ω	ohm	electric resistance	Ohm, Georg Simon	Germany	1787–1854
Pa	pascal	pressure/stress	Pascal, Blaise	France	1623–1662
S	siemens	electric conductance	Siemens, Carl Wilhelm (Sir William)	Germany (England)	1823–1883
T	tesla	magnetic flux density	Tesla, Nikola	Croatia (US)	1856–1943
V	volt	electrical potential	Volta, Count Alessandro	Italy	1745–1827
W	watt	power	Watt, James	Scotland	1736–1819
Wb	weber	magnetic flux	Weber, Wilhelm Eduard	Germany	1804–1891

The unit of pressure and stress, the pascal, was named after the French physicist, mathematician, and philosopher Blaise Pascal (1623–1662). Pascal conducted important investigations on the characteristics of vacuums and barometers and also invented a machine that would make mathematical calculations. His work in the area of hydrostatics helped lay the foundation for the later development of these scientific fields. In addition to his scientific pursuits, Pascal was passionately interested in religion and philosophy and thus wrote extensively on a wide range of subjects.

The base unit of temperature, the kelvin, was named in honor of Lord William Thomson Kelvin (1824–1907). Named William Thomson, he was educated at the University of Glasgow and Cambridge University. Early in his career, Thomson investigated the thermal properties of steam at a scientific laboratory in Paris. At the age of 32, he returned to Glasgow to accept the chair of Natural Philosophy. His meeting with James Joule in 1847 stimulated interesting discussions on the nature of heat, which eventually led to the establishment of Thomson's absolute scale of temperature, the Kelvin scale. In recognition of Thomson's contributions to the field of thermodynamics, King Edward VII conferred on him the title of Lord Kelvin.

The commonly used unit of temperature, the degree Celsius, was named after the Swedish astronomer and inventor Anders Celsius (1701–1744). Celsius was appointed professor of astronomy at the University of Uppsala at the age of 29 and remained at the university until his death 14 years later. In 1742, he described the centigrade thermometer in a paper prepared for the Swedish Academy of Sciences. The name of the centigrade temperature scale was officially changed to Celsius in 1948.

Converting to SI from Other Units of Measurement

Box A-1 contains the formulae for the conversion of measurements expressed in English and non-SI metric units into SI units. One fundamental source of confusion in converting from one system to another is

---

BOX A-1 Conversion of Units

Length
1 centimeter (cm) = 0.01 meter (m)
1 inch (in) = 0.0254 m
1 foot (ft) = 0.3048 m
1 yard (yd) = 0.9144 m
1 mile = 1,609 m
1 angstrom (Å) = 10^{-10} m

Time
1 minute (min) = 60 seconds (s)
1 hour (h) = 3,600 s
1 day (d) = 86,400 s

Mass
1 pound mass (1bm) = 0.4536 kilogram (kg)
1 slug = 14.59 kg

Force
1 kilogram force (kgf) = 9.807 newtons (N)
1 pound force (lbf) = 4.448 N
1 dyne (dyn) = 10^{-5} N

Pressure and Stress
1 kg/m-s^2 = 1 N/m^2 = 1 Pascal (Pa)
1 lbf/in^2 (psi) = 6,896 Pa
1 lbf/ft^2 (psf) = 992,966 Pa
1 dyn/cm^2 = 0.1 Pa

Moment (Torque)
1 dyn-cm = 10^{-7} N-m
1 lbf-ft = 1.356 N-m

Work and Energy
1 kg-m^2/s^2 = 1 N-m = 1 Joule (J)
1 dyn-cm = 1 erg = 10^{-7} J
1 lbf-ft = 1.356 J

Power
1 kg-m^2/s^2 = 1 J/s = 1 Watt (W)
1 horsepower (hp) = 550 lbf-ft/s = 746 W

Plane Angle
1 degree (°) = $\pi/180$ radian (rad)
1 revolution (rev) = 360°
1 rev = 2π rad = 6.283 rad

Temperature
°C = °K − 273.2
°C = 5 (°F − 32)/9

From Özkaya, N., Nordin, M. (1999). Fundamentals of Biomechanics: Equilibrium, Motion, and Deformation (2nd ed.) New York: Springer-Verlag, 11.

that two basic types of measurement systems exist. In the "physical" system (such as SI), the units of length, time, and *mass* are arbitrarily defined, and other units (including force) are derived from these base units. In "technical" or "gravitational" systems (such as the English system), the units of length, time, and *force* are arbitrarily defined, and other units (including mass) are derived from these base units. Since the units of force in gravitational systems are in fact the *weights* of standard masses, conversion to SI depends on the acceleration of mass due to the Earth's gravity. By international agreement, the acceleration due to gravity is 9.806650 m/s^2. This value has been used in establishing some of the conversion factors in Box A-1.

SUGGESTED READING

Feirer, J.L. (1977). *SI Metric Handbook.* New York: Charles Scribner's Sons.

Özkaya, N., Nordin, M. (1999). *Fundamentals of Biomechanics: Equilibrium, Motion, and Deformation (2nd ed.).* New York: Springer-Verlag.

Pennychuick, C.J. (1974). *Handy Matrices of Unit Conversion Factors for Biology and Mechanics.* New York: John Wiley and Sons.

World Health Organization. (1977). *The SI for the Health Professions.* Geneva, Switzerland: WHO.

PART **1**

Biomechanics of Tissues and Structures of the Musculoskeletal System

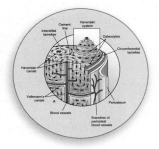

Biomechanics of Bone

Victor H. Frankel and Margareta Nordin

Introduction

The purpose of the skeletal system is to protect internal organs, provide rigid kinematic links and muscle attachment sites, and facilitate muscle action and body movement. Bone has unique structural and mechanical properties that allow it to carry out these roles. Bone is among the body's hardest structures; only dentin and enamel in the teeth are harder. It is one of the most dynamic and metabolically active tissues in the body and remains active throughout life. A highly vascular tissue, it has an excellent capacity for self-repair and can alter its properties and configuration in response to changes in mechanical demand. For example, changes in bone density are commonly observed after periods of disuse and of greatly increased use; changes in bone shape are noted during fracture healing and after certain surgeries. Thus, bone adapts to the mechanical demands placed on it.

This chapter describes the composition and structure of bone tissue, the mechanical properties of bone, and the behavior of bone under different loading conditions. Various factors that affect the mechanical behavior of bone in vitro and in vivo are also discussed.

Bone Composition and Structure

Bone tissue is a specialized connective tissue whose solid composition suits it for its supportive and protective roles. Like other connective tissues, it consists of cells and an organic extracellular matrix of fibers and ground substance produced by the cells. The distinguishing feature of bone is its high content of inorganic materials, in the form of mineral salts, which combine intimately with the organic matrix (Buckwalter et al., 1995). The inorganic component of bone makes the tissue hard and rigid, while the organic component gives bone its flexibility and resilience. The composition of bone differs depending on site, age, dietary history, and the presence of disease (Kaplan et al., 1994).

In normal human bone, the mineral or inorganic portion of bone consists primarily of calcium and phosphate, in the form of small crystals resembling synthetic hydroxyapatite crystals with the composition $Ca_{10}(PO_4)_6(OH)_2$, although impurities such as carbonate, fluoride, and other molecules can be found throughout (Currey, 2002). These minerals, which account for 60% of its weight, give bone its solid consistency, while water accounts for 10% of its weight. The organic matrix, which is predominantly type I collagen, makes up the remaining 30%. The proportions of these substances in terms of volume are approximately 40% mineral, 25% water, and 35% collagen. Bone serves as a reservoir for essential minerals in the body, particularly calcium. Most of the water in bone is found in the organic matrix, around the collagen fibers and ground substance, and in the hydration shells surrounding the bone crystals, although a small amount is located in the canals and cavities that house bone cells and carry nutrients to the bone tissue.

Bone mineral is embedded in variously oriented fibers of the protein collagen (predominantly type I), the fibrous portion of the extracellular matrix—the organic matrix. These collagen fibers are tough and pliable, yet resist stretching and have little extensibility (Bartel et al., 2006). Ninety percent of the extracellular matrix is composed of type I collagen, with some other minor collagen types (III and IV) and a mix of noncollagenous proteins constituting the remaining portion. A universal building block of the body, collagen is also the chief fibrous component of other skeletal structures. (A short discussion of the biomechanical importance of type I collagen can be found later in this chapter, and a detailed description of its microstructure and mechanical behavior is provided in Chapter 3.)

The gelatinous ground substance surrounding the mineralized collagen fibers consists mainly of protein polysaccharides, or glycosaminoglycans (GAGs), primarily in the form of complex macromolecules called proteoglycans (PGs). The GAGs serve as a cementing substance between layers of mineralized collagen fibers. These GAGs, along with various noncollagenous glycoproteins, constitute approximately 5% of the extracellular matrix. (The structure of PGs, which are vital components of articular cartilage, is described in detail in Chapter 3.)

At the microscopic level, the fundamental structural unit of bone is the osteon, or haversian system (Fig. 2-1). At the center of each osteon is a small channel—a haversian canal—that contains blood vessels and nerve fibers. The osteon itself consists of a concentric series of layers (lamellae) of mineralized matrix surrounding the central canal, a configuration similar to growth rings in a tree trunk.

Along the boundaries of each layer, or lamella, are small cavities known as lacunae, each containing a bone cell—an osteocyte—that has entombed itself within the bony matrix (Fig. 2-1C). Numerous small channels, called canaliculi, radiate from each lacuna, connecting the lacunae of adjacent lamellae and ultimately reaching the haversian canal. These allow for cell-to-cell communication, much like a spider's web alerts a spider to

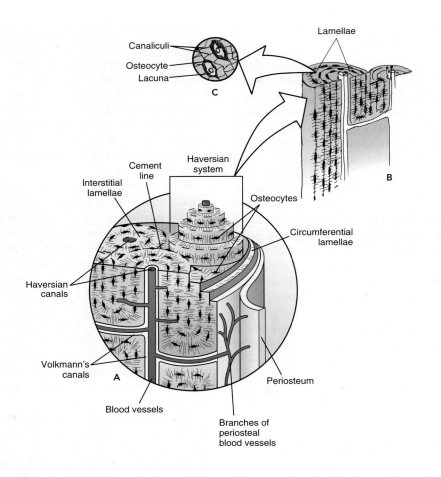

FIG. 2-1

A. The fine structure of bone is illustrated schematically in a section of the shaft of a long bone depicted without inner marrow. The osteons, or haversian systems, are apparent as the structural units of bone. The haversian canals are in the center of the osteons, which form the main branches of the circulatory network in bone. Each osteon is bounded by a cement line. One osteon is shown extending from the bone (×20). *Adapted from Smeltzer SC, Bare BG. (2000). Textbook of Medical-Surgical Nursing (9th ed.). Philadelphia: Lippincott Williams & Wilkins.* **B.** Each osteon consists of lamellae, concentric rings composed of a mineral matrix surrounding the haversian canal. *Adapted from Tortora G J., Anagnostakos, N.P. (1984). Principles of Anatomy and Physiology (4th ed.). New York: Harper & Row.* **C.** Along the boundaries of the lamellae are small cavities known as lacunae, each of which contains a single bone cell, or osteocyte. Radiating from the lacunae are tiny canals, or canaliculi, into which the cytoplasmic processes of the osteocytes extend. *Adapted from Tortora G.J., Anagnostakos, N.P. (1984). Principles of Anatomy and Physiology (4th ed.). New York: Harper & Row.*

movement anywhere in its vicinity (Fig. 2-2). In return, cell processes extend from the osteocytes into the canaliculi, allowing nutrients from the blood vessels in the haversian canal to reach the osteocytes.

At the periphery of each osteon is a cement line, a narrow area of cement-like ground substance composed primarily of GAGs. The canaliculi of the osteon do not pass this cement line. Like the canaliculi, the collagen fibers in the bone matrix interconnect from one lamella to another within an osteon but do not cross the cement line. Effective cross-linking of collagen fibers within the osteon strongly increases the bone's resistance to

mechanical stress and probably explains why the cement line is the weakest portion of the bone's microstructure. The cement line also improves the fatigue properties of cortical bone by dissipating energy through crack propagation, allowing cracks and microdamage to be confined to older, more densely mineralized interstitial bone between osteons (Hernandez and Keveaney, 2006).

A typical osteon is approximately 200 micrometers (μm) in diameter. Hence, every point in the osteon is no more than 100 μm from the centrally located blood supply. In the long bones, the osteons usually run longitudinally, but they branch frequently and anastomose

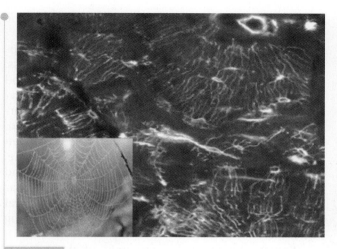

FIG. 2-2

Canaliculi spread like a spider's web **(bottom left)** from the lacunae of an osteon, connecting osteocytes to one another. A crack is visible in the interstitial lamellae between osteons. *Reprinted with permission from Seeman, E. (2006). Osteocytes— martyrs for the integrity of bone strength. Osteoporos Int, 17, 1444.*

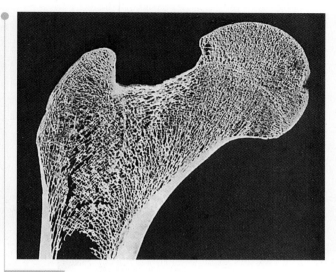

FIG. 2-3

Frontal longitudinal section through the head, neck, greater trochanter, and proximal shaft of an adult femur. Cancellous bone, with its trabeculae oriented in a lattice, lies within the shell of cortical bone. *Reprinted with permission from Gray, H. (1985). Anatomy of the Human Body (13th American ed.). Philadelphia: Lea & Febiger.*

extensively with each other. This interconnection, through the osteon network of canaliculi, allows the osteocytes to detect strain and send signals to each other, thus facilitating bone remodeling.

Interstitial lamellae span the regions between complete osteons (Fig. 2-1A). They are continuous with the osteons and consist of the same material in a different geometric configuration. As in the osteons, no point in the interstitial lamellae is farther than 100 μm from its blood supply, although many of the lacunae in the interstitial lamellae are not inhabited by osteocytes. Because of this, interstitial lamellae tend to be areas of dead bone with increased mineralization and fragility.

At the macroscopic level, all bones are composed of two types of osseous tissue: cortical (compact) bone and cancellous (trabecular) bone (Fig. 2-3). Cortical bone forms the outer shell, or cortex, of the bone and has a dense structure similar to that of ivory. Cancellous bone within this shell is composed of thin rods or plates, called trabeculae, in a loose mesh structure; the interstices between the trabeculae are filled with red marrow (Fig. 2-4). Cancellous bone tissue is arranged in concentric lacunae-containing lamellae but does not contain haversian canals. The osteocytes receive nutrients through canaliculi from blood vessels passing through the red marrow. Cortical bone always surrounds cancellous bone, but the relative quantity of each type varies among bones and within individual bones according to functional requirements.

Bone is found in two forms at the microscopic level: woven and lamellar bone (Fig. 2-5). Woven bone is considered immature bone. This type of bone is found in the embryo, in the newborn, in the fracture callus, and in the metaphysial region of growing bone as well as in tumors, osteogenesis imperfecta, and pagetic bone. Lamellar bone begins to form one month after birth and actively replaces woven bone, meaning it is a more mature form of bone.

All bones are surrounded by a dense fibrous membrane called the periosteum (Fig. 2-1A). The outer periosteal layer is permeated with blood vessels (Fig. 2-6) and nerve fibers that pass into the cortex via Volkmann canals, connecting with haversian canals and extending to the cancellous bone. An inner, osteogenic layer contains bone cells responsible for generating new bone during growth and repair (osteoblasts). The periosteum covers the entire bone except for the joint surfaces, which are covered with articular cartilage. In the long bones, a thinner membrane, the endosteum, lines the central (medullary) cavity, which is filled with yellow fatty marrow. The endosteum contains osteoblasts and giant multinucleated bone cells called osteoclasts, both of which are important in the remodeling and resorption of bone.

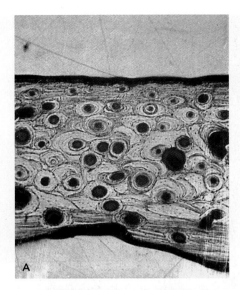

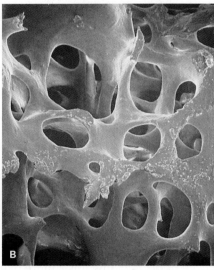

FIG. 2-4

A. Reflected-light photomicrograph of cortical bone from a human tibia (×40). **B.** Scanning electron photomicrograph of cancellous bone from a human tibia (x30). *Reprinted with permission from Carter, D.R., Hayes, W.C. (1977). Compact bone fatigue damage. A microscopic examination. Clin Orthop 127, 265.*

Biomechanical Properties of Bone

Biomechanically, bone tissue may be regarded as a two-phase (biphasic) composite material, with the mineral as one phase and the collagen and ground substance as the other. In such materials (a nonbiologic example is fiberglass) in which a strong, brittle material is embedded in a weaker, more flexible one, the combined substances are stronger for their weight than either substance is alone (Bassett, 1965).

Functionally, the most important mechanical properties of bone are its strength, stiffness, and toughness. These and other characteristics can best be understood for bone, or any other structure, by examining its behavior under loading, that is, under the influence of externally applied forces. Loading causes a deformation, or a change in the dimensions, of the structure. When a load in a known direction is imposed on a structure, the deformation of that structure can be measured and plotted on a load-deformation curve. Much information about the strength, stiffness, and other mechanical properties of the structure can be gained by examining this curve.

Although the mineral component of the bone is thought to give strength and stiffness to the bone, it has been shown that type I collagen is most important in conferring the fundamental toughness and postyield properties to bone tissue (Burr, 2002). Research shows that denaturing collagen decreases bone's toughness and overall strength by up to 60% as shown in Figure 2-7 (Wang et al., 2002). These studies also show that total collagen content is strongly related to failure energy and fracture toughness of bone tissue, suggesting that type I collagen is a primary arrestor of cracks. Type I collagen is a vital element relating to the energy required for matrix failure, independent of size or geometry. It is thus the main determinant of bone toughness, which is defined by the area under the stress-strain curve, known as the modulus of toughness (Fig. 2-8).

Whole Bone

A hypothetical load-deformation curve for a somewhat pliable fibrous structure, such as a long bone, is shown in Figure 2-9. The initial (straight line) portion of the curve, the elastic region, reveals the elasticity of the structure, that is, its capacity to return to its original shape after the load is removed. As the load is applied, deformation occurs but is not permanent; the structure recovers its original shape when unloaded. As loading continues, the outermost fibers of the structure begin to yield. This yield point signals the elastic limit of the structure. As the load exceeds this limit, the structure exhibits plastic behavior, reflected in the second (curved) portion of the curve, the plastic region. The structure will no longer return to its original dimensions when the load has been released; some residual deformation will be permanent. If loading is progressively increased, the structure will fail at some point (bone will fracture). This point is indicated by the ultimate failure point on the curve.

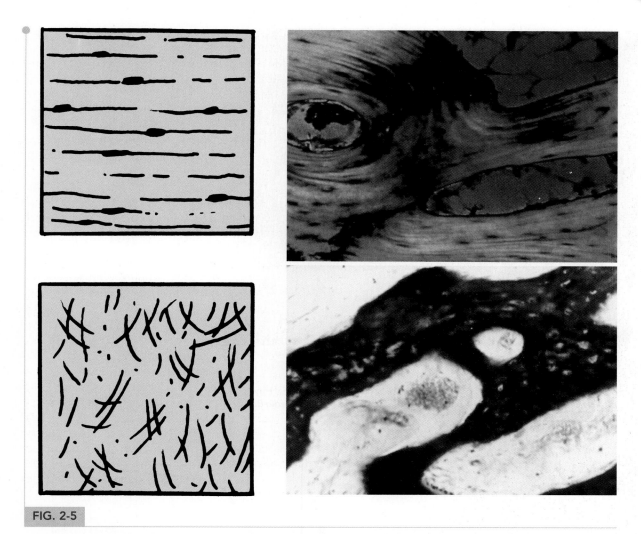

FIG. 2-5

Schematic drawing and photomicrographs of lamellar and woven bone. *Adapted from Kaplan, F.S., Hayes, W.C., Keaveny, T.M., et al. (1994). Form and function of bone. In S.R. Simon (Ed.). Orthopaedic Basic Science. Rosemont, IL: AAOS, 129, 130.*

Three parameters for determining the strength of a structure are reflected on the load-deformation curve: (1) the load the structure can sustain before failing, (2) the deformation it can sustain before failing, and (3) the energy it can store before failing. The strength in terms of load and deformation, or ultimate strength, is indicated on the curve by the ultimate failure point. The strength in terms of energy storage is indicated by the size of the area under the entire curve. The larger the area, the greater the energy that builds up in the structure as the load is applied. The stiffness of the structure is indicated by the slope of the curve in the elastic region. The steeper the slope, the stiffer the material.

The load-deformation curve is useful for determining the mechanical properties of whole structures such as a whole bone, an entire ligament or tendon, or a metal implant. This knowledge is helpful in studying fracture behavior and repair, the response of a structure to physical stress, and the effect of various treatment programs.

Bone Material

Characterizing a bone or other structure in terms of the material of which it is composed, independent of its geometry, requires standardization of the testing conditions, as well as the size and shape of the test specimens. Such standardized testing is useful for comparing the mechanical properties of two or more materials, such as the relative strength of bone and tendon tissue or the relative stiffness of various materials used in prosthetic implants. More

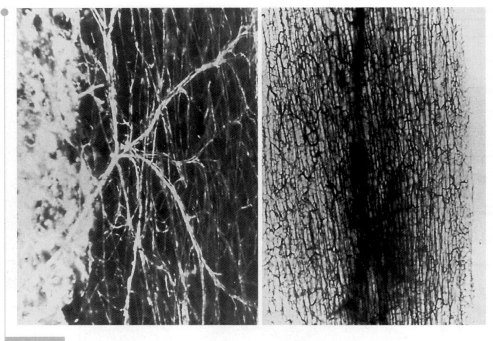

FIG. 2-6

Photomicrograph showing the vasculature of cortical bone. *Adapted from Kaplan, F.S., Hayes, W.C., Keaveny, T.M., et al. (1994). Form and function of bone. In S.R. Simon (Ed.). Orthopaedic Basic Science. Rosemont, IL: AAOS, 131.*

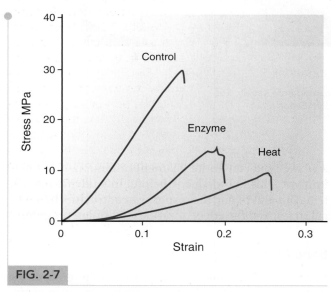

FIG. 2-7

The stress-strain curves of the collagen network, with and without treatments. Collagen denaturation induced by heating or enzymatic cleavage makes the collagen network weaker, more compliant, and less tough. *Adapted from Wang, X., Li, X, Bank, R.A., et al. (2002). Effects of collagen unwinding and cleavage on the mechanical integrity of the collagen network in bone. Calcif Tissue Int, 71, 188.*

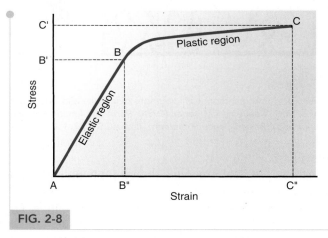

FIG. 2-8

Stress-strain curve for a cortical bone sample tested in tension (pulled). Yield point (B): point past which some permanent deformation of the bone sample occurred. Yield stress (B'): load per unit area sustained by the bone sample before plastic deformation took place. Yield strain (B''): amount of deformation withstood by the sample before plastic deformation occurred. The strain at any point in the elastic region of the curve is proportional to the stress at that point. Ultimate failure point (C): the point past which failure of the sample occurred. Ultimate stress (C'): load per unit area sustained by the sample before failure. Ultimate strain (C''): amount of deformation sustained by the sample before failure. As labeled, the area under the curve is known as the modulus of toughness.

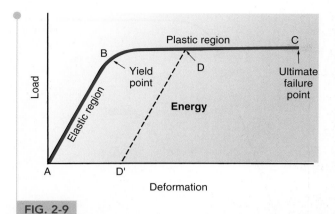

FIG. 2-9

Load-deformation curve for a structure composed of a somewhat pliable material. If a load is applied within the elastic range of the structure (A to B on the curve) and is then released, no permanent deformation occurs. If loading is continued past the yield point (B) and into the structure's plastic range (B to C on the curve) and the load is then released, permanent deformation results. The amount of permanent deformation that occurs if the structure is loaded to point D in the plastic region and then unloaded is represented by the distance between A and D. If loading continues within the plastic range, an ultimate failure point (C) is reached.

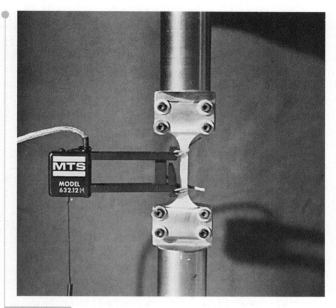

FIG. 2-10

Standardized bone specimen in a testing machine. The strain in the segment of bone between the two gauge arms is measured with a strain gauge. The stress is calculated from the total load measured. *Courtesy of Dennis R. Carter, Ph.D.*

precise units of measurement can be used when standardized samples are tested—that is, the load per unit area of the sample (stress) and the amount of deformation in terms of the percentage of change in the sample's dimensions (strain). The curve generated is a stress-strain curve.

Stress is considered the intensity of the load, or force, per unit area that develops on a plane surface within a structure in response to externally applied loads. The units most commonly used for measuring stress in standardized samples of bone are newtons per centimeter squared (N/cm^2); newtons per meter squared, or Pascals (N/m^2, Pa); and meganewtons per meter squared, or megapascals (MN/m^2, MPa).

Strain is the deformation (change in dimension) that develops within a structure in response to externally applied loads. The two basic types of strain are linear strain, which causes a change in the length of the specimen, and shear strain, which causes a change in the angular relationships between imaginary lines within the structure. Linear strain is measured as the amount of linear deformation (lengthening or shortening) of the sample divided by the sample's original length. It is a non-dimensional parameter expressed as a percentage (e.g., centimeter per centimeter). Shear strain is measured as the amount of angular change (γ) in a right angle lying in the plane of interest in the sample. It is expressed in radians (one radian equals approximately 57.3°) (International Society of Biomechanics, 1987).

Stress and strain values can be obtained for bone by placing a standardized specimen of bone tissue in a testing jig and loading it to failure (Fig. 2-10). These values can then be plotted on a stress-strain curve (Fig. 2-8). The regions of this curve are similar to those of the load-deformation curve. Loads in the elastic region do not cause permanent deformation, but once the yield point is exceeded, some deformation is permanent. As noted earlier, the strength of the material in terms of energy storage is known as the modulus of toughness and is represented by the area under the entire curve (Fig. 2-9). The stiffness is represented by the slope of the curve in the elastic region. A value for stiffness is obtained by dividing the stress at a point in the elastic (straight line) portion of the curve by the strain at that point. This value is called the modulus of elasticity (Young's modulus). Young's modulus (E) is derived from the relationship between stress (σ) and strain (ε):

$$E = \sigma/\varepsilon$$

The elasticity of a material or the Young's modulus E is equal to the slope of the stress (σ) and strain (ε) diagram in the elastic linear region. E represents the stiffness of the material, meaning the higher the elastic modulus or Young's modulus, the stiffer the material (Özkaya and Nordin, 1999).

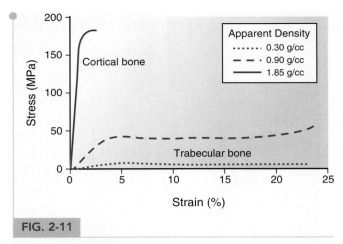

FIG. 2-11

Example of stress-strain curves of cortical and trabecular bone with different apparent densities. Testing was performed in compression. The figure depicts the difference in mechanical behavior for the two bone structures. The trabecular bone has a substantially greater modulus of toughness (area under the curve) than does the cortical bone. *Reprinted with permission from Keaveny, T.M., Hayes, W.C. (1993). Mechanical properties of cortical and trabecular bone. Bone, 7, 285–344.*

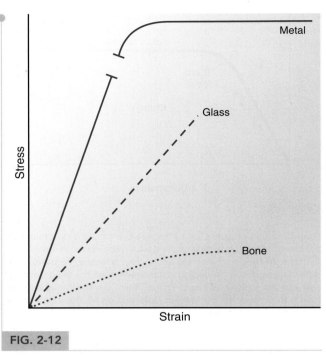

FIG. 2-12

Schematic stress-strain curves for three materials. Metal has the steepest slope in the elastic region and is thus the stiffest material. The elastic portion of the curve for metal is a straight line, indicating linearly elastic behavior. The fact that metal has a long plastic region indicates that this typical ductile material deforms extensively before failure. Glass, a brittle material, exhibits linearly elastic behavior but fails abruptly with little deformation, as indicated by the lack of a plastic region on the stress-strain curve. Bone possesses both ductile and brittle qualities demonstrated by a slight curve in the elastic region, which indicates some yielding during loading within this region.

Mechanical properties differ in the two bone types. Cortical bone is stiffer than cancellous bone, withstanding greater stress but less strain before failure. Cancellous bone in vitro may sustain up to 50% of strains before yielding, whereas cortical bone yields and fractures when the strain exceeds 1.5% to 2.0%. Because of its porous structure, cancellous bone has a large capacity for energy storage (Keaveny and Hayes, 1993). The physical difference between the two bone tissues is quantified in terms of the apparent density of bone, which is defined as the mass of bone tissue present in a unit of bone volume (gram per cubic centimeter [g/cc]). Figure 2-11 depicts typical stress-strain qualities of cortical and trabecular bone with different bone densities tested under similar conditions. In general, it is not enough to describe bone strength with a single number. It is better to examine the stress-strain curve for the bone tissue under the circumstances tested.

Schematic stress-strain curves for bone, metal, and glass are useful in illustrating the relationship between their various mechanical behaviors (Fig. 2-12). The disparities in stiffness are reflected in the different slopes of the curves in the elastic region. The steep slope of metal denotes that it is the stiffest of the three.

The elastic portion of the curve for glass and metal is a straight line, indicating linearly elastic behavior; virtually no yielding takes place before the yield point is reached. By comparison, precise testing of cortical bone has shown that the elastic portion of the curve is not straight but is instead slightly curved, indicating that bone yields somewhat during loading in the elastic region and is therefore not linearly elastic in its behavior (Bonefield and Li, 1967). Table 2-1 depicts the mechanical properties of selected biomaterials for comparison. Materials are classified as brittle or ductile depending on the extent of deformation before failure. Glass is a typical brittle material, and soft metal is a typical ductile material. The difference in the amount of deformation is reflected in the fracture surfaces of the two materials (Fig. 2-13). When pieced together after fracture, the ductile material will not conform to its original shape whereas the brittle material will. Bone exhibits more brittle or more ductile behavior depending on its age (younger bone being more ductile) and the rate at which it is loaded (bone being more brittle at a higher loading rate).

After the yield point is reached, glass deforms very little before failing, as indicated by the absence of a plastic region on the stress-strain curve (Fig. 2-10). By contrast, metal exhibits extensive deformation before failing, as

TABLE 2-1			
Mechanical Properties of Selected Biomaterials			
	Ultimate Strength (MPa)	*Modulus (GPa)*	*Elongation (%)*
Metals			
Co-Cr alloy			
Cast	600	220	8
Forged	950	220	15
Stainless steel	850	210	10
Titanium	900	110	15
Polymers			
Bone cement	20	2.0	2–4
Ceramics			
Alumina	300	350	<2
Biologicals			
Cortical bone	100–150	10–15	1–3
Trabecular bone	8–50		2–4
Tendon, ligament	20–35	2.0–4.0	10–25

Adapted from Kummer, J.K. (1999). Implant biomaterials. In J.M. Spivak, P.E. DiCesare, D.S. Feldman, et al. (Eds.). Orthopaedics: A Study Guide. New York: McGraw-Hill, 45–48.

indicated by the long plastic region on the curve. Bone also deforms before failing, but to a much lesser extent than metal. The difference in the plastic behavior of metal and bone is the result of differences in micromechanical events at yield. Yielding in metal (tested in tension, or pulled) is caused by plastic flow and the formation of plastic slip lines; slip lines are formed when the molecules of the lattice structure of metal dislocate. Yielding in bone (tested in tension) is caused by debonding of the osteons at the cement lines and microfracture (Fig. 2-14), whereas yielding in bone as a result of compression is indicated by cracking of the osteons or interstitial lamellae (Fig. 2-15).

BIOMECHANICAL BEHAVIOR OF BONE

The mechanical behavior of bone—its behavior under the influence of forces and moments—is affected by its mechanical properties, its geometric characteristics, the loading mode applied, direction of loading, rate of loading, and frequency of loading.

Anisotropy

Bone has a grain like wood in the form of its lamellae and therefore behaves anisotropically. This means it exhibits

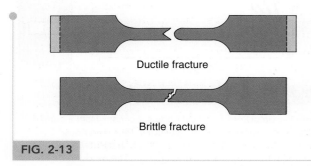

Ductile fracture

Brittle fracture

FIG. 2-13

Fracture surfaces of samples of a ductile and a brittle material. The broken lines on the ductile material indicate the original length of the sample, before it deformed. The brittle material deformed very little before fracture.

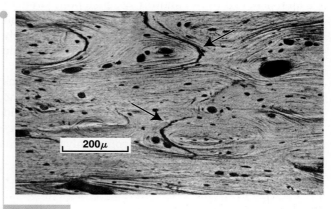

FIG. 2-14

Reflected-light photomicrograph of a human cortical bone specimen tested in tension (×30). *Arrows* indicate debonding at the cement lines and pulling out of the osteons. *Courtesy of Dennis R. Carter, Ph.D.*

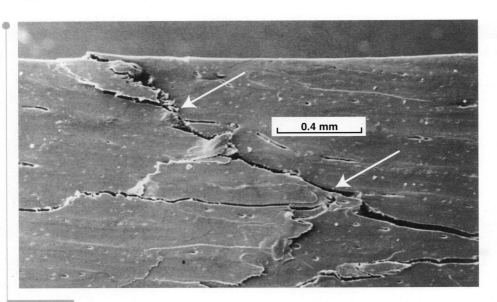

FIG. 2-15

Scanning electron photomicrograph of a human cortical bone specimen tested in compression (×30). *Arrows* indicate oblique cracking of the osteons. *Courtesy of Dennis R. Carter, Ph.D.*

distinct mechanical properties when loaded along various axes because its structure differs in the transverse and longitudinal directions. Isotropic materials like metal, on the other hand, have the same properties when loaded in any direction.

Figure 2-16 shows the variations in strength and stiffness for cortical bone samples from a human femoral shaft, tested in tension in four directions (Carter, 1978; Frankel and Burstein, 1970). The values for both parameters are highest for the samples loaded in the longitudinal direction. Figure 2-11 shows trabecular bone strength and stiffness tested in two directions: compression and tension. Trabecular or cancellous bone is approximately 25% as dense, 5% to 10% as stiff, and five times as ductile as cortical bone.

Although the relationship between loading patterns and the mechanical properties of bone throughout the skeleton is extremely complex, it can generally be said that bone strength and stiffness are greatest in the direction in which daily loads are most commonly imposed.

BONE BEHAVIOR UNDER VARIOUS LOADING MODES

Forces and moments can be applied to a structure in various directions, producing tension, compression, bending, shear, torsion, and combined loading (Fig. 2-17). Bone in vivo is subjected to all these loading modes. The following descriptions of these modes

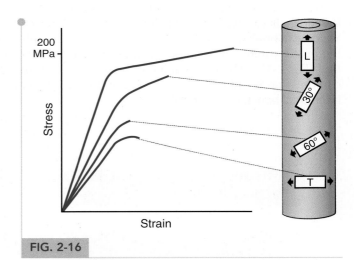

FIG. 2-16

Anisotropic behavior of cortical bone specimens from a human femoral shaft tested in tension (pulled) in four directions: longitudinal (*L*), tilted 30° with respect to the neutral axis of the bone, tilted 60°, and transverse (*T*). The modulus of toughness is clearly also anisotropic in bone and is greatest when tension is applied longitudinally. *Data from Frankel, V.H., Burstein, A.H. (1970). Orthopaedic Biomechanics. Philadelphia: Lea & Febiger.*

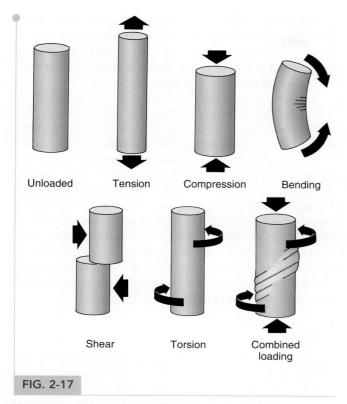

FIG. 2-17

Schematic representation of various loading modes.

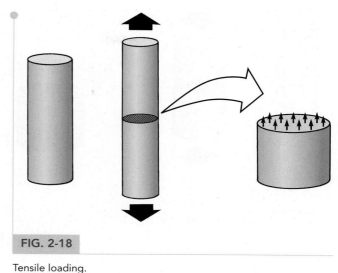

FIG. 2-18

Tensile loading.

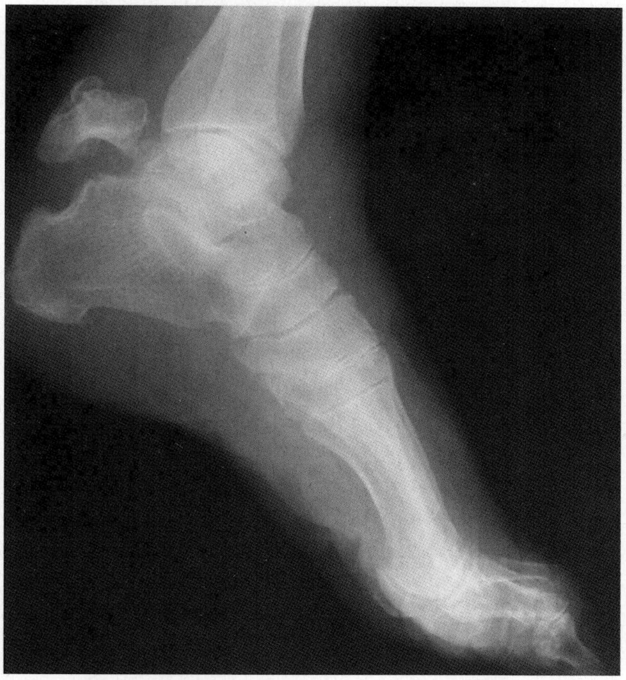

apply to structures in equilibrium (at rest or moving at a constant speed); loading produces an internal, deforming effect on the structure.

Tension

During tensile loading, equal and opposite loads are applied outward from the surface of the structure, and tensile stress and strain result inside the structure. Tensile stress can be thought of as many small forces directed away from the surface of the structure. Maximal tensile stress occurs on a plane perpendicular to the applied load (Fig. 2-18). Under tensile loading, the structure lengthens and narrows.

Clinically, fractures produced by tensile loading are usually seen in bones with a large proportion of cancellous bone. Examples are fractures of the base of the fifth metatarsal adjacent to the attachment of the peroneus brevis tendon and fractures of the calcaneus adjacent to the attachment of the Achilles tendon. Figure 2-19 shows a tensile fracture through the calcaneus; intense contraction of the triceps surae muscle produces abnormally high tensile loads on the bone, which is problematic because bone is usually weaker in tension than in compression (Currey, 2002).

FIG. 2-19

Tensile fracture through the calcaneus produced by strong contraction of the triceps surae muscle during a tennis match. *Courtesy of Robert A. Winquist, M.D.*

Compression

During compressive loading, equal and opposite loads are applied toward the surface of the structure and compressive stress and strain result inside the structure.

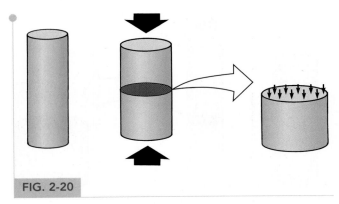

FIG. 2-20

Compressive loading.

Compressive stress can be thought of as many small forces directed into the surface of the structure. Maximal compressive stress occurs on a plane perpendicular to the applied load (Fig. 2-20). Under compressive loading, the structure shortens and widens.

Clinically, compression fractures are commonly found in the vertebrae, which are subjected to high compressive loads. These fractures are most often seen in the elderly with osteoporotic bone tissue. Figure 2-21 shows the shortening and widening that takes place in a human vertebra subjected to a high compressive load. In a joint, compressive loading to failure can be produced by abnormally strong contraction of the surrounding muscles. An example of this effect is presented in Figure 2-22: Bilateral subcapital fractures of the femoral neck were sustained by a patient undergoing electroconvulsive therapy. Strong contractions of the muscles around the hip joint compressed the femoral head against the acetabulum, causing injury.

Shear

During shear loading, a load is applied parallel to the surface of the structure, and shear stress and strain result inside the structure. Shear stress can be thought of as many small forces acting on the surface of the structure on a plane parallel to the applied load (Fig. 2-23). A structure subjected to a shear load deforms internally in an angular manner; right angles on a plane surface within the structure become obtuse or acute (Fig. 2-24). Whenever a structure is subjected to tensile or compressive

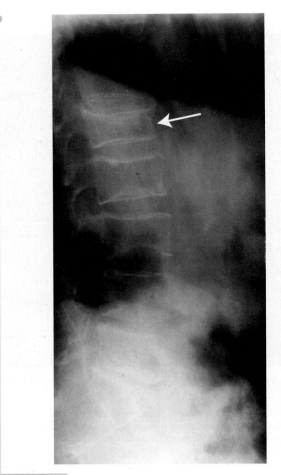

FIG. 2-21

Compression fracture of a human first lumbar vertebra. The vertebra has shortened and widened.

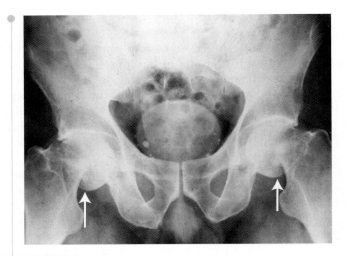

FIG. 2-22

Bilateral subcapital compression fractures of the femoral neck in a patient who underwent electroconvulsive therapy.

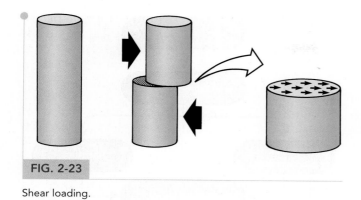

FIG. 2-23

Shear loading.

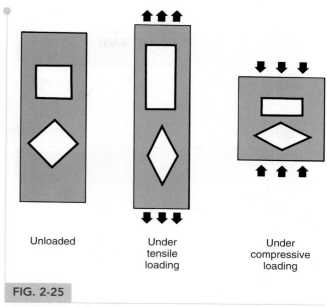

| Unloaded | Under tensile loading | Under compressive loading |

FIG. 2-25

The presence of shear strain in a structure loaded in tension and in compression is indicated by angular deformation.

loading, shear stress is produced. Figure 2-25 illustrates angular deformation in structures subjected to these loading modes. Clinically, shear fractures are most often seen in cancellous bone. Shear stress is greatest when the angle of applied force is equal to 45°, where it has a value that is half that of the maximum normal stress.

Bending

In bending, loads are applied to a structure in a manner that causes it to bend about an axis. When a bone is loaded in bending, it is subjected to a combination of tension and compression. Tensile stresses and strains act on one side of the neutral axis, and compressive stresses and strains act on the other side (Fig. 2-26); there are no stresses and strains along the neutral axis. The magnitude of the stresses is proportional to their distance from the neutral axis of the bone. The farther the stresses are from the neutral axis, the higher their magnitude.

Because a bone structure is asymmetrical, the stresses may not be equally distributed.

Bending may be produced by three forces (three-point bending) or four forces (four-point bending) (Fig. 2-27). Fractures produced by both types of bending are commonly observed clinically, particularly in the long bones.

Three-point bending takes place when three forces acting on a structure produce two equal moments, each the product of one of the two peripheral forces and its

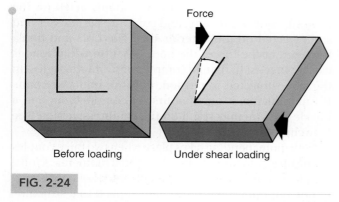

Force

| Before loading | Under shear loading |

FIG. 2-24

When a structure is loaded in shear, lines originally at right angles on a plane surface within the structure change their orientation and the angle becomes obtuse or acute. This angular deformation indicates shear strain.

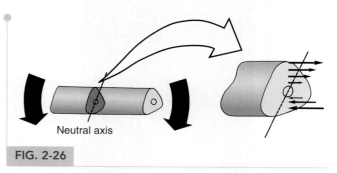

Neutral axis

FIG. 2-26

Cross-section of a bone subjected to bending, showing distribution of stresses around the neutral axis. Tensile stresses act on the superior side, and compressive stresses act on the inferior side. The stresses are highest at the periphery of the bone and lowest near the neutral axis. The tensile and compressive stresses are unequal because the bone is asymmetric.

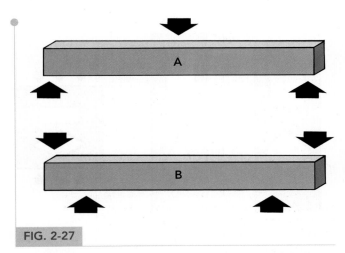

FIG. 2-27

Two types of bending. **A.** Three-point bending. **B.** Four-point bending.

perpendicular distance from the axis of rotation (the point at which the middle force is applied) (Fig. 2-27A). If loading continues to the yield point, the structure, if homogeneous, symmetrical, and with no structural or tissue defect, will break at the point of application of the middle force.

The "boot top" fracture sustained by skiers is a typical three-point bending fracture. In this type of fracture, as shown in Figure 2-28, one bending moment acted on the proximal tibia as the skier fell forward over the top of the ski boot. An equal moment, produced by the fixed foot and ski, acted on the distal tibia. As the proximal tibia

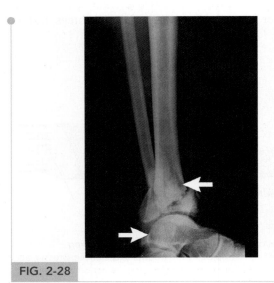

FIG. 2-28

Lateral roentgenogram of a "boot top" fracture produced by three-point bending. *Courtesy of Robert A. Winquist, M.D.*

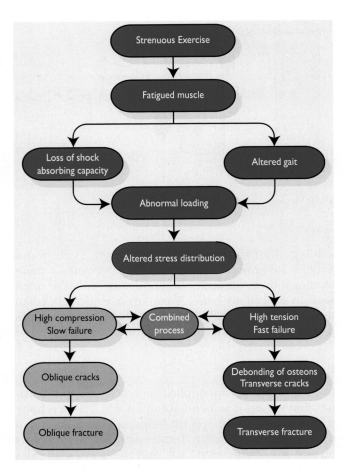

FLOW CHART 2-1

Effects of fatigue and stress on bone.*

*This flow chart is designed for classroom or group discussion. Flow chart is not meant to be exhaustive.

bent forward, tensile stresses and strains acted on the posterior side of the bone and compressive stresses and strains acted on the anterior side. The tibia and fibula fractured at the top of the boot. Because adult bone is weaker in tension than in compression, failure begins on the side subjected to tension. Because immature bone is more ductile, it may fail first in compression, and a buckle fracture may result on the compressive side (See Flowchart 2-1).

Four-point bending takes place when two force couples acting on a structure produce two equal moments. A force couple is formed when two parallel forces of equal magnitude but opposite direction are applied to a structure. Because the magnitude of the bending moment is the same throughout the area between the two force couples, the structure breaks at its weakest point. An example of a four-point bending fracture is shown in Case Study 2-1.

Bone Failure

A stiff knee joint was manipulated incorrectly during rehabilitation of a patient with a postsurgical infected femoral fracture. During the manipulation, the posterior knee joint capsule and tibia formed one force couple and the femoral head and hip joint capsule formed the other. As a bending moment was applied to the femur, the bone failed at its weakest point, the original but now infected fracture site (Case Study Fig. 2-1).

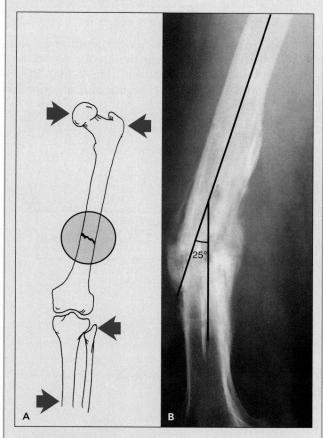

Case Study Figure 2-1 **A.** During manipulation of a stiff knee joint during fracture rehabilitation, four-point bending caused the femur to refracture at its weakest point, the original fracture site. **B.** Lateral radiograph of the fractured femur. *Courtesy of Kaj Lundborg, M.D.*

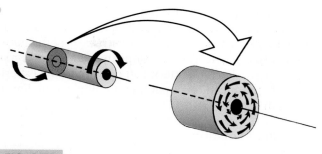

FIG. 2-29

Cross-section of a cylinder loaded in torsion, showing the distribution of shear stresses around the neutral axis. The magnitude of the stresses is highest at the periphery of the cylinder and lowest near the neutral axis.

Torsion

In torsion, a load is applied to a structure in a manner that causes it to twist about an axis, and a torque (or moment) is produced within the structure. When a structure is loaded in torsion, shear stresses are distributed over the entire structure. As in bending, the magnitude of these stresses is proportional to their distance from the neutral axis (Fig. 2-29). The farther the stresses are from the neutral axis, the higher their magnitude.

Under torsional loading, maximal shear stresses act on planes parallel and perpendicular to the neutral axis of the structure. In addition, maximal tensile and compressive stresses act on a plane diagonal to the neutral axis of the structure. Figure 2-30 illustrates these planes in a small segment of bone loaded in torsion.

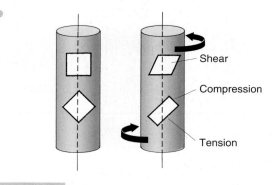

FIG. 2-30

Schematic representation of a small segment of bone loaded in torsion. Maximal shear stresses act on planes parallel and perpendicular to the neutral axis. Maximal tensile and compressive stresses act on planes diagonal to this axis.

The fracture pattern for bone loaded in torsion suggests that the bone fails first in shear, with the formation of an initial crack parallel to the neutral axis of the bone. A second crack usually forms along the plane of maximal tensile stress. Such a pattern can be seen in the experimentally produced torsional fracture of a canine femur shown in Figure 2-31.

Combined Loading

Although each loading mode has been considered separately, living bone is seldom loaded in only one mode. Loading of bone in vivo is complex for two principal reasons: Bones are constantly subjected to multiple indeterminate loads, and their geometric structure is irregular. In vivo measurement of the strains on the anteromedial surface of a human adult tibia during walking and jogging demonstrates the complexity of the loading patterns during these common physiologic activities (Lanyon et al., 1975). Stress values calculated from these strain measurements by Carter (1978) showed that dur-

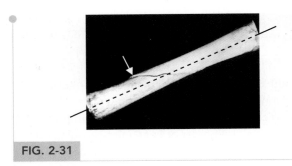

FIG. 2-31

Experimentally produced torsional fracture of a canine femur. The short crack (*arrow*) parallel to the neutral axis represents shear failure; the fracture line at a 30° angle to the neutral axis represents the plane of maximal tensile stress.

ing normal walking, the stresses were compressive during heel strike, tensile during the stance phase, and again compressive during push-off (Fig. 2-32A). Values for shear stress were relatively high in the later portion

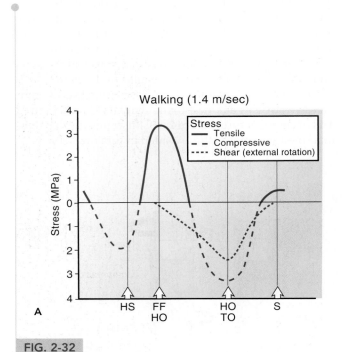

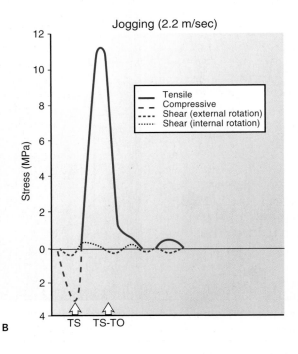

FIG. 2-32

A. Calculated stresses on the anterolateral cortex of a human tibia during walking. *HS,* heel strike; *FF,* foot flat; *HO,* heel-off; *TO,* toe-off; *S,* swing. *Calculated from Lanyon, L.E., Hampson, W.G.J., Goodship, A.E., et al. (1975). Bone deformation recorded in vivo from strain gauges attached to the human tibial shaft. Acta Orthop Scand, 46, 256. Courtesy of Dennis R. Carter, Ph.D.*

B. Calculated stresses on the anterolateral cortex of a human tibia during jogging. *TS,* toe strike; *TO,* toe-off. *Calculated from Lanyon, L.E., Hampson, W.G.J., Goodship, A.E., et al. (1975). Bone deformation recorded in vivo from strain gauges attached to the human tibial shaft. Acta Orthop Scand, 46, 256. Courtesy of Dennis R. Carter, Ph.D.*

of the gait cycle, denoting significant torsional loading. This torsional loading was associated with external rotation of the tibia during stance and push-off.

During jogging, the stress pattern was quite different (Fig. 2-32B). The compressive stress predominating at toe strike was followed by high tensile stress during push-off. The shear stress was low throughout the stride, denoting minimal torsional loading produced by slight external and internal rotation of the tibia in an alternating pattern. The increase in speed from slow walking to jogging increased both the stress and the strain on the tibia (Lanyon et al., 1975). This increase in strain with greater speed was confirmed in studies of locomotion in sheep, which demonstrated a fivefold increase in strain values from slow walking to fast trotting (Lanyon and Bourn, 1979).

Fracture

Because of its anisotropic properties, human adult cortical bone exhibits different values for ultimate stress under compressive, tensile, and shear loading. As shown in Table 2-2, cortical bone can withstand greater stress in compression (approximately 190 MPa) than in tension (approximately 130 MPa) and greater stress in tension than in shear (70 MPa). The elasticity (Young's modulus) is approximately 17,000 MPa in longitudinal or axial loading and approximately 11,000 MPa in transverse loading, as shown in Table 2-3. Human trabecular bone values for testing in compression are approximately 50 MPa and are reduced to approximately 8 MPa if loaded in tension. The modulus of elasticity is low (0–400 MPa) and depends on the apparent density of the trabecular bone and direction of loading. The clinical biomechanical consequence is that the direction of compression failure results in general in a stable fracture, whereas a fracture initiated by tension or shear may have catastrophic consequences.

TABLE 2-2

Average Anisotropic and Asymmetric Ultimate Stress Properties of Human Femoral Cortical Bone*

Longitudinal (MPa)	Tension	133
	Compression	193
Transverse (MPa)	Tension	51
	Compression	133
Shear (MPa)		68

*These properties refer to the principal material coordinate system.
Reproduced with permission from Reilly, D.T., Burstein, A.H. (1975). J Biomech, 8(6), 393–405.

TABLE 2-3

Average Anisotropic and Elastic Properties of Human Femoral Cortical Bone*

Longitudinal Modulus (MPa)	17,000
Transverse Modulus (MPa)	11,500
Shear Modulus (MPa)	3,300

*These properties refer to the principal material coordinate system.
Reproduced with permission from Reilly, D.T., Burstein, A.H. (1975). J Biomech, 8(6), 393–405.

When bone begins to heal after fracture, blood vessels and connective tissue from the periosteum migrate into the region of the fracture, forming a cuff of dense fibrous tissue, or callus (woven bone), around the fracture site, stabilizing that area (Fig. 2-33A). The callus significantly increases the area and polar moments of inertia, thereby increasing the strength and stiffness of the bone in bending and torsion during the healing period. As the fracture heals and the bone gradually regains its normal strength, the callus cuff is progressively resorbed and the bone returns to as near its normal size and shape as possible (Fig. 2-33B).

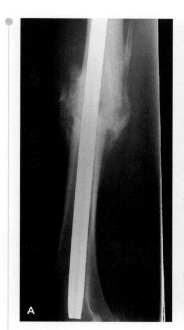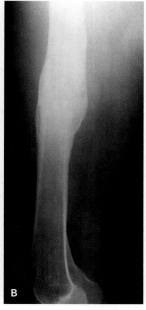

FIG. 2-33

A. Early callus formation in a femoral fracture fixed with an intramedullary nail. **B.** Nine months after injury, the fracture has healed and most of the callus cuff has been resorbed. *Courtesy of Robert A. Winquist, M.D.*

STRAIN RATE DEPENDENCY IN BONE: VISCOELASTICITY

Because bone is a viscoelastic material, its biomechanical behavior varies with the rate at which it is loaded (i.e., the rate at which the load is applied and removed). Bone is stiffer and sustains a higher load to failure when loads are applied at higher rates. Bone also stores more energy before failure at higher loading rates, provided that these rates are within the physiologic range.

The in vivo daily strain can vary considerably. The calculated strain rate for slow walking is 0.001 per second, whereas slow running displays a strain rate of 0.03 per second.

In general, when activities become more strenuous, the strain rate increases (Keaveny and Hayes, 1993). Figure 2-34 shows cortical bone behavior in tensile testing at different physiologic strain rates. As shown, the same change in strain rate produces a larger change in ultimate stress (strength) than in elasticity (Young's modulus). The data indicate that the bone is approximately 30% stronger for brisk walking than for slow walking. At very high strain rates (>1 per second), representing impact trauma, the bone becomes more brittle. In a full range of experimental testing for ultimate tensile strength and elasticity of cortical bone, the strength increases by a factor of three and the modulus by a factor of two (Keaveny and Hayes, 1993).

The loading rate is clinically significant because it influences both the fracture pattern and the amount of soft tissue damage at fracture. When a bone fractures, the stored energy is released. At a low loading rate, the energy can dissipate through the formation of a single crack; the bone and soft tissues remain relatively intact, with little or no displacement of the bone fragments. At a high loading rate, however, the greater energy stored cannot dissipate rapidly enough through a single crack, and comminution of bone and extensive soft tissue damage result. Figure 2-35 shows a human tibia tested in vitro in torsion at a high loading rate; numerous bone fragments were produced, and displacement of the fragments was pronounced.

Clinically, bone fractures fall into three general categories based on the amount of energy released at fracture: low-energy, high-energy, and very high-energy. A low-energy fracture is exemplified by the simple torsional ski fracture, a high-energy fracture is often sustained during automobile accidents, and a very high-energy fracture is produced by very high muzzle-velocity gunshot.

INFLUENCE OF MUSCLE ACTIVITY ON STRESS DISTRIBUTION IN BONE

When bone is loaded in vivo, the contraction of the muscles attached to the bone alters the stress distribution in the bone. This muscle contraction decreases or eliminates tensile stress on the bone by producing compressive stress that neutralizes it either partially or totally.

The effect of muscle contraction can be illustrated in a tibia subjected to three-point bending. Figure 2-36A represents the leg of a skier who is falling forward, subjecting the tibia to a bending moment. High tensile stress is produced on the posterior aspect of the tibia, and high compressive stress acts on the anterior aspect. Contraction of the triceps surae muscle produces great compressive stress on the posterior aspect (Fig. 2-36B), neutralizing the great tensile stress and thereby protecting the tibia from failure in tension. This muscle contraction may result in higher compressive stress on the anterior surface of the tibia and thus protect the bone from failure. Adult bone usually withstands this stress, but immature bone, which is weaker, may fail in compression.

Muscle contraction produces a similar effect in the hip joint (Fig. 2-37). During locomotion, bending

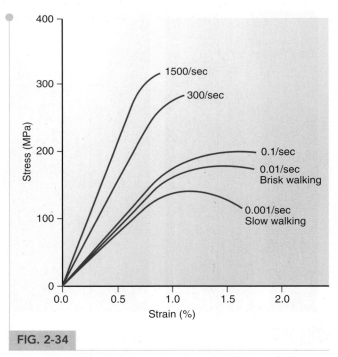

FIG. 2-34

Rate dependency of cortical bone is demonstrated at five strain rates. Both stiffness (modulus) and strength increase considerably at increased strain rates. *Adapted from McElhaney, J.H. (1966). Dynamic response of bone and muscle tissue. J Appl Physiol, 21, 1231–1236.*

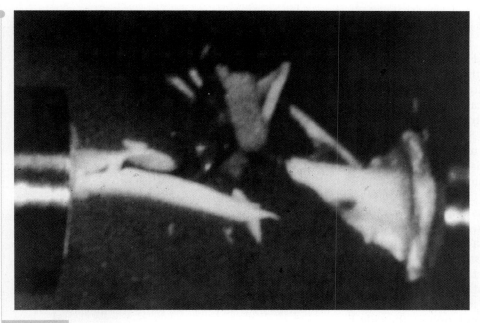

FIG. 2-35

Human tibia experimentally tested to failure in torsion at a high loading rate. Displacement of the numerous fragments was pronounced.

moments are applied to the femoral neck and tensile stress is produced on the superior cortex. Contraction of the gluteus medius muscle produces compressive stress that neutralizes this tensile stress, with the net result that neither compressive nor tensile stress acts on the superior cortex. Thus, the muscle contraction allows the femoral neck to sustain higher loads than would otherwise be possible.

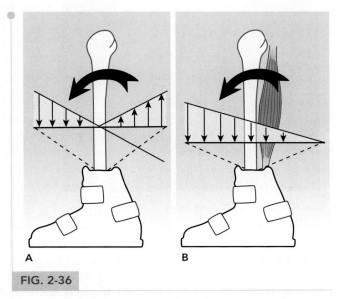

FIG. 2-36

A. Distribution of compressive and tensile stresses in a tibia subjected to three-point bending. **B.** Contraction of the triceps surae muscle produces high compressive stress on the posterior aspect, neutralizing the high tensile stress.

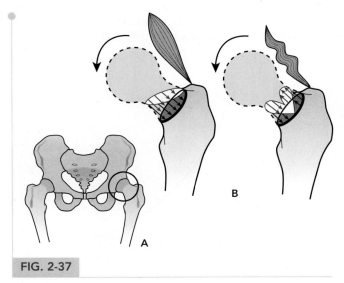

FIG. 2-37

Stress distribution in a femoral neck subjected to bending. When the gluteus medius muscle is relaxed (**A**), tensile stress acts on the superior cortex and compressive stress acts on the inferior cortex. Contraction of this muscle (**B**) neutralizes the tensile stress.

FATIGUE OF BONE UNDER REPETITIVE LOADING

Bone fractures can be produced by a single load that exceeds the ultimate strength of the bone or by repeated applications of a lower-magnitude load. A fracture caused by a repeated load application is called a fatigue fracture. Case Study 2-1 demonstrates the impact of few repetitions of a high load. Case Study 2-3 displays the effect of many repetitions of a relatively normal load. Strain leading to microdamage is between usual strain (400–1,500 $\mu\varepsilon$) and strain causing failure (10,000 $\mu\varepsilon$) (Warden et al., 2006).

This theory of muscle fatigue as a cause of fatigue fracture in the lower extremities is outlined in the schema in Flowchart 2-1. Fatigue fractures fall into two main subcategories: Fatigue-type stress fractures are seen in normal bone after excessive activity, whereas insufficiency-type fractures are brought on by normal activity and are frequently caused by osteoporosis and osteomalacia, thus appearing most commonly in the elderly (Case Study 2-4). Each of these forms of stress fractures falls into two further subcategories: tension and compression fatigue fractures. The more dangerous of the two is a tension fracture, which is caused by debonding of osteons and appears as a transverse crack—"the dreaded black line"—that will continue to completion and ultimately displacement if activity is continued. In compression fractures, on the other hand, bone fails through the formation of oblique cracks, which isolate areas of bone, leading to devascularization. Compression fractures often appear more slowly and most can heal on their own (Egol and Frankel, 2001).

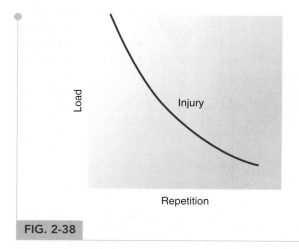

FIG. 2-38

The interplay of load and repetition is represented on a fatigue curve.

Fatigue Fracture

A 21-year-old military recruit was told to run 100 yards with another recruit on his back. After running approximately 60 yards, the man collapsed with a fatigue fracture in his left femur. Although muscle fatigue clearly played a role in this case, the intensity of the load, which was close to the ultimate failure point, is likely the main cause, leading to a fatigue fracture after a few applications (low repetition). After several applications of such high force, initial tensile microdamage quickly advanced to a complete fracture (Case Study Fig. 2-2).

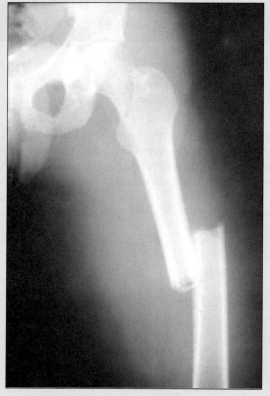

Case Study Figure 2-2 Radiograph of a fatigue fracture.

The interplay of load and repetition for any material can be plotted on a fatigue curve (Fig. 2-38). For some materials (some metals, for example), the fatigue curve is asymptotic, indicating that if the load is kept below a certain level, the material theoretically will remain intact no matter how many repetitions are applied. For bone tested in vitro, the curve is not asymptotic. When

Case Study 2-3

Bone Overloading

A 23-year-old military recruit was exposed to an intensive heavy physical training regime that included repetitive continuous crawling in an awkward position for several weeks (A) (Case Study Fig. 2-3). The repeated application of loads (high repetitions) and the number of applications of a load during a short period of time (high frequency of loading) surpassed the time for the bone remodeling process to prevent failure. Muscle fatigue occurred as a result of the abnormal loading pattern and the intensive training. It affected the muscle function in the neutralization of the stress imposed, leading to abnormal loading and altered stress distribution (B).

After four weeks of strenuous physical activity, the damage accumulation from fatigue at the femoral shaft led to an oblique fracture.

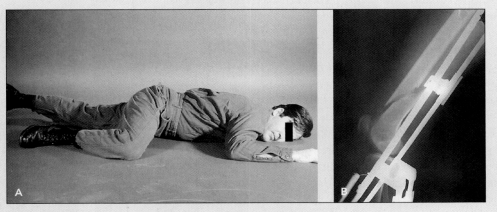

Case Study Figure 2-3 Abnormal loads at the femoral shaft occurred.

bone is subjected to repetitive low loads, it may sustain microfractures. Testing of bone in vitro also reveals that bone fatigues rapidly when the load or deformation approaches its yield strength; that is, the number of repetitions needed to produce a fracture diminishes rapidly. Figure 2-39 shows bone's nonlinear dependence on load intensity and cycle number in the maintenance of bone mass and morphology. It has been hypothesized that any point above the optimal ratio of load intensity to cycle number is anabolic (increasing fracture risk) while any point below will stimulate resorption.

In repetitive loading of living bone, the fatigue process is affected not only by the amount of load and the number of repetitions but also by the number of applications of the load within a given time (frequency of loading), as shown in Figure 2-40 (Burr et al., 1985). Because living bone is self-repairing, a fatigue fracture can result when the remodeling process is outpaced by the fatigue process—that is, when loading is so frequent that it precludes the remodeling necessary to prevent failure.

Remodeling time required to reach a new equilibrium following a change in activity routine is approximately one remodeling period, which is around three to four months. Increasing the number of active bone remodeling units also provisionally removes bone, leading to reduced bone mass and increased fracture risk (Warden et al., 2006).

Muscle fatigue, leading to its inability to contract effectively, can lead to fatigue fractures in cases of sustained strenuous physical activity. As a result, they are less able to store energy and ultimately to neutralize the stresses imposed on the bone. The resulting alteration of the stress distribution in the bone causes abnormally high loads to be imposed, and a fatigue damage accumulation occurs that may lead to a fracture.

Resistance to fatigue behavior is greater in compression than in tension (Keaveny and Hayes, 1993). On average, approximately 5,000 cycles of experimental loading correspond to the number of steps in 10 miles of running. One million cycles corresponds to approximately

Case Study 2-4

Bone Loss and Aging

A woman in her late 60s with no history of fractures or osteoporosis presented with bilateral ankle fractures, reporting no increase in activity around the onset of symptoms. A radiograph of the right ankle showed a trimalleolar fracture with medial displacement of the medial malleolus due to an oblique fracture line, as well as a malunion of the fibula (Case Study Figs. 2-4 A and B). A radiograph of the left ankle showed an oblique fracture line and a minimally displaced fracture of the medial malleolus (Case Study Figs. 2-4 A and C).

The only identifiable risk factor was the diffuse osteopenia noted in both radiographs. Previous dual-energy x-ray absorptiometry (DXA) excluded osteoporosis and bone; CT and positron emission tomography revealed no signs of pathology or malignancy. The rheumatologist also found no evidence of systemic inflammatory arthritis. The most likely causes of these insufficiency fractures are the diffuse osteopenia and possibly a decrease in collagen cross-linking, which would have led to lower bone toughness.

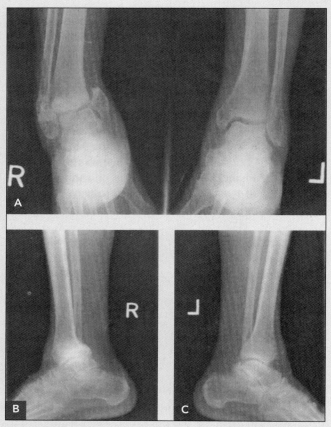

Case Study Figure 2-4 Preoperative anteroposterior radiograph of left and right ankles (**A**), lateral radiograph of right ankle (**B**), and lateral radiograph of left ankle (**C**) show bilateral medial malleoli fractures with fracture of lateral malleolus on right.

1,000 miles. A total distance of less than 1,000 miles can cause a fracture of the *cortical* bone tissue.

This is consistent with stress fractures reported among military recruits undergoing strenuous training of marching and running over a short period of time (6 weeks) as well as by collegiate runners who run up to 120 miles per week. Fractures of individual trabeculae in *cancellous* bone have been observed in postmortem human specimens and may be caused by fatigue accumulation. Common sites are the lumbar vertebrae, the femoral head, and the proximal tibia.

Research has shown that these fractures play a role in bone remodeling, as well as in age-related fractures, collapse of subchondral bone, degenerative joint diseases, and other bone disorders. Bone's ability to sustain this type of microdamage is important because it is a vital means of energy dissipation as a defense against complete fracture, the only alternative means of energy release (Seeman and Delmas, 2006).

BONE REMODELING

Bone has the ability to remodel, by altering its size, shape, and structure, to meet the mechanical demands placed on it (Buckwalter et al., 1995). This phenomenon, in which bone gains or loses cancellous and/or cortical bone in response to the level of stress sustained, is summarized as Wolff's law, which states that the remodeling of bone is influenced and modulated by mechanical stresses (Wolff, 1892).

Load on the skeleton can be accomplished by either muscle activity or gravity. A positive correlation exists between bone mass and body weight. A greater body weight has been associated with a larger bone mass

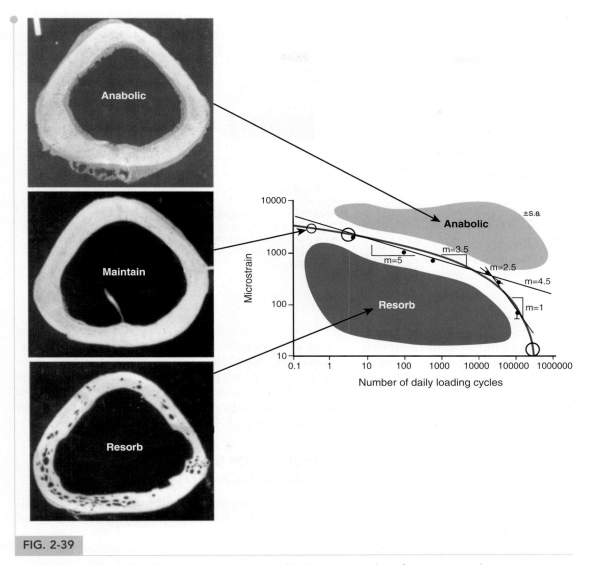

FIG. 2-39

The fatigue process is affected not only by the amount of load and the number of repetitions (cycles), but also by the number of applications of the load within a given time (frequency of loading).

(Exner et al., 1979). Conversely, a prolonged condition of weightlessness, such as that experienced during space travel, has been found to result in decreased bone mass in weight-bearing bones. Astronauts experience a fast loss of calcium and consequent bone loss (Rambaut and Johnston, 1979; Whedon, 1984). These changes are not completely reversible.

Disuse or inactivity has deleterious effects on the skeleton. Bed rest induces a bone mass decrease of approximately 1% per week (Jenkins and Cochran, 1969; Krolner and Toft, 1983). In partial or total immobilization, bone is not subjected to the usual mechanical stresses, which leads to resorption of the periosteal and subperiosteal

bone and a decrease in the mechanical properties of bone (i.e., strength and stiffness). This decrease in bone strength and stiffness was shown by Kazarian and Von Gierke (1969), who immobilized rhesus monkeys in full-body casts for 60 days. Subsequent compressive testing in vitro of the vertebrae from the immobilized monkeys and from controls showed up to a threefold decrease in load to failure and toughness in the vertebrae that were immobilized; stiffness was also significantly decreased (Fig. 2-41).

Current research has shown that microdamage is an important part of bone remodeling because it leads to two major processes that assist the body in identifying where bone resorption and thus remodeling are required. These

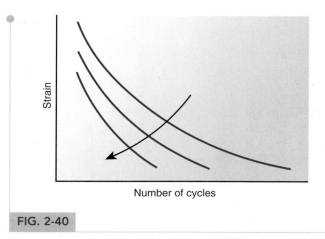

FIG. 2-40

The interplay of strain and repetition is represented in this fatigue curve. The higher the strain (Y-axis) and the higher the repetitions (X-axis), the more damage to the bone. The time (*arrow*) indicates a higher strain rate.

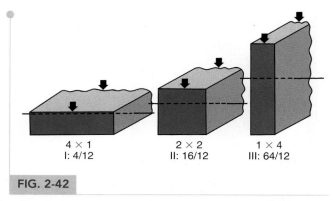

FIG. 2-42

Three beams of equal area but different shapes subjected to bending. The area moment of inertia for beam I is 4/12; for beam II, 16/12; and for beam III, 64/12. *Adapted from Frankel, V.H., Burstein, A.H. (1970). Orthopaedic Biomechanics. Philadelphia: Lea & Febiger.*

are osteocyte apoptosis, which is important because it is believed that live osteocytes inhibit bone resorption, and a decrease in the cross-linking of collagen (Seeman, 2006). Though the effects of aging on bone will be discussed later, it is important to note that the remodeling process itself is strongly impacted by aging; where young bone goes through resorption and remodeling at areas of high stress, osteopenia is frequently symptomatic at sites subject to the greatest mechanical demand (Rubin et al., 2001), as seen in Case Study 2-4.

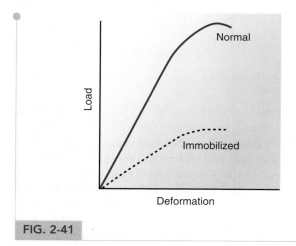

FIG. 2-41

Load-deformation curves for vertebral segments L5–L7 from normal and immobilized rhesus monkeys. Note the extensive loss of strength and stiffness in the immobilized specimens. *Adapted from Kazarian, L.L., Von Gierke, H.E. (1969). Bone loss as a result of immobilization and chelation. Preliminary results in Macaca mulatta. Clin Orthop, 65, 67.*

Several techniques inspired by an understanding of Wolff's law, including low-intensity ultrasound and low-magnitude, high-frequency mechanical stimuli, have been shown to support the body in quickening bone remodeling and to support the bone remodeling process itself in older individuals and in those unable to bear weight (Rubin et al., 2001).

INFLUENCE OF BONE GEOMETRY ON BIOMECHANICAL BEHAVIOR

The geometry of a bone greatly influences its mechanical behavior (Wright and Maher, 2008). In tension and compression, the load to failure and the stiffness are proportional to the cross-sectional area of the bone. The larger the area, the stronger and stiffer the bone. In bending, both the cross-sectional area and the distribution of bone tissue around a neutral axis affect the bone's mechanical behavior. The quantity that takes these two factors into account in bending is called the area moment of inertia. A larger moment of inertia results in a stronger and stiffer bone. Figure 2-42 shows the influence of the area moment of inertia on the load to failure and the stiffness of three rectangular structures that have the same area but different shapes. In bending, beam III is the stiffest of the three and can withstand the highest load because the greatest amount of material is distributed at a distance from the neutral axis. For rectangular cross-sections, the formula for the area moment of inertia is the width (B) multiplied by the cube of the height (H^3) divided by 12:

$$(B \times H^3)/12$$

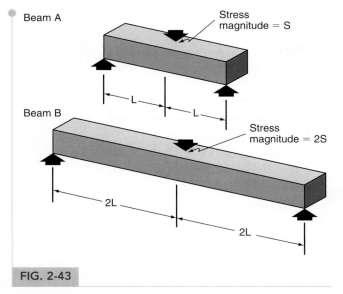

FIG. 2-43

Beam B is twice as long as Beam A and sustains twice the bending moment. Hence, the stress magnitude throughout beam B is twice as high. *Adapted from Frankel, V.H., Burstein, A.H. (1970). Orthopaedic Biomechanics. Philadelphia: Lea & Febiger.*

Because of its large area moment of inertia, beam III can withstand four times more load in bending than beam I.

The length of a bone also influences its strength and stiffness in bending. The longer the bone, the greater the magnitude of the bending moment caused by the application of a force. In a rectangular structure, the magnitude of the stresses produced at the point of application of the bending moment is proportional to the length of the structure. Figure 2-43 depicts the forces acting on two beams with the same width and height but different lengths: Beam B is twice as long as Beam A. The bending moment for the longer beam is twice that for the shorter beam; consequently, the stress magnitude throughout the beam is twice as high. Because of their length, the long bones of the skeleton are subjected to high bending moments and, in turn, to high tensile and compressive stresses, yet their tubular shape gives them the ability to resist bending moments in all directions.

These bones also have large area moments of inertia because much of their bone tissue is distributed at a distance from the neutral axis. Periosteal apposition—endosteal bone resorption and periosteal bone formation—is the natural remodeling process that increases a bone's moment of inertia by thickening the cortical bone and displacing it further from the neutral axis. This is especially important in terms of bone strength

in men versus women since estrogen inhibits periosteal bone formation while promoting bone formation on the endocortical surface in women, which narrows the inner diameter of the bone (Fig. 2-44). After menopause and, in turn, estrogen deficiency, endosteal remodeling increases, leading to a greater quantity of endocortical bone removal and thus bone fragility (Seeman, 2003).

The factors that affect bone strength and stiffness in torsion are the same that operate in bending: the cross-sectional area and the distribution of bone tissue around a neutral axis. The quantity that takes into account these two factors in torsional loading is the polar moment of inertia. The larger the polar moment of inertia, the stronger and stiffer the bone.

Figure 2-45 shows distal and proximal cross-sections of a tibia subjected to torsional loading. Although the proximal section has a slightly smaller bony area than does the distal section, it has a much higher polar moment of inertia because much of the bone tissue is distributed at a distance from the neutral axis. The distal section, although it has a larger bony area, is subjected to much higher shear stress because much of the bone tissue is distributed close to the neutral axis. The magnitude of the shear stress in the distal section is approximately double that in the proximal section. For this reason, torsional fractures of the tibia commonly occur distally.

Surgical Effects

Certain surgical procedures produce defects that greatly weaken the bone, particularly in torsion. These defects fall into two categories: those whose length is less than the diameter of the bone (stress raisers) and those whose length exceeds the bone diameter (open section defects).

STRESS RAISERS

A stress raiser is produced surgically when a small piece of bone is removed or a screw is inserted. Bone strength is reduced because the stresses imposed during loading are prevented from being distributed evenly throughout the bone and instead become concentrated around the defect. This defect is analogous to a rock in a stream, which diverts the water, producing high water turbulence around it. The weakening effect of a stress raiser is particularly marked under torsional loading; the total decrease in bone strength in this loading mode can reach 60%.

Burstein et al. (1972) showed the effect of stress raisers produced by screws and by empty screw holes on the energy storage capacity of rabbit bones tested in torsion at a high loading rate. The immediate effect of

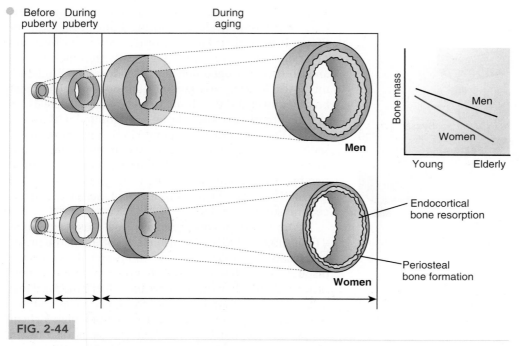

FIG. 2-44

Variances between men's and women's bones in periosteal apposition and net bone loss in the aging process. *Adapted from Seeman, E. (2003). Periosteal bone formation—a neglected determinant of bone strength. N Engl J Med, 349(4), 322.*

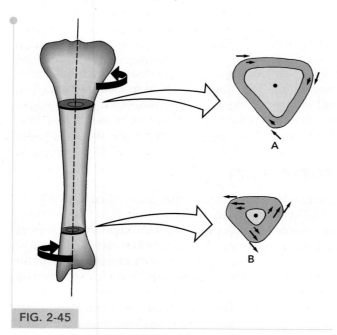

FIG. 2-45

Distribution of shear stress in two cross-sections of a tibia subjected to torsional loading. The proximal section (**A**) has a higher moment of inertia than does the distal section (**B**) because more bony material is distributed away from the neutral axis. *Adapted from Frankel, V.H., Burstein, A.H. (1970). Orthopaedic Biomechanics. Philadelphia: Lea & Febiger.*

drilling a hole and inserting a screw in a rabbit femur was a 74% decrease in energy storage capacity. After 8 weeks, the stress raiser effect produced by the screws and by the holes without screws had disappeared completely because the bone had remodeled: Bone had been laid down around the screws to stabilize them, and the empty screw holes had been filled in with bone. In femora from which the screws had been removed immediately before testing, however, the energy storage capacity of the bone decreased by 50%, mainly because the bone tissue around the screw sustained microdamage during screw removal (Fig. 2-46).

Surgical insertion of large intramedullary implants can also lead to fracture directly due to the anisotropic nature of bone. For instance, tapered, uncemented hip stems that are driven too far into the femoral diaphysis during surgery can produce circumferential or hoop stresses, ultimately leading to fracture (Bartel et al., 2006).

OPEN SECTION DEFECTS

An open section defect is a discontinuity in the bone caused by the surgical removal of a piece of bone longer than the bone's diameter (e.g., by the cutting of a slot

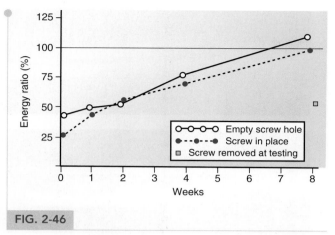

FIG. 2-46

Effect of screws and of empty screw holes on the energy storage capacity of rabbit femora. The energy storage for experimental animals is expressed as a percentage of the total energy storage capacity for control animals. When screws were removed immediately before testing, the energy storage capacity decreased by 50%. *Adapted from Burstein, A.H., Currey J., Frankel, V.H., et al. (1972). Bone strength: The effect of screw holes. J Bone Joint Surg, 54A, 1143.*

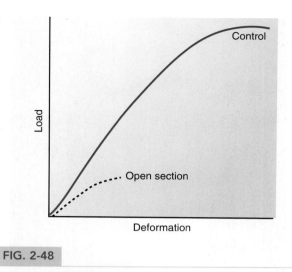

FIG. 2-48

Load-deformation curves for human adult tibiae tested in vitro under torsional loading. The control curve represents a tibia with no defect; the open section curve represents a tibia with an open section defect. *Data from Frankel, V.H., Burstein, A.H. (1970). Orthopaedic Biomechanics. Philadelphia: Lea & Febiger.*

during a bone biopsy). Because the outer surface of the bone's cross-section is no longer continuous, its ability to resist loads is altered, particularly in torsion.

In a normal bone subjected to torsion, the shear stress is distributed throughout the bone and acts to resist the torque. This stress pattern is illustrated in the cross-section of a long bone shown in Figure 2-47A. (A cross-section with a continuous outer surface is called a closed section.) In a bone with an open section defect, only the shear stress at the periphery of the bone resists the applied torque. As the shear stress encounters the discontinuity, it is forced to change direction (Fig. 2-47B). Throughout the interior of the bone, the stress runs

parallel to the applied torque, and the amount of bone tissue resisting the load is greatly decreased.

In torsion tests in vitro of human adult tibiae, an open section defect reduced the load to failure and energy storage to failure by as much as 90%. The deformation to failure was diminished by approximately 70% (Frankel and Burstein, 1970) (Fig. 2-48).

Clinically, the surgical removal of a piece of bone can greatly weaken the bone, particularly in torsion. Figure 2-49 is a radiograph of a tibia from which a graft was removed for use in an arthrodesis of the hip. A few weeks after operation, the patient tripped while twisting and the bone fractured through the defect.

BONE DEPOSITION AND RESORPTION

In the case of a 30-year-old man who had a surgical removal of an ulna plate after stabilization of a displaced ulnar fracture, Figure 2-50 shows anteroposterior (A) and lateral (B) roentgenograms of the ulna after late plate removal.

The implant was used to stabilize the fracture for rapid healing. However, in situations such as this, the late plate removal decreased the amount of mechanical stresses necessary for bone remodeling. It is of concern when the plate carries most or all of the mechanical load and remains after fracture healing. Thus, according to Wolff's law, it will promote localized osseous resorption as a result of decreased mechanical stress and stimulus

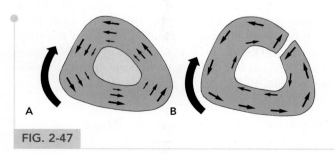

FIG. 2-47

Stress pattern in an open and closed section under torsional loading. **A.** In the closed section, all the shear stress resists the applied torque. **B.** In the open section, only the shear stress at the periphery of the bone resists the applied torque.

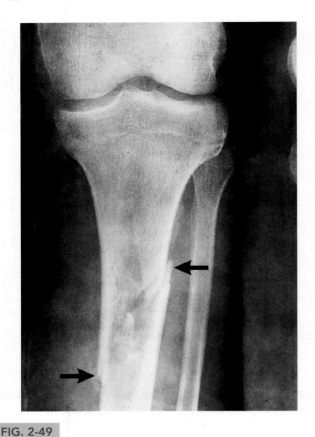

FIG. 2-49

A patient sustained a tibial fracture through a surgically produced open section defect when she tripped a few weeks after the biopsy.

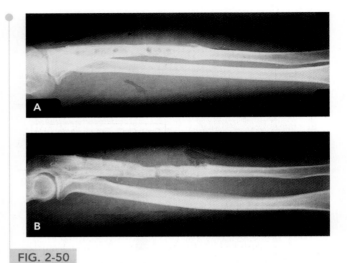

FIG. 2-50

Anteroposterior (A) and lateral (B) roentgenograms of an ulna after plate removal show a decreased bone diameter caused by resorption of the bone under the plate. Cancellization of the cortex and the presence of screw holes also weaken the bone. *Courtesy of Marc Martens, M.D.*

of the bone under the plate, resulting in a decrease in strength and stiffness of the bone.

An implant that remains firmly attached to a bone after a fracture has healed may also diminish the strength and stiffness of the bone. In the case of a plate fixed to the bone with screws, the plate and the bone share the load in proportions determined by the geometry and material properties of each structure. A large plate, carrying high loads, unloads the bone to a great extent; the bone then atrophies in response to this diminished load. (The bone may hypertrophy at the bone-screw interface in an attempt to reduce the micromotion of the screws.)

Bone resorption under a plate is illustrated in Figure 2-50. A compression plate made of a material approximately 10 times stiffer than the bone was applied to a fractured ulna and remained after the fracture had healed. The bone under the plate carried a lower load than normal; it was partially resorbed, and the diameter of the diaphysis became markedly smaller. A reduction in the size of the bone diameter greatly decreases bone strength, particularly in bending and torsion, because it

reduces the area and polar moments of inertia. A 20% decrease in bone diameter may reduce the strength in torsion by 60%. The changes in bone size and shape illustrated in Figure 2-50 suggest that rigid plates should be removed shortly after a fracture has healed and before the bone has markedly diminished in size. Such a decrease in bone size is usually accompanied by secondary osteoporosis, which further weakens the bone (Slätis et al., 1980).

An implant may cause bone hypertrophy at its attachment sites. An example of bone hypertrophy around screws is illustrated in Figure 2-51. A nail plate was applied to a femoral neck fracture and the bone hypertrophied around the screws in response to the increased load at these sites.

Hypertrophy may also result if bone is repeatedly subjected to high mechanical stresses within the normal physiologic range. Hypertrophy of normal adult bone in response to strenuous exercise has been observed (Dalén and Olsson, 1974; Huddleston et al., 1980; Jones et al., 1977), as has an increase in bone density (Nilsson and Westlin, 1971).

Degenerative Changes in Bone Associated with Aging

A progressive loss of bone density has been observed as part of the normal aging process. The longitudinal trabeculae become thinner, and some of the transverse trabeculae

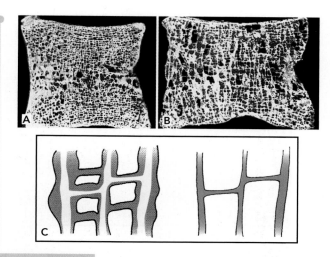

FIG. 2-52

Vertebral cross-sections from autopsy specimens of young **(A)** and old **(B)** bone show a marked reduction in cancellous bone in the latter. *Reprinted with permission from Nordin, B.E.C. (1973). Metabolic Bone and Stone Disease. Edinburgh, Scotland: Churchill Livingstone.* **C.** Bone reduction with aging is schematically depicted. As normal bone **(top)** is subjected to absorption **(shaded area)** during the aging process, the longitudinal trabeculae become thinner and some transverse trabeculae disappear **(bottom).** *Adapted from Siffert, R.S Levy, R.N. (1981). Trabecular patterns and the internal architecture of bone. Mt. Sinai J Med, 48, 221.*

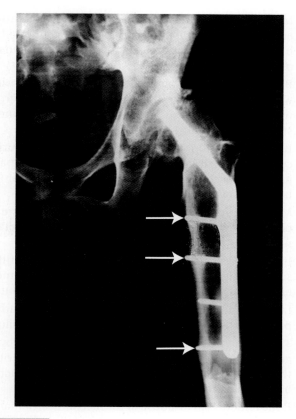

FIG. 2-51

Roentgenogram of a fractured femoral neck to which a nail plate was applied. Loads are transmitted from the plate to the bone via the screws. Bone has been laid down around the screws to bear these loads due to local increased stress.

are resorbed (Siffert and Levy, 1981) (Fig. 2-52). The result is a marked reduction in the amount of cancellous bone and a thinning of cortical bone. The relationship between bone mass, age, and gender is shown in Figure 2-53. The decrease in bone tissue and the slight decrease in the size of the bone reduce bone strength and stiffness.

Stress-strain curves for specimens from human adult tibiae of two widely differing ages tested in tension are shown in Figure 2-54. The ultimate stress was approximately the same for the young and the old bone; however, the old bone specimen could withstand only half the strain that the young bone could, indicating greater brittleness and a reduction in bone toughness (Wang and Qian, 2006). The reduction in collagen cross-linking, bone density, strength, stiffness, and toughness results in increased bone fragility. Age-related bone loss depends on several factors, including gender, age, postmenopausal estrogen deficiency, endocrine

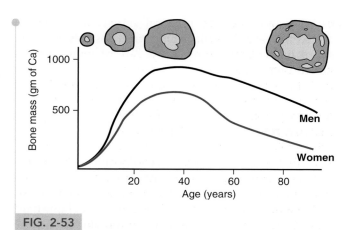

FIG. 2-53

Graph showing the relationship between bone mass, age, and gender. On the **(top)** of the figure, a cross-section of the diaphysis of the femur and the bone mass configuration is shown. *Reprinted with permission from Kaplan, F.S., Hayes, W.C., Keaveny, T.M., et al. (1994). Form and function of bone. In S.R. Simon (Ed.). Orthopaedic Basic Science. Rosemont, IL: American Academy of Orthopaedic Surgeons, 167.*

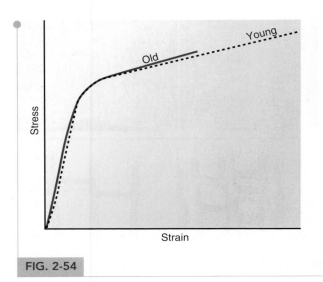

FIG. 2-54

Stress-strain curves for samples of adult young and old human tibiae tested in tension. Note that the bone strength is comparable but that the old bone is more brittle and has lost its ability to deform. This portrays the substantial decrease in the modulus of toughness in the old bone sample. *Adapted from Burstein, A.H., Reilly, D.T., Martens, M. (1976). Aging of bone tissue: Mechanical properties. J Bone Joint Surg, 58A, 82.*

abnormality, inactivity, disuse, and calcium deficiency. Over several decades, the skeletal mass may be reduced to 50% of original trabecular and 25% of cortical mass. In the fourth decade, women lose approximately 1.5% to 2% a year, due in large part to endocortical resorption, whereas men lose at less than half that rate (0.5%–0.75%) yearly. Regular physical activity and exercise (Zetterberg et al., 1990), calcium, and possibly estrogen supplements may decrease the rate of bone mineral loss during aging.

Summary

- Bone is a complex two-phase composite material. One phase is composed of inorganic mineral salts and the other is an organic matrix of type I collagen and ground substance. The inorganic component makes bone hard and rigid, whereas the organic component gives bone its flexibility and toughness.
- Microscopically, the fundamental structural unit of bone is the osteon, or haversian system, composed of concentric layers of a mineralized matrix surrounding a central canal containing blood vessels and nerve fibers.

- Macroscopically, the skeleton is composed of cortical and cancellous (trabecular) bone. Cortical bone has high density, whereas trabecular bone varies in density over a wide range.
- Bone is an anisotropic material, having a grain like wood and exhibiting different mechanical properties when loaded in different directions. Mature bone is strongest and stiffest in compression.
- Bone is subjected to complex loading patterns during common physiologic activities such as walking and jogging. Most bone fractures are produced by a combination of several loading modes.
- Muscle contraction affects stress patterns in bone by producing compressive stress that partially or totally neutralizes the tensile stress acting on the bone.
- Bone is a viscoelastic material. It is therefore stiffer, sustains higher loads before failing, and stores more energy when loaded at higher physiologic strain rates.
- Living bone fatigues when the frequency of loading precludes the remodeling necessary to prevent failure.
- The mechanical behavior of a bone is influenced by its geometry (length, cross-sectional area, and distribution of bone tissue around the neutral axis, which is influenced by periosteal apposition).
- Bone remodels in response to the mechanical demands placed on it; it is laid down where needed and resorbed where not needed, as defined by Wolff's law.
- With aging comes a marked reduction in the amount of cancellous bone and a decrease in the thickness of cortical bone. These changes diminish bone toughness, as well as its strength and stiffness.

Practice Questions

1. The skeleton is made up of cortical and trabecular bone. A. Describe the physiologic and anatomic differences between cortical and trabecular bone. B. Explain the biomechanical properties of the two types of bone.

2. Bone is anisotropic. What does that mean?

3. Bone can remodel. Discuss when, how, and where bone is remodeling.

4. With age, bone density changes. How does this change affect the biomechanical properties of cortical and trabecular bone?

5. What are the best ways to maintain and increase bone density? Mention both negative and positive factors for bone health.

REFERENCES

Bartel, D.L, Davy, D.T., Keaveny, T.M. (2006). *Orthopaedic Biomechanics: Mechanics and Design in Musculoskeletal Systems.* New York: Pearson Prentice Hall.

Bassett, C.A.L. (1965). Electrical effects in bone. *Sci Am, 213,* 18.

Bonefield, W., Li, C.H. (1967). Anisotropy of nonelastic flow in bone. *J Appl Phys, 38,* 2450.

Buckwalter, J.A., Glimcher, M.J., Cooper, R.R., et al. (1995). Bone biology. Part I: Structure, blood supply, cells, matrix and mineralization. Part II: Formation, form, remodelling and regulation of cell function. (Instructional Course Lecture). *J Bone Joint Surg, 77A,* 1256–1289.

Burr, D.B. (2002). The contribution of the organic matrix to bone's material properties. *Bone, 31*(1), 8–11.

Burr, D.B., Martin, R.B., Schaffler, M.B., et al. (1985). Bone remodeling in response to in vivo fatigue microdamage. *J Biomech, 18*(3), 189–200.

Burstein, A.H., Reilly, D.T., Martens, M. (1976). Aging of bone tissue: Mechanical properties. *J Bone Joint Surg, 58A,* 82.

Burstein, A.H., Currey, J., Frankel, V.H., et al. (1972). Bone strength: The effect of screw holes. *J Bone Joint Surg, 54A,* 1143.

Carter, D.R. (1978). Anisotropic analysis of strain rosette information from cortical bone. *J Biomech, 11,* 199.

Carter, D.R., Hayes, W.C. (1977). Compact bone fatigue damage: A microscopic examination. *Clin Orthop, 127,* 265.

Currey, J.D. (2002). *Bones: Structure and Mechanics.* Princeton, NJ: Princeton University Press.

Dalén, N., Olsson, K.E. (1974). Bone mineral content and physical activity. *Acta Orthop Scand, 45,* 170.

Egol, K.A., Frankel, V.H. (2001). Problematic stress fractures. In D.B. Burr, C.M. Milgrom (Eds.). *Musculoskeletal Fatigue and Stress Fractures.* Boca Raton, FL: Lewis Publishers.

Exner, G.U., Prader, A., Elsasser, U., et al. (1979). Bone densitometry using computed tomography. Part 1: Selective determination of trabecular bone density and other bone mineral parameters. Normal values in children and adults. *Br J Radiol, 52,* 14.

Frankel, V.H., Burstein, A.H. (1970). *Orthopaedic Biomechanics.* Philadelphia: Lea & Febiger.

Hernandez, C.J., Keaveny, T.M. (2006). A biomechanical perspective on bone quality. *Bone, 39*(6), 1173–1181.

Huddleston, A. L., Rockwell, D., Kulund, D. N., et al. (1980). Bone mass in lifetime tennis athletes. *JAMA, 244,* 1107.

International Society of Biomechanics (1987). *Quantities and Units of Measurements in Biomechanics* (unpublished).

Jenkins, D.P., Cochran, T.H. (1969). Osteoporosis: The dramatic effect of disuse of an extremity. *Clin Orthop, 64,* 128.

Jones, H., Priest, J., Hayes, W., et al. (1977). Humeral hypertrophy in response to exercise. *J Bone Joint Surg, 59A,* 204.

Kaplan, F.S., Hayes, W.C., Keaveny, T.M., et al. (1994). Form and function of bone. In S.R. Simon (Ed.): *Orthopaedic Basic Science.* Rosemont, IL: American Academy of Orthopaedic Surgeons, 127–184.

Kazarian, L.L., Von Gierke, H.E. (1969). Bone loss as a result of immobilization and chelation. Preliminary results in *Macaca mulatta. Clin Orthop, 65,* 67.

Keaveny, T.M., Hayes, W.C. (1993). Mechanical properties of cortical and trabecular bone. *Bone, 7,* 285–344.

Krolner, B., Toft, B. (1983). Vertebral bone loss: An unheeded side effect of bedrest. *Clin Sci, 64,* 537–540.

Kummer, J.K. (1999). Implant biomaterials. In J M. Spivak, P.E. DiCesare, D.S., Feldman, et al. (Eds.). *Orthopaedics: A Study Guide.* New York: McGraw-Hill, 45–48.

Lanyon, L.E., Bourn, S. (1979). The influence of mechanical function on the development and remodeling of the tibia: An experimental study in sheep. *J Bone Joint Surg, 61A,* 263.

Lanyon, L.E., Hampson, W.G.J., Goodship, A.E., et al. (1975). Bone deformation recorded in vivo from strain gauges attached to the human tibial shaft. *Acta Orthop Scand, 46,* 256.

Nilsson, B.E., Westlin, N.E. (1971). Bone density in athletes. *Clin Orthop, 77,* 179.

Özkaya, N., Nordin, M. (1999). *Fundamentals of Biomechanics: Equilibrium, Motion, and Deformation (2nd ed.).* New York: Springer-Verlag.

Rambaut, P.C., Johnston, R.S. (1979). Prolonged weightlessness and calcium loss in man. *Acta Astronautica, 6,* 1113.

Rubin, C., Bolander, M., Ryaby, J.P., et al. (2001). The use of low-intensity ultrasound to accelerate the healing of fractures. *J Bone Joint Surg Am, 83A*(2), 259–270.

Rubin, C.T., Sommerfeldt, D.W., Judex, S., et al. (2001). Inhibition of osteopenia by low magnitude, high-frequency mechanical stimuli. *Drug Discov Today, 6*(16), 848–858.

Seeman, E. (2003). Periosteal bone formation—a neglected determinant of bone strength. *N Engl J Med, 349*(4), 320–323.

Seeman, E. (2006). Osteocytes—martyrs for integrity of bone strength. *Osteoporos Int, 17*(10), 1443–1448.

Seeman, E., Delmas, P.D. (2006). Bone quality—the material and structural basis of bone strength and fragility. *N Engl J Med, 354*(21), 2250–2261.

Siffert, R.S., Levy, R.N. (1981). Trabecular patterns and the internal architecture of bone. *Mt Sinai J Med, 48,* 221.

Slätis, P., Paavolainen, P., Karaharju, E., et al. (1980). Structural and biomechanical changes in bone after rigid plate fixation. *Can J Surg, 23,* 247.

Wang, X., Qian, C. (2006). Prediction of microdamage formation using a mineral-collagen composite model of bone. *J Biomech, 39,* 595–602.

Wang, X., Shen, X., Li, X., et al. (2002). Age-related changes in the collagen network and toughness of bone. *Bone, 31*(1), 1–7.

Warden, S.J., Burr, D.B., Brukner, P.D. (2006). Stress fractures: Pathophysiology, epidemiology, and risk factors. *Curr Osteoporos Rep, 4*(3), 103–109.

Whedon, G.D. (1984). Disuse osteoporosis: Physiological aspects. *Calcif Tissue Int, 36*, 146–150.

Wolff, J. (1892). *Das Gesetz der Transformation der Knochen.* Berlin, Germany: Hirschwald.

Wright, T.M., Maher, S.A. (2008). Musculoskeletal biomechanics. In J.D. Fischgrund (Ed.): *Orthopaedic Knowledge Update 9.* New York: American Academy of Orthopaedic Surgeons.

Zetterberg, C., Nordin, M., Skovron, M.L., et al. (1990). Skeletal effects of physical activity. *Geri-Topics, 13*(4), 17–24.

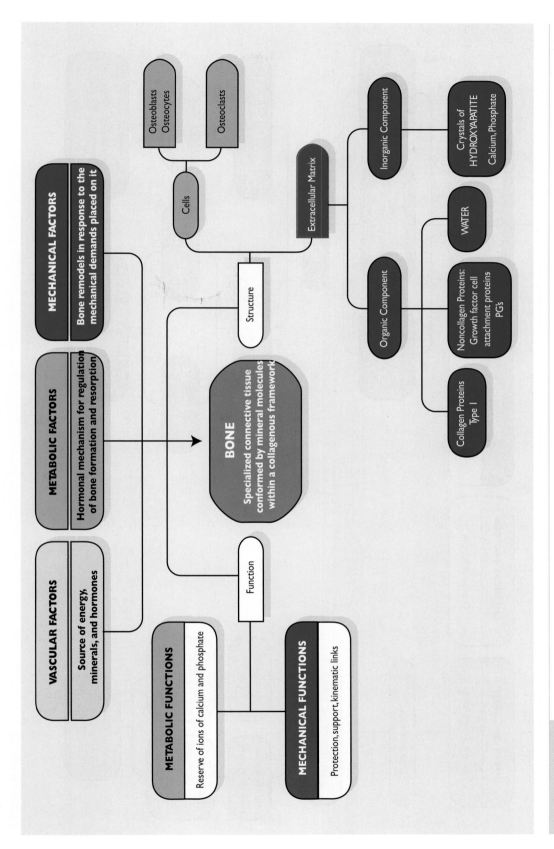

Bone composition, structure, and functions.*

*This flow chart is designed for classroom or group discussion. Flow chart is not meant to be exhaustive.
PGs, proteoglycans.

VASCULAR FACTORS

Source of energy, minerals, and hormones

METABOLIC FACTORS

Hormonal mechanism for regulation of bone formation and resorption

MECHANICAL FACTORS

Bone remodels in response to the mechanical demands placed on it

Osteoblasts Osteocytes

Osteoclasts

Cells

Extracellular Matrix

Inorganic Component

Crystals of HYDROXYAPATITE Calcium, Phosphate

WATER

Organic Component

Noncollagen Proteins: Growth factor cell attachment proteins PG's

Collagen Proteins Type I

Structure

BONE

Specialized connective tissue conformed by mineral molecules within a collagenous framework

Function

METABOLIC FUNCTIONS

Reserve of ions of calcium and phosphate

MECHANICAL FUNCTIONS

Protection, support, kinematic links

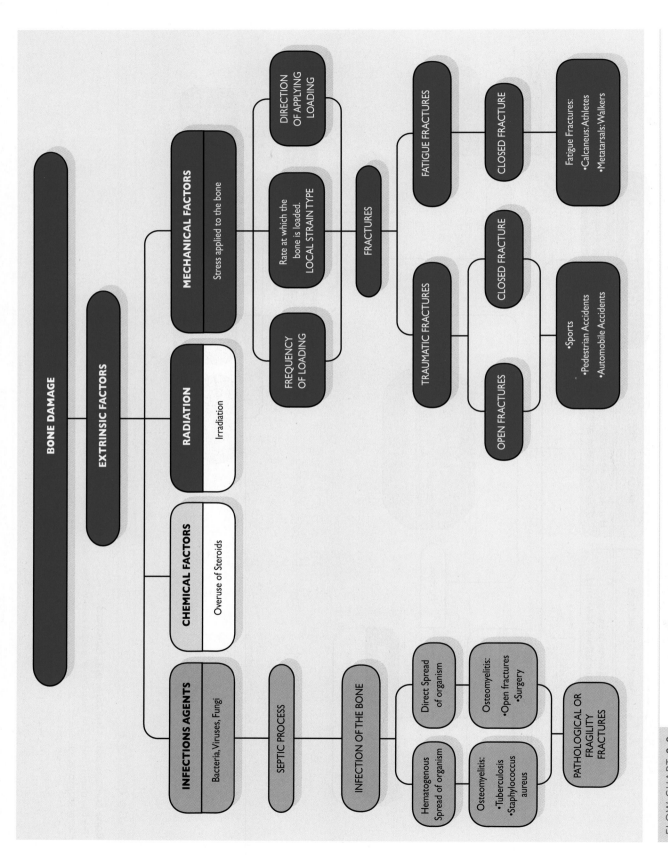

FLOW CHART 2-3

Extrinsic factors associated with bone damage. Clinical examples.*

*This flow chart is designed for classroom or group discussion. Flow chart is not meant to be exhaustive.

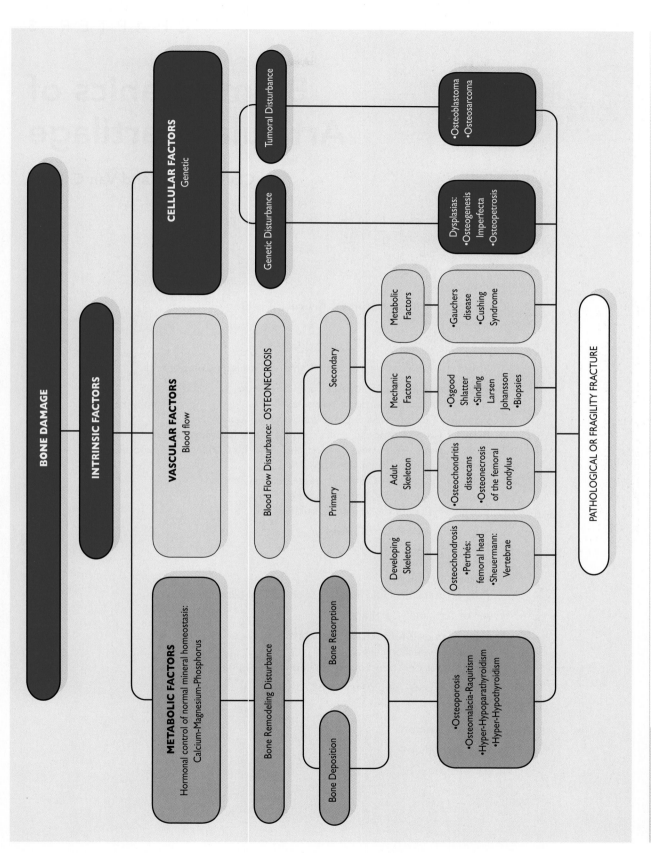

Intrinsic factors associated with bone damage. Clinical examples.*

*This flow chart is designed for classroom or group discussion. Flow chart is not meant to be exhaustive.

CHAPTER 3

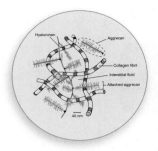

Biomechanics of Articular Cartilage

Clark T. Hung and Van C. Mow

Introduction

Three types of joints exist in the human body: fibrous, cartilaginous, and synovial. Only one of these, the synovial, or diarthrodial, joint, allows a large degree of motion. In young normal joints, the articulating bone ends of diarthrodial joints are covered by a thin (1–6 mm), dense, translucent, white connective tissue called hyaline articular cartilage (Box 3-1). Articular cartilage is a very specialized tissue precisely suited to withstand the highly loaded joint environment without failure during an average individual's lifetime. Physiologically, however, it is virtually an isolated tissue, devoid of blood vessels, lymphatic channels, and neurologic innervation. Furthermore, its cellular density is less than that of any other tissue (Stockwell, 1979).

In diarthrodial joints, articular cartilage has two primary functions: (1) to distribute joint loads over a wide area, thus decreasing the stresses sustained by the contacting joint surfaces (Ateshian et al., 1995; Helminen et al., 1987) and (2) to allow relative movement of the opposing joint surfaces with minimal friction and wear (Mow and Ateshian, 1997). In this chapter, we will describe how the biomechanical properties of articular cartilage, as determined by its composition and structure, allow for the optimal performance of these functions.

Composition and Structure of Articular Cartilage

Chondrocytes, the sparsely distributed cells in articular cartilage, account for less than 10% of the tissue's volume

(Stockwell, 1979). Schematically, the zonal arrangement of chondrocytes is shown in Figure 3-1. Despite their sparse distribution, chondrocytes manufacture, secrete, organize, and maintain the organic component of the extracellular matrix (ECM) (Fosang and Hardingham, 1996; Muir, 1983). The organic matrix is composed of a dense network of fine collagen fibrils (mostly type II collagen, with minor amounts of types V, VI, IX, and XI) that are enmeshed in a concentrated solution of proteoglycans (PGs) (Bateman et al., 1996; Eyre, 1980; Muir, 1983). In normal articular cartilage the collagen content ranges from 15% to 22% by wet weight and the

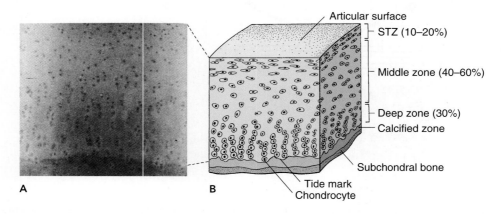

FIG. 3-1

Photomicrograph **(A)** and schematic representation **(B)** of the chondrocyte arrangement throughout the depth of noncalcified articular cartilage. In the superficial tangential zone, chondrocytes are oblong with their long axes aligned parallel to the articular surface. In the middle zone, the chondrocytes are "round" and randomly distributed. Chondrocytes in the deep zone are arranged in a columnar fashion oriented perpendicular to the tidemark, the demarcation between the calcified and noncalcified tissue.

PG content from 4% to 7% by wet weight; the remaining 60% to 85% is water, inorganic salts, and small amounts of other matrix proteins, glycoproteins, and lipids (Mow and Ratcliffe, 1997). Collagen fibrils and PGs, each being capable of forming structural networks of significant strength (Broom and Silyn-Roberts, 1990; Kempson et al., 1976; Schmidt et al., 1990; Zhu et al., 1991, 1993), are the structural components supporting the internal mechanical stresses that result from loads being applied to the articular cartilage. Moreover, these structural components, together with water, determine the biomechanical behavior of this tissue (Ateshian et al., 1997; Maroudas, 1979; Mow et al., 1980, 1984; Mow and Ateshian, 1997).

COLLAGEN

Collagen is the most abundant protein in the body (Bateman et al., 1996; Eyre, 1980). In articular cartilage, collagen has a high level of structural organization that provides a fibrous ultrastructure (Clark, 1985; Clarke, 1971; Mow and Ratcliffe, 1997). The basic biologic unit of collagen is tropocollagen, a structure composed of three procollagen polypeptide chains (alpha chains) coiled into left-handed helixes (Fig. 3-2A) that are further coiled about each other into a right-handed triple helix (Fig. 3-2B). These rod-like tropocollagen molecules, 1.4 nanometers (nm) in diameter and 300 nm long (Figs. 3-2C and D), polymerize into larger collagen fibrils (Bateman et al., 1996; Eyre, 1980). In articular cartilage, these fibrils have an average diameter of 25 to 40 nm (Fig. 3-2E, Box 3-2); however, this is highly variable.

Scanning electron microscopic studies, for instance, have described fibers with diameters ranging up to 200 nm (Clarke, 1971). Covalent cross-links form between these tropocollagen molecules, adding to the fibrils' high tensile strength (Bateman et al., 1996).

The collagen in articular cartilage is inhomogeneously distributed, giving the tissue a layered character (Lane and Weiss, 1975; Mow and Ratcliffe, 1997). Numerous investigations using light, transmission electron, and scanning electron microscopy have identified

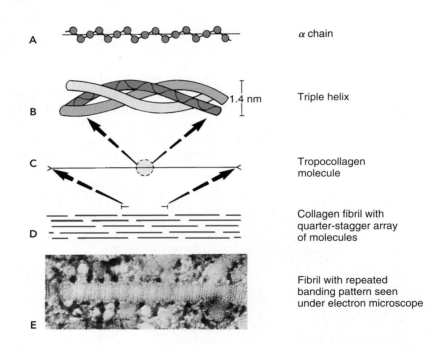

A — α chain

B — 1.4 nm — Triple helix

C — Tropocollagen molecule

D — Collagen fibril with quarter-stagger array of molecules

E — Fibril with repeated banding pattern seen under electron microscope

FIG. 3-2

Molecular features of collagen structure from the alpha chain (α) to the fibril. The flexible amino acid sequence in the alpha chain **(A)** allows these chains to wind tightly into a right-handed triple helix configuration **(B)**, thus forming the tropocollagen molecule **(C)**. This tight triple helical arrangement of the chains contributes to the high tensile strength of the collagen fibril. The parallel alignment of the individual tropocollagen molecules, in which each molecule overlaps the other by about one quarter of its length **(D)**, results in a repeating banded pattern of the collagen fibril seen by electron microscopy (×20,000) **(E)**. *Reprinted with permission from Donohue, J.M., Buss, D., Oegema, T.R., et al. (1983). The effects of indirect blunt trauma on adult canine articular cartilage. J Bone Joint Surg, 65A, 948.*

three separate structural zones. For example, Mow et al. (1974) proposed a zonal arrangement for the collagen network shown schematically in Figure 3-3A. In the superficial tangential zone, which represents 10% to 20% of the total thickness, sheets of fine, densely packed fibers are randomly woven in planes parallel to the articular surface (Clarke, 1971; Redler and Zimny, 1970; Weiss et al., 1968). In the middle zone (40%–60% of

the total thickness), there are greater distances between the randomly oriented and homogeneously dispersed fibers. Below this, in the deep zone (approximately 30% of the total thickness), the fibers come together, forming larger, radially oriented fiber bundles (Redler et al., 1975). These bundles then cross the tidemark, the interface between articular cartilage and the calcified cartilage beneath it, to enter the calcified cartilage, thus forming an interlocking "root" system anchoring the cartilage to the underlying bone (Bullough and Jagannath, 1983; Redler et al., 1975). This anisotropic fiber orientation is mirrored by the inhomogeneous zonal variations in the collagen content, which is highest at the surface and then remains relatively constant throughout the deeper zones (Lipshitz et al., 1975). This compositional layering appears to provide an important biomechanical function by enhancing superficial interstitial fluid support and frictional properties (Krishnan et al., 2003).

Cartilage is composed primarily of type II collagen. In addition, an array of different collagen (types V, VI, IX,

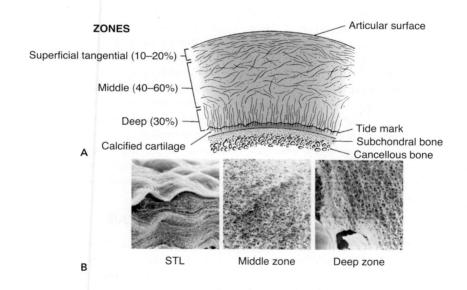

ZONES

Superficial tangential (10–20%)

Middle (40–60%)

Deep (30%)

Calcified cartilage

Articular surface

Tide mark
Subchondral bone
Cancellous bone

A

B   STL   Middle zone   Deep zone

FIG. 3-3

A. Schematic representation. *Reprinted with permission from Mow, V.C., Lai, W.M., Redler, I. (1974). Some surface characteristics of articular cartilages. A scanning electron microscopy study and a theoretical model for the dynamic interaction of synovial fluid and articular cartilage. J Biomech, 7, 449.* **B.** Photomicrographs (×3000; courtesy of Dr. T. Takei, Nagano, Japan) of the ultrastructural arrangement of the collagen network throughout the depth of articular cartilage. In the superficial tangential zone (STZ), collagen fibrils are tightly woven into sheets arranged parallel to the articular

surface. In the middle zone, randomly arrayed fibrils are less densely packed to accommodate the high concentration of proteoglycans and water. The collagen fibrils of the deep zone form larger radially oriented fiber bundles that cross the tidemark, enter the calcified zone, and anchor the tissue to the underlying bone. Note the correspondence between this collagen fiber architecture and the spatial arrangement of the chondrocytes shown in Figure 3-1. In photomicrographs **B**, the STZ is shown under compressive loading while the middle and deep zones are unloaded.

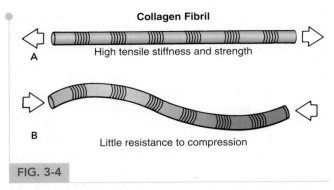

Collagen Fibril

A — High tensile stiffness and strength

B — Little resistance to compression

FIG. 3-4

Illustration of the mechanical properties of collagen fibrils: stiff and strong in tension **(A),** but weak and buckling easily with compression **(B).** *Adapted from Myers, E.R., Lai, W.M., Mow, V.C. (1984). A continuum theory and an experiment for the ion-induced swelling behavior cartilage. J Biomech Eng, 106(2), 151–158.*

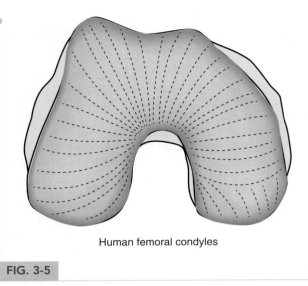

Human femoral condyles

FIG. 3-5

Diagrammatic representation of a split line pattern on the surface of human femoral condyles.

XI) can be found in quantitatively minor amounts within articular cartilage. Type II collagen is present primarily in articular cartilage, the nasal septum, and sternal cartilage, as well as in the inner regions of the intervertebral disc and meniscus. For reference, type I is the most abundant collagen in the human body and can be found in bone and soft tissues such as intervertebral discs (mainly in the annulus fibrosis), skin, meniscus, tendons, and ligaments. The most important mechanical properties of collagen fibers are their tensile stiffness and their strength (Fig. 3-4A). The properties of a single type I collagen fibril have been tested in tension and measured to be 860 MPa (Shen et al., 2008). Collagen fibril interactions with other fibrils as well as with other extracellular matrix components yield much lower effective tissue mechanical properties. For example, tendons are about 80% collagen (dry weight) and have a tensile stiffness of 10^3 MPa and a tensile strength of 50 MPa (Akizuki et al., 1986; Kempson, 1979; Kempson et al., 1976; Woo et al., 1987, 1997). Steel, by comparison, has a tensile stiffness of approximately 220×10^3 MPa. Although strong in tension, collagen fibrils offer little resistance to compression because their large slenderness ratio, the ratio of length to thickness, makes it easy for them to buckle under compressive loads (Fig. 3-4B).

Like bone, articular cartilage is anisotropic; its material properties differ with the direction of loading (Akizuki et al., 1986; Huang et al., 2005; Kempson, 1979; Mow and Ratcliffe, 1997; Roth and Mow, 1980; Wang et al., 2003; Woo et al., 1987). It is thought that this anisotropy is related to the varying collagen fiber arrangements within the planes parallel to the articular surface. It is also thought, however, that variations in collagen fiber cross-link density, as well as variations in collagen-PG interactions, also contribute to articular cartilage tensile anisotropy. In tension, this anisotropy is usually described with respect to the direction of the articular surface split lines. These split lines are elongated fissures produced by piercing the articular surface with a small round awl (Fig. 3-5; Hultkrantz, 1898). The origin of the pattern is related to the directional variation of the tensile stiffness and strength characteristics of articular cartilage described previously. To date, however, the exact reasons as to why articular cartilage exhibits such pronounced anisotropies in tension is not known, nor is the functional significance of this tensile anisotropy.

PROTEOGLYCAN

Many types of PGs are found in cartilage. Fundamentally, it is a large protein-polysaccharide molecule composed of a protein core to which one or more glycosaminoglycans (GAGs) are attached (Fosang and Hardingham, 1996; Muir, 1983; Ratcliffe and Mow, 1996). Even the smallest of these molecules, biglycan and decorin, are quite large (approximately 1×10^4 mw), but they comprise less than 10% of all PGs present in the tissue. Aggrecans are much larger ($1–4 \times 10^6$ mw), and they have the remarkable capability to attach to a hyaluronan molecule (HA: 5×10^5 mw) via a specific HA-binding region (HABR). This binding is stabilized by a link protein (LP) ($40–48 \times 10^3$ mw). Stabilization is crucial to the function of normal cartilage; without it, the components of the PG molecule would rapidly escape from the tissue (Hardingham and Muir, 1974; Hascall, 1977; Muir, 1983).

Two types of GAGs comprise aggrecan: chondroitin sulfate (CS) and keratan sulfate (KS). Each CS chain contains 25 to 30 disaccharide units, whereas the shorter KS chain contains 13 disaccharide units (Muir, 1983). Aggrecans (previously referred to as subunits in the American literature or as monomers in the UK and European literature) consist of an approximately 200-nanometer-long protein core to which approximately 150 GAG chains, and both O-linked and N-linked oligosaccharides, are covalently attached (Fosang and Hardingham, 1996; Muir, 1983). Furthermore, the distribution of GAGs along the protein core is heterogeneous; there is a region rich in KS and O-linked oligosaccharides and a region rich in CS (Fig. 3-6A). Figure 3-6A depicts the famous "bottle-brush" model for an aggrecan (Muir, 1983). Also shown in Figure 3-6A is the heterogeneity of the protein core that contains three globular regions: G_1, the HABR located at the N-terminus that contains a small amount of KS (Poole, 1986) and a few N-linked oligosaccharides, G_2, located between the HABR- and the KS-rich region (Hardingham et al., 1987), and G_3, the core protein C-terminus. A 1:1 stoichiometry exists between the LP and the G1 binding region in cartilage. More recently, the other two globular regions have been extensively studied (Fosang and Hardingham, 1996), but their functional significance has not yet been elucidated. Figure 3-6B is the accepted molecular conformation of a PG aggregate; Rosenberg et al. (1975) were the first to obtain an electron micrograph of this molecule (Fig. 3-6C). Atomic force microscopy (AFM) techniques have been used to image (Seog et al., 2005) as well as to measure the net intermolecular interaction forces (Ng et al., 2003) of proteoglycans.

In native cartilage, most aggrecans are associated with HA to form the large PG aggregates (Fig. 3-6C). These aggregates may have up to several hundred aggrecans noncovalently attached to a central HA core via their HABR, and each site is stabilized by an LP. The filamentous HA core molecule is a nonsulfated disaccharide chain that may be as long as 4 mm in length. PG biochemists have dubbed the HA an "honorary" PG because it is so intimately involved in the structure of the PG aggregate in articular cartilage. The stability afforded by the PG aggregates has a major functional significance. It is accepted now that PG aggregation promotes immobilization of the PGs within the fine collagen meshwork, adding structural stability and rigidity to the ECM (Mow et al., 1989b; Muir, 1983; Ratcliffe et al., 1986). Furthermore, two additional forms of dermatan sulfate PG have been identified in the ECM of articular cartilage (Rosenberg et al., 1985). In tendons, dermatan sulfate PGs have been shown to bind noncovalently to the surfaces of collagen fibrils (Scott and Orford, 1981); however, the role of dermatan sulfate in articular cartilage is unknown, biologically and functionally.

Although aggrecans generally have the basic structure as described previously, they are not structurally identical (Fosang and Hardingham, 1996). Aggrecans vary in length, molecular weight, and composition in various ways; in other words, they are polydisperse. Studies have demonstrated two distinct populations of aggrecans (Buckwalter et al., 1985; Heinegard et al., 1985). The first population is present throughout life and is rich in CS; the second contains PGs rich in KS and is present only in adult cartilage. As articular cartilage matures, other age-related changes in PG composition and structure occur. With cartilage maturation, the water content (Armstrong and Mow, 1982; Bollet and Nance, 1965; Linn and Sokoloff, 1965; Maroudas, 1979; Venn, 1978) and the carbohydrate/protein ratio progressively decrease (Garg and Swann, 1981; Roughley and White, 1980). This decrease is mirrored by a decrease in the CS content. Conversely, KS, which is present only in small amounts at birth, increases throughout development and aging. Thus, the CS/KS ratio, which is approximately 10:1 at birth, is only approximately 2:1 in adult cartilage (Roughley and White, 1980; Sweet et al., 1979; Thonar et al., 1986). Furthermore, sulfation of the CS molecules, which can occur at either the 6 or the 4 position, also undergoes age-related changes. In utero, chondroitin-6-sulfate and chondroitin-4-sulfate are present in equal molar amounts; however, by maturity, the chondroitin-6-sulfate:chondroitin-4-sulfate ratio has increased to approximately 25:1 (Roughley et al., 1981). Other studies have also documented an age-related decrease in the hydrodynamic size of the aggrecan. Many of these early changes seen in articular cartilage may reflect cartilage maturation, possibly as a result of increased functional demand with increased weight-bearing. However, the functional significance of these changes, as well as those occurring later in life, is undetermined.

WATER

Water, the most abundant component of articular cartilage, is most concentrated near the articular surface (~80%) and decreases in a near-linear fashion with increasing depth to a concentration of approximately 65% in the deep zone (Lipshitz et al., 1976; Maroudas, 1979). This fluid contains many free mobile cations (e.g., Na^+, K^+, and Ca^{2+}) that greatly influence the mechanical and physicochemical behaviors of cartilage (Gu et al.,

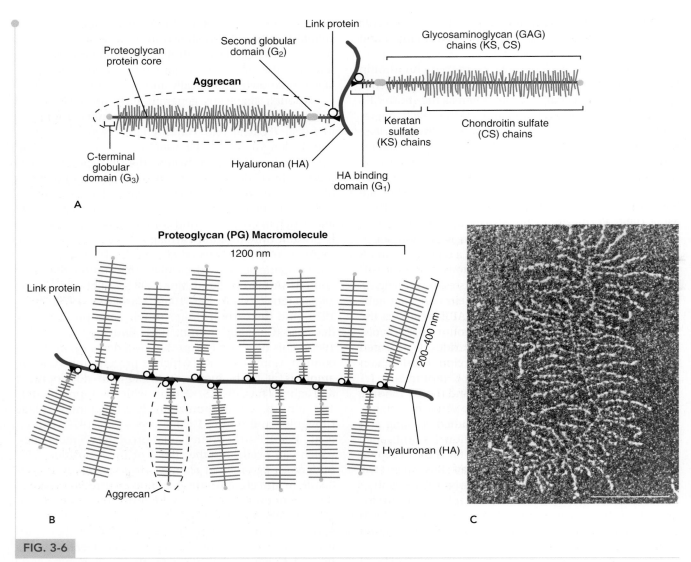

FIG. 3-6

A. Schematic depiction of aggrecan, which is composed of keratan sulfate and chondroitin sulfate chains bound covalently to a protein core molecule. The proteoglycan protein core has three globular regions as well as keratan sulfate-rich and chondroitin sulfate-rich regions. **B.** Schematic representation of a proteoglycan macromolecule. In the matrix, aggrecan noncovalently binds to HA to form a macromolecule with a molecular weight of approximately 200×10^6. Link protein stabilizes this interaction between the binding region of the aggrecan and the HA core molecule. **C.** Dark field electron micrograph of a proteoglycan aggregate from bovine humeral articular cartilage (×120,000). *Horizontal line at lower right represents 0.5 mm. Reprinted with permission from Rosenberg, L., Hellmann, W., Kleinschmidt, A.K. (1975). Electron microscopic studies of proteoglycan aggregates from bovine articular cartilage. J Biol Chem, 250, 1877.*

1998; Lai et al., 1991; Linn and Sokoloff, 1965; Maroudas, 1979). The fluid component of articular cartilage is also essential to the health of this avascular tissue because it permits gas, nutrient, and waste product movement back and forth between chondrocytes and the surrounding nutrient-rich synovial fluid (Bollet and Nance, 1965; Linn and Sokoloff, 1965; Mankin and Thrasher, 1975; Maroudas, 1975, 1979).

A small percentage of the water in cartilage resides intracellularly, and approximately 30% is strongly associated with the collagen fibrils (Maroudas et al., 1991; Torzilli et al., 1982). The interaction between collagen,

PG, and water, via Donnan osmotic pressure, is believed to have an important function in regulating the structural organization of the ECM and its swelling properties (Donnan, 1924; Maroudas, 1968, 1975). Most of the water thus occupies the interfibrillar space of the ECM and is free to move when a load or pressure gradient or other electrochemical motive forces are applied to the tissue (Gu et al., 1998; Maroudas, 1979). When loaded by a compressive force, approximately 70% of the water may be moved. This interstitial fluid movement is important in controlling cartilage mechanical behavior and joint lubrication (Ateshian et al., 1997, 1998; Hlavacek, 1995; Hou et al., 1992; Mow and Ateshian, 1997; Mow et al., 1980).

STRUCTURAL AND PHYSICAL INTERACTION AMONG CARTILAGE COMPONENTS

The chemical structure and physical interactions of the PG aggregates influence the properties of the ECM (Guterl et al., 2010; Ratcliffe and Mow, 1996). The closely spaced (5–15 angstroms) sulfate and carboxyl charge groups on the CS and KS chains dissociate in solution at physiologic pH (Fig. 3-7), leaving a high concentration of fixed negative charges that create strong intramolecular and intermolecular charge-charge repulsive forces (Seog et al., 2005); the colligative sum of these forces (when the tissue is immersed in a physiologic saline solution) is equivalent to the Donnan osmotic pressure (Buschmann and Grodzinsky, 1995; Donnan, 1924; Gu et al., 1998; Lai et al., 1991). Structurally, these charge-charge repulsive forces tend to extend and stiffen the PG macromolecules into the interfibrillar space formed by the surrounding collagen network. To appreciate the magnitude of this force, according to Stephen Hawking (1988), this electrical repulsion is one million, million, million, million, million, million, million times (42 zeros) greater than gravitational forces.

In nature, a charged body cannot persist long without discharging or attracting counter-ions to maintain electroneutrality. Thus, the charged sulfate and carboxyl groups fixed along the PGs in articular cartilage must attract various counter-ions and co-ions (mainly Na^+, Ca^{2+}, and Cl^{1-}) into the tissue to maintain electroneutrality. The total concentration of these counter-ions and co-ions is given by the well-known Donnan equilibrium ion distribution law (Donnan, 1924). Inside the tissue, the mobile counter-ions and co-ions form a cloud surrounding the fixed sulfate and carboxyl charges, thus shielding these charges from each other. This charge shielding acts to diminish the very large electrical repulsive forces that otherwise would exist. The net result is a swelling pressure given by the Donnan osmotic pressure law (Buschmann and Grodzinsky, 1995; Donnan, 1924; Gu et al., 1998; Lai

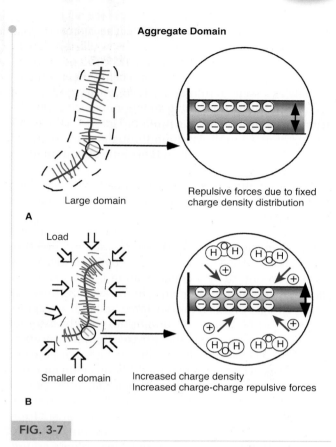

Aggregate Domain

Large domain

Repulsive forces due to fixed charge density distribution

A

Load

Smaller domain

Increased charge density
Increased charge-charge repulsive forces

B

FIG. 3-7

A. Schematic representation of a proteoglycan aggregate solution domain **(left)** and the repelling forces associated with the fixed negative charge groups on the GAGs of aggrecan **(right)**. These repulsive forces cause the aggregate to assume a stiffly extended conformation, occupying a large solution domain. **B.** Applied compressive stress decreases the aggregate solution domain **(left)**, which in turn increases the charge density and thus the intermolecular charge repulsive forces **(right)**.

et al., 1991; Schubert and Hamerman, 1968). The Donnan osmotic pressure theory has been extensively used to calculate the swelling pressures of articular cartilage and the intervertebral disc (Maroudas, 1979; Urban and McMullin, 1985). By Starling's law, this swelling pressure is, in turn, resisted and balanced by tension developed in the collagen network, confining the PGs to only 20% of their free solution domain (Maroudas, 1976; Mow and Ratcliffe, 1997; Setton et al., 1995). Consequently, this swelling pressure subjects the collagen network to a "prestress" of significant magnitude even in the absence of external loads (Setton et al., 1995, 1998).

Cartilage PGs are inhomogeneously distributed throughout the matrix, with their concentration generally being highest in the middle zone and lowest in the superficial and deep zones (Lipshitz et al., 1976;

Maroudas, 1968, 1979; Venn, 1978). The biomechanical consequence of this inhomogeneous swelling behavior of cartilage (caused by the varying PG content throughout the depth of the tissue) has recently been quantitatively assessed (Setton et al., 1998). Also, results from recent models incorporating an inhomogeneous PG distribution show that it has a profound effect on the interstitial counter-ion distribution throughout the depth of the tissue (Sun et al., 1998) and osmotic environment in cartilage (Oswald et al., 2008).

When a compressive stress is applied to the cartilage surface, there is an instantaneous deformation caused primarily by a change in the PG molecular domain, Figure 3-7B. This external stress causes the internal pressure in the matrix to exceed the swelling pressure and thus liquid will begin to flow out of the tissue. As the fluid flows out, the PG concentration increases, which in turn increases the Donnan osmotic swelling pressure or the charge-charge repulsive force and bulk compressive stress until they are in equilibrium with the external stress. In this manner, the physicochemical properties of the PG gel trapped within the collagen network enable it to resist compression. This mechanism complements the role played by collagen that, as previously described, is strong in tension but weak in compression. The ability of PGs to resist compression thus arises from two sources: (1) the Donnan osmotic swelling pressure associated with the tightly packed fixed anionic groups on the GAGS and (2) the bulk compressive stiffness of the collagen-PG solid matrix. Experimentally, the Donnan osmotic pressure ranges from 0.05 to 0.35 MPa (Maroudas, 1979), whereas the elastic modulus of the collagen-PG solid matrix ranges from 0.5 to 1.5 MPa (Armstrong and Mow, 1982; Athanasiou et al., 1991; Mow and Ratcliffe, 1997).

It is now apparent that collagen and PGs also interact and that these interactions are of great functional importance. A small portion of the PGs have been shown to be closely associated with collagen and may serve as a bonding agent between the collagen fibrils, spanning distances that are too great for collagen cross-links to develop (Bateman et al., 1996; Mow and Ratcliffe, 1997; Muir, 1983).

PGs are also thought to play an important role in maintaining the ordered structure and mechanical properties of the collagen fibrils (Muir, 1983; Scott and Orford, 1981). Recent investigations show that in concentrated solutions, PGs interact with each other to form networks of significant strength (Mow et al., 1989b; Zhu et al., 1991, 1996). Moreover, the density and strength of the interaction sites forming the network were shown to depend on the presence of LP between aggrecans and aggregates, as well as collagen. Evidence suggests that there are fewer aggregates, and more biglycans and decorins than aggrecans, in the superficial zone of articular cartilage. Thus, there must be a difference in the interaction between these PGs and the collagen fibrils from the superficial zone than from those of the deeper zones (Poole et al., 1986). Indeed, the interaction between PG and collagen not only plays a direct role in the organization of the ECM but also contributes directly to the mechanical properties of the tissue (Kempson et al., 1976; Schmidt et al., 1990; Zhu et al., 1993).

The specific characteristics of the physical, chemical, and mechanical interactions between collagen and PG have not yet been fully determined. Nevertheless, as discussed previously, we know that these structural macromolecules interact to form a porous-permeable, fiber-reinforced composite matrix possessing all the essential mechanical characteristics of a solid that is swollen with water and ions and that is able to resist the high stresses and strains of joint articulation (Andriacchi et al., 1997; Hodge et al., 1986; Mow and Ateshian, 1997; Paul, 1976). It has been demonstrated that these collagen-PG interactions involve an aggrecan, an HA filament, type II collagen, other minor collagen types, an unknown bonding agent, and possibly smaller cartilage components such as collagen type IX, recently identified glycoproteins, and/or polymeric HA (Poole et al., 1986). A schematic diagram depicting the structural arrangement within a small volume of articular cartilage is shown in Figure 3-8.

When articular cartilage is subjected to external loads, the collagen-PG solid matrix and interstitial fluid function together in a unique way to protect against high levels of stress and strain developing in the ECM. Furthermore, changes to the biochemical composition and structural organization of the ECM, such as during osteoarthritis (OA), are paralleled by changes to the biomechanical properties of cartilage. In the following section, the behavior of articular cartilage under loading and the mechanisms of cartilage fluid flow will be discussed in detail.

Biomechanical Behavior of Articular Cartilage

The biomechanical behavior of articular cartilage can best be understood when the tissue is viewed as a multiphasic medium. In the present context, articular cartilage will be treated as biphasic material consisting of two intrinsically incompressible, immiscible, and distinct phases (Bachrach et al., 1998; Mow et al., 1980): an interstitial fluid phase and a porous-permeable solid phase (i.e., the ECM). For explicit analysis of the

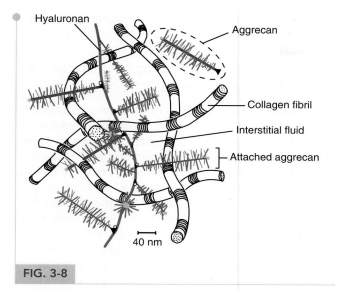

Hyaluronan

Aggrecan

Collagen fibril

Interstitial fluid

Attached aggrecan

40 nm

FIG. 3-8

Schematic representation of the molecular organization of cartilage. The structural components of cartilage, collagen, and proteoglycans interact to form a porous composite fiber-reinforced organic solid matrix that is swollen with water. Aggrecans bind covalently to HA to form large proteoglycan macromolecules.

contribution of the PG charges and ions, one would have to consider three distinct phases: a fluid phase, an ion phase, and a charged solid phase (Gu et al., 1998; Lai et al., 1991). For understanding how the water contributes to its mechanical properties, in the present context, articular cartilage may be considered as a fluid-filled porous-permeable (uncharged) biphasic medium, with each constituent playing a role in the functional behavior of cartilage.

During joint articulation, forces at the joint surface may vary from almost zero to more than ten times body weight (Andriacchi et al., 1997; Paul, 1976). The contact areas also vary in a complex manner, and typically they are only of the order of several square centimeters (Ahmed and Burke, 1983; Ateshian et al., 1994). It is estimated that the peak contact stress may reach 20 MPa in the hip while rising from a chair and 10 MPa during stair climbing (Hodge et al., 1986; Newberry et al., 1997). Thus, articular cartilage, under physiologic loading conditions, is a highly stressed material. To understand how this tissue responds under these high physiologic loading conditions, its intrinsic mechanical properties in compression, tension, and shear must be determined. From these properties, one can understand the load-carrying mechanisms within the ECM. Accordingly, the following subsections will characterize the tissue behavior under these loading modalities.

NATURE OF ARTICULAR CARTILAGE VISCOELASTICITY

If a material is subjected to the action of a constant (time-independent) load or a constant deformation and its response varies with time, then the mechanical behavior of the material is said to be viscoelastic. In general, the response of such a material can be theoretically modeled as a combination of the response of a viscous fluid (dash-pot) and an elastic solid (spring), hence viscoelastic.

The two fundamental responses of a viscoelastic material are creep and stress relaxation. Creep occurs when a viscoelastic solid is subjected to the action of a constant load. Typically, a viscoelastic solid responds with a rapid initial deformation followed by a slow (time-dependent), progressively increasing deformation known as creep until an equilibrium state is reached. Stress relaxation occurs when a viscoelastic solid is subjected to the action of a constant deformation. Typically, a viscoelastic solid responds with a rapid, high initial stress followed by a slow (time-dependent), progressively decreasing stress required to maintain the deformation; this phenomenon is known as stress relaxation.

Creep and stress relaxation phenomena may be caused by different mechanisms. For single-phase solid polymeric materials, these phenomena are the result of internal friction caused by the motion of the long polymeric chains sliding over each other within the stressed material (Fung, 1981). The viscoelastic behavior of tendons and ligaments is primarily caused by this mechanism (Woo et al., 1987, 1997). For bone, the long-term viscoelastic behavior is thought to be caused by a relative slip of lamellae within the osteons along with the flow of the interstitial fluid (Lakes and Saha, 1979). For articular cartilage, the compressive viscoelastic behavior is primarily caused by the flow of the interstitial fluid and the frictional drag associated with this flow (Ateshian et al., 1997; Mow et al., 1980, 1984). In shear, as in single-phase viscoelastic polymers, it is primarily caused by the motion of long polymer chains such as collagen and PGs (Zhu et al., 1993, 1996). The component of articular cartilage viscoelasticity caused by interstitial fluid flow is known as the biphasic viscoelastic behavior (Mow et al., 1980), and the component of viscoelasticity caused by macromolecular motion is known as the flow-independent (Hayes and Bodine, 1978) or the intrinsic viscoelastic behavior of the collagen-PG solid matrix.

Although the deformational behavior has been described in terms of a linear elastic solid (Hirsch, 1944) or viscoelastic solid (Hayes and Mockros, 1971), these models fail to recognize the role of water in the viscoelastic behavior of and the significant contribution that fluid pressurization plays in joint load support and cartilage lubrication (Ateshian et al., 1998; Elmore et al., 1963, Mow and Ratcliffe, 1997; Sokoloff, 1963). Recently, experimental measurements have determined that interstitial fluid pressurization supports more than 90% of the applied load to the cartilage surface (Soltz and Ateshian, 1998) immediately following loading. This effect can persist for more than one thousand seconds and thus shields the ECM and chondrocytes from the crushing deformations of the high stresses (20 MPa) resulting from joint loading. The shielding effects of interstitial fluid pressure are also observed for physiologic loading conditions (Ateshian, 2009; Caligaris and Ateshian, 2008; Park et al., 2003, 2004).

CONFINED COMPRESSION EXPLANT LOADING CONFIGURATION

The loading of cartilage in vivo is extremely complex. To achieve a better understanding of the deformational behavior of the tissue under load, an explant loading configuration known as confined compression (Mow et al., 1980) has been adopted by researchers. In this configuration, a cylindrical cartilage specimen is fitted snugly into a cylindrical, smooth-walled (ideally frictionless) confining ring that prohibits motion and fluid loss in the radial direction. Under an axial loading condition via a rigid porous-permeable loading platen (Fig. 3-9A), fluid will flow from the tissue into the porous-permeable platen, and, as this occurs, the cartilage sample will compress in creep. At any time the amount of compression equals the volume of fluid loss because both the water and the ECM are each intrinsically incompressible (Bachrach et al., 1998). The advantage of the confined compression test is that it creates a uniaxial, one-dimensional flow and deformational field within the tissue, which does not depend on tissue anisotropy or properties in the radial direction. This greatly simplifies the mathematics needed to solve the problem.

It should be emphasized that the stress-strain, pressure, fluid, and ion flow fields generated within the tissue during loading can only be calculated; however, these calculations are of idealized models and testing conditions. There are many confounding factors, such as the time-dependent nature and magnitude of loading and alterations in the natural state of pre-stress (acting within the tissue) (Mow et al., 1999), that arise from disruption of the collagen network during specimen harvesting. Despite limitations in determining the natural physiologic states of stress and strain within the tissue in vivo, several researchers have made gains toward an understanding of potential mechanosignal transduction mechanisms in cartilage through the use of explant loading studies (Bachrach et al., 1995; Buschmann et al., 1992; Kim et al., 1994; Valhmu et al., 1998) based on the

biphasic constitutive law for soft hydrated tissues (Mow et al., 1980).

BIPHASIC CREEP RESPONSE OF ARTICULAR CARTILAGE IN COMPRESSION

The biphasic creep response of articular cartilage in a one-dimensional confined compression experiment is depicted in Figure 3-9. In this case, a constant compressive stress (σ_0) is applied to the tissue at time t_0 (point A in Fig. 3-9B) and the tissue is allowed to creep to its final equilibrium strain ($\epsilon\infty$). For articular cartilage, as illustrated in the top diagrams, creep is caused by the exudation of the interstitial fluid. Exudation is most rapid initially, as evidenced by the early rapid rate of increased deformation, and it diminishes gradually until flow cessation occurs. During creep, the load applied at the surface is balanced by the compressive stress developed within the collagen-PG solid matrix and the frictional drag generated by the flow of the interstitial fluid during exudation. Creep ceases when the compressive stress developed within the solid matrix is sufficient to balance the applied stress alone; at this point no fluid flows and the equilibrium strain $\epsilon\infty$ is reached.

Typically, for relatively thick human and bovine articular cartilages, 2 to 4 mm, it takes four to sixteen hours to reach creep equilibrium. For rabbit cartilage, which is generally less than 1.0 mm thick, it takes approximately one hour to reach creep equilibrium. Theoretically, it can be shown that the time it takes to reach creep equilibrium varies inversely with the square of the thickness of the tissue (Mow et al., 1980). Under relatively high loading conditions, >1.0 MPa, 50% of the total fluid content may be squeezed from the tissue (Edwards, 1967). Furthermore, in vitro studies demonstrate that if the tissue is immersed in physiologic saline, this exuded fluid is fully recoverable when the load is removed (Elmore et al., 1963; Sokoloff, 1963).

Because the rate of creep is governed by the rate of fluid exudation, it can be used to determine the permeability coefficient of the tissue (Mow et al., 1980, 1989a). This is known as the indirect measurement for tissue permeability (k). Average values of normal human, bovine, and canine patellar groove articular cartilage permeability k obtained in this manner are 2.17×10^{-15} m^4/N·s, 1.42×10^{-15} m^4/N·s, and 0.9342×10^{-15} m^4/N·s, respectively (Athanasiou et al., 1991). At equilibrium no fluid flow occurs, and thus the equilibrium deformation can be used to measure the intrinsic compressive modulus (H_A) of the collagen-PG solid matrix (Armstrong and Mow, 1982; Mow et al., 1980). Average values of normal human, bovine, and canine patellar groove articular cartilage compressive modulus H_A are 0.53, 0.47, and 0.55 megapascal

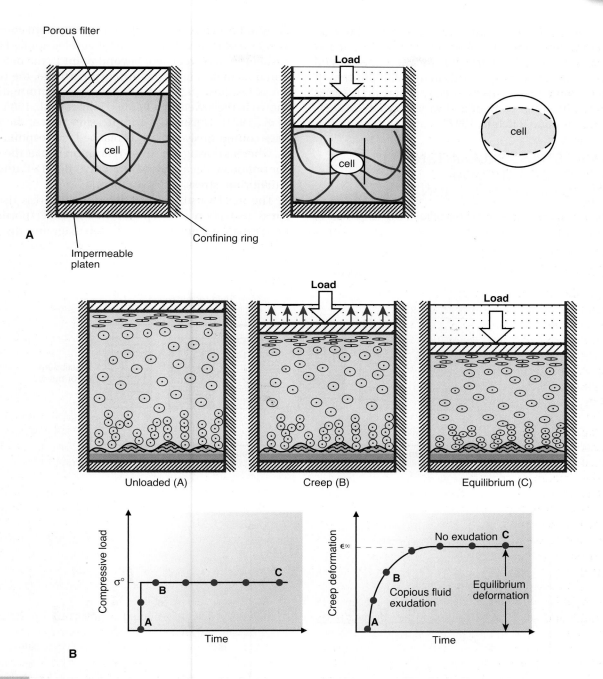

FIG. 3-9

A. A schematic of the confined compression loading configuration. A cylindrical tissue specimen is positioned tightly into an impermeable confining ring that does not permit deformation (or fluid flow) in the radial direction. Under loading, fluid exudation occurs through the porous platen in the vertical direction.
B. A constant stress σ^o applied to a sample of articular cartilage **(bottom left)** and creep response of the sample under the constant applied stress **(bottom right).** The drawings of a block of tissue above the curves illustrate that creep is accompanied by copious exudation of fluid from the sample and that the rate of exudation decreases over time from points A to B to C. At equilibrium (ε_∞), fluid flow ceases and the load is borne entirely by the solid matrix (point C). *Adapted from Mow, V.C., Kuei, S.C., Lai, W.M., et al. (1980). Biphasic creep and stress relaxation of articular cartilage in compression: Theory and experiments. J Biomech Eng, 102, 73–84.*

(MPa; note: 1.0 MPa = 145 lb/in²), respectively. Because these coefficients are a measure of the intrinsic material properties of the solid matrix, it is therefore meaningful to determine how they vary with matrix composition. It was determined that k varies directly, whereas H_A varies inversely with water content and varies directly with PG content (Mow and Ratcliffe, 1997).

BIPHASIC STRESS-RELAXATION RESPONSE OF ARTICULAR CARTILAGE IN COMPRESSION

The biphasic viscoelastic stress-relaxation response of articular cartilage in a 1D compression experiment is depicted in Figure 3-10. In this case, a constant compression rate (line 0-A-B of lower left figure) is applied to the tissue until u_0 is reached; beyond point B, the deformation u_0 is maintained. For articular cartilage, the typical stress response caused by this imposed deformation is shown in the lower right figure (Holmes et al., 1985; Mow et al., 1984). During the compression phase, the stress rises continuously until σ_0 is reached, corresponding to u_0, whereas during the stress-relaxation phase, the stress continuously decays along the curve B-C-D-E until the equilibrium stress (σ_∞) is reached.

The mechanisms responsible for the stress rise and stress relaxation are depicted in the lower portion of Figure 3-10. As illustrated in the top diagrams, the stress

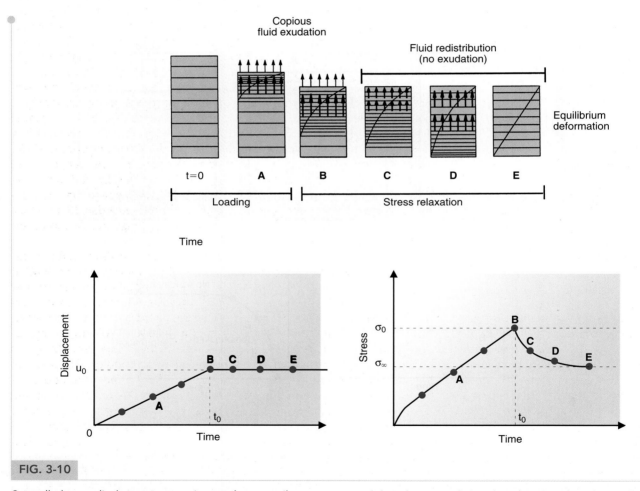

FIG. 3-10

Controlled ramp displacement curve imposed on a cartilage specimen commencing at t_0 **(bottom left)** and the stress-response curve of the cartilage in this uniaxial confined-compression experiment **(bottom right)**. The sample is compressed to point B and maintained over time (points B to E). The history of the stress and response shows a characteristic stress that rises during the compressive phase (points t_0 to B) and then decreases during the relaxation phase (points B to D) until an equilibrium is reached (point E). Above these two curves, schematics illustrate interstitial fluid flow (represented by arrows) and solid matrix deformation during this compressive process. Fluid exudation gives rise to the peak stress (point B), and fluid redistribution gives rise to the stress-relaxation phenomena.

rise in the compression phase is associated with fluid exudation, whereas stress relaxation is associated with fluid redistribution within the porous solid matrix. During the compressive phase, the high stress is generated by forced exudation of the interstitial fluid and the compaction of the solid matrix near the surface. Stress relaxation is in turn caused by the relief or rebound of the high compaction region near the surface of the solid matrix. This stress-relaxation process will cease when the compressive stress developed within the solid matrix reaches the stress generated by the intrinsic compressive modulus of the solid matrix corresponding to u_0 (Holmes et al., 1985; Mow et al., 1980, 1984). Analysis of this stress-relaxation process leads to the conclusion that under physiologic loading conditions, excessive stress levels are difficult to maintain because stress relaxation will rapidly attenuate the stress developed within the tissue; this must necessarily lead to the rapid spreading of the contact area in the joint during articulation (Ateshian et al., 1995, 1998; Mow and Ateshian, 1997).

Much focus has been on the inhomogeneity of H_A with cartilage depth (Schinagl et al., 1996, 1997; Wang et al., 2002a). Based on these data, from an analysis of the stress-relaxation experiment it was found that an inhomogeneous tissue would relax at a faster rate than would the uniform tissue (Wang and Mow, 1998). Moreover, the stress, strain, pressure, and fluid flow fields within the tissue were significantly altered as well. Thus it seems that the variations in biochemical and structural composition in the layers of cartilage provide another challenge to understanding the environment of chondrocytes in situ (Mow et al., 1999).

PERMEABILITY OF ARTICULAR CARTILAGE

Fluid-filled porous materials may or may not be permeable. The ratio of fluid volume (V^f) to the total volume (V^T) of the porous material is known as the porosity ($\beta = V^f/V^T$); thus, porosity is a geometric concept. Articular cartilage is therefore a material of high porosity (approximately 80%). If the pores are interconnected, the porous material is permeable. Permeability is a measure of the ease with which fluid can flow through a porous material, and it is inversely proportional to the frictional drag exerted by the fluid flowing through the porous-permeable material. Thus, permeability is a physical concept; it is a measure of the resistive force that is required to cause the fluid to flow at a given speed through the porous-permeable material. This frictional resistive force is generated by the interaction of the interstitial fluid and the pore walls of the porous-permeable material. The permeability coefficient k is related to the frictional drag coefficient K by the relationship $k = \beta^2/K$

(Lai and Mow, 1980). Articular cartilage has a very low permeability and thus high frictional resistive forces are generated when fluid is caused to flow through the porous solid matrix.

In the previous sections on cartilage viscoelasticity, we discussed the process of fluid flow through articular cartilage induced by solid matrix compression and how this process influences the viscoelastic behavior of the tissue. This process also provides an indirect method to determine the permeability of the tissue. In this section, we discuss the experimental method used to directly measure the permeability coefficient. Such an experiment is depicted in Figure 3-11A. Here, a specimen of the tissue is held fixed in a chamber subjected to the action of a pressure gradient; the imposed upstream pressure P_1 is greater than the downstream pressure P_2. The thickness of the specimen is denoted by h and the cross-sectional area of permeation is defined by A. Darcy's law, used to determine the permeability k from this simple experimental setup, yields $k = Qh/A(P_1 - P_2)$, where Q is the volumetric discharge per unit time through the specimen whose area of permeation is A (Mow and Ratcliffe, 1997). Using low pressures, approximately 0.1 MPa, this method was first used to determine the permeability of articular cartilage (Edwards, 1967; Maroudas, 1975). The value of k obtained in this manner ranged from 1.1×10^{-15} m^4/N·s to 7.6×10^{-15} m^4/N·s. In addition, using a uniform straight tube model, the average "pore diameter" has been estimated at 6 nm (Maroudas, 1979). Thus, the "pores" within articular cartilage are of molecular size.

The permeability of articular cartilage under compressive strain and at high physiologic pressures (3 MPa) was first obtained by Mansour and Mow (1976) and later analyzed by Lai and Mow (1980). The high pressure and compressive strain conditions examined in these studies more closely resemble those conditions found in diarthrodial joint loading. In these experiments, k was measured as a function of two variables: the pressure gradient across the specimen and the axial compressive strain applied to the sample. The results from these experiments are shown in Figure 3-11B. Permeability decreased exponentially as a function of both increasing compressive strain and increasing applied fluid pressure. It was later shown, however, that the dependence of k on the applied fluid pressure derives from compaction of the solid matrix that, in turn, results from the frictional drag caused by the permeating fluid (Lai and Mow, 1980). From the point of view of pore structure, compaction of the solid matrix decreases the porosity and hence the average "pore diameter" within the solid matrix; thus, solid matrix compaction increases frictional resistance (Mow et al., 1984).

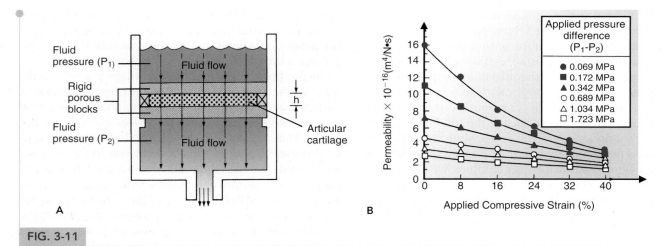

FIG. 3-11

A. Experimental configuration used in measuring the permeability of articular cartilage, involving the application of a pressure gradient $(P_1 - P_2)/h$ across a sample of the tissue (h = tissue thickness). Because the fluid pressure (P_1) above the sample is greater than that beneath it (P_2), fluid will flow through the tissue. The permeability coefficient k in this experiment is given by the expression $Qh/A(P_1 - P_2)$, where Q is the volumetric discharge per unit time and A is the area of permeation. *Adapted from Torzilli, P.A., Mow, V.C. (1976). On the fundamental fluid transport mechanisms through normal and pathologic cartilage* *during function. I. The formulation. J Biomech, 9(8), 541–552.* **B.** Experimental curves for articular cartilage permeability show its strong dependence on compressive strain and applied pressure. Measurements were taken at applied pressure differential $(P_1 - P_2)$ and applied strains. The permeability decreased in an exponential manner as a function of both increasing applied compressive strain and increasing applied pressure. *Adapted from Lai, W.M., Mow, V.C. (1980). Drag-induced compression of articular cartilage during a permeation experiment. Biorheology 17, 111.*

The nonlinear permeability of articular cartilage demonstrated in Figure 3-11B suggests that the tissue has a mechanical feedback system that may serve important purposes under physiologic conditions. When subjected to high loads through the mechanism of increased frictional drag against interstitial fluid flow, the tissue will appear stiffer and it will be more difficult to cause fluid exudation. Moreover, this mechanism also is important in joint lubrication.

BEHAVIOR OF ARTICULAR CARTILAGE UNDER UNIAXIAL TENSION

The mechanical behavior of articular cartilage in tension is highly complex. In tension, the tissue is strongly anisotropic (being stiffer and stronger for specimens harvested in the direction parallel to the split line pattern than those harvested perpendicular to the split line pattern) and strongly inhomogeneous (for mature animals, being stiffer and stronger for specimens harvested from the superficial regions than those harvested deeper in the tissue) (Kempson, 1979; Roth and Mow, 1980). Interestingly, articular cartilage from immature bovine knee joints does not exhibit these layered inhomogeneous variations; however, the superficial zones of both mature and immature bovine cartilage appear to have the same tensile stiffness (Roth and Mow, 1980). These anisotropic and inhomogeneous characteristics in mature joints are believed to be caused by the varying collagen and PG structural organization of the joint surface and the layering structural arrangements found within the tissue. Thus, the collagen-rich superficial zone appears to provide the joint cartilage with a tough wear-resistant protective skin (Setton et al., 1993) (Fig. 3-3A).

Articular cartilage also exhibits viscoelastic behavior in tension (Woo et al., 1987). This viscoelastic behavior is attributable to both the internal friction associated with polymeric motion and the flow of the interstitial fluid. To examine the intrinsic mechanical response of the collagen-PG solid matrix in tension, it is necessary to negate the biphasic fluid flow effects. To do this, one must perform slow, low–strain-rate experiments (Akizuki et al., 1986; Roth and Mow, 1980; Woo et al., 1987) or perform an incremental strain experiment in which stress relaxation is allowed to progress toward equilibration at each increment of strain (Akizuki et al., 1986). Typically, in a low–strain-rate (or near-equilibrium tensile) experiment, a displacement rate of 0.5 cm/minute is used and the specimens usually are pulled to failure. Unfortunately, using these procedures to negate the effect of interstitial

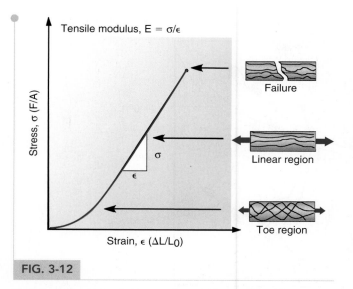

FIG. 3-12

Typical tensile stress-strain curve for articular cartilage. The drawings on the right of the curve show the configuration of the collagen fibrils at various stages loading. In the toe region, collagen fibril pull-out occurs as the fibrils align themselves in the direction of the tensile load. In the linear region, the aligned collagen fibers are stretched until failure occurs.

fluid flow also negates the manifestation of the intrinsic viscoelastic behavior of the solid matrix. Thus, only the equilibrium intrinsic mechanical properties of the solid matrix may be determined from these tensile tests. The intrinsic viscoelastic properties of the solid matrix must be determined from a pure shear study.

The "equilibrium" stress-strain curve for a specimen of articular cartilage tested under a constant low–strain-rate condition is shown in Figure 3-12. Like other fibrous biologic tissues (tendons and ligaments), articular cartilage tends to stiffen with increasing strain when the strain becomes large. Thus, over the entire range of strain (up to 60%) in tension, articular cartilage cannot be described by a single Young's modulus. Rather, a tangent modulus, defined by the tangent to the stress-strain curve, must be used to describe the tensile stiffness of the tissue. This fundamental result has given rise to the wide range of Young's modulus, 3 to 100 MPa, reported for articular cartilage in tension (Akizuki et al., 1986; Kempson, 1979; Roth and Mow, 1980; Woo et al., 1987). At physiologic strain levels, however, less than 15% (Armstrong et al., 1979) of the linear Young's modulus of articular cartilage ranges between 5 and 10 MPa (Akizuki et al., 1986).

Morphologically, the cause for the shape of the tensile stress-strain curve for large strains is depicted in the diagrams on the right of Figure 3-12. The initial toe region is caused by collagen fiber pull-out and realignment during the initial portion of the tensile experiment, and the final linear region is caused by the stretching of the straightened-aligned collagen fibers. Failure occurs when all the collagen fibers contained within the specimen are ruptured. Figure 3-13A depicts an unstretched articular cartilage specimen, while Figure 3-13B depicts a stretched specimen. Figures 3-14A and B show scanning electron micrographs of cartilage blocks under 0% and 30% stretch (right) and the corresponding histograms of collagen fiber orientation determined from the scanning electron micrograph pictures (left). Clearly it can be seen that the collagen network within cartilage responds to tensile stress and strain (Wada and Akizuki, 1987).

If the molecular structure of collagen, the organization of the collagen fibers within the collagenous network, or the collagen fiber cross-linking is altered (such as that occurring in mild fibrillation or OA), the tensile properties of the network will change. Schmidt et al. (1990) have shown a definitive relationship between collagen hydroxypyridinium cross-linking and tensile stiffness and strength of normal bovine cartilage. Akizuki et al. (1986) showed that progressive degradation of human knee joint cartilage, from mild fibrillation to OA, yields a progressive deterioration of the intrinsic tensile properties of the collagen-PG solid matrix. Similar results have been observed recently in animal models of OA (Guilak et al., 1994; Setton et al., 1994). Together, these observations support the belief that disruption of the collagen network is a key factor in the initial events leading to the development of OA. Also, loosening of the collagen network is generally believed to be responsible for the increased swelling, hence the water content, of osteoarthritic cartilage (Mankin and Thrasher, 1975; Maroudas, 1979). Moreover, experimental culture studies have confirmed that collagen contributes significantly to the dynamic compressive properties of cartilage, by demonstrating that collagenase digestion impairs these properties under stress amplitudes and frequencies that are representative of physiologic loading conditions (Park et al., 2008). We have already discussed how increased water content leads to decreased compressive stiffness and increased permeability of articular cartilage.

BEHAVIOR OF ARTICULAR CARTILAGE IN PURE SHEAR

In tension and compression, only the equilibrium intrinsic properties of the collagen-PG solid matrix can be determined. This is because a volumetric change always occurs within a material when it is subjected to uniaxial

Unloaded

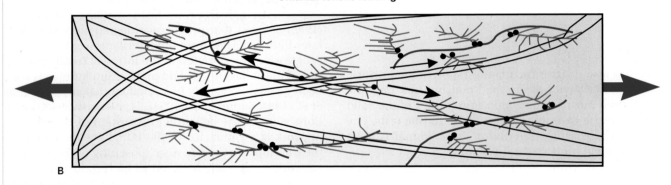

Collagen

Water

Proteoglycan

A

Uniaxial tensile loading

B

FIG. 3-13

Schematic depiction of the main components of articular cartilage when the tissue is unloaded **(A)** and when a tensile load is applied **(B)**. Loading would result in an alignment of collagen fibrils along the axis of tension. *Adapted from Myers, E.R., Lai, W.M., Mow, V.C. (1984). A continuum theory and an experiment for the ion-induced swelling behavior cartilage. J Biomech Eng, 106(2), 151–158.*

tension or compression. This volumetric change causes interstitial fluid flow and induces biphasic viscoelastic effects within the tissue. If, however, articular cartilage is tested in pure shear under infinitesimal strain conditions, no pressure gradients or volumetric changes will be produced within the material; hence, no interstitial fluid flow will occur (Hayes and Bodine, 1978; Zhu et al. 1993) (Fig. 3-15). Thus, a steady dynamic pure shear experiment can be used to assess the intrinsic viscoelastic properties of the collagen-PG solid matrix.

In a steady dynamic shear experiment, the viscoelastic properties of the collagen-PG solid matrix are determined by subjecting a thin circular wafer of tissue to a steady sinusoidal torsional shear, shown in Figure 3-16. In an experiment of this type, the tissue specimen is held by a precise amount of compression between two rough porous platens. The lower platen is attached to a sensitive torque transducer and the upper platen is attached to a precision mechanical spectrometer with a servo-controlled dc motor. A sinusoidal excitation signal may

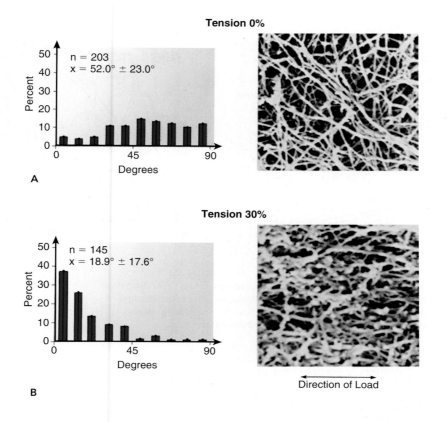

Tension 0%

n = 203
x = 52.0° ± 23.0°

A

Tension 30%

n = 145
x = 18.9° ± 17.6°

B

Direction of Load

FIG. 3-14

Collagen fibril alignment is clearly demonstrated by the scanning electron micrographs (X10,000) **(right)** of cartilage blocks under 0% stretch **(A)** and 30% stretch **(B)**. The histograms **(left)**, calculated from the micrographs, represent the percent of collagen fibers oriented in the direction of the applied tension.

At 0% stretch the fibers have a random orientation; however, at 30% they are aligned in the direction of the applied tension. *Reprinted with permission from Wada, T., Akizuki, S. (1987). An ultrastructural study of solid matrix in articular cartilage under uniaxial tensile stress. J Jpn Orthop Assoc, 61.*

be provided by the motor in a frequency of excitation range of 0.01 to 20 hertz (Hz). For shear strain magnitudes ranging from 0.2% to 2.0%, the viscoelastic properties are equivalently defined by the elastic storage modulus G′; the viscous loss modulus G″ of the collagen-PG solid matrix may be determined as a function of frequency (Fung, 1981; Zhu et al., 1993).

Sometimes it is more convenient to determine the magnitude of the dynamic shear modulus $|G^*|$ given by

$$|G^*|^2 = (G')^2 + (G'')^2$$

and the phase shift angle given by

$$\delta = \tan^{-1}(G''/G').$$

The magnitude of the dynamic shear modulus is a measure of the total resistance offered by the viscoelastic material. The value of δ, the angle between the steady applied sinusoidal strain and the steady sinusoidal

torque response, is a measure of the total frictional energy dissipation within the material. For a pure elastic material with no internal frictional dissipation, the phase shift angle δ is zero; for a pure viscous fluid, the phase shift angle δ is 90°.

The magnitude of the dynamic shear modulus for normal bovine articular cartilage has been measured to range from 1 to 3 MPa, whereas the phase shift angle has been measured to range from 9° to 20° (Hayes and Bodine, 1978; Zhu et al., 1993). The intrinsic transient shear stress-relaxation behavior of the collagen-PG solid matrix along with the steady dynamic shear properties also has been measured (Zhu et al., 1986). With both the steady dynamic and the transient results, the latter investigators showed that the quasi-linear viscoelasticity theory proposed by Fung (1981) for biologic materials provides an accurate description of the flow-independent viscoelastic behavior of the collagen-PG

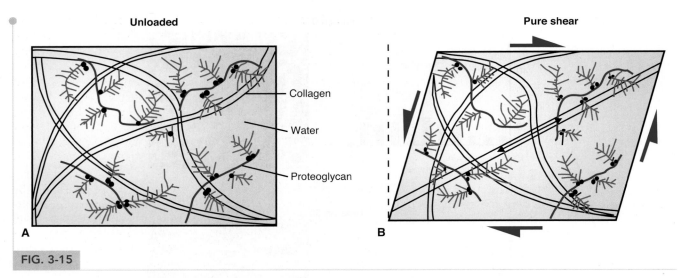

FIG. 3-15

Schematic depiction of unloaded cartilage (**A**) and cartilage subjected to pure shear (**B**). When cartilage is tested in pure shear under infinitesimal strain conditions, no volumetric changes or pressure gradients are produced; hence, no interstitial fluid flow occurs. This figure also demonstrates the functional role of collagen fibrils in resisting shear deformation.

solid matrix. Figure 3-17 depicts a comparison of the theoretical prediction of the transient stress-relaxation phenomenon in shear with the results from Fung's 1981 quasi-linear viscoelasticity theory.

From these shear studies, it is possible to obtain some insight as to how the collagen-PG solid matrix functions.

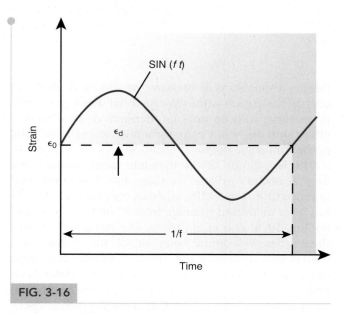

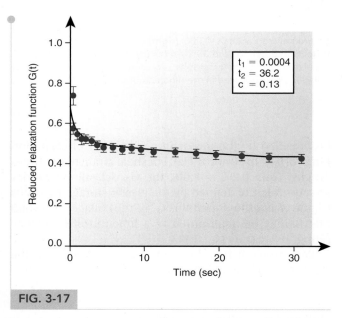

FIG. 3-17

Typical stress-relaxation curve after a step change in shear strain, expressed in terms of the mean of ten cycles of stress relaxation normalized by the initial stress. The *solid line* represents the theoretical prediction of the quasilinear viscoelasticity theory. *Adapted from Zhu, W.B., Lai, W.M., Mow, V.C. (1986). Intrinsic quasi-linear viscoelastic behavior of the extracellular matrix of cartilage. Trans Orthop Res Soc, 11, 407.*

FIG. 3-16

Steady sinusoidal torsional shear imposed on a specimen in pure shear. The fluctuating strain in the form of a sine wave with a strain amplitude ε_d and frequency f.

First, we note that measurements of PG solutions at concentrations similar to those found in articular cartilage in situ yield a magnitude of shear modulus to be of the order of 10 Pa and phase shift angle ranging up to 70° (Mow et al., 1989b; Zhu et al., 1991, 1996). Therefore, it appears that the magnitude of the shear modulus of concentrated PG solution is one hundred thousand times less and the phase angle is six to seven times greater than that of articular cartilage solid matrix. This suggests that PGs do not function in situ to provide shear stiffness for articular cartilage. The shear stiffness of articular cartilage must therefore derive from its collagen content, or from the collagen-PG interaction (Mow and Ratcliffe, 1997). From this interpretation, an increase in collagen, which is a much more elastic element than PG and the predominant load-carrying element of the tissue in shear, would decrease the frictional dissipation and hence the observed phase angle.

SWELLING BEHAVIOR OF ARTICULAR CARTILAGE

The Donnan osmotic swelling pressure, associated with the densely packed fixed anionic groups (SO_3^- and COO^-) on the GAG chains as well as the bulk compressive stiffness of the PG aggregates entangled in the collagen network, permits the PG gel in the collagen network to resist compression (Donnan, 1924; Maroudas, 1979; Mow and Ratcliffe, 1997). To account for such fixed charge density (FCD) effects in cartilage, a triphasic mechano-electrochemical, multi-electrolyte theory was developed that models cartilage as a mixture of three miscible phases: a charged solid phase representing the collagen-PG network, a fluid phase representing the interstitial water, and an ion phase comprising the monovalent cation Na^+ and anion Cl^- as well as other multivalent species such as Ca^{2+} (Gu et al., 1998; Lai et al., 1991). In this theory, the total stress is given by the sum of two terms: $\sigma^{total} = \sigma^{solid} + \sigma^{fluid}$, where σ^{solid} and σ^{fluid} are the solid matrix stress and interstitial fluid pressure, respectively. At equilibrium, σ^{fluid} is given by the Donnan osmotic pressure, π (see discussion later). Derived from all of the fundamental laws of mechanics and thermodynamics rather than through the ad hoc combination of existing specialized theories (e.g., Frank and Grodzinsky, 1987a,b), this triphasic theory provides a set of thermodynamically permissible constitutive laws to describe the time-dependent physico-chemical, mechanical, and electrical properties of charged-hydrated soft tissues. Moreover, the triphasic multi-electrolyte theory has been shown to be entirely consistent with the specialized classical osmotic pressure theory for charged polymeric solutions, phenomenologic transport theories, and the biphasic theory (Donnan, 1924; Katchalsky and Curran, 1975; Mow et al., 1980; Onsager, 1931), all of which have been frequently used to study specific facets of articular cartilage.

The triphasic theory has been used successfully to describe many of the mechano-electrochemical behaviors of articular cartilage. These include the prediction of free swelling under chemical load; nonlinear dependence of hydraulic permeability with FCD; nonlinear dependence of streaming potentials with FCD; curling of cartilage layers; pre-stress; osmotic and negative osmotic flows; swelling and electrical responses of cells to osmotic shock loading; and the influence of inhomogeneous fixed charge density (Gu et al., 1993, 1997, 1998; Lai et al., 1991; Mow et al., 1998; Setton et al., 1998; Sun et al., 1998). Providing more versatility, the triphasic theory has been generalized to include multi-electrolytes in the tissue (Gu et al., 1998).

From analysis using the triphasic theory, it becomes clear that the swelling behavior of the tissue can be responsible for a significant fraction of the compressive load-bearing capacity of articular cartilage at equilibrium (Mow and Ratcliffe, 1997). For example, the triphasic theory predicts for confined-compression at equilibrium that the total stress (σ^{total}) acting on the cartilage specimen is the sum of the stress in the solid matrix (σ^{solid}) and the Donnan osmotic pressure ($\sigma^{fluid} = \pi$). The Donnan osmotic pressure is the swelling pressure caused by the ions in association with the FCD and represents the physicochemical motive force for cartilage swelling (Fig. 3-18). From the classical theory for osmotic pressure, the Donnan osmotic pressure caused by the excess of ion particles inside the tissue is given as follows:

$$\pi = RT[\varphi(2c + c^F) - 2\varphi^*c^*] + P\infty$$

where c is the interstitial ion concentration, c^* is the external ion concentration, c^F is the FCD, R is the universal gas constant, T is the absolute temperature, φ and φ^* are osmotic coefficients, and $P\infty$ is the osmotic pressure caused by the concentration of PG particles in the tissue, usually assumed to be negligible (Lai et al., 1991). For a lightly loaded tissue, the swelling pressure may contribute significantly to the load support. But for highly loaded tissues, such as those found under physiologic conditions and certainly for dynamically loaded tissues, the interstitial fluid pressurization (σ^{fluid}) would dominate; the contribution of this swelling pressure to load support would be less than 5% (Soltz and Ateshian, 1998).

As with the biphasic theory, the triphasic mechano-electrochemical theory can be used to elucidate potential mechanosignal transduction mechanisms in cartilage. For example, because of their potential effects

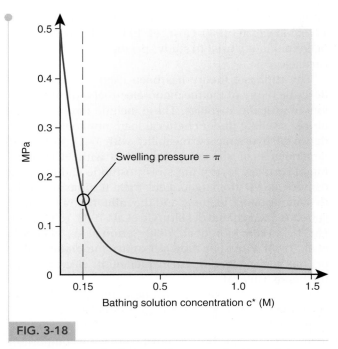

FIG. 3-18

Swelling pressure of articular cartilage versus bathing solution concentration (c^*). At equilibrium, the interstitial fluid pressure is equal to the swelling pressure, which is defined by the tissue Donnan osmotic pressure (π).

on chondrocyte function, it is important to describe and predict electrokinetic phenomena such as streaming potentials and streaming currents (Gu et al., 1993, 1998; Katchalsky and Curran, 1975; Kim et al., 1994) that arise from ion movement caused by the convection of interstitial fluid flow past the FCD of the solid matrix. As a second example, the pressure produced in the interstitial fluid by polyethylene glycol-induced osmotic loading of cartilage explants (Schneiderman et al., 1986) was recently shown to be theoretically nonequivalent to the pressure produced in any other commonly used mechanically loaded explant experiment or by hydrostatic loading (Lai et al., 1998). In light of this finding, earlier interpretations of biologic data from studies making such an assumption of equivalency should be revisited.

Lubrication of Articular Cartilage

As already discussed, synovial joints are subjected to an enormous range of loading conditions, and under normal circumstances the cartilage surface sustains little wear. The minimal wear of normal cartilage associated with such varied loads indicates that sophisticated lubrication processes are at work within the joint

and within and on the surface of the tissue. These processes have been attributed to a lubricating fluid-film forming between the articular cartilage surface and to an adsorbed boundary lubricant on the surface during motion and loading. The variety of joint demands also suggests that several mechanisms are responsible for diarthrodial joint lubrication. To understand diarthrodial joint lubrication, one should use basic engineering lubrication concepts.

From an engineering perspective, there are two fundamental types of lubrication. One is boundary lubrication, which involves a single monolayer of lubricant molecules adsorbed on each bearing surface. The other is fluid-film lubrication, in which a thin fluid-film provides greater surface-to-surface separation (Bowden and Tabor, 1967). Both lubrication types appear to occur in articular cartilage under varying circumstances. Intact synovial joints have an extremely low coefficient of friction, approximately 0.02 (Dowson, 1966/1967; Linn, 1968; McCutchen, 1962; Mow and Ateshian, 1997). Boundary-lubricated surfaces typically have coefficients of friction one or two orders of magnitude higher than surfaces lubricated by a fluid-film, suggesting that synovial joints are lubricated, at least in part, by the fluid-film mechanism. It is quite possible that synovial joints use the mechanism that will most effectively provide lubrication at a given loading condition. Unresolved, though, is the manner by which synovial joints generate the fluid lubricant film.

FLUID-FILM LUBRICATION

Fluid-film lubrication uses a thin film of lubricant that causes a bearing surface separation. The load on the bearing is then supported by the pressure that is developed in this fluid-film. The fluid-film thickness associated with engineering bearings is usually less than 20 μm. Fluid-film lubrication requires a minimum fluid-film thickness (as predicted by a specific lubrication theory) to exceed three times the combined statistical surface roughness of cartilage (e.g., 4–25 μm; Clarke, 1971; Walker et al., 1970). If fluid-film lubrication is unachievable because of heavy and prolonged loading, incongruent gap geometry, slow reciprocating-grinding motion, or low synovial fluid viscosity, boundary lubrication must exist (Mow and Ateshian, 1997).

The two classical modes of fluid-film lubrication defined in engineering are hydrodynamic and squeeze-film lubrication (Figs. 3-19A and B). These modes apply to rigid bearings composed of relatively undeformable material such as stainless steel. Hydrodynamic lubrication occurs when nonparallel rigid bearing surfaces lubricated by a fluid-film move tangentially with

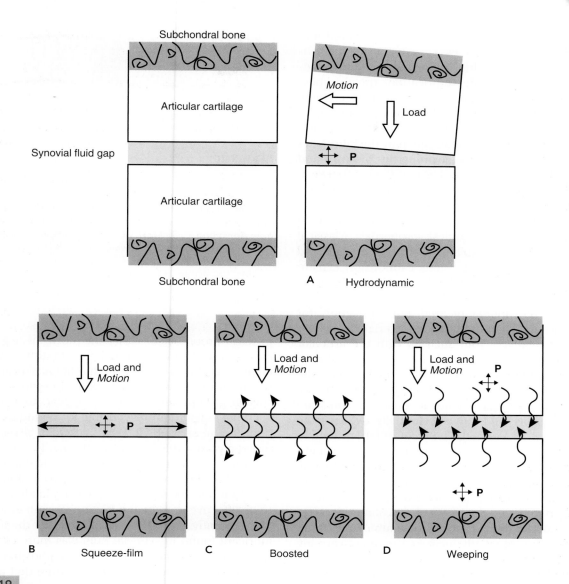

FIG. 3-19

A. In hydrodynamic lubrication, viscous fluid is dragged into a convergent channel, causing a pressure field to be generated in the lubricant. Fluid viscosity, gap geometry, and relative sliding speed determine the load-bearing capacity. **B.** As the bearing surfaces are squeezed together, the viscous fluid is forced from the gap into the transverse direction. This squeeze action generates a hydrodynamic pressure in the fluid for load support. The load-bearing capacity depends on the size of the surfaces, velocity of approach, and fluid viscosity. **C.** The direction of fluid flow under squeeze-film lubrication in the boosted mode for joint lubrication. **D.** Depicts the weeping lubrication hypothesis for the uniform exudation of interstitial fluid from the cartilage. The driving mechanism is a self-pressurization of the interstitial fluid when the tissue is compressed.

respect to each other (i.e., slide on each other), forming a converging wedge of fluid. A lifting pressure is generated in this wedge by the fluid viscosity as the bearing motion drags the fluid into the gap between the surfaces, as shown in Figure 3-19A. In contrast, squeeze-film lubrication occurs when the bearing surfaces move perpendicularly toward each other. A pressure is generated in the fluid-film as a result of the viscous resistance of the fluid that acts to impede its escape from the gap (Fig. 3-19B). The squeeze-film mechanism is sufficient to carry high loads for short durations. Eventually, however, the fluid-film becomes so thin that contact between the asperities (peaks) on the two bearing surfaces occurs.

Calculations of the relative thickness of the fluid-film layer and the surface roughness are valuable in establishing when hydrodynamic lubrication may exist. In hydrodynamic and squeeze-film lubrication, the thickness and extent of the fluid-film, as well as its load-carrying capacity, are characteristics independent of the (rigid) bearing surface material properties. These lubrication characteristics are instead determined by the lubricant's properties, such as its rheologic properties, viscosity and elasticity, film geometry, the shape of the gap between the two bearing surfaces, and the speed of the relative surface motion.

Cartilage is unlike any man-made material with respect to its near frictionless properties. Classical theories developed to explain lubrication of rigid and impermeable bearings (e.g., steel) cannot fully explain the mechanisms responsible for lubrication of the natural diarthrodial joint. A variation of the hydrodynamic and squeeze-film modes of fluid-film lubrication, for example, occurs when the bearing material is not rigid but instead relatively soft, such as with the articular cartilage covering the joint surface. This type of lubrication, termed elastohydrodynamic, operates when the relatively soft bearing surfaces undergo either a sliding (hydrodynamic) or squeeze-film action and the pressure generated in the fluid-film substantially deforms the surfaces (Fig. 3-19 A and B). These deformations tend to increase the surface area and congruency, thus beneficially altering film geometry. By increasing the bearing contact area, the lubricant is less able to escape from between the bearing surfaces, a longer-lasting lubricant film is generated, and the stress of articulation is lower and more sustainable. Elastohydrodynamic lubrication enables bearings to greatly increase their load-carrying capacity (Dowson, 1966/1967, 1990).

Note that several studies have shown that hyaluronidase treatment of synovial fluid, which decreases its viscosity (to that of saline) by causing depolymerization of HA, has little effect on lubrication (Linn, 1968; Linn and Radin, 1968). Because fluid-film lubrication is highly dependent on lubricant viscosity, these results strongly suggest that an alternative mode of lubrication is the primary mechanism responsible for the low frictional coefficient of joints.

BOUNDARY LUBRICATION

During diarthrodial joint function, relative motion of the articulating surfaces occurs. In boundary lubrication, the surfaces are protected by an adsorbed layer of boundary lubricant, which prevents direct, surface-to-surface contact and eliminates most of the surface wear. Boundary lubrication is essentially independent of the

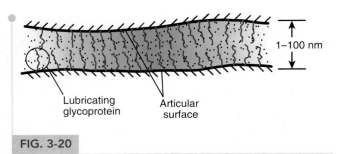

FIG. 3-20

Boundary lubrication of articular cartilage. The load is carried by a monolayer of the lubricating glycoprotein (LGP), which is adsorbed onto the articular surfaces. The monolayer effectively serves to reduce friction and helps to prevent cartilaginous wear. *Adapted from Armstrong, C.G., Mow, V.C. (1980). Friction, lubrication and wear of synovial joints. In R. Owen, J. Goodfellow, P. Bullough (Eds.). Scientific Foundations of Orthopaedics and Traumatology. London, UK: William Heinemann, 223–232.*

physical properties of either the lubricant (e.g., its viscosity) or the bearing material (e.g., its stiffness), instead depending almost entirely on the chemical properties of the lubricant (Dowson, 1966/1967). In synovial joints, a specific glycoprotein, "lubricin," appears to be the synovial fluid constituent responsible for boundary lubrication (Swann et al., 1979, 1985). Lubricin, also known as proteoglycan-4 (PRG-4), $(25 \times 10^4$ mw) is adsorbed as a macromolecular monolayer to each articulating surface (Fig. 3-20). These two layers, ranging in combined thickness from 1 to 100 nm, are able to carry loads and appear to be effective in reducing friction (Swann et al., 1979). More recently, Hills (1989) suggested that the boundary lubricant found in synovial fluid was more likely to be a phospholipid named dipalmitoyl phosphatidylcholine. Although experiments demonstrate that a boundary lubricant can account for a reduction of the friction coefficient by a factor of threefold to sixfold (Swann et al., 1985; Williams et al., 1993), this reduction is quite modest compared with the much greater range (e.g., up to 60-fold) reported earlier (McCutchen, 1962). Moreover, removing the superficial zone of cartilage (with its adsorbed lubricin layer) does not affect its friction coefficient (Krishnan et al., 2004). Even so, these results do suggest that boundary lubrication exists as a complementary mode of lubrication (Gleghorn et al., 2009; Jay et al., 2007; Schmidt et al., 2007).

MIXED LUBRICATION

There are two joint lubrication scenarios that can be considered a combination of fluid-film and boundary lubrication or simply mixed lubrication (Dowson,

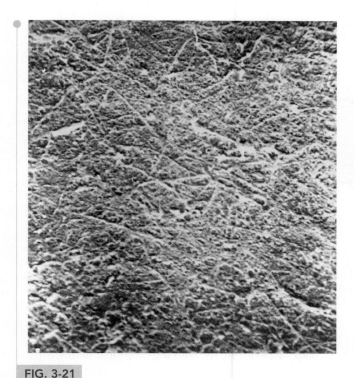

FIG. 3-21

Scanning electron micrograph of the surface of human articular cartilage from a normal young adult showing the typical irregularities characteristic of this tissue (×3,000). *Adapted from Armstrong, C.G., Mow, V.C. (1980). Friction, lubrication and wear of synovial joints. In R. Owen, J. Goodfellow, P. Bullough (Eds.). Scientific Foundations of Orthopaedics and Traumatology. London, UK: William Heinemann, 223–232.*

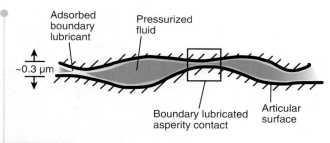

FIG. 3-22

Schematic depiction of mixed lubrication operating in articular cartilage. Boundary lubrication occurs when the thickness of the fluid-film is on the same order as the roughness of the bearing surfaces. Fluid-film lubrication takes place in areas with more widely separated surfaces. *Adapted from Armstrong, C.G., Mow, V.C. (1980). Friction, lubrication and wear of synovial joints. In R. Owen, J. Goodfellow, P. Bullough (Eds.). Scientific Foundations of Orthopaedics and Traumatology. London, UK: William Heinemann, 223–232.*

1966/1967). The first case refers to the temporal coexistence of fluid-film and boundary lubrication at spatially distinct locations, whereas the second case, termed "boosted lubrication," is characterized by a shift of fluid-film to boundary lubrication with time over the same location (Walker et al., 1970).

The articular cartilage surface, like all surfaces, is not perfectly smooth; asperities project out from the surface (Clarke, 1971; Gardner and McGillivray, 1971; Redler and Zimny, 1970) (Figs. 3-3B and 3-21). In synovial joints, situations may occur in which the fluid-film thickness is of the same order as the mean articular surface asperity (Walker et al., 1970). During such instances, boundary lubrication between the asperities may come into play. If this occurs, a mixed mode of lubrication is operating, with the joint surface load sustained by both the fluid-film pressure in areas of noncontact and by the boundary lubricant lubricin in the areas of asperity contact (shown in Fig. 3-22). In this mode of mixed lubrication,

it is probable that most of the friction (which is still extremely low) is generated in the boundary lubricated areas while most of the load is carried by the fluid-film (Dowson, 1966/1967, 1990).

The second mode of mixed lubrication (boosted lubrication) proposed by Walker et al. (1968, 1970) and Maroudas (1966/1967) is based on the movement of fluid from the gap between the approaching articular surfaces into the articular cartilage (Fig. 3-19C). Specifically, in boosted lubrication, articular surfaces are believed to be protected during joint loading by the ultrafiltration of the synovial fluid through the collagen-PG matrix. This ultrafiltration permits the solvent component of the synovial fluid (water and small electrolytes) to pass into the articular cartilage during squeeze-film action, yielding a concentrated gel of HA protein complex that coats and lubricates the bearing surfaces (Lai and Mow, 1978). According to this theory, it becomes progressively more difficult, as the two articular surfaces approach each other, for the HA macromolecules in the synovial fluid to escape from the gap between the bearing surfaces because they are physically too large (0.22–0.65 µm), as shown in Figure 3-23. The water and small solute molecules can still escape into the articular cartilage through the cartilage surface and/or laterally into the joint space at the periphery of the joint. Theoretical results by Hou et al. (1992) predict that fluid entry into the cartilage-bearing surface is possible, leading them to suggest that boosted lubrication may occur. The role of this HA gel in joint lubrication remains unclear, however, particularly in view of the findings by Linn (1968), which demonstrated that purified HA acts as a poor lubricant.

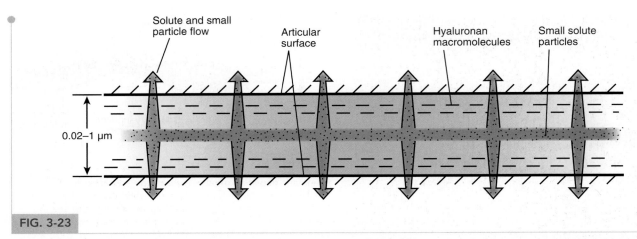

FIG. 3-23

Ultrafiltration of the synovial fluid into a highly viscous gel. As the articular surfaces come together, the small solute molecules escape into the articular cartilage and into the lateral joint space, leaving the large HA macromolecules that, because of their size, are unable to escape. These HA macromolecules form a concentrated gel less than 1 μm thick that lubricates the articular surfaces. This hypothesized lubrication mode is termed "boosted lubrication."

To summarize, in any bearing, the effective mode of lubrication depends on the applied loads and on the relative velocity (speed and direction of motion) of the bearing surfaces. Adsorption of the synovial fluid glycoprotein, lubricin, to articular surfaces seems to be most important under severe loading conditions, that is, contact surfaces with high loads, low relative speeds, and long duration. Under these conditions, as the surfaces are pressed together, the boundary lubricant monolayers interact to prevent direct contact between the articular surfaces. Conversely, fluid-film lubrication operates under less severe conditions, when loads are low and/or oscillate in magnitude and when the contacting surfaces are moving at high relative speeds. In light of the varied demands on diarthrodial joints during normal function, it is unlikely that only a single mode of lubrication exists. As yet, it is impossible to state definitely under which conditions a particular lubrication mechanism may operate. Nevertheless, using the human hip as an example, some general statements are possible:

1. Elastohydrodynamic fluid-films of both the sliding (hydrodynamic) and the squeeze type probably play an important role in lubricating the joint. During the swing phase of walking, when loads on the joint are minimal, a substantial layer of synovial fluid-film is probably maintained. After the first peak force, at heel strike, a supply of fluid lubricant is generated by articular cartilage. However, this fluid-film thickness will begin to decrease under the high load of stance phase; as a result, squeeze-film action occurs. The second peak force during the walking cycle, just before the toe leaves the ground, occurs when the joint is swinging in the opposite direction. Thus, it is possible that a fresh supply of fluid-film could be generated at toe-off, thereby providing the lubricant during the next swing phase.

2. With high loads and low speeds of relative motion, such as during standing, the fluid-film will decrease in thickness as the fluid is squeezed out from between the surfaces (fluid film). Under these conditions, the fluid exuded from the compressed articular cartilage could become the main contributor to the lubricating film.

3. Under extreme loading conditions, such as during an extended period of standing following impact, the fluid-film may be eliminated, allowing surface-to-surface contact. The surfaces, however, will probably still be protected, either by a thin layer of ultrafiltrated synovial fluid gel (boosted lubrication) or by the adsorbed lubricin monolayer (boundary lubrication).

ROLE OF INTERSTITIAL FLUID PRESSURIZATION IN JOINT LUBRICATION

During joint articulation, loads transmitted across a joint may be supported by the opposing joint surfaces via solid-to-solid contact, through a fluid-film layer, or by a mixture of both. Although fluid-film lubrication is achievable, its contribution to joint lubrication is transient, a consequence of the rapid dissipation of the fluid-film thickness by joint loads. With this caveat, Ateshian (1997), adopting the theoretical framework of the biphasic theory (Mow et al., 1980), proposed a mathematical formulation of a boundary friction model of articular cartilage to describe the underlying mechanism behind diarthrodial joint lubrication, in particular, the time-dependence of the friction coefficient for cartilage reported during creep and stress-relaxation experiments (Malcolm, 1976; McCutchen, 1962).

Although load is partitioned between the solid and fluid phases of a biphasic material (Mow et al., 1980), Ateshian (1997) derived an expression for the effective (or measured) coefficient of friction that was dependent solely on the proportion of the load supported by the solid matrix (e.g., the difference between total load and that supported by hydrostatic pressure in the fluid). The implication of such an expression is that the frictional properties of cartilage vary with time during applied loading, a reflection of the interstitial fluid and collagen-PG matrix interactions that give rise to the flow-dependent viscoelastic properties of the tissue described earlier and shown in Figures 3-9 and 3-10. Degradation of cartilage with enzymes for collagen (Park et al., 2008) and proteoglycan (Basalo et al., 2004) alters the mechanical and friction properties of cartilage.

To validate his model, Ateshian developed a novel loading experiment that superimposed a frictional torque load on a cartilage explant undergoing creep loading in a confined compression configuration (Fig. 3-24A) (Ateshian et al., 1998). More specifically, a cylindrical biphasic cartilage plug was compressed in a confining ring (e.g., prohibiting radial motion and fluid exudation) under a constant applied load generated by an impermeable rigid platen that was rotating at a prescribed angular speed. The surface of the plug opposite the platen was pressed against a fixed rigid porous filter whereby the interdigitation of the cartilage with the rough surface of the porous filter prevented it from rotating. In this manner, a frictional torque was developed in the tissue. Because the application of a torque load that yields pure shear, under infinitesimal deformations, induces no volume change in the tissue or associated fluid exudation, the load generated by the frictional torque is independent of the biphasic creep behavior of the tissue.

Theoretical predictions, which closely match experimental results, show that during initial loading, when interstitial pressurization is high, the friction coefficient can be very low (Fig. 3-24B). As creep equilibrium is reached and the load is transferred to the solid matrix, the friction coefficient becomes high (e.g., 0.15). The time constant for this transient response is in excellent agreement with observed experimental results

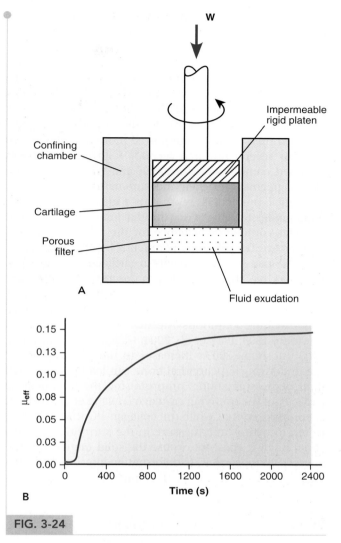

FIG. 3-24

Experimental configuration superimposing a frictional torque with creep loading of an articular cartilage explant in confined compression. **A.** Note that fluid exudation occurs on the opposite face of the tissue exposed to the frictional load, indicating that the frictional properties of cartilage are not dependent on the weeping of interstitial fluid to the lubricating boundary. **B.** Note that effective friction coefficient (μ_{eff}) varies with increasing proportion of load on the solid matrix, as can be seen from the theoretical curve for μ_{eff} as a function of time during the experiment. *Adapted from Mow, V.C., Ateshian, G.A. (1997). Lubrication and wear of diarthrodial joints. In V.C. Mow, W.C. Hayes (Eds.). Basic Biomechanics (2nd ed.). Philadelphia: Lippincott–Raven Publishers, 275–315.*

(Malcolm, 1976; McCutchen, 1962). Another important result of this work is that fluid pressurization can function in joint lubrication without concomitant fluid exudation to the lubricating boundary as is proposed for weeping lubrication (McCutchen, 1962) (Fig. 3-19D).

Equally significant, this lubrication theory is capable of explaining the observed decrease of the effective friction coefficient with increasing rolling and sliding joint velocities and with increasing joint load (Linn, 1968).

The interstitial fluid pressurization within cartilage during uniaxial creep and stress relaxation experiments was successfully measured (Soltz and Ateshian, 1998). As predicted by the biphasic theory, they found that interstitial fluid pressurization supported more that 90% of the load for several hundred seconds following loading in confined compression (Ateshian and Wang, 1995). The close agreement of their measurements with biphasic theoretical predictions represents a major advancement in the understanding of diarthrodial joint lubrication and provides compelling evidence for the role of interstitial fluid pressurization as a fundamental mechanism underlying the load-bearing capacity in cartilage. Subsequently, this work has been extended to describe the important role of interstitial fluid pressurization in cartilage load-bearing and lubrication properties under more physiologic loading conditions (Ateshian et al., 1998; Soltz and Ateshian, 1998; Park et al., 2004; Caligaris and Ateshian, 2008; Ateshian, 2009). Increases to the permeability of the underlying subchondral bone can lead to interstitial fluid depressurization, compromising the load-bearing capacity of the overlying cartilage (Hwang et al., 2008). It is emphasized that while the collagen-PG matrix is subjected to hydrostatic pressure in the surrounding interstitial fluid, it does not expose the solid matrix (nor the encased chondrocytes) to deformation, presumably causing no mechanical damage (Bachrach et al., 1998).

Wear of Articular Cartilage

Wear is the unwanted removal of material from solid surfaces by mechanical action. There are two components of wear: interfacial wear resulting from the interaction of bearing surfaces and fatigue wear resulting from bearing deformation under load.

Interfacial wear occurs when bearing surfaces come into direct contact with no lubricant film (boundary or fluid) separating them. This type of wear can take place in either of two ways: adhesion or abrasion. Adhesive wear arises when, as the bearings come into contact, surface fragments adhere to each other and are torn off from the surface during sliding. Abrasive wear, conversely, occurs when a soft material is scraped by a harder one; the harder material can be either an opposing bearing or loose particles between the bearings. The low rates of interfacial wear observed in articular cartilage tested in vitro (Lipshitz and Glimcher, 1979) suggest that direct surface-to-surface contact between the asperities of the two cartilage surfaces rarely occurs. Abrasive wear in these experiments, however, was not ruled out. The multiple modes of effective lubrication working in concert are the mechanisms making interfacial wear of articular cartilage unlikely. Nevertheless, adhesive and abrasive wear may take place in an impaired or degenerated synovial joint. Once the cartilage surface sustains ultrastructural defects and/or decreases in mass, it becomes softer and more permeable (Akizuki et al., 1986; Armstrong and Mow, 1982; Setton et al., 1994). Thus, fluid from the lubricant film separating the bearing surfaces may leak away more easily through the cartilage surface. This loss of lubricating fluid from between the surfaces increases the probability of direct contact between the asperities and exacerbates the abrasion process.

Fatigue wear of bearing surfaces results not from surface-to-surface contact but from the accumulation of microscopic damage within the bearing material under repetitive stressing. Bearing surface failure may occur with the repeated application of high loads over a relatively short period or with the repetition of low loads over an extended period even though the magnitude of those loads may be much lower than the material's ultimate strength. This fatigue wear, resulting from cyclically repeated deformation of the bearing materials, can take place even in well-lubricated bearings.

In synovial joints, the cyclical variation in total joint load during most physiologic activities causes repetitive articular cartilage stressing (deformation). In addition, during rotation and sliding, a specific region of the articular surface "moves in and out" of the loaded contact area, repetitively stressing that articular region. Loads imposed on articular cartilage are supported by the collagen-PG matrix and by the resistance generated by fluid movement throughout the matrix. Thus, repetitive joint movement and loading will cause repetitive stressing of the solid matrix and repeated exudation and imbibition of the tissue's interstitial fluid (Mow and Ateshian, 1997). These processes give rise to two possible mechanisms by which fatigue damage may accumulate in articular cartilage: disruption of the collagen-PG solid matrix and PG "washout."

First, repetitive collagen-PG matrix stressing could disrupt the collagen fibers, the PG macromolecules, and/or the interface between these two components. A popular hypothesis is that cartilage fatigue is the result of a tensile failure of the collagen fiber network (Freeman, 1975). Also, as discussed previously, pronounced changes in the articular cartilage PG population have been observed with age and disease (Buckwalter et al., 1985; Muir, 1983; Roughley et al., 1980; Sweet et al., 1979). These PG changes could be considered as part of the accumulated tissue damage. These molecular structural changes would result in lower PG-PG interaction sites and thus lower network strength (Mow et al., 1989b; Zhu et al., 1991, 1996). Second, repetitive and massive exudation and imbibition of the interstitial fluid may cause the degraded PGs to "wash out" from the ECM, with a resultant decrease in stiffness and increase in permeability of the tissue that in turn defeats the stress-shielding mechanism of interstitial fluid-load support and establishes a vicious cycle of cartilage degeneration.

A third mechanism of damage and resultant articular wear is associated with synovial joint impact loading—that is, the rapid application of a high load. With normal physiologic loading, articular cartilage undergoes surface compaction during compression with the lubricating fluid being exuded through this compacted region, as shown in Figure 3-10. As described previously, however, fluid redistribution within the articular cartilage occurs over time, which relieves the stress in this compacted region. This process of stress relaxation takes place quickly; the stress may decrease by 63% within two to five seconds (Ateshian et al., 1998; Mow et al., 1980). If, however, loads are supplied so quickly that there is insufficient time for internal fluid redistribution to relieve the compacted region, the high stresses produced in the collagen-PG matrix may induce damage (Newberry et al., 1997, Thompson et al., 1991). This phenomenon could well explain why Radin and Paul (1971) found dramatic articular cartilage damage with repeated impact loads.

These mechanisms of wear and damage may be the cause of the commonly observed large range of structural defects observed in articular cartilage (Bullough and Goodfellow, 1968; Meachim and Fergie, 1975) (Figs. 3-25A–C). One such defect is the splitting of the cartilage surface. Examination of vertical sections of cartilage exhibiting these lesions, known as fibrillation, show that they eventually extend through the full depth of the

articular cartilage. In other specimens, the cartilage layer appears to be eroded rather than split. This erosion is known as smooth-surfaced destructive thinning.

Considering the variety of defects noted in articular cartilage, it is unlikely that a single wear mechanism is responsible for all of them. At any given site, the stress history may be such that fatigue is the initiating failure mechanism. At another, the lubrication conditions may be so unfavorable that interfacial wear dominates the progression of cartilage failure. As yet, there is little experimental information on the type of defect produced by any given wear mechanism.

Once the collagen-PG matrix of cartilage is disrupted, damage resulting from any of the three wear mechanisms mentioned becomes possible: (1) further disruption of the collagen-PG matrix as a result of repetitive matrix stressing; (2) an increased "washing out" of the PGs as a result of violent fluid movement and thus impairment of articular cartilage's interstitial fluid load support capacity; and (3) gross alteration of the normal load carriage mechanism in cartilage, thus increasing frictional shear loading on the articular surface.

All these processes will accelerate the rate of interfacial and fatigue wear of the already disrupted cartilage microstructure.

Hypotheses on the Biomechanics of Cartilage Degeneration

ROLE OF BIOMECHANICAL FACTORS

Articular cartilage has only a limited capacity for repair and regeneration, and if subjected to an abnormal range of stresses can quickly undergo total failure (Fig. 3-26). It

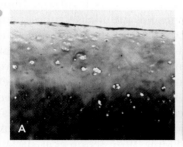

FIG. 3-25

Photomicrographs of vertical sections through the surface of articular cartilage showing a normal intact surface (**A**), an eroded articular surface (**B**), and a vertical split or fibrillation of the articular surface that will eventually extend through the full depth of the cartilage (**C**). Photomicrographs courtesy of Dr. S. Akizuki, Nagano, Japan.

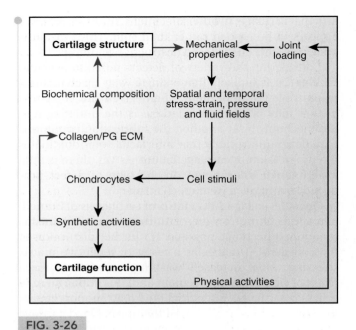

FIG. 3-26

Flow diagram of the events mediating the structure and function of articular cartilage. Physical activities result in joint loads that are transmitted to the chondrocyte via the extracellular matrix (ECM). The chondrocyte varies its cellular activities in response to the mechano-electrochemical stimuli generated by loading of its environment. The etiology of osteoarthritis is unclear but may be traced to intrinsic changes to the chondrocyte or to an altered ECM (e.g., resulting from injury or gradual wear) that leads to abnormal chondrocyte stimuli and cell activities.

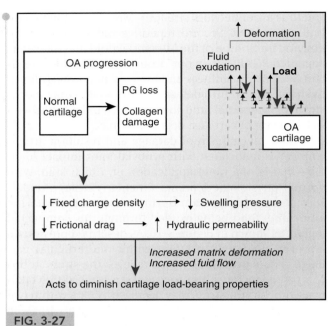

FIG. 3-27

A figure illustrating how osteoarthritic changes to the collagen-PG network can compromise the ability of articular cartilage to maintain interstitial fluid pressurization, which underlies the tissue's load-bearing and joint lubrication capacity. Loss of PG and damage to the collagen fibers result in an increased hydraulic permeability (decreased resistance to fluid flow) and supranormal loads and strains on the solid matrix (and chondrocyte).

has been hypothesized that failure progression relates to the following: (1) the magnitude of the imposed stresses; (2) the total number of sustained stress peaks; (3) the changes in the intrinsic molecular and microscopic structure of the collagen-PG matrix; and (4) the changes in the intrinsic mechanical property of the tissue. The most important failure-initiating factor appears to be the "loosening" of the collagen network that allows abnormal PG expansion and thus tissue swelling (Maroudas, 1976; McDevitt and Muir, 1976). Associated with this change is a decrease in cartilage stiffness and an increase in cartilage permeability (Altman et al., 1984; Armstrong and Mow, 1982; Guilak et al., 1994, Setton et al., 1994), both of which alter cartilage function in a diarthrodial joint during joint motion, as shown in Figure 3-27 (Mow and Ateshian, 1997).

The magnitude of the stress sustained by the articular cartilage is determined by both the total load on the joint and how that load is distributed over the articular surface contact area (Ahmed and Burke, 1983; Armstrong et al., 1979; Paul, 1976). Any intense stress concentration in the contact area will play a primary role in tissue degeneration. A large number of well-known conditions cause excessive stress concentrations in articular cartilage and result in cartilage failure. Most of these stress concentrations are caused by joint surface incongruity, resulting in an abnormally small contact area. Examples of conditions causing such joint incongruities include OA subsequent to congenital acetabular dysplasia, a slipped capital femoral epiphysis, and intra-articular fractures. Two further examples are knee joint meniscectomy, which eliminates the load-distributing function of the meniscus (Mow et al., 1992), and ligament rupture, which allows excessive movement and the generation of abnormal mechanical stresses in the affected joint (Altman et al., 1984; Guilak et al, 1994; McDevitt and Muir, 1976; Setton et al., 1994). In all these cases, abnormal joint articulation increases the stress acting on the joint surface, which appears to predispose the cartilage to failure.

Macroscopically, stress localization and concentration at the joint surfaces have a further effect. High contact pressures between the articular surfaces decrease the probability of fluid-film lubrication (Mow and Ateshian, 1997). Subsequent actual surface-to-surface contact of asperities will cause microscopic stress concentrations that are responsible for further tissue damage (Ateshian et al., 1995, 1998; Ateshian and Wang, 1995) (Case Study 3-1).

The high incidence of specific joint degeneration in individuals with certain occupations, such as foot-ball players' knees and ballet dancers' ankles, can be explained by the increase in high and abnormal load frequency and magnitude sustained by the joints of these individuals. It has been suggested that, in some cases, OA may be caused by deficiencies in the mechanisms that act to minimize peak forces on the joints. Examples of these mechanisms include the active processes of joint flexion and muscle lengthening and the passive absorption of shocks by the subchondral bone (Radin, 1976) and meniscus (Mow et al., 1992).

Degenerative changes to the structure and composition of articular cartilage could lead to abnormal tissue swelling and functionally inferior biomechanical properties. In this weakened state, the cartilage ultrastructure will then be gradually destroyed by stresses of normal joint articulation (Fig. 3-27). OA may also arise secondarily from insult to the intrinsic molecular and microscopic structure of the collagen-PG matrix. Many conditions may promote such a breakdown in matrix integrity; these include degeneration associated with rheumatoid arthritis, joint space hemorrhage associated with hemophilia, various collagen metabolism disorders, and tissue degradation by proteolytic enzymes. The presence of soluble mediators such as cytokines (e.g., interleukin-1) (Ratcliffe et al., 1986) and growth factors (e.g., transforming growth factor-beta 1) also appear to play an important role in OA. Another contributing factor to the etiology of OA may be age-related changes to the chondrocyte (Case Study 3-2).

IMPLICATIONS OF CHONDROCYTE FUNCTION

The ECM modulates the transmission of joint loads to the chondrocyte, acting as a transducer that converts mechanical loading to a plethora of environmental cues that mediate chondrocyte function. In healthy articular cartilage, loads from normal joint function motion result in the generation of mechano-electrochemical stimuli (e.g., hydrostatic pressure, stress and strain fields, streaming potentials) that promote normal cartilage maintenance (by the chondrocytes) and normal tissue function (Fig. 3-26). However, when the integrity of the collagen-PG network (the transducer) of articular cartilage is compromised, such as from trauma or disease, normal joint articulation leads to abnormal mechano-electrochemical stimuli, with ensuing abnormal ECM remodeling by the chondrocytes and debilitated tissue function.

In the absence of joint loading, the normal environment of the chondrocyte is characterized by the pre-stress established by the balance between tension in

---

Case Study 3-1

Knee Meniscectomy

Consider the case of a 40-year-old man who had a meniscectomy 10 years ago in his right knee. Currently, he is suffering pain associated with movement, swelling, and limitations of knee motion (Case Study Fig. 3-1).

The history of knee meniscectomy not only implies an alteration in joint surface congruence but also the elimination of the load-distribution function of the meniscus. The effect is an abnormal joint, characterized by an increase in the stress acting on the joint surface, which results in cartilage failure. Most of these stress concentrations are caused by joint surface incongruity, resulting in an abnormally small contact area. This small contact area will suffer high contact pressure, decreasing the probability of fluid-film lubrication, and thus the actual surface-to-surface contact will cause microscopic stress concentrations that lead to damage.

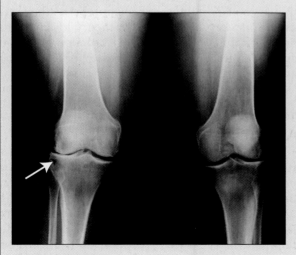

Case Study Figure 3-1

Case Study 3-2

Osteoarthritis

A 70-year-old woman, overweight, presents with OA of the right hip joint with associated symptoms of pain, limitation of motion, joint deformity, and abnormal gait (Case Study Fig. 3-2).

OA is characterized by erosive cartilage lesions, cartilage loss and destruction, subchondral bone sclerosis and cysts, and large osteophyte formation at the margins of the joint (Mow and Ratcliffe, 1997). In this case, roentgenograms of the right hip of the patient show a decrease in the interarticular space and changes in bone surfaces as sclerotic and osteophyte formations. The most severe alterations are found at the point of maximum pressure against the opposing cartilage surface, in this case at the superior aspect of the femoral head.

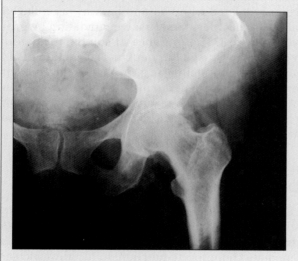

Case Study Figure 3-2

the collagen fibers and the Donnan osmotic pressure. During joint loading, by virtue of the tissue's low permeability, the normal environment of the chondrocyte is dominated by hydrostatic pressure in the interstitial fluid. Various phenomena originating from interstitial fluid flow exist as well. Implicated in enhancing nutrient diffusion, interstitial fluid flow (i.e., of unbound water) gives rise to cellular stimuli of an electrical nature, namely streaming potentials and currents (Frank and Grodzinsky, 1987a, 1987b; Gu et al., 1993, 1998). In addition, interstitial fluid flow through the small pores associated with the solid matrix (~50 nm) of normal cartilage, which offer considerable resistance to fluid flow (Maroudas, 1979; McCutchen, 1962; Mow et al., 1984), can give rise to a mechanical phenomenon termed fluid-induced matrix compaction (Lai and Mow, 1980). The frictional interaction between interstitial fluid and solid are a result of drag resistance to forced flow through the porous-permeable cartilage matrix and a viscous shear stress exerted by the interstitial fluid. Given the nominal flow rates of the interstitial fluid mentioned earlier and the low permeability of the cartilage matrix, chondrocyte perception of this frictional interaction force is likely to be dominated by the drag resistance of flow through the matrix rather than by direct viscous shear stress on the cell (Ateshian et al., 2007). This frictional drag force can produce solid matrix deformation on the order of 15% to 30%.

From the previous discussion, chondrocyte deformation can be considered to be governed by three coupled loading mechanisms: direct ECM deformation; flow-induced compaction; and fluid pressurization (Mow et al., 1999). In OA, the increased tissue permeability diminishes the cartilage's normal fluid pressure load-support mechanism. Thus, there is a shift of load support onto the solid matrix, causing supranormal stresses and strains to be imposed on the chondrocytes (Fig. 3-27). These abnormally high stress and strain levels, and other mechano-electrochemical changes that are manifested with OA, can trigger an imbalance of chondrocyte anabolic and catabolic activities, further contributing to a vicious cycle of progressive cartilage degeneration. Indeed, changes to the biochemical composition and structure of cartilage can have a profound impact on tissue and chondrocyte function. With multidisciplinary collaborations and an appropriate theoretical framework, such as the biphasic theory, insights into the factors that govern chondrocyte function, cartilage structure and function, and the etiology of OA can be obtained.

Functional Tissue Engineering of Articular Cartilage

Because articular cartilage is avascular, it is unable to mount a typical wound healing response associated with the chemical factors and cells in blood. This poor healing capacity and the 15- to 20-year lifespan of orthopaedic implants has generated significant research toward the development of cell-based therapies and engineered cartilage tissues for joint repair. Tissue engineering strategies typically incorporate an appropriate cell source (e.g., stem cells, chondrocytes), scaffold material, and culture environment (chemical and biophysical stimuli) to grow fabricated tissues for repair and replacement of damaged or diseased tissues and organs.

Rooted in the field of biomechanics, functional tissue engineering (FTE) refers to the incorporation of physiologic loading during in vitro cultivation of engineered tissues to promote development of tissue substitutes with functional mechanical properties capable of sustaining the biomechanical demands imposed on them after in vivo implantation (Butler et al., 2000). Tissue engineering of cartilage has been performed in 3D scaffold and scaffold-free systems, which maintain the normal chondrocyte phenotype (e.g., type II collagen and aggrecan expression). The physical stimuli arising from joint loading can be described from material and biochemical properties of cartilage coupled with an appropriate constitutive framework such as the triphasic theory (Lai et al., 1991). Together, they allow description and prediction of the tissue loading–induced spatiotemporal stimuli that arise in cartilage and that modulate chondrocyte activities (Wang et al., 2002b).

Many of the testing devices described earlier to measure the material properties of soft hydrated tissues such as cartilage have inspired designs for physiologic loading bioreactor systems to foster tissue engineered cartilage growth in culture. Confined-compression permeation bioreactors (Dunkelman et al., 1995; Pazzano et al., 2000) and sliding bioreactors (Grad et al., 2005), arising from classical material testing configurations to measure tissue permeability and friction/wear properties, respectively, represent two examples. With increasing tissue maturation in culture, engineered tissues are reported to develop many of the important structure-function relationships that govern the behavior of cartilage in applied loading (e.g., Mauck et al., 2002; Vunjak-Novakovic et al., 1999).

Tissue deformation during joint loading is critical for providing nutrients to chondrocytes in avascular cartilage by augmenting diffusional transport of nutrients from the synovial fluid bathing the joint into the tissue (Albro et al., 2008; Mauck et al., 2003a). As such, joint loading provides nutrients as well as a plethora of complex multidimensional physical stimuli to chondrocytes that are important to the maintenance of cartilage. Accounting for bioengineering considerations for cartilage tissue engineering

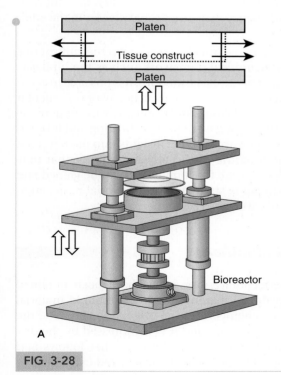

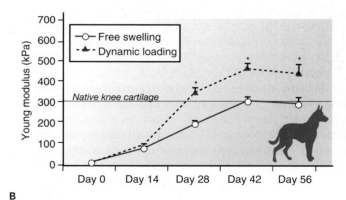

FIG. 3-28

Adult canine chondrocytes were expanded in 2D culture using a growth factor cocktail for 2 passages and then seeded in 2% w/v agarose hydrogel. **(A)** Constructs were then cultured in serum-free defined medium supplemented with transforming growth factor-β3 and subjected to applied deformational loading (uniaxial unconfined compression between smooth impermeable platens depicted in the **top figure**, 10% peak-to-peak strain, 3 hours/day) using a custom bioreactor depicted in the **bottom figure** or maintained in free swelling (or unloaded). **(B)** Application of loading expedited tissue development and led to significantly stiffer tissue properties versus free-swelling (Hung et al., 2009).

(Mow and Wang, 1999), the first study to demonstrate that long-term physiologic deformational loading to engineered constructs can promote improved tissue development was reported by Mauck et al. (2000). In the laboratory, deformation and loading bioreactors have been shown to promote development of functional engineered cartilage (Hung et al., 2004), via mechanisms that are likely to include mechanotransduction and increased nutrient availability (Mauck et al., 2003b).

In the laboratory, tissues with near native cartilage properties have been generated in eight weeks or less using juvenile (Lima et al., 2007) and adult (Hung et al., 2009) chondrocytes and applied uniaxial, unconfined compression deformational loading. Unconfined compression loading between smooth impermeable platens (another material testing configuration used to determine the compressive Young's modulus of cartilage) subjects constructs to both compressive strains, normal to the surface (i.e., parallel to the loading direction) and tensile strains, tangential to the surface (i.e., perpendicular to the loading direction), much like what native cartilage experiences in situ under physiologic conditions of loading (Park et al., 2003). In contrast to confined compression, it provides for nutrient access and fluid exudation-imbibition from the peripheral surfaces of the cylindrical tissues during loading (Fig. 3-28, top). Using unconfined compression loading bioreactors (Fig. 3-28, bottom), expedited tissue growth and resulting significant increases to construct properties (Fig. 3-28, right) compared with free swelling control constructs have been reported. The enhanced Young's modulus can be attributed in part to development of radial tensile properties (i.e., tangential to the surface) that effectively restrain lateral tissue expansion to axially applied loads (i.e., normal to the surface) (Kelly et al., 2006).

It remains to be seen if engineered cartilage cultivated using a functional tissue engineering paradigm and with mechanical "preconditioning" results in a significantly better clinical outcome when compared with repair with unloaded engineered tissue constructs. In addition to loading-induced differences in tissue development and properties, it is anticipated that subjecting chondrocytes to a regimen of in vitro loading may serve as training (such as for athletes for competition) for the dynamic in vivo joint loading environment.

Summary

- The function of articular cartilage in diarthrodial joints is to increase the area of load distribution (thereby reducing the stress) and provide a smooth, wear-resistant bearing surface.

- Biomechanically, articular cartilage should be viewed as a multiphasic material. In terms of a biphasic material, articular cartilage is composed of a porous-permeable collagen-PG solid matrix (approximately 25% by wet weight) filled by the freely movable interstitial fluid (approximately 75% by wet weight). In addition to solid and fluid there exists an additional ion phase when considering articular cartilage as a triphasic medium. The ion phase is necessary to describe swelling and other electromechanical behaviors of the tissue.

- Important biomechanical properties of articular cartilage are the intrinsic material properties of the solid matrix and the frictional resistance to the flow of interstitial fluid through the porous-permeable solid matrix (a parameter inversely proportional to the tissue permeability). Together, these parameters define the level of interstitial fluid pressurization, a major determinant of the load-bearing and lubrication capacity of the tissue, which can be generated in cartilage.

- Damage to articular cartilage, from whatever cause, can disrupt the normal interstitial fluid load-bearing capacity of the tissue and thus the normal lubrication process operating within the joint. Therefore, lubrication insufficiency may be a primary factor in the etiology of OA.

- When describing articular cartilage in the context of a rigorous theoretical framework such as the biphasic, triphasic, or multiphasic theories, it is possible to accurately predict the biomechanical behaviors of articular cartilage under loading and to elucidate the underlying mechanisms that govern its load-bearing and lubrication function. Furthermore, insights into the temporal and spatial nature of the physical stimuli that may affect chondrocyte function in situ can be gained and be used to guide strategies for functional tissue engineering of cartilage.

Practice Questions

1. The permeability data below have been obtained for a hydrogel material used as a scaffold material for tissue engineering. The original thickness of the specimen is 3 mm and it is compressed by the displacements provided in the table that follows and the respective permeability calculated using Darcy's Law.

    (a) Plot the permeability data and determine the intrinsic permeability for the hydrogel.
    (b) An exponential function can be used to describe the strain dependence of the permeability.

Determine the coefficient M that signifies the strain dependence of k.

(c) Why might strain-dependent permeability benefit the load support mechanisms of a hydrated tissue?

Displacement (mm)	Permeability m⁴/N·s
−0.1	2.39×10^{-14}
−0.2	1.96×10^{-14}
−0.3	1.62×10^{-14}
−0.4	1.48×10^{-14}

(d) From a practical perspective, why is a tare load needed to perform the experiment?

(e) Why is it important that the platens in the permeability device are free-flowing?

2. Given the material testing data curves,

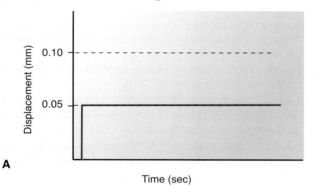

A

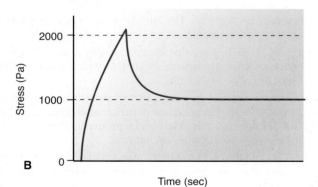

B

(a) Calculate the equilibrium modulus of a cylindrical specimen (thickness: 5 mm) that is axially loaded in unconfined compression. The x-axis is time (seconds).

(b) For a material composed of fluid and solid phases, what mechanism of load support exists for most of the peak load? Explain.

(c) What type of test is this?

Acknowledgements

This work was sponsored by the National Institutes of Health grants AR41913, AR42850, AR52871, and AR46568.

REFERENCES

Ahmed, A.M., Burke, D.L. (1983). In vitro measurement of static pressure distribution in synovial joints—Part 1: Tibial surface of the knee. *J Biomech Eng, 105,* 216.

Akizuki, S., Mow, V.C., Muller, F., et al. (1986). Tensile properties of knee joint cartilage: 1. Influence of ionic condition, weight bearing, and fibrillation on the tensile modulus. *J Orthop Res, 4,* 379.

Albro, M.B., Chahine N.O., Li, R., et al. (2008). Dynamic loading of deformable porous media can induce active solute transport. *J Biomech, 41*(15): 3152–3157.

Altman, R.D., Tenebaum, J., Latta, L., et al. (1984). Biomechanical and biochemical properties of dog cartilage in experimentally induced osteoarthritis. *Ann Rheum Dis, 43,* 83.

Andriacchi, T.P., Natarajan, R.N., Hurwitz, D.E. (1997). Musculoskeletal dynamics, locomotion, and clinical application. In V.C. Mow, W.C. Hayes (Eds.). *Basic Orthopaedic Biomechanics (2nd ed.).* Philadelphia: Lippincott–Raven Publishers, 31–68.

Armstrong, C.G., Bahrani, A.S., Bardner, D L. (1979). In vitro measurement of articular cartilage deformations in the intact human hip joint under load. *J Bone Joint Surg, 61A,* 744.

Armstrong, C.G., Mow, V.C. (1980). Friction, lubrication and wear of synovial joints. In R. Owen, P. Goodfellow, P. Bullough (Eds.). *Scientific Foundations of Orthopaedics and Traumatology.* London: William Heinemann, 223–232.

Armstrong, C.G., Mow, V.C. (1982). Variations in the intrinsic mechanical properties of human articular cartilage with age, degeneration, and water content. *J Bone Joint Surg, 64A,* 88.

Ateshian, G.A. (1997). Theoretical formulation for boundary friction in articular cartilage. *J Biomech Eng, 119,* 81.

Ateshian, G.A. (2009). The role of interstitial fluid pressurization in articular cartilage lubrication. *J Biomech, 42*(9), 1163–1176.

Ateshian, G.A., Costa K.D., Hung, C.T. (2007). A theoretical analysis of water transport through chondrocytes. *Biomech Model Mechanobiol, 6*(1–2), 91–101.

Ateshian, G.A., Kwak, S.D., Soslowsky, L.J., et al. (1994). A new stereophotogrammetry method for determining in situ contact areas in diarthrodial joints: A comparison study. *J Biomech, 27,* 111.

Ateshian, G.A., Lai, W.M., Zhu, W.B., et al. (1995). An asymptotic solution for the contact of two biphasic cartilage layers. *J Biomech, 27,* 1347.

Ateshian, G.A., Wang, H. (1995). A theoretical solution for the frictionless rolling contact of cylindrical biphasic articular cartilage layers. *J Biomech, 28,* 1341.

Ateshian, G.A., Wang, H., Lai, W.M. (1998). The role of interstitial fluid in pressurization and surface porosities on the boundary friction of articular cartilage. *J Tribol, 120,* 241.

Ateshian, G.A., Warden, W.H., Kim, J.J., et al. (1997). Finite deformation biphasic material properties of bovine articular cartilage from confined compression experiments. *J Biomech, 30*, 1157.

Athanasiou, K.A., Rosenwasser, M.P., Buckwalter, J.A., et al. (1991). Interspecies comparison of in situ mechanical properties of distal femoral cartilage. *J Orthop Res, 9*, 330.

Bachrach, N.M., Mow, V.C., Guilak, F. (1998). Incompressibility of the solid matrix of articular cartilage under high hydrostatic pressure. *J Biomech, 31*, 445.

Bachrach, N.M., Valhmu, W.B., Stazzone, E J., et al. (1995). Changes in proteoglycan synthesis rates of chondrocytes in articular cartilage are associated with the time dependent changes in the mechanical environment. *J Biomech, 28*, 1561.

Basalo, I.M., Mauck, R.L., Kelly, T.A., et al. (2004). Cartilage interstitial fluid load support in unconfined compression following enzymatic digestion. *J Biomech Eng, 126*(6), 779–786.

Bateman, J F., Lamande, S.R., Ramshaw, J.A.M. (1996). Collagen superfamily. In W.D. Comper (Ed.): *Extracellular Matrix* (Vol. 2). Amsterdam, Netherlands: Harwood Academic Publishers, 2267.

Bollet, A.J., Nance, J.L. (1965). Biochemical findings in normal and osteoarthritic articular cartilage, II: Chondroitin sulfate concentration and chain length, and water and ash content. *J Clin Invest, 45*, 1170.

Bowden, F.P., Tabor, D. (1967). *Friction and Lubrication*. London, UK: Methuen.

Broom, N D., Silyn-Roberts, H. (1990). Collagen-collagen versus collagen-proteoglycan interactions in the determination of cartilage strength. *Arthritis Rheum, 33*, 1512.

Buckwalter, J.A., Kuettner, K.E., Thonar, E.J.-M.A. (1985). Age-related changes in articular cartilage proteoglycans: Electron microscopic studies. *J Orthop Res, 3*, 251.

Bullough, P.G., Goodfellow, J. (1968). The significance of the fine structures of articular cartilage. *J Bone Joint Surg, 50B*, 852.

Bullough, P.G., Jagannath, A. (1983). The morphology of the calcification front in articular cartilage. *J Bone Joint Surg, 65B*, 72.

Buschmann, M. D., Gluzband, Y.A., Grodzinsky, A.J., et al. (1992). Chondrocytes in agarose culture synthesize a mechanically functional extracellular matrix. *J Orthop Res, 10*, 745.

Buschmann, M.D., Grodzinsky, A.J. (1995). A molecular model of proteoglycan-associated electrostatic forces in cartilage mechanics. *J Biomech Eng, 117*, 170.

Butler, D.L., Goldstein, S.A., Guilak, F. (2000). Functional tissue engineering: The role of biomechanics. *J Biomech Eng, 122*(6), 570–575.

Caligaris, M., Ateshian, G.A. (2008). Effects of sustained interstitial fluid pressurization under migrating contact area, and boundary lubrication by synovial fluid, on friction. *Osteoarthritis Cartilage, 16*(10), 1220–1227.

Clark, J. M. (1985). The organization of collagen in cryofractured rabbit articular cartilage: A scanning electron microscopy study. *J Orthop Res, 3*, 17.

Clarke, I C. (1971). Articular cartilage: A review and scanning electron microscope study—l. The interterritorial fibrillar architecture. *J Bone Joint Surg, 53B*, 732.

Donnan, F.G. (1924). The theory of membrane equilibria. *Chem Rev, 1*, 73.

Donohue, J.M., Buss, D., Oegema, T.R., et al. (1983). The effects of indirect blunt trauma on adult canine articular cartilage. *J Bone Joint Surg, 65A*, 948.

Dowson, D. (1966/1967). Modes of lubrication in human joints. *Proc Inst Mech Eng, 181J*, 45.

Dowson, D. (1990). Bio-tribology of natural and replacement joints. In V.C. Mow, A. Ratcliffe, S-L.Y. Woo (Eds.). *Biomechanics of Diarthrodial Joints*. New York: Springer-Verlag, 305–345.

Dunkelman, N.S., Zimber, M.P., LeBaron, R.G., et al. (1995). Cartilage production by rabbit articular chondrocytes on polyglycolic acid scaffolds in a closed bioreactor system. *Biotech Bioeng, 46*, 299–305.

Edwards, J. (1967). Physical characteristics of articular cartilage. *Proc Inst Mech Eng, 181J*, 16.

Elmore, S.M., Sokoloff, L., Norris, G., et al. (1963). Nature of "imperfect" elasticity of articular cartilage. *J Applied Physiol, 18*, 393.

Eyre, D.R. (1980). Collagen: Molecular diversity in the body's protein scaffold. *Science, 207*, 1315.

Fosang, A.J., Hardingham, T.E. (1996). Matrix proteoglycans. In W.D. Comper (Ed.) *Extracellular Matrix* (Vol. 2). Amsterdam, Netherlands: Harwood Academic Publishers, 200–229.

Frank, E.H., Grodzinsky, A.J. (1987a). Cartilage electromechanics—I. Electrokinetic transduction and effects of pH and ionic strength. *J Biomech, 30*, 615.

Frank, E.H., Grodzinsky, A.J. (1987b). Cartilage electromechanics—II. A continuum model of cartilage electrokinetics and correlation with experiments. *J Biomech, 20*, 629.

Freeman, M.A.R. (1975). The fatigue of cartilage in the pathogenesis of osteoarthrosis. *Acta Orthop Scand, 46*, 323.

Fung, Y.C. (1981). Quasi-linear viscoelasticity of soft tissues. In *Biomechanics: Mechanical Properties of Living Tissues*. New York: Springer-Verlag, 226.

Gardner, S.L., McGillivray, D.C. (1971). Living articular cartilage is not smooth. The structure of mammalian and avian joint surfaces demonstrated in vivo by immersion incident light microscopy. *Ann Rheum Dis, 30*, 3.

Garg, H.G., Swann, D.A. (1981). Age-related changes in the chemical composition of bovine articular cartilage. *Biochem J, 193*, 459.

Gleghorn, J.P., Jones, A.R., Flannery, C.R., et al. (2009). Boundary mode lubrication of articular cartilage by recombinant human lubricin. *J Orthop Res, 27*(6), 771–777.

Grad, S., Lee, C.R., Goma, K., et al. (2005). Surface motion upregulates superficial zone protein and hyaluronan production in chondrocyte-seeded three-dimensional scaffolds. *Tissue Eng, 11*(1–2), 249–256.

Gu, W.Y., Lai, W.M., Mow, V.C. (1993). Transport of fluid and ions through a porous-permeable charged-hydrated tissue, and streaming potential data on normal bovine articular cartilage. *J Biomech, 26*, 709.

Gu, W.Y., Lai, W.M., Mow, V.C. (1997). A triphasic analysis of negative osmotic flows through charged hydrated soft tissues. *J Biomech, 30*, 71.

Gu, W.Y., Lai, W.M., Mow, V.C. (1998). A mixture theory for charged hydrated soft tissues containing multi-electrolytes: Passive transport and swelling behaviors. *J Biomech Eng, 102*, 169.

Guilak, F., Ratcliffe, A., Lane, N., et al. (1994). Mechanical and biochemical changes in the superficial zone of articular cartilage in a canine model of osteoarthritis. *J Orthop Res, 12*, 474.

Guterl, C., Hung, C.T., Ateshian, G.A. (2010). Electrostatic and non-electrostatic contributions of proteoglycans to the compressive equilibrium modulus of bovine articular cartilage. *J Biomech, 43*(7), 1343–1350.

Hardingham, T.E., Beardmore-Garg, M., Dunham, D.G. (1987). Protein domain structure of the aggregating proteoglycan from cartilage. *Trans Orthop Res Soc, 12*, 61.

Hardingham, T.E., Muir, H. (1974). Hyaluronic acid in cartilage and proteoglycan aggregation. *Biochem J, 139*, 565.

Hascall, V.C. (1977). Interactions of cartilage proteoglycans with hyaluronic acid. *J Supramol Struct, 7*, 101.

Hayes, W.C., Bodine, A.J. (1978). Flow-independent viscoelastic properties of articular cartilage matrix. *J Biomech, 11*, 407.

Hayes, W.C., Mockros, L.F. (1971). Viscoelastic properties of human articular cartilage. *J Appl Physiol 31*, 562.

Hawking, S.W. (1988). *A Brief History of Time: From the Big Bang to Black Holes.* New York: Bantam Books.

Heinegard, D., Wieslander, J., Sheehan, J., et al. (1985). Separation and characterization of two populations of aggregating proteoglycans from cartilage. *Biochem J, 225*, 95.

Helminen, H.J., Kiviranta, I., Tammi, M., et al. (Eds.) (1987). *Joint Loading: Biology and Health of Articular Structures.* Bristol, UK: Wright and Sons, Publishers.

Hills, B.A. (1989). Oligolamellar lubrication of joints by surface active phospholipid. *J Rheum, 1–6*, 82–91.

Hirsch, C. (1944). The pathogenesis of chondromalacia of the patella. *Acta Chir Scand, 83*(Suppl), 1.

Hlavacek, M. (1995). The role of synovial fluid filtration by cartilage in lubrication of synovial joints: IV. Squeeze-film lubrication for axial symmetry under high loading conditions. *J Biomech, 28*, 1199.

Hodge, W.A., Fijan, R.S., Carlson, K., et al. (1986). Contact pressure in the human hip joint measured in vivo. *Proc Natl Acad Sci U S A, 83*, 2879.

Holmes, M.H., Lai, W.M., Mow, V.C. (1985). Singular perturbation analysis on the nonlinear, flow-dependent, compressive stress-relaxation behavior of articular cartilage. *J Biomech Eng, 107*, 206.

Hou, J.S., Mow, V.C., Lai, W.M., et al. (1992). An analysis of the squeeze-film lubrication mechanism for articular cartilage. *J Biomech, 25*, 247.

Huang, C.Y., Stankiewicz, A., Ateshian, G.A., et al. (2005). Anisotropy, inhomogeneity, and tension-compression nonlinearity of human glenohumeral cartilage in finite deformation. *J Biomech, 38*(4), 799–809.

Hultkrantz, W. (1898). Ueber die Spaltrichtungen der Gelenkknorpel. *Verh Anat Ges, 12*, 248.

Hung, C.T., Bian, L., Stoker, A.M., et al. (2009). Functional tissue engineering of articular cartilage using adult chondrocytes. In *Transactions of the International Cartilage Repair Society,* Miami, Florida, May 23–26, 2009, 8th World Congress of the International Cartilage Repair Society, 169.

Hung, C.T., Mauck, R.L., Wang, C.-C.B., et al. (2004). A paradigm for functional tissue engineering of articular cartilage via applied physiologic deformational loading. *Ann Biomed Eng, 32*(1), 35–49.

Hwang, J., Bae, W.C., Shieu, W., et al. (2008), Increased hydraulic conductance of human articular cartilage and subchondral bone plate with progression of osteoarthritis. *Arthritis Rheum, 58*(12), 3831–3842.

Jay, G.D., Torres, J.R., Rhee, D.K., et al. (2007). Association between friction and wear in diarthrodial joints lacking lubricin. *Arthritis Rheum, 56*(11), 3662–3669.

Katchalsky, A., Curran, P.F. (1975). *Nonequilibrium Thermodynamics in Biophysics (4th ed.).* Cambridge, MA: Harvard University Press.

Kelly, T.N., Ng, K.W., Wang, C.-C.B., et al. (2006). Spatial and temporal development of chondroctye-seeded agarose constructs in free-swelling and dynamically loaded cultures. *J Biomech, 39*(8), 1489–1497.

Kempson, G.E. (1979). Mechanical properties of articular cartilage. In M.A.R. Freeman (Ed.). *Adult Articular Cartilage (2nd ed).* Tunbridge Wells, UK: Pitman Medical, 333–414.

Kempson, G.E., Tuke, M.A., Dingle, J.T., et al. (1976). The effects of proteolytic enzymes on the mechanical properties of adult human articular cartilage. *Biochem Biophys Acta, 428*, 741.

Kim, Y.J., Sah, R.L., Grodzinsky, A.J., et al. (1994). Mechanical regulation of cartilage biosynthetic behavior: Physical stimuli. *Arch Biochem Biophys, 311*, 1.

Krishnan, R., Caligaris, M., Mauck, R.L., et al. (2004). Removal of the superficial zone of bovine articular cartilage does not increase its frictional coefficient. *Osteoarthritis Cartilage, 12*(12), 947–955.

Krishnan, R., Park, S., Eckstein, F., et al. (2003). Inhomogeneous cartilage properties enhance superficial interstitial fluid support and frictional properties, but do not provide a homogeneous state of stress. *J Biomech Eng, 125*(5), 569–577.

Lai, W.M., Gu, W.Y., Mow, V.C. (1998). On the conditional equivalence of chemical loading and mechanical loading on articular cartilage. *J Biomech, 31*(12), 1181–1185.

Lai, W.M., Hou, J.S., Mow, V.C. (1991). A triphasic theory for the swelling and deformation behaviors of articular cartilage. *J Biomech Eng, 113*, 245.

Lai, W.M., Mow, V.C. (1978). Ultrafiltration of synovial fluid by cartilage. *J Eng Mech Div ASCE, 104*, 79.

Lai, W.M., Mow, V.C. (1980). Drag-induced compression of articular cartilage during a permeation experiment. *Biorheology 17*, 111.

Lakes, R., Saha, S. (1979). Cement line motion in bone. *Science, 204*, 501.

Lane, J.M., Weiss, C. (1975). Review of articular cartilage collagen research. *Arthritis Rheum, 18*, 553.

Lima, E.G., Bian, L., Ng, K.W., et al. (2007). The beneficial effect of delayed compressive loading on tissue-engineered cartilage constructs cultured with TGF-B3. *Osteoarthritis Cartilage, 15*(9), 1025–1033.

Linn, F.C. (1968). Lubrication of animal joints: 1. The mechanism. *J Biomech, 1*, 193.

Linn, F.C., Radin, E.L. (1968). Lubrication of animal joints: III. The effect of certain chemical alterations of the cartilage and lubricant. *Arthritis Rheum, 11*, 674.

Linn, F.C., Sokoloff, L. (1965). Movement and composition of interstitial fluid of cartilage. *Arthritis Rheum, 8*, 481.

Lipshitz, H., Etheredge, R., Glimcher, M.J. (1975). In vitro wear of articular cartilage: I. Hydroxyproline, hexosamine, and amino acid composition of bovine articular cartilage as a function of depth from the surface; hydroxyproline content of the lubricant and the wear debris as a measure of wear. *J Bone Joint Surg, 57A*, 527.

Lipshitz, H., Etheredge, R., Glimcher, M.J. (1976). Changes in the hexosamine content and swelling ratio of articular cartilage as functions of depth from the surface. *J Bone Joint Surg, 58A*, 1149.

Lipshitz, H., Glimcher, M.J. (1979). In vitro studies of the wear of articular cartilage. *Wear, 52*, 297.

Malcolm, L.L. (1976). *An experimental investigation of the frictional and deformational responses of articular cartilage interfaces to static and dynamic loading.* Doctoral thesis, University of California, San Diego.

Mankin, H.A., Thrasher, A.Z. (1975). Water content and binding in normal and osteoarthritic human cartilage. *J Bone Joint Surg, 57A*, 76.

Mansour, J.M., Mow, V.C. (1976). The permeability of articular cartilage under compressive strain and at high pressures. *J Bone Joint Surg, 58A*, 509.

Maroudas, A. (1966/1967). Hyaluronic acid films. *Proc Inst Mech Eng London, 181J*, 122.

Maroudas, A. (1968). Physicochemical properties of cartilage in light of ion-exchange theory. *Biophys J, 8*, 575.

Maroudas, A. (1975). Biophysical chemistry of cartilaginous tissues with special reference to solute and fluid transport. *Biorheology, 12*, 233.

Maroudas, A. (1976). Balance between swelling pressure and collagen tension in normal and degenerate cartilage. *Nature, 260*, 808.

Maroudas, A. (1979). Physicochemical properties of articular cartilage. In M.A.R. Freeman (Ed.), *Adult Articular Cartilage (2nd ed)*. Tunbridge Wells, UK: Pitman Medical, 215–290.

Maroudas, A., Wachtel, E., Grushko, G., et al. (1991). The effect of osmotic and mechanical pressures on water partitioning in articular cartilage. *Biochem Biophys Acta, 1073*, 285.

Mauck, R.L., Hung, C.T., Ateshian, G.A. (2003a). Modeling of neutral solute transport in a dynamically loaded porous permeable gel: Implications for articular cartilage biosynthesis and tissue. *J Biomech Eng, 125*(5), 602–614.

Mauck, R.L., Nicoll, S.B., Seyhan, S.L., et al. (2003b). Synergistic effects of growth factors and dynamic loading for cartilage tissue engineering. *Tissue Eng, 9*(4), 597–611.

Mauck, R.L., Seyhan, S.L., Ateshian, G.A., et al. (2002). The influence of seeding density and dynamic deformational loading on the developing structure/function relationships of chondrocyte-seeded agarose hydrogels. *Ann Biomed Eng, 30*, 1046–1056.

Mauck, R.L., Soltz, M.A., Wang, C.C.-B., et al. (2000). Functional tissue engineering of articular cartilage through dynamic loading of chondrocyte-seeded agarose gels. *J Biomech Eng, 122*, 252–260.

McCutchen, C.W. (1962). The frictional properties of animal joints. *Wear, 5*, 1.

McDevitt, C.A., Muir, H. (1976). Biochemical changes in the cartilage of the knee in experimental and natural osteoarthritis in the dog. *J Bone Joint Surg, 58B*, 94.

Meachim, G., Fergie, I.A. (1975). Morphological patterns of articular cartilage fibrillation. *J Pathol, 115*, 231.

Mow, V.C., Amoczky, S.P., Jackson, D.W. (1992). *Knee Meniscus: Basic and Clinical Foundations.* New York: Raven Press.

Mow, V.C., Ateshian, G.A. (1997). Lubrication and wear of diarthrodial joints. In V.C. Mow, W.C. Hayes (Eds.). *Basic Biomechanics (2nd ed.)*. Philadelphia: Lippincott–Raven Publishers, 275–315.

Mow, V.C., Ateshian, G.A., Lai, W.M., et al. (1998). Effects of fixed charges on the stress-relaxation behavior of hydrated soft tissues in a confined compression problem. *Int J Solids Struct, 35*, 4945–4962.

Mow, V.C., Gibbs, M.C., Lai, W.M., et al. (1989a). Biphasic indentation of articular cartilage–Part II. A numerical algorithm and an experimental study. *J Biomech, 22*, 853.

Mow, V.C., Holmes, M.H., Lai, W.M. (1984). Fluid transport and mechanical properties of articular cartilage: A review. *J Biomech, 17*, 377.

Mow, V.C., Kuei, S.C., Lai, W.M., et al. (1980). Biphasic creep and stress relaxation of articular cartilage in compression: Theory and experiments. *J Biomech Eng, 102*, 73.

Mow, V.C., Lai, W.M., Redler, I. (1974). Some surface characteristics of articular cartilages. A scanning electron microscopy study and a theoretical model for the dynamic interaction of synovial fluid and articular cartilage. *J Biomech, 7*, 449.

Mow, V.C., Ratcliffe, A. (1997). Structure and function of articular cartilage and meniscus. In V.C. Mow, W.C. Hayes (Eds.). *Basic Orthopaedic Biomechanics (2nd ed.)*. Philadelphia: Lippincott–Raven Publishers, 113–177.

Mow, V.C., Wang, C.C.-B. (1999). Some bioengineering considerations for tissue engineering of articular cartilage. *Clin Orthop Rel Res, 367* (Suppl), S204–S223.

Mow, V.C., Wang, C.C., Hung, C.T. (1999). The extracellular matrix, interstitial fluid and ions as a mechanical signal transducer in articular cartilage. *Osteoarthritis Cartilage, 7*(1), 41–58.

Mow, V.C., Zhu, W.B., Lai, W.M., et al. (1989b). The influence of link protein stabilization on the viscoelastic properties of proteoglycan aggregates. *Biochem Biophys Acta, 992*, 201.

Muir, H. (1983). Proteoglycans as organizers of the extracellular matrix. *Biochem Soc Trans, 11*, 613.

Myers, E.R., Lai, W.M., Mow, V.C. (1984). A continuum theory and an experiment for the ion-induced swelling behavior cartilage. *J Biomech Eng, 106*(2), 151–158.

Newberry, W.N., Zukosky, D.K., Haut, R.C. (1997). Subfracture insult to a knee joint causes alterations in the bone and in the functional stiffness of overlying cartilage. *J Orthop Res, 15*, 450.

Ng, L., Grodzinsky, A.J., Patwari, P., et al. (2003). Individual cartilage aggrecan macromolecules and their constituent glycosaminoglycans visualized via atomic force microscopy. *J Struct Biol, 143*(3), 242–257.

Onsager, L. (1931). Reciprocal relations in irreversible processes. *I Phys Rev, 37*, 405.

Oswald, E.S., Chao, P.H., Bulinski, J.C., et al. (2008). Dependence of zonal chondrocyte water transport properties on osmotic environment. *Cell Mol Bioeng, 1*(4), 339–348.

Park, S., Hung, C.T., Ateshian, G.A. (2004). Mechanical response of bovine articular cartilage under dynamic unconfined compression loading at physiological stress levels. *Osteoarthritis Cartilage, 12*(1), 65–73.

Park, S., Krishnan, R., Nicoll, S.B., et al. (2003). Cartilage interstitial fluid load support in unconfined compression. *J Biomech, 36*(12), 1785–1796.

Park S., Nicoll, S.B., Mauck, R.L., et al. (2008). Cartilage mechanical response under dynamic compression at physiological stress levels following collagenase digestion. *Ann Biomed Eng, 36*(3):425–434.

Paul, J.P. (1976). Force actions transmitted by joints in the human body. *Proc Roy Soc Lond, 192B*, 163.

Pazzano, D., Mercier, K.A., Moran, J.M., et al. (2000). Comparison of chondrogensis in static and perfused bioreactor culture. *Biotechnol Prog, 16*(5), 893–896.

Poole, A.R. (1986). Proteoglycans in health and disease: Structure and function. *Biochem J, 236*, 1.

Radin, E.L. (1976). Aetiology of osteoarthrosis. *Clin Rheum Dis, 2*, 509.

Radin, E.L., Paul, I.L. (1971). Response of joints to impact loading. I. In vitro wear. *Arthritis Rheum, 14*, 356.

Ratcliffe, A., Mow, V.C. (1996). Articular cartilage. In W.D. Comper (Ed.), *Extracellular Matrix* (Vol. 1). Amsterdam, Netherlands: Harwood Academic Publishers, 234–302.

Ratcliffe, A., Tyler, J., Hardingham, T.E. (1986). Articular cartilage culture with interleukin 1: Increased release of link protein, hyaluronate-binding region and other proteoglycan fragments. *Biochem J*, 238, 571.

Redler, I., Zimny, M.L. (1970). Scanning electron microscopy of normal and abnormal articular cartilage and synovium. *J Bone Joint Surg, 52A*, 1395.

Redler, I., Zimny, M.L., Mansell, J., et al. (1975). The ultrastructure and biomechanical significance of the tidemark of articular cartilage. *Clin Orthop Rel Res, 112*, 357.

Rosenberg, L., Choi, H.U., Tang, L.-H., et al. (1985). Isolation of dermatan sulfate proteoglycans from mature bovine articular cartilage. *J Biol Chem, 260*, 6304.

Rosenberg, L., Hellmann, W., Kleinschmidt, A.K. (1975). Electron microscopic studies of proteoglycan aggregates from bovine articular cartilage. *J Biol Chem, 250*, 1877.

Roth, V., Mow, V.C. (1980). The intrinsic tensile behavior of the matrix of bovine articular cartilage and its variation with age. *J Bone Joint Surg, 62A*, 1102.

Roughley, P.J., White, R.J. (1980). Age-related changes in the structure of the proteoglycan subunits from human articular cartilage. *J Biol Chem, 255*, 217.

Roughley, P.J., White, R.J., Santer, V. (1981). Comparison of proteoglycans extracted from high and low-weight-bearing human articular cartilage, with particular reference to sialic acid content. *J Biol Chem, 256*, 12699.

Schinagl, R.M., Gurskis, D., Chen, A.C., et al. (1997). Depth-dependent confined compression modulus of full-thickness bovine articular cartilage. *J Orthop Res, 15*, 499.

Schinagl, R.M., Ting, M.K., Price, J.H., et al. (1996). Video microscopy to quantitate the inhomogeneous equilibrium strain within articular cartilage during confined compression. *Ann Biomed Eng, 24*, 500.

Schmidt, M.B., Mow, V.C., Chun, L.E., et al. (1990). Effects of proteoglycan extraction on the tensile behavior of articular cartilage. *J Orthop Res, 8*, 353.

Schmidt, T.A., Gastelum, N.S., Nguyen, Q.T., et al. (2007). Boundary lubrication of articular cartilage: Role of synovial fluid constituents. *Arthritis Rheum, 56*(3), 882–891.

Schneiderman, R., Keret, D., Maroudas, A. (1986). Effects of mechanical and osmotic pressure on the rate of glycosaminoglycan synthesis in adult femoral head cartilage: An in vitro study. *J Orthop Res, 4*, 393.

Schubert, M., Hamerman, D. (1968). *A Primer on Connective Tissue Biochemistry*. Philadelphia: Lea & Febiger.

Scott, J.E., Orford, C.R. (1981). Dermatan sulphate-rich proteoglycan associates with rat tail-tendon collagen at the d band in the gap region. *Biochem J, 197*, 213.

Seog, J., Dean D., Rolauffs, B., et al, C. (2005). Nanomechanics of opposing glycosaminoglycan macromolecules. *J Biomech, 38*(9), 1789–1797.

Setton, L.A., Gu, W.Y., Lai, W.M., et al. (1995). Predictions of the swelling induced pre-stress in articular cartilage. In A.P.S. Selvadurai (Ed.): *Mechanics of Porous Media*. Kluwer Academic Publishers. Dordrecht, Netherlands, 299–322.

Setton, L.A., Mow, V.C., Muller, F.J., et al. (1994). Mechanical properties of canine articular cartilage are significantly altered following transection of the anterior cruciate ligament. *J Orthop Res, 12*, 451.

Setton, L.A., Tohyama, H., Mow, V.C. (1998). Swelling and curling behavior of articular cartilage. *J Biomech Eng, 120*, 355.

Setton, L.A., Zhu, W.B., Mow, V.C. (1993). The biphasic porovis-coelastic behavior of articular cartilage in compression: Role of the surface zone. *J Biomech, 26*, 581.

Shen, Z.L., Dodge, M.R., Kahn, H., et al J. (2008). Stress-strain experiments on individual collagen fibrils. *Biophys J, 95*(8), 3956–3963.

Sokoloff, L. (1963). Elasticity of articular cartilage: Effect of ions and viscous solutions. *Science, 141*, 1055.

Soltz, M.A., Ateshian, G.A. (1998). Experimental verification and theoretical prediction of cartilage interstitial fluid pressurization at an impermeable contact interface in confined compression. *J Biomech, Oct 31* (10), 927–934.

Stockwell, R.S. (1979). *Biology of Cartilage Cells.* Cambridge, UK: Cambridge University Press.

Sun, D.N., Gu, W.Y., Guo, X.E., et al. (1998). The influence of inhomogeneous fixed charge density on cartilage mechano-electrochemical behaviors. *Trans Orthop Res Soc, 23*, 484.

Swann, D.A., Radin, E.L., Hendren, R.B. (1979). The lubrication of articular cartilage by synovial fluid glycoproteins. *Arthritis Rheum, 22*, 665.

Swann, D.A., Silver, F.H., Slayter, H.S., et al. (1985). The molecular structure and lubricating activity of lubricin from bovine and human synovial fluids. *Biochem J, 225*, 195.

Sweet, M.B.E., Thonar, E.J.-M.A., Marsh, J. (1979). Age-related changes in proteoglycan structure. *Arch Biochem Biophys, 198*, 439–448.

Thompson, R.C., Oegema, T.R., Lewis, J.L., et al. (1991). Osteo-arthrotic changes after acute transarticular load. An animal model. *J Bone Joint Surg, 73A*, 990.

Thonar, E.J.-M.A., Bjornsson, S., Kuettner, K.E. (1986). Age-related changes in cartilage proteoglycans. In K. Kuettner, R.S. Schleyerbach, V.C. Hascall (Eds.). *Articular Cartilage Biochemistry.* New York: Raven Press, 273–287.

Torzilli, P.A., Mow, V.C. (1976). On the fundamental fluid transport mechanisms through normal and pathologic cartilage during function. I. The formulation. *J Biomech, 9*(8), 541–552.

Torzilli, P.A., Rose, D.E., Dethemers, S.A. (1982). Equilibrium water partition in articular cartilage. *Biorheology, 19*, 519.

Urban, J.P.G., McMullin, J.F. (1985). Swelling pressure of the intervertebral disc: Influence of collagen and proteoglycan content. *Biorheology, 22*, 145.

Valhmu, W.B., Stazzone, E.J., Bachrach, N.M., et al. (1998). Load-controlled compression of articular cartilage induces a transient stimulation of aggrecan gene expression. *Arch Biochem Biophys, 353*, 29.

Venn, M.F. (1978). Variation of chemical composition with age in human femoral head cartilage. *Ann Rheum Dis, 37*, 168.

Vunjak-Novakovic, G., Martin, I., Obradovic, B., et al. (1999). Bioreactor cultivation conditions modulate the composition and mechanical properties of tissue-engineered cartilage. *J Orthop Res, 17*, 130–138.

Wada, T., Akizuki, S. (1987). An ultrastructural study of solid matrix in articular cartilage under uniaxial tensile stress. *J Jpn Orthop Assoc, 61*.

Walker, P.S., Dowson, D., Longfeild, M.D., et al. (1968). 'Boosted lubrication' in synovial joints by fluid entrapment and enrichment. *Ann Rheum Dis, 27*, 512.

Walker, P.S., Unsworth, A., Dowson, D., et al. (1970). Mode of aggregation of hyaluronic acid protein complex on the surface of articular cartilage. *Ann Rheum Dis, 29*, 591.

Wang, C.B., Mow, V.C. (1998). Inhomogeneity of aggregate modulus affects cartilage compressive stress-relaxation behavior. *Trans Orthop Res Soc, 23*(1), 481.

Wang, C.C., Chahine, N.O., Hung, C.T., et al. (2003) Optical determination of anisotropic material properties of bovine articular cartilage in compression. *J Biomech, 36*(3), 339–353.

Wang, C.C.-B., Deng, J.M., Ateshian, G.A., et al. (2002a). An automated approach for direct measurement of strain distributions within articular cartilage under unconfined compression. *J Biomech Eng, 124*, 557–567.

Wang, C C.-B., Guo, X.E., Sun, D., et al. (2002b). The functional environment of chondrocytes within cartilage subjected to compressive loading: Theoretical and experimental approach. *Biorheology, 39*(1–2), 39–45.

Weiss, C., Rosenberg, L., Helfet, A.J. (1968). An ultrastructural study of normal young adult human articular cartilage. *J Bone Joint Surg, 50A*, 663.

Williams, P.F., Powell, G.L., Laberge, M. (1993). Sliding friction analysis of phosphatidylcholine as a boundary lubricant for articular cartilage. *Proc Inst Mech Eng H, 207*, 59.

Woo, S.L.-Y., Mow, V.C., Lai, W.M. (1987). Biomechanical properties of articular cartilage. In *Handbook of Bioengineering.* New York: McGraw-Hill, 4.1–4.44.

Woo, S.L.Y., Levesay G.A., Runco, T.J., et al. (1997). Structure and function of tendons and ligaments. In V.C. Mow, W.C. Hayes (Eds.). *Basic Orthopaedic Biomechanics* (2nd ed.). Philadelphia: Lippincott–Raven Publishers, 209–251.

Zhu, W.B., Iatridis, J.C., Hlibczjk, V., et al. (1996). Determination of collagen-proteoglycan interactions in vitro. *J Biomech, 29*, 773.

Zhu, W.B., Lai, W.M., Mow, V.C. (1986). Intrinsic quasi-linear viscoelastic behavior of the extracellular matrix of cartilage. *Trans Orthop Res Soc, 11*, 407.

Zhu, W.B., Lai, W.M., Mow, V.C. (1991). The density and strength of proteoglycan-proteoglycan interaction sites in concentrated solutions. *J Biomech, 24*, 1007.

Zhu, W.B., Mow, V.C., Koob, T.J., et al. (1993). Viscoelastic shear properties of articular cartilage and the effects of glycosidase treatments. *J Orthop Res, 11*, 771.

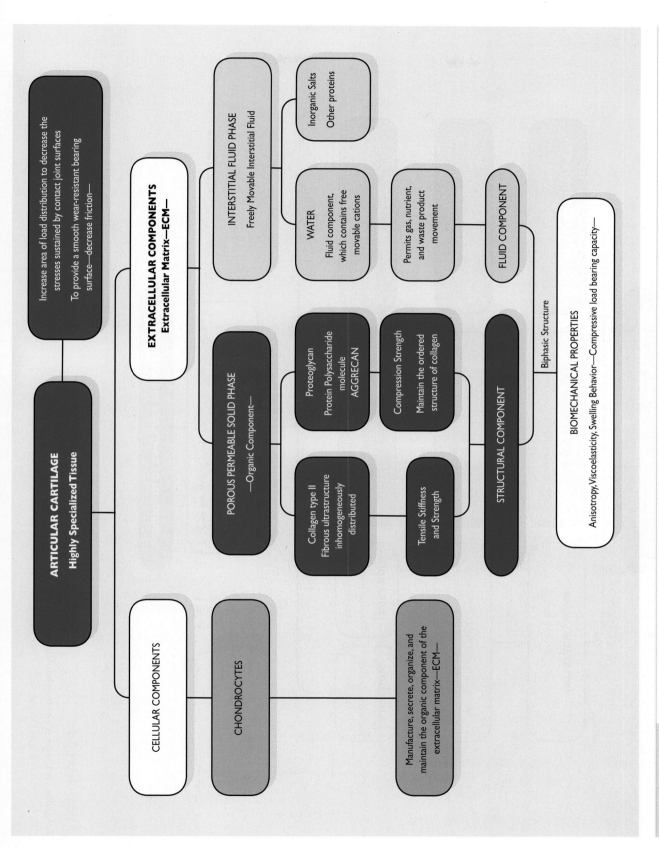

Articular cartilage structure and biomechanical properties.*

*This flow chart is designed for classroom or group discussion. Flow chart is not meant to be exhaustive.

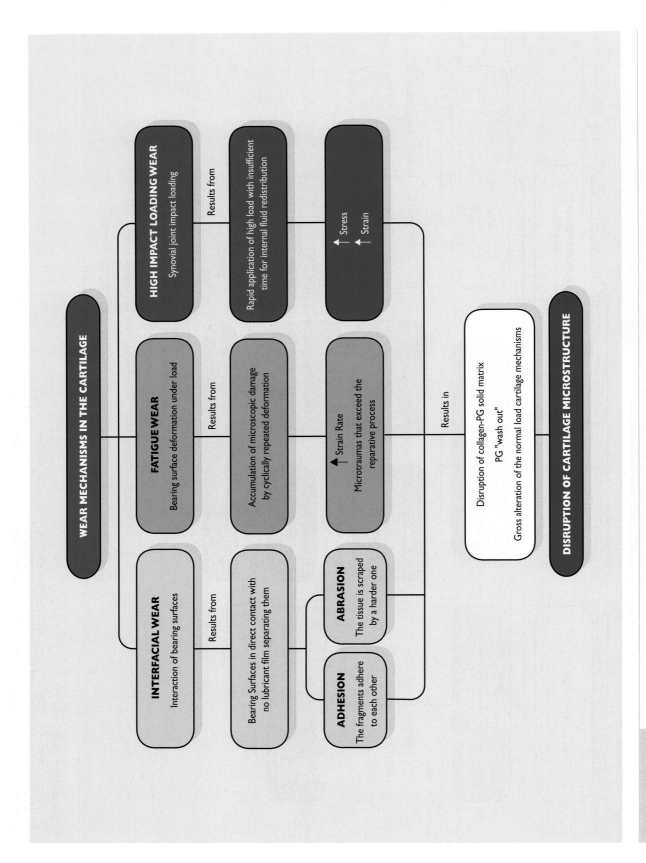

FLOW CHART 3-2

Articular cartilage wear mechanisms.*

*This flow chart is designed for classroom or group discussion. Flow chart is not meant to be exhaustive.

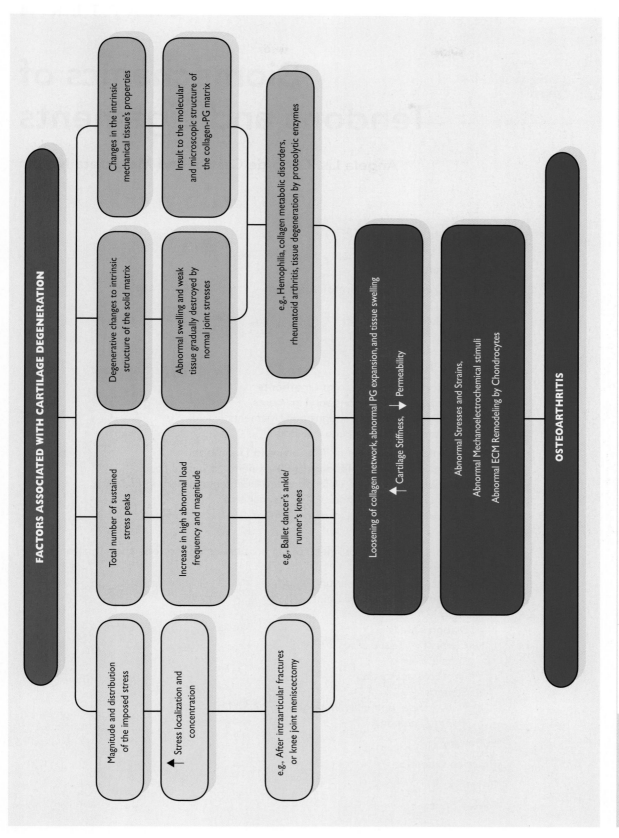

FLOW CHART 3-3

Factors associated with cartilage degeneration.*

*This flow chart is designed for classroom or group discussion. Flow chart is not meant to be exhaustive.
PG, proteoglycan; ECM, extracellular matrix.

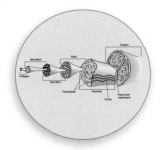

CHAPTER 4

Biomechanics of Tendons and Ligaments

Angela Lis, Carlo de Castro, and Margareta Nordin

Introduction

The three principal structures that closely surround and connect the joints of the skeletal system are tendons, ligaments, and joint capsules. Although these structures are mechanically passive (i.e., they do not contract and produce motion as do the muscles), each plays an essential role in joint motion and stability.

The role of the ligaments and joint capsules, which connect bone to bone, is to augment the mechanical stability of the joints, to guide joint motion, to prevent excessive motion, and to contribute to proprioception or position sense. Ligaments and joint capsules act as static restraints. The function of the tendons is to attach muscle and transmit the tensile load it generates to bone to produce joint motion or promote joint stability, and to contribute in maintaining body posture. The tendons and the muscles compose the muscle-tendon unit, which acts as a dynamic restraint, and allow muscles to be pre-positioned at an optimal distance from joints without need for increased muscular length.

Tendons, ligaments, and joint capsules also have a critical role in motor control because they contain neural structures that provide constant feedback regarding joint position in space. Tendon and ligament injuries and derangements are common. Proper management of these disorders requires an understanding of the biomechanical properties and functions of tendons and ligaments and their healing mechanisms. This chapter discusses the following aspects of tendons and ligaments:

1. Composition and structure
2. Normal biomechanical properties and behavior
3. Biomechanical properties and behavior following injury
4. Factors that affect the biomechanical function of tendons and ligaments

Composition and Structure of Tendons

Tendons connect muscle to bone as they extend from the muscle to the bony insertion, allowing it to transmit the tensile load generated through muscular contraction or passive elongation. They also protect joints from instability. To transmit the tensile load and to protect the joint from unstable positions, tendons are composed of dense connective tissue. They are composed of an extracellular matrix dominated by a parallel-fibered collagenous network and by metabolically active fibroblastic cells called tenocytes.

Like other connective tissues, tendons have relatively few cells (tenocytes) and an abundant extracellular matrix. In general, the cellular material occupies approximately 20% of the total tissue volume, while the extracellular matrix accounts for the remaining 80%. Approximately 55% to 70% of the matrix consists of water, and a substantial part of this is associated with the proteoglycans in the extracellular matrix (ECM). The remaining percentage are solids composed mainly of collagen (60%–85%), an inorganic substance such as proteoglycan (<0.2%), a small amount of elastin (~2%), and other proteins (~4.5%) (Kjaer, 2004).

TENDON CELLS (TENOCYTES)

Cells within the tendon substance are specialized fibroblasts called tenocytes. The primary role of these cells is to control tendon metabolism (production and degradation of the extracellular matrix) and to respond to the mechanical stimuli applied to the tendon, particularly tensile loads that serve as signals for collagen production in a process known as mechanotransduction. These cells lie in longitudinal rows along the collagen fibrils, following the tensile load at which they are stressed. It has been found that tenocytes have multiple extensions that stretch extensively within the extracellular matrix, allowing for three-dimensional intercellular communication via gap junctions (Kjaer, 2004). Blocking these gap junctions in vitro resulted in the cessation of collagen production in response to tensile loads (Benjamin et al., 2008).

EXTRACELLULAR MATRIX (ECM)

The extracellular matrix of tendons is largely composed of a network of collagen fibers and of a smaller percentage of proteoglycans, elastin, and other proteins. These components' primary function is to maintain the tendon's structure and facilitate the biomechanical response to mechanical loading.

COLLAGEN

The collagen network is dominated by type I fibers (~60%), but other types (e. g., III, IV, V, VI) are also present (Benjamin et al., 2008; Kjaer, 2004). Collagen type I fibers are characterized by their capacity to sustain large tensile loads while allowing for some level of compliance or mechanical deformation. Collagen is synthesized by tenocytes in an intricate process that ultimately contributes to the quality and stability of the collagen molecule (Fig. 4-1). This process of collagen synthesis is similar for all of the connective tissues, with some differences based on the type of collagen produced. Therefore tendons, ligaments, and bone that share collagen type I have similar synthesis and degradation processes.

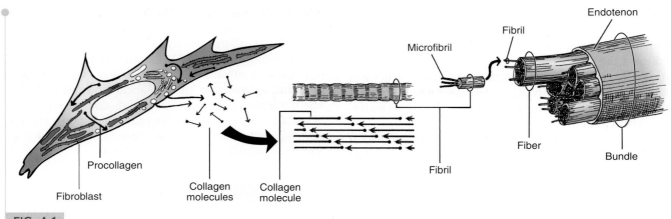

FIG. 4-1

Schematic representation of collagen fibrils, fibers, and bundles in tendons and collagenous ligaments (not drawn to scale). Collagen molecules, triple helices of coiled polypeptide chains, are synthesized and secreted by the fibroblasts. These molecules (depicted with "heads" and "tails" to represent positive and negative polar charges) aggregate in the extracellular matrix in a parallel arrangement to form microfibrils and then fibrils. The staggered array of the molecules, in which each overlaps the other, gives a banded appearance to the collagen fibrils under the electron microscope. The fibrils aggregate further into fibers, which come together into densely packed bundles.

The synthesis process starts at the membrane of the fibroblasts (in tendons: tenocytes) (Fig. 4-2). At this level there are *integrin* molecules that serve as a direct link between the cytoskeleton and the ECM. Integrin has a key role in the production of collagen because these molecules are sensitive to the transfer of mechanical loading from the outside to the inside of the cell and vice versa. It is hypothesized that integrins are sensors, a bridge through which forces are transmitted. It is believed that they are sensitive to the tensile strain at the cell membrane (Kjaer, 2004) and have the capacity to transform these mechanical stimuli into adaptive responses of the cell. Thus, this mechanotransduction along with the presence of growth factors such as TGF-β (transforming growth factor-β), IGF (insulin-like growth factor), IGF-BP (insulin-like growth factor and its binding proteins), FGB (fibroblast growth factor), and VEGF (vasoactive endothelial growth factor) have been presumed to be the main regulators of collagen production. Interleukins (IL-1, IL-6) and prostaglandins (PGs) are also involved in this process.

Several pathways of mechanotransduction signaling between the previously mentioned regulators and the cell nucleus have been suggested. The most crucial among this is MAPK (mitogen-activated protein kinase), which is an enzyme that induces signaling from the cytosol to the nucleus. This information mediates gene expression and the activation of protein synthesis to

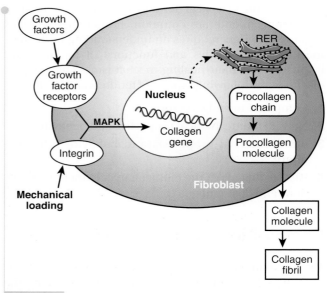

FIG. 4-2

Schematic diagram simplifying the process of mechanotransduction. In the presence of mechanical loading and key growth factors, a fibroblast responds in a series of events that involves signaling of integrin, mitogen-activated protein kinase (MAPK), and the cell nucleus to trigger production by the Rough Endoplasmic Reticulum (RER) of procollagen fibrils that become cleaved extracellularly to form collagen. *Adapted from Kjaer M. (2004). Role of extracellular matrix in adaptation of tendon and skeletal muscle to mechanical loading. Physiol Rev, 84(2), 658.*

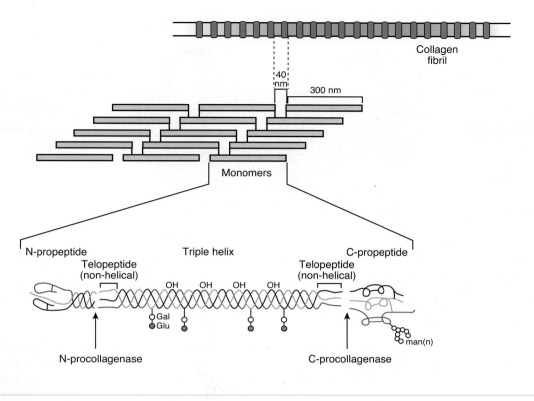

FIG. 4-3

Schematic drawing of collagen microstructure. The collagen molecule consists of three alpha chains in a triple helix **(bottom)**. Several collagen molecules are aggregated into a staggered parallel array. This staggering, which creates hole zones and overlap zones, causes the cross-striation (banding pattern) visible in the collagen fibril under the electron microscope. Gal and Glu (aminoacids); OH represents the hydrogen-bonded bridges.

initiate procollagen production. The synthesis of collagen fiber occurs first at the intracellular level with the assembly and secretion of procollagen within the rough endoplasmic reticulum (RER). Procollagen fibrils are then secreted and cleaved extracellularly to form collagen. The collagen molecule consists of three polypeptide chains (α chains), each coiled in a left-handed helix with approximately 100 amino acids (Fig. 4-3). Two of the peptide chains (called α-1 chains) are identical, and one differs slightly (the α-2 chain). The three α-chains are combined in a right-handed triple helix, which is a unique and characteristic feature and gives this molecule its rod-like shape (Prockop, 1990). The length of the molecule is approximately 280 nanometers (nm), and its diameter is approximately 1.5 nm.

About 300 repeating sequences of amino acids (glycine, proline, and hydroxyproline), not usually found in other proteins, characterize collagen (Prockop, 1990). Every third amino acid in each chain is glycine, and this repetitive sequence is essential for the proper formation of the triple helix. The small size of this amino acid allows the tight helical packing of the collagen molecule. Moreover, glycine enhances the stability of the molecule by forming hydrogen bonds among the three chains of the superhelix. Hydroxyproline and proline form hydrogen bonds, or hydrogen-bonded water bridges, within each chain. The intrachain and interchain bonding, or cross-linking, between specific groups on the chains is essential to the stability of the molecule.

Cross-links are also formed between collagen molecules and are essential to aggregation at the fibril level. It is the cross-linked character of the collagen fibrils that gives strength to the tissues they compose and that allows these tissues to function under mechanical stress. Within the fibrils, the molecules are apparently cross-linked by "head-to-tail" interactions (Fig. 4-1), but interfibrillar cross-linking of a more complex nature also may occur.

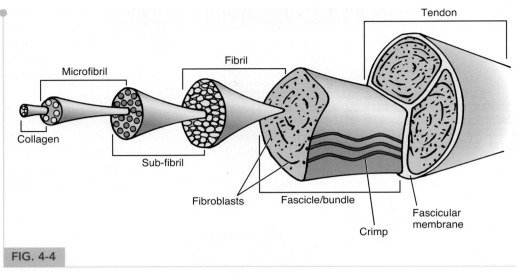

FIG. 4-4

Schematic representation of the microarchitecture of a tendon.

In newly formed collagen, the cross-links are relatively few and are reducible; the collagen is soluble in neutral salt solutions and in acid solutions, and the cross-links are fairly easily denatured by heat. As collagen ages, the total number of reducible cross-links decreases to a minimum and many stable, nonreducible cross-links are formed due to glycation. Mature collagen is not soluble in neutral salt solutions or in acid solutions, and it survives a higher denaturation temperature. This is thought to be associated with an accumulation of advanced glycation products (Riley, 2004).

A fibril is formed by the aggregation of several collagen molecules in a quaternary structure. This structure, in which each molecule overlaps the other, is responsible for the repeating bands observed on the fibrils under the electron microscope (Fig. 4-3). The quaternary structure of collagen relates to the organization of collagen molecules into a stable, low energetic biologic unit. By arranging adjacent collagen molecules in a quarter-stagger, oppositely charged amino acids are aligned. This stable structure will require a great amount of energy and force to separate its molecules, thus contributing to the strength of the structure. In this way, organized collagen molecules (five) form units of microfibrils, subfibrils, and fibrils (Fig. 4-4) (Simon, 1994). The fibrils aggregate further to form collagen fibers, which are visible under the light microscope. The fibers aggregate further into bundles and tenocytes are elongated in rows between these bundles, which are aligned in the direction of the mechanical load (Fig. 4-5).

Fibril segments can range in length from a few microns to approximately 100 µm with diameters that vary with the shape of the fibril (Kjaer, 2004). They have a characteristic undulated form referred to as *crimp* (Fig. 4-4), a characteristic that serves an important biomechanical function as it allows room for the straightening of this wavy configuration during tensile loading

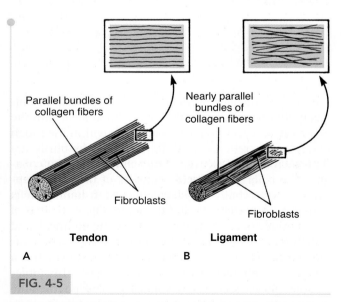

FIG. 4-5

Schematic diagram of the structural orientation of the fibers of tendon **(A)** and ligament **(B); insets** show longitudinal sections. In both structures the fibroblasts are elongated along an axis in the direction of function. *Adapted from Snell, R.S. (1984). Clinical and Functional Histology for Medical Students. Boston: Little, Brown and Company.*

(Benjamin, 2008; Kjaer, 2004). It has been suggested that the interplay between the crimp and the fibrous extracellular matrix may be responsible for the nonlinear viscoelastic properties of these tissues (e.g., the toe region of deformation) (Dourte et al., 2008; Riley, 2004).

The metabolic turnover of collagen fibers can occur intracellularly and extracellularly. Intracellular degradation occurs via phagocytosis. Extracellularly, particularly in type I collagen, matrix metalloproteinases (MMPs) and tissue inhibitors of matrix metalloproteinases (TIMPs) have been found to facilitate the breakdown of collagen in the presence of injury and inflammation. However, it is not clear if MMPs and TIMPs play a role during normal physiologic turnover of collagen. Collagen in mature animals tends to have a very long half-life, and most molecules last through a lifetime (Kjaer, 2004).

ELASTIN

The mechanical properties of tendons and ligaments depend not only on the architecture and properties of the collagen fibers but also on the proportion of elastin that these structures contain. The protein elastin is scarcely present in tendons and extremity ligaments composing approximately 2% of the dry weight (Kjaer, 2004), but in elastic ligaments such as the ligamentum flavum, the proportion of elastic fibers is substantial. Nachemson and Evans (1968) found a 2-to-1 ratio of elastic to collagen fibers in the flavum ligament. This ligament, which connects the laminae of adjacent vertebrae, appears to have a specialized role, which is to protect the spinal nerve roots from mechanical impingement, to pre-stress (preload) the motion segment (the functional unit of the spine), and to provide some intrinsic stability to the spine.

GROUND SUBSTANCE

The ground substance in tendons and ligaments is composed mainly of inorganic substances and other proteins, accounting for <0.2% and approximately 4.5% of the weight, respectively. Of the inorganic substances, the most dominant molecules are proteoglycans (PGs), which are macromolecules composed of various sulfated polysaccharide chains (glycosaminoglycans), bonded to a core protein linked to a long hyaluronic acid (HA) chain that forms an extremely high molecular weight PG aggregate such as that found in the ground substance of articular cartilage. Only a few PGs are found in the ground substance, the most common of which, and those that have been found to contribute to biomechanical and viscoelastic properties, are decorin and cartilage oligomatrix protein (COMP) (Kjaer, 2004).

The PG aggregates bind most of the extracellular water of the ligament and tendon, making the matrix a highly structured gel-like material rather than an amorphous solution. This combination allows for spacing and lubrication between the collagen microfibrils while at the same time acts as a cement-like substance that may help stabilize the collagenous skeleton of tendons and ligaments, and contributes to the overall strength of these composite structures. Apart from decorin and COMP, other PGs are found in the substance of tendons and ligaments. However, because these molecules are nonaggregated to HA chains, their function remains unclear (Kjaer, 2004).

Composition and Structure of Ligaments

Ligaments have the same general composition as tendons with a few key differences. Similar to tenocytes, ligaments have fibroblasts that are found within the ligament substance aligned with the collagen fibrils. Like tenocytes, ligamentous fibroblasts also form an extensive network with other cells via cytoplasmic extensions that are linked by gap junctions. The extracellular matrix is also composed mainly of type I collagen, although in contrast to tendons, the fibers are not parallel and are multidirectional (Fig. 4-5). Most are in line with the axis of the ligament. Although ligaments generally sustain tensile loads in one predominant direction, they may also bear smaller tensile loads in other directions, which suggests that the fibers are interlaced even if they are not completely parallel (Fig. 4-5B). Thus, the specific orientation of the fiber bundles varies to some extent among the ligaments and depends on the function of the ligament (Amiel et al., 1984).

VASCULAR SUPPLY OF TENDONS AND LIGAMENTS

Tendons and ligaments have a limited vascularization. This directly affects their metabolic activity, most critically during healing and repair. Blood vessels in tendons represent only about 1% to 2% of the extracellular matrix (Kjaer, 2004). Thus they appear white compared with the highly vascular red-colored muscles from which they originate. Besides their limited vascularity, several factors also contribute to their blood supply, such as their anatomic location, morphology, prior injury, and levels of physical activity. For instance, there is evidence that some ligaments and tendons are more vascular due to their anatomic location and attachments, or their shape and function (e.g., avascular areas of long flexor tendons overlaying bony pulleys). In addition, there is also evidence that blood flow is increased in tendons and ligaments and in surrounding tissues following periods of increased physical activity (Benjamin et al., 2008) without evidence of tissue ischemia, even with intense loading (Kjaer, 2004) and after an injury that seems to trigger revascularization and neovascularization in previously avascular areas.

Tendons receive their blood supply directly from vessels in the perimysium, the periosteal insertion, and the surrounding tissue via vessels in the paratenon or mesotenon. Tendons surrounded by paratenon have been referred to as vascular tendons, and those surrounded by tendon sheaths as avascular tendons. This is a misnomer because these "avascular" tendons possess blood vessels that run through the mesotenon. In vascular tendons, vessels enter from many points on the periphery, anastomose with a longitudinal system of capillaries, and pass through the endotenon surrounding the fiber bundles (fascicles) (Fig. 4-6).

Tendons that are wrapped in sheaths have a different blood supply mechanism and are also bathed in synovial fluid within the sheaths. Blood vessels run through mesotenons, which can appear as folds or elongated structures known as vinculae (Fig. 4-7). The latter are found in digital tendons. Vinculae connect the tendon sheath to the tendon at regular intervals, creating areas of increased and decreased vascularity (Benjamin et al., 2008). The hypovascular regions led various researchers to propose a dual pathway for tendon nutrition: a

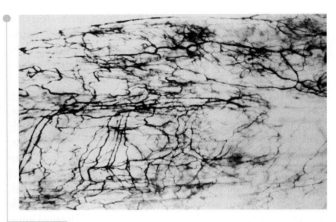

FIG. 4-6

India ink–injected (Spälteholz technique) into the calcaneal tendon of a rabbit, illustrating the vasculature of a paratenon-covered tendon. Vessels enter from many points on the periphery and anastomose with a longitudinal system of capillaries. *Reprinted with permission from Woo, S.L.Y., An, K.N., Arnoczky, D.V.M., et al. (1994). Anatomy, biology, and biomechanics of the tendon, ligament, and meniscus. In S.R. Simon (Ed.). Orthopaedic Basic Science. Rosemont, IL: American Academy of Orthopaedic Surgeons.*

vascular pathway, and, for the hypovascular regions, a synovial (diffusion) pathway. The concept of diffusional nutrition is of primary clinical significance because it implies that for tendons enclosed in a sheath, healing and repair can occur in the absence of adhesions.

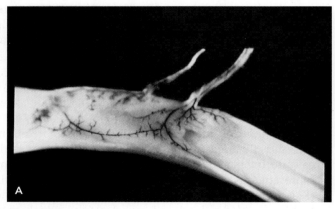

FIG. 4-7

A. India ink-injected specimen illustrating the vascular supply of the flexor digitorum profundus in a human through the vinculum longus. **B.** Close-up specimen (Spälteholz technique) showing the extent of the blood supply from the vinculum longus. The vessels in the vinculum divide into the dorsal, proximal, and distal branches, giving off vascular loops into the tendon substance. *Reprinted with permission from Woo, S.L.Y., An, K.N., Arnoczky, D.V.M., et al. (1994). Anatomy, biology, and biomechanics of the tendon, ligament, and meniscus. In S.R. Simon (Ed.). Orthopaedic Basic Science. Rosemont, IL: American Academy of Orthopaedic Surgeons.*

In contrast to tendons, ligaments have an outer, often indistinguishable layer called the epiligament that connects directly to the periosteum of the adjacent bones and primarily contains most of its sparse vascular supply. In intra-articular ligaments, the epiligament is replaced by synovium (Frank et al., 1999; Frank, 2004). Nevertheless, ligaments are generally hypovascular in comparison with surrounding tissues. Despite the small size and limited blood flow of this vascular system, it is of primary importance in the maintenance of the ligament. Specifically, this occurs by providing nutrition and maintaining the continuous process of matrix synthesis and repair. In its absence, damage from normal activities accumulates (fatigue) and the ligament may be at risk for rupture (Woo et al., 1994).

NEURAL COMPONENTS OF TENDONS AND LIGAMENTS

Ligaments and tendons have been shown in both human and animal studies to have various specialized nerve endings and mechanoreceptors. They play an important role in joint proprioception and nociception, which is directly related to the functionality of joints and possibly in the regulation of blood flow of tendons and ligaments. In the presence of tendon healing, studies have found that in disorders such as chronic tendonitis there is evidence of neurovascular in-growth, which presumably has a role in the presence of chronic pain. The vascular in-growth seems to be an attempt of the tendon to heal, but along with this growth may be nerves that sensitize previously less pain-sensitive areas (Benjamin et al., 2008). This evidence strongly suggests the importance of the neural components not only in proprioception but also in nociception.

OUTER STRUCTURE AND INSERTION INTO BONE

Certain similarities are found in the outer structure of tendons and ligaments, but there are also important differences related to function. Both tendons and ligaments are surrounded by a loose areolar connective tissue. In ligaments, this tissue is called the epiligament and in tendons it is referred to as the paratenon. More structured than the epiligament, the paratenon forms a sheath that protects the tendon and enhances gliding. Tendon sheaths have two continuous layers: the outer parietal and the inner visceral. The visceral layer is surrounded by synovial cells that produce synovial fluid. In some tendons, such as the flexor tendons of the digits, the sheath runs the length of the tendons, and in others

the sheath is found only at the point where the tendon bends in concert with a joint. If adhesions should occur within the sheath, such as during chronic inflammation, the tendons are rendered unable to glide and mobility is severely impaired (Benjamin et al., 2008).

In locations where the tendons are subjected to particularly high friction forces (e.g., in the palm, in the digits, and at the level of the wrist joint), a parietal synovial layer is found just beneath the paratenon; this synovium-like membrane, called the epitenon, surrounds several fiber bundles. The synovial fluid produced by the synovial cells of the epitenon facilitates gliding of the tendon. In locations where tendons are subjected to less friction, they are surrounded by paratenon only.

Each fiber bundle is bound together by the endotenon (Fig. 4-1), which continues at the musculotendinous junction into the perimysium. At the tendo-osseous junction, the collagen fibers of the endotenon continue into the bone as Sharpey perforating fibers and become continuous with the periosteum (Woo et al., 1988).

The structure of the insertions into bone is similar in ligaments and tendons and consists of four zones; Figure 4-8 illustrates these zones in a tendon. At the end of the tendon (zone 1), the collagen fibers intermesh with fibrocartilage (zone 2). This fibrocartilage gradually becomes mineralized fibrocartilage (zone 3) and then merges into cortical bone (zone 4). The change from more tendinous to more bony material produces a gradual alteration in the mechanical properties of the tissue (i.e., increased stiffness), which results in a decreased stress concentration effect at the insertion of the tendon into the stiffer bone (Cooper and Misol, 1970).

It is also important to add that tendons and ligaments are intimately connected to fascia. It has been suggested that this interconnection serves three important biomechanical functions: dissipation of loads to reduce wear and tear, facilitate linkage to form mechanical chains, and in the case of tendons, improve muscular efficiency via force transmission to noncontractile tissues (Benjamin et al., 2008).

Biomechanical Properties of Tendons and Ligaments

Tendons and ligaments are viscoelastic structures with unique mechanical properties. Tendons are strong enough to sustain the high tensile forces that result from muscle contraction during joint motion, yet they are sufficiently flexible to angulate around bone surfaces and to deflect beneath retinacula to change the final direction of muscle pull. The ligaments are pliant and flexible, allowing natural movement of the bones to which

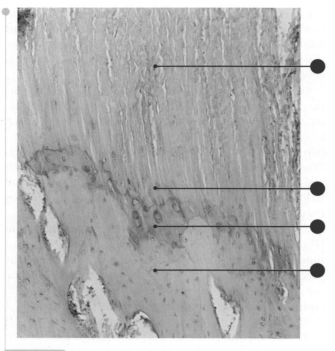

FIG. 4-8

Electron micrograph of a patellar tendon insertion from a dog, showing four zones (×25,000): zone 1, parallel collagen fibers; zone 2, unmineralized fibrocartilage; zone 3, mineralized fibrocartilage; zone 4, cortical bone. The ligament-bone junction (not pictured) has a similar appearance. *Reprinted with permission from Cooper, R.R., Misol, S. (1970). Tendon and ligament insertion. A light and electron microscopic study. J Bone Joint Surg, 52A, l.*

they attach, but are strong and inextensible so as to offer suitable resistance to applied forces. Both structures sustain chiefly tensile loads during normal and excessive loading. When injury happens, the degree of damage is related to the rate of loading as well as the amount of load.

The biomechanical properties of tendons and ligaments are often evaluated by using mounted specimens such as a bone-ligament-bone specimen. However, recent advances have allowed for some instrumentation to be used in the measurement of in situ forces in humans. These include the use of buckle transducers, instrumentation at insertion sites, magnetic resonance imaging, kinematic linkage measurements, and implantable transducers (Woo et al., 2000). Advances in other techniques such as finite element modeling (FEM), elastographic imaging, and robotic/universal force-moment sensor (UFS) testing, are contributing invaluable information to the understanding of tissue biomechanics (Dourte et al., 2008). Structural and mechanical

properties for tendons and ligaments are then analyzed by using the above techniques to obtain load-elongation curves and stress-strain diagrams.

A load-elongation curve offers information regarding the tensile capacity of a tendon-ligament structure after loading a tendon or a ligament to failure. In a load-elongation curve (Fig. 4-9A) the stiffness of the structure (N/nm) is the slope of the curve between two limits of elongation. It represents how much load and or elongation the structure can sustain before it fails. The ultimate load (N) is the highest load placed on the structure before failure. The ultimate elongation (mm) is the maximum elongation of the complex at failure. Finally, the energy absorbed at failure (N/mm) is the area under the entire curve, which represents the maximum energy stored by the complex (Woo et al., 2000).

Load-elongation curves have several regions that characterize the behavior of the tissue (Fig. 4-9A). The first region of the load-elongation curve is called the "toe" region. The elongation reflected in this region is believed to be the result of a change in the wavy pattern or crimp (Fig. 4-4) of the relaxed collagen fibers. In this region, the tissue stretches easily without much force, the collagen fibers become straight and lose their wavy appearance, and sliding occurs between fibrils and fascicles as the loading progresses (Woo et al., 1994). Figure 4-10 shows the appearance of relaxed and loaded collagen fibers under an electron microscope.

As loading continues, the stiffness of the tissue increases and there is a resultant change in the tissue elongation. This region is called the elastic or linear region of the curve. It follows the toe region and it is observed as a sudden increase in the slope of the curve. When the linear region is surpassed, major failure of fiber bundles occurs in an unpredictable manner. The curve can end abruptly or curve downward as a result of irreversible changes (failure) (Woo et al., 1994). With the attainment of maximum load that reflects the ultimate tensile strength of the specimen, complete failure occurs rapidly, and the load-supporting ability of the tendon or ligament is substantially reduced (complete failure). The tissue is elongated then until it ruptures, and the resulting force, or load (P), is plotted.

Where the curve levels off toward the elongation axis, the load value is designated as P_{lin}. The point at which this value is reached is the yield point for the tissue. The energy uptake to P_{lin} is represented by the area under the curve up to the end of the linear region.

To further test tendon and ligament specimen to tensile deformation, stress-strain curves are also generated (Fig. 4-9B). In a stress-strain diagram, the elongation is often expressed as strain (ε), which is the deformation of the tissue calculated as a percentage of the original length

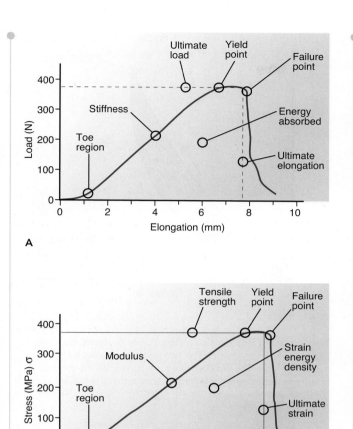

A

A

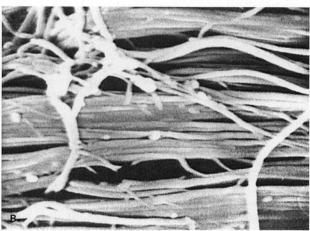

B

FIG. 4-9

FIG. 4-10

A. Load-elongation curve of a tendon-ligament structure after loading to failure. The X-axis is the elongation the structure suffers as a result of loading, measured in mm. The Y-axis is the tensile load applied to the tissue measured in newtons (N). The stiffness of the structure (N/nm) is the slope of the curve between two limits of elongation. The ultimate load (N) is the highest load placed on the structure before failure. The ultimate elongation (mm) is the maximum elongation of the complex at failure. The energy absorbed at failure (N/mm) is the area under the entire curve, which represents the maximum energy stored by the complex. **B.** A stress-strain curve of a tendon/ligament structure under tensile loading. The X-axis is the percentage of deformation (elongation) expressed as strain (ε) and the Y-axis is the stress or load per unit of area (MPa) expressed as stress (σ), which refers to the tensile strength of the tissue. A modulus of elasticity (N/mm² or MPa) is obtained from the linear slope of the stress-strain curve between two limits of strain (deformation). The tensile strength (N/mm²) is the maximum stress achieved, the ultimate strain (in percentage) is the strain at failure, and the strain energy density (MPa) is the area under the stress-strain curve. *Adapted from Woo, S.L.Y., et al. (2000). Injury and repair of ligaments and tendons. Ann Rev Biomed Eng, 2, 86.*

Scanning electron micrographs of unloaded (relaxed) and loaded collagen fibers of human knee ligaments (×10,000). **A.** The unloaded collagen fibers have a wavy configuration. **B.** The collagen fibers have straightened out under load. *Reprinted with permission from Kennedy, J.C., Hawkins, R.J., Willis, R.B., et al. (1976). Tension studies of human knee ligaments. Yield point, ultimate failure, and disruption of the cruciate and tibial collateral ligaments. J Bone Joint Surg, 58A, 350.*

of the specimen. The force per unit of area (in this case, the total tensile load per unit by the cross-sectional area of the tendon or ligament under analysis) is expressed as the stress (σ). From stress-strain curves (Fig. 4-9B), a modulus (N/mm² or MPa) is obtained from the linear slope of the stress-strain curve between two limits of strain (deformation) where the tensile strength (N/mm²) is the maximum stress achieved, the ultimate strain (in percentage) is the strain at failure, and the strain energy density (MPa) is the area under the stress-strain curve.

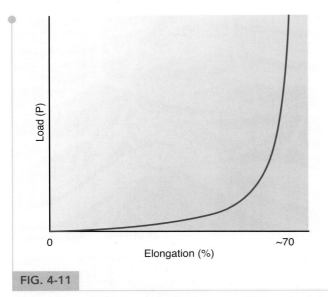

FIG. 4-11

Load-elongation curve for a human ligamentum flavum (60%–70% elastic fibers) tested in tension to failure. At 70% elongation the ligament exhibited a great increase in stiffness with additional loading and failed abruptly without further deformation. *Adapted from Nachemson, A.L., Evans, J.H. (1968). Some mechanical properties of the third human lumbar interlaminar ligament (ligamentum flavum). J Biomech, 1, 211–220.*

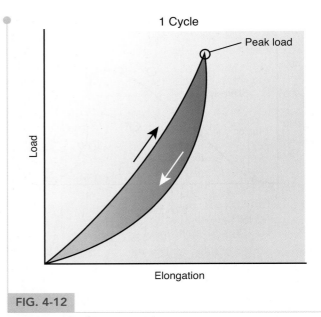

FIG. 4-12

Typical loading **(top)** and unloading curves **(bottom)** from tensile testing of knee ligaments. The two nonlinear curves form a hysteresis loop. The area between the curves, called the area of hysteresis, represents the energy losses within the tissue.

This modulus of elasticity for tendons and ligaments has been determined in several investigations (Fung, 1967, 1972; Viidik, 1968), and it represents a linear and proportional relationship between load and deformation or stress and strain:

$E = \sigma/\varepsilon$ (where E = modulus of elasticity, σ = stress, ε = strain)

The load-elongation curves analyzed previously generally apply to tendons and extremity ligaments. The curve for the ligamentum flavum, with its high proportion of elastic fibers, is different (Fig. 4-11). In tensile testing of a human ligamentum flavum, elongation of the specimen reached 50% before the stiffness increased appreciably. Beyond this point, the stiffness increased greatly with additional loading and the ligament failed abruptly (reached P_{max}), with little further deformation (Nachemson and Evans, 1968). The greater proportion of elastic proteins and the resultant elastic capacity of the ligament flavum results then in a larger capacity to elongate before failure (large strain to failure).

The proportion of elastic proteins in ligaments and capsules is extremely important for the small elastic deformation that they endure under tensile strain and the storage and loss of energy. During the loading and unloading of a ligament between two limits of elongation, the elastic fibers allow the material to return to its original shape and size after being deformed. Meanwhile, part of the energy spent is stored. What remains represents the energy loss during the cycle and is called hysteresis. The area enclosed by the loop represents the energy loss (Fig. 4-12).

VISCOELASTIC BEHAVIOR IN TENDONS AND LIGAMENTS TO TENSILE LOADS

Biologic materials under loading such as ligaments and tendons exhibit time-dependent viscoelastic behaviors, and their mechanical properties change with different rates of loading. Both ligaments and tendons display this viscoelastic behavior that is assumed to result from the complex interaction of its constituents (i.e., collagen, water, surrounding protein, and ground substance) (Woo et al., 2000). When ligament and tendon specimens are subjected to increased loading rates, the linear portion of the stress-strain curve becomes steeper, indicating greater stiffness of the tissue at higher strain rates. With higher strain rates, ligaments and tendons in isolation store more energy, require more force to rupture, and undergo greater elongation (Kennedy et al., 1976).

During cyclic testing of ligaments and tendons, where loads are applied and released at specific intervals, the stress-strain curve may be displaced to the right along the elongation (strain) axis with each loading cycle. This reveals the presence of a nonelastic (plastic) component characterized by the permanent deformation of the tissue that is progressively greater with every loading cycle. As repetitive loading progresses, the specimen also shows an increase in elastic stiffness as a result of plastic deformation (molecular displacement). Microfailure can then occur within the physiologic range if frequent loading is imposed on an already damaged structure where the stiffness has decreased. This phenomenon is illustrated later in Case Study 4-2.

Two standard experimental tests are used to illustrate the viscoelastic, time-dependent, nonlinear behavior of ligaments and tendons. These are the stress-relaxation and the creep-deformation tests (Fig. 4-13). In a stress-relaxation test (load relaxation test, Fig. 4-13A), the specimen is stretched (deformed) to a constant length, so the strain is kept constant over an extended period allowing the stress to vary with time. As observed in Figure 4-13A, when the length is held constant, the stress decreases rapidly at first and then more slowly. When the stress-relaxation test is repeated cyclically, the decrease in stress gradually becomes less pronounced.

A creep-deformation test, on the other hand, involves subjecting the specimen to a constant load (the stress is kept constant over an extended period) while the length (deformation) gradually increases with time (Fig. 4-13B). The strain increases relatively quickly at first and then more and more slowly. When this test is performed cyclically, the increase in strain gradually becomes less pronounced.

The clinical application of a constant low load to the soft tissues over a prolonged period, which takes advantage of the creep response, is a useful treatment for several types of deformities. One example is the manipulation of a child's clubfoot by subjecting it to constant loads by means of a plaster cast, or the treatment of idiopathic scoliosis with a brace, whereby constant loads are applied to the spinal area to elongate the soft tissues surrounding the abnormally curved spine (Fig. 4-14).

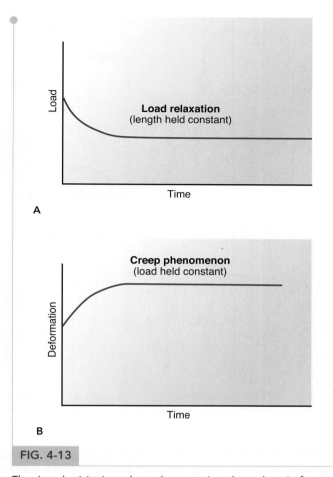

FIG. 4-13

The viscoelasticity (rate dependency, or time dependency) of ligaments and tendons can be demonstrated by two standard tests: the load-relaxation test and the creep test. **A.** Load relaxation is demonstrated when the loading of a specimen is halted safely below the linear region of the load-deformation curve and the specimen is maintained at a constant length over an extended period (i.e., the amount of elongation is constant). The load decreases rapidly at first (i.e., during the first six to eight hours of loading) and then gradually more slowly, but the phenomenon may continue at a low rate for months. **B.** The creep response takes place when loading of a specimen is halted safely below the linear region of the load-deformation curve and the amount of load remains constant over an extended period. The deformation increases relatively quickly at first (within the first six to eight hours of loading) but then progressively more slowly, continuing at a low rate for months.

BIOMECHANICAL RESPONSE OF TENDONS AND LIGAMENTS TO NONTENSILE LOADS

Tendons and ligaments can also be subjected to compression and shear. Although few researchers have investigated the mechanical properties of these tissues under these loading conditions, it has been found that adaptations to these forces are evident in the structure of tendons and ligaments. In the case of the long flexor tendons of the digits, compressive loads are present on the side of the tendons closest to the bony pulleys where fibrocartilage is found along these points as a mechanical adaptation to these loads when tensile loads are present on the side away from it (Benjamin et al., 2008).

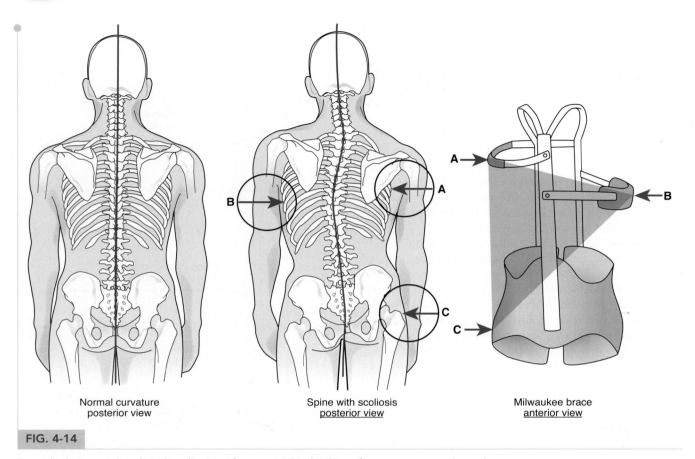

Normal curvature
posterior view

Spine with scoliosis
posterior view

Milwaukee brace
anterior view

FIG. 4-14

Example demonstrating clinical application of a constant low load to soft tissues over a prolonged period. The figure illustrates a normal spinal curve, a spine with scoliosis, and a typical brace used for correction of scoliosis (the Milwaukee brace). The *arrows* depict the areas where the brace places the corrective forces.

Frictional shear also occurs in tendons that rub against bony prominences or other muscles. Ligaments, in performing their function to guide joint motion and stabilize segments also frequently encounter shear, compression, and torsional forces. In particular, it has been found that ligament response to shear forces is nonlinear and independent of the rate of loading (Weiss et al., 2002).

LIGAMENT FAILURE AND TENDON INJURY MECHANISMS

Injury mechanisms are similar for ligaments and tendons. These fall into two general categories or a combination of both: high levels of stress or load (such as those cases where an external violence occurs), high rates of strain (such as those in which an overuse injury or repetitive microtrauma surpasses the reparative process), or high levels of both stress and strain (such as in a ligament injury in contact-collision sports).

When a ligament in vivo is subjected to loading that exceeds the physiologic range (injury due to high levels of stress), either microfailure takes place even before the yield point (P_{lin}) is reached (i.e., partial rupture of a ligament) or if the P_{lin} is exceeded, the ligament will undergo gross failure (complete rupture). When this occurs, the joint will simultaneously begin to displace abnormally and show signs of instability. This displacement can also result in damage to the surrounding structures, such as the joint capsule, the adjacent ligaments, and the blood vessels that supply these structures. Noyes (1976) demonstrated the progressive failure of the anterior cruciate ligament (ACL) and displacement of the tibiofemoral joint by applying a clinical test, the anterior drawer test, to a cadaver knee up to the point of ACL failure (Fig. 4-15).

At maximum load, the joint had displaced several millimeters, generating a progressive increase in the ligament elongation beyond its elastic region. Thus, although the ligament was still in continuity, it had undergone

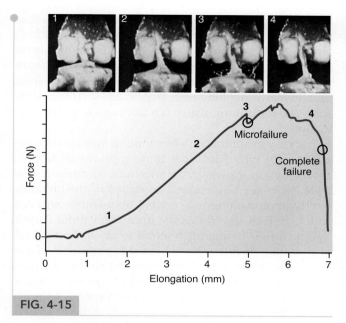

FIG. 4-15

Progressive failure of the anterior cruciate ligament from a cadaver knee tested in tension to failure at a physiologic strain rate. The joint was displaced 7 mm before the ligament failed completely. The force-elongation curve generated during this experiment is correlated with various degrees of joint displacement recorded photographically; photos correspond to similarly numbered points on the curve. *Reprinted with permission from Noyes, F.R., Grood, E.S. (1976). The strength of the anterior cruciate ligament in humans and rhesus monkeys. Age-related and species-related changes. J Bone Joint Surg, 58A, 1074–1082.*

extensive macrofailure and microfailure and extensive elongation with a resultant structural and mechanical damage. As shown in Figure 4-15, the force-elongation curve generated during the experiment indicated when the microfailure of the ligament began compared with various stages of joint displacement recorded photographically.

Correlation of the results of this test in vitro with clinical findings sheds light on the microevents that take place in the ACL during normal daily activity and during injuries of various degrees of severity. In Figure 4-15, the curve for the experimental study on cadaver knees has been divided into four regions, corresponding respectively to (1) the load placed on the ACL during tests of knee joint stability performed clinically, (2) the load placed on this ligament during physiologic activity, (3) the load imposed on the ligament and its resultant permanent deformation from the beginning of microfailure to (4) complete rupture (Case Study 4-1).

Ligament injuries are categorized clinically in three ways according to degree of severity (Magee, 2007).

Case Study 4-1

ACL Failure

A 25-year-old male occasional soccer player injured his ACL as a result of an abnormal torque in rotation of the knee. The player locked his foot on the ground and pivoted on his lower limb, which produced a high rotational torque on the knee and increased tensile loads on the ACL.

If a load-joint displacement curve is generated, the first region of the curve will show a normal physiologic loading response. However, the mechanism of injury (in this case, the abnormally high rotational torque) increased a strain deformation leading to high internal stress that led to a complete rupture.

A knee with an ACL injury will undergo an abnormal intra-articular joint motion as a result of the lack of such an important stabilizer. A change in the pathway of the joint center of rotation could be observed and a resultant change in the distribution of the loads with abnormally high stresses on other joint surfaces and structures such as cartilage. This can lead to generative joint diseases. In addition, a deficiency in joint stability that results from ACL impairments can increase the likelihood of experiencing the "giving way" sensation or functional instability, affecting activities of daily living such as gait, jogging, and squatting (Case Study Fig. 4-1).

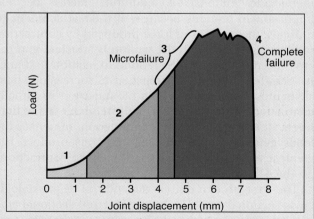

Case Study Figure 4-1 This is a representation of a load (Y axis)-elongation (X axis) curve of a progressive failure of an Anterior Cruciate Ligament of the knee. Zone 1 represents the TOE region. Zones 2 and 3 represent the elastic and plastic regions, respectively. The microfailure starts as permanent deformation that begins in the plastic region. As the load/deformation continues, a complete failure is reached. The injury region represents microfailure and complete failure within the ligament.

Injuries in the first category (first-degree sprain) produce negligible symptoms; some pain is felt, but no joint instability can be detected even though there may be some microfailure of the collagen fibers. There is no macroscopically observable disruption of the fibers of the ligament.

Injuries in the second category (second-degree sprain) produce severe pain, and some joint instability can be detected or experienced. Progressive failure of the collagen fibers has taken place, resulting in partial ligament rupture. The strength and stiffness of the ligament may have decreased by 50% or more, corresponding to the percentage of fibrous disruption, mainly because the amount of undamaged tissue has been reduced. As a result, there may be joint instability that in some cases may or may not be symptomatic depending on the functional stability imparted by the muscles.

Injuries in the third category (third-degree sprain) produce severe pain during the course of trauma with less pain after injury. The joint is found to be completely unstable. Either a total rupture has occurred or most collagen fibers have ruptured with a few left intact, giving the ligament the appearance of continuity even though it is unable to perform its function.

Loading a joint that is unstable as a result of a ligament injury has been associated with abnormally high stresses on the articular cartilage. Cartilage wear mechanisms are then generated by this abnormally increased loading and changes in the loading pattern, leading to early onset of degenerative joint disease. Although injury mechanisms are generally comparable in ligaments and tendons, two additional factors become important in tendons because of their attachment to muscles: the amount of force produced by contraction of the muscle to which the tendon is attached and the cross-sectional area of the tendon in relation to that of its muscle. A tendon is subjected to increasing stress as its muscle contracts. When the muscle is maximally contracted, the tensile stress on the tendon reaches high levels. The type of muscular contraction influences the loads generated. Eccentric contractions produce the greatest tensile load whereas concentric contractions produce the least.

The strength of a muscle depends on its physiologic cross-sectional area. The larger the cross-sectional area of the muscle, the higher the magnitude of the force produced by the contraction and thus the greater the tensile loads transmitted through the tendon. Similarly, the larger the cross-sectional area of the tendon, the greater the loads it can bear. Although the maximal stress to failure for a muscle has been difficult to compute accurately, such measurements have shown that the tensile strength of a healthy tendon may be more than twice that of its muscle (Elliott, 1967). This finding is supported clinically by the fact that muscle ruptures are more common than ruptures through a tendon. Large muscles usually have tendons with large cross-sectional areas. Examples are the quadriceps muscle with its patellar tendon and the triceps surae muscle with its Achilles tendon. Some small muscles have tendons with large cross-sectional areas such as the plantaris, which is a tiny muscle with a large tendon.

It is easy to appreciate when tendon injuries are associated with high loads like acute ruptures such as in sports injuries. However, the two most common pathologic issues seen in tendons described in the literature (tendinitis and tendinosis) are not frequently associated with high loads but rather the opposite. Painful tendon dysfunction is frequently referred to as *tendinitis*. Lack of evidence of inflammation in this condition makes the use of this term inappropriate. An alternative term is *tendinosis*, suggesting a degenerative condition that is asymptomatic and can lead to tendon rupture. A limitation of this term is that it is extremely difficult to look at the state of the tendon prior to the development of symptoms to verify the existence of a degenerative process. *Tendinopathy* is the most appropriate term to use because it encompasses all conditions related to tendon pathology (Riley, 2004). Although multiple factors have been implicated, there is agreement that a common mechanism of injury in tendinopathy is related to high rates of strain. (Case Study 4-2).

HEALING OF TENDONS AND LIGAMENTS

Tendons and ligaments heal in the same manner as other tissues following injury with the same three phases occurring in succession: the inflammatory phase, the proliferative phase or fibroplasia, and the remodeling and maturation phase. However, this process is slower for tendons and ligaments because of the limited vascularity of these tissues. The end result is the creation of scar tissue that is characterized by persistent flaws, abnormal ECM components, and the abnormally reduced diameter of collagen fibrils, which makes it biomechanically inferior to normal tissue. Although scar tissue can respond and adapt to load, it can improve to within only 10% to 20% of normal tissue properties (Frank, 2004).

There is also variation in the ability of tendons and ligaments to heal. A good example is the medial collateral ligament of the knee, which can heal fully after total rupture without the need for surgical intervention (Woo et al., 2000); other ligaments, such as the ACL, require grafting and reattachment following total rupture. Similarly, in the case of tendons, incomplete tears can heal, particularly in the case of the Achilles tendon. However, it is nearly impossible for the long flexor tendons of the

Case Study 4-2

de Quervain Tenosynovitis

A 45-year-old female graphic designer who has recently been working long hours due to increased workload developed pain and discomfort along the dorsolateral aspect of the dominant wrist. Increased symptoms with active abduction of the thumb or passive ulnar deviation with adduction of the thumb confirmed presence of De Quervain tenosynovitis, which is a painful condition affecting the abductor pollicis longus and extensor pollicis brevis tendons and their common sheath as it passes through the wrist. (Case Study Fig. 4-2)

This is a good example of a repetitive strain injury where there is a cumulative exposure in this situation to a high strain rate (a combination of high-repetition, low-force task). In this case, the mechanical tolerance of the affected structures is exceeded and microtrauma occurs, generating an inflammatory response and subsequent pain and altered function. If a stress-strain curve is plotted, a displacement toward the right would be observed as cyclic loading progresses, indicating greater deformation and possible molecular displacement (plastic deformation). As time passes, due to the repetitive nature of the injury, the injury ends up overpassing the healing processes. This leads to a chronic inflammatory response that makes the tissue even more susceptible to further injuries as it surpasses its ultimate tensile strength and increases its susceptibility to complete failure. This may be less frequently observed in patients with De Quervain syndrome but is more common in individuals with rotator cuff tendinopathy.

The main risk factor in this case seems to be the repetition of the activity. However, current evidence has suggested that a combination of factors is what ultimately contributes to the onset of symptoms. These include age, force, posture, vibration, temperature, state of the tendon prior to increased frequency of deformation, work organization, and individual predisposition (comorbidities). Pain may be explained by the presence of inflammation, but in some cases the absence of inflammatory markers has suggested that the attempts of tendons to adapt to increasing load and/or deformation triggers neurovascular in-growth. This leads not only to an increase in blood vessels in the area but also in nerves, which make the tendons more sensitive to pain Barbe and Barr, 2006; (Barr and Barbe, 2002; Benjamin et al., 2008).

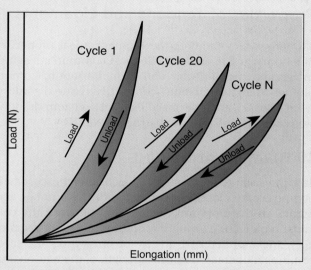

Case Study Figure 4-2 This curvature represents a typical loading and unloading cycle. The y axis corresponds to the load applied and the x axis to the elongation within the tissue. The area between the curves represents the energy losses within the tissue (hysteresis). As the number of cycles increases, the percentage of deformation increases, indicating a greater deformation and possible molecular disruption.

fingers to heal without surgical reattachment following complete disruption.

GRAFTS

Reconstruction of torn ligaments, especially of the anterior and posterior cruciate ligament, is now a frequent procedure. The need for reconstruction is related to age, activity level, and associated injuries. Grafts derived from different individuals of the same species are called allografts; grafts derived from the same individual are called autografts. Allograft tissue preservation is done through freeze-drying and low-dose irradiation to reduce rates of rejection and infection and to limit effects on the structural properties. Bone-patellar, tendon-bone, and Achilles tendon are usually used as allograft tissue, whereas the central tissue of the patellar tendon is commonly used as autograft tissue.

Several authors have described allogenous and autograft procedures (Shino et al., 1995; Strocchi et al., 1992) for ACL reconstruction in humans. Currently there is evidence of the advantage of using autograft tissues rather than allograft tissues, after extended follow-up allografts showed collagen fibril profiles that do not resemble normal tendon grafts or normal ACL.

Biopsies of patellar tendons that had been autografted to reconstruct torn ACLs have shown that the autograft underwent considerable changes over time, and after 24 months it had the appearance of normal ligament tissue. Strocchi et al. (1992) then suggested that the patellar tendon autograft is a valid functional ACL substitution for patients who desire to perform normal mechanical activity. Still, even a fully incorporated graft will never duplicate the native ACL but works instead as a check rein that increases knee function (Corsetti and Jackson, 1996).

Factors That Affect the Biomechanical Properties of Tendons and Ligaments

Numerous factors affect the biomechanical properties of tendons and ligaments. The most common are aging, pregnancy, mobilization and immobilization, comorbidities (diabetes mellitus, connective tissue disorders, renal disease), and pharmacologic agents (steroids, nonsteroidal anti-inflammatory drugs or NSAIDs).

MATURATION AND AGING

The physical properties of collagen and the tissues it composes are closely associated with the number and quality of the cross-links within and between the collagen molecules. During maturation (up to 20 years of age), the number and quality of cross-links increase, resulting in increased tensile strength of the tendon and ligament (Viidik et al., 1982). Although cross-linking has been correlated with tissue strength and maturation, it has been found that there is an optimum level of permanent cross-linking. It appears that beyond this ideal level of cross-linking, a reduction in the biomechanical properties of the tissue has been observed, thus explaining the deterioration of biomechanical properties with aging (Dressler et al., 2002). After maturation, as aging progresses, collagen reaches a plateau with respect to its mechanical properties, after which the tensile strength and stiffness of the tissue begin to decrease.

An increase in collagen fibril diameter has been observed (Parry et al., 1978) with high variability in size (range 20–180 nm) (Strocchi et al., 1996) noted in the young (<20 years). The diameter in adults (20–60 years) and in the elderly (>60 years) decreases remarkably (120 and 110 nm, respectively) but with a more even distribution.

Strocchi et al. (1996) investigated age-related changes in human ACL collagen fibrils and report an increase of fibril concentration from 68 fibrils/mu^2 in the young to 140 fibrils/mu^2 in the elderly. However, Amiel et al. (1991) report that the water content and the collagen concentration decreases significantly in the medial cruciate ligament of 2-, 12-, and 39-month-old rabbits. Dressler et al. (2002) also found that in aging rabbits, fibril diameter and the general biomechanical properties of patellar tendons were decreased and an increase in type V collagen was found, which was associated with increased tissue stiffness.

It has also been found that with aging, tenocytes decrease in number and tend to be flattened and with less cytoplasmic extensions, suggesting reduced intercellular communication and reduced capability for the mechanotransduction needed for collagen production and maturation (Benjamin et al., 2008). Skeletal maturity has also been found to reduce the ability of growth factors to induce fibroblastic activity (Woo et al., 2000).

PREGNANCY AND THE POSTPARTUM PERIOD

A common clinical observation is the increased laxity of the tendons and ligaments in the pubic area during later stages of pregnancy and the postpartum period that is attributed to the hormone relaxin. Relaxin is an insulin-like growth factor that is capable of altering gene expression in fibroblasts, which ultimately results in temporary changes in the biomechanical properties of tissue (Frank et al., 1999). This observation has been confirmed in animal studies. Rundgren (1974) found that the tensile strength of the tendons and the pubic symphysis in rats decreased at the end of pregnancy and during the postpartum period. Stiffness of these structures decreased in the early postpartum period but was later restored.

MOBILIZATION AND IMMOBILIZATION

Living tissues are dynamic and change their mechanical properties in response to stress, which leads to functional adaptation and optimal operation of the tissue.

Like bone, ligaments and tendons appear to remodel in response to the mechanical demands placed on them; they become stronger and stiffer when subjected to increased stress, and weaker and less stiff when the stress is reduced (Noyes et al., 1977).

Exercise has been found to increase the tensile strength of tendons and of the ligament-bone interface (Woo et al., 1981). The cross-sectional area of tendons was increased with continuous training that contributed to an increase in the resilience to overload and rupture

(Kjaer, 2004). Initially training induced a temporary net loss of collagen presumed to be consistent with a process of restructuring and adaptation to increasing loads. With continuous training, there is a net gain consistent with a concurrent reparative process and biomechanical adaptation to load, which ultimately may lead to stronger and more resilient tissues (Benjamin et al., 2008). Conversely, stress deprivation associated with immobilization has been found to be associated with a decrease in collagen synthesis and an increase in matrix metalloproteinases (MMPs), which are enzymes linked to collagen breakdown that ultimately reduces the mechanical properties of the tissue (Dourte et al., 2008; Kjaer, 2004).

Immobilization has been found to decrease the tensile strength of ligaments (Newton et al., 1995; Walsh et al., 1993). Amiel et al. (1982) found a decrease in the strength and stiffness of lateral collateral ligaments in rabbits immobilized for nine weeks. Because the cross-sectional area of the specimens did not change significantly, the degeneration of mechanical properties was attributed to changes in the ligament substance itself. The tissue metabolism was noted to increase, leading to proportionally more immature collagen with a decrease in the amount and quality of the cross-links between collagen molecules. Newton et al. (1995) also reported that the cross-sectional area of ligaments in immobilized rabbit knees was 74% of the control value.

COMORBIDITIES

There are many conditions that contribute to musculoskeletal disorders involving tendons and ligaments, either directly or indirectly. Several of these are described in this section.

Diabetes Mellitus

Diabetes mellitus, more frequently the insulin-dependent form of the disease (type I diabetes), is known to be correlated with musculoskeletal disorders due to its association with changes in connective tissue and the metabolic fluctuations that directly affect its microvascularity and promote collagen accumulation in periarticular tissues (Riley, 2004). Specific tendon and ligament pathologies associated with diabetes include diabetic cheiroarthropathy (stiff hand syndrome), flexor tenosynovitis (De Quervain syndrome), Dupuytren contracture, adhesive capsulitis (frozen shoulder), and calcific periarthritis (Kim, 2001).

Connective Tissue Disorders

Connective tissue disorders such as rheumatologic conditions (e.g., rheumatoid arthritis, spondyloarthropathies) have been associated with inflammatory infiltrate that promotes the destruction of collagenous tissue. In addition, genetically inherited conditions (e.g., Marfan syndrome, Ehlers-Danlos syndrome) can result in deficiencies in the amount or the type of collagen present in tendons and ligaments, abnormalities in fibril structure, and an imbalance in the amount of elastin and other proteins altering its biomechanical properties (Riley, 2004).

Renal Disease

Tendinous failure resulting from chronic renal failure does occur, with tendon rupture reaching 36% among individuals receiving hemodialysis. Hyperlaxity of tendons and ligaments was found in 74%, patellar tendon elongation in 49%, and articular hypermobility in 51% of individuals receiving long-term hemodialysis (Rillo et al., 1991). Dialysis-related amyloidosis may cause the deposit of amyloid in the synovium of tendons. The major constituent of the amyloid fibrils is the beta 2-microglobulin (Bardin et al., 1985; Honda et al., 1990; Morita et al., 1995). An increase in the amount of elastin and collagen destruction has also been suggested (Riley, 2004).

PHARMACOLOGIC AGENTS

Systemic intake of medications that include steroids, nonsteroidal anti-inflammatory drugs, and other medications such as fluoroquinolones can influence tendon and ligament metabolism.

Steroids

There is conflicting evidence regarding the short- and long-term effect of steroids on tendons and ligaments. However, they have been associated with the inhibition of collagen synthesis and the subsequent altered healing and a decrease on the peak load of these tissues (Campbell et al., 1996; Liu et al. 1997; Oxlund 1980; Walsh et al., 1995; Wiggins et al., 1994, 1995).

Nonsteroidal Anti-inflammatory Drugs (NSAIDs)

NSAIDs are widely used for the control of inflammatory conditions. There is some evidence supporting their effectiveness in the treatment of inflammatory disorders of the tendons and ligaments (indomethacin and diclofenac in particular). Even though there is some evidence to support the enhancement of the biomechanical properties of tendons and ligaments with use of NSAIDs, there is little use of these medications for this purpose (Carlstedt et al., 1986a, 1986b; Marsolais et al., 2003; Vogel, 1977).

Fluoroquinolones

Intake of fluoroquinolones, chemotherapeutic antibiotic medications, has been found to cause an increase in the activity of matrix metalloproteinases (MMPs), which therefore promotes the increased degradation of collagen (Kjaer, 2004).

Summary

- Tendons and extremity ligaments are composed largely of collagen, whose mechanical stability gives these structures their characteristic strength and flexibility. The proportion of elastin accounts for variations in extensibility.
- The arrangement of the collagen fibers is nearly parallel in tendons, equipping them to withstand high unidirectional loads. The less parallel arrangement in ligaments allows these structures to sustain predominantly tensile stresses in one direction and smaller stresses in other directions.
- Tendons and ligaments have an intricate neurovascular supply that plays a significant role in metabolism, healing, proprioception, and pain generation.
- At the insertion of ligament and tendon into stiffer bone, the gradual change from a more fibrous to more bony material results in a reduced stress concentration.
- Tendons and ligaments undergo deformation before failure. When the ultimate tensile strength of these structures is surpassed, complete failure occurs rapidly, and their load-bearing ability is substantially decreased.
- Injury mechanisms in a tendon are influenced by the amount of force produced by the contraction of the muscle to which the tendon is attached and the cross-sectional area of the tendon in relation to that of its muscle.
- The biomechanical behavior of ligaments and tendons is viscoelastic, or rate-dependent, so that these structures display an increase in strength and stiffness with an increased loading rate.
- An additional effect of rate dependency is the slow deformation, or creep, that occurs when tendons and ligaments are subjected to a constant low load over an extended period; stress relaxation takes place when these structures sustain a constant elongation over time.
- Ligaments and tendons remodel in response to the mechanical demands placed on them.

- Allografts and autografts are useful in ligament reconstruction, but material properties do not return completely to normal levels.
- Aging results in a decline in the mechanical properties of tendons and ligaments (their strength, stiffness, and ability to withstand deformation).
- Pregnancy, immobilization, systemic conditions, and certain pharmacologic agents influence the biomechanical properties of ligaments and tendons.

Practice Questions

1. What are the structural differences in collagen fiber orientation between tendons and ligaments? What is the relationship between collagen fiber orientation and its respective function?

2. Draw a hypothetical load-elongation graph (curve) for a tendon-ligament living tissue, display all the regions on the curve, and label the axes.

3. Explain the difference between the yield point and the ultimate failure point.

4. Ligaments and tendons are tissues whose response varies with time and that exhibit a rate-dependent behavior under loading. What is the biomechanical property of this response? Explain and try to clarify this property using stress-relaxation and creep deformation curves.

5. Illustrate three real-life scenarios that demonstrate the mechanisms of injury in tendons and ligaments where there is a high external load, high strain rate with a low load, and a combined high load with a high strain rate.

REFERENCES

Amiel, D., Frank, C., Harwood, F., et al. (1984). Tendons and ligaments: A morphological and biochemical comparison. *J Orthop Res, 1*, 257.

Amiel, D., Kuiper, S.D., Wallace, C.D., et al. (1991). Age-related properties of medial collateral ligament and anterior cruciate ligament: A morphologic and collagen maturation study in the rabbit. *J Gerontol, 46*(4), B156–B165.

Amiel, D., Woo, S.L.Y., Harwood, F.L., et al. (1982). The effect of immobilization on collagen turnover in connective tissue. A biochemical-biomechanical correlation. *Acta Orthop Scand, 53*, 325.

Barbe, M.F., Barr, A.E. (2006). Inflammation and the pathophysiology of work-related musculoskeletal disorders. *Brain Behav Immun, 20*(5), 423–429.

Bardin, T., Kuntz, S., Zingraff, J., et al. (1985). Synovial amyloidosis in patients undergoing long-term hemodialysis. *Arthritis Rheum, 28*(9), 1052–1058.

Barr, A.E., Barbe, M.F. (2002). Pathophysiological tissue changes associated with repetitive movement: A review of the evidence. *Phys Ther, 82*(2), 173–187.

Benjamin, M., Kaiser, E., Milz, S. (2008). Structure-function relationships in tendons: A review. *J Anat, 212*(3), 211–228.

Campbell, R.B., Wiggins, M.E., Cannistra, L.M., et al. (1996). Influence of steroid injection in ligament healing in the rat. *Clin Orthop, 332*, 242–253.

Carlstedt, C.A., Madson, K., Wredmark, T. (1986a). The influence of indomethacin on collagen synthesis during tendon healing in the rabbit. *Prostaglandins, 32*, 353.

Carlstedt, C.A., Madson, K., Wredmark, T. (1986b). The influence of indomethacin on tendon healing. A biomechanical and biochemical study. *Arch Orthop Trauma Surg, 105*, 332.

Cooper, R.R., Misol, S. (1970). Tendon and ligament insertion. A light and electron microscopic study. *J Bone Joint Surg, 52A*, l.

Corsetti, J.R., Jackson, D.W. (1996). Failure of anterior cruciate ligament reconstruction: The biologic basis. *Clin Orthop, 325*, 42–49.

Dourte, L.M., Kuntz, A.F., Soslowsky, L.J. (2008). Twenty-five years of tendon and ligament research. *J Orthop Res, 26*(10), 1297–1305.

Dressler, M.R., Butler, D.L., Wenstrup, R., et al. (2002). A potential mechanism for age-related declines in patellar tendon biomechanics. *J Orthop Res, 20*(6), 1315–1322.

Elliott, D.H. (1967). The biomechanical properties of tendon in relation to muscular strength. *Ann Phys Med, 9*, 1.

Frank, C.B. (2004). Ligament structure, physiology and function. *J Musculoskelet Neuronal Interact, 4*(2), 199–201.

Frank, C.B., David, A.H., Shrive, N.G. (1999). Molecular biology and biomechanics of normal and healing ligaments—a review. *Osteoarthritis Cartilage, 7*(1), 130–140.

Fung, Y.C.B. (1967). Elasticity of soft tissues in simple elongation. *Am J Physiol, 213*, 1532.

Fung, Y.C.B. (1972). Stress-strain-history relations of soft tissues in simple elongation. In Y.C. Fung, N. Perrone, M. Anliker (Eds.). *Biomechanics: Its Foundations and Objectives.* Englewood Cliffs, NJ: Prentice-Hall, 181–208.

Honda, K., Hara, M., Ogura, Y., et al. (1990). Beta-2-microglobulin amyloidosis in hemodialysis patients. An autopsy study of intervertebral disks and posterior ligaments. *Acta Pathol Jpn, 40*(11), 820–826.

Kennedy, J.C., Hawkins, R.J., Willis, R.B., et al. (1976). Tension studies of human knee ligaments. Yield point, ultimate failure, and disruption of the cruciate and tibial collateral ligaments. *J Bone Joint Surg, 58A*, 350.

Kim, R.P., Edelman, S.V., Kim, D.D. (2001). Musculoskeletal complications of diabetes mellitus. *Clin Diabetes, 19*(3), 132–135.

Kjaer, M. (2004). Role of extracellular matrix in adaptation of tendon and skeletal muscle to mechanical loading. *Physiol Rev, 84*(2), 649–698.

Liu, S.H., Al-Shaikh, R.A., Panossian, V., et al. (1997). Estrogen affects the cellular metabolism of the anterior cruciate ligament. A potential explanation for female athletic injury. *Am J Sports Med, 25*(5), 704–709.

Magee, D.J. (2007). *Orthopedic Physical Assessment.* Saint Louis, MO: Saunders Elsevier, 859.

Marsolais, D., Cote, C.H., Frenette, J. (2003). Nonsteroidal antiinflammatory drug reduces neutrophil and macrophage accumulation but does not improve tendon regeneration. *Lab Invest, 83*(7), 991–999.

Morita, H., Shinzaato, T., Cai, Z., et al. (1995). Basic fibroblast growth factor-hepatin sulphate complex in the human dialysis-related amyloidosis. *Virchows Arch, 427*(4), 395–400.

Nachemson, A.L., Evans, J.H. (1968). Some mechanical properties of the third human lumbar interlaminar ligament (ligamentum flavum). *J Biomech, 1*, 211–220.

Newton, P.O., Woo, S.L., Mackenna, D.A., et al. (1995). Immobilization of the knee joint alters the mechanical and ultrastructural properties of the rabbit anterior cruciate ligament. *J Orthop Res, 13*(2), 191–200.

Noyes, F.R., et al. (1977). Functional properties of knee ligaments and alterations induced by immobilization. *Clin Orthop, 123*, 210–242.

Noyes, F.R., Grood, E.S. (1976). The strength of the anterior cruciate ligament in humans and rhesus monkeys. Age-related and species-related changes. *J Bone Joint Surg, 58A*, 1074–1082.

Oxlund, H. (1980). The influence of a local injection of cortisol on the mechanical properties of tendons and ligaments and the indirect effect on skin. *Acta Orthop Scand, 51*(2), 231–238.

Parry, D.A.D., Barnes, G.R.G., Craig, A.S. (1978). A comparison of the size distribution of collagen fibrils in connective tissues as a function of age and possible relation between fibril size and mechanical properties. *Proc R Soc Lond, 203*, 305–321.

Prockop, D.J. (1990). Mutations that alter the primary structure of collagen. *J Biol Chem, 265*(26), 15349–15352.

Prockop, D.J., Guzman, N.A. (1977). Collagen diseases and the biosynthesis of collagen. *Hosp Pract, 12*(12), 61–68.

Riley, G. (2004). The pathogenesis of tendinopathy. A molecular perspective. *Rheumatology, 43*(2), 131–142.

Rillo, O.L., Babini, S.M., Basnak, A., et al. (1991). Tendinous and ligamentous hyperlaxity in patients receiving long-term hemodialysis. *J Rheumatol, 18*(8), 1227–1231.

Rundgren, A. (1974). Physical properties of connective tissue as influenced by single and repeated pregnancies in the rat. *Acta Physiol Scan Suppl, 417*, 1–138.

Shino, K., Oakes, B.W., Horibe, S., et al. (1995). Collagen fibril populations in human anterior cruciate ligament allografts. Electron microscopic analysis. *Am J Sports Med, 23*(2), 203–208.

Simon, S.R. (1994). *Orthopedic Basic Science.* Rosemont, IL: American Academy of Orthopaedic Surgeons.

Snell, R.S. (1984). *Clinical and Functional Histology for Medical Students.* Boston: Little, Brown and Company.

Strocchi, R., De Pasquale, V., Facchini, A., et al. (1996). Age-related changes in human anterior cruciate ligament (ACL) collagen fibrils. *Ital Anat Embryol, 101*(4), 213–220.

Strocchi, R., De Pasquale, V., Guizzardi, S., et al. (1992). Ultrastructural modifications of patellar tendon fibres used as anterior cruciate ligament (ACL) replacement. *Ital J Anat Embryol, 97*(4), 221–228.

Viidik, A. (1968). Elasticity and tensile strength of the anterior cruciate ligament in rabbits as influenced by training. *Acta Physiol Scand, 74,* 372.

Viidik, A., Danielsen, C.C., Oxlund, H. (1982). Fourth International Congress of Biorheology Symposium on Mechanical Properties of Living Tissues: On fundamental and phenomenological models, structure and mechanical properties of collagen, elastin and glycosaminoglycan complexes. *Biorheology, 19,* 437.

Vogel, H.C. (1977). Mechanical and chemical properties of various connective tissue organs in rats as influenced by nonsteroidal antirheumatic drugs. *Connect Tissue Res, 5,* 91.

Walsh, S., Frank, C., Shrive, N., et al. (1993). Knee immobilization inhibits biomechanical maturation of the rabbit medial collateral ligament. *Clin Orthop, 297,* 253–261.

Walsh, W.R., Wiggins, M.E., Fadale, P.D., et al. (1995). Effects of delayed steroid injection on ligament healing using a rabbit medial collateral ligament model. *Biomaterials, 16*(12), 905–910.

Weiss, J.A., Gardiner, J.C., Bonifasi-Lista, C. (2002). Ligament material behavior is nonlinear, viscoelastic and rate-independent under shear loading. *J Biomech, 35*(7), 943–950.

Wiggins, M.E., Fadale, P.D., Barrach, H., et al. (1994). Healing characteristics of a type I collagenous structure treated with corticosteroids. *Am J Sports Med, 22*(2), 279–288.

Wiggins, M.E., Fadale, P.D., Barrach, H., et al. (1995). Effects of local injection of corticosteroids on the healing of ligaments. A follow-up report. *J Bone Joint Surg Am, 77*(11), 1682–1691.

Woo, S.L.Y. (1988). Ligament, tendon, and joint capsule insertions to bone. In Woo, S.L.Y., Buckwalter, J. (Eds.). *Injury and Repair of the Musculoskeletal Soft Tissues.* Park Ridge, IL: American Academy of Orthopaedic Surgeons, 133–166.

Woo, S.L.Y., An, K.N., Arnoczky, D.V.M., et al. (1994). Anatomy, biology, and biomechanics of the tendon, ligament, and meniscus. In S.R. Simon (Ed.): *Orthopaedic Basic Science.* Rosemont, IL: American Academy of Orthopaedic Surgeons, 52.

Woo, S.L.Y., Debski, R.E., Zeminski, J., et al. (2000). Injury and repair of ligaments and tendons. *Ann Rev Biomed Eng, 2,* 83–118.

Woo, S.L.Y., Gomez, M.A., Amiel, D., et al. (1981). The effects of exercise on the biomechanical and biochemical properties of swine digital flexor tendons. *J Biomech Eng, 103,* 51.

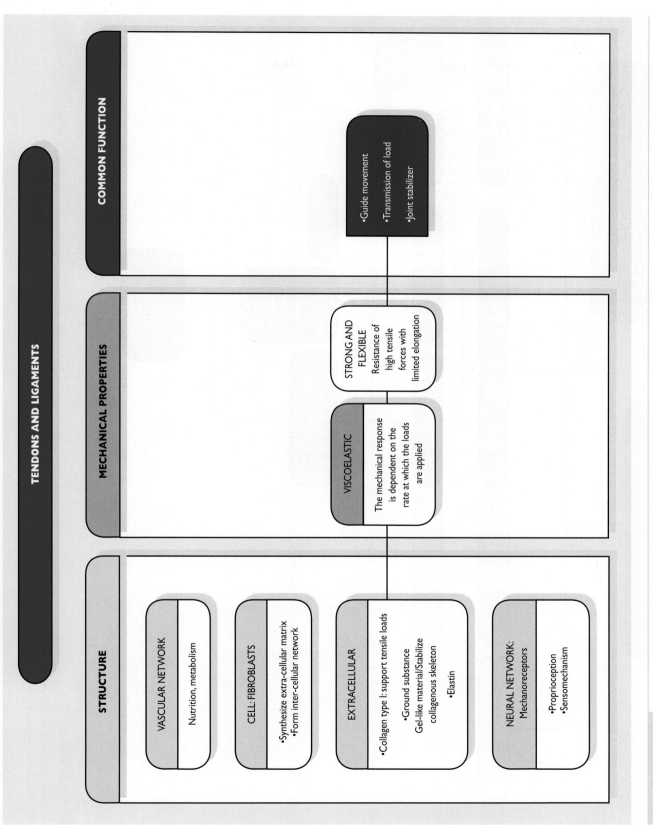

FLOW CHART 4-1

Common structure and mechanical properties of tendons and ligaments*

*This flow chart is designed for classroom or group discussion. Flow chart is not meant to be exhaustive.

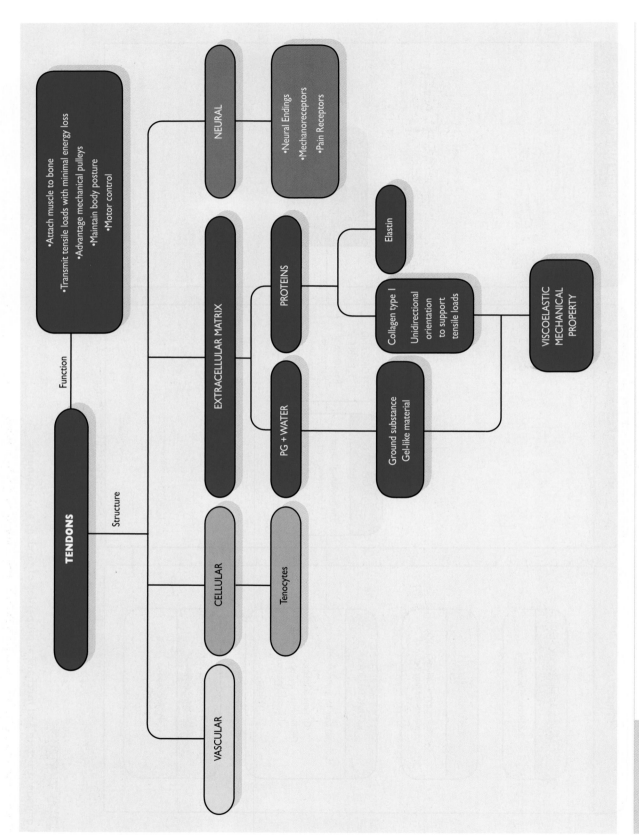

FLOW CHART 4-2

Tendon structure and mechanical properties*

*This flow chart is designed for classroom or group discussion. Flow chart is not meant to be exhaustive.

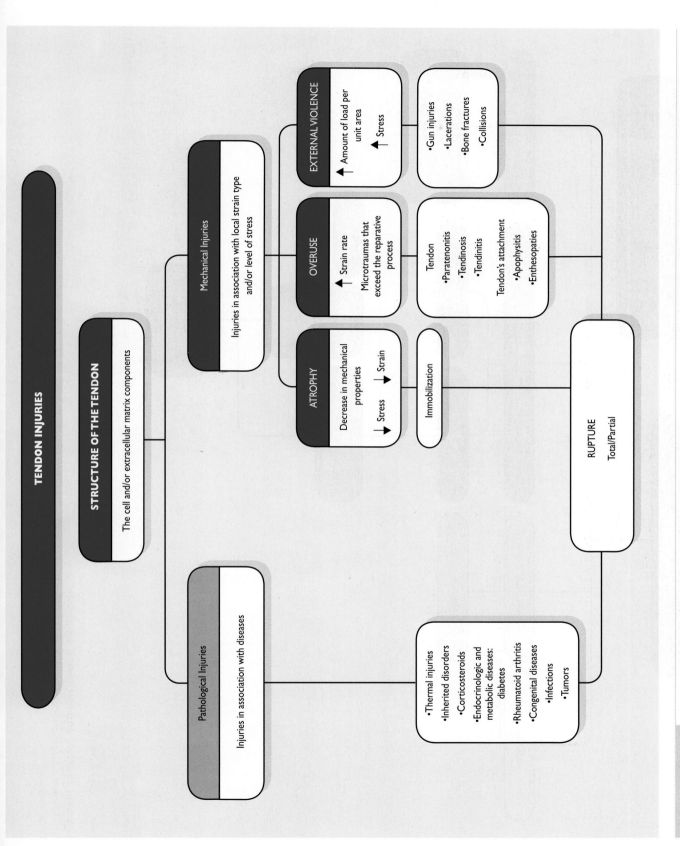

TENDON INJURIES

STRUCTURE OF THE TENDON

The cell and/or extracellular matrix components

Mechanical Injuries

Injuries in association with local strain type and/or level of stress

ATROPHY

Decrease in mechanical properties

Stress → → Strain

Immobilization

OVERUSE

Strain rate →

Microtraumas that exceed the reparative process

Tendon
•Paratenonitis
•Tendinosis
•Tendinitis

Tendon's attachment
•Apophysitis
•Enthesopaties

EXTERNAL VIOLENCE

→ Amount of load per unit area

← Stress

•Gun injuries
•Lacerations
•Bone fractures
•Collisions

Pathological Injuries

Injuries in association with diseases

•Thermal injuries
•Inherited disorders
•Corticosteroids
•Endocrinologic and metabolic diseases: diabetes
•Rheumatoid arthritis
•Congenital diseases
•Infections
•Tumors

RUPTURE

Total/Partial

FLOW CHART 4-3

Tendon injuries. Clinical examples*

*This flow chart is designed for classroom or group discussion. Flow chart is not meant to be exhaustive.

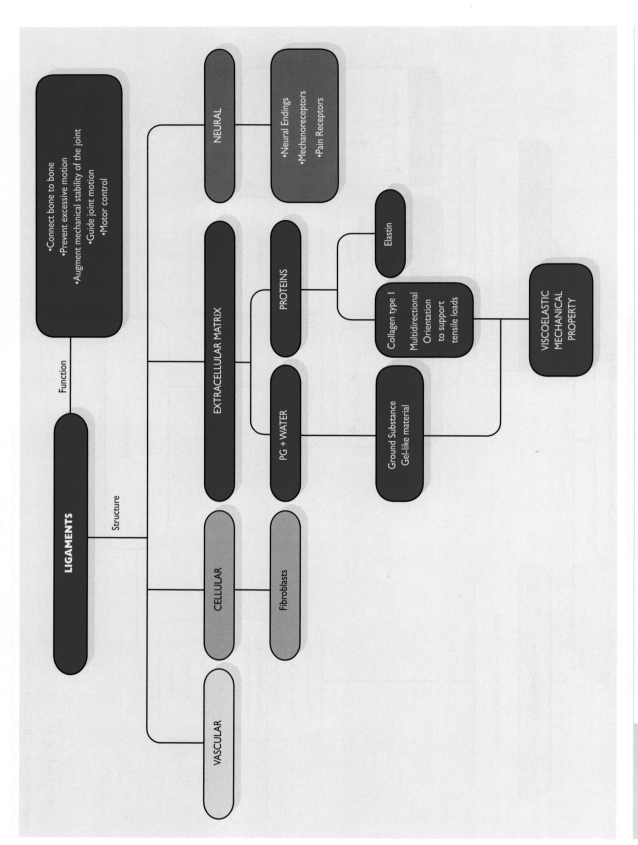

FLOW CHART 4-4

Ligament structure and mechanical properties*

*This flow chart is designed for classroom or group discussion. Flow chart is not meant to be exhaustive.
PG, proteoglycan.

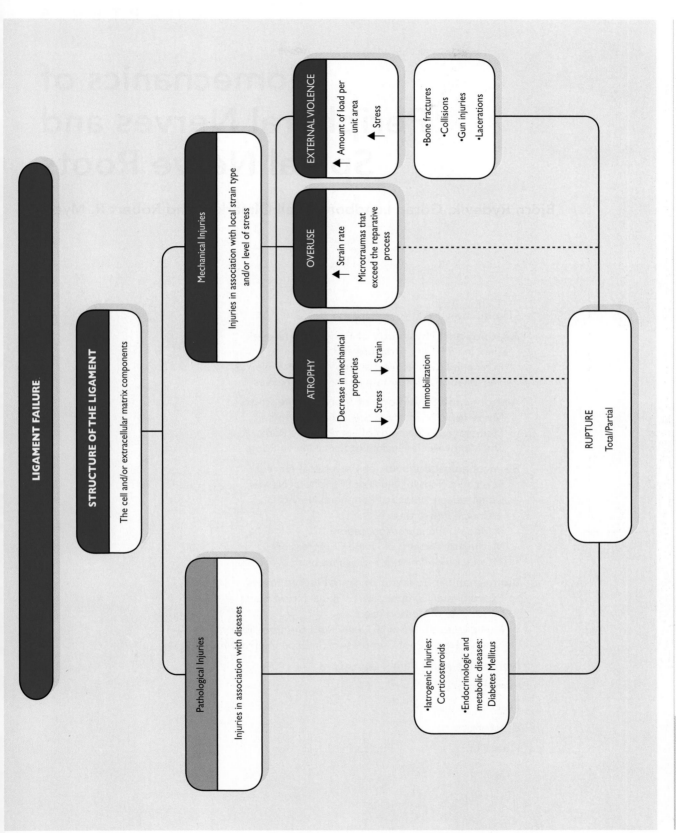

Ligament failure. Clinical examples*

*This flow chart is designed for classroom or group discussion. Flow chart is not meant to be exhaustive.

LIGAMENT FAILURE

STRUCTURE OF THE LIGAMENT

The cell and/or extracellular matrix components

Mechanical Injuries

Injuries in association with local strain type and/or level of stress

ATROPHY

Decrease in mechanical properties

↓ Stress → ↓ Strain

Immobilization

OVERUSE

↑ Strain rate

Microtraumas that exceed the reparative process

EXTERNAL VIOLENCE

↑ Amount of load per unit area

↓ Stress

• Bone fractures
• Collisions
• Gun injuries
• Lacerations

Pathological Injuries

Injuries in association with diseases

• Iatrogenic Injuries: Corticosteroids
• Endocrinologic and metabolic diseases: Diabetes Mellitus

RUPTURE

Total/Partial

CHAPTER 5

Biomechanics of Peripheral Nerves and Spinal Nerve Roots

Björn Rydevik, Göran Lundborg, Kjell Olmarker, and Robert R. Myers

Introduction

The nervous system serves as the body's control center and communications network. As such, it has three broad roles: It senses changes in the body and in the external environment, it interprets these changes, and it responds to this interpretation by initiating action in the form of muscle contraction or gland secretion.

For descriptive purposes, the nervous system can be divided into two parts: the central nervous system, consisting of the brain and spinal cord, and the peripheral nervous system, composed of the various nerve processes that extend from the brain and spinal cord. These peripheral nerve processes provide input to the central nervous system from sensory receptors in skin, joints, muscles, tendons, viscera, and sense organs and provide output from it to effectors (muscles and glands). The peripheral nervous system includes 12 pairs of cranial nerves and their branches and 31 pairs of spinal nerves and their branches (Fig. 5-1A). These branches are called peripheral nerves.

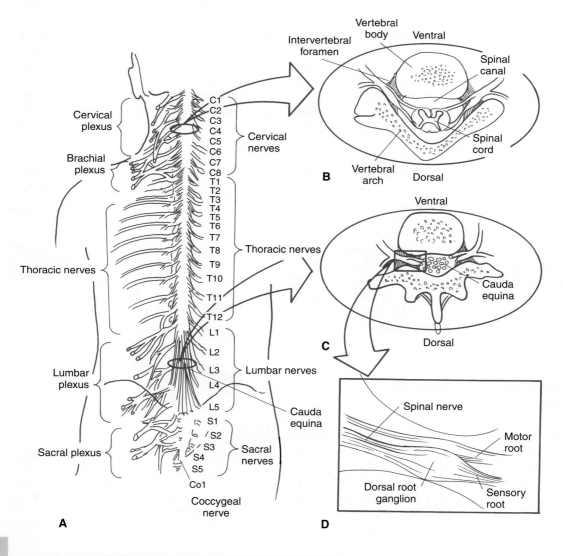

FIG. 5-1

A. Schematic drawing of the spinal cord and the spinal nerves (posterior view). The spinal nerves emerge from the spinal canal through the intervertebral foramina. There are 8 pairs of cervical nerves, 12 pairs of thoracic nerves, 5 pairs of lumbar nerves, 5 pairs of sacral nerves, and 1 pair of coccygeal nerves. Except in the region of the 2nd to the 11th thoracic vertebrae (T2–T11), the nerves form complex networks called plexuses after exiting the intervertebral foramina. Only the main branch of each nerve, the ventral ramus, is depicted. **B.** Cross-section of the cervical spine showing the spinal cord in the spinal canal and the nerve roots exiting through the intervertebral foramina. **C.** Cross-section of the lumbar spine showing the nerve roots of the cauda equina in the spinal canal. **D.** Each exiting nerve root complex in the intervertebral foramen consists of a motor root, a sensory root, and a dorsal root ganglion.

Each spinal nerve is connected to the spinal cord through a posterior (dorsal) root and an anterior (ventral) root, which unite to form the spinal nerve at the intervertebral foramen (Fig. 5-1B–D). The posterior roots contain fibers of sensory neurons (those conducting sensory information from receptors in the skin, muscles, tendons, and joints to the central nervous system), and the anterior roots contain mainly fibers of motor neurons (those that conduct impulses from the central nervous system to distal targets such as muscle fibers).

Shortly after the spinal nerves leave their intervertebral foramina, they divide into two main branches: the dorsal rami, which innervate the muscles and skin of the head, neck, and back, and the generally larger and more important ventral rami, which innervate the ventral and lateral parts of these structures as well as the upper and lower extremities. Except in the thoracic region, the ventral rami do not run directly to the structures that they innervate but first form interlacing networks, or plexuses, with adjacent nerves (Fig. 5-1A).

This chapter focuses on both the peripheral nerves and spinal nerve roots, which contain not only nerve fibers but also connective tissue elements and vascular structures that encompass the nerve fibers. The nerves possess some special anatomic properties that may serve to protect the nerve from mechanical damage, for instance, stretching (tension) and compression. In this chapter, the basic microanatomy of the peripheral nerves and the spinal nerve roots is reviewed with special reference to these built-in mechanisms of protection. The mechanical behavior of peripheral nerves that are subjected to tension and compression is also described in some detail.

Anatomy and Physiology of Peripheral Nerves

The peripheral nerves are complex composite structures consisting of nerve fibers, connective tissue, and blood vessels. Because the three tissue elements that make up these nerves react to trauma in different ways and may each play distinct roles in the functional deterioration of the nerve after injury, each element is described separately.

THE NERVE FIBERS: STRUCTURE AND FUNCTION

The term nerve fiber refers to the elongated process (axon) extending from the nerve cell body along with its myelin sheath and Schwann cells (Figs. 5-2 and 5-3).

The nerve fibers of sensory neurons conduct impulses from the skin, skeletal muscles, and joints to the central nervous system. The nerve fibers of the motor neurons convey impulses from the central nervous system to the skeletal muscles, causing muscle contraction. (A detailed description of the mechanics of muscle contraction is given in Chapter 6.)

The nerve fibers not only transmit impulses but also serve as an anatomic connection between the nerve cell body and its end organs. This connection is maintained by axonal transport systems, through which various substances synthesized within the cell body (e.g., proteins) are transported from the cell body to the periphery and in the opposite direction. The axonal transport takes place at speeds that vary from approximately 1 to approximately 400 mm per day.

Most axons of the peripheral nervous system are surrounded by multilayered, segmented coverings known

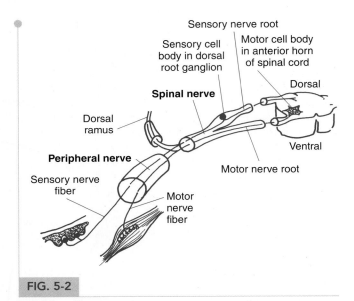

FIG. 5-2

Schematic representation of the arrangement of a typical spinal nerve as it emerges from its dorsal and ventral nerve roots. The peripheral nerve begins after the dorsal ramus branches off. (For the sake of simplicity, the nerve is not shown entering a plexus.) Spinal nerves and most peripheral nerves are mixed nerves: They contain both sensory (afferent) and motor (efferent) nerve fibers. The cell body and its nerve fibers make up the neuron. The cell bodies of the motor neurons are located in the anterior horn of the spinal cord, and those of the sensory neurons are found in the dorsal root ganglia. Here, a motor nerve fiber is shown innervating muscle and a sensory nerve fiber is depicted innervating skin. *Adapted from Rydevik, B., Brown, M.D., Lundborg, G. (1984). Pathoanatomy and pathophysiology of nerve root compression. Spine, 9, 7.*

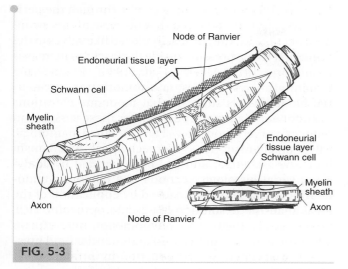

FIG. 5-3

Schematic drawings of the structural features of a myelinated nerve fiber. *Adapted from Sunderland, S. (1978). Nerves and Nerve Injuries (2nd ed.). Edinburgh, Scotland: Churchill Livingstone.*

as myelin sheaths (Fig. 5-3). Fibers with this covering are said to be myelinated, whereas those without it (mainly small sensory fibers conducting impulses for pain from the skin) are unmyelinated. The myelin sheath of the axons of the peripheral nerves is produced by flattened cells called Schwann cells arranged along the axon (Fig. 5-3). A sheath is formed as the Schwann cell encircles the axon and winds around it many times, pushing its cytoplasm and nucleus to the outside layer. Unmyelinated gaps called nodes of Ranvier lie between the segments of the myelin sheath at approximately 1 to 2 mm apart.

The myelin sheath increases the speed of the conduction of nerve impulses and insulates and maintains the axon. Impulses are propagated along the unmyelinated nerve fibers in a slow, continuous way, whereas in the myelinated nerve fibers the impulses "jump" at a higher speed from one node of Ranvier to the next in a process called saltatory conduction. The conduction velocity of a myelinated nerve is directly proportional to the diameter of the fiber, which usually ranges from 2 to 20 mm. Motor fibers that innervate skeletal muscle have large diameters, as do sensory fibers that relay impulses associated with touch, pressure, heat, cold, and kinesthetic sense, such as skeletal muscle tension and joint position. Sensory fibers that conduct impulses for dull, diffuse pain (as opposed to sharp, immediate pain) have the smallest diameters.

Nerve fibers are packed closely in fascicles, which are further arranged into bundles that make up the nerve itself. The fascicles are the functional subunits of the nerve.

INTRANEURAL CONNECTIVE TISSUE OF PERIPHERAL NERVES

Successive layers of connective tissue surround the nerve fibers—called the endoneurium, perineurium, and epineurium—and protect the fibers' continuity (Fig. 5-4). The protective function of these connective tissue layers is essential because nerve fibers are extremely susceptible to stretching and compression.

The outermost layer, the epineurium, is located between the fascicles and superficially in the nerve. This rather loose connective tissue layer serves as a cushion during movements of the nerve, protecting the fascicles from external trauma and maintaining the oxygen supply system via the epineural blood vessels. The amount of epineural connective tissue varies among nerves and at different levels within the same nerve. Where the nerves lie close to bone or pass joints, the epineurium is often more abundant than elsewhere, as the need for protection may be greater in these locations. The spinal nerve roots are devoid of both epineurium and perineurium, and the nerve fibers in the nerve root may therefore be more susceptible to trauma (Rydevik et al., 1984).

The perineurium is a lamellar sheath that encompasses each fascicle. This sheath has great mechanical strength as well as a specific biochemical barrier. Its strength is demonstrated by the fact that the fascicles can be inflated by fluid to a pressure of approximately l,000 mm of mercury (Hg) before the perineurium ruptures.

The barrier function of the perineurium chemically isolates the nerve fibers from their surroundings, thus preserving an ionic environment of the interior of the fascicles, a special milieu intérieur. The endoneurium, the connective tissue inside the fascicles, is composed principally of fibroblasts and collagen.

The interstitial tissue pressure in the fascicles, the endoneurial fluid pressure, is normally slightly elevated ($+1.5 \pm 0.7$ mm Hg [Myers and Powell, 1981]) compared with the pressure in surrounding tissues such as subcutaneous tissue (-4.7 ± 0.8 mm Hg) and muscle tissue (-2 ± 2 mm Hg). The elevated endoneurial fluid pressure is illustrated by the phenomenon whereby incision of the perineurium results in herniation of nerve fibers. The endoneurial fluid pressure may increase further as a result of trauma to the nerve, with subsequent edema. Such a pressure increase may affect the microcirculation and the function of the nerve.

THE MICROVASCULAR SYSTEM OF PERIPHERAL NERVES

The peripheral nerve is a well-vascularized structure containing vascular networks in the epineurium, the perineurium, and the endoneurium. Because both impulse propagation and axonal transport depend on a local oxygen supply, it is natural that the microvascular system has a large reserve capacity.

The blood supply to the peripheral nerve as a whole is provided by large vessels that approach the nerve segmentally along its course. When these local nutrient vessels reach the nerve, they divide into ascending and descending branches. These vessels run longitudinally and frequently anastomose with the vessels in the perineurium and endoneurium. Within the epineurium, large arterioles and venules, 50 to 100 mm in diameter, constitute a longitudinal vascular system (Fig. 5-4).

Within each fascicle lies a longitudinally oriented capillary plexus with loop formations at various levels. The capillary system is fed by arterioles 25 to 150 mm in diameter that penetrate the perineurial membrane.

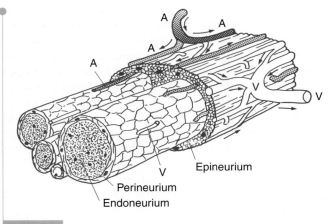

FIG. 5-4

Schematic drawing of a segment of a peripheral nerve. The individual nerve fibers are located within the endoneurium. They are closely packed in fascicles, each of which is surrounded by a strong sheath, the perineurium. A bundle of fascicles is embedded in a loose connective tissue, the epineurium. Blood vessels are present in all layers of the nerve. *A,* arterioles *(shaded); V,* venules *(unshaded).* The *arrows* indicate the direction of blood flow. *Adapted from Rydevik B (1979). Compression injury of peripheral nerve, University of Gothenburg, Sweden, 1979.*

These vessels run an oblique course through the perineurium, and it is believed that because of this structural peculiarity, they are easily closed like valves in the event that tissue pressure inside the fascicles increases (Lundborg, 1975; Myers et al., 1986). Engineering structural analysis of these anastomotic transperineurial blood vessels suggests that moderate elevations in endoneurial pressure compresses these vessels and reduces nerve blood flow. This so-called pathologic valve mechanism is a biomechanical mechanism for nerve ischemia, and has been verified by analysis of serial histologic sections in a moderate endoneurial edematous state caused by application of the local anaesthetic solution 3% 2-chloroprocaine HCL (Myers et al., 1986). This phenomenon may explain why even a limited increase in endoneurial fluid pressure is associated with a reduction in intrafascicular blood flow.

The built-in safety system of longitudinal anastomoses provides a wide margin of safety if the regional segmental vessels are transected. In an experimental animal in vivo model, it is extremely difficult to induce complete ischemia to a nerve by local surgical procedures. For example, if the whole sciatic-tibial nerve complex of a rabbit (15 cm long) is surgically separated from its surrounding structures and the regional nutrient vessels are cut, there is no detectable reduction in the intrafascicular blood flow as studied by intravital microscopic techniques. Even if such a mobilized nerve is cut distally or proximally, the intraneural longitudinal vascular systems can maintain the microcirculation at least 7 to 8 cm from the cut end. If a nonmobilized nerve is cut, there is still perfect microcirculation even at the very tip of the nerve; this phenomenon demonstrates the sufficiency of the intraneural vascular collaterals. However, other studies in rats indicate that stripping the epineural circulation from nerve bundles causes demyelination of subperineural nerve fibers.

Anatomy and Physiology of Spinal Nerve Roots

In the early embryologic developmental stages, the spinal cord has the same length as the spinal column. However, in the fully grown individual, the spinal cord ends as the conus medullaris, approximately at the level of the first lumbar vertebra. A nerve root that leaves the spinal canal through an intervertebral foramen in the lumbar or sacral spine therefore has to pass from the point where it leaves the spinal cord, which is in the lower thoracic spine, to the point of exit from the spine

(Fig. 5-5). Because the spinal cord is not present below the first lumbar vertebra, the nervous content of the spinal canal is composed of only the lumbosacral nerve roots. This bundle of nerve roots within the lumbar and sacral part of the spinal canal has been suggested to resemble the tail of a horse and is therefore often called the cauda equina, that is, tail of horse.

Two different types of nerve roots are found within the lumbosacral spine, ventral/motor roots and dorsal/sensory roots. The cell bodies of the motor axons are located in the anterior horns of the gray matter in the spinal cord, and because these nerve roots leave the spinal cord from the ventral aspect, they are also called ventral roots. The other type of nerve root is the sensory, or dorsal, root. As the name suggests, these nerve roots mainly comprise sensory (i.e., afferent) axons and reach the spinal cord at the dorsal region of the spinal cord. The cell bodies of the sensory axons are located in a swelling of the most caudal part of the respective dorsal nerve root, called the dorsal root ganglion. The dorsal root ganglia are located in or close to the intervertebral foramen. Unlike the nerve roots, the dorsal root ganglia are not enclosed by cerebrospinal fluid and the meninges. Instead, they are enclosed by both a multilayered connective tissue sheath, similar to the perineurium of the peripheral nerve, and a loose connective tissue layer called epineurium.

When the nerve root approaches the intervertebral foramen, the root sleeve gradually encloses the nerve tissue more tightly. The subarachnoid space and the amount of cerebrospinal fluid surrounding each nerve root pair will thus become gradually reduced in the caudal direction. Compression injury of a nerve root may induce an increase in the permeability of the endoneurial capillaries, resulting in edema formation (Olmarker et al., 1989b; Rydevik and Lundborg, 1977). This can lead to an increase of the intraneural fluid and subsequent impairment of the nutritional transport to the nerve (Myers, 1998; Myers and Powell, 1981). Such a mechanism might be particularly important at locations where the nerve roots are tightly enclosed by connective tissue. Thus there is a more pronounced risk for an "entrapment syndrome" within the nerve roots at the intervertebral foramen than more central in the cauda equina (Rydevik et al., 1984). The dorsal root ganglion, with its content of sensory nerve cell bodies, tightly enclosed by meninges, might be particularly susceptible to edema formation.

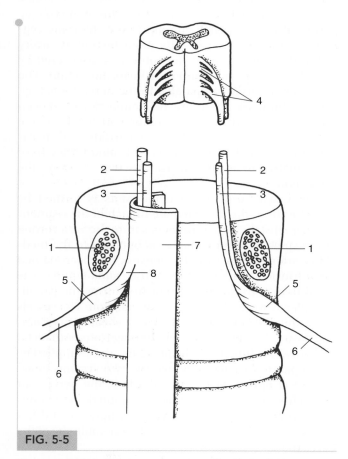

FIG. 5-5

The intraspinal nervous structures as seen from behind. The vertebral arches are removed by cutting the pedicles (1). A ventral (2) and a dorsal (3) nerve root leave the spinal cord as small rootlets (4). Before leaving the spinal canal, the dorsal root forms a swelling called the dorsal root ganglion (5), which contains the sensory cell bodies, before forming the spinal nerve (6) together with the ventral nerve root. The nerve roots are covered by a central dural sac (7) or with extensions of this sac called nerve root sleeves (8). *Reproduced with permission from Olmarker, K. (1991). Spinal nerve root compression. Nutrition and function of the porcine cauda equina compressed in vivo. Acta Orthop Scand Suppl, 242, 1–27.*

MICROSCOPIC ANATOMY OF SPINAL NERVE ROOTS

There are two microscopically different regions of the nerve roots. Closest to the spinal cord is a central glial segment comprised of glial cells that resembles the microscopic organization of central nervous structures at the spinal cord or the brain. This glial segment is transferred to a nonglial segment in a dome-shaped junction a few millimeters from the spinal cord. This nonglial segment is organized in the same manner as the endoneurium of the peripheral nerves, that is, with Schwann cells instead of glia cells. However, some small islets of glia cells also are found in this otherwise peripherally organized endoneurium.

MEMBRANOUS COVERINGS OF SPINAL NERVE ROOTS

The axons in the endoneurium are separated from the cerebrospinal fluid by a thin layer of connective tissue called the root sheath. This root sheath is the structural analogue to the pia mater that covers the spinal cord. There are usually 2 to 5 cellular layers in the root sheath, but as many as 12 layers have been identified. The cells of the proximal part of the outer layers of the root sheath are similar to the pia cells of the spinal cord, and the cells in the distal part are more similar to the arachnoid cells of the spinal dura. The inner layers of the root sheath are composed of cells that show similarities to the cells of the perineurium of peripheral nerves. An interrupted basement membrane encloses these cells separately. The inner layers of the root sheath constitute a diffusion barrier between the endoneurium of the nerve roots and the cerebrospinal fluid. This barrier is considered to be relatively weak and may prevent only the passage of macromolecules.

The spinal dura encloses the nerve roots and the cerebrospinal fluid. When the two layers of the cranial dura enter the spinal canal, the outer layer blends with the periosteum of the part of the laminae of the cervical vertebrae facing the spinal canal. The inner layers join the arachnoid and become the spinal dura. In contrast to the root sheath, the spinal dura is an effective diffusion barrier. The barrier properties are located in a connective tissue sheath between the dura and the arachnoid called the neurothelium. Similar to the inner layer of the root sheath, this neurothelium resembles the perineurium of the peripheral nerves. It is suggested that these two layers in fact form the perineurium when the nerve root is transformed to a peripheral nerve on leaving the spine.

THE MICROVASCULAR SYSTEM OF SPINAL NERVE ROOTS

Information about the vascular anatomy of the nerve roots has mainly been derived from studies on the vascularization of the spinal cord. Therefore, the nomenclature of the various vessels has been somewhat confusing. A summary of the existing knowledge on nerve root vasculature will be presented in the paragraphs that follow.

The segmental arteries generally divide into three branches when approaching the intervertebral foramen: (1) an anterior branch that supplies the posterior abdominal wall and lumbar plexus, (2) a posterior branch that supplies the paraspinal muscles and facet joints, and (3) an intermediate branch that supplies the contents of the spinal canal. A branch of the intermediate branch joins the nerve root at the level of the dorsal root ganglion. There are usually three branches from this vessel: one to the ventral root, one to the dorsal root, and one to the vasa corona of the spinal cord.

The branches to the vasa corona of the spinal cord, called medullary arteries, are inconsistent. In adults, only 7 to 8 remain of the 128 from the embryologic period of life, and each supplies more than one segment of the spinal cord. The main medullary artery in the thoracic region of the spine was discovered by Adamkiewicz in 1881 and still bears his name. The medullary arteries run parallel to the nerve roots (Fig. 5-6). In humans, there are no connections between these vessels and the vascular network of the nerve roots. Because the medullary feeder arteries only occasionally supply the nerve roots with blood, they have been referred to as the extrinsic vascular system of the cauda equina.

The vasculature of the nerve roots is formed by branches from the intermediate branch of the segmental artery distally and by branches from the vasa corona of the spinal cord proximally. As opposed to the medullary arteries, this vascular network has been named the intrinsic vascular system of the cauda equina. The distal branch to the dorsal root first forms the ganglionic plexus within the dorsal root ganglion. The vessels run within the outer layers of the root sheath, called epipial tissue. As there are vessels coming from both distal and proximal directions, the nerve roots are supplied by two separate vascular systems. The two systems anastomose at approximately two thirds of the nerve root length from the spinal cord. This location demonstrates a region of a less-developed vascular network and has been suggested to be a particularly vulnerable site of the nerve roots.

The arteries of the intrinsic system send branches down to the deeper parts of the nerve tissue in a T-like manner. To compensate for elongation of the nerve roots, the arteries are coiled both longitudinally and in the steep running branches between the different fascicles (Fig. 5-6). Unlike peripheral nerves, the venules do not course together with the arteries in the nerve roots but instead usually have a spiraling course in the deeper parts of the nerve.

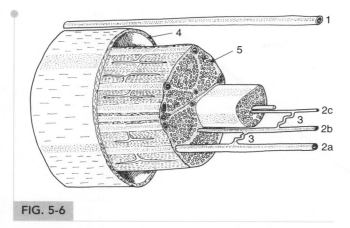

FIG. 5-6

Schematic presentation of some anatomic features of the intrinsic arteries of the spinal nerve roots. The arterioles within the cauda equina may refer to either the extrinsic (1) or the intrinsic (2) vascular system. From the superficial intrinsic arterioles are branches that continue almost at right angles down between the fascicles. These vessels often run in a spiraling course, thus forming vascular "coils" (3). When reaching a specific fascicle they branch in a T-like manner, with one branch running cranially and one caudally, forming interfascicular arterioles (2b). From these interfascicular arterioles are small branches that enter the fascicles, where they supply the endoneurial capillary networks (2c). Arterioles of the extrinsic vascular system run outside the spinal dura (4) and have no connections with the intrinsic system by local vascular branches. The superficial intrinsic arterioles (2a) are located within the root sheath (5). *Reproduced with permission from Olmarker, K. (1991). Spinal nerve root compression. Nutrition and function of the porcine cauda equina compressed in vivo. Acta Orthop Scand Suppl, 242, 1–27.*

There is a barrier of the endoneurial capillaries in peripheral nerves called the blood–nerve barrier, which is similar to the blood–brain barrier of the central nervous system (Lundborg, 1975; Rydevik and Lundborg, 1977). The presence of a corresponding barrier in nerve roots has been questioned. If present, a blood–nerve barrier in nerve roots does not seem to be as well developed as in endoneurial capillaries of peripheral nerve, which implies that edema may be formed more easily in nerve roots than in peripheral nerves (Rydevik et al., 1984).

Biomechanical Behavior of Peripheral Nerves

External trauma to the extremities and nerve entrapment may produce mechanical deformation of the peripheral nerves that results in the deterioration of nerve function. If the mechanical trauma exceeds a certain degree, the nerves' built-in mechanisms of protection may not be sufficient, resulting in changes in nerve structure and function. Common modes of nerve injury are stretching and compression, which may be inflicted, respectively, by rapid extension and crushing.

STRETCHING (TENSILE) INJURIES OF PERIPHERAL NERVES

Nerves are strong structures with considerable tensile strength. The maximal load that can be sustained by the median and ulnar nerves is in the range of 70 to 220 newtons (N) and 60 to 150 N, respectively. These figures are of academic interest only because severe intraneural tissue damage is produced by tension long before a nerve breaks.

A discussion of the elasticity and biomechanical properties of nerves is complicated by the fact that nerves are not homogeneous isotropic materials but, instead, composite structures, with each tissue component having its own biomechanical properties. The connective tissues of the epineurium and perineurium are primarily longitudinal structures.

When tension is applied to a nerve, initial elongation of the nerve under a very small load is followed by an interval in which stress and elongation show a linear relationship characteristic of an elastic material (Fig. 5-7). As the limit of the linear region is approached, the nerve fibers start to rupture inside the endoneurial tubes and inside the intact perineurium. The perineurial sheaths rupture at approximately 25% to 30% elongation (ultimate strain) above in vivo length (Rydevik et al., 1990). After this point, there is a disintegration of the elastic properties, and the nerve behaves more like a plastic material (i.e., its response to the release of loads is incomplete recovery).

Although variations exist in the tensile strength of various nerves, the maximal elongation at the elastic limit is approximately 20%, and complete structural failure seems to occur at a maximum elongation of approximately 25% to 30%. These values are for normal nerves; injury to a nerve may induce changes in its mechanical properties, namely increased stiffness and decreased elasticity.

Stretching, or tensile, injuries of peripheral nerves are usually associated with severe accidents, such as when high-energy tension is applied to the brachial plexus in association with a birth-related injury, as a result of high-speed vehicular collision, or after a fall from

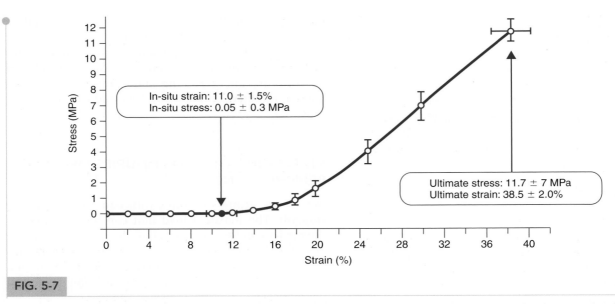

FIG. 5-7

The stress-strain behavior of a rabbit tibial nerve. The nerve exhibits a low stiffness toe region of approximately 15% and begins to retain significant tension as the strain increases beyond 20%. *Reproduced with permission from Rydevik, B.L., Kwan, M K., Myers, R.R., et al. (1990). An in vitro mechanical and histologic study of acute stretching on rabbit tibial nerve. J Orthop Res, 8, 694–701.*

a height. Such plexus injuries may result in partial or total functional loss of some or all of the nerves in the upper extremity, and the consequent functional deficits represent a considerable disability in terms of sensory and motor loss. The outcome depends on which tissue components of the nerves are damaged as well as on the extent of the tissue injury. Of clinical importance is the observation that there can be considerable structural damage (perineurial sheath injuries) induced by stretching with no visible injury on the surface of the nerve (Case Study 5-1).

High-energy plexus injuries represent an extreme type of stretching lesion caused by sudden violent trauma. A different stretching situation of considerable clinical interest is the suturing of the two ends of a cut nerve under moderate tension. This situation occurs when a substantial gap exists in the continuity of a nerve trunk and the restoration of the continuity requires the application of tension to bring the nerve ends back together. The moderate, gradual tension applied to the nerve in these cases may stretch and angulate local feeding vessels. It may also be sufficient to reduce the transverse fascicular cross-sectional area and impair the intraneural nutritive capillary flow (Fig. 5-8).

As the sutured nerve is stretched, the perineurium tightens; as a result, the endoneurial fluid pressure is increased and the intrafascicular capillaries may be obliterated. Also, the flow is impaired in the segmental, feeding, and draining vessels, as it is in larger vessels in the epineurium, and at a certain stage the intraneural microcirculation ceases. Intravital observations of intraneural blood flow in rabbit tibial nerves (Lundborg and Rydevik, 1973) showed that an elongation of 8% induced impaired venular flow and that even greater tension produced continuous impairment of capillary and arteriolar flow until, at 15% elongation, all intraneural microcirculation ceased completely. For the same nerve, an elongation (strain) of 6% induced a reduction of nerve action potential amplitude by 70% at one hour with recovery to normal values during one-hour restitution. At 12% elongation, conduction was completely blocked by one hour and showed minimal recovery (Wall et al., 1992). Such data have clinical implications in nerve repair, limb trauma, and limb lengthening.

A situation of even more gradual stretching, applied over a long time, is the growth of intraneural tumors such as schwannomas. In this situation, the nerve fibers are forced into a circumferential course around the gradually expanding tumor. Functional changes in cases of such very gradual stretching are often minimal or nonexistent.

Brachial Plexus Palsy

During the birth process, a newborn suffered a traction injury in his left brachial plexus. A few months later, he presents with the upper left arm in a static position of adduction, internal rotation of the shoulder, extension of the elbow, pronation of the forearm, and flexion of the wrist. He does not respond to sensory stimulus in his shoulder and presents biceps and brachioradialis areflexia. A sudden deformation and high tensile stress injuries in the C5–C6 nerve roots affected the mixed (motor and sensory) neural functions, mainly the muscles responsible for the scapulohumeral rhythm (see Chapter 12).

An Erb palsy is diagnosed. The sudden elongation suffered during the traction can lead to structural damage and reduction in the transverse fascicular cross-sectional area, producing impairment of the intraneural vascular flow and impulse transmission.

In less severe cases, functional restoration may occur within weeks or months. In more severe cases, healing may take place during the first two to three years, but if the structural nerve injury is severe, considerable long-term functional disability can result. If structural derangement of the nerve trunk has taken place, nerve grafting may be required.

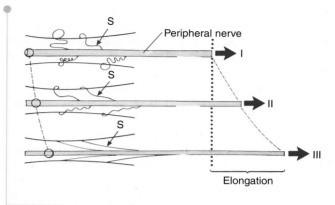

FIG. 5-8

Schematic representation of a peripheral nerve and its blood supply at three stages during stretching. **I.** The segmental blood vessels (S) are normally coiled to allow for the physiologic movements of the nerve. **II.** Under gradually increasing elongation, these regional vessels become stretched and the blood flow in them is impaired. **III.** The cross-sectional area of the nerve (represented within the *circle*) is reduced during stretching and the intraneural blood flow is further impaired. Complete cessation of all blood flow in the nerve usually occurs at approximately 15% elongation. *Adapted from Lundborg, G., Rydevik, B. (1973). Effects of stretching the tibial nerve of the rabbit: A preliminary study of the intraneural circulation and the barrier function of the perineurium. J Bone Joint Surg, 55B, 390.*

COMPRESSION INJURIES OF PERIPHERAL NERVES

It has long been known that compression of a nerve can induce symptoms such as numbness, pain, and muscle weakness. The biologic basis for the functional changes has been investigated extensively (Rydevik and Lundborg, 1977; Rydevik et al., 1981). In these investigations (Fig. 5-9), even mild compression was observed to induce structural and functional changes, and the significance of mechanical factors such as pressure level and mode of compression became apparent.

Critical Pressure Levels

Experimental and clinical observations have revealed some data on the critical pressure levels at which disturbances occur in intraneural blood flow, axonal

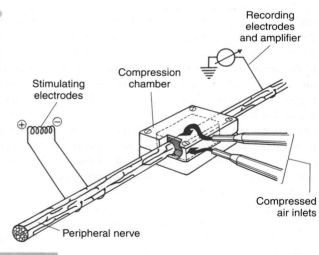

FIG. 5-9

Schematic drawing of an experimental setup for studying deterioration of nerve function during compression. *Adapted from Dahlin, L.B., Danielsen, N., Ehira, T., Lundborg, G., Rydevik, B. (1968). Mechanical effects of compression of peripheral nerves. J Biomech Eng, 108, 120–122.*

transport, and nerve function. Certain pressure levels seem to be well defined with respect to structural and functional changes induced in the nerve. The duration of the compression also influences the development of these changes.

At 30 mm Hg of local compression, functional changes may occur in the nerve, and its viability may be jeopardized during prolonged compression (four to six hours) at this pressure level (Lundborg et al., 1982). Such changes appear to be caused by impairment of the blood flow in the compressed part of the nerve (Rydevik et al., 1981). Corresponding pressure levels (approximately 32 mm Hg) were recorded close to the median nerve in the carpal tunnel in patients with carpal tunnel syndrome, whereas in a group of control subjects the pressure in the carpal tunnel averaged only 2 mm Hg. Long-standing or intermittent compression at low pressure levels (approximately 30–80 mm Hg) may induce intraneural edema, which in turn may become organized into a fibrotic scar in the nerve (Rydevik and Lundborg, 1977).

Compression at approximately 30 mm Hg also brings about changes in the axonal transport systems, and long-standing compression may thus lead to depletion of axonally transported proteins distal to the compression site. Such blockage of axonal transport induced by local compression (pinching) may cause the axons to be more susceptible to additional compression distally, the so-called double-crush syndrome.

Slightly higher pressure (80 mm Hg, for example) causes complete cessation of intraneural blood flow; the nerve in the locally compressed segment becomes completely ischemic. Yet, even after two hours or more of compression, blood flow is rapidly restored when the pressure is released (Rydevik et al., 1981). Even higher levels of pressure (200–400 mm Hg, for example) applied directly to a nerve can induce structural nerve fiber damage and rapid deterioration of nerve function, with incomplete recovery after even shorter periods of compression. Hence, the magnitude of the applied pressure and the severity of the induced compression lesion appear to be correlated.

Mode of Pressure Application

The pressure level is not the only factor that influences the severity of nerve injury brought about by compression. Experimental and clinical evidence indicates that the mode of pressure application is also of major significance. Its importance is illustrated by the fact that direct compression of a nerve at 400 mm Hg by means of a small inflatable cuff around the nerve induces a more severe nerve injury than does indirect compression of the nerve at 1,000 mm Hg via a tourniquet applied around the extremity. Even though the hydrostatic pressure acting on the nerve in the former situation is less than half that in the latter, the nerve lesion is more severe, probably because direct compression causes a more pronounced deformation of the nerve (especially at its edges) than does indirect compression, in which the tissue layers between the compression device and the nerve "bolster" the nerve. One may also conclude that the nerve injury caused by compression is not directly related to the high hydrostatic pressure in the center of the compressed nerve segment but instead is more dependent on the specific mechanical deformation induced by the applied pressure.

MECHANICAL ASPECTS OF NERVE COMPRESSION

Electron microscopic analysis of the deformation of the nerve fibers in the peroneal nerve of the baboon hind limb induced by tourniquet compression demonstrated the so-called edge effect; that is, a specific lesion was induced in the nerve fibers at both edges of the compressed nerve segment: The nodes of Ranvier were displaced toward the noncompressed parts of the nerve. The nerve fibers in the center of the compressed segment, where the hydrostatic pressure is highest, generally were not affected acutely. The large-diameter nerve fibers were usually affected, but the thinner fibers were spared. This finding confirms theoretical calculations that indicate larger nerve fibers undergo a relatively greater deformation than do thinner fibers at a given pressure. It is also known clinically that a compression lesion of a nerve first affects the large fibers (e.g., those that carry motor function), whereas the thin fibers (e.g., those that mediate pain sensation) are often preserved. The intraneural blood vessels have also been shown to be injured at the edges of the compressed segment (Rydevik and Lundborg, 1977). Basically, the lesions of nerve fibers and blood vessels seem to be consequences of the pressure gradient, which is maximal just at the edges of the compressed segment.

In considering the mechanical effects on nerve compression, keep in mind that the effect of a given pressure depends on the way in which it is applied, its magnitude, and its duration. Although pressure may be applied with various spatial distributions, two basic types of pressure applications are generally encountered in experimental settings and in pathologic conditions. One type is uniform pressure applied around the entire circumference of a longitudinal segment of a nerve or extremity. This is the kind of purely radial pressure that is applied by the common pneumatic tourniquet. It has also been used in miniature apparatus to produce controlled compression of individual nerves (Rydevik and Lundborg, 1977) (Fig. 5-9). Clinically, this type of loading on a nerve probably occurs when the pressure on the median nerve is elevated in the carpal tunnel, producing a characteristic syndrome.

Another type of mechanical action takes place when the nerve is compressed laterally. This is the kind of deformation that occurs if a nerve or extremity is placed between two parallel flat rigid surfaces that are then moved toward each other, squeezing the nerve or extremity. This type of deformation occurs if a sudden blow by a rigid object squeezes a nerve against the surface of an underlying bone. It may also occur when a spinal nerve is compressed by a herniated disc (Case Study 5-2).

The details of the deformation of a nerve may be quite different in these two cases of loading. In uniform circumferential compression like that applied by a pneumatic tourniquet (Fig. 5-10), the cross-section of the nerve or extremity tends to remain circular but decreases in diameter in the loaded region. Because the material of the tissues is relatively incompressible, this radial compression requires a squeezing out of the tissue under the tourniquet, moving it outward from the centerline toward each of the free edges. It can be seen readily that the displacement of the tissue builds up from zero at the centerline to a maximum at the edge of the tourniquet. It is this large displacement, along with accompanying shear stresses, that is believed to cause the edge effect mentioned, that is observed in experiments in vivo. This region sustains both the maximum pressure gradient and the maximum displacement.

Lateral compression does not necessarily produce any axial motion of material, but it may simply deform the cross-section from nearly circular to more elliptical, as shown in Figure 5-11. In this kind of compression it is clear that in the direction perpendicular to the direction of compression (x), the nerve must be extended. This extension is illustrated by the movement of point G to G′ during compression. At the same time point A moves to A′, indicating shortening or compression, in the direction of loading. The degree of compression can be measured by the maximum extension ratio (l), which is defined as the maximum diameter divided by the initial diameter of the nerve. The theoretically computed shapes are shown for l values of 1.1, 1.3, and 1.5. The theoretical results shown in the figure are based on the theory of elasticity. Point B moves to B′, C moves to C′, and so on during the deformation.

Case Study 5-2

Sciatic Pain

A 35-year-old male construction worker has chronic low back pain radiating below the left knee that is more severe with lifting activities and prolonged positions. After a careful examination, certain neurologic signs were found. Positive straight leg raising and L5 motor and sensory functions were affected.

A MRI shows a herniated disc at level L4–L5 with posterolateral protrusion, which laterally compresses the left L5 nerve root. Compression of the nerve deforms it toward a more elliptical shape, increasing strain and stress loads. The effects of the pressure and mechanical deformation resultant from the load affects the nerve tissue, its nutrition, and the transmission function. Inflammation of the nerve root, induced by the nucleus pulposus, may sensitize the nerve root so that mechanical nerve root deformation causes sciatic pain (Case Study Fig. 5-2).

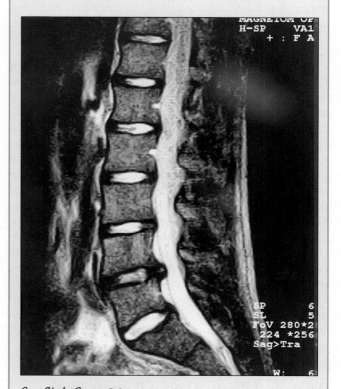

Case Study Figure 5-2

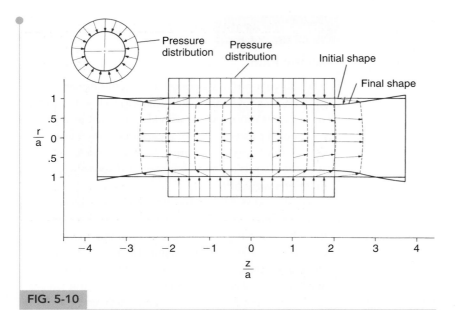

FIG. 5-10

Theoretical displacement field under a pressure cuff applied to a model cylindrical nerve of radius (*a*). Displacements in the radial (*r*) and longitudinal (*z*) directions are computed on the basis of isotropic material properties and elastic theory, so the deformation is proportional to the pressure. The *arrows* represent displacement vectors.

The effects of a deformation such as that shown in Figure 5-11 on the functioning of the axoplasm and the neural membrane are not known. It seems likely that the initial degeneration of function would be associated with damage to the membrane. It can be shown that if the cross-sectional area of the nerve shown in Figure 5-11 remains constant during the deformation, the perimeter must increase in moving from the initial circular shape to the final elliptical shape. This increase indicates that there must be stretching of the membrane, which is likely to affect its permeability and electrical properties. This deformation is similar to that of a pacinian corpuscle, which senses pressure applied to the skin. It may be that this kind of deformation can trigger firing of nerves, resulting in a sensation of pain when the nerve fibers are laterally compressed. The details of such deformation of nerves and their functional consequences have not been studied extensively and require further research for their elucidation.

Duration of Pressure versus Pressure Level

Knowledge is limited regarding the relative importance of pressure and time, respectively, in the production of nerve compression lesions. Mechanical factors seem to be relatively more important at higher than at lower pressures. Time is a significant factor at both high and low pressures, but ischemia plays a dominant role in longer-duration compression. This phenomenon is illustrated by the fact that direct nerve compression at 30 mm Hg for two to four hours produces reversible changes, whereas prolonged compression greater than this time period at this pressure level may cause irreversible damage to the nerve (Lundborg et al., 1982; Rydevik et al., 1981). Compression at 400 mm Hg causes a much more severe nerve injury after two hours than after fifteen minutes. Such information indicates that even high pressure has to "act" for a certain period of time for injury to occur. These data also give some information about the viscoelastic (time-dependent) properties of peripheral nerve tissue. Sufficient time must elapse for permanent deformation to develop.

Biomechanical Behavior of Spinal Nerve Roots

The nerve roots in the thecal sac lack epineurium and perineurium, but under tensile loading they exhibit both elasticity and tensile strength. The ultimate load for

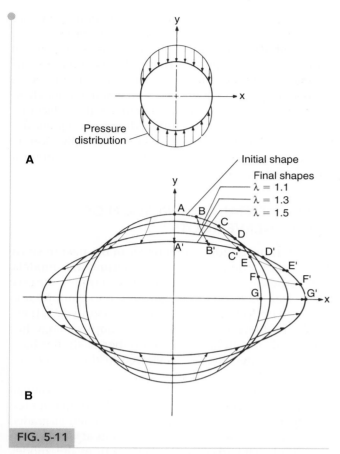

A. Theoretical displacement field under lateral compression as a result of uniform clamping pressure. **B.** The original and deformed cross-sections are shown for maximum elongation in the x direction of 10%, 30%, and 50%. The vectors shown from A to A', B to B', and so forth, indicate the paths followed by the particular points A, B, and so forth during the deformation.

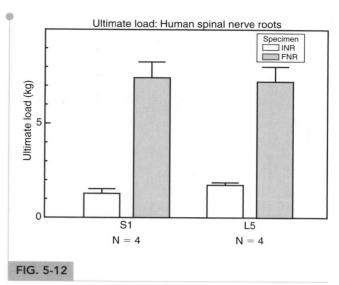

FIG. 5-12

Diagram illustrating values for ultimate load obtained for human spinal nerve roots under tensile loading. *INR*, intrathecal nerve roots; *FNR*, foraminal nerve root. Note the marked difference in ultimate load for the intrathecal and the foraminal portions of the nerve roots. *Error bars* indicate standard deviation. *Reproduced with permission from Weinstein, J.N., LaMotte, R., Rydevik, B., et al. (1989). Nerve. In J.W. Frymoyer, S.L. Gordon (Eds.). New Perspectives on Low Back Pain. Park Ridge, IL: American Academy of Orthopaedic Surgeons, 35–130.* (Based on a workshop arranged by the National Institutes of Health [NIH] in Airlie, Virginia, May 1988.)

ventral spinal nerve roots from the thecal sac is between 2 and 22 N, and for dorsal nerve roots from the thecal sac the load is between 5 and 33 N. The length of the nerve roots from the spinal cord to the foramina varies from approximately 60 mm at the L1 level to approximately 170 mm at the S1 level. The mechanical properties of human spinal nerve roots are different for any given nerve root at its location in the central spinal canal and in the lateral intervertebral foramina. The ultimate load for the intrathecal portion of human S1 nerve roots at the S1 level is approximately 13 N, and that for the foraminal portion is approximately 73 N. For human nerve roots at the L5 level, the corresponding values are 16 N and 71 N, respectively (Fig. 5-12). Thus, the values for

ultimate load are approximately five times higher for the foraminal segment of the spinal nerve roots than for the intrathecal portion of the same nerve roots under tensile loading. However, the cross-sectional area of the nerve root in the intervertebral foramen is significantly larger than that of the same nerve root in the thecal sac; thus, the ultimate tensile stress is more comparable for the two locations. The ultimate strain under tensile loading is 13% to 19% for the human nerve root at the L5 to S1 level (Fig. 5-13).

The nerve roots in the spine are not static structures; they move relative to the surrounding tissues with every spinal motion. To allow for such motion the nerve roots in the intervertebral foramina, for example, must have the capacity to glide. Chronic irritation with subsequent fibrosis around the nerve roots, in association with conditions such as disc herniation and/or foraminal stenosis, can thus impair the gliding capacity of the nerve roots. This produces repeated "microstretching" injuries of the nerve roots even during normal spinal movements, which might be speculated to induce yet further tissue

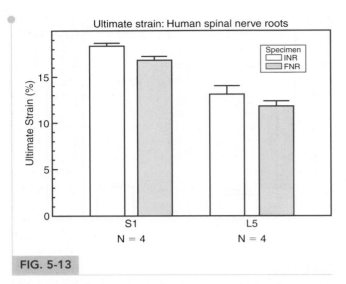

FIG. 5-13

Ultimate strain for human spinal nerve roots under tensile loading. *INR*, intrathecal nerve root; *FNR*, foraminal nerve root. *Reproduced with permission from Weinstein, J.N., LaMotte, R., Rydevik, B., et al. (1989). Nerve. In J.W. Frymoyer, S.L. Gordon (Eds.). New Perspectives on Low Back Pain. Park Ridge, IL: American Academy of Orthopaedic Surgeons, 35–130.* (Based on a workshop arranged by the National Institutes of Health [NIH] in Airlie, Virginia, May 1988.)

irritation in the nerve root components. The normal range of movements of nerve roots in the human lumbar spine has been measured in cadaver experiments. It was found that straight leg raising moved the nerve roots at the level of the intervertebral foramina approximately 2 to 5 mm.

Certain biomechanical factors are obviously involved in the pathogenesis of various symptoms induced by nerve root deformation in association with disc herniation and spinal stenosis and resulting in radiating pain. In disc herniation, only one nerve root is usually compressed. Because individual nerve roots normally adhere to the surrounding tissues above and below the intervertebral disc they traverse, compression may give rise to intraneural tension. Spencer et al. (1984) measured the contact force between a simulated disc herniation and a deformed nerve root in cadavers. Taking the area of contact into account, they assumed a contact pressure of approximately 400 mm Hg. With reduced disc height, the contact force and pressure between the experimental disc herniation and the nerve root was reduced. They suggested that these findings may explain in part why sciatic pain is relieved after

chemonucleolysis, and as disc degeneration progresses over time and the disc height thereby decreases.

In central spinal stenosis, the mechanics of nerve root compression are different. Under these conditions, the pressure is applied circumferentially around the nerve roots in the cauda equina at a slow, gradual rate. These different deformation factors, together with the fact that the nerve roots centrally within the cauda equina differ from the nerve roots located more laterally, close to the discs, may explain some of the different symptoms found in spinal stenosis and disc herniation.

EXPERIMENTAL COMPRESSION OF SPINAL NERVE ROOTS

There has been moderate interest in the past to study nerve root compression in experimental models. Early studies in the 1950s and 1970s found that nerve roots seemed to be more susceptible to compression than did peripheral nerves. During recent years, however, the interest in nerve root pathophysiology has increased considerably and a number of studies have been performed that are reviewed in the paragraphs that follow.

Some years ago, a model was presented to evaluate the effects of compression of the cauda equina in pigs, which for the first time allowed for experimental, graded compression of cauda equina nerve roots at known pressure levels (Olmarker, 1991) (Fig. 5-14). In this model, the cauda equina was compressed by an inflatable balloon that was fixed to the spine. The cauda equina could also be observed through the translucent balloon. This model made it possible to study the flow in the intrinsic nerve root blood vessels at various pressure levels (Olmarker et al., 1989a). The experiment was designed in a way that the pressure in the compression balloon was increased by 5 mm Hg every twenty seconds. Blood flow and vessel diameters of the intrinsic vessels could simultaneously be observed through the balloon using a vital microscope. The average occlusion pressure for the arterioles was found to be slightly below and directly related to the systolic blood pressure, and the blood flow in the capillary networks was intimately dependent on the blood flow of the adjacent venules. This corroborates the assumption that venular stasis may induce capillary stasis and thus changes in the microcirculation of the nerve tissue and is in accordance with previous studies in which such a mechanism has been suggested as involved in carpal tunnel syndrome. The mean occlusion pressures for the venules demonstrated large variations. However, a pressure of 5 to 10 mm Hg was found to be sufficient for inducing venular occlusion. Because

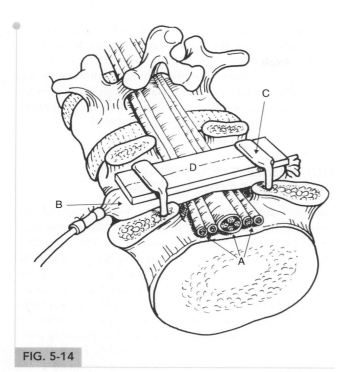

FIG. 5-14

Schematic drawing of an experimental model. The cauda equina **(A)** is compressed by an inflatable balloon **(B)** that is fixed to the spine by two L-shaped pins **(C)** and a plexiglas plate **(D)**. *Reproduced with permission from Olmarker, K., Rydevik, B., Holm, S. (1989a). Edema formation in spinal nerve roots induced by experimental, graded compression. An experimental study on the pig cauda equina with special reference to differences in effects between rapid and slow onset of compression. Spine, 14, 579.*

of retrograde stasis, it is not unlikely to assume that the capillary blood flow will be affected as well in such situations.

In the same experimental set-up, the effects of gradual decompression, after initial acute compression was maintained for only a short while, were studied. The average pressure for starting the blood flow was slightly lower at decompression than at compression for arterioles, capillaries, and venules. However, with this protocol, a full restoration of the blood flow did not occur until the compression was lowered from 5 to 0 mm Hg. This observation further supports the previous hypothesis that vascular impairment is present even at low pressure levels.

A compression-induced impairment of the vasculature may thus be one mechanism for nerve root dysfunction because the nutrition of the nerve root will be affected. However, the nerve roots will also derive a considerable nutritional supply via dif-

fusion from the cerebrospinal fluid. To assess the compression-induced effects on the total contribution to the nerve roots, an experiment was designed in which 3H-labeled methyl glucose was allowed to be transported to the nerve tissue in the compressed segment via both the blood vessels and the cerebrospinal fluid diffusion after systemic injection. The results showed that no compensatory mechanism from cerebrospinal fluid diffusion could be expected at the low pressure levels. On the contrary, 10 mm Hg compression was sufficient to induce a 20% to 30% reduction of the transport of methyl glucose to the nerve roots, as compared with the control.

We know from experimental studies on peripheral nerves that compression may also induce an increase in the vascular permeability, leading to an intraneural edema formation. Such edema may increase the endoneurial fluid pressure, which in turn may impair the endoneurial capillary blood flow and jeopardize the nutrition of the nerve roots. Because the edema usually persists for some time after the removal of a compressive agent, edema may negatively affect the nerve root for a longer period than the compression itself. The presence of intraneural edema is also related to the subsequent formation of intraneural fibrosis and may therefore contribute to the slow recovery seen in some patients with nerve compression disorders. To assess if intraneural edema also may form in nerve roots as the result of compression, the distribution of Evans blue-labeled albumin in the nerve tissue was analyzed after compression at various pressures and at various durations (Olmarker et al., 1989b). The study showed that edema was formed even at low pressure levels. The predominant location was at the edges of the compression zone.

The function of the nerve roots has been studied by direct electrical stimulation and recordings either on the nerve itself or in the corresponding muscular segments. During a two-hour compression period, a critical pressure level for inducing a reduction of muscle action potential (MAP) amplitude seems to be located between 50 and 75 mm Hg. Higher pressure levels (100–200 mm Hg) may induce a total conduction block with varying degrees of recovery after compression release. To study the effects of compression on sensory nerve fibers, electrodes in the sacrum were used to record a compound nerve action potential after stimulating the sensory nerves in the tail, that is, distal to the compression zone. The results showed that the sensory fibers are slightly more susceptible to compression than are the motor fibers. Also, the nerve roots are more susceptible to compression injury if the blood pressure is lowered pharmacologically. This further indicates the importance of the blood supply to maintain the functional properties of the nerve roots.

ONSET RATE OF COMPRESSION

One factor that has not been fully recognized in compression trauma of nerve tissue is the onset rate of the compression. The onset rate, that is, the time from start to full compression, may vary clinically from fractions of a second in traumatic conditions to months or years in association with degenerative processes. Even in the clinically rapid-onset rates, there may be a wide variation of onset rates. With the presented model, it was possible to vary the onset time of the applied compression. Two onset rates have been investigated. Either the pressure is present and compression is started by flipping the switch of the compressed-air system used to inflate the balloon, or the compression pressure level is slowly increased during twenty seconds. The first onset rate was measured at 0.05 to 0.1 second, thus providing a rapid inflation of the balloon and a rapid compression onset.

Such a rapid-onset rate has been found to induce more pronounced effects on edema formation, methyl glucose transport, and impulse propagation than the slow-onset rate (Olmarker, 1991). Regarding methyl glucose transport, the results show that the levels within the compression zone are more pronounced at the rapid than at the slow onset rate at corresponding pressure levels. There was also a striking difference between the two onset rates when considering the segments outside the compression zones. In the slow-onset series, the levels approached baseline values closer to the compression zone than in the rapid-onset series. This may indicate the presence of a more pronounced edge-zone edema in the rapid-onset series, with a subsequent reduction of the nutritional transport in the nerve tissue adjacent to the compression zone.

For the rapid-onset compression, which is likely to be more closely related to spine trauma or disc herniation than to spinal stenosis, a pressure of 600 mm Hg maintained for only one second is sufficient to induce a gradual impairment of nerve conduction during the two hours studied after the compression was ended. Overall, the mechanisms for these pronounced differences between the different onset rates are not clear but may be related to differences in the displacement rates of the compressed nerve tissue toward the uncompressed parts as a result of the viscoelastic properties of the nerve tissue. Such phenomena may lead not only to structural damage to the nerve fibers but also to structural changes in the blood vessels with subsequent edema formation.

The gradual formation of intraneural edema may also be closely related to observations of a gradually increasing difference in nerve conduction impairment between the two onset rates (Olmarker et al., 1989b).

MULTIPLE LEVELS OF SPINAL NERVE ROOT COMPRESSION

Patients with double or multiple levels of spinal stenosis seem to have more pronounced symptoms than patients with stenosis at only one level. The presented model was modified to address this interesting clinical question. Using two balloons at two adjacent disc levels, which resulted in a 10-mm uncompressed nerve segment between the balloons, induced a much more pronounced impairment of nerve impulse conduction than previously had been found at corresponding pressure levels (Olmarker and Rydevik, 1992). For instance, a pressure of 10 mm Hg in two balloons induced a 60% reduction of nerve impulse amplitude during two hours of compression, whereas 50 mm Hg in one balloon showed no reduction.

The mechanism for the difference between single and double compression may not simply be based on the fact that the nerve impulses have to pass more than one compression zone at double-level compression. There may also be a mechanism based on the local vascular anatomy of the nerve roots. Unlike for peripheral nerves, there are no regional nutritive arteries from the surrounding structures to the intraneural vascular system in spinal nerve roots. Compression at two levels might therefore induce a nutritionally impaired region between the two compression sites. In this way, the segment affected by the compression would be widened from one balloon diameter (10 mm) to two balloon diameters including the interjacent nerve segment (30 mm).

This hypothesis was partly confirmed in an experiment on continuous analyses of the total blood flow in the uncompressed nerve segment located between two compression balloons (Takahashi et al., 1993). The results showed that a 64% reduction of total blood flow in the uncompressed segment was induced when both balloons were inflated to 10 mm Hg. At a pressure close to the systemic blood pressure there was complete ischemia in the nerve segment. Thus, experimental evidence shows that the blood supply to the nerve segment located between two compression sites in nerve roots is severely impaired although this nerve segment itself is uncompressed. Regarding nerve conduction, the effects were much enhanced if the distance between the compression balloons was increased from one vertebral segment to two vertebral segments (Olmarker and Rydevik, 1992). This indicates that the functional impairment may be directly related to the distance between the two compression sites.

CHRONIC NERVE ROOT COMPRESSION IN EXPERIMENTAL MODELS

The discussion of compression-induced effects on nerve roots has dealt primarily with acute compression, that is, compression that lasts for some hours and with no survival of the animal. To better mimic various clinical situations, compression must be applied over longer periods of time. There are probably many changes in the nerve tissue, such as adaptation of axons and vasculature, that will occur in patients but cannot be studied in experimental models using only one to six hours of compression. Another important factor in this context is the onset rate that was discussed previously. In clinical syndromes with nerve root compression, the onset time may in many cases be quite slow. For instance, a gradual remodeling of the vertebrae to induce a spinal stenosis probably leads to an onset time of many years. It will of course be difficult to mimic such a situation in an experimental model.

It will also be impossible to have control over the pressure acting on the nerve roots in chronic models because of the remodeling and adaptation of the nerve tissue to the applied pressure. However, knowledge of the exact pressures is probably of less importance in chronic than in acute compression situations. Instead, chronic models should induce a controlled compression with a slow onset time that is easily reproducible. Such models may be well suited for studies on pathophysiologic events as well as intervention by surgery or drugs. Some attempts have been made to induce such compression.

Delamarter et al. (1990) presented a model on dog cauda equina in which they applied a constricting plastic band. The band was tightened around the thecal sac to induce a 25%, 50%, or 75% reduction of the cross-sectional area. The band was left in place for various times. Analyses were performed and showed both structural and functional changes that were proportional to the degree of constriction.

To induce a slower onset and more controlled compression, Cornefjord et al. (1997) used a constrictor

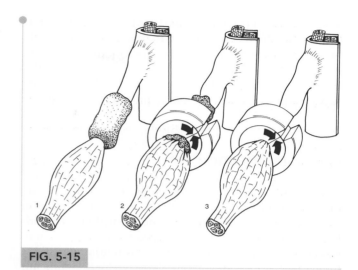

FIG. 5-15

Experimental study to analyze the effects on nerve conduction velocity of nucleus pulposus (1), the combination of nucleus pulposus and compression (2), and compression only (3). The nucleus pulposus and the constrictor were applied to the first sacral nerve root in pigs. The contralateral nerve root served as a control. *Reproduced with permission from Cornefjord, M., Sato, K., Olmarker, K., et al. (1997). A model for chronic nerve root compression studies. Presentation of a porcine model for controlled slow-onset compression with analyses of anatomic aspects, compression onset rate, and morphologic and neurophysiologic effects. Spine, 22, 946–957.*

to compress the nerve roots in the pig (Fig. 5-15). The constrictor was initially intended for inducing vascular occlusion in experimental ischemic conditions in dogs. The constrictor consists of an outer metal shell that on the inside is covered with a material called amaroid that expands when in contact with fluids. Because of the metal shell, the amaroid expands inwards with a maximum expansion after two weeks, resulting in compression of a nerve root placed in the central opening of the constrictor. Compression of the first sacral nerve root in the pig resulted in a significant reduction of nerve conduction velocity and axonal injuries using a constrictor with a defined original diameter. An increase in substance P in the nerve root and the dorsal root ganglion following such compression also has been found. Substance P is a neurotransmitter that is related to pain transmission, and the study may thus provide experimental evidence that the compression of nerve roots can produce pain.

The constrictor model has also been used to study blood flow changes in the nerve root vasculature. It could then be observed that the blood flow is not reduced just outside the compression zone but significantly reduced in parts of the nerve roots located inside the constrictor. In this context, it should be noted that in case of disc herniation, the nerve root may become sensitized by substances from the disc tissue (nucleus pulposus) so that mechanical root deformation can induce sciatic pain (Olmarker et al., 1993, 2002).

Nerve Injury and Pain

The mechanical factors discussed previously can be important causes of painful nerve injury, as can metabolic and toxic insults to nerve fibers or their cell bodies. During the past two decades extensive use of rodent models of nerve injury, coupled with quantitative assessments of pain-like behaviors, have yielded significant insight into the neuropathology and pathophysiology of pain. In particular, two compression models of nerve injury have become standard.

The first, by Bennett and Xie (1988), was developed to mimic a low-grade compression neuropathy like peripheral nerve entrapments. This model loosely ties chromic gut ligatures around the sciatic nerve and produces a robust thermal hyperalgesia lasting several weeks. The second model (Kim and Chung, 1992) is based on tight ligature of (typically) the L5 spinal nerve and is noted to produce a prolonged mechanical allodynia in the distribution of the sciatic nerve.

Both of these models have a similar basic pathologic change: wallerian degeneration, a cytokine-mediated process that begins with disintegration and degeneration of the axoplasm and axolemma followed by macrophage phagocytosis of axonal and Schwann cell debris (Stoll et al., 2002). In fact, it has been shown that both the magnitude of the hyperalgesic response to nerve injury and its persistence are directly related to the magnitude of wallerian degeneration, that is, the percentage of fibers undergoing wallerian degeneration (Myers et al., 1996).

Subsequent research has linked wallerian degeneration and pain to the expression of tumor necrosis factor alpha, which is liberated immediately after nerve injury by Schwann cells and then significantly increased by macrophages invading the injured nerve from the circulation (Myers et al., 2006). This neuroinflammatory state is potentially amenable to anticytokine therapy, and clinical trials provide hope for future success in treating neuropathic pain associated with compression and other mechanical forms of injury to peripheral nerves and nerve roots.

Summary

- The peripheral nerves are composed of nerve fibers, layers of connective tissue, and blood vessels.
- The nerve fibers are extremely susceptible to trauma, but because they are surrounded by successive layers of connective tissue (the epineurium and perineurium), they are mechanically protected.
- Stretching induces changes in intraneural blood flow and nerve fiber structure before the nerve trunk ruptures.
- Compression of a nerve can cause injury to both nerve fibers and blood vessels in the nerve, mainly at the edges of the compressed nerve segment, but also by ischemic mechanisms.
- Pressure level, duration of compression, and mode of pressure application are significant variables in the development of nerve injury.
- Spinal nerve roots are anatomically different from peripheral nerves and therefore react differently to mechanical deformation.
- Spinal nerve roots are more susceptible than peripheral nerves to mechanical deformation, mainly because of the lack of protective connective tissue layers in nerve roots.
- Mechanical injury to nerve roots or peripheral nerves can cause nerve degeneration and a neuroinflammatory state that may lead to neuropathic pain.

Practice Questions

1. Describe the anatomic structures in a peripheral nerve that protect the nerve against the effects of mechanical loading. Explain how these structures protect the nerve.

2. Explain the differences in terms of anatomic structure and biomechanical properties between a peripheral nerve and a spinal nerve root.

3. What type of mechanical loading can injure peripheral nerves? What type of mechanical loading can injure spinal nerve roots? Discuss similarities and differences between the mechanisms behind injury to a spinal nerve root and a peripheral nerve.

4. Describe symptoms and/or signs indicating that there is peripheral nerve injury. Describe symptoms and/or signs indicating that there is spinal nerve root injury.

5. Discuss biologic and biomechanical factors that can be involved in painful peripheral nerve and spinal nerve root injuries. Illustrate examples with real case studies.

REFERENCES

Bennett, G.J., Xie, Y.K. (1988) A peripheral neuropathy in rat that produces disorders of pain sensation like those seen in man. *Pain, 33,* 87–107.

Cornefjord, M., Sato, K., Olmarker, K., et al. (1997). A model for chronic nerve root compression studies. Presentation of a porcine model for controlled slow-onset compression with analyses of anatomic aspects, compression onset rate, and morphologic and neurophysiologic effects. *Spine, 22,* 946–957.

Dahlin, L.B., Rydevik, B., Lundborg, G. (1986). The pathophysiology of nerve entrapments and nerve compression injuries. In A.R. Hargens (Ed.). *Effects of Mechanical Stress on Tissue Viability.* New York: Springer-Verlag.

Delamarter, R.B., Bohlman, H.H., Dodge, L.D., et al. (1990). Experimental lumbar spinal stenosis. Analysis of the cortical evoked potentials, microvasculature and histopathology. *J Bone Joint Surg, 72A,* 110–120.

Kim, S.H., Chung, J.M. (1992) An experimental model for peripheral neuropathy produced by segmental spinal nerve ligation in the rat. *Pain, 50,* 355–363.

Lundborg, G. (1975). Structure and function of the intraneural microvessels as related to trauma, edema formation and nerve function. *J Bone Joint Surg, 57A,* 938.

Lundborg, G., Gelberman, R.H., Minteer-Convery, M., et al. (1982). Median nerve compression in the carpal tunnel: The functional response to experimentally induced controlled pressure. *J Hand Surg Am, 7,* 252.

Lundborg, G., Rydevik, B. (1973). Effects of stretching the tibial nerve of the rabbit: A preliminary study of the intraneural circulation and the barrier function of the perineurium. *J Bone Joint Surg, 55B,* 390.

Myers, R.R. (1998). Morphology of the peripheral nervous system and its relationship to neuropathic pain. In T.L. Yaksh, C. Lynch III, W.M. Zapol, et al., (Eds.). *Anesthesia: Biologic Foundations.* Philadelphia: Lippincott–Raven Publishers, 483–514.

Myers, R.R., Campana, W.M., Shubayev, V.I. (2006) The role of neuroinflammation in neuropathic pain: Mechanisms and therapeutic targets. *Drug Discov Today, 11,* 8–20.

Myers, R.R., Heckman, H.M., Powell, H.C. (1996) Axonal viability and the persistence of thermal hyperalgesia after partial freeze lesions of nerve. *J Neurol Sci, 139,* 28–38.

Myers, R.R., Murakami, H., Powell, H.C. (1986). Reduced nerve blood flow in edematous neuropathies—A biomechanical mechanism. *Microvascular Res, 32,* 145–151.

Myers, R.R., Powell, H.C. (1981). Endoneurial fluid pressure in peripheral neuropathies. In A.R. Hargens (Ed.). *Tissue Fluid Pressure and Composition.* Baltimore: Williams & Wilkins, 193.

Olmarker, K. (1991). Spinal nerve root compression. Nutrition and function of the porcine cauda equina compressed in vivo. *Acta Orthop Scand Suppl, 242,* 1–27.

Olmarker, K., Rydevik, B. (1992). Single versus double level compression. An experimental study on the porcine cauda equina with analyses of nerve impulse conduction properties. *Clin Orthop, 279,* 3539.

Olmarker, K., Rydevik, B., Holm, S. (1989a). Edema formation in spinal nerve roots induced by experimental, graded compression. An experimental study on the pig cauda equina with special reference to differences in effects between rapid and slow onset of compression. *Spine, 14,* 579.

Olmarker, K., Rydevik, B., Holm, S., et al. (1989b). Effects of experimental graded compression on blood flow in spinal nerve roots. A vital microscopic study on the porcine cauda equina. *J Orthop Res, 7,* 817.

Olmarker, K., Rydevik, B., Nordborg, C. (1993). Autologous nucleus pulposus induces neurophysiologic and histologic changes in porcine cauda equina nerve roots. *Spine, 18,* 1425–1432.

Olmarker, K., Størkson, R., Berge, O.G. (2002). Pathogenesis of sciatic pain: A study of spontaneous behavior in rats exposed to experimental disc herniation. *Spine, 27,* 1312–1317.

Rydevik, B., Brown, M.D., Lundborg, G. (1984). Pathoanatomy and pathophysiology of nerve root compression. *Spine, 9,* 7.

Rydevik, B.L., Kwan, M.K., Myers, R.R., et al. (1990). An in vitro mechanical and histological study of acute stretching on rabbit tibial nerve. *J Orthop Res, 8,* 694–701.

Rydevik, B., Lundborg, G. (1977). Permeability of intraneural microvessels and perineurium following acute, graded experimental nerve compression. *Scand J Plast Reconstr Surg, 11,* 179.

Rydevik, B., Lundborg, G., Bagge, U. (1981). Effects of graded compression on intraneural blood flow. An in vivo study on rabbit tibial nerve. *J Hand Surg, 6.*

Spencer, D.L., Miller, J.A., Bertolini, J.E. (1984). The effects of intervertebral disc space narrowing on the contact force between the nerve root and a simulated disc protrusion. *Spine, 9,* 422.

Stoll, G., Jander, S., Myers, R.R. (2002). Degeneration and regeneration of the peripheral nervous system: From Augustus Waller's observations to neuroinflammation. *J Peripher Nerv Syst, 7,* 13–27.

Sunderland, S. (1978). *Nerves and Nerve Injuries (2nd ed.).* Edinburgh, Scotland: Churchill Livingstone.

Takahashi, K., Olmarker, K., Holm, S., et al. (1993). Double-level cauda equina compression. An experimental study with continuous monitoring of intraneural blood flow. *J Orthop Res, 11,* 104.

Tortora, G.J., Anagnostakos, N.P. (1984). *Principles of Anatomy and Physiology* (4th ed.). New York: Harper & Row.

Wall, E.J., Massie, J.B., Kwan, M.K., et al. (1992). Experimental stretch neuropathy. Changes in nerve conduction under tension. *J Bone Joint Surg, 74-B*, 126.

Weinstein, J.N., LaMotte, R., Rydevik, B. et al.,(1989). Nerve. In J.W. Frymoyer, S.L. Gordon (Eds.). *New Perspectives on Low Back Pain.* Park Ridge, IL: American Academy of Orthopaedic Surgeons, 35–130. (Based on a workshop arranged by the National Institutes of Health [NIH] in Airlie, Virginia, May 1988.)

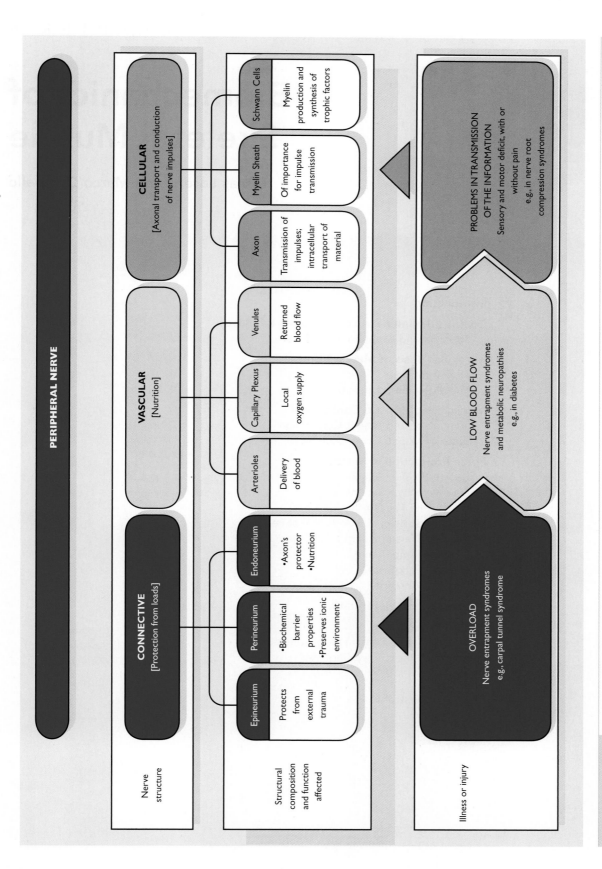

FLOW CHART 5-1

Peripheral nerve structure and alteration. Clinical examples.*

*This flow chart is designed for classroom or group discussion. It is not meant to be exhaustive.

149

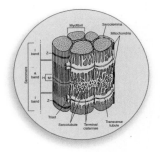

CHAPTER 6

Biomechanics of Skeletal Muscle

Tobias Lorenz and Marco Campello

Introduction

The muscular system consists of three muscle types: the cardiac muscle, which composes the heart; the smooth (nonstriated or involuntary) muscle, which lines the hollow internal organs; and the skeletal (striated or voluntary) muscle, which attaches to the skeleton via the tendons. The focus of this chapter is on the role and function of the third type, skeletal muscle.

Skeletal muscle is the most abundant tissue in the human body, accounting for 40 to 45% of the total body weight. There are more than 430 skeletal muscles, found in pairs on the right and left sides of the body. The most vigorous movements are produced by fewer than 80 pairs. The muscles provide strength and protection to the skeleton by distributing loads and absorbing shock; they enable the bones to move at the joints, and maintain body posture against force. Such abilities usually represent the action of muscle groups rather than of individual muscles.

The skeletal muscles perform both dynamic or isokinetic and static or isometric work. Dynamic work permits locomotion and the positioning of the body segments in space. Static work maintains body posture.

This chapter gives a description of the following: the composition and structure of skeletal muscle, the molecular basis of muscle contraction, the mechanics of muscle contraction, force production in muscle, muscle fiber differentiation, and muscle remodeling.

Composition and Structure of Skeletal Muscle

An understanding of the biomechanics of muscle function requires knowledge of the gross anatomic structure and function of the musculotendinous unit, and the basic microscopic structure and chemical composition of the muscle fiber.

STRUCTURE AND ORGANIZATION OF MUSCLE

The structural unit of skeletal muscle is the muscle fiber, a long cylindrical cell with many hundreds of nuclei. Muscle fibers range in thickness from about 10 to 100 μm and in length from about 1 to 30 cm. A muscle fiber consists of many myofibrils, which are invested by a delicate plasma membrane, the sarcolemma. The sarcolemma is connected via vinculin- and dystrophin-rich costameres with the sarcomeric Z lines, which represent a part of the extramyofibrillar cytoskeleton. The myofibril is made up of several sarcomeres, which contain thin (actin), thick (myosin), elastic (titin), and inelastic (nebulin) filaments. Each fiber is encompassed by a loose connective tissue called the endomysium, and the fibers are organized into various-sized bundles, or fascicles (Figs. 6-1A and B), which are in turn encased in a dense connective tissue sheath known as the perimysium. The muscle is composed of several fascicles, surrounded by a fascia of fibrous connective tissue called the epimysium.

In general, each end of a muscle is attached to bone by tendons, which have no active contractile properties. The muscles form the contractile component and the tendons the series elastic component. The collagen fibers in the epimysium and perimysium are continuous with those in the tendons, and together these fibers act as a structural framework for the attachment of bones and muscle fibers. The epimysium, perimysium, the endomysium, and the sarcolemma act as parallel elastic components. The forces produced by the contracting muscles are transmitted to bone through these connective tissues and tendons (Beason et al., 2007).

Each muscle fiber is composed of many delicate strands, the myofibrils. Their structure and function have been studied exhaustively by light and electron microscopy, and their histochemistry and biochemistry have been explained elsewhere (Arvidson et al., 1984; Guyton, 1986). About 1 μm in diameter, the myofibrils lie parallel to each other within the cytoplasm (sarcoplasm) of the muscle fiber and extend throughout its length. They vary in number from a few to several thousand, depending on the diameter of the muscle fiber, which depends in turn on the type of muscle fiber.

The transverse banding pattern in striated muscles repeats itself along the length of the muscle fiber, each repeat being known as a sarcomere (Fig. 6-lC). These striations are caused by the individual myofibrils, which are aligned continuously throughout the muscle fiber. The sarcomere is the functional unit of the contractile system in muscle, and the events that take place in one sarcomere are duplicated in the others. Various sarcomere build a myofibril, various myofibrils build the muscle fiber, and various muscle fibers build the muscle.

Each sarcomere is composed of the following:

1. The thin filaments (approximately 5 nm in diameter) composed of the protein actin
2. The thick filaments (about 15 nm in diameter) composed of the protein myosin (Figs. 6-lD and E)
3. The elastic filaments composed of the protein titin (Fig. 6-2)
4. The inelastic filaments composed of the proteins nebulin and titin

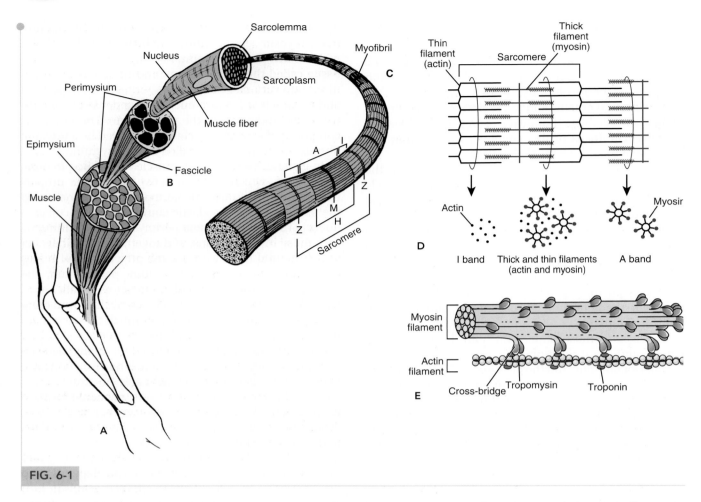

FIG. 6-1

Schematic drawings of the structural organization of muscle. **A.** A fibrous connective tissue fascia, the epimysium, surrounds the muscle, which is composed of many bundles, or fascicles. The fascicles are encased in a dense connective tissue sheath, the perimysium. **B.** The fascicles are composed of muscle fibers, which are long, cylindrical, multinucleated cells. Between the individual muscle fibers are capillary blood vessels. Each muscle fiber is surrounded by a loose connective tissue called the endomysium. Just beneath the endomysium lies the sarcolemma, a thin elastic sheath with infoldings that invaginate the fiber interior. Each muscle fiber is composed of numerous delicate strands, myofibrils, the contractile elements of muscle. **C.** Myofibrils consist of smaller filaments, which form a repeating banding pattern along the length of the myofibril. One unit of this serially repeating pattern is called a sarcomere. The sarcomere is the functional unit of the contractile system of muscle. **D.** The banding pattern of the sarcomere is formed by the organization of thick and thin filaments, composed of the proteins myosin and actin, respectively. The actin filaments are attached at one end but are free along their length to interdigitate with the myosin filaments. The thick filaments are arranged in a hexagonal fashion. A cross-section through the area of overlap shows the thick filaments surrounded by six equally spaced thin filaments. **E.** The lollipop-shaped molecules of each myosin filament are arranged so that the long tails form a sheaf with the heads, or cross-bridges, projecting from it. The cross-bridges point in one direction along half of the filament and in the other direction along the other half. Only a portion of one-half of a filament is shown here. The cross-bridges are an essential element in the mechanism of muscle contraction, extending outward to interdigitate with receptor sites on the actin filaments. Each actin filament is a double helix, appearing as two strands of beads spiraling around each other. Two additional proteins, tropomyosin and troponin, are associated with the actin helix and play an important role in regulating the interdigitation of the actin and myosin filaments. Tropomyosin is a long polypeptide chain that lies in the grooves between the helices of actin. Troponin is a globular molecule attached at regular intervals to the tropomyosin.

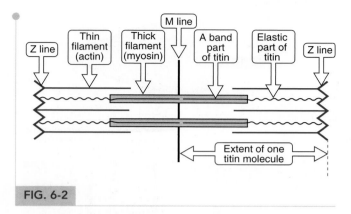

FIG. 6-2

The arrangement of titin molecules within the sarcomere.

Actin and myosin are the contractile part of the myofibrils, whereas titin and nebulin are part of the intramyofibrillar cytoskeleton (Stromer, 1998). The myofibrils are the basic unit of contraction.

Actin, the chief component of the thin filament, has the shape of a double helix and appears as two strands of beads spiraling around each other. Two additional proteins, troponin and tropomyosin, are important constituents of the actin helix, because they appear to regulate the making and breaking of contacts between the actin and myosin filaments during contraction. Tropomyosin is a long polypeptide chain that lies in the grooves between the helices of actin. Troponin is a globular molecule attached at regular intervals to the tropomyosin. (Figs. 6-1D and E)

The thick filaments are located in the central region of the sarcomere, where their orderly, parallel arrangement gives rise to dark bands, known as A bands because they are strongly anisotropic. The thin filaments are attached at either end of the sarcomere to a structure known as the Z line, which consists of short elements that link the thin filaments of adjacent sarcomeres, defining the limits of each sarcomere. The thin filaments extend from the Z line toward the center of the sarcomere, where they overlap with the thick filaments. Recently it was shown that there is a third set of myofibril filaments in the vertebrate striated muscles. This connecting filament, named titin, links the thick filaments with the Z line (elastic I band region of titin) and is part of the thick filaments (A band region of titin). This filament maintains the central position of the A band throughout contraction and relaxation and might act as a template during myosin assembly.

Myosin, the thicker filament, is composed of individual molecules each of which resembles a lollipop with a globular "head" projecting from a long shaft, or "tail." Several hundred such molecules are packed tail to tail in a sheaf with their heads pointed in one direction along half of the filament and in the opposite direction along the other half, leaving a head-free region (the H zone) in between. The globular heads spiral about the myosin filament in the region where actin and myosin overlap (the A band) and extend as cross-bridges to interdigitate with sites on the actin filaments, thus forming the structural and functional link between the two filament types.

The intramyofibrillar cytoskeleton includes inelastic nebulin filaments, (not depicted in Figure 6-2) which span from the Z line to the actin filaments. Nebulin might also act as a template for the thin filament assembly.

Titin is 1 μm long. It is the largest polypeptide and spans from the Z line to the M line. Titin is an elastic filament. The part between the Z line and myosin has a stringlike appearance. It has been suggested that titin contributes greatly to the passive force development of muscle during stretching (Fig. 6-2). However, recent studies have demonstrated that titin has little or no contribution to passive force development (Reisman et al., 2009). It also might act as a template for the thick filament assembly (Linke et al., 1998; Squire et al., 1997; Stromer 1998).

The I band is bisected by the Z lines, which contain the portion of the thin filaments that does not overlap with the thick filaments and the elastic part of titin. In the center of the A band, in the gap between the ends of the thin filaments, is the H zone, a light band containing only thick filaments and that part of titin that is integrated in the thick filaments. A narrow dark area in the center of the H zone is the M line, produced by transversely and longitudinally-oriented proteins that link adjacent thick filaments, maintaining their parallel arrangement. The various areas of the banding pattern are apparent in the photomicrograph of human skeletal muscle shown in Figure 6-3B.

Closely correlated with the repeating pattern of the sarcomeres is an organized network of tubules and sacs known as the sarcoplasmic reticulum. The tubules of the sarcoplasmic reticulum lie parallel to the myofibrils and tend to enlarge and fuse at the level of the junctions between the A and I bands, forming transverse sacs—the terminal cisternae—that surround the individual myofibril completely.

The terminal cisternae enclose a smaller tubule that is separated from them by its own membrane.

The smaller tubule and the terminal cisternae above and below it are known as a triad. The enclosed tubule is part of the transverse tubule system, or T system, which are invaginations of the surface membrane of the fiber. This membrane, the sarcolemma, is a plasma membrane that invests every striated muscle (Fig. 6-4).

Molecular Basis of Muscle Contraction

The most widely held theory of muscle contraction is the sliding filament theory, proposed simultaneously by A.F. Huxley and H.E. Huxley in 1964 and subsequently refined (Huxley, 1974). According to this theory, active shortening of the sarcomere, and hence of the muscle, results from the relative movement of the actin and myosin filaments past one another while each retains its original length. The force of contraction is developed by the myosin heads, or cross-bridges, in the region of overlap between actin and myosin (the A band). These cross-bridges swivel in an arc around their fixed positions on the surface of the myosin filament, much like the oars of a boat. This movement of the cross-bridges in contact with the actin filaments produces the sliding of the actin filaments toward the center of the sarcomere. A muscle fiber contracts when all sarcomere shorten simultaneously in an all-or-nothing fashion, which is called a twitch.

Since a single movement of a cross-bridge produces only a small displacement of the actin filament relative to the myosin filament, each individual cross-bridge detaches itself from one receptor site on the actin filament and reattaches itself to another site farther along, repeating the process five or six times, "with an action similar to a man pulling on a rope hand over hand" (Wilkie, 1968). The cross-bridges do not act in a synchronized manner; each acts independently. Thus, at any given moment only about half of the cross-bridges actively generate force and displacement, and when these detach, others take up the task so that shortening is maintained. The shortening is reflected in the sarcomere as a decrease in the I band and a decrease in the H zone as the Z lines move closer together; the width of the A band remains constant.

A key to the sliding mechanism is the calcium ion (Ca^{2+}), which turns the contractile activity on and off. Muscle contraction is initiated when calcium is made available to the contractile elements and ceases when calcium is removed. The mechanisms that regulate the availability of calcium ions to the contractile machinery are coupled to electric events occurring in the muscle

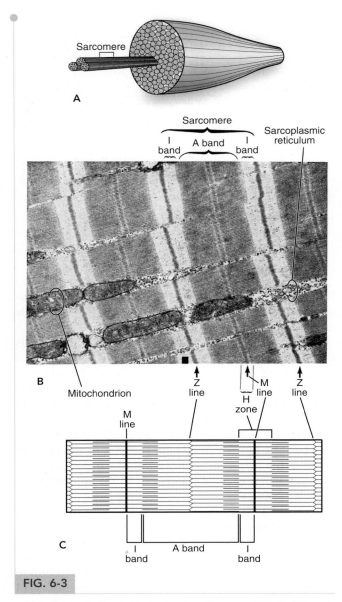

FIG. 6-3

A. Single muscle fiber with three protruding myofibrils. **B.** Electron photomicrograph of a cross-section of human skeletal muscle. The sarcomeres are apparent along the myofibrils. Characteristic regions of the sarcomere are indicated. *Adapted from Craig, R.W., Padrón R. Molecular Structure of the Sarcomere. In Engel A.G., Franzini-Armstrong C., eds (1994). Myology (2nd ed.). New York: McGraw-Hill, 135.* **C.** Schematic representation of B depicting the contraction mechanism.

membrane (sarcolemma). An action potential in the sarcolemma provides the electric signal for the initiation of contractile activity. The mechanism by which the electric signal triggers the chemical events of contraction is known as excitation-contraction coupling.

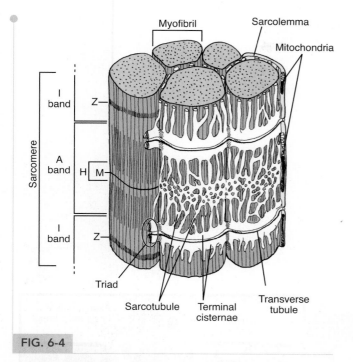

Myofibril   Sarcolemma

Mitochondria

Sarcomere

I band   Z

A band   H   M

I band   Z

Triad

Sarcotubule   Terminal cisternae   Transverse tubule

FIG. 6-4

Diagram of a portion of a skeletal muscle fiber illustrating the sarcoplasmic reticulum that surrounds each myofibril. The various regions of the sarcomere are indicated on the left myofibril to show the correlation of these regions with the sarcoplasmic reticulum, shown surrounding the middle and right myofibrils. The transverse tubules represent an infolding of the sarcolemma, the plasma membrane that encompasses the entire muscle fiber. Two transverse tubules supply each sarcomere at the level of the junctions of the A band and I bands. Terminal cisternae are located on each side of the transverse tubule, and together these structures constitute a triad. The terminal cisternae connect with a longitudinal network of sarcotubules spanning the region of the A band. *Adapted from Ham, A.W., Cormack, D.H. (1979). Histology (8th ed.). Philadelphia: JB Lippincott Co.*

When the motor neuron stimulates the muscle at the neuromuscular junction (Fig. 6-5A), and the propagated action potential depolarizes the muscle cell membrane (sarcolemma), there is an inward spread of the action potential along the T system. (Details of this process are given in Figures 6-5A–C and Box 6-1, which summarizes the events during the excitation, contraction, and relaxation of muscle. Figure 6-5D shows the structural features between actin and the cross-bridges of myosin.)

THE MOTOR UNIT

The functional unit of skeletal muscle is the motor unit, which includes a single motor neuron, and all of the muscle fibers innervated by it. This unit is the smallest part of the muscle that can be made to contract independently. When stimulated, all muscle fibers in the motor unit respond as one. The fibers of a motor unit are said to show an all-or-none response to stimulation: They either contract maximally or not at all.

The number of muscle fibers forming a motor unit is closely related to the degree of control required of the muscle. In small muscles that perform very fine movements, such as the extraocular muscles, each motor unit may contain less than a dozen muscle fibers, whereas in large muscles that perform coarse movements, such as the gastrocnemius, the motor unit may contain 1,000 to 2,000 muscle fibers.

The fibers of each motor unit are not contiguous but are dispersed throughout the muscle with fibers of other units. Thus, if a single motor unit is stimulated, a large portion of the muscle appears to contract. If additional motor units of the nerve innervating the muscle are stimulated, the muscle contracts with greater force. The calling in of additional motor units in response to greater stimulation of the motor nerve is called recruitment.

THE MUSCULOTENDINOUS UNIT

The tendons and the connective tissues in and around the muscle belly are viscoelastic structures that help determine the mechanical characteristics of entire muscle during contraction and passive extension. Hill (1970) showed that the tendons represent a springlike elastic component located in a series with the contractile component (the contractile proteins of the myofibril, actin, and myosin), while the epimysium, perimysium, endomysium, and sarcolemma represent a second elastic component located parallel to the contractile component (Fig. 6-6).

When the parallel and series elastic components stretch during active contraction or passive extension of a muscle, tension is produced and energy is stored. When they recoil with muscle relaxation, this energy is released. The series elastic fibers are more important in the production of tension than are the parallel elastic fibers (Wilkie, 1956). Several investigators have suggested that the cross-bridges of the myosin filaments have a springlike property and also contribute to the elastic properties of muscle (Hill, 1968).

The distensibility and elasticity of the elastic components are valuable to the muscle in several ways:

1. They tend to keep the muscle in readiness for contraction and ensure that muscle tension is produced and transmitted smoothly during contraction.

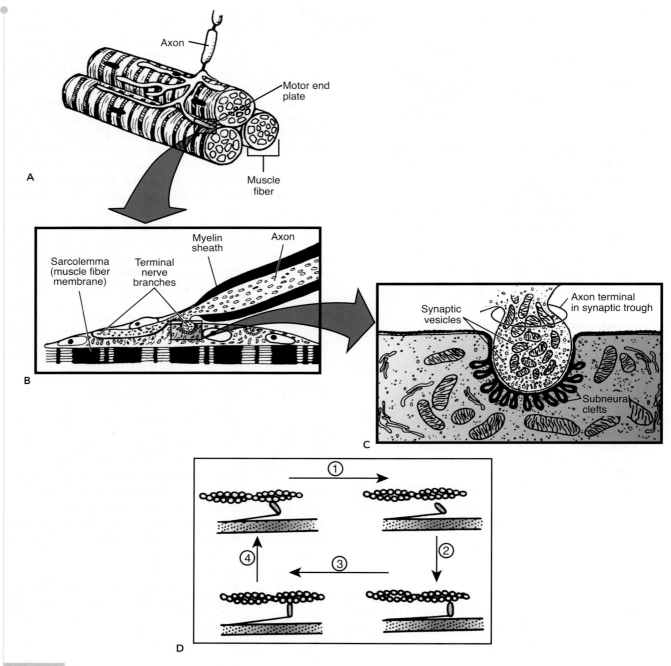

FIG. 6-5

Schematic representation of the innervation of muscle fibers.
A. An axon of a motor neuron (originating from the cell body in the anterior horn of the spinal cord) branches near its end to innervate several skeletal muscle fibers, forming a neuromuscular junction with each fiber. The region of the muscle membrane (sarcolemma) lying directly under the terminal branches of the axon has special properties and is known as the motor end plate, or motor end-plate membrane. **B.** The rectangular area is shown in detail. The fine terminal branches of the nerve (axon terminals), devoid of myelin sheaths, lie in grooves on the sarcolemma. **C.** The rectangular area in this section is shown in detail depicting the ultrastructure of the junction of an axon terminal and the sarcolemma. The invagination of the sarcolemma forms the synaptic trough into which the axon terminal protrudes. The invaginated sarcolemma has many folds, or subneural clefts, which greatly increase its surface area. Acetylcholine is stored in synaptic vesicles in the axon terminal. **B** and **C** *adapted from Brobeck, J.R. (Ed.). (1979). Best and Taylor's Physiological Basis of Medical Practice (10th ed.). Baltimore: Williams & Wilkins, 59–113.* **D.** Cross-bridge cycle of muscle contraction.

BOX 6-1 Events During Excitation, Contraction, and Relaxation of Muscle Fiber

1. An action potential is initiated and propagated in a motor axon.

2. This action potential causes the release of acetylcholine from the axon terminals at the neuromuscular junction.

3. Acetylcholine is bound to receptor sites on the motor end plate membrane.

4. Acetylcholine increases the permeability of the motor end plate to sodium and potassium ions, producing an end-plate potential.

5. The end-plate potential depolarizes the muscle membrane (sarcolemma), generating a muscle action potential that is propagated over the membrane surface.

6. Acetylcholine is rapidly destroyed by acetylcholinesterase on the end plate membrane.

7. The muscle action potential depolarizes the transverse tubules.

8. Depolarization of the transverse tubules leads to the release of calcium ions from the terminal cisternae of the sarcoplasmic reticulum surrounding the myofibrils. These ions are released into the sarcoplasm in the direct vicinity of the regulatory proteins tropomyosin and troponin.

9. Calcium ions bind to troponin, allowing movement of the tropomyosin molecule away from the myosin receptor sites on the actin filament that it had been blocking and releasing the inhibition that had prevented actin from combining with myosin.

10. Actin (A) combines with myosin ATP (M-ATP). In this state, ATP has been hydrolyzed to ADP and phosphate but the products are still attached to myosin (receptor sites on the myosin cross-bridges bind to receptor sites on the actin chain):

$$A + M \cdot ATP \rightarrow A \cdot M \cdot ATP$$

11. Actin activates the myosin ATPase found on the myosin cross-bridge, enabling ATP to be split (hydrolyzed). This process releases energy used to produce movement of the myosin cross-bridges:

$$A \cdot M \cdot ATP \rightarrow A \cdot M + ADP + P_1$$

12. Oar-like movements of the cross-bridges produce relative sliding of the thick and thin filaments past each other.

13. Fresh ATP binds to the myosin cross-bridge, breaking the actin-myosin bond and allowing the cross-bridge to dissociate from actin:

$$A \cdot M + ATP \rightarrow A + M \cdot ATP$$

14. The ATPase hydrolyzes the myosin ATP complex to the M · ATP complex, which represents the relaxed state of the sarcomere:

$$M \cdot ATP \rightarrow M \cdot ATP$$

15. Cycles of binding and unbinding of actin with the myosin cross-bridges at successive sites along the actin filament (steps 11, 12, 13, and 14) continue as long as the concentration of calcium remains high enough to inhibit the action of the troponin-tropomyosin system.

16. Concentration of calcium ions falls as they are pumped into the terminal cisternae of the sarcoplasmic reticulum by an energy-requiring process that splits ATP.

17. Calcium dissociates from troponin, restoring the inhibitory action of troponin-tropomyosin. The actin filament slides back and the muscle lengthens. In the presence of ATP, actin and myosin remain in the dissociated, relaxed state.

Modified from Luciano, Vander, A.J., Sherman, J.H. (1978). Human Function and Structure. New York: McGraw-Hill, Fig 5.5D; and adapted from Craig, R.W., Padrón R. Molecular Structure of the Sarcomere. In Engel A.G., Franzini-Armstrong C., eds (1994). Myology (2nd ed.). New York: McGraw-Hill, 162.

2. They ensure that the contractile elements return to their original (resting) positions when contraction is terminated.

3. They may help prevent the passive overstretch of the contractile elements when these elements are relaxed, thereby lessening the danger of muscle injury.

4. The viscous property of the series and parallel elastic components allows them to absorb energy proportional to the rate of force application and to dissipate energy in a time-dependent manner (for a discussion of viscoelasticity, see Chapter 4).

This viscous property, combined with the elastic properties of the musculotendinous unit, is demonstrated in everyday activities. For example, when a person attempts to stretch and touch the toes, the stretch is initially elastic. As the stretch is held, however, further

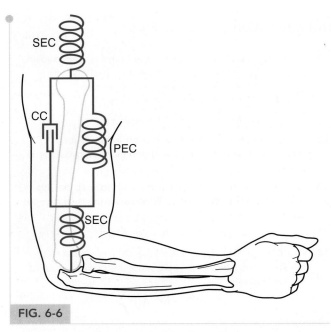

FIG. 6-6

The musculotendinous unit may be depicted as consisting of a contractile component (*CC*) in parallel with an elastic component (*PEC*) and in series with another elastic component (*SEC*). The contractile component is represented by the contractile proteins of the myofibril, actin and myosin. (The myosin cross-bridges may also exhibit some elasticity.) The parallel elastic component comprises the connective tissue surrounding the muscle fibers (the epimysium, perimysium, and endomysium) and the sarcolemma. The series elastic component is represented by the tendons.

elongation of the muscle results from the viscosity of the muscle-tendon structure, and the fingers slowly reach closer to the floor.

Mechanics of Muscle Contraction

Electromyography provides a mechanism for evaluating and comparing neural effects on muscle and the contractile activity of the muscle itself in vivo and in vitro. Much has been learned by using electromyography to study various aspects of the contractile process, particularly the time relationship between the onset of electrical activity in the muscle and of the actual contraction of the muscle or muscle fiber. The following sections discuss the mechanical response of a muscle to electrical (neural) stimulation and the various ways in which the muscle contracts to move a joint, control its motion, or maintain its position.

SUMMATION AND TETANIC CONTRACTION

The mechanical response of a muscle to a single stimulus of its motor nerve is known as a twitch, which is the fundamental unit of recordable muscle activity. Following stimulation there is an interval of a few milliseconds known as the latency period before the tension in the muscle fibers begins to rise. This period represents the time required for the slack in the elastic components to be taken up. The time from the start of tension development to peak tension is the contraction time, and the time from peak tension until the tension drops to zero is the relaxation time. The contraction time and relaxation time vary among muscles, because they depend largely on the muscle fiber makeup (described later). Some muscle fibers contract with a speed of 10 msec, whereas others may take 100 msec or longer.

An action potential lasts only about 1 to 2 msec. This is a small fraction of the time taken for the subsequent mechanical response, or twitch, even in muscles that contract quickly; so it is possible for a series of action potentials to be initiated before the first twitch is completed if the activity of the motor axon is maintained. When mechanical responses to successive stimuli are added to an initial response, the result is known as summation (Fig. 6-7). If a second stimulus occurs during the latency period of the first muscle twitch, it produces no additional response, and the muscle is said to be completely refractory.

The frequency of stimulation is variable and is modulated by individual motor units. The greater the frequency of stimulation of the muscle fibers, the greater will be the tension produced in the muscle as a whole. However, a maximal frequency will be reached beyond which the tension of the muscle no longer increases. When this maximal tension is sustained as a result of summation, the muscle is said to contract tetanically. In this case, the rapidity of stimulation outstrips the contraction-relaxation time of the muscle so that little or no relaxation can occur before the next contraction is initiated (Fig. 6-8).

The considerable gradation of contraction exhibited by whole muscles is achieved by the differential activity of their motor units, in both stimulation frequency and the number of units activated. The repetitive twitching of all recruited motor units of a muscle in an asynchronous manner results in brief summations or more prolonged subtetanic or tetanic contractions of the muscle as a whole and is a principal factor responsible for the smooth movements produced by the skeletal muscles.

TYPES OF MUSCLE CONTRACTION

During contraction, the force exerted by a contracting muscle on the bony lever(s) to which it is attached is

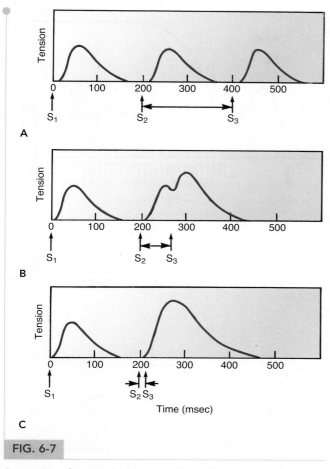

FIG. 6-7

Summation of contractions in a muscle held at a constant length. **A.** An initial stimulus (S_1) is applied to the muscle, and the resulting twitch lasts 150 msec. The second (S_2) and third (S_3) stimuli are applied to the muscle after 200 msec intervals when the muscle has relaxed completely, and thus no summation occurs. **B.** S_3 is applied 60 msec after S_2, when the mechanical response from S_2 is beginning to decrease. The resulting peak tension is greater than that of the single twitch. **C.** The interval between S_2 and S_3 is further reduced to 10 msec. The resulting peak tension is even greater than in B, and the increase in tension produces a smooth curve. The mechanical response evoked by S_3 appears as a continuation of that evoked by S_2. Adapted from Luciano, D.S., Vander, A.J., Sherman, J.H. (1978). Human Function and Structure. New York: McGraw-Hill, 113–136.

known as the muscle tension, and the external force exerted on the muscle is known as the resistance, or load. As the muscle exerts its force, it generates a turning effect, or moment (torque), on the involved joint, as the line of application of the muscle force usually lies at a distance from the center of motion of the joint. The moment is calculated as the product of the muscle force and the perpendicular distance between

its point of application and the center of motion (this distance is known as the lever arm, or moment arm, of the force).

Muscle contractions and the resulting muscle work can be classified according to the relationship between either the muscle tension and the resistance to be overcome or the muscle moment generated and the resistance to be overcome, as shown in Box 6-2.

Although no motion is accomplished and no mechanical work is performed during an isometric contraction, muscle work (physiologic work) is performed: Energy is expended and is mostly dissipated as heat, which is also called the isometric heat production. All dynamic contractions involve what may be considered an initial static (isometric) phase as the muscle first develops tension equal to the load it is expected to overcome.

The tension in a muscle varies with the type of contraction. Isometric contractions produce greater tension than concentric contractions. Studies suggest that the tension developed in an eccentric contraction may even exceed that developed during an isometric contraction. These differences are thought to be due in large part to the varying amounts of supplemental tension produced in the series elastic component of the muscle and to differences in contraction time. The longer contraction time of the isometric and eccentric contractions allows greater cross-bridge formation by the contractile components, thus permitting greater tension to be generated (Kroll, 1987). More time is also available for this tension to be transmitted to the series elastic component as the muscle-tendon unit is stretched. Furthermore, the longer contraction time allows the recruitment of additional motor units.

Komi (1986) has pointed out that concentric, isometric, and eccentric muscle contractions seldom occur alone in normal human movement. Rather, one type of contraction or load is preceded by a different type. An example is the eccentric loading prior to the concentric contraction that occurs at the ankle from midstance to toe-off during gait.

Since muscles normally shorten or lengthen at varying velocities and with varying amounts of tension, the performance and measurement of isokinetic work require the use of an isokinetic dynamometer. This device provides constant velocity of joint motion and maximum external resistance throughout the range of motion of the involved joint, thereby requiring maximal muscle torque. The use of the isokinetic dynamometer provides a method of selective training and measurement, but physiologic movement is not simulated.

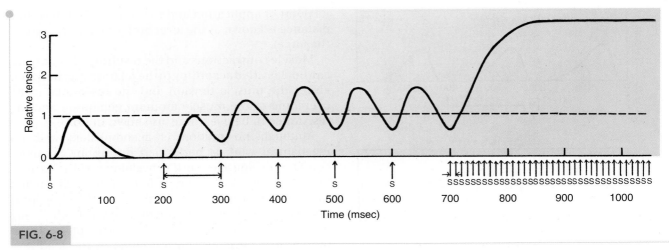

FIG. 6-8

Generation of muscle tetanus. As the frequency of stimulation (S) increases (i.e., the intervals shorten from 200 to 100 msec), the muscle tension rises as a result of summation. When the frequency is increased to 100 per second, summation becomes maximal and the muscle contracts tetanically, exerting sustained peak tension. *Adapted from Luciano, D.S., Vander, A.J., Sherman, J.H. (1978). Human Function and Structure. New York: McGraw-Hill, 113–136.*

Force Production in Muscle

The total force that a muscle can produce is influenced by its mechanical properties, which can be described by examining the length-tension, load-velocity, and force-time relationships of the muscle and the skeletal muscle architecture such as the fiber angle. Other principal factors in force production are muscle temperature, muscle fatigue, and pre-stretching.

LENGTH-TENSION RELATIONSHIP

The force, or tension, that a muscle exerts varies with the length at which it is held when stimulated. This relationship can be observed in a single fiber contracting isometrically and tetanically, as illustrated by the length-tension curve in Figure 6-9. Maximal tension is produced when the muscle fiber is approximately at its slack, or resting, length. If the fiber is held at shorter lengths, the tension falls off slowly at first and then rapidly. If the fiber is lengthened beyond the resting length, tension progressively decreases.

The changes in tension when the fiber is stretched or shortened are due primarily to structural alterations in the sarcomere. Maximal isometric tension can be exerted when the sarcomeres are at their resting length (2.0 to 2.25 μm), because the actin and myosin filaments overlap along their entire length and the number of cross-bridges is maximal. If the sarcomeres are lengthened, there are fewer junctions between

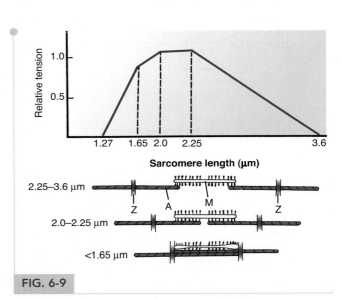

FIG. 6-9

Tension-length curve from part of an isolated muscle fiber stimulated at different lengths. The isometric tetanic tension is closely related to the number of cross-bridges on the myosin filament overlapped by the actin filament. The tension is maximal at the slack length, or resting length, of the sarcomere (2 μm), where the overlap is greatest, and falls to zero at the length where overlap no longer occurs (3.6 μm). The tension also decreases when the sarcomere length is reduced below the resting length, falling sharply at 1.65 μm and reaching zero at 1.27 μm as the extensive overlap interferes with cross-bridge formation. The structural relationship of the actin and myosin filaments at various stages of sarcomere shortening and lengthening is portrayed below the curve. *A,* actin filaments; *M,* myosin filaments; *Z,* Z lines. *Adapted from Crawford, C.N.C., James, N.T. (1980). The design of muscles. In R. Owen, J. Goodfellow, P. Bullough (Eds.). Scientific Foundations of Orthopaedics and Traumatology. London: William Heinemann, 67–74.*

BOX 6-2 Types of Muscle Work and Contraction

Dynamic work: Mechanical work is performed and joint motion is produced through the following forms of muscle contraction:

1. Concentric (con, together; centrum, center) contraction: When muscles develop sufficient tension to overcome the resistance of the body segment, the muscles shorten and cause joint movement. The net moment generated by the muscle is in the same direction as the change in joint angle. An example of a concentric contraction is the action of the quadriceps in extending the knee when ascending stairs.

2. Eccentric (ec, out of, centrum, center) contraction: When a muscle cannot develop sufficient tension and is overcome by the external load, it progressively lengthens instead of shortening. The net muscle moment is in the opposite direction from the change in joint angle. One purpose of eccentric contraction is to decelerate the motion of a joint. For example, when one descends stairs, the quadriceps works eccentrically to decelerate flexion of the knee, thus decelerating the limb. The tension that it applies is less than the force of gravity pulling the body downward, but it is sufficient to allow controlled lowering of the body.

3. Isokinetic (iso, constant; kinetic, motion) contraction: This is a type of dynamic muscle work in which movement of the joint is kept at a constant velocity, and hence the velocity of shortening or lengthening of the muscle is constant. Because velocity is held constant, muscle energy cannot be dissipated through acceleration of the body part and is entirely converted to a resisting moment. The muscle force varies with changes in its lever arm throughout the range of joint motion (Hislop and Perrine, 1967). The muscle contracts concentrically and eccentrically with different directions of joint motion. For example, the flexor muscles of a joint contract concentrically during flexion and eccentrically during extension, acting as decelerators during the latter.

4. Isoinertial (iso, constant; inertial, resistance) contraction: This is a type of dynamic muscle work wherein the resistance against which the muscle must contract remains constant. If the moment (torque) produced by the muscle is equal to or less than the resistance to be overcome, the muscle length remains unchanged and the muscle contracts isometrically. If the moment is greater than the resistance, the muscle shortens (contracts concentrically) and causes acceleration of the body part. Isoinertial contraction occurs, for example, when a constant external load is lifted. At the extremes of motion, the inertia of the load must be overcome; the involved muscles contract isometrically and muscle torque is maximal. In the midrange of the motion, with the inertia overcome, the muscles contract concentrically and the torque is submaximal.

5. Isotonic (iso, constant; tonic, force) contraction: This term is commonly used to define muscle contraction in which the tension is constant throughout a range of joint motion. This term does not take into account the leverage effects at the joint. However, because the muscle force moment arm changes throughout the range of joint motion, the muscle tension must also change. Thus, isotonic muscle contraction in the truest sense does not exist in the production of joint motion (Kroll, 1987).

Static work: No mechanical work is performed and posture or joint position is maintained through the following form of muscle contraction:

1. Isometric (iso, constant; metric, length) contraction: Muscles are not always directly involved in the production of joint movements. They may exercise either a restraining or a holding action, such as that needed to maintain the body in an upright position in opposing the force of gravity. In this case the muscle attempts to shorten (i.e., the myofibrils shorten and in doing so stretch the series elastic component, thereby producing tension), but it does not overcome the load and cause movement; instead, it produces a moment that supports the load in a fixed position (e.g., maintains posture) because no change takes place in the distance between the muscle's points of attachment.

the filaments, and the active tension decreases. At a sarcomere length of about 3.6 μm, there is no overlap and hence no active tension. Sarcomere shortening to less than resting length decreases the active tension because it allows overlapping of the thin filaments at opposite ends of the sarcomere, which are functionally polarized in opposite directions. At a sarcomere length of less than 1.65 μm, the thick filaments completely overlap the Z line and the tension diminishes sharply.

The length-tension relationship illustrated in Figure 6-9 is for an individual muscle fiber. If this relationship is measured in a whole muscle contracting isometrically and tetanically, the tension produced by both active components and passive components must be taken into account (Fig. 6-10).

The curve labeled *Active tension* in Figure 6-10 represents the tension developed by the contractile elements of the muscle, and it resembles the curve for the individual fiber. The curve labeled *Passive tension*

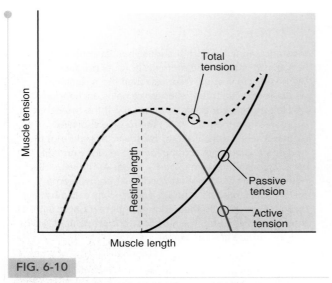

FIG. 6-10

The active and passive tension exerted by a whole muscle contracting isometrically and tetanically is plotted against the muscle's length. The active tension is produced by the contractile muscle components and the passive tension by the series and parallel elastic components, which develop stress when the muscle is stretched beyond its resting length. The greater the amount of stretching, the larger the contribution of the elastic component to the total tension. The shape of the active curve is generally the same in different muscles, but the passive curve, and hence the total curve, varies depending on how much connective tissue (elastic component) the muscle contains.

LOAD-VELOCITY RELATIONSHIP

The relationship between the velocity of shortening or eccentric lengthening of a muscle and different constant loads can be determined by plotting the velocity of motion of the muscle lever arm at various external loads, thereby generating a load-velocity curve (Fig. 6-11). The velocity of shortening of a muscle contracting concentrically is inversely related to the external load applied (Guyton, 1986). The velocity of shortening is greatest when the external load is zero, but as the load increases the muscle shortens more and more slowly. When the external load equals the maximal force that the muscle can exert, the velocity of shortening becomes zero and the muscle contracts isometrically. When the load is increased still further, the muscle contracts eccentrically: It elongates during contraction. The load-velocity relationship is reversed from that of the concentrically contracting muscle; the muscle eccentrically lengthens more quickly with increasing load (Kroll, 1987) (Case Study 6-1).

reflects the tension developed when the muscle surpasses its resting length and the noncontractile muscle belly is stretched. This passive tension is mainly developed in the parallel and series elastic components (see Fig. 6-6). When the belly contracts, the combined active and passive tensions produce the total tension exerted. The curve demonstrates that as a muscle is progressively stretched beyond its resting length, the passive tension rises and the active tension decreases.

Most muscles that cross only one joint normally are not stretched enough for the passive tension to play an important role, but the case is different for two-joint muscles, in which the extremes of the length-tension relationship may be functioning (Crawford and James, 1980). For example, the hamstrings shorten so much when the knee is fully flexed that the tension that they can exert decreases considerably. Conversely, when the hip is flexed and the knee extended, the muscles are so stretched that it is the magnitude of their passive tension that prevents further elongation and causes the knee to flex if hip flexion is increased.

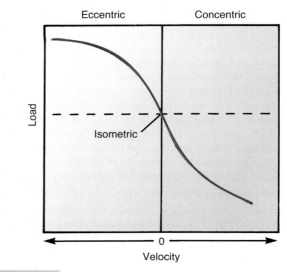

FIG. 6-11

Load-velocity curve generated by plotting the velocity of motion of the muscle lever arm against the external load. When the external load imposed on the muscle is negligible, the muscle contracts concentrically with maximal speed. With increasing loads the muscle shortens more slowly. When the external load equals the maximum force that the muscle can exert, the muscle fails to shorten (i.e., has zero velocity) and contracts isometrically. When the load is increased further, the muscle lengthens eccentrically. This lengthening is more rapid with greater load.

Case Study 6-1

Gastrocnemius Muscle Tear

A 22-year-old male professional athlete tears his gastrocnemius during a race (Case Study Fig. 6-1). The tensile overload that happens during strenuous eccentric and concentric contractions increases the risk of injury, especially when the forces involve biarticular muscles such as the gastrocnemius. This indirect trauma is associated with high tensile forces during rapid contraction (high velocity) and continued changes in muscle length. The status of muscle contraction at the time of overload is usually eccentric, and failure most often occurs at or near the myotendinous junction unless the muscle has been previously injured (Kasser, 1996). Swelling from hemorrhage occurs initially in the inflammatory phase. The cellular response is more rapid and repair is more complete if the vascular channels are not disrupted and the nutrition of the tissue is not disturbed. The degree of injury from a tensile overload will dictate the potential host response and the time needed for repair.

Case Study Figure 6-1

FORCE-TIME RELATIONSHIP

The force, or tension, generated by a muscle is proportional to the contraction time: The longer the contraction time, the greater the force developed, up to the point of maximum tension. In Figure 6-12, this relationship is illustrated by a force-time curve for a whole muscle contracting isometrically. Slower contraction leads to greater force production because time is allowed for the tension produced by the contractile elements to be transmitted through the parallel elastic components to the tendon. Although tension production in the contractile component can reach a maximum in as little as 10 msec, up to 300 msec may be needed for that tension to be transferred to the elastic components. The tension in the tendon will reach the maximum tension developed by the contractile element only if the active contraction process is of sufficient duration (Ottoson, 1983).

EFFECT OF SKELETAL MUSCLE ARCHITECTURE

The muscle's architecture regarding the lengths and angles of its fibers and fascicles has a great impact on its biomechanics, such as the force production (Blazevich, 2006).

The muscles consist of the contractile component, the sarcomere, which produces active tension. The arrangement of the contractile components affects the contractile properties of the muscle dramatically. The more sarcomere lie in series, the longer the myofibril; the more sarcomere lie parallel, the larger the cross-sectional area of the myofibril. These two basic architectural patterns

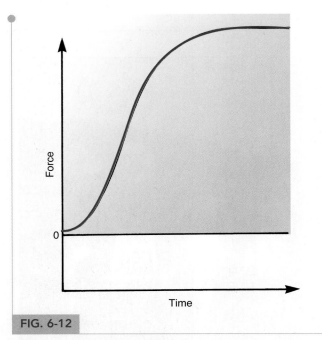

FIG. 6-12

Force-time curve for a whole muscle contracting isometrically. The force exerted by the muscle is greater when the contraction time is longer because time is required for the tension created by the contractile components to be transferred to the parallel elastic component and then to series elastic component as the musculotendinous unit is stretched.

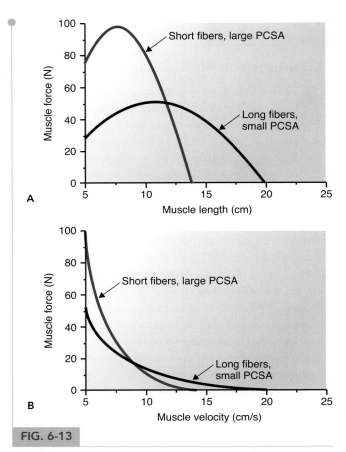

FIG. 6-13

Isometric and isotonic properties of muscles with different architecture. **A.** Force-length relationship. **B.** Force-velocity relationship. *PSCA,* physiologic cross-sectional area. *Reprinted with permission of the American Physical Therapy Association from Lieber, R.L. (1993). Skeletal muscle mechanics: Implications for rehabilitation. Phys Ther 73(12), 852.*

of myofibrils (long or thick) affect the contractile properties of the muscles through the velocity and the excursion (working range) the muscle can produce, which are proportional to the length of the myofibril (Fig. 6-13A), and the force the muscle can produce, which is proportional to the cross-section of the myofibril (Fig. 6-13B).

Muscles with shorter fibers and a larger cross-sectional area are designed to produce force, whereas muscles with long fibers are designed for excursion and velocity. The quadriceps muscle contains shorter myofibrils and appears to be specialized for force production. The sartorius muscle has longer fibers and a smaller cross-sectional area and is better suited for high excursion (Baratta et al., 1998; Lieber and Bodine-Fowler, 1993).

In some human muscles, the fascicles run directly from origin to insertion. However, usually within a muscle, the fascicles are connected to the aponeuroses of that muscle. The fascicle lie in an angle relative to the aponeurosis (fascicle angle), and the aponeuroses lie in an angle relative to the tendon (aponeurosis angle). Fascicle angle minus aponeurosis angle forms the pennation angle, which is described to have an impact on general muscle function. The pennation angle influences the force production in three ways:

1. A larger fascicle angle results in a greater physiologic cross-sectional area and in a greater force production (Fig. 6-14).

2. A larger fascicle angle results in a possible muscle operation closer to that fiber length that allows maximal force production by means of the length-tension relationship) (Fig. 6-15).

3. A larger fascicle angle results in shortening of the fibers. This shortening increases the force production by means of the force-velocity relationship. The velocity of shortening decreases if fiber displacement is reduced and contraction time is kept the same ($v = d/t$, where v = velocity of shortening, d = fiber displacement, and t = time). This leads to a slower contraction resulting in greater force production.

In essence, muscles with a larger pennation angle, such as the vastus lateralis of the quadriceps muscle, do have

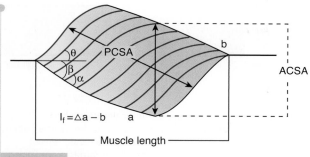

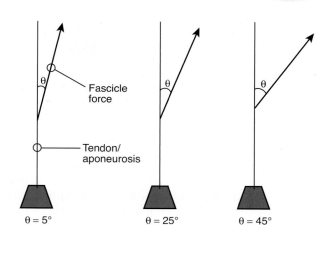

FIG. 6-14

Muscle architectural parameters include the following: fiber length, distance between ends of a fiber (*a* to *b*); pennation angle (θ), fascicle angle (relative to the aponeurosis [α]) minus the aponeurosis angle (relative to the tendon [β]); muscle length; and anatomical (ACSA) or physiologic (PCSA) cross-sectional area. The PCSA can be calculates as (V/t) × $\sin\theta$ for a simple, unipenate muscle, where V is the muscle volume, t is the muscle thickness from one aponeurosis to the other, and θ is the pennation angle. In more complex muscles, PCSA is calculated as V/l_f × $\cos\theta$ where l_f is the mean fiber/fascicle length. *Used with permission from Blazevich AJ (2006). Effects of physical training and detraining, immobilisation, growth and aging on human fascicle geometry [Review]. Sports Med, 36(12), 1003–1017.*

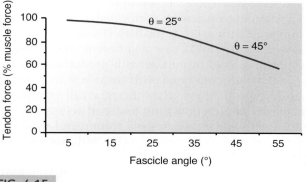

FIG. 6-15

The effect of the fascicle angle on the quantity of force directed along the tendon axis. As fascicle angle (θ) increases, the proportion of fiber force directed along the tendon decreases (tendon force = sum of fiber forces × cos [fiber angle]), where the fiber force is represented by the arrow attached to the tendon/aponeurosis. The tendon is attached to a mass (m) representing the inertia of the system on which the muscle-tendon complex does work. The effect of fascicle angulation on the proportion of force directed along the tendon is minimal when fascicle angle is moderate (e.g. <25°), but increases non-linearly as fascicle angle increases, as shown in the graph. *Used with permission from Blazevich AJ (2006). Effects of physical training and detraining, immobilisation, growth and aging on human fascicle geometry [Review] Sports Medicine 36(12):1003–1017.*

shorter fascicles. This results in increased force production due to length-force and force-velocity relationship, and through a greater physiologic cross-sectional area (CSA) (Blazevich, 2006; Fukunaga et al., 2001). Furthermore, these muscles have a short range of motion. In contrast, muscles with a lower pennation angle, such as the adductor magnus and longus, have longer fascicles, a decreased force production, but high shortening velocity over a long range of motion (Burkholder et al., 1994; Lieber et al., 2001).

EFFECT OF PRE-STRETCHING

It has been demonstrated in amphibians and in humans (Ciullo and Zarins, 1983) that a muscle performs more work when it shortens immediately after being stretched in the concentrically contracted state than when it shortens from a state of isometric contraction. This phenomenon is not entirely accounted for by the elastic energy stored in the series elastic component during stretching but must also be due to energy stored in the contractile component. It has been suggested that changes in the intrinsic mechanical properties of myofibrils are important in the stretch-induced enhancement of work production (Takarada et al., 1997).

EFFECT OF TEMPERATURE

Temperature has been shown to have an effect on muscle performance. Changes in the temperature will have

an effect on the contractile properties of the skeletal muscles. Extreme environmental conditions will produce changes in the rate of enzymatic activity within the muscle.

A rise in muscle temperature causes an increase in conduction velocity across the sarcolemma (Phillips and Petrofsky, 1983), increasing the frequency of stimulation and hence the production of muscle force. Increasing the muscle temperature from 6°C to 34°C (42.8°F to 93.2°F, respectively) results in an almost linear increase of the tension/stiffness ratio (Galler et al., 1998). A rise in temperature also causes greater enzymatic activity of muscle metabolism, thus increasing the efficiency of muscle contraction. A further effect of a rise in temperature is the increased elasticity of the collagen in the series and parallel elastic components, which enhances the extensibility of the muscle-tendon unit. This pre-stretch thus increases the force production of the muscle.

Conversely, studies have shown that with a decrease in temperature there will be a decrease in the production or utilization of adenosine 5'-triphosphate (ATP) and depletion of intracellular glycogen (Hong et al., 2008); consequently it will affect the muscle performance power. A study conducted in males showed that there was a 23% depletion of intracellular glycogen when exercises were performed at 9°C (48.2°F) as compared with at 21°C (69.8°F) (Jacobs et al., 1985). Gossen et al. (2001) saw a different mechanism of response to cold in muscle with predominantly type I fibers when compared with muscle with predominantly type II fibers. Hypothermia will produce an increased contraction time and half-relaxation time and it will reduce twitch peak torque and maximum rate torque development (Gossen, 2001; Kimura et al., 2003; Nomura et al., 2002).

In another study conducted in rats, it was found that under prolonged exposure to cold, the muscle configuration changed from fast-twitch fibers into slow-twitch fibers (Nomura et al., 2002).

EFFECT OF FATIGUE

The ability of a muscle to contract and relax depends on the availability of ATP (see Box 6-1). If a muscle has an adequate supply of oxygen and nutrients that can be broken down to provide ATP, it can sustain a series of low-frequency twitch responses for a long time. The frequency must be low enough to allow the muscle to synthesize ATP at a rate sufficient to keep up with the rate of ATP breakdown during contraction. If the frequency of stimulation increases and outstrips the rate of replacement of ATP, the twitch responses soon grow progressively weaker and eventually fall to zero (Fig. 6-16). This drop in tension following prolonged stimulation is muscle fatigue. If the frequency is high enough to produce tetanic contractions, fatigue occurs even sooner. If a period of rest is allowed before stimulation is continued, the ATP concentration rises and the muscle briefly recovers its contractile ability before again undergoing fatigue.

Three sources supply ATP in muscle: creatine phosphate, oxidative phosphorylation in the mitochondria,

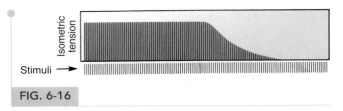

FIG. 6-16

Fatigue in a muscle contracting isometrically. Prolonged stimulation occurs at a frequency that outstrips the muscle's ability to produce sufficient ATP for contraction. As a result, tension production declines and eventually ceases. *Adapted from Luciano, D.S., Vander, A.J., Sherman, J.H. (1978). Human Function and Structure. New York: McGraw-Hill, 113–136.*

and substrate phosphorylation during anaerobic glycolysis. When contraction begins, the myosin ATPase breaks down ATP very rapidly. The increase in ADP and Pi concentrations resulting from this breakdown ultimately leads to increased rates of oxidative phosphorylation and glycolysis. However, after a short lapse, these metabolic pathways begin to deliver ATP at a high rate. During this interval, the energy for ATP formation is provided by creatine phosphate, which offers the most rapid means of forming ATP in the muscle cell.

At moderate rates of muscle activity, most of the required ATP can be formed by the process of oxidative phosphorylation. During very intense exercise, when ATP is being broken down very rapidly, the cell's ability to replace ATP by oxidative phosphorylation may be limited, primarily by inadequate delivery of oxygen to the muscle by the circulatory system.

Even when oxygen delivery is adequate, the rate at which oxidative phosphorylation can produce ATP may be insufficient to sustain very intense exercise, since the enzymatic machinery of this pathway is relatively slow. Anaerobic glycolysis then begins to contribute an increasing portion of the ATP. The glycolytic pathway, although it produces much smaller amounts of ATP from the breakdown of glucose, operates at a much faster rate. It can also proceed in the absence of oxygen, with the formation of lactic acid as its end product. Thus, during intense exercise, anaerobic glycolysis becomes an additional source for rapidly supplying the muscle with ATP.

The glycolytic pathway has the disadvantage of requiring large amounts of glucose for the production of small amounts of ATP. Thus, even though muscle stores glucose in the form of glycogen, existing glycogen supplies may be depleted quickly when muscle activity is intense. Finally, myosin ATPase may break down ATP faster than even glycolysis can replace it, and fatigue occurs rapidly as ATP concentrations drop.

After a period of intense exercise, creatine phosphate levels drop and much of the muscle glycogen may have been converted to lactic acid. For the muscle to be returned to its original state, creatine phosphate must be resynthesized and the glycogen stores must be replaced. Since both processes require energy, the muscle will continue to consume oxygen at a rapid rate even though it has stopped contracting. This sustained high oxygen uptake is demonstrated by the fact that a person continues to breathe heavily and rapidly after a period of strenuous exercise.

When the energy necessary to return glycogen and creatine phosphate to their original levels is taken into account, the efficiency with which muscle converts chemical energy to work (movement) is usually no more than 20% to 25%, most of the energy being dissipated as heat. Even when muscle is operating in its most efficient state, a maximum of only about 45% of the energy is used for contraction (Arvidson et al., 1984; Guyton, 1986). Training has an effect in improving metabolic oxidative production and this may increase time to fatigue when involve in certain sports activity (Harmer et al., 2000).

Although the phenomena of fatigue or exhaustion may overlap with muscle injury or damage, it is important to point out that they are two distinct entities. Both will cause a decline in muscle performance but fatigue does not necessarily involve structural damage (Allen et al., 2008).

The cellular mechanism of muscle damage is not well understood (Clarkson et al., 2002). Currently, there are two models proposed to study the mechanism, one a mechanical model and the other a metabolic model (Tee et al., 2007). The mechanical model demonstrates that eccentric contraction produces a greater amount of force that will increase force per cross bridges and predispose the contractile proteins to fail. This is primarily true with exercise-induced injuries that involve eccentric muscle contraction. The metabolic model mechanism proposes that deficiencies occur within the stressed muscle, increasing the presence of Ca^{2+} and possibly resulting in muscle fiber degeneration (Allen et al., 2008). This may be an explanation for the occurrence of muscle damage resulting from activities that involve primarily concentric muscle contraction such as when subjects were exposed to long cycling events or run marathons (Warhol et al., 1985).

EFFECT OF VIBRATION

There have been some laboratory studies looking at the effect of exposure to vibration and changes in force output, power, and velocity. A study conducted on female professional volleyball players in which subjects were exposed to vibration training showed a significant increase in their average velocity, power, and strength when compared with a control group (Bosco et al., 1999).

A study conducted of healthy subjects subjected to bed rest for almost two months showed that those subjects exposed to a vibration plate were more likely to maintain muscle structure and force production in calf muscle compared with the control. The training did not have the same effect on the thigh muscles (Blottner et al., 2006).

Another study in young untrained individuals exposed to whole body vibration (wbv) training showed an increase in torque production in knee extension after the training (Jacobs and Burns, 2009).

Bogaerts et al. (2007) studied elderly men and compared three groups: fitness, wbv training (at 30–45 hz), and a control. The fitness group and the wbv training group showed a significant increase in muscle strength when compared with the control group. Pietrangelo et al. (2009) conducted a study in an elderly population in which both genders were submitted to local vibration training at 300 hz. They concluded that the training was effective in decreasing muscle loss due to sarcopenia.

Marin and Rhea (2010) conducted a metanalysis review of different types of vibration training in humans. They concluded that the type of vibration equipment used would have different results in muscle strength. Vertical vibration produced higher effects on muscle strength in chronic training when compared with oscillatory vibration equipment. On the other hand, they concluded that oscillatory vibration produced a higher effect on acute training.

Muscle Fiber Differentiation

In the preceding section, the major factors that determine the total tension developed by the whole muscle when it contracts were described. In addition, individual muscle fibers display distinct differences in their rates of contraction, development of tension, and susceptibility to fatigue.

Many methods of classifying muscle fibers have been devised. As early as 1678, Lorenzini observed anatomically the gross difference between red and white muscle, and in 1873, Ranvier typed muscle on the basis of speed of contractility and fatigability. Although considerable confusion has existed concerning the method and terminology for classifying skeletal muscle, recent histologic and histochemical observations have led to the identification of three distinct types of muscle fibers on the basis

TABLE 6-1			
Properties of Three Types of Skeletal Muscle Fibers			
	Type I Slow-Twitch Oxidative (SO)	Type IIA Fast-Twitch Oxidative-Glycolytic (FOG)	Type IIB Fast-Twitch Glycolytic (FG)
Speed of contraction	Slow	Fast	Fast
Primary source of ATP production	Oxidative phosphorylation	Oxidative phosphorylation	Anaerobic glycolysis
Glycolytic enzyme activity	Low	Intermediate	High
Capillaries	Many	Many	Few
Myoglobin content	High	High	Low
Glycogen content	Low	Intermediate	High
Fiber diameter	Small	Intermediate	Large
Rate of fatigue	Slow	Intermediate	Fast

of differing contractile and metabolic properties (Brandstater and Lambert, 1969; Buchtahl and Sohmalburch, 1980) (Table 6-1).

The fiber types are distinguished mainly by the metabolic pathways by which they can generate ATP and the rate at which its energy is made available to the contractile system of the sarcomere, which determines the speed of contraction. The three fiber types are termed type I, slow-twitch oxidative (SO) red fibers; type IIA, fast-twitch oxidative-glycolytic (FOG) red fibers; and type IIB, fast-twitch glycolytic (FG) white fibers.

Type I (SO) fibers are characterized by a low activity of myosin ATPase in the muscle fiber and, therefore, a relatively slow contraction time. The glycolytic (anaerobic) activity is low in this fiber type, but a high content of mitochondria produces a high potential for oxidative (aerobic) activity. Type I fibers are very difficult to fatigue, because the high rate of blood flow to these fibers delivers oxygen and nutrients at a sufficient rate to keep up with the relatively slow rate of ATP breakdown by myosin ATPase. Thus, the fibers are well suited for prolonged, low-intensity work. These fibers are relatively small in diameter and thus produce relatively little tension. The high myoglobin content of type I fibers gives the muscle a red color.

Type II muscle fibers have been divided into two main subgroups, IIA and IIB, on the basis of differing susceptibility to treatment with different buffers prior to incubation (Brooke and Kaiser, 1970). A third subgroup, the type IIC fibers, are rare, undifferentiated fibers, which are usually seen before the 30th week of gestation. This fiber type is infrequent in human muscle (Banker, 1994). Type IIA and IIB fibers are characterized by a high activity of myosin ATPase, which results in relatively fast contraction.

Type IIA (FOG) fibers are considered intermediate between type I and type IIB, because their fast contraction time is combined with a moderately well-developed capacity for both aerobic (oxidative) and anaerobic (glycolytic) activity. These fibers also have a well-developed blood supply. They can maintain their contractile activity for relatively long periods; however, at high rates of activity, the high rate of ATP splitting exceeds the capacity of both oxidative phosphorylation and glycolysis to supply ATP and these fibers eventually fatigue. Because the myoglobin content of this muscle type is quite high, the muscle is often categorized as red muscle.

Type IIB (FG) fibers rely primarily on glycolytic (anaerobic) activity for ATP production. Very few capillaries are found in the vicinity of these fibers, and because they contain little myoglobin they are often referred to as white muscle. Although type IIB fibers are able to produce ATP rapidly, they fatigue very easily, because their high rate of ATP splitting quickly depletes the glycogen needed for glycolysis. These fibers generally are of large diameter and are thus able to produce great tension but for only short periods before they fatigue.

It has been well demonstrated that the nerves innervating the muscle fiber determines its type (Burke et al., 1971); thus, the muscle fibers of each motor unit are of a single type. In humans and other species, electrical stimulation was found to change the fiber type (Munsat et al., 1976). In animal studies, transacting the nerves that innervate slow-twitch and fast-twitch muscle fibers and then crossing these nerves was noted to reverse the fiber types. After recovery from the cross-innervation, the slow-twitch fibers became fast in their contractile and histochemical properties and the fast-twitch fibers became slow.

The fiber composition of a given muscle depends on the function of that muscle. Some muscles perform predominantly one form of contractile activity and are often composed mostly of one muscle fiber type. An example is the soleus muscle in the calf, which primarily maintains posture and is composed of a high percentage of type II fibers. More commonly, however, a muscle is required to perform endurance-type activity under some circumstances and high-intensity strength activity under others. These muscles generally contain a mixture of the three muscle fiber types.

In a typical mixed muscle exerting low tension, some of the small motor units, composed of type I fibers, contract. As the muscle force increases, more motor units are recruited and their frequency of stimulation increases. As the frequency becomes maximal, greater muscle force is achieved by recruitment of larger motor units composed of type IIA (FOG) fibers and eventually type IIB (FG) fibers. As the peak muscle force decreases, the larger units are the first to cease activity (Guyton, 1986; Luciano et al., 1978).

It is generally, but not universally, accepted that fiber types are genetically determined (Costill et al., 1979; Gollnick, 1982). In the average population, about 50% to 55% of muscle fibers are type I, about 30% to 35% are type IIA, and about 15% are type IIB, but these percentages vary greatly among individuals.

In elite athletes, the relative percentage of fiber types differs from that in the general population and appears to depend on whether the athlete's principal activity requires a short, explosive, maximal effort or involves submaximal endurance. Sprinters and shot putters, for example, have a high percentage of type II fibers, whereas distance runners and cross-country skiers have a higher percentage of type I fibers. Endurance athletes may have as many as 80% type I fibers, and those engaged in short, explosive efforts as few as 30% of these fibers (Saltin et al., 1977).

The genetically determined fiber typing may be responsible for the natural selective process by which athletes are drawn to the type of sport for which they are most suited. Since fiber types are determined by the nerve that innervates the muscle fiber, there may be some cortical control of this innervation that influences an athlete to choose the sport in which he or she is genetically able to excel.

Muscle Remodeling

The remodeling of muscle tissue is similar to that of other skeletal tissues such as bone, articular cartilage, and ligaments. As in these other tissues, muscle atrophies in response to disuse and immobilization and hypertrophies when subjected to greater use than usual.

EFFECTS OF DISUSE AND IMMOBILIZATION

Disuse and immobilization have detrimental effects on muscle fibers. These include loss of endurance and force production, and muscle atrophies on microstructural and macrostructural levels. These effects are dependent on fiber type and muscle length during immobilization and may be dependent on cause of immobilization. Immobilization in a lengthened position has a less deleterious effect (Appell, 1997; Kasser, 1996; Ohira et al., 1997; Sandmann et al., 1998).

It has been described that loss of muscle force production may predominantly be due to reduced physical activity and biologic aging. From human and animal studies including bed rest, immobilization, and space flight trials, it has been well documented that disuse and immobilization have a deteriorating affect on muscle force. It seems that due to immobilization, loss of strength is greater in the lower limbs than in the upper limbs. This may be due to the fact that legs and arms are exposed to different tasks such as moving the whole body weight (lower limbs) or handling small goods (upper limbs).

Cast immobilization has been shown to reduce the cross-sectional area (CSA) of the human triceps surae muscle by 8.6%, and to reduce the isometric strength by 14.2% after four weeks (Clark et al., 2006), with 15% reduction of the CSA and 54% reduction of isometric strength after seven weeks of immobilization, respectively (Christensen et al., 2008).

In the case of disuse due to pain, as in the case of osteoarthritis, these changes are less pronounced. Suetta et al. (2007) found a reduction of the CSA of the quadriceps muscle in humans with symptomatic osteoarthritis of the hip for more than one year. The CSA was reduced by 7% in men and 8.7% in women, and the isometric force was reduced by 19.8% and 20.3%, respectively. However, to date it remains unclear if these differences are due to the different causes of immobilization (pain vs. no pain).

The reduction of the CSA of the affected muscles is believed to be due to different mechanisms. Jones et al. (2004) undertook a study including healthy human subjects who were exposed to two weeks of lower limb immobilization. They found a reduction of muscle mass of 4.7%. This was explained by an increased expression of genes, which are linked to the atrophy of muscle tissue and result in an increase of protein degradation and a decrease in protein synthesis. Interestingly, only one gene linked to hypertrophy of muscle tissue was affected by immobilization.

Powers et al. (2007) found that reactive oxygen species (ROS) such as hydrogen peroxide are produced in active and inactive skeletal muscles. The ROS regulate physiologic and pathologic signaling, playing a role in regulating protein degradation and cell death. Low levels of ROS may promote the survival of the cell. The production of ROS in periods of muscle disuse may exceed the cell's antioxidant capacity. In that state, the ROS may serve as second messengers inducing intracellular pathways leading to protein degradation, cell death, and muscle atrophy, and thus to a reduction of the CSA.

The reduction of force production may be linked to other factors also. Udaka et al. (2008) found in an animal study a shortening of thick and thin filaments, which changes the biomechanical properties of the muscle, due to the length dependent force production (see Fig. 6-13A). The authors also found a reduction of Ca^{2+} sensitivity after six weeks of immobilization. Both changes may explain in part the reduced force production that was found in this study.

It has also been described that immobilization affects energy production. Studies of human and animal space flight showed that immobilization decreases fat oxidation and increases glycolysis as the energy resources of the muscles. This favors short and high intensity activities (fiber type IIA) and disfavors sustained activities (fiber type I), which may alter the biomechanical properties of the muscle, namely the reduced capacity of postural maintenance (Stein and Wade, 2005).

It has been widely accepted that disuse and immobilization also have an effect on muscle composition. Human muscle biopsy studies have shown that it is mainly the type I fibers that atrophy with immobilization; their cross-sectional area decreases, and their potential for oxidative enzyme activity is reduced (Kannus et al., 1998b). Prolonged bed rest results in a reduced demand on postural maintenance and leads to atrophy of slow twitch fibers resulting in difficulties in postural maintenance (Fitts et al., 2000). Rat hindlimb unloading, which is widely accepted as a model for muscle disuse, revealed that immobilization of a postural slow twitch (fiber type I) muscle such as the soleus results not only in an atrophy of the muscle but also in a transition from slow to fast twitch fibers within the muscle. The soleus muscle in healthy individuals usually consists of less than 15% fast twitch fibers (type II) which increases to 40% after fifteen days of immobilization (Pierno et al., 2007). In this case, there is not only an atrophy of the muscle but also a shift in the composition of the muscle, which alters the biomechanical properties of a given muscle. Different muscle fibers show different properties in force production (see Box 6-2). The composition regarding the muscle fiber mix leads to specific biomechanical properties of that given muscle, which may change if the fiber composition of the muscle changes. However, the results from animal stud-

ies are not observed consistently in humans. The results of a human study including patients with osteoarthritis of the knee, revealed a selective atrophy of the type II fibers in the vastus medialis muscle (Fink et al., 2007).

Early motion may prevent this atrophy. It appears that if the muscle is placed under tension when the body segment moves, afferent (sensory) impulses from the intrafusal muscle spindles will increase, leading to increased stimulation of the type I fiber. While intermittent isometric exercise may be sufficient to maintain the metabolic capacity of the type II fiber, the type I fiber (the postural fiber) requires a more continuous impulse. There is also evidence that electric stimulation may prevent the decrease in type I fiber size and the decline in its oxidative enzyme activity caused by immobilization (Eriksson et al., 1981).

In elite athletes, inactivity following injury, surgery, or immobilization rapidly decreases the size and aerobic capability of muscle fibers, particularly in the fiber type affected by the chosen sport. In endurance athletes, type I fibers are affected, whereas in athletes engaged in an explosive activity such as sprinting, type II fibers are affected (see Case Study 6-2).

Clinical and laboratory studies of human and animal muscle tissue suggest that a program of immediate or early motion may prevent muscle atrophy after injury or surgery. In a study of crush injuries to rat muscle, the effect of immobilization of the crushed limb was compared with that of immediate motion. The muscle fibers were found to regenerate in a more parallel orientation in the mobilized animal than in the immobilized animal, capillarization occurred more rapidly, and tensile strength returned more quickly. Similar results were found in a later study on the effect of immobilization on the morphology of rat calf muscles (Kannus et al., 1998a).

It has been found clinically that atrophy of the quadriceps muscle that develops while the limb is immobilized in a rigid plaster cast cannot be reversed through the use of isometric exercises. Atrophy may be limited by allowing early motion such as that permitted by a partly mobile cast brace. In this case, dynamic exercises can be performed.

Finally, it remains unclear whether the reason for immobilization (pain vs no pain) or the different biomechanic properties of the immobilized muscle (force vs velocity) or both play a role regarding the deteriorating effects of immobilization on different muscle fibers. The above shows, however, that disuse changes the biomechanical properties of muscles on many different levels.

EFFECTS OF PHYSICAL TRAINING

Physical activity influences the architecture of the muscles (Blazevich, 2006) and force production. Postural muscles of the spine need different metabolic and contractile

Case Study 6-2

Ruptured Left Anterior Cruciate Ligament

A 25-year-old male, status postsurgical repair of the ruptured left anterior cruciate ligament, had torque measurements taken from the involved and uninvolved limb ten weeks after the surgical procedure (Case Study Fig. 6-2A) and repeated six weeks after the training began (Case Study Fig. 6-2B). An increase of muscle torque is shown in the repeated isokinetic test. The initial deficit of the involved side was approximately 63% when compared with the uninvolved side. After six weeks of training, the deficit of the involved side compared with the uninvolved side decreased to 43%.

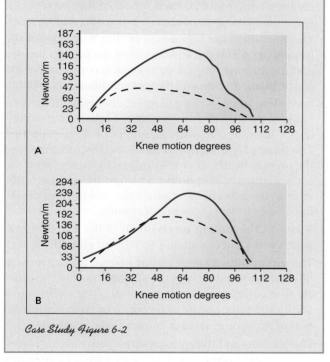

Case Study Figure 6-2

properties (long-term activity for postural maintenance) than the extensor of the leg (explosive activity for locomotion). This is believed to be related to the motor nerve-dependent gene expression during physical activity, which is shown to be applicable in development and adulthood (Schiaffino and Serrano, 2002). Research has shown that the architecture regarding fascicle length, CSA, and pennation angle differs in highly and lesser trained athletes and untrained controls. An unique muscle geometry can also be found in athletes from different

sports (e.g. sprinter vs endurance runner vs weight lifter). Interestingly, there was no relative or absolute difference of the fascicle length between gender (Abe et al., 2001; Kumagai, 2000). However, there is conflicting evidence regarding the effect of training in nonathletes. Gondin et al. (2005) described an increase of the fascicle angle of 14% in the vastus medialis after eight weeks of strengthening with a 27% increase of maximum force and a 6% increase of the CSA. Blazevich and Giorgi (2001) did not find such changes in fascicle length or angle related to physical training.

Due to different populations (untrained vs well trained), different muscles (triceps brachii vs quadriceps), and different protocols (five to sixteen weeks with or without testosterone injections), no final conclusions regarding change in muscle architecture due to training in non-athletes can be made. To date, no studies concerning endurance training and its effects on the geometry of muscles have been conducted.

Physical training has been found to increase the cross-sectional area of all muscle fibers, accounting for the increase in muscle bulk and strength. Physical training is described to influence the neural activation of the muscle resulting in an increased force production within the first six weeks of training. Ongoing training is believed to result in an increase of contractile tissue and thus an increase of the CSA of the trained muscle. Long-term training may then lead to further, rather small increases of force production via the neural mechanisms (Chilibeck et al., 1998; Enoka et al., 1997).

Some evidence suggests that the relative percentage of fiber types in a person's muscles may also change with physical training (Arvidson et al., 1984). Besides genetic factors, physical activity influences the differentiation into slow-twitch and fast-twitch fiber, which may change the biomechanical properties of force production. The latter may be activity-specific. Furthermore, skeletal muscles do adapt to changing demands, which is in part due to the plasticity of the muscle tissue.

Labarque et al. (2002) performed a study on healthy humans that involved cast immobilization of a lower limb for two weeks and retraining for ten weeks. Dynamic and isometric torque was reduced significantly for the knee extensor and flexor muscles after immobilization. However, after three weeks of isokinetic training, the dynamic and isometric torque showed no significant difference from the pre-immobilization state. During immobilization, fiber types I and IIA did show a more pronounced but not significant reduction compared with fiber type IIB. However, after retraining, the fiber distribution returned to its pre-immobilization state.

The effects that different exercises have on different types of muscle fiber have also been demonstrated.

Hortobagyi et al. (2000) performed a human study on muscle composition after immobilization and retraining. They found that fiber types IIA and B showed greater enlargement after eccentric training than after concentric training, which is in line with increased isometric force production.

The different studies use different muscles (upper extremity vs. lower extremity) and different training protocols (concentric vs. eccentric), which prevents comparison of the results. However, these studies suggest that physical training does have an effect on the fiber type distribution. The cross-sectional area of the fibers affected by the athlete's principal activity increases. For example, in endurance athletes, the area of muscle taken up by type I and type IIA fibers increases at the expense of the total area of type IIB fibers.

Aging results in sarcopenia (reduction of muscle mass) and reduction of CSA, which are recognized as factors in muscle weakness (Frontera et al., 2000; Narici et al., 2003). These reductions usually start in the late 50s and are found to be greater in the lower limbs. The resulting reduction of the CSA is as high as 25% to 33% of the quadriceps between younger (aged 20–29 years) and older (aged 70–81 years) individuals (Narici et al., 2003). It is believed that the reduction of the CSA does not explain the force reduction solely. Thom et al. (2007) found that in older individuals (aged 69–82 years), torque velocity was 38.2% lower and power velocity 48.5% lower than in the younger control group (aged 19–35 years). Peak power was 80% less. CSA and fascicle length were reduced by 14.3% and 19.3%, respectively, in the older individuals. However, these changes were found to be insufficient to explain the differences in torque and power.

D'Antona et al. (2003) performed an in vitro study on human muscles. The authors found that even if the fiber composition of a muscle remains unchanged in aging muscles, the force production per muscle unit is reduced. This may be due to changing intrinsic factors. They found a lower myosin concentration in type I and IIA fibers in elderly individuals. This may then result in reduction of actomyosin interaction and reduced force production. Furthermore, Conley et al. (2000) described that the loss of aerobic performance with aging may be due to the reduced maximal O^2 uptake rate (minus 45%) and the reduced oxidative capacity of the investigated quadriceps muscle (minus 36%).

It remains unclear if aging changes the pennation angle. Binzoni et al. (2001) investigated the effects of aging on the pennation angle. They scanned individuals from birth to 70 years of age. The authors found an increase of the pennation angle from birth to the adolescent growth spurt, which remains stable afterwards. It is believed that the pennation angle does not change with aging and therefore does not contribute to the reduction of force production in elderly individuals. However, Narici et al. (2003) compared young adults (age 27–42 years) and older individuals (age 71–80 years). Both groups were described to be similarly physically active. The authors found a significant reduction of the pennation angle of 13.2% in the older group. A reduction of the pennation angle reduces the force production (see Fig. 6-15).

Summary

- The structural unit of skeletal muscle is the fiber, which is encompassed by the endomysium and organized into fascicles encased in the perimysium. The epimysium surrounds the entire muscle.
- The fibers are composed of myofibrils, aligned so as to create a band pattern. Each repeat of this pattern is a sarcomere, the functional unit of the contractile system.
- The myofibrils are composed of thin filaments of the protein actin and thick filaments of the protein myosin and the intramyofibrillar cytoskeleton composed of the elastic filaments titin and the inelastic filaments nebulin.
- According to the sliding filament theory, active shortening of the muscle results from the relative movement of the actin and myosin filaments past one another. The force of contraction is developed by movement of the myosin heads, or cross-bridges, in contact with the actin filaments. Troponin and tropomyosin, two proteins in the actin helix, regulate the making and breaking of the contacts between filaments.
- A key to the sliding mechanism is the calcium ion, which turns the contractile activity on and off.
- The motor unit, a single motor neuron and all muscle fibers innervated by it, is the smallest part of the muscle that can contract independently. The calling in of additional motor units in response to greater stimulation of the motor nerve is known as recruitment.
- The tendons and the endomysium, perimysium, sarcolemma, and epimysium represent parallel and series elastic components that stretch with active contraction or passive muscle extension and recoil with muscle relaxation.
- Summation occurs when mechanical responses of the muscle to successive stimuli are added to an initial response. When maximal tension is sustained as a result of summation, the muscle contracts tetanically. The muscle fiber contracts in an all-or-nothing fashion.
- Muscles may contract concentrically, eccentrically, or isometrically depending on the relationship between the muscle tension and the resistance to be overcome. Concentric and eccentric contractions involve dynamic

work, in which the muscle moves a joint or controls its movement.

- Force production in muscle is influenced by the length-tension, load-velocity, and force-time relationships of the muscle. The length-tension relationship in a whole muscle is influenced by both active (contractile) and passive (series and parallel elastic) components.

- Two other factors that increase force production are pre-stretching of the muscle and a rise in muscle temperature.

- The energy for muscle contraction and its release is provided by the hydrolytic splitting of ATP. Muscle fatigue occurs when the ability of the muscle to synthesize ATP is insufficient to keep up with the rate of ATP breakdown during contraction.

- Three main fiber types have been identified: type I, slow-twitch oxidative; type IIA, fast-twitch oxidative-glycolytic; and type IIB, fast-twitch glycolytic fibers. Most muscles contain a mixture of these types.

- Muscle atrophies occur under disuse and immobilization and muscle trophism can be restored through early and active remobilization.

- Aging changes the muscle architecture and force production per muscle unit and reduces the muscle mass. This results in a reduction of the force production of the muscular system in older age.

Practice Questions

1. Name and explain the different forms of muscle work and their different forms of muscle contraction.

2. Name and explain the functions of the different components of a sarcomere.

3. What is the pennation angle and in what ways does it influence the force production of a muscle?

4. How can muscle remodeling be influenced?

5. Name and explain the different types of muscle fibers.

REFERENCES

Abe, T., Fukashiro, S., Harada, Y., et al. (2001). Relationship between sprint performance and muscle fascicle length in female sprinters. *J Physiol Anthropol Appl Human Sci, 20*(2), 141–147.

Allen, D.G., Lamb, G.D., Westerblad, H. (2008). Skeletal muscle fatigue: Cellular mechanisms. *Physiol Rev, 88*, 287–332.

Appell, H.J. (1997). The muscle in the rehabilitation process. *Orthopade, 26*(11), 930–934.

Arvidson, I., Eriksson, E., Pitman, M. (1984). Neuromuscular basis of rehabilitation. In E. Hunter, J. Funk (Eds.). *Rehabilitation of the Injured Knee*. St Louis, MO: Mosby, 210–234.

Banker, B.Q. (1994). Basic reaction of muscle. In A.G. Engel, C. Franzini-Armstrong (Eds.). *Myology (2nd ed.)*. New York: McGraw-Hill.

Baratta, R.V., Solomonow, M., Zhou, B. H. (1998). Frequency domain-based models of skeletal muscle. *J Electromyogr Kinesiol, 8*(2), 79–91.

Beason, D., Soslowsky, L., Karthikeyan, T., et al. (2007). Muscle, tendon and ligament. In J.S. Fischgrund (Ed.): *Orthopedic Update 9*. Rosemont, IL: American Academy of Orthopedic Surgeons, Chapter 4.

Binzoni, T., Bianch, S., Hanguinet, S., et al. (2001). Human gastrocnemius medialis pennation angle as a function of age: From newborn to the elderly. *J Physiol Anthropol, 20*(5), 293–298.

Blazevich, A.J. (2006). Effects of physical training and detraining, immobilization, growth and aging on human fascicle geometry. *Sports Med, 36*(12), 1003–1017.

Blazevich, A.J., Giorgi, A. (2001). Effect of testosterone administration and weight training on muscle architecture. *Med Sci Sports Exerc, 33*(10), 1688–1693.

Blottner, D., Salanova, M., Puttmann, B., et al. (2006). Human skeletal muscle structure and function preserved by vibration muscle exercise following 55 days of bed rest. *Eur J Appl Physiol, 97*, 261–271.

Bogaerts, A., Delecluse, C., Claessens, A.L., et al. (2007). Impact of whole-body vibration training versus fitness training on muscle strength and muscle mass in older men: A 1-year randomized controlled trial. *J Gerontol A Biol Sci Med Sci, 62*(6), 630–635.

Bosco, C., Colli, R., Introini, E., et al. (1999). Adaptive responses of human skeletal muscle to vibration exposure. *Clinical Physiol, 19*(2), 183–187.

Brandstater, M.E., Lambert, E.H. (1969). A histologic study of the spatial arrangements of muscle fibers in single motor units within rat tibialis anterior muscle. *Bull Am Assoc Electromyog Electro Diag, 82*, 15.

Brooke, M.H., Kaiser, K.K. (1970). Three myosin adenosine triphosphatase systems: The nature of their pH liability and sulfhydryl dependence. *J Histochem Cytochem, 18*, 670.

Buchtahl, F., Sohmalburch, H. (1980). Motor units of mammalian muscle. *Physiol Rev, 60*, 90.

Burke, R.E., Levine, D.N., Zajac, F. E. (1971). Mammalian motor units: Physiological histochemical correlation in three types of motor units in cat gastrocnemius. *Science, 174*, 709.

Burkholder, T.J., Fingado, B., Baron, S., et al. (1994). Relationship between muscle fiber types and sizes and muscle architectural properties in the mouse hindlimb. *J Morphol, 221*, 177–190.

Chilibeck, P.D., Calder, A.W., Sale, D.G., et al. (1998). A comparison of strength and muscle mass increases during resistance training in young women. *Eur J Appl Physiol Occup Physiol, 77*(1–2), 170–175.

Christensen, B., Dyrberg, E., Aagaard, P., et al. (2008). Effects of long-term immobilization and recovery on human triceps surae and collagen turnover in the Achilles tendon in patients with healing ankle fracture. *J Appl Physiol, 105*, 420–426.

Ciullo, J.V., Zarins, B. (1983). Biomechanics of the musculotendinous unit: Relation to athletic performance and injury. *Clin Sports Med, 2*, 71.

Clark, B.C., Fernhall, B., Ploutz-Snyder, L.L. (2006). Adaptations in human neuromuscular function following prolonged unweighting: I. Skeletal muscle contractile properties and applied ischemia efficacy. *J Appl Physiol, 101*, 256–263.

Clarkson, P.M., Hubal, M.J. (2002). Exercise-induced muscle damage in humans. *Am J Phys Med Rehabil, 81, 11*(Suppl), S52–S69.

Conley, K.E., Esselman, P.C., Jubrias, S.A., et al. (2000). Ageing, muscle properties and maximal O(2) uptake rate in humans. *J Physiol*. Jul 1;526 Pt 1:211–7.

Costill, P.L., Coyle, E.F., Fink, W.F., et al. (1979). Adaptations in skeletal muscles following strength training. *J Appl Physiol, 46*, 96.

Crawford, C.N.C., James, N.T. (1980). The design of muscles. In R. Owen, J. Goodfellow, P. Bullough (Eds.). *Scientific Foundations of Orthopaedics and Traumatology.* London: William Heinemann, 67–74.

D'Antona, G., Pellegrino, M.A., Adami, R., et al. (2003). The effect of ageing and immobilization on structure and function of human skeletal muscle fibres. *J Physiol, 552*(2), 499–511.

Enoka, R.M. (1997). Neural adaptations with chronic physical activity. *J Biomech, 30*(5), 447–455.

Eriksson, E., Haggmark, T., Kiessling, K.H., et al. (1981). Effect of electrical stimulation on human skeletal muscle. *Int Sports Med, 2*, 18.

Fink, B., Egl, M., Singer, J., et al. (2007). Morphologic changes in the vastus medialis muscle in patients with osteoarthritis of the knee. *Arthritis Rheum, 56*(11), 3626–3633.

Fitts, R.H., Romatowski, J.G., Blaser, C., et al. (2000). Effect of spaceflight on the isotonic contractile properties of single skeletal muscle fibers in the rhesus monkey. *J Gravit Physiol, 7*(1), S53–S54.

Frontera, W.R., Suh, D., Krivickas, L.S., et al. (2000). Skeletal muscle fiber quality in older men and women. *Am J Physiol Cell Physiol, 279*, 611–618.

Fukunaga, T., Miyatani, M., Tachi, M., et al. (2001). Muscle volume is a major determinant of joint torque in humans. *Acta Physiol Scand, 172*, 240–255.

Galler, S., Hilber, K. (1998). Tension/stiffness ratio of skinned rat skeletal muscle fiber types at various temperatures. *Act Physiol Scand, 162*(2), 119–126.

Gollnick, P.D. (1982). Relationship of strength and endurance with skeletal muscle structure and metabolic potential. *Int J Sports Med, 3*(Suppl 1), 26.

Gondin J., Guette, M., Ballay, Y., et al. (2005). Electromyostimulation training effects on neural drive and muscle architecture. *Med Sci Sports Exerc, 37*(8), 1291–1299.

Gossen, E.R., Allingham, K., Sale, D.G. (2001). Effect of temperature on post-tetanic potentiation in human dorsiflexor muscles. *Can J Physiol Pharmacol, 79*, 49–58.

Guyton, A.C. (1986). *Textbook of Medical Physiology* (7th ed.). Philadelphia: WB Saunders.

Harmer, A., McKenna, M., Sutton, J., et al. (2000). Skeletal muscle metabolic and ionic adaptations during intense exercise following sprint training in humans. *J Appl Physiol, 89*, 1793–1803.

Hill, A.V. (1970). *First and Last Experiments in Muscle Mechanics.* Cambridge: Cambridge University Press.

Hill, D.K. (1968). Tension due to interaction between the sliding filaments of resting striated muscle. The effect of stimulation. *J Physiol, 199*, 637.

Hislop, H.J., Perrine, J. (1967). The isokinetic concept of exercise. *Phys Ther, 47*, 114.

Hong, J.H., Kim, H.J., Kim, K.J., et al. (2008). Comparison of metabolic substrates between exercise and cold exposure in skaters. *J Physiol Anthropol, 27*, 273–281.

Hortobagyi, T., Dempsey, L., Fraser, D., et al. (2000). Changes in muscle strength, muscle fibre size and myofibrillar gene expression after immobilization and retraining in humans. *J Physiol, 524*(1), 293–304.

Huxley, A.F. (1974). Muscular contraction. *J Physiol, 243*, 1.

Huxley, A.F., Huxley, H.E. (1964). A discussion on the physical and chemical basis of muscular contraction. Introductory remarks. *Proc R Soc Lond B Biol Sci, 160*, 433.

Jacobs, P.L., Burns, P. (2009). Acute enhancement of lower-extremity dynamic strength and flexibility with whole-body vibration. *J Strength Cond Res,23*(1):51–75.

Jacobs, I., Romet, T., Kerrigan-Brown, D. (1985). Muscle glycogen depletion during exercise at 9 degrees Celsius and 21 degrees Celsius. *Eur J Appl Physiol, 54*, 35–39.

Jones, S.W., Hill, R.J., Krasney, P.A., et al. (2004). Disuse atrophy and exercise rehabilitation in humans profoundly affects the expression of genes associated with the regulation of skeletal muscle mass. *FASEB J, 18*(9), 1025–1027.

Kannus, P., Jozsa, L., Järvinen, T.L., et al. (1998b). Free mobilization and low- to high-intensity exercise in immobilization-induced muscle atrophy. *J Appl Physiol, 84*(4), 1418–1424.

Kannus, P., Jozsa, L., Kvist, M., et al. (1998a). Effects of immobilization and subsequent low- and high-intensity exercise on morphology of rat calf muscles. *Scand J Med Sci Sports, 8*(3), 160–171.

Kasser, J.R. (1996). General knowledge. In J.R. Kasser (Ed.). *Orthopaedic Knowledge Update 5: Home study syllabus.* Rosemont, IL: American Academy of Orthopaedic Surgeons.

Kimura, T., Hamada, T., Ueno, L., et al. (2003). Changes in contractile properties and neuromuscular propagation evaluated by simultaneous mechanomyogram and electromyogram during experimentally induced hypothermia. *J Electromyog Kinesiol, 13*, 433–440.

Komi, P.V. (1986). The stretch-shortening cycle and human power output. In N.L. Jones, N. McCartney, A.J. McConas (Eds.). *Human Muscle Power.* Champaign, IL: Human Kinetics Publishers, 27–39.

Kroll, P.G. (1987). *The effect of previous contraction condition on subsequent eccentric power production in elbow flexor muscles.* Doctoral dissertation, New York University, New York.

Kumagai, K. (2000). Sprint performance is related to muscle fascicle length in male 100-m sprinters. *J Appl Physiol, 88,* 811–816.

Labarque, V.L., Op't Eijnde B., Van Leemputte, M. (2002). Effect of immobilization and retraining on torque-velocity relationship of human knee flexor and extensor muscles. *Eur J Appl Physiol, 86,* 251–257.

Lieber, R.L., Bodine-Fowler, S.C. (1993). Skeletal muscle mechanics. Implications for rehabilitation. *Phys Ther, 73*(12), 844–856.

Lieber R.L., Jacobson, M.D., Fazeli, B.M., et al. 2001). Architecture of selected muscles of the arm and forearm: Anatomy and implications for tendon transfer. *J Hand Surg, 17A,* 787–798.

Linke, W.A., Ivemeyer, M., Mundel, P., et al. (1998). Nature of PEVK-titin elasticity in skeletal muscle. *Proc Natl Acad Sci U S A, 95*(14), 8052–8057.

Luciano, D.S., Vander, A.J., Sherman, J.H. (1978). *Human Function and Structure.* New York: McGraw-Hill, 113–136.

Marin, P., Rhea, M. (2010). Effects of vibration training in muscle strength. A meta-analysis. *J Strength Cond Res, 24*(2), 548–556.

Munsat, T.L., NcNeal, D., Waters, R. (1976). Effects of nerve stimulation on human muscle. *Arch Neurol, 33,* 608.

Narici M.V., Maganaris, C.N., Reeves, N.D., et al. (2003). Effect of aging on human muscle architecture. *J Appl Physiol, 95,* 2229–2234.

Nomura, T., Kawano, F., Kang, M.S., et al. (2002). Effects of long-term cold exposure on contractile properties in slow- and fast-twitch muscles of rats. *Japanese J Physiol, 52,* 85–93.

Ohira, Y., Yasui, W., Roy, R.R., et al. (1997). Effects of muscle length on the response to unloading. *Acta Anat (Basel), 159*(2–3), 90–98.

Ottoson, D. (1983). *Physiology of the Nervous System.* New York: Oxford University Press, 78–116.

Phillips, C.A., Petrofsky, J.S. (1983). *Mechanics of Skeletal and Cardiac Muscle.* Springfield, IL: Charles C Thomas Publishers.

Pierno, S., Desaphy, J.F., Liantonio, A., et al. (2007). Disuse of rat muscle in vivo reduces protein kinase C activity controlling the sarcolemma chloride conductance. *J Physiol, 584*(3), 983–995.

Pietrangelo, T., Mancinelli, R., Toniolo, L., et al. (2009). Effects of local vibrations on skeletal muscle trophism in elderly people: Mechanical, cellular, and molecular events. *Int J Mol Med, 24,* 503–512.

Powers, S.K., Kavazis, A.N., McClung, J.M. (2007). Oxidative stress and disuse muscle atrophy. *J Appl Physiol, 102*(6), 2389–2397.

Reisman, S., Allen. T., Proske, U. (2009). Changes in passive tension after stretch of unexercised and eccentrically exercised human plantar flexor muscles. *Exp Brain Res, 193,* 545–554.

Saltin, B., Henriksson, J., Nygaard, E., et al. (1977). Fiber types and metabolic potentials of skeletal muscles in sedentary man and endurance runners. *Ann NY Acad Sci, 301,* 3.

Sandmann, M.E., Shoeman, J.A., Thompson, L.V. (1998). The fiber-type-specific effect of inactivity and intermittent weight-bearing on the gastrocnemius of 30-month-old rats. *Arch Phys Med Rehabil, 79*(6), 658–662.

Schiaffino, S., Serrano, A. (2002). Calcineurin signaling and neural control of skeletal muscle fiber type and size. *Trends Pharmacol Sci, 23*(12), 569–575.

Squire, J.M. (1997). Architecture and function in the muscle sarcomere. *Curr Opin Struct Biol, 7*(2), 247–257.

Stein, T.P., Wade, C.E. (2005). Metabolic consequences of muscle disuse atrophy. *J Nutr, 135*(7), 1824S–1828S.

Stromer, M.H. (1998). The cytoskeleton in skeletal, cardiac and smooth muscle cells. *Histol Histopathol, 13*(1), 283–291.

Suetta, C., Aagaard, P., Magnusson, S.P., et al. (2007). Muscle size, neuromuscular activation, and rapid force characteristics in elderly men and women: Effects of unilateral long-term disuse due to hip-osteoarthritis. *J Appl Physiol, 102,* 942–948.

Takarada, Y., Iwamoto, H., Sugi, H., et al. (1997). Stretch-induced enhancement of mechanical work production in long frog single fibers and human muscle. *J Appl Physiol, 83*(5), 1741–1748.

Tee, J.C., Bosch, A.N., Lambert, M.I. (2007). Metabolic consequences of exercise-induced muscle damage. *Sports Med, 37*(10), 827–836.

Thom J.M., Morse, C.I., Birch, K.M., et al. (2007). Influence of muscle architecture on the torque and power-velocity characteristics of young and elderly men. *Eur J Appl Physiol,100,* 613–619.

Udaka, J., Ohmori, S., Terui, T., et al. (2008). Disuse-induced preferential loss of the giant protein titin depresses muscle performance via abnormal sarcomeric organization. *J Gen Physiol, 131*(1), 33–41.

Warhol, M.J., Siegel, A.J., Evans, W., et al. (1985). Skeletal muscle injury and repair in marathon runners after competition. *Am J Pathol, 118,* 331–339.

Wilkie, D.R. (1956). The mechanical properties of muscle. *Br Med Bull, 12,*177.

Wilkie, D.R. (1968). *Muscle.* London, UK: Edward Arnold.

FLOW CHART 6-1

Structure and organization of the skeletal muscle.*

*This flowchart is designed for classroom or group discussion. Flow chart is not meant to be exhaustive.

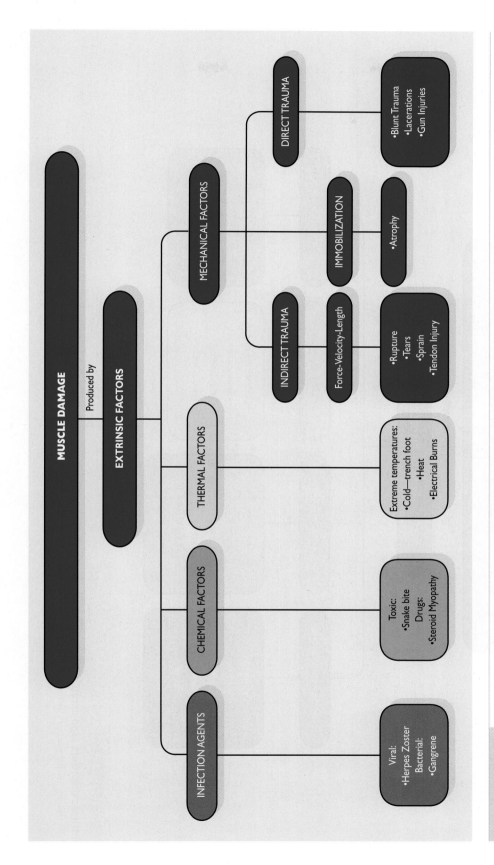

Extrinsic factors associated with muscle damage. Clinical examples.*

*This flowchart is designed for classroom or group discussion. Flow chart is not meant to be exhaustive.

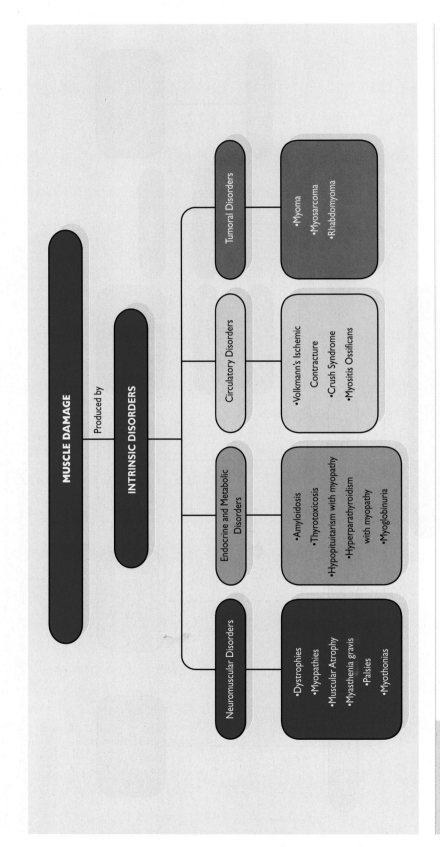

FLOW CHART 6-3

Intrinsic disorders associated with muscle damage. Clinical examples.*

*This flowchart is designed for classroom or group discussion. Flow chart is not meant to be exhaustive.

PART 2

Biomechanics of Joints

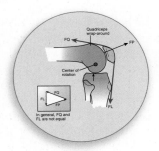

CHAPTER 7

Biomechanics of the Knee

Victor H. Frankel, Margareta Nordin, and Peter S. Walker

Introduction

The knee joint transmits loads, facilitates positions and movements of the body, aids in the conservation of momentum, and provides the necessary moments for activities involving the leg. The human knee, the largest and perhaps most complex joint in the body, is principally a two-joint structure composed of the tibiofemoral joint and the patellofemoral joint (Fig. 7-1). The tibiofibular joint has a valuable role but does not participate in motion. The knee sustains high forces and moments and is situated between the body's two longest lever arms, the femur and the tibia, making it particularly susceptible to injury. As well as dealing with the knee specifically, this chapter also introduces basic terms, explains the methods, and demonstrates the calculations necessary for analyzing joint motion and the forces and moments acting on a joint. This methodology is applied to other joints in subsequent chapters.

The knee is particularly well suited for demonstrating biomechanical analyses of joints because these analyses can be simplified in the knee and still yield useful data. Although knee motion occurs simultaneously in three planes, the motion in the sagittal plane dominates so that it accounts for nearly all of the motion. Also, although many muscles produce forces on the knee, at any particular instant the quadriceps muscle group predominates, generating a force that accounts for most of the muscle force acting on the knee. Thus, basic biomechanical analyses can be limited to motion in one plane and to the force produced by a single muscle group and still give an understanding of knee motion and an estimation of the magnitude of the principal forces and moments on the knee. Advanced biomechanical dynamic analyses of the knee joint that include all soft tissue structures are complex and continue to be investigated.

This chapter will be divided into two parts: kinematics and kinetics. Kinematics is the branch of mechanics

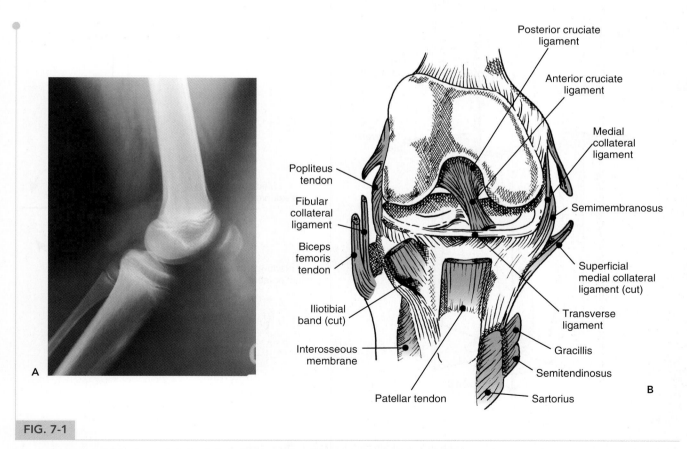

FIG. 7-1

Two-joint structure of the knee. **A.** Lateral view of a knee joint with open growth plates, the femur at the top, the tibia beneath. The fibula can be seen posterior to the tibia. **B.** Anterior view without patella. The lateral and medial menisci are located on the upper surface of the tibia.

that deals with the motion of a body without reference to force or mass. However, a basic understanding of the kinematics requires consideration of the different structures of the joint, and hence this will be included in this section. Kinetics is the branch of mechanics that deals with the motion of a body under the action of forces and moments. This section will deal with these forces and the methods by which these data were determined. In addition, some of the effects of the forces on the structures of the knee will be discussed.

Kinematics

Kinematics describes the motion of a joint in three planes: frontal (coronal or longitudinal), sagittal, and transverse (horizontal) (Figs. 7-2A and B). Clinical measurements of joint range of motion define the anatomic position where the knee is in a relaxed standing position as a zero position for measurement. This taxonomy regarding reference axes and positions will be used for joint motion throughout this book. Other taxonomies and reference

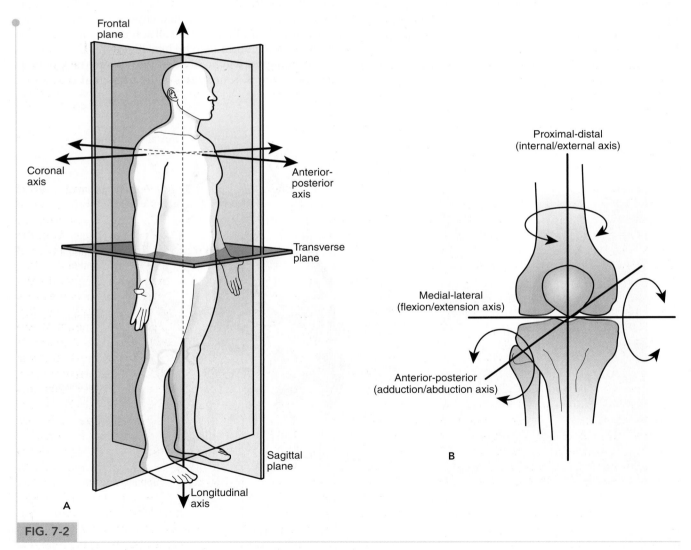

FIG. 7-2

A. Frontal (coronal or longitudinal), sagittal, and transverse (horizontal) planes in the human body.
B. Depiction and nomenclature of the six degrees of freedom of knee motion: anterior posterior translation, medial-lateral translation, proximal-distal translation, flexion-extension rotation, internal-external rotation, and varus-valgus rotation.

systems exist (Andriacchi et al., 1979; Grood and Suntay 1983; Kroemer et al., 1990; Özkaya and Nordin, 1999), but the anatomic reference system by far is the most commonly used among clinicians. Of the two joints composing the knee, the tibiofemoral joint lends itself particularly well to an analysis of joint motion. Analysis of the relative rolling and sliding at the joint surfaces can be performed from the overall motion and the geometry of the surfaces. Any impediment to the range of motion or the surface joint motion will disturb the normal loading pattern of a joint and result in adverse consequences. For example, a torn meniscus will cause abnormalities in both tibiofemoral and patellofemoral motions and can lead to later joint degeneration.

RANGE OF MOTION

The range of motion of any joint can be measured in any plane. Approximate measurements can be made with a goniometer, where the arms of the goniometer are lined up with the estimated long axes of the femur and tibia. More exact measurements require the use of methods such as electrogoniometry, roentgenography, fluoroscopy, stereophotogrammetry, or photographic and video techniques using skin markers. The femoral axis is defined as a line between the center of the femoral head and the center of the distal condyles: The tibial axis joins the center of the proximal tibia to the center of the ankle. It is noted that it is difficult to define the axes of long bones as the center lines of the shafts, due to the bowing that is generally present in the sagittal and frontal planes.

In the tibiofemoral joint, motion takes place in all three planes, but the range of motion is greatest in the sagittal plane. Motion in this plane from full extension to full flexion of the knee is typically from 3° of hyperextension (–3° flexion) to 155° of flexion. At maximum flexion, the medial posterior femoral cortex impacts the posterior horn of the meniscus. Thigh-calf contact is usually the major factor in limiting flexion. On the other hand, in cultures where deep kneeling is common, flexion angles can reach beyond 155°, at which point the femoral condyles actually lever over the posterior tibial condyles. In measuring flexion angle at the extremes, it is important to distinguish between active motion and passive motion (Dennis et al., 1998). Active motion is when the subject actively applies muscle forces to reach the extremes of motion, while lying at the side or seated with the leg hanging down. Passive motion is that achieved by the measurer applying slight force at the extremes of flexion and extension. Passive motion is usually 5° to 10° more than active, hence studies involving flexion angles should carefully define the method of measurement.

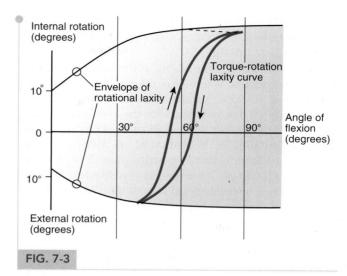

FIG. 7-3

The torque-rotation laxity curve describes the freedom of internal-external rotation about the long axis of the tibia at a particular flexion angle, 60° in the curve shown. Connecting the laxity values for the full flexion range gives the envelope of rotational laxity of the knee.

Motion in the transverse plane, internal and external rotation, is variable during function and can be thought of as freedom of rotation. This term is scientifically described as laxity (Fig. 7-3). Each laxity curve is in the form of a hysteresis loop that reflects the properties of the soft tissues, such as ligaments, capsule, and menisci, which restrict the motion at the extremes. First, the tissues are strain stiffening, such that at higher elongation, their stiffness is higher. Second, the tissues are viscoelastic, meaning that the elongation is time dependent and there is a time lag in recovering their original shape. At any angle of flexion, if an internal and then an external torque is applied to the tibia, there is a rotational laxity in each direction with a limit at each extreme within normal physiologic torques. Relative to the neutral position, the angles of rotation at the range of flexion angles are called internal and external rotational laxity. With the knee in full extension (or hyperextension), the rotational laxity is restricted by the interlocking of the femoral and tibial condyles. This occurs mainly because the medial femoral condyle is longer than the lateral condyle; it also occurs when the collateral ligaments, the anterior cruciate, and the posterior capsule are tightened. The range of rotational laxity increases as the knee is flexed, reaching a maximum at 30° to 40° of flexion; with the knee in this position, external tibial rotation is approximately 18° and internal rotation is approximately 25° (Blankevoort et al., 1988). Beyond 40° of flexion, the range of internal and external rotation remains constant up to about 120° flexion and then diminishes again up to full flexion due to soft tissue tightening.

Motion in the frontal plane, abduction (varus) and adduction (valgus), is similarly affected by the amount of joint flexion. Note that the varus and valgus motions of the femur are defined relative to the axes in the frontal plane of the tibia. Full extension of the knee precludes almost all motion in the frontal plane. Passive abduction and adduction increase with knee flexion up to 30°, but each reaches a maximum of only a few degrees. With the knee flexed beyond 30°, motion in the frontal plane again decreases because of the limiting function of the soft tissues. Varus rotation is greater than valgus, especially in flexion because of the higher stiffness of the medial collateral ligament than the lateral collateral. However, during function, axial forces and muscle action around the knee usually prevent varus and valgus rotations, although short periods of lift-off of the lateral or medial femoral condyle can occur.

The range of tibiofemoral joint motion required for the performance of various physical activities can be determined from kinematic analysis. Motion in the knee during walking has been measured in all planes. The range of motion in the sagittal plane during level walking was measured with an electrogoniometer (Kettelkamp et al., 1970; Lamoreaux, 1971; Murray et al., 1964) (Fig. 7-4). Full or nearly full extension was noted at the beginning of the stance phase (0% of cycle) at heel strike. As weight-bearing was applied, the angle of flexion increased to approximately

TABLE 7-1
Range of Tibiofemoral Joint Motion in the Sagittal Plane during Common Activities

Activity	*Range of Motion from Knee Extension to Knee Flexion (Degrees)*
Walking	0–67[a]
Climbing stairs	0–83[b]
Descending stairs	0–90
Sitting down	0–93
Tying a shoe	0–106
Lifting an object	0–117

[a]Mean for 22 subjects. A slight difference was found between right and left knees (mean for right knee 68.1°, mean for left knee 66.7°). Data are from Kettelkamp, D.B., Johnson, R.J., Smidt, G.L., et al. (1970). An electrogoniometric study of knee motion in normal gait. *J Bone Joint Surg Am, 52,* 775.

[b]Mean for 30 subjects. These and subsequent data are from Laubenthal, K.N., Smidt, G.L., Kettelkamp, D.B. (1972). A quantitative analysis of knee motion during activities of daily living. *Phys Ther, 52,* 34.

15°, followed by extension back to almost 0°. Flexion then increased rapidly to begin the swing phase. Maximum flexion, approximately 60°, was observed during the first part of the swing phase (see Chapter 17, Biomechanics of Gait, for more detailed information).

Values for the range of motion of the tibiofemoral joint in the sagittal plane during several common activities are presented in Table 7-1. Maximal knee flexion occurs during lifting an object from the ground. A range of motion from full extension to at least 117° of flexion appears to be required for an individual to carry out the activities of daily living in a normal manner. However, there are additional activities that can be considered biomechanically demanding, such as squatting and kneeling, which require even higher flexion angles. In studying the range of tibiofemoral joint motion during walking and other activities, researchers found that an increased speed of movement requires a greater range of motion in the tibiofemoral joint (Holden et al., 1997; Perry et al., 1977). As the pace accelerates from walking slowly to running, progressively more knee flexion is needed during the stance phase (Table 7-2).

SURFACE JOINT MOTION

Surface joint motion, which is the motion between the articulating surfaces of a joint, can be described for any joint in any plane with the use of stereophotogrammetric methods (Selvik, 1978, 1983). Because these methods are highly technical and complex, a simpler method that

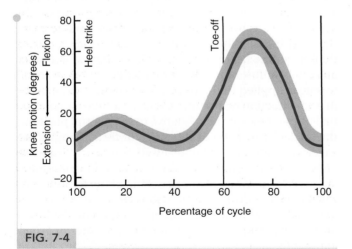

FIG. 7-4

Range of motion of the tibiofemoral joint in the sagittal plane during level walking in one gait cycle. The shaded area indicates variation among 60 subjects (age range 20–65 years). *Adapted from Murray, M.P., Drought, A.B., Kory, R.C. (1964). Walking patterns of normal men. J Bone Joint Surg, 46A, 335.*

TABLE 7-2

Amount of Knee Flexion during Stance Phase of Walking and Running

Activity	Range (Degrees)
Walking	
Slow	0–6
Free	6–12
Fast	12–18
Running	18–30

Range for seven subjects. Data are from Perry, J., Norwood, L., House, K. (1977). Knee posture and biceps and semimembranosus muscle action in running and cutting (an EMG study). *Trans Orthop Res Soc, 2, 258.*

evolved in the nineteenth century is still used (Reuleaux, 1876). This method, called the instant center technique, allows surface joint motion to be analyzed in the sagittal and frontal planes but not in the transverse plane. The instant center technique provides a description of the relative uniplanar motion of two adjacent segments of a body and the direction of displacement of the contact points between these segments.

The skeletal portion of a body segment is called a link. As one link rotates about the other, at any instant there is a point that does not move, that is, a point that has zero velocity. As an example, consider rising from a chair. Consider the tibia as the fixed link, and the femur the moving link. For an arc of motion seen in the sagittal plane, there is a point in the femur that remains fixed. This point constitutes an instantaneous center of motion, or instant center. The instant center is found by identifying the displacement of two points on the moving link as the link moves from one position to another in relation to the adjacent fixed link. The points on the moving link in its original position and in its displaced position are designated on a graph, and lines are drawn connecting the two pairs of points. The perpendicular bisectors of these two lines are then drawn. The intersection of the perpendicular bisectors is the instant center.

Clinically, a pathway of the instant center for a joint can be determined by taking successive roentgenograms of the joint in a sequence of positions, such as 10° apart, throughout the range of motion in one plane and applying the Reuleaux method for locating the instant center for each interval of motion.

When the instant center pathway has been determined for joint motion in one plane, the surface joint motion can be described. For each interval of motion, the point at which the joint surfaces make contact is located on the roentgenograms used for the instant

center analysis, and a line is drawn from the instant center to the contact point. A second line drawn at right angles to this line indicates the instantaneous direction of displacement of the contact point. The direction of displacement of these points throughout the range of motion describes the surface motion in the joint. In most joints, the instant centers lie at a distance from the joint surface, and the line indicating the direction of displacement of the contact points is tangential to the load-bearing surface, demonstrating that one joint surface is sliding on the other surface. In the case in which the instant center is located on the surface, the joint has a rolling motion and there is no sliding. Because the instant center technique allows a description of motion in one plane only, it loses accuracy if there is significant rotation in another plane. For example, if there is as much as 15° to 20° of internal-external rotation during the flexion range, the instant center data would be suspect. However this can be countered by choosing reference points on the roentgenograms that are not sensitive to this rotation as seen in the sagittal plane.

In the knee, surface joint motion occurs between the tibial and femoral condyles and between the femoral condyles and the patella. In the tibiofemoral joint, surface motion takes place primarily in the anterior-posterior direction. Surface motion in the patellofemoral joint takes place in two planes simultaneously, the frontal and transverse, but is far greater in the frontal plane.

Tibiofemoral Joint

An example will illustrate the use of the instant center technique to describe the surface motion of the tibiofemoral joint in the sagittal plane. To determine the pathway of the instant center of this joint during flexion, a lateral roentgenogram is taken of the knee in full extension and successive films are taken at 10° intervals of increased flexion. Care is taken to keep the tibia parallel to the x-ray table and to prevent rotation about the femur.

Two points on the femur that are easily identified on all roentgenograms are selected and designated on each roentgenogram (Fig. 7-5A). The films are then compared in pairs, with the images of the tibiae superimposed on each other. Lines are drawn between the points on the femur in the two positions, and the perpendicular bisectors of these lines are then drawn. The point at which these perpendicular bisectors intersect is the instant center of the tibiofemoral joint for each 10° interval of motion (Fig. 7-5B). It will be noted that the movement of the point close to the instant center moves only a small distance, and the opposite for the point distant from the center. The movement is given by the equation

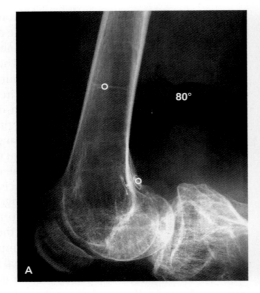 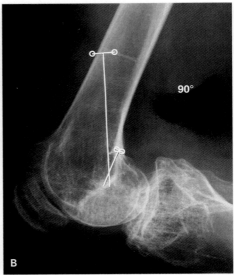

FIG. 7-5

Locating the instant center. **A.** Two easily identifiable points on the femur, shown by *circles*, are designated on a roentgenogram of a knee flexed 80°. **B.** This roentgenogram is compared with a roentgenogram of the knee flexed 90°, on which the same two points have been indicated. The images of the tibia are superimposed, and lines are drawn connecting each pair of points. The perpendicular bisectors of these two lines are then drawn. The point at which these perpendicular bisectors intersect is the instant center of the tibiofemoral joint for the motion between 80° and 90° of flexion. Courtesy of Ian Goldie, M.D., University of Gothenburg, Gothenburg, Sweden.

$s = r\,\theta$ where s is the distance, r the radius to the instant center, and θ the range of flexion in radians (1 radian = 57.3°). The instant center pathway throughout the entire range of knee flexion and extension can then be plotted. In a normal knee, the instant center pathway for the tibiofemoral joint is semicircular. The reason is that the radius of curvature of the femoral condyles gradually reduces from the distal end, which articulates at low flexion angles, to the posterior-superior, which articulates in high flexion. In addition, the motion at the articulating surfaces is a combination of rolling and sliding, lowering the instant centers in the femur toward the contact point.

After the instant center pathway has been determined for the tibiofemoral joint, the surface motion can be described. On each set of superimposed roentgenograms, the point of contact of the tibiofemoral joint surfaces is determined and a line is drawn connecting this point with the instant center. A second line drawn at right angles to this line indicates the direction of displacement at the contact points. In a normal knee, this line is tangential to the surface of the tibia for each interval of motion from full extension to full flexion, demonstrating that the femur is sliding on the tibial condyles (Frankel et al., 1971) (Fig. 7-6). During normal knee motion in

the sagittal plane from full extension to full flexion, the instant center pathway of the midsagittal plane moves posteriorly, indicating a combination of rolling and sliding between the articular surfaces (Figs. 7-6A and B).

The unique mechanism prevents the femur from rolling off the posterior aspect of the tibia plateau as the knee goes into increased flexion (Draganich et al., 1987; Fu et al., 1994; Kapandji, 1970). The motion shown in Figure 7-6B is characteristic of the medial side of the knee where the anterior-posterior displacement of the femur on the tibia is small, and there is almost complete sliding of the femur on the tibia. If there were pure rolling, the femoral condyle would displace off the posterior of the tibial plateau (Fig. 7-6C). Figure 7-6D represents the lateral side where the contact point displaces to the very posterior of the tibia by a combination of rolling and sliding. The mechanism that prevents complete roll-off is the link formed between the tibial and femoral attachment sites of the anterior and posterior cruciate ligaments and the geometry of the femoral condyles (Fu et al., 1994).

A sagittal plane model that has been used to explain the function of the cruciate ligaments is the four-bar linkage (Fig. 7-7) (O'Connor et al., 1989; Zavatsky and O'Connor, 1992). The four bars are the line PA on the tibia, the anterior cruciate AA, the posterior cruciate

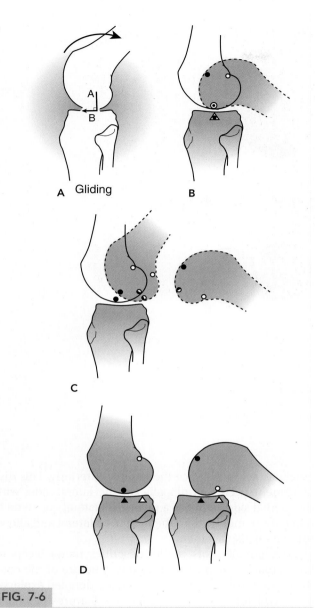

FIG. 7-6

A. In a normal knee, a line drawn from the instant center of the tibiofemoral joint to the tibiofemoral contact point (line A) forms a right angle with a line tangential to the tibial surface (line B). The *arrow* indicates the direction of displacement of the contact points. Line B is tangential to the tibial surface, indicating that the femur slides on the tibial condyles during the measured interval of motion. **B.** Pure sliding of the femur on the tibia with knee extension. Note that the contact point of the tibia does not change as the femur slides over it. Eventually impingement would occur if all surface motion were restricted to sliding. *Round points* delineate contact points at the femur and *triangles* delineate contact points at the tibia. **C.** Pure rolling of the femur on the tibia with knee flexion. Note that both the tibia and the femoral contact points change as the femur rolls on the tibia. Also note that with moderate flexion, the femur will begin to roll off the tibia if surface motion is restricted to rolling. **D.** Actual knee motion including both sliding and rolling.

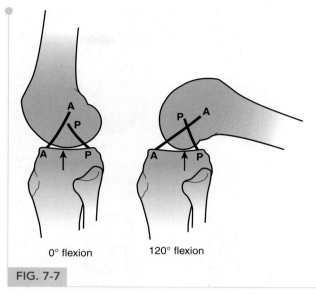

FIG. 7-7

The four-bar linkage model for knee motion in the sagittal plane. The model predicts posterior femoral displacement with flexion (*arrows*) and shows how the cruciates provide anterior and posterior stability throughout flexion. *AA*, anterior cruciate; *PP*, posterior cruciate.

PP, and the line PA on the femur. For the purpose of the model, all bars are assumed to be in a single plane and to be of constant length throughout the motion. This is a reasonable approximation for the cruciates even though they are made up from bands of fibers rather that straight lines and their lengths change by approximately 5% during flexion-extension (Girgis et al., 1975). Nevertheless, as shown in the figure, as the knee is flexed from 0° to 120°, the model predicts that the contact point between the femur and the tibia, shown by arrows, will displace posteriorly. If the average of the motion of the lateral and medial femoral condyles is taken, this prediction is correct. In reality, the medial condyle does not displace, while the lateral condyle displaces to the very posterior.

The model also shows that at all angles of flexion, the posterior cruciate ligament (PCL) limits anterior displacement of the femur on the tibia, while the anterior cruciate ligament (ACL) prevents posterior displacement. Both of the cruciates will prevent distraction between the femur and tibia. Although the four-bar linkage model provides an understanding of the control of the anterior-posterior displacements between the femur and the tibia, a more elaborate model is needed to explain three-dimensional phenomena.

Such a three-dimensional kinematic model is shown in Figure 7-8. The femoral condyles are shown as spherical surfaces that have been shown to be valid from approximately 0° to 120° flexion (Kurosawa et al.,

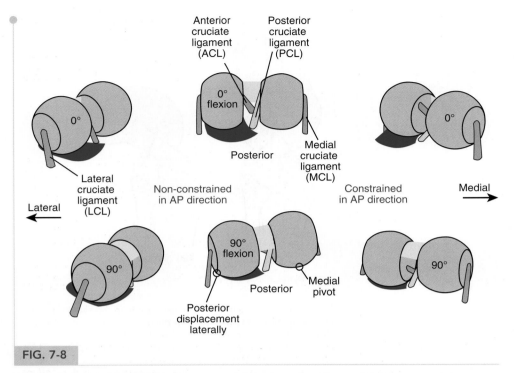

FIG. 7-8

A three-dimensional model of the knee to explain knee motion. The medial side acts as a pivot. The lateral femoral condyle displaces posteriorly with flexion, guided by the cruciate ligaments. *ACL*, anterior cruciate ligament; *PCL*, posterior cruciate ligament; *LCL*, lateral collateral ligament; *MCL*, medial collateral ligament.

1985). The medial tibial condyle is shown dished with a radius only a few millimeters larger than that of the femoral condyle. This tibial dishing represents the shallow dishing of the tibial condyle itself in combination with a relatively immobile medial meniscus. The lateral tibial condyle has a curved trough to allow unrestricted anteroposterior (AP) displacement of the femoral condyle as the femur pivots about the medial side. Although there is a lateral meniscus in close conformity with the femoral condyle, in terms of kinematics it has little influence because of its AP mobility on the upper tibia.

The collateral ligaments are shown as flat bars, the medial having a much larger cross-section than the lateral. Their general orientation is vertical, in the sagittal plane. The cruciate ligaments are shown as rod-shaped and are angled to the horizontal from 30° to 60° in the sagittal plane, but are also angled in the frontal plane (Girgis et al., 1975). When the knee is at 0° flexion, the lateral femoral condyle is anterior to the center of the tibia, the position being determined mainly by the cruciates. The four-bar linkage model described previously is also helpful in relation to the three-dimensional model. As the knee is flexed, the medial constraint and the stiff medial collateral ligament limit AP displacement. However, the

cruciates act to displace the femur posteriorly. This displacement occurs principally on the lateral side, with the medial side acting as a pivot. The lateral collateral is much more mobile than the medial collateral and allows this lateral displacement.

Conceptually, this model describes many facets of knee kinematics in relation to the structure of the condyles and the ligaments. A much more detailed explanation involving many more anatomic features has been described (Freeman and Pinskerova, 2005). The kinematics in extension are complex and are characterized by a "screw-home" mechanism, where the tibia rotates externally (Fig. 7-9) and the contact points shift anteriorly, acting as a brake to further extension and providing a stable position of the knee. This action is caused by the large sagittal radii of the distal-anterior femoral condyles, the shallow anterior lateral tibia, and the upsweep on the anterior medial tibia. All of the ligaments tighten in this position.

The model shown in Figure 7-8 shows the ligaments as rigid links. If they were, it would probably be impossible to obtain kinematic compatibility between the ligaments and the joint surfaces throughout a flexion range. However, the ligaments are composed of bands

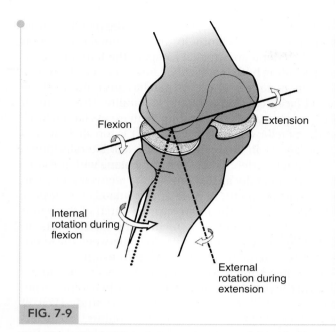

FIG. 7-9

Screw-home mechanism of the tibiofemoral joint. During knee extension, the tibia rotates externally. This motion is reversed as the knee is flexed. This figure shows an oblique view of the femur and tibia. The shaded area indicates the tibial plateau. *Adapted from Helfet, A.J. (1974). Anatomy and mechanics of movement of the knee joint. In A. Helfet (Ed.). Disorders of the Knee. Philadelphia: JB Lippincott Co, 1–17.*

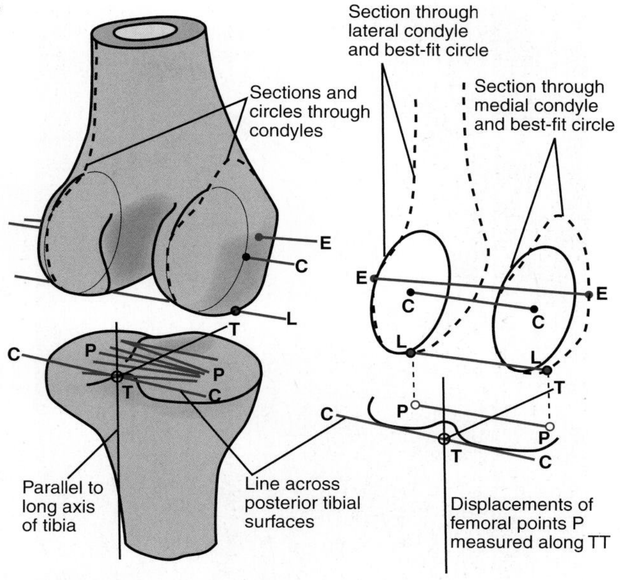

FIG. 7-10

Different possible lines or axes, fixed in the femur and in the tibia. Tracking the movement of the femoral lines (EE, CC, LL) on the horizontal tibial plane allows for an understanding of the motions of the femoral condyles on the tibial surface.

or bundles of fibers with large areas of attachment to the femur and tibia. Therefore during flexion, different parts of the ligaments become tighter or looser. The ligaments themselves can elongate a few millimeters under tension, whereas the cartilage surfaces can deform up to 0.5 to 1.0 mm. These factors allow for guided motion of the knee but with a considerable amount of laxity, especially in axial rotation. Hence, although the knee has a characteristic neutral path of motion, the actual motion depends on the external forces and the activity being performed.

A complete description of three-dimensional knee motion involves axes embedded in the femur and the tibia and the six degrees of freedom determined in a succession of positions. Specific motions can be chosen for special study (Fig. 7-10), however. The initial problem is to select axes based on anatomic landmarks. For the femur, two possible transverse axes are the circular axis CC (Eckoff et al., 2007; Kurosawa et al., 1985,) and the epicondylar axis EE. Studies have shown that the circular axis corresponds better to the contact points on the tibial surface. At any flexion angle, points CC are transferred down to the plane through the tibial surface PP. The axis system in the tibia can be defined as a line across the

posterior condyles CC, an anterior perpendicular line TT, and a vertical line through T. AP coordinates of points PP are then measured along TT while the axial rotation angle is the inclination of PP to CC. The image on the left of Figure 7-10 shows a group of lines PP representing different activities. The anterior line shows how the medial contact has moved anteriorly in hyperextension. The other lines show that during flexion, there is a medial pivot action. In high flexion, however, the medial side displaces posteriorly, levering over the posterior horn of the meniscus. Recent studies of the kinematics of the knee using fluoroscopy have shown this pattern of kinematics for several different activities (Dennis et al., 2001; Komistek et al., 2003; Li et al., 2005). The method involves taking a succession of fluoroscopic images during the activity. A CT (or MRI) scan is then taken and a solid model is generated using special software such as Mimics (Materialise, Leuven, Belgium). The models for the femur and tibia are then rotated in the computer until they match each fluoroscopic image. The relative 3D positions of the femur and tibia are then calculated. This technique is called shape matching, which is also used for the determination of the 3D motion of artificial knee joints.

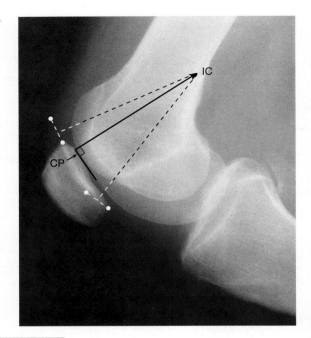

FIG. 7-11

After the instant center (IC) is determined for the patellofemoral joint for the motion from 75% to 90° of knee flexion, a line is drawn from the instant center to the contact point (CP) between the patella and the femoral condyle. A line drawn at right angles to this line is tangential to the surface of the patella, indicating sliding.

Patellofemoral Joint

The surface motion of the patellofemoral joint can be described by the instant center technique (Fig. 7-11). For a 15° range of flexion, the displacements of a superior and inferior point on the patella are marked. The perpendicular bisector of the two lines is drawn, and the intersection point IC is the instant center of rotation. IC is then connected to the contact point CP. The perpendicular line (arrowed) represents the sliding direction. It is noted that for this range of flexion, 75° to 90°, the contact point is superior on the patella. In contrast, in early flexion, the contact point is distal on the patella. This transference distributes the contact areas over the entire patella surface during flexion, and also acts to modify the lever arm of the quadriceps, which is essential for the efficient control of knee motion.

Sectional views through the patella and the femoral condyles are shown in Figure 7-12. In early flexion, the patella usually contacts the femur on the periosteum above its cartilage-bearing surface. This causes no problem because even if the quadriceps force is high, the component of that force compressing the patella against the femur is small. From 30° to 90°, contact occurs on both the lateral and medial facets of the femoral trochlea, providing both medial and lateral stability. However, there are instances where the patella can sublux over the lateral facet, causing pain and instability. This occurs if the Q-angle is too large, leading to an excess lateral force on the patella, a lateral trochlea that is too shallow, and an imbalance in the vastus medialis and lateralis forces. The Q-angle is defined in the frontal plane with the knee in extension as the angle between the rectus femoris and the patella ligament. Beyond approximately 90° flexion, the patella strides the intercondylar notch of the femur and the contact splits into medial and lateral parts. In higher flexion, the patella sinks down between the femoral condyles reducing the quadriceps tension.

It is interesting to study the positions of the contact areas at the patellofemoral and femorotibial joints, especially in high flexion (Fig. 7-13). In this study (Walker et al., 2006), the centers of the contact points were determined for six different knees. Knees were placed in a test machine, and 3D positions were recorded for the full flexion range, and AP and rotational laxities, to encompass a complete spectrum of possible activities. Reconstructions of the shapes of the femur, tibia, and patella in software enabled the centers of the contact areas, the contact points, to be determined. On the medial side of the femur, the tibiofemoral contacts in early flexion are down the centers of the condyle, whereas the patella contacts in high flexion are toward the intercondylar notch, essentially separating these areas. The opposite is the case on the lateral side where there is an overlap between the contacts in high flexion and low flexion. However in both cases, the areas of overlap (or potential overlap) occur only after a flexion angle of approximately 135°. The contacts on the tibial surface are seen to cover almost the entire cartilage surface due to large amounts of AP laxity of about ±2 to 4 mm, rotational laxity of about ±20°, and the lateral femoral rollback in high flexion. The latter accounts for the extreme posterior contacts seen on the lateral side. It will be appreciated that the actual contact areas will need to be accounted for in a full analysis, but these will depend on many factors such as the effect of the menisci and the loads acting across the joint.

Kinetics

Kinetics involves both static and dynamic analysis of the forces and moments acting on a joint. Statics is the study of the forces and moments acting on a body in equilibrium, meaning that the body is at rest or moving

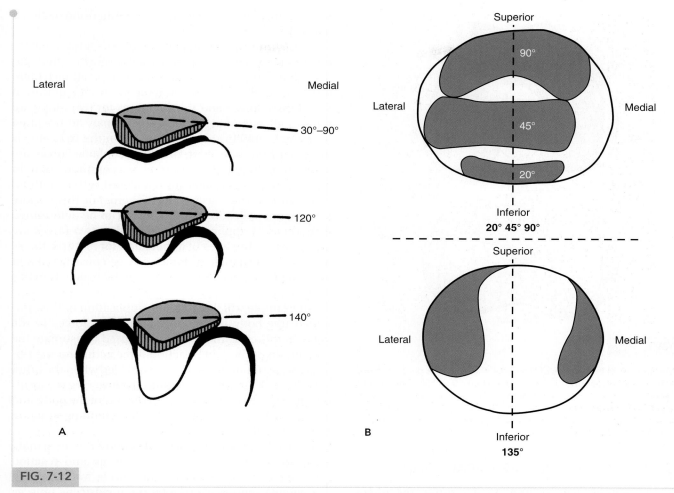

FIG. 7-12

A. The position of the patella at different ranges of knee flexion. Until 90° flexion, contact is on the lateral and medial sides of the femoral trochlea. In high flexion the contact splits into two distinct areas. *Adapted from Goodfellow, J., Hungerford, D.S., Zindel, M. (1976). Patellofemoral joint mechanics and pathology. 1. Functional anatomy of the patellofemoral joint. J Bone Joint Surg, 58B, 287; and Hehne, H.J. (1990). Biomechanics of the patellofemoral joint and its clinical relevance. Clin Orthop, 258, 73–85.* **B.** The contact areas at different flexion angles. Note the gradual superior shift of the contacts with the flexion angle. The split contact in high flexion is evident.

at a constant speed. For a body to be in equilibrium, two conditions must be met: The sum of the forces in any direction must be zero, and the sum of the moments about any point or axis must be zero. These conditions are expressed as $\Sigma F = 0$ and $\Sigma M = 0$.

Dynamics is the study of the forces and moments acting on a body when it is accelerating or decelerating. If the resultant force on the body is not equal to zero, there will be acceleration in the direction of the force: Newton's second law expresses this as $F = ma$ where F is the force, m is the mass, and a is the acceleration. Similarly, a resultant torque will produce an angular acceleration. Kinetic analysis allows the determination of the magnitude of the

moments and forces on a joint produced by body weight, muscle action, soft tissue resistance, and externally applied loads in any situation, either static or dynamic: It identifies those situations that produce excessively high moments or forces.

In this and subsequent chapters, the discussion of statics and dynamics of the joints of the skeletal system concerns the magnitude of the forces and moments acting to move a joint about an axis or to maintain its position. This is called rigid body analysis in that it does not take into account the deforming effect of these forces and moments on the joint structures. Such effects will however be discussed at various points in this chapter.

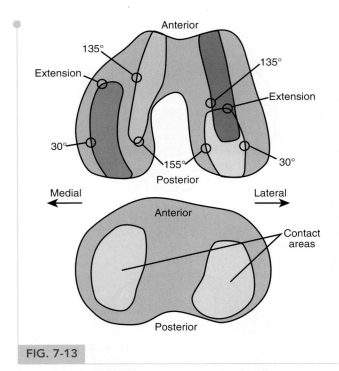

FIG. 7-13

Composites of the centers of the contact areas of the patellofemoral and tibiofemoral joints, for six different knees, 0 to 155° flexion, and for maximum laxity positions in anterior-posterior displacement and internal-external rotation.

STATICS OF THE TIBIOFEMORAL JOINT

Static analysis may be used to determine the forces and moments acting on a joint when no motion takes place or at one instant in time during a dynamic activity such as walking, running, or lifting an object. It can be performed for any joint in any position and under any loading configuration. In such analyses, either graphic or mathematical methods may be used to solve for the unknown forces or moments.

A complete static analysis involving all moments and all forces imposed on a joint in three dimensions is complicated. For this reason, a simplified technique can be used. One such technique is to use a free-body diagram and limit the analysis to one plane, to the main coplanar forces acting on the free body, and to the main moments acting about the joint under consideration. The magnitudes of the forces acting at the joint surfaces or in the muscles, for example, can then be calculated.

When the free-body technique is used to analyze coplanar forces, one portion of the body is isolated from the entire body, and all forces acting on this free body are identified. A diagram is drawn of the free body in the loading situation to be analyzed. The principal coplanar forces acting on the free body are identified and depicted on the free-body diagram.

These forces are designated as vectors if four characteristics are known: magnitude, sense (positive or negative), line of application, and point of application. If there are a total of three forces and if the points of application for all three forces and the directions for two forces are known, all remaining characteristics can be obtained for a force equilibrium situation. When the free body is in equilibrium, the three principal coplanar forces are concurrent; that is, they intersect at a common point. In other words, these forces form a closed system with no resultant and their vector sum is zero. For this reason, the line of application for one force can be determined if the lines of application for the other two forces are known. Once the lines of application for all three forces are known, a triangle of forces can be constructed and the magnitudes of all three forces can be scaled from this triangle.

An example will illustrate the application of this simplified free-body technique for coplanar forces to the knee. In this case, the technique is used to estimate the magnitude of the joint reaction force acting on the tibiofemoral joint of the weight-bearing leg when the other leg is lifted during stair climbing. The lower leg is considered as a free body, distinct from the rest of the body, and a diagram of this free body in the stair-climbing situation is drawn (Calculation Box 7-1).

From all forces acting on the free body, the three main coplanar forces are identified as the ground reaction force, equal to body weight, the tensile force through the patellar tendon exerted by the quadriceps muscle, and the joint reaction force on the tibial plateau. (Note that the latter is the sum of the forces on the lateral and medial plateaus.) The ground reaction force (W) has a known magnitude (equal to body weight), sense, line of application, and point of application (point of contact between the foot and the ground). The patellar tendon force (P) has a known sense (away from the knee joint), line of application (along the patellar tendon), and point of application (point of insertion of the patellar tendon on the tibial tuberosity), but an unknown magnitude. The joint reaction force (J) has a known point of application on the surface of the tibia (the contact point of the joint surfaces between the tibial and femoral condyles, estimated from a roentgenogram of the joint in the proper loading configuration), but an unknown magnitude, sense, and line of application. Using vector calculations, the joint reaction force (J) and the patellar tendon force (P) can be calculated.

It can be seen that the quadriceps force has a much greater influence on the magnitude of the joint reaction force than does the ground reaction force produced by

CALCULATION BOX 7-1

Free-Body Diagram of the Knee Joint

The three main coplanar forces acting on the lower leg (ground reaction force [*W*], patellar tendon force [*P*], and joint reaction force [*J*]) are designated on a free-body diagram of the lower leg while climbing stairs (Calculation Box Fig. 7-1-1).

Because the lower leg is in equilibrium, the lines of application for all three forces intersect at one point. Because the lines of application for two forces (*W* and *P*) are known, the line of application for the third force (*J*) can be determined. The lines of application for forces *W* and *P* are extended until they intersect. The line of application for *J* can then be drawn from its point of application on the tibial surface through the intersection point (Calculation Box Fig. 7-1-2).

Now that the line of application for *J* has been determined, it is possible to construct a triangle of forces (Calculation Box Fig. 7-1-3). First, a vector representing *W* is drawn. Next, *P* is drawn from the head of vector

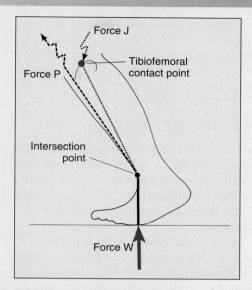

Calculation Box Figure 7-1-2

W. Then, to close the triangle, force *J* is drawn from the head of vector *W*. The point at which forces *P* and *J* intersect defines the length of these vectors. Now that the length of all three vectors is known, the magnitude of forces *P* and *J* can be scaled from force *W*, which is equal to body weight. In this case, force *P* is 3.2 times body weight and force *J* is 4.1 times body weight.

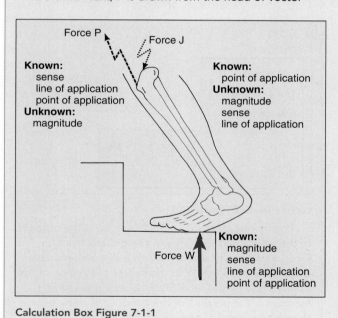

Calculation Box Figure 7-1-1

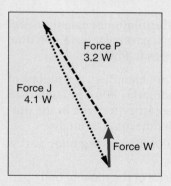

Calculation Box Figure 7-1-3

body weight. Note that, in this example, only the minimum magnitude of the joint reaction force has been calculated. If other muscle forces are considered, such as the force produced by the contraction of the hamstring muscles in stabilizing the knee, the joint reaction force increases.

Even with only the quadriceps acting, the joint force is calculated to be more than four times body weight. This is because the lever arm of the muscle about the center of rotation of the knee joint is small, whereas the

moment of the ground-to-foot force is much larger. This is an important principle that applies to any joint in the body. Physiologically, however, muscles are capable of generating large forces, whereas joint surfaces and the menisci distribute the forces over large areas of contact, producing acceptable contact stresses.

The next step in the static analysis is calculation of the moments acting around the center of rotation of the tibiofemoral joint with the knee in the same position and the loading configuration shown in Calculation Box Figure 7-1-1. The moment analysis is used to estimate the minimum magnitude of the moment produced through the patellar tendon, which counterbalances the moment on the lower leg produced by the weight of the body as the subject ascends stairs (Calculation Box 7-2).

DYNAMICS OF THE TIBIOFEMORAL JOINT

Although estimations of the magnitude of the forces and moments imposed on a joint in static situations are useful, most of our activities are of a dynamic nature. Analysis of the forces and moments acting on a joint during motion requires the use of a different technique for solving dynamic problems.

As in static analysis, the main forces considered in dynamic analysis are those produced by body weight, muscles, other soft tissues, and externally applied loads. Friction forces are negligible in a normal joint and are not considered here. In dynamic analysis, two factors in addition to those in static analysis must be taken into account: the acceleration of the body part under consideration and the mass moment of inertia of the body part. The mass moment of inertia is the unit used to express the amount of force needed to accelerate a body and depends on the shape of the body and the mass distribution. (For more in-depth studies of dynamics, see Özkaya and Nordin, 1999.)

The steps for calculating the minimum magnitudes of the forces acting on a joint at a particular instant in time during a dynamic activity are as follows:

1. The anatomic structures are identified: definitions of structures, anatomic landmarks, point of contact of articular surface, and lever arms involved in the production of forces for the biomechanical analyses.
2. The angular acceleration of the moving body part is determined.
3. The mass moment of inertia of the moving body part is determined.
4. The torque (moment) acting about the joint is calculated.
5. The magnitude of the main muscle force accelerating the body part is calculated.
6. The magnitude of the joint reaction force at a particular instant in time is calculated by static analysis.

In the first step, the structures of the body involved in producing forces on the joint are identified. These are the moving body part and the main muscles in that body part that are involved in the production of the motion. Care must be taken in applying this first step.

Calculation Box 7-2

Free-Body Diagram of the Lower Leg During Stair Climbing

The two main moments acting around the center of motion of the tibiofemoral joint (solid dot) are designated on the free-body diagram of the lower leg during stair climbing (Calculation Box Fig. 7-2-1).

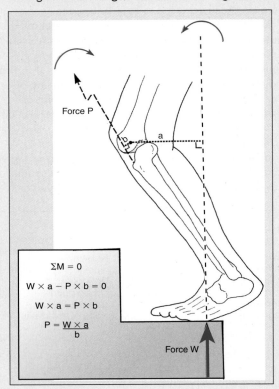

$$\Sigma M = 0$$
$$W \times a - P \times b = 0$$
$$W \times a = P \times b$$
$$P = \frac{W \times a}{b}$$

Calculation Box Figure 7-2-1

The flexing moment on the lower leg is the product of the weight of the body (W, the ground reaction force) and its lever arm (a), which is the perpendicular distance of the force W to the center of rotation of the tibiofemoral joint. The counterbalancing extending moment is the product of the quadriceps muscle force through the patellar tendon (P) and its lever arm (b). Because the lower leg is in equilibrium, the sum of these two moments must equal zero ($\Sigma M = 0$).

In this example, the counterclockwise moment is arbitrarily designated as positive [W × (a – P) × b = 0]. Values for lever arms a and b can be measured from anatomic specimens or on soft tissue imaging or fluoroscopy (Kellis and Baltzopoulos, 1999; Wretenberg et al., 1996), and the magnitude of W can be determined from the body weight of the individual. The magnitude of P can then be found from the moment equilibrium equation:

$$P = (W \times a)/b$$

For example, the lever arms for all major knee muscles change according to the degree of knee flexion and gender (Wretenberg et al., 1996).

In joints of the extremities, acceleration of the body part involves a change in joint angle. To determine this angular acceleration of the moving body part, the entire movement of the body part is recorded photographically. Recording can be done with a stroboscopic light and movie camera, with video photogrammetry, with Selspot systems, with stereophotogrammetry, or with other methods (Gardner et al., 1994; Ramsey and Wretenberg, 1999; Winter, 1990). The maximum angular acceleration for a particular motion is calculated.

Next, the mass moment of inertia for the moving body part is determined. Anthropometric data on the body part can be used for this determination. Because calculating these data is a complicated procedure, tables are commonly used (Drillis et al., 1964).

The torque about the joint can now be calculated using Newton's second law of motion, which states that when motion is angular, the torque is a product of the mass moment of inertia of the body part and the angular acceleration of that part:

$$T = Ia$$

where

T is the torque expressed in newton meters (Nm)

I is the mass moment of inertia expressed in newton meters seconds squared (Nm sec^2)

a is the angular acceleration expressed in radians per second squared (r/sec^2).

The torque is not only a product of the mass moment of inertia and the angular acceleration of the body part, but also a product of the main muscle force accelerating the body part and the perpendicular distance of the force from the center of motion of the joint (lever arm). Thus,

$$T = Fd$$

where

F is the force expressed in newtons (N)

d is the perpendicular distance expressed in meters (m).

Because T is known and d can be measured on the body part from the line of application of the force to the center of motion of the joint, the equation can be solved for F. When F has been calculated, the remaining problem can be solved like a static problem using the simplified free-body technique to determine the minimum magnitude of the joint reaction force acting on the joint at a certain instant in time.

A classic example will illustrate the use of dynamic analysis in calculating the joint reaction force on the tibiofemoral joint at a particular instant during a dynamic activity, namely kicking a football (Frankel and Burstein, 1970). A stroboscopic film of the knee and lower leg was taken, and the angular acceleration was found to be maximum at the instant the foot struck the ball; the lower leg was almost vertical at this instant. From the film, the maximal angular acceleration was computed to be 453 r/sec^2. From anthropometric data tables (Drillis et al., 1964), the mass moment of inertia for the lower leg was determined to be 0.35 Nm sec^2. The torque about the tibiofemoral joint was calculated according to the equation: Torque equals mass moment of inertia times angular acceleration ($T = Ia$),

$$0.35 \text{ Nm sec}^2 \times 453 \text{ r/sec}^2 = 158.5 \text{ Nm}$$

After the torque had been determined to be 158.5 Nm and the perpendicular distance from the subject's patellar tendon to the instant center for the tibiofemoral joint had been found to be 0.05 m, the muscle force acting on the joint through the patellar tendon was calculated using the equation torque equals force times distance ($T = Fd$),

$$158.5 \text{ Nm} = F \times 0.05 \text{ m}$$
$$F = 158.5 \text{ Nm} / 0.05 \text{ m}$$
$$F = 3{,}170 \text{ N}$$

Thus, 3,170 N was the maximal force exerted by the quadriceps muscle during the kicking motion. This is equal to approximately two times body weight for an average person.

Static analysis can now be performed to determine the minimum magnitude of the joint reaction force on the tibiofemoral joint. The main forces on this joint are identified as the patellar tendon force (P), the gravitational force of the lower leg (T), and the joint reaction force (J). P and T are known vectors. J has an unknown magnitude, sense, and line of application. The free-body technique for three coplanar forces is used to solve for J, which is found to be only slightly lower than P.

As is evident from the calculations, the two main factors that influence the magnitude of the forces on a joint in dynamic situations are the acceleration of the body part and its mass moment of inertia. An increase in angular acceleration of the body part will produce a proportional increase in the torque about the joint. Although in the body the mass moment of inertia is anatomically set, it can be manipulated externally. For example, it is increased when a weight of the boot is applied to the foot during rehabilitative exercises of the extensor muscles of the knee. Normally, a joint reaction force of approximately 50% of body weight results when the knee is slowly (with no acceleration forces) extended from 90° of flexion to full extension. In a person weighing 70 kg, this force is approximately 350 N. If a 10-kg weight boot is placed on the foot, it will exert a gravitational force of 100 N. This will increase the joint reaction force by 1,000 N, making the joint force almost four times greater than it would be without the boot.

FORCES IN THE KNEE IN VIVO

In recent years, two groups, one at the Scripps Clinic in San Diego (D'Lima et al., 2006, 2007, 2008; Munderman, 2008; Zhao et al., 2007) and the other at the University of Berlin (Heinlein et al., 2009), have developed telemetric instrumentation for the tibial component of an artificial knee joint, to measure the forces in vivo for a range of activities. The following example refers to the Berlin study where two patients were implanted with a moderately conforming condylar resurfacing device. The cruciate ligaments were resected, such that the tibial surface would carry compressive forces on the lateral and medial condyles, as well as AP shear forces and internal-external

torques. The varus-valgus moment would be represented by unequal forces on the lateral and medial condyles. The coordinate system for measuring and describing the forces and moments was at the center of the bearing surfaces on the tibial component, as shown in Figure 7-14.

The figure shows the force vectors, which are composed of the resultant of the vertical, AP, and mediolateral (ML) forces. All of the peak forces occur at heel-strike and toe-off, which is when all the body weight is carried on one leg. The vertical force component clearly predominates in all activities. To determine the other force components from this figure, assume the resultant R is an angle θ to the vertical. The vertical component will be $R \cos \theta$ and the horizontal component $R \sin \theta$. For angles of 3°, 5°, and 8°, which cover all of the charts, the force components are calculated to two decimal places: $1.00 R$ and $0.05 R$; $1.00 R$ and $0.08 R$; $0.99 R$ and $0.13 R$. In other words, the vertical force components are close to the resultants, whereas the shear force components are much smaller but still significant.

For level walking, the maximum compressive force was 2.65 times body weight (BW) for one of the two patients. These forces are the sum of the lateral and medial joint forces, the relative amounts being discussed

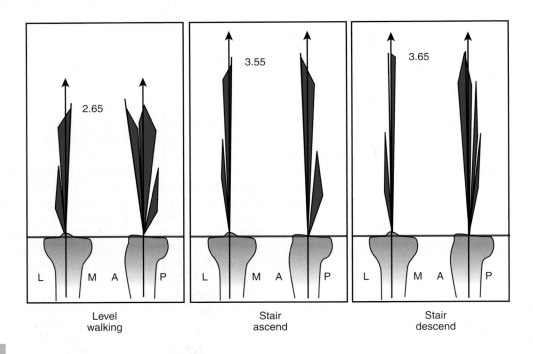

FIG. 7-14

The resultant forces on the tibia for three activities. The *vertical arrowed lines* are the vertical reference axes. The *tall narrow triangles* represent the boundaries of the force vectors during the stance phase of the activity. The *shorter triangles* are the force vectors in the swing phase. The *heavy vertical lines* are the vectors of maximum force, represented by the numerals that are units of body weight. *L*, lateral; *M*, medial; *A*, anterior; *P*, posterior. *Adapted from Heinlein, B., Kutzner, I., Graichen, F., et al. (2009). Complete data of total knee replacement loading for level walking and stair climbing measured in vivo with a follow-up of 6–10 months. Clin Biomech, 24(4), 315–326.*

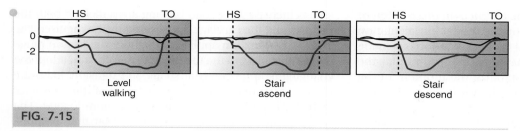

FIG. 7-15

The axial torque (curves of smaller magnitude, in *black*) and the varus moment (curves of larger magnitude, in *blue*) for the three activities. Positive values are internal rotation and valgus moment. *HS,* heel-strike; *TO,* toe-off; 0, −2 are in units of % BW × m.

later. In the frontal view, the shear forces in the lateral medial directions were present but of small value. However, there were significant peak anterior and posterior shear forces in early and late stance of 0.3 to 0.4 BW.

In ascending stairs, the frontal plane resultant forces were concentrated in a vertical direction with very small lateral and medial force components. The peak value was 3.55 BW, 34% higher than for level walking. Even in the sagittal plane, the forces were more concentrated than for level walking. Here the peak AP shear forces were in the range of 0.2 to 0.3 BW. It was notable that in the swing phase, where the overall resultant was much smaller than in stance, the anterior shear force was still 0.3 BW. In descending stairs, the forces in stance were again concentrated vertically, with a peak of 3.65 BW. The anterior and posterior shear forces were similar to those for ascending stairs.

Again considering the frontal plane, the distribution of the resultant force between the lateral and medial sides can be determined from Figure 7-15, which shows the frontal plane moments. The peak values for all three activities are in the range 3% to 4% BW × m of varus. Forces *FL* and *FM* can be calculated by assuming a spacing apart of the lateral and medial contacts of 48 mm and a body weight of 750 newtons (Fig. 7-16). The equations are, in appropriate units: $FL + FM = FR$, $(FM − FL) \times 24 = M$. For the 4% value, the medial/lateral ratio is 2.7. In other words, the force on the medial side is much greater than on the lateral for this activity at the instants of the peak forces.

Comparisons of the forces in different activities were given by Mundermann et al. (2008). They defined three categories: category one, high cycle loading with a moderate load (2–3 BW), many cycles, and peak load at low flexion angles: This is represented by level walking; category two, high loading in midflexion (3–4 BW) including stair ascent and descent and golf swing; category three is moderate loads at high flexion angles (2–3 BW) including sit-to-stand and squatting. The forces on a stationary bicycle were measured by the Berlin group (Kutzner et al., 2008), finding up to 1.4 BW depending on speed and effort. In almost all activities, the medial load was greater than the lateral, by a factor of up to eight times in squatting.

These data from instrumental knee replacements, which will be expanded on in further studies, have provided invaluable information on the forces in the knee in a range of activities. Biomechanical principles can be applied in the interpretation. These data were obtained from total knee patients, who have reduced quadriceps strength even up to one year after surgery and beyond (Mizner et al., 2005), so in a normal individual the forces could be higher. The gait patterns are not entirely normal, and the frontal plane alignment, and hence the mediolateral force distribution, depends on the surgical placement of the total knee. Also, due to the expense of the technology, data are likely to be obtained for only a few subjects.

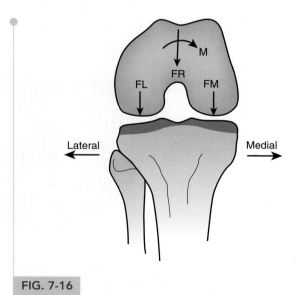

FIG. 7-16

A front view of the knee with an instrumental total knee replacement. The resultant force and moment acting on the tibia are *FR* and *M*. These are equivalent to the separate lateral and medial forces *FL* and *FM* where in general *FL < FM*.

Overall, the magnitudes of the forces in these different activities can be explained by the model in Calculation Boxes 7-1 and 7-2. The axial compressive force is made up of the direct weight of the body on the joint plus the muscle force necessary to stabilize the moment of the external ground-to-foot force about the center of rotation of the joint. In general, the lever arm increases with the angle of flexion, which explains why stair activities generate higher forces. Sit-to-stand uses two legs simultaneously, requiring less force, whereas in cycling the full body weight is not supported on the foot. The effect of the knee forces on the various structures in the knee will now be discussed.

The anteroposterior shear forces acting on the tibia of up to approximately 0.4 times body weight are the result of the body weight and muscle forces. An anterior shear force on the tibia would be carried by the PCL, the medial meniscus, and the upsweep of the tibial surface. A posterior shear force would be carried by the ACL and the medial meniscus. In both directions the MCL will also carry some of the force. The contributions of the cruciates (ACL and PCL) can be appreciated from Figure 7-7.

In a normal knee, the vertical force component is carried by the menisci as well as by the articular cartilage not covered by the meniscus as determined by several investigators (Ahmed and Burke, 1983; Kurosawa et al., 1980; Walker and Erkman, 1975). Under low loads, the femoral condyles are supported substantially by the meniscus. As the load is increased, the load-bearing areas include both the menisci and the cartilage not covered by the menisci (Fig. 7-17). This situation applies throughout the flexion range due to the ability of the meniscus to deform and adopt the changing femoral condylar shape. In extension, loads are more concentrated on the anterior horn of the meniscus; in high flexion, the posterior horn is more loaded. The wedge shape of the meniscus means that the compressive force on the proximal surface has a radial component, expanding the meniscus outward and causing tensile stresses (or hoop stresses) in the meniscus itself. The tensile stresses act circumferentially around the entire meniscus. These stresses can be related to the ultrastructure of the meniscus, which consists largely of collagen fiber bundles oriented circumferentially (Bullough et al., 1970).

The circumferential stress can be elevated by sudden twisting or shearing actions, which can produce meniscal tears. The catching of loosely held parts of meniscus is usually treated by arthroscopic removal of the parts, but certain types of tears are repairable. Horizontal cleavage lesions, parallel to the base of the meniscus, can also be produced by the radial force component. Preservation of the meniscus is important due to its role in distributing the loads among other functions.

In the normal knee and even in the arthritic knee, the tibial cartilage beneath the menisci is usually intact with preservation of the original surface. However, the exposed cartilage not covered by the menisci is usually soft and

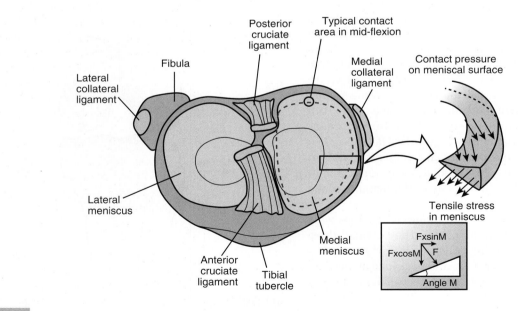

FIG. 7-17

A top view of the tibia showing the various structures, excluding muscles. On the medial side (and on the lateral side) the femoral-tibial force acts over the *dotted area*. The resultant force F produces a radial force $F \times \sin M$ and a compressive force $F \times \cos M$. The radial force component produces tensile (hoop) stresses in the meniscus.

fibrillated. These conditions might be explained by the difference in the sliding velocities and the efficiency of lubrication. If the menisci are removed, the stresses are no longer distributed over such a large area but instead are limited to a contact area in the center of the plateau (Walker and Hajek, 1972). Thus, removal of the menisci not only increases the magnitude of the stresses on the cartilage and subchondral bone at the center of the tibial plateau but also diminishes the size and changes the location of the contact areas. Over the long term, the high stresses placed on these smaller contact areas may be harmful to the exposed cartilage, which is already soft and fibrillated. It is well known that removal of a meniscus will result in osteoarthritis in only a few years in many cases (McDermott and Amis, 2006).

STABILITY OF THE KNEE JOINT

One of the most important keys to a healthy knee joint is stability in response to forces and moments in all planes. The osseous configuration, the menisci, the ligaments, and the capsule provide static stability. The muscles surrounding the knee joint produce dynamic stability. As described previously, the cruciate ligaments primarily provide AP stability, but they can also become taut in internal and external rotation. The collateral ligaments primarily provide varus-valgus stability, but in rotation the medial collateral ligament (MCL) carries some force. When there is excessive displacement or rotation in any direction, the meniscus can contribute to stability. If any of these structures is injured, knee joint instability will occur (Case Study 7-1).

Today, after ACL rupture, immediate repair is often carried out if the patient expects to return to an active lifestyle. On the other hand, for a sedentary person, a ruptured ACL can be tolerated, although there is evidence that there is a higher risk of developing osteoarthritis, especially if there was a meniscal injury at the time of the ACL rupture (Neuman et al., 2008). The options for repair are to use a patella tendon graft (as previously described) or a tendon graft from hamstring tendons. The grafts are threaded through tunnels drilled through the desired attachment points in the femur and tibia and held using special grommets.

Numerous studies have been carried out to determine the specific roles of the different ligaments in providing static stability. In general, the conclusion has been that one ligament is the primary stabilizer, whereas others are secondary stabilizers.

Fu et al. (1994) summarized the overall situation. The ACL is the predominant restraint to anterior tibial displacement. The ligament accepts 75% of the anterior force at full extension and an additional 10% up to 90° of knee flexion. The posterior cruciate ligament is the primary restraint to posterior tibial translation; it sustains 85% to 100% of the posterior force at both 30° and 90° of knee flexion. The lateral collateral ligament is the primary restraint to varus angulation and it resists approximately 55% of the applied load at full extension. The medial collateral ligament (superficial portion) is the primary restraint to valgus (adduction) angulation and resists 50% of the external valgus load. The capsule, the anterior cruciate ligament, and the posterior cruciate ligament share the remaining valgus load. Internal rotational laxity seen in the 20° to 40° range of knee flexion is restrained by the medial collateral ligament and the ACL. Finally, the posterior cruciate is a secondary stabilizer, especially at higher flexion angles.

Considering that varus moments and opening of the lateral side of the knee can occur, much more so than on the medial side, it seems illogical that the MCL should be much stronger than the lateral collateral ligament (LCL). However, on the lateral side, this can be explained by the action of the iliotibial band (ITB), which is a dynamic stabilizer of the knee. The MCL, as well as restraining valgus moments, is important in limiting AP displacement of the medial femoral condyle and providing the medial pivot action in function.

Focusing on the ACL, Beynnon et al. (1992) have performed in vivo measurements. They placed a strain transducer arthroscopically in the ACL. The results showed that strain in the ACL was related to knee flexion (with the most strain occurring near full extension) and increased with quadriceps contraction. Less strain occurred with co-contractions of both the quadriceps and the hamstring muscle groups and at greater degrees of knee flexion. This indicates that muscle contraction and co-contraction contribute to the stability of the knee joint by increasing the stiffness of the joint. Kwak et al. (2000) studied the effect of hamstrings and iliotibial band forces on the kinematics of the knee in vitro. At various knee flexion angles, human knee specimens were tested with different muscle-loading patterns. The quadriceps muscle force was always present, and the test was performed with and without hamstring muscle force and with and without iliotibial band force. With loading of simultaneous quadriceps and hamstring muscle force, the tibia translated posteriorly and rotated externally. The effect was similar for the iliotibial band simulated forces, but the effect was smaller.

Many in vitro studies suggest that the hamstrings are important anterior and rotational stabilizers of the tibia. In vivo studies have shown that co-contractions of the quadriceps and hamstring muscles are present in normal knee joints in daily activities (Baratta et al., 1988; Solomonow and D'Ambrosia, 1994). The co-contraction mechanisms also increase the knee joint stability in vivo (Åagaard et al., 2000; Markolf et al., 1978; Solomonow and D'Ambrosia, 1994).

Case Study 7-1

ACL Injury

A 30-year-old man suffered an internal tibial rotation trauma in his right knee while downhill skiing. Following the trauma, he experienced sharp pain, progressive joint effusion, and subjective instability. During examination by a sports medicine specialist, an anterior positive drawer test was diagnosed, and the Lachman test and pivot shift test were found positive. An MRI confirmed the ACL rupture (Case Study Fig. 7-1-1).

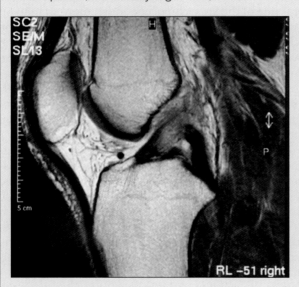

Case Study Figure 7-1-1

The rupture of a primary stabilizer of the knee joint, the anterior cruciate ligament, can lead to a progressive structural alteration of the knee if untreated. A primary objective of the treatment is the prevention of reinjury of the knee in the hope of preventing additional ligamentous injuries, meniscal injuries, and possible cartilage degeneration. In this case, the patient first completed a course of conservative treatment with physical therapy. However, after six months, the subjective instability was present during sports and daily activities such as gait and stair climbing. To compensate for the ACL deficiency, the patient altered his gait patterns, including quadriceps avoidance gait to prevent the anterior translation of the tibia when the quadriceps contracts at the midstance phase of the gait (Andriacchi and Birac, 1993; Berchuck et al., 1990). However, this is not a situation to be tolerated for the long term and hence the patient opted for surgical treatment. The MRI (Case Study Fig. 7-1-2) shows the ACL status after patella bone-tendon-bone autograft was performed 10 months post-trauma.

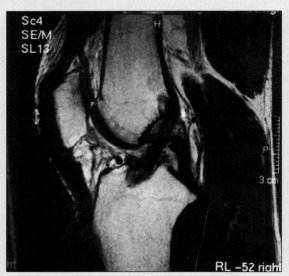

Case Study Figure 7-1-2

FUNCTION OF THE PATELLA

The patella provides an important biomechanical function in the knee by lengthening the lever arm of the quadriceps muscle force about the center of rotation of the knee and thereby increasing the mechanics and efficiency of the quadriceps in all (Hehne, 1990). Figure 7-18 shows the lines of actions of the three forces acting on the patella at 90° of flexion. The lever arms can be considered for the femur and tibia separately depending on which bone is actively rotating. The lever arms are the perpendicular distances from the center of rotation to the lines of actions of the forces. It can be seen that these distances are increased due to the patella. It is often assumed that the quadriceps and ligament forces FQ and FL will be equal, using the analogy of a rope around a smooth pulley. However, both experiment and analysis have shown that this is not the case, due to the patellofemoral contact geometry (Ellis et al., 1980; Huberti and Hayes, 1984).

In the patellofemoral joint, the quadriceps muscle force generally increases with knee flexion. During relaxed upright standing, minimal quadriceps muscle forces are required to counterbalance the small flexion moments about the center of the joint because the center of gravity of the body above the knee is almost directly above the center of rotation. As knee flexion increases, the external forces move farther away from the center of rotation, thereby greatly increasing the flexion moments to be counterbalanced by the quadriceps muscle force. As the quadriceps muscle force increases, so does the patellofemoral joint reaction force (Hungerford and Barry, 1979; Reilly and Martens, 1972).

In Figure 7-18, the reaction force of the patella on the femur is shown as a single resultant force. In the isometric view of the transverse section shown in Figure 7-19, the resultant forces at the lateral and medial sides are treated separately. To calculate these, it is first necessary to consider the quadriceps and patella ligament forces. The components of these forces in the frontal plane XY are QS and TS, respectively.

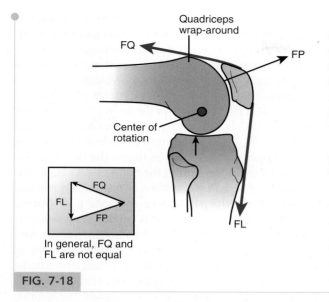

FIG. 7-18

Sagittal section of the knee at 90° flexion showing the quadriceps force GQ, the patella reaction force FP, and the patella ligament force FL. The force triangle shows the relative values.

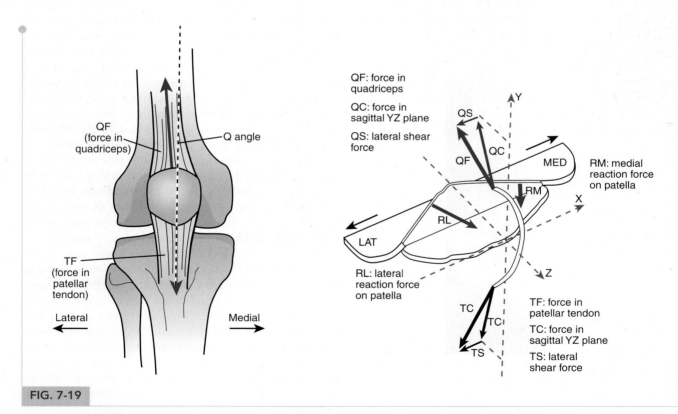

FIG. 7-19

Forces on the patella mechanism. **Left diagram.** The quadriceps and patella tendon forces are not collinear due to the Q-angle. **Right diagram.** A transverse (horizontal) plane section through the patella and femoral trochlea at approximately 60° flexion.

Case Study 7-2

Extensor Mechanism Injury

A 30-year-old basketball player had a forceful knee flexion while coming down from a jump. A strong eccentric contraction of the quadriceps produced abnormally high tensile loads in the patella, leading to a fracture at the inferior pole. In this case, the patella fracture occurred because the muscle forces of the quadriceps overcame the osseous strength of the patella. The weakest link was the patella.

The picture shows a fracture at the patella accompanied by a significant fracture separation that resulted from the quadriceps traction force (Case Study Figure 7-2). Because of the fracture, the extensor mechanism is unable to function and extend the knee. It will directly affect the stability of the patellofemoral joint and the distribution of the compressive stresses on the femur. At the same time, the impaired function of the quadriceps decreases the dynamic stability at the knee joint (patellofemoral and tibiofemoral joints) that is necessary for daily activities such as gait and stair climbing.

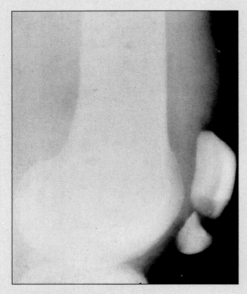

Case Study Figure 7-2

Hence $QS + QT = (RL - RM) \cos G$ where G is the angle of the lateral and medial facets of the patella groove (considered equal) in the transverse XZ plane. The angle G is approximately 25° in most knees. It can be seen that if the Q-angle is zero, QS and QT are zero and $RL = RM$, indicating equal lateral and medial patella facet forces. For $QS > 0$ and $TS > 0$, then $RL > RM$, a greater force on the lateral force, the most common condition. Solving for the magnitudes of RL and RM in terms of QF and TF requires resolution of the forces in all three planes. In general, the higher the flexion angle, the higher all of the forces, and the higher the Q-angle, the higher the RL/RM ratio. A condition is theoretically reached where $RM = 0$ and the patella is on the point of displacing laterally (Hvid, 1983). However, this condition applies only under extreme valgus or under a traumatic dynamic condition. Many patella subluxations occur in flexing from an extended position, due to the patella traversing lateral to the patella groove and not entering the groove at all.

In Figure 7-19, as the frontal view shows, the quadriceps and ligament forces are not collinear, producing a laterally directed force on the patella. In general, the higher the valgus angle of the knee, the higher the Q-angle. The Q-angle is also higher with the knee in extension when the screw-home mechanism causes the tibia to rotate externally. However, the exact direction of the resultant force will depend on the relative forces in the vastus lateralis, vastus medialis, and rectus femoris. In any case, the lateral force component puts the patella at risk for subluxing laterally. This is prevented by the slope and height of the patella groove on the lateral side. In contrast, the medial groove is shallow. When the knee flexes beyond approximately 90°, the patella begins to sink into the intercondylar notch, which has high slopes on both lateral and medial sides. In this situation, the internal-external rotational position of the tibia on the femur is highly variable, requiring high stability under all conditions. However, each of these structures can carry forces of many times body weight, but in a traumatic injury, the bone of the patella itself can fracture, as shown in Case Study 7-2.

When the knee is extended, the lower part of the patella rests against the femur. As the knee is flexed to 90°, the contact surface between the patella and femur shifts cranially in vivo and under weight-bearing conditions (Komistek et al., 2000). The contact surface area increases in size (Goodfellow et al., 1976). To some extent, this increase in the contact surface with knee flexion compensates for the larger patellofemoral joint reaction force. If a tight iliotibial band is present, the patellofemoral joint force may shift laterally, causing abnormal patellar kinematics and load bearing (Kwak et al., 2000).

The quadriceps muscle force and the torque around the patellofemoral joint can be extremely high under certain circumstances, particularly when the knee is flexed—for instance, when a basketball player suffers a patella fracture as a result of indirect forces played by an eccentric contraction of the quadriceps (Case Study 7-2). Another extreme situation was observed during a study of the external torque on the knee produced by weight lifting: One subject ruptured his patellar tendon when he lifted a barbell weighing 175 kg (Zernicke et al., 1977). At the instant of tendon rupture, the knee was flexed 90°, the torque on the knee joint was 550 Nm, and the quadriceps muscle force was approximately 10,330 N.

Because of the high magnitude of quadriceps muscle force and joint reaction force during activities requiring a large amount of knee flexion, patients with patellofemoral joint derangements experience increased pain when performing these activities. An effective mechanism for reducing these forces is to limit the amount of knee flexion.

Summary

- The knee is a two-joint structure that is composed of the tibiofemoral joint and the patellofemoral joint.
- In the tibiofemoral joint, surface motion occurs in three planes simultaneously with the largest motions occurring in the sagittal plane. In the patellofemoral joint, the surface motion occurs primarily in the sagittal plane with respect to axes fixed in the femur.
- The surface joint motion can be described with the use of an instant center technique. When performed on a normal knee, the technique reveals the following: The instant center for successive intervals of motion of the tibiofemoral joint in the sagittal plane follows a curved pathway reflecting the changing femoral radii of curvature in the sagittal plane. The direction of displacement of the tibiofemoral contact points is tangential to the surface of the tibia, indicating sliding throughout the range of motion.
- On the medial side, the contact point on the tibia is close to constant, indicating sliding motion. On the lateral side, the contact point displaces posteriorly with flexion, indicating a combination of rolling and sliding.
- The screw-home mechanism of the tibiofemoral joint in extension adds stability to the joint in full extension. Additional passive stability to the knee is given by the ligamentous structures and menisci, and the dynamic stability by the muscles around the knee.
- The kinematics and stability of the knee can be modeled with either 2D or 3D models incorporating the joint surfaces and the major ligaments.
- The tibiofemoral and patellofemoral joints are subjected to high forces. Muscle forces have the greatest influence on the magnitude of the joint reaction force, which can reach several times body weight in both joints. In the patellofemoral joint, knee flexion also affects the joint reaction force, with greater knee flexion resulting in a higher joint reaction force.
- The total compressive forces on the knee are in the range of 2 to 4 BW, with higher flexion activities having the highest forces, and with the medial side carrying higher forces than the lateral.
- Although the tibial plateaus are the main load-bearing structures in the knee, the cartilage, menisci, and ligaments also carry forces. The menisci aid in distributing the stresses and reducing the pressures imposed on the tibial plateaus.
- The patella aids knee extension by lengthening the lever arm of the quadriceps muscle force throughout the entire range of motion and allows a wider distribution of compressive stress on the femur.

Practice Questions

1. What is meant by the laxity of the knee joint? Approximately how many degrees of rotational laxity occur in the knee, and what are the active and passive restraints to this laxity?

2. Explain the four-bar linkage as a model for the flexion-extension motion of the knee in the sagittal plane. In what ways is this model a simplification of reality?

3. Define the neutral path of motion of the knee during flexion. What are the structures that define this motion, and what occurs when there are superimposed anterior-posterior shear forces or torques applied externally or from muscle action?

4. Why are there contact areas on the femorotibial surfaces when a force is acting across the knee? What is the role of the menisci in relation to the contact areas?

5. What are the resultant forces across the knee joint during level walking, stair ascend, and stair descend? How large are the anteroposterior shear forces in relation to the axial compressive forces?

6. How are the forces across the knee joint generated? What are the relative forces on the lateral and medial sides of the joint? What can happen if the medial force component becomes abnormally large over time due to a gait abnormality or bone misalignment?

7. Describe the force transmission between the patella and the femur. When is the resultant force the greatest? Which muscles generate this force?

8. What are the main symptoms of osteoarthritis of the knee, and what are the main treatment options?

REFERENCES

Åagaard, P., Simonsen, E.B., Andersen, J.L., et al. (2000). Antagonist muscle co-activation during isokinetic extension. *Scan J Med Sci Sports, 10*(2), 58–67.

Ahmed, A.M., Burke, D.L. (1983). In-vitro measurement of static pressure distribution in synovial joints—Part 1: Tibial surface of the knee. *J Biomech Eng, 105*, 216–225.

Andriacchi, T.P., Birac, D. (1993). Functional ligament testing in the anterior cruciate ligament deficient knee. *Clin Orthop Rel Res, 288* (March), 40–47.

Andriacchi, T.P., Kramer, G.M., Landon, G.C. (1979). Three-dimensional coordinate data processing in human motion analysis. *J Biomech Eng, 101*, 279–283.

Baratta, R., Solomonow, M., Zhou, B.H., et al. (1988). Muscular co-activation: The role of the antagonist musculature in maintaining knee stability. *Am J Sports Med, 16*, 113–122.

Berchuck, M., Andriacchi, T.P., Bach, B.R., et al. (1990). Gait adaptations by patients who have a deficient anterior cruciate ligament. *J Bone Joint Surg Am, 72*, 871–877.

Beynnon, B., Howe, J.G., Pope, M.H. (1992). The measurements of anterior cruciate ligament strain in vivo. *Int Orthop, 16*, 1–12.

Blankevoort, L., Huiskes, R., De Lange, A. (1988). The envelope of passive knee joint motion. *J Biomech, 21*(9), 705–720.

Bullough, P.G., Munuera, L., Murphy, J., et al. (1970). The strength of the menisci of the knee as it relates to their fine structure. *J Bone Joint Surg Am, 52*, 564–570.

Dennis, D., Komistek, R., Scuderi, G., et al. (2001). In vivo three-dimensional determination of kinematics for subjects with a normal knee or a unicompartmental or total knee replacement. *J Bone Joint Surg Am, 83*, S104–115.

Dennis, D.A., Komistek, R.D., Stiehl, J.B., et al. (1998). Range of motion after total knee arthroplasty: The effect of implant design and weight-bearing conditions. *J Arthroplasty, 13*(7), 748.

D'Lima, D.D., Patil, S., Steklov, N., et al. (2006). Tibial forces measured in vivo after total knee arthroplasty. *J Arthroplasty, 21*(2), 255–262.

D'Lima, D.D., Patil, S., Steklov, N., et al. (2007). In vivo knee moments and shear after total knee arthroplasty. *J Biomech, 41*(10), 2332–2335.

D'Lima, D.D., Steklov, N., Patil, S., et al. (2008). The Mark Coventry Award: In vivo knee forces during recreation and exercise after knee arthroplasty. *Clin Orthop Rel Res, 466*(11), 2605–2611.

Draganich, L.D., Andriacchi, T.P., Andersson, G.B.J. (1987). Interaction between intrinsic knee mechanism and the knee extensor mechanism. *J Orthop Res, 5*, 539–547.

Drillis, R., Contini, R., Bluestein, M. (1964). Body segment parameters. A survey of measurement techniques. *Artif Limbs, 8*, 44.

Eckhoff, D., Hogan, C., DeMatteo, L., et al. (2007). Difference between the epicondylar and cylindrical axis of the knee. *Clin Orthop, 461*, 238–244.

Ellis, M.I., Seedhom, B.B., Wright, V., et al. (1980). An evaluation of the ratio between the tensions along the quadriceps tendon and the patellar ligament. *Eng Med, 9*(4), 189–194.

Frankel, V.H., Burstein, A.H. (1970). *Orthopaedic Biomechanics.* Philadelphia: Lea & Febiger.

Frankel, V.H., Burstein, A.H., Brooks, D.B. (1971). Biomechanics of internal derangement of the knee. Pathomechanics as determined by analysis of the instant centers of motion. *J Bone Joint Surg Am, 53*, 945.

Freeman, M.A.R., Pinskerova, V. (2005). The movement of the normal tibio-femoral joint. *J Biomech, 38*, 197–208.

Fu, F.H., Harner, C.D., Johnson, D.L., et al. (1994). Biomechanics of the knee ligaments: Basic concepts and clinical application. *Instr Course Lecture, 43*, 137–148.

Gardner, T.R., Ateshian, G.A., Grelsamer, R.P., et al. (1994). A 6 DOF knee testing device to determine patellar tracking and patellofemoral joint contact area via stereophotogrammetry. *Adv Bioeng ASME BED, 28*, 279–280.

Girgis, F.G., Marshall, J.L., Al Monajema, R.M. (1975). The cruciate ligaments of the knee joint. Anatomical, functional and experimental analysis. *Clin Orthop, 106*, 216–231.

Goodfellow, J., Hungerford, D.S., Zindel, M. (1976). Patellofemoral joint mechanics and pathology. 1. Functional anatomy of the patellofemoral joint. *J Bone Joint Surg Br, 58*, 287.

Grood, E.S., Suntay, W.J. (1983). A coordinate system for clinical description of three-dimensional motions: Application to the knee. *J Biomech Eng, 105*, 136–144.

Hehne, H.J. (1990). Biomechanics of the patellofemoral joint and its clinical relevance. *Clin Orthop, 258*, 73–85.

Heinlein, B., Kutzner, I., Graichen, F., et al. (2009). Complete data of total knee replacement loading for level walking and stair climbing measured in vivo with a follow-up of 6–10 months. *Clin Biomech, 24*(4), 315–326.

Helfet, A.J. (1974). Anatomy and mechanics of movement of the knee joint. In A. Helfet (Ed.). *Disorders of the Knee.* Philadelphia: JB Lippincott Co, 1–17.

Holden, J.P., Chou, G., Stanhope, S.J. (1997). Changes in knee joint function over a wide range of walking speeds. *Clin Biomech, 12*(6), 375–382.

Huberti, H.H., Hayes, W.C. (1984). Patellofemoral contact pressures. The influence of Q-angle and tendofemoral contact. *J Bone Joint Surg, 66-A*(5), 715–724.

Hungerford, D.S., Barry, M. (1979). Biomechanics of the patellofemoral joint. *Clin Orthop, 144*, 9–15.

Hvid, I. (1983). The stability of the human patello-femoral joint. *Eng Med, 12*(2), 55–59.

Kapandji, I.A. (1970). The knee. In I.A. Kapandji (Ed.). *The Physiology of the Joints,* vol 2. Paris, France: Editions Maloine, 72–135.

Kellis, E., Baltzopoulus, V. (1999). In vivo determination of the patella and hamstrings moment arms in adult males using videofluoroscopy during submaximal knee extension and flexion. *Clin Biomech, 14*, 118–124.

Kettelkamp, D.B., Johnson, R.J., Smidt, G.L., et al. (1970). An electrogoniometric study of knee motion in normal gait. *J Bone Joint Surg Am, 52,* 775.

Komistek, R.D., Dennis, D.A., Mabe, J.A., et al. (2000). An in vivo determination of patellofemoral contact positions. *Clin Biomech, 15,* 29–36.

Komistek, R.D., Dennis, D.A., Mahfoua, M. (2003). In vivo fluoroscopic analysis of the normal human knee. *Clin Orthop Relat Res, 410,* 69–81.

Kroemer, K.H., Marras, W.S., McGlothin, J.D., et al. (1990). On the measurements of human strength. *Int J Ind Ergonomics, 6,* 199–210.

Kurosawa, H., Fukubayashi, T., Nakajima, H. (1980). Load-bearing mode of the knee joint: Physical behavior of the knee joint with or without menisci. *Clin Ortho, 49,* 283–290.

Kurosawa, H., Walker, P.S., Abe, S., et al. (1985). Geometry and motion of the knee for implant and orthotic design. *J Biomech, 18*(7), 487–499.

Kutzner, I., Heinlein, B., Graichen, F., et al. (2008). In vivo measurements of loads during ergometer cycling 6 months post-operatively. 16th ESB Congress Short Talks. *J Biomech, 41*(S1), S323.

Kwak, S.D., Ahmad, C.S., Gardner, T.R., et al. (2000). Hamstrings and iliotibial forces affect knee ligaments and contact pattern. *J Orthop Res, 18,* 101–108.

Lamoreux, L. (1971). Kinematic measurements in the study of human walking. Biomechanics Lab, University of California, San Francisco. *Bull Prosthet Res, 10*(15), 3–84.

Laubenthal, K.N., Smidt, G.L., Kettelkamp, D.B. (1972). A quantitative analysis of knee motion during activities of daily living. *Phys Ther, 52,* 34.

Li, G., DeFrate, L.E., Park, S.E., et al. (2005). An investigation using dual-orthogonal fluoroscopy and magnetic resonance image-based computer models. *Am J Sports Med, 33*(1), 102–107.

Markholf, K., Graff-Radford, A., Amstutz, H. (1978). In vivo knee stability. *J Bone Joint Surg Am, 60,* 664–674.

McDermott, I.D., Amis, A.A. (2006). The consequences of meniscectomy. *J Bone Joint Surg Br, 88,* 1549–1556.

Mizner, R.L., Petterson, S.C., Stevens, J.E., et al. (2005). Early quadriceps strength loss after total knee arthroplasty. The contributions of muscle atrophy and failure of voluntary muscle activation. *J Bone Jt Surg Am, 87,* 1047–1053.

Mundermann, A., Dyrby, C.O., D'Lima, D.D., et al. (2008). In vivo knee loading characteristics during activities of daily living as measured by an instrumented total knee replacement. *J Ortho Res,* 1167–1172.

Murray, M.P., Drought, A.B., Kory, R.C. (1964). Walking patterns of normal men. *J Bone Joint Surg Am, 46,* 335.

Neuman, P., Englund, M., Kostogiannis, I., et al. (2008). Prevalence of tibiofemoral osteoarthritis 15 years after nonoperative treatment of anterior cruciate ligament injury: A prospective cohort study. *Am J Sports Med, 36,* 1717–1725.

O'Connor, J.J., Shercliff, T.L., Biden, E., et al. (1989). The geometry of the knee in the sagittal plane. *Proc Inst Mech Eng H, 203*(4), 223–233.

Özkaya, N., Nordin, M. (1999). *Fundamentals of Biomechanics: Equilibrium, Motion, and Deformation (2nd ed.).* New York: Springer-Verlag.

Perry, J., Norwood, L., House, K. (1977). Knee posture and biceps and semimembranosus muscle action in running and cutting (an EMG study). *Trans Orthop Res Soc, 2,* 258.

Ramsey, D.K., Wretenberg, P.F. (1999). Biomechanics of the knee: Methodological considerations in the in vivo kinematic analysis of the tibiofemoral and patellofemoral joint. Review paper. *Clin Biomech, 14,* 595–611.

Reilly, D.T., Martens, M. (1972). Experimental analysis of the quadriceps muscle force and patellofemoral joint reaction force for various activities. *Acta Orthop Scand, 43,* 126.

Reuleaux, F. (1876). *The Kinematics of Machinery: Outline of a Theory of Machines.* London, UK: Macmillan.

Selvik, G. (1978). Roentgen stereophotogrammetry in Lund, Sweden. In A.M. Coblenz, R.E. Herron (Eds.). *Applications of Human Biostereometrics. Proc. SPIE (166),* 184–189.

Selvik, G. (1983). Roentgen stereophotogrammetry in orthopaedics. In R.E. Herron (Ed.). *Biostereometrics '82. Proc SPIE (361),* 178–185.

Solomonow, M., D'Ambrosia, R. (1994). Neural reflex arcs and muscle control of knee stability and motion. In W.N. Scott (Ed.). *The Knee.* New York: Mosby Elsevier, 107–120.

Walker, P.S., Erkman, M.J. (1975). The role of the menisci in force transmission across the knee. *Clin Orthop Relat Res, 109,* 184–192.

Walker, P.S., Hajek, J.V. (1972). The load-bearing area in the knee joint. *J Biomech, 5*(6), 581–589.

Walker, P.S., Yildirim, G., Sussman-Fort, J., et al. (2006). Relative positions of the contacts on the cartilage surfaces of the knee joint. *Knee, 13,* 382–388.

Wilson, S.A., Vigorita, V.J., Scott, W.N. (1994). Anatomy. In N. Scott (Ed.). *The Knee.* Philadelphia: Mosby Elsevier, 17.

Winter, D.A. (1990). *Biomechanics and Motor Control of Human Behaviour (2nd ed.).* New York: John Wiley and Sons.

Wretenberg, P., Németh, G., Lamontagne, M., et al. (1996). Passive knee muscle moment arms measured in vivo with MRI. *Clinical Biomech, 11*(8), 439–446.

Zavatsky, A.B., O'Connor, J.J. (1992). A model of the human knee ligaments in the sagittal plane. Part 1: Response to passive flexion. *Proc Inst Mech Eng H, 206*(3), 125–134.

Zernicke, R.F., Garhammer, J., Jobe, F.W. (1977). Human patellar tendon rupture. *J Bone Joint Surg Am, 59,* 179–183.

Zhao, D., Banks, S.A., D'Lima, D.D., et al. (2007). In vivo medial and lateral tibial loads during dynamic and high flexion activities. *J Orthop Res,* 593–602.

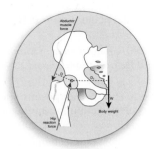

C H A P T E R **8**

Biomechanics of the Hip

Ali Sheikhzadeh, Owen Kendall, and Victor Frankel

Introduction

The primary function of the hip joint is to support the weight of the head, arms, and trunk during daily activities such as walking, running, and climbing stairs. Its ability to transmit forces between the torso and the lower extremities is vital to the normal functioning of the human body. The joint's ball-and-socket configuration provides it with inherent stability while supporting substantial mobility. Injury to and diseases of the hip are quite common, and derangement of the hip can produce altered stress distribution in the joint cartilage and bone. This can lead to degenerative arthritis, as well as substantial functional limitations including difficulty walking, dressing, driving, and lifting and carrying objects. This chapter will address the functional anatomy of the hip joint and then deal with fundamental kinetics and kinematics considerations such as range of motion and static and dynamic forces on the hip joint.

Anatomic Considerations

The hip joint is composed of the pelvic acetabulum, the head of the femur, and the femoral neck, and is controlled and protected by the acetabular labrum, the joint capsule, and many powerful muscles (Fig. 8-1). When these fundamental structures work in tandem, the hip joint has substantial stability, flexibility, and strength.

THE ACETABULUM

The acetabulum is the concave portion of the ball-and-socket hip joint structure. The acetabulum is not completely spherical due to the acetabular notch in its inferior region, which makes it fundamentally horseshoe shaped (Fig. 8-2). Articular cartilage, covering the surface of the acetabulum, thickens peripherally and laterally, though most notably in the superior-anterior region of the dome. Articulation in the acetabulum occurs only on the horseshoe-shaped hyaline cartilage on the periphery of the lunate surface. The acetabular labrum, a fibrocartilaginous lip encircling and deepening the acetabulum, blends with the transverse acetabular ligament, which spans the acetabular notch and prevents inferior dislocation of the femoral head (Ferguson et al., 2003).

The cavity of the acetabulum faces obliquely forward, outward, and downward, and a malaligned acetabulum does not adequately cover the femoral head, often causing chronic dislocation and osteoarthritis. The center edge angle (the angle of Wiberg) and the angle of acetabular anteversion are the angles that describe how much coverage the acetabulum provides the femoral head (Fig. 8-3). The center edge angle denotes the extent to which the acetabulum covers the femoral head in the frontal plane; it is highly variable, but measures 35° to 40° on average in radiographs of adults. A normal center edge angle provides a protective shelf over the femoral head, whereas a more vertical configuration (a smaller angle) contains the femoral head less, increasing the risk of dislocation. The acetabular anteversion angle describes how much the acetabulum surrounds the femoral head within

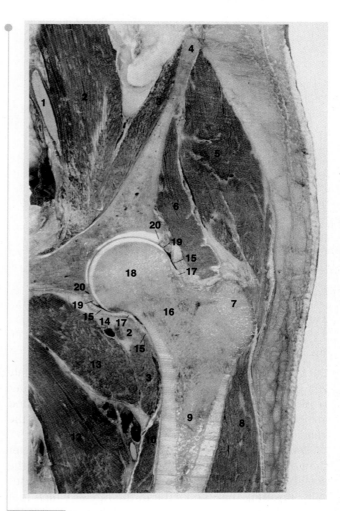

FIG. 8-1

The hip joint (front view): 1. External iliac artery. 2. Psoas major muscle. 3. Iliacus muscle. 4. Iliac crest. 5. Gluteus medius muscle. 6. Gluteus minimus muscle. 7. Greater trochanter. 8. Vastus lateralis muscle. 9. Shaft of femur. 10. Vastus medialis muscle. 11. Profunda femoris vessels. 12. Adductor longus muscle. 13. Pectineus muscle. 14. Medial circumflex femoral vessels. 15. Capsule of the hip joint. 16. Neck of femur. 17. Zona orbicularis of capsule. 18. Head of femur. 19. Acetabular labrum. 20. Rim of acetabulum. *Reprinted with permission from McMinn, R.H., Huchings, R.H.R. (1988). Color Atlas of Human Anatomy (2nd ed.). Chicago: Year Book Medical, 302.*

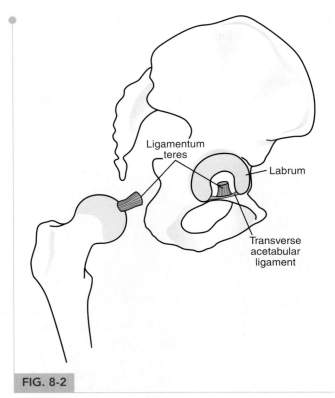

FIG. 8-2

Schematic drawing showing the lateral view of the acetabulum with the labrum and the transverse acetabular ligament intact. *Reprinted with permission from Kelly, B.T., Williams, R.J. III, Philippon, M.J. (2003). Hip arthroscopy: Current indications, treatment options, and management issues. Am J Sports Med, 31(6), 1020–1037.*

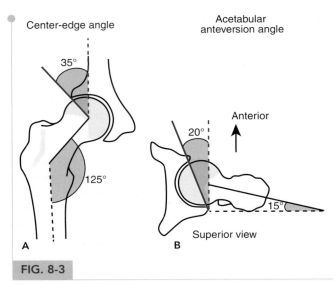

FIG. 8-3

The center edge angle (the angle of Wiberg) (**A**) and the angle of acetabular anteversion (**B**) are the angles that describe how much coverage the acetabulum provides the femoral head. The center edge angle denotes the extent to which the acetabulum covers the femoral head in the frontal plane; it is highly variable but measures 35° to 40° on average in radiographs of adults. The acetabular anteversion angle relates to how much the acetabulum surrounds the femoral head within the horizontal plane. The average value is approximate 20°. *Adapted from Neumann, D.H. (2002). Kinesiology of the Musculoskeletal System: Foundations for Physical Rehabilitation. St Louis, MO: Mosby.*

the horizontal plane. The average value is approximately 20°; pathologic increases in the angle of acetabular anteversion are associated with decreased joint stability and an increased likelihood of anterior dislocation of the head of the femur (Barrack, 2003).

The unloaded acetabulum has a smaller diameter than the femoral head. When the hip joint is loaded, the acetabulum deforms about the femoral head viscoelastically, meaning that the load and the rate at which that load is applied to the hip are inversely proportionate to the amount of deformation that occurs.

The acetabular labrum itself is extremely important to the proper functioning of the hip joint (Fig. 8-2). Unlike capsular tissue, labral tissue is made up predominantly of fibrocartilage. Arthroscopic visualization of damaged labral tissue has shown more extensive penetration of vascular tissue throughout the entire labral structure, which suggests more healing potential than previously believed (Ranawat et al., 2005). The labrum plays a role in containing the femoral head in extremes of motion, particularly in flexion. In conjunction with the joint

capsule, the labrum also acts as a load-bearing structure during flexion, meaning that subjects with a deficient labrum experience instability and capsular laxity. There is a potential for rotational instability and hypermobility of a hip joint with labral deficiencies due to the role the labrum plays in stabilizing and maintaining the congruity of the joint. Such instability can result in redundant joint capsule tissue and abnormal load distribution.

It is believed that there is interaction between the intra-articular fluid and the labrum, which decreases peak pressures in the joint (Ferguson et al., 2003). The labrum also plays a role in maintaining a vacuum within the joint space, which is produced by the acetabular fossa, a depression in the center of the acetabulum (Fig. 8-2). The vacuum produced by the acetabular fossa and maintained by the labrum has been shown to play a larger role than the capsuloligamentous structure in stabilizing the hip joint (Wingstrand et al., 1990).

THE FEMORAL HEAD

The femoral head, the convex component of the ball-and-socket configuration of the hip joint, forms two-thirds of a sphere. The articular cartilage covering the femoral head is thickest on the medial-central surface surrounding the fovea into which the ligamentum teres attaches and thinnest toward the periphery. Variations in thickness result in differences in strength and stiffness in various regions of the femoral head. As noted previously, the cartilage in the hip joint is viscoelastic, which influences the loading pattern on the femoral head based on the magnitude of the load applied.

The load-bearing area is concentrated at the periphery of the lunate surface of the femoral head at smaller loads, but it shifts to the center of the lunate surface and the anterior and posterior horns as loads increase (Von Eisenhart-Rothe et al., 1997). Studies done by Bergman et al. (1993, 1995), using an instrumented prosthetic head, show that the anterior and medial lunate surfaces transmit most of the load during daily activity, though direct measurements such as these are extremely difficult to attain due to various restrictions and considerations. Improper formation of the femoral head has also been shown to lead to osteoarthritis, which develops into further changes in load distribution during activity.

THE FEMORAL NECK

The femoral neck's structure also plays a role in the proper functioning of the hip joint, especially in terms of its angular relationships with the femoral shaft. The two most relevant angles are the neck-to-shaft angle, known as the angle of inclination (Fig. 8-4), and the torsion angle, which is the angle between the axis through the femoral head and neck and the axis through the femoral condyles (Fig. 8-5).

The inclination angle is about 140° to 150° at birth and gradually reduces to approximately 125°, with a range of 45° (90°–135°) in adulthood (Ogus, 1996). An angle greater than 125° produces a condition known as coxa valga, while an angle less than 125° is known as coxa vara (Figs. 8-4 and 8-6). These abnormal angles shift the alignment between the acetabulum and femoral head, altering the hip moments by changing the lever arm and the impact of forces applied to the joint by the upper body. Although there are some benefits to both coxa valga and coxa vara, the negative effects, as shown in Figure 8-6, outweigh these benefits and attest to the value and importance of the median (125°) angle of inclination.

The torsion angle reflects medial rotary migration of the lower limb bud that occurs in fetal development; it is commonly estimated at 40° in newborns, but decreases substantially in the first two years of life. A torsion angle of between 10° and 20° is considered normal. Angles greater than 12°, known as anteversion, cause a portion of the femoral head to be uncovered and create a tendency toward internal rotation of the leg during gait to keep the femoral head in the acetabular cavity (Fig. 8-5B). Retroversion, an angle of less than 12°, produces a tendency toward external rotation of the leg during gait (Fig. 8-5C). Both are fairly common during childhood and are usually outgrown.

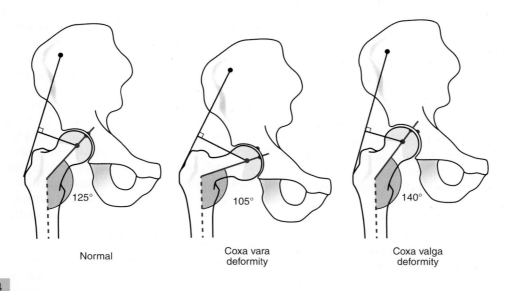

Normal   Coxa vara deformity   Coxa valga deformity

FIG. 8-4

The normal neck-to-shaft angle (angle of inclination of the femoral neck to the shaft in the frontal plane) is approximately 125°. The condition in which this angle is less than 125° is called coxa vara. If the angle is greater than 125°, the condition is called coxa valga.

The angle of inclination is about 140° to 150° at birth and gradually reduces to approximately 125°, with a range of 45° (90°–135°) in adulthood. *Modified from Callaghan, J.J., Rosenburg, A.G., Rubash, H.E. (2007). The Adult Hip. Philadelphia: Lippincott Williams & Wilkins.*

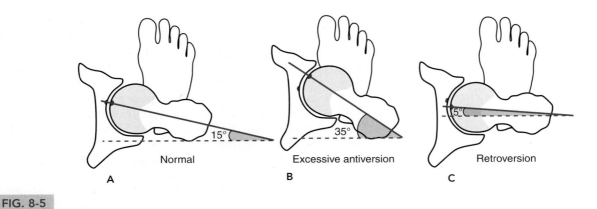

FIG. 8-5

This image shows the effect of various torsion angles, the torsion angle being that between the axis through the femoral head and neck and the axis through the femoral condyles. The torsion angle reflects medial rotary migration of the lower limb bud that occurred in fetal development; it is commonly estimated at 40° in newborns but decreases substantially in the first two years of life. A torsion angle of between 10° and 20° is considered normal, though angles greater than 12° (known as anteversion) and angles less than 12° (known as retroversion) are moderately common and impact internal and external rotation, respectively. *Modified from Neumann, D.H. (2002). Kinesiology of the Musculoskeletal System: Foundations for Physical Rehabilitation. St. Louis, MO: Mosby.*

Abnormal inclination and torsion angles, as well as structural deformities, can also lead to femoroacetabular impingement, which is caused by bone of the femoral neck abutting that of the acetabulum. Impingement, caused by structural deformities, is currently considered the single most common cause of osteoarthritis (Maheshwari et al., 2007).

The interior of the femoral head and neck are composed of cancellous bone with trabeculae organized into medial and lateral trabecular systems (Figs. 8-7A and B). The forces and stresses on the femoral head—most specifically the joint reaction force—parallel the trabeculae of the medial system (Frankel, 1960), which underscores their importance in supporting this force. The epiphyseal plates are at right angles to the trabeculae of the medial system and are believed to be perpendicular to the joint reaction force (Inman, 1947). It is likely that the lateral trabecular system resists the compressive force on the femoral head produced by contraction of the abductor muscles—the gluteus medius, the gluteus minimus, and the tensor fasciae latae. The thin shell of cortical bone around the superior femoral neck progressively thickens in the inferior region.

With aging, the femoral neck gradually undergoes degenerative changes: the cortical bone thins and is cancellated, and the trabeculae are gradually resorbed (see Fig. 2-54). These changes may predispose the femoral neck to fracture, which is discussed in greater detail in Chapter 2, Biomechanics of Bone. It is worth noting that the femoral neck is the most common fracture site in the elderly (Case Study 8-1).

THE HIP CAPSULE AND MUSCLES SURROUNDING THE HIP JOINT

The hip capsule, composed of three capsular ligaments, is an important stabilizer of the hip joint, especially in extremes of motion, where it acts as a check rein to prevent dislocation (Johnston et al., 2007). The capsular ligament is made up of three reinforcing ligaments—two found anteriorly and one posteriorly (Fig. 8-8). The capsule is thickened anterosuperiorly, where the predominant stresses occur, and is relatively thin and loosely attached posteroinferiorly (Lavigne et al., 2005). Due to the rotation that occurs in fetal development of the hip joint, the capsular ligaments are coiled around the femoral neck in a clockwise direction, meaning that they are tightest in combined extension and medial rotation of the hip joint, which further coils the ligaments, and loosest in flexion and lateral rotation, which uncoils the ligaments. Understanding how this influences hip stability in various positions, such as when a person sits cross-legged, further destabilizing the joint through adduction, can inform us of why forces applied up the femoral shaft can push the femoral head out of the acetabulum, causing dislocation (Barrack, 2003).

The strength and flexibility of the more than 27 separate musculotendinous units that cross the hip joint are vital to the joint's proper functioning. To achieve a realistic estimate of joint forces, a biomechanical model of the hip joint should include the agonist-antagonist muscle forces in three-dimensional dynamic environments, though such three-dimensional modeling is too complex for this

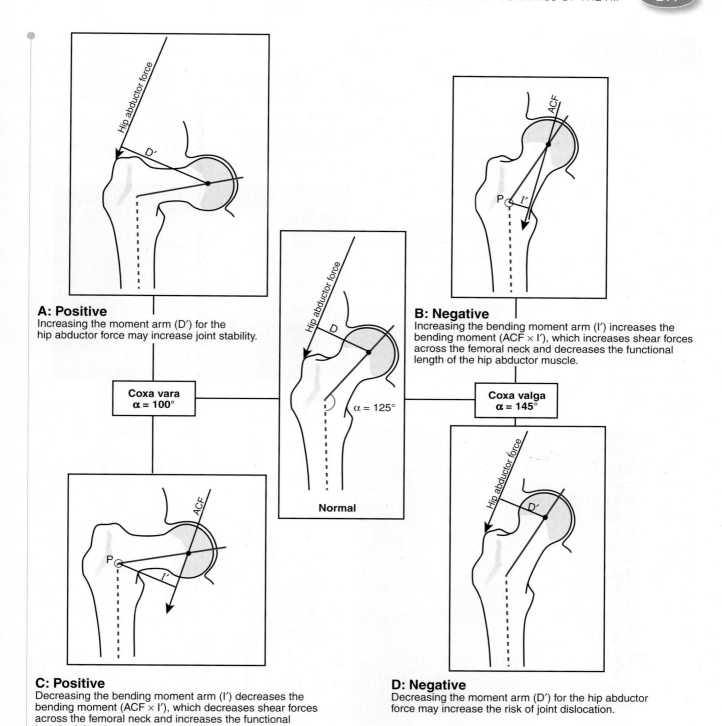

A: Positive
Increasing the moment arm (D′) for the
hip abductor force may increase joint stability.

Coxa vara
α = 100°

Normal

B: Negative
Increasing the bending moment arm (I′) increases the
bending moment (ACF × I′), which increases shear forces
across the femoral neck and decreases the functional
length of the hip abductor muscle.

Coxa valga
α = 145°

C: Positive
Decreasing the bending moment arm (I′) decreases the
bending moment (ACF × I′), which decreases shear forces
across the femoral neck and increases the functional
length of the hip abductor muscle.

D: Negative
Decreasing the moment arm (D′) for the hip abductor
force may increase the risk of joint dislocation.

FIG. 8-6

The negative and positive biomechanical effects of coxa vara and
coxa valga are contrasted. As a reference, a hip with a normal
angle of inclination (α = 125°) is shown in the center of the
display. D is the internal moment arm used by hip abductor force;

I is the bending moment arm across the femoral neck. *Modified
from Neumann, D.H. (2002). Kinesiology of the Musculoskeletal
System: Foundations for Physical Rehabilitation. St Louis, MO:
Mosby.*

Femoral Intertrochanteric Fractures

An 80-year-old woman fell from a standing position after losing her balance. She presented with sharp pain in her hip and an inability to stand or walk by herself. She was transported to the ER and, after careful examination and x-ray evaluation, a right intertrochanteric fracture was diagnosed (Case Study Fig. 8-1).

The radiograph illustrates an unstable right femoral intertrochanteric fracture with separation of the lesser trochanter. The image shows osteoporotic changes characteristic of the aging process. The decrease in the bone mass at the femoral neck leads to reduced bone strength and stiffness as a result of diminution in the amount of cancellous bone and thinning of cortical bone, increasing the likelihood of fracture at the weakest level.

In the fall, the magnitude of the compressive forces at the femoral neck overcame its stiffness and strength. In addition, the tensile forces produced by protective contraction of muscles such as the iliopsoas generated a traction fracture at the lesser trochanter level.

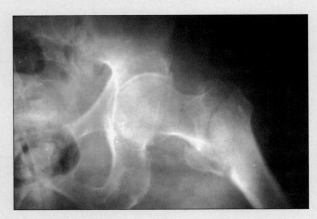

Case Study Figure 8-1

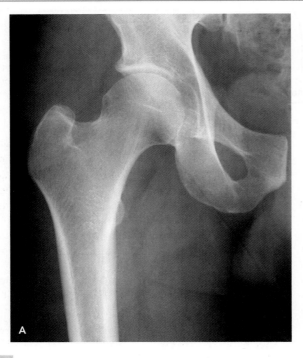

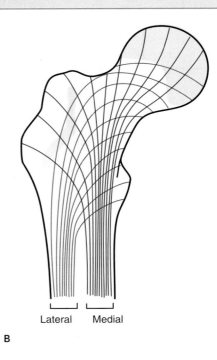

Lateral   Medial

B

FIG. 8-7

A. The interior of the femoral head and neck are composed of cancellous bone with trabeculae organized into medial and lateral trabecular systems. The roentgenogram of a femoral neck shows these trabecular systems. The thin shell of cortical bone around the superior femoral neck progressively thickens in the inferior region. **B.** A simplified drawing of the medial and lateral trabecular systems. *Modified from Brody, L.T., Hall, C.M., (2005). Therapeutic Exercise: Moving Toward Function (2nd ed.). Baltimore: Lippincott Williams & Wilkins.*

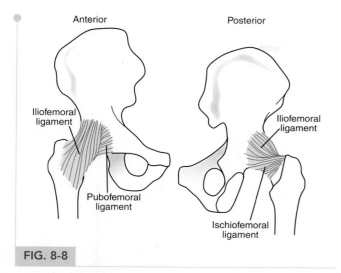

FIG. 8-8

Shown here are the primary structures of the capsular ligaments, which act as a check rein in extremes of motion to prevent dislocation. Due to the rotation that occurs in the fetal development of the hip joint, the capsular ligaments are coiled around the femoral neck in a clockwise direction, meaning that they are tightest in combined extension and medial rotation of the hip joint, which further coils the ligaments, and loosest in flexion and lateral rotation, which uncoils the ligaments. *Modified from Kelly, B.T., Williams, R.J. III, Philippon, M.J. (2003). Hip arthroscopy: Current indications, treatment options, and management issues. Am J Sports Med, 31(6), 1020–1037.*

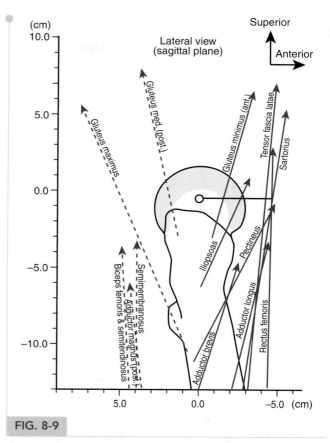

FIG. 8-9

This image uses two-dimensional vectors to describe the line of action of individual muscular forces acting on the sagittal plane of the hip joint. *Modified from Neumann, D.H. (2002). Kinesiology of the Musculoskeletal System: Foundations for Physical Rehabilitation. St Louis, MO: Mosby.*

chapter. Figures 8-9 and 8-10 represent the line of action of the joint musculature, a complex modeling of muscular forces on the hip joint. Simplification of these forces can be accomplished by combining vectors, though further analysis of these musculoskeletal forces will not be done here.

Musculature in other portions of the lower limbs, from the knee through the ankle and foot, has also been shown to influence the functioning of the hip and vice versa. Chronic hyperextension of the knee due to weak quadriceps and short ankle plantar flexors, for instance, transmits an anterior force to the head of the femur. This may contribute to anterior compression of the head of the femur in the acetabulum. Stretched and weak lateral hip rotators can also cause problems, leading the hip to function in chronic medial rotation and resulting in excessive pronation of the foot. Sufficient strength and flexibility in each of the 27 musculotendinous units, as well as the musculature throughout the lower limbs, is vital to the effective functioning of the hip joint.

A background in functional anatomy is fundamental to understanding the biomechanics of the hip joint by allowing us to comprehend how force is transferred from the acetabulum to the femoral head, through the femoral neck, and ultimately to the femur. The effectiveness of this force transmission, which can be understood as the joint reaction force between the acetabulum and femoral head, depends on the anatomic configuration of the femoral neck and head in relation to the acetabulum (Fig. 8-11). Due to the three-dimensional nature of this structure, the contact and force transmissions can be radically different from one individual to the next. Force transmission can also be influenced by changes in activity, various arthritic diseases, and other pathology.

Kinematics

Hip motion takes place in all three planes: sagittal (flexion-extension), frontal (abduction-adduction), and transverse (internal-external rotation) (Fig. 8-12). Understanding the range of motion of the hip joint is important in understanding an individual's motion requirements for returning to work and other activities.

Motion is greatest in the sagittal plane, where the range of flexion is from 0° to approximately 140° and the range of extension is from 0° to 15°. The range of abduction is from 0° to 30°, whereas that of adduction is somewhat less, from 0° to 25°. External rotation ranges from 0° to 90° degrees and internal rotation from 0° to 70° when the hip joint is

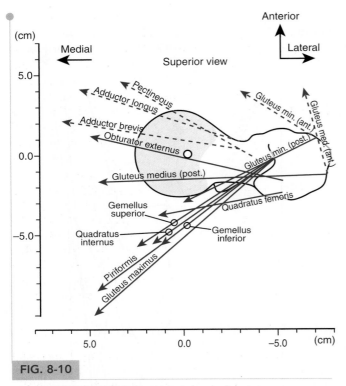

FIG. 8-10

This image, similar to Figure 8-9, uses two-dimensional vectors to describe the line of action of individual muscular forces acting on the transverse plane of the hip joint. *Modified from Neumann, D.H. (2002). Kinesiology of the Musculoskeletal System: Foundations for Physical Rehabilitation. St Louis, MO: Mosby.*

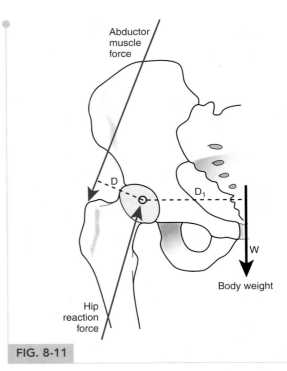

FIG. 8-11

This image displays the joint reaction force between the acetabulum and femoral head, which depends on the anatomic configuration of the femoral neck and head in relation to the acetabulum. Due to the three-dimensional nature of this structure, the contact and force transmissions can be radically different from one individual to another. Force transmission can also be influenced by changes due to activity, various arthritic diseases, and other pathology. *Modified from Oatis, C.A. (2009). Kinesiology (2nd Ed.). Baltimore: Lippincott Williams & Wilkins.*

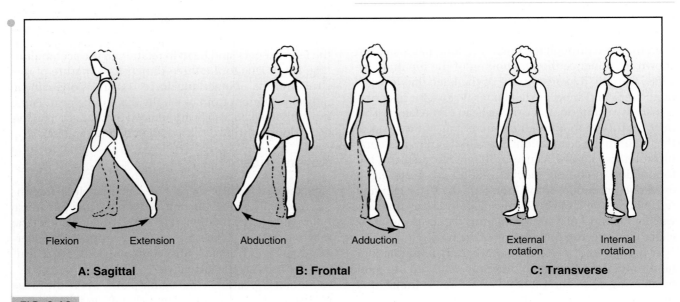

FIG. 8-12

Movements of the hip joint. **A.** Sagittal: Flexion-extension. **B.** Frontal: Abduction-adduction. **C.** Transverse: External rotation-internal rotation.

TABLE 8-1

Mean Values for Maximum Hip Motion in Three Planes during Common Activities

Activity	Plane of Motion	Recorded Value (Degrees)
Tying shoe with foot on floor	Sagittal	124
	Frontal	19
	Transverse	15
Tying shoe with foot across opposite thigh	Sagittal	110
	Frontal	23
	Transverse	33
Sitting down on chair and rising from seated position	Sagittal	104
	Frontal	20
	Transverse	17
Stooping to obtain object from floor	Sagittal	117
	Frontal	21
	Transverse	18
Squatting	Sagittal	122
	Frontal	28
	Transverse	26
Ascending stairs	Sagittal	67
	Frontal	16
	Transverse	18
Descending stairs	Sagittal	36

Mean values for 33 normal men. *Data are from Johnston, R.C., Smidt, G.L. (1970). Hip motion measurements for selected activities of daily living. Clin Orthop, 72, 205.*

flexed. Less rotation occurs when the hip joint is extended because of the restricting function of the soft tissues.

The clinical goal is commonly to restore a patient's functional movements. One indicator for evaluating a patient's ability to engage in physical activities is to assess his or her range of motion during activities involved in daily living. The range of motion in three planes during common daily activities—tying a shoe, sitting down on a chair and rising from it, picking up an object from the floor, and climbing stairs—was measured electrogoniometrically in 33 normal men by Johnston and Smidt (1970). The mean motion during these activities is shown in Table 8-1. Maximal motion in the sagittal plane (hip flexion) was needed for tying the shoe and bending down to squat to pick up an object from the floor. The greatest motion in the frontal and transverse planes was recorded during squatting and during shoe tying with the foot across the opposite thigh. The values obtained for these common activities indicate that hip flexion of at

least 120° and external rotation of at least 20° are necessary for carrying out daily activities in a normal manner.

Range of motion during daily activities should be interpreted with caution. Reported ranges of motion have been shown to be influenced by age, speed of movement, and environmental task constraints such as chair and stair height. Researchers such as Mulholland and Wyss (2001) have discussed the implications of cultural insensitivity regarding daily activities. In many parts of the world a chair is not commonly used at home or at work and sitting on the floor without support, sitting cross-legged, or kneeling are more common than they are in western countries. It has even been suggested that rural and urban lifestyles in varying geographic locations might require divergent approaches when considering daily physical activities (Mulholland and Wyss, 2001). The range of motion requirements for certain activities related in this chapter are based on western standards and are therefore unlikely to be culturally transferable.

RANGE OF MOTION IN WALKING

The range of motion of the hip joint during gait has been measured electrogoniometrically in all three planes. Measurements in the sagittal plane during level walking (Murray, 1967) showed that the joint was maximally flexed during the late swing phase of gait, as the limb moved forward for heel-strike. The joint extended as the body moved forward at the beginning of stance phase. Maximum extension was reached at heel-off. The joint reversed into flexion during swing phase and again reached maximal flexion, 35° to 40°, prior to heel-strike. Figure 8-13A shows the pattern of hip joint motion in the sagittal plane during a gait cycle and allows a comparison of this motion with that of the knee and ankle.

Motion in the frontal plane (abduction-adduction) and transverse plane (internal-external rotation) during gait (Johnston and Smidt, 1969) is illustrated in Figure 8-13B. Abduction occurred during swing phase, reaching its maximum just after toe-off; at heel-strike, the hip joint reversed into adduction, which continued until late stance phase. The hip joint was externally rotated throughout the swing phase, rotating internally just before heel-strike. The joint remained internally rotated until late stance phase, at which point it again rotated externally. The average ranges of motion recorded for the 33 normal men in this study were 12° for the frontal plane and 13° for the transverse plane.

IMPACT OF AGE ON RANGE OF MOTION

As people age, they use a progressively smaller portion of the range of motion of the lower extremity joints during ambulation. Murray et al. (1969) studied the

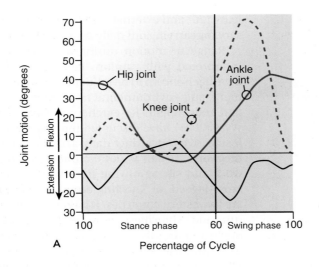

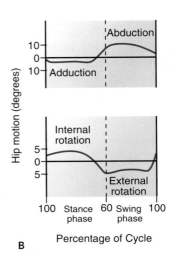

FIG. 8-13

A. Range of hip joint motion in the sagittal plane for 30 normal men during level walking, one gait cycle. The ranges of motion for the knee and ankle joints are shown for comparison. *Adapted from Murray, M.P. (1967). Gait as a total pattern of movement. Am J Phys Med, 46, 290.* **B.** A typical pattern for range of motion in the frontal plane **(top)** and transverse plane **(bottom)** during level walking, one gait cycle. *Adapted from Johnston, R.C., Smidt, G.L. (1969). Measurement of hip-joint motion during walking. Evaluation of an electrogoniometric method. J Bone Joint Surg, 51A, 1083.*

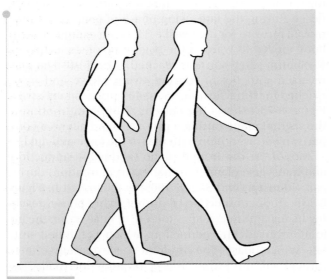

FIG. 8-14

Differences in the sagittal body positions of older men **(left)** and younger men **(right)** at the instant of heel-strike. The older men showed shorter strides, a decreased range of hip flexion and extension, decreased plantar flexion of the ankle, and a decreased heel-to-floor angle of the tracking limb; they also showed less dorsiflexion of the ankle and less elevation of the toe of the forward limb. *Reprinted with permission from Murray, M.P., Kory, R.C., Clarkson, B.H. (1969). Walking patterns in healthy old men. J Gerontol, 24, 169–178.*

walking patterns of 67 normal men of similar weight and height ranging in age from 20 to 87 years and compared the gait patterns of the older and younger men. The differences in the sagittal body positions of the two groups at the instant of heel-strike are illustrated in Figure 8-14. The older men had shorter strides, a decreased range of hip flexion and extension, decreased plantar flexion of the ankle, and a decreased heel-to-floor angle of the tracking limb; they also showed reduced dorsiflexion of the ankle and diminished elevation of the toe of the forward limb.

One of the fundamental reasons for modifications in joint kinetics that come with aging involves changes in motor control, loss of motor units, and decreases in fast twitch muscle fibers. Aging muscles, unable to produce as much force per equivalent time period in relation to young muscles, lead to substantial force production modifications. Large deficits have been shown in the plantar flexor groups compared with other lower extremity muscle groups, which likely play a role in age-related changes in the joint ranges of motion and gait mechanics (Boyer et al., 2008).

Kinetics

Kinetics studies have shown that substantial forces act on the hip joint during simple activities (Hurwitz and Andriacchi, 1997, 1998). A biomechanical analysis of the hip joint can address either the forces acting on the joint as a simple static snapshot while standing on one or both legs, or the forces acting on the joint during a dynamic task (e.g., climbing stairs, walking, or running). The main objectives of these biomechanical analyses are as follows:

1. To provide an understanding of the factors involved in producing the total forces acting on the joint, as well as their magnitude

2. To provide a better understanding of activities that may well be harmful to joints and the surrounding soft tissue

3. To understand the functioning of a healthy versus a diseased joint during various activities

4. To design treatment and evaluation plans for patients with hip problems or total joint replacements

5. To understand the structure of the hip joint for optimal performance

Formulating a comprehensive, dynamic model of the hip joint would provide an understanding of the actual joint forces produced by activity, but is a challenge due to the complexity of the internal forces acting on the joint, as well as the difficulties intrinsic in measuring the precise anatomic parameters. A comprehensive model of the hip joint, for instance, should include the line of action of muscles crossing it with respect to its axes of rotation in the sagittal, frontal, and transverse planes, as described previously and shown in Figures 8-9 and 8-10. Additionally, such a model must take into consideration the dynamic changes of these parameters that occur during joint motion.

INDIRECT MEASUREMENT OF JOINT FORCES

Frequently, even the simplest models of external forces generated by gravity acting on the body provide crucial functional and clinical information about a joint. A simplified free-body diagram of the hip joint of a patient standing on a single leg can provide an indirect measurement of the joint forces involved. Though this does not take into account the influence of the abductor muscle group (gluteus medius and minimus muscles) on the hip joint as the main stabilizer during single-legged stance, it can give a relatively good estimate of joint forces (Kumagai et al., 1997). Calculating the joint force during two-legged stance is substantially easier and is considered more accurate than the one-leg stance due to the simplified nature of the musculature involved and the inherent stability of the hip joint in that position.

During two-legged stance, the line of gravity of the superincumbent body passes posterior to the pubic symphysis; an erect stance can thus be achieved through the stabilizing effect of the joint capsule and capsular ligaments alone, without muscle contraction. With no muscle activity to produce moments around the hip joint,

calculation of the joint reaction force becomes simple: the magnitude of the force on each femoral head during upright two-legged stance is one-half the weight of the superincumbent body. Each lower extremity is one-sixth body weight, meaning the reaction force on each hip joint will be one-third of body weight, which is one-half of the remaining two-thirds. When the muscles surrounding the hip joint contract to prevent swaying and to maintain an upright position of the body, as during prolonged standing, this force changes and becomes difficult to measure indirectly, though the force does tend to increase in proportion to the level of muscle activity.

The line of gravity of the superincumbent body shifts in all three planes when going from a two-legged stance to a single-leg stance. This produces moments around the hip joint that must be counteracted by muscle forces, which increases the joint reaction force. The magnitude of the moments, and hence the magnitude of the joint reaction force, depends on the position of the trunk and head, upper extremities, and position of the non-weight-bearing leg. Figure 8-15 demonstrates how the line of gravity in the frontal plane shifts with different positions of the upper body and inclinations of the pelvis. The shift in the line of gravity, and hence in the length of the lever arm of the gravitational force (the perpendicular distance between the line of gravity and the center of rotation in the femoral head), influences the magnitude of the moments about the hip joint and, consequently, the joint reaction force. Though most changes increase the joint reaction force, when the trunk is tilted over the hip joint, the gravitational force lever arm and the joint reaction force are minimized since the load line is in a more vertical position (Fig. 8-15). Nonetheless, following arthroplasty for arthritis, the abductor muscles are weak and atrophic as a result of the disease process and the surgery, meaning that external support such as a cane should be used until the abductor muscles are rehabilitated—best indicated by a lack of limping.

Figure 8-11 is an example of a simplified free-body diagram that could be used to calculate the joint reaction force on the hip joint during a one-legged stance. It is important to include the force's direction, magnitude, and distance from the center of rotation of the hip joint. Using free-body diagram methods allows one to compare the effects of certain conditions on the hip's joint reaction force such as bilateral stance, one-legged stance, and carrying external loads.

DIRECT MEASUREMENT: USING SURGICAL IMPLANTS

Although biomechanical models can produce indirect estimates of internal forces, the real-time continuous signal from an instrumented telemetric prosthesis can

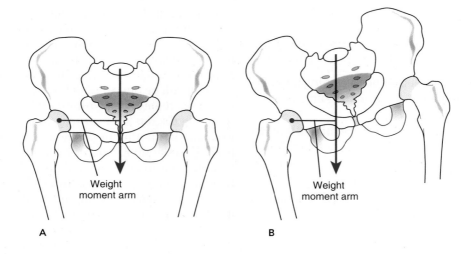

FIG. 8-15

These simplified drawings of the pelvis and lower body show that the line of gravity shifts in the frontal plane with different positions of the upper body and inclinations of the pelvis. **A.** The pelvis is in a neutral position. The gravity line falls approximately through the pubic symphysis. The lever arm for the force produced by body weight (the perpendicular distance between the gravity line and the center of rotation in the femoral head) influences the moment about the hip joint and hence the joint reaction force. **B.** The shoulders are maximally tilted to the left, supporting the hip joint. The gravity line has shifted and is now nearest the supporting hip. Because this shift minimizes the lever arm, the moment about the hip joint, as well as the joint reaction force are minimized.

give direct measurements of internal forces acting on the distal or proximal femur during daily activities such as walking and stair climbing.

Though implants historically used to determine in vivo load measurements were attached to hip replacements or directly to the bone as nail plates during hip fracture fixation, they are now intrinsically combined with hip replacements, mounted within ceramic enclosures. One of the more recent variations contains a low-power circuit that measures six load components, as well as the temperature and supplied voltage. It also includes a programmable memory, a pulse interval modulator, and a radio-frequency transmitter to deliver results of these in vivo measurements to researchers. These measurement-equipped hip replacements are then cemented into the femur to create stable hip replacements, producing improved in vivo measurements (Fig. 8-16).

Direct measurement produces more realistic estimates of internal forces and can be used to validate biomechanical models, providing useful insight into wear, strength, and fixation stability. Direct measurement is nonetheless difficult due to technological restriction, ethical considerations, and because of the practical problem that only a limited number of subjects can be studied. It should also be noted that the validity of the collected information depends on the extent to which joint mechanics and surrounding tissues have been altered, meaning that generalizations are often imprecise and cannot adequately be made.

Combining the information provided by both direct and indirect methods of force measurement can provide comprehensive and ideal methods with which to study joint forces, since both methods of measurement alone have intrinsic qualities and shortcomings. The pattern of loading for walking is similar in every study that has been done, but the magnitude of joint peak load differs. External measurements generally yield higher calculated peak force on the hip joint, whereas in vivo measurements using instrumented implants yield lower peak forces. There are many reasons for this, such as the method and instrumentation, the normal hip versus the "abnormal" instrumented implant, the gait velocity, and the age of the individual tested (Brand et al., 1994). Nonetheless, all studies comparing direct and indirect measurements present calculated results that can be considered to be in reasonable agreement with those measured using instrumented implants (Stansfield et al., 2003).

JOINT REACTION FORCE DURING ACTIVITIES

The loads on the hip joint during dynamic activities have been studied by many investigators (Draganich et al., 1980; English and Kilvington, 1979; Heller et al., 2005; Hurwitz et al., 2003; Rydell, 1966; Wang et al., 2006). It has been shown that peak hip forces vary during gait from 1.8 to 4.3 times body weight, with peak pressure occurring during heel-strike and early midstance (Andriacchi et al., 1980). These hip forces are related to the ground

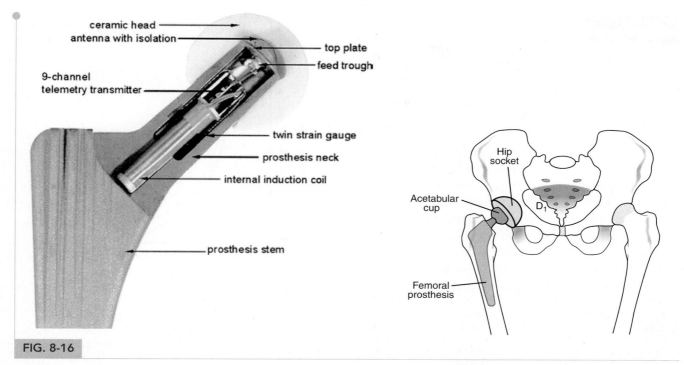

FIG. 8-16

The two images shown—the one on the **left** detailing the circuitry a hip prosthesis that is used to measure in vivo forces on the hip joint and the one on the **right** displaying the hip prosthetic as it would look cemented into the shaft of the femur, with its large ceramic head settled into the acetabulum, show how modern prostheses are attached and used for in vivo measurements. *Modified from Damm, P., Graichen, F., Rohlmann, A., et al. (2010). Total hip joint prosthesis for in vivo measurement of forces and moments. Med Eng Phys, 32, 95–100.* Drawing from Anatomical Chart Company.

reaction forces acting on the superior anterior acetabulum. For patients measured 11 to 31 months postoperatively, the average hip forces during fast walking and climbing stairs were about 250% body weight and slightly less going down stairs (Bergmann et al., 2001).

Taylor and Walker (2001) studied two patients over two-and-a-half years during various daily activities. The average peak distal femoral forces for one patient during various activities were as follows: for jogging, 3.6 times body weight; for stair descending, 3.1 times body weight; for walking, 2.8 times body weight; for treadmill walking, 2.75 times body weight; and for stair ascending, 2.8 times body weight. Bending moments about the mediolateral axis (flexion-extension) and anteroposterior axis (varus-valgus) peaked in the range of 4.7 to 7.6 body weight/cm^2 and 8.5 to 9.8 body weight/cm^2, respectively, over the follow-up period. During similar activities, however, forces and moments for the second subject were generally 45% to 70% less than those for the first subject due to the first subject's inadequate musculature around the knee.

Activities other than walking, such as stair ascending and descending, have yielded loads of around 2.6 to 5.5 times body weight when measured with an instrumented hip implant (Bergmann et al., 1995; Kotzar et al., 1991). The highest load magnitudes, as well as the most substantial hip contact forces, that occur during daily activities are measured during stair climbing and getting up from a low chair when the hip is flexed more than 100° (Bergmann et al., 2001; Catani et al., 1995; Johnston et al., 1979). Co-contraction of the biarticular muscles was evident during these activities, though analysis of the effect of that contraction is beyond the scope of this chapter. Beyond this, activities such as running and skiing yield forces determined to be up to eight times body weight in middle-aged and older people when measured with an accelerometer (van den Bogert et al., 1999) (Case Study 8-2).

IMPACT OF GENDER ON HIP KINETICS

In men, two peak forces were produced during the stance phase when the abductor muscles contracted to stabilize the pelvis. One peak of approximately four times body weight occurred just after heel-strike, and a large peak of approximately seven times body weight was reached just before toe-off (Fig. 8-17A). During foot flat, the joint reaction force decreased to approximately body weight because of the rapid deceleration of the body's center of

Case Study 8-2

Fatigue Fracture of the Hip

A 64-year-old, very active retired man experienced a femoral neck fracture after changing his training regimes to prepare for a marathon. The fracture was classified as a fatigue fracture caused by overload of the hip joint.

Case Study Figure 8-2 shows an MRI (frontal view) of the pelvis and both hip joints. The fracture is seen in the left femoral neck distal to the femoral head. The fracture is believed to have occurred during running and after an extensive change of training program. Due to high repetitive loading, muscle fatigue, and the change in the load pattern on the hip joint and femoral neck, the bone fractured.

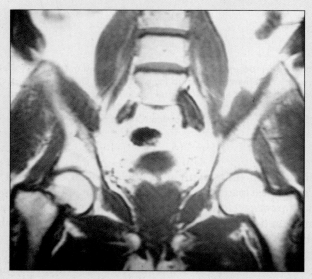

Case Study Figure. 8-2

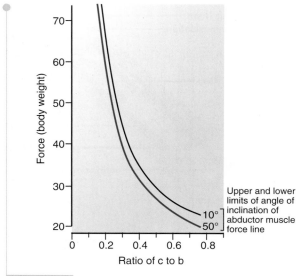

FIG. 8-17

The ratio of the abductor muscle force lever arm (*c*) to the gravitational force lever arm (*b*) is plotted against the joint reaction force on the femoral head in units of body weight. Because the line of application of the abductor muscle force (its angle of inclination in the frontal plane) has finite upper and lower limits (10° and 50°), the force envelope is plotted. The curve can be used to determine the minimal force acting on the femoral head during a one-leg stance if the ratio of *c* to *b* is known. *Courtesy of Dr. V.H. Frankel.*

In a recent study, it was shown that the greatest difference in gait between genders was found for the extension and adduction joint moments ($\cong$14.5). It was found that females walked with greater adduction angles at the hip, which contributed to the greater adduction moment, suggesting a narrower step width relative to pelvic width (Boyer et al., 2008). This implies that the hip joint stress for the female population is higher, not only in static situations, but also in dynamic activities, as compared with males.

IMPLANTS

As noted earlier, the ratio of the abductor muscle force lever arm to the gravitational force lever arm is a key factor influencing the magnitude of the joint reaction force on the femoral head (Fig. 8-11). Many researchers have addressed the importance of this ratio in respect to prosthetic replacements of the hip joint (Delp and Maloney, 1993; Free and Delp, 1996; Heller et al., 2005; Lim et al., 1999; Sutherland et al., 1999; Vasavada et al., 1994). Joint reaction forces can be decreased in several ways: (1) by altering the center of motion in the prosthetic design,

gravity. During the swing phase, the joint reaction force was influenced by contraction of the extensor muscles in decelerating the thigh, and the magnitude remained relatively low, approximately equal to body weight.

In women, the force pattern was the same but the magnitude was somewhat lower, reaching a maximum of only approximately four times body weight at late stance phase (Fig. 8-17B). The lower magnitude of the joint reaction force in the women may have been the result of several factors: a wider female pelvis, a difference in the inclination of the femoral neck-to-shaft angle, a difference in footwear, and differences in the general pattern of gait.

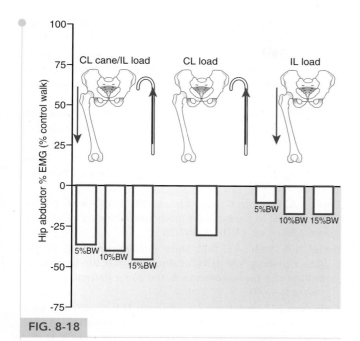

FIG. 8-18

The mean of normalized electromyography produced by the hip abductor muscles during three walking conditions: walking with contralateral (*CL*) cane and ipsilateral (*IL*) load, with contralateral cane and with ipsilateral load. Loads are 5%, 10%, and 15% BW. The hip abductor electromyogram is normalized to normal control walking conditions. *Reprinted with permission from Neumann, D.A. (1999). An electromyographic study of the hip abductor muscles as subjects with a hip prosthesis walked with different methods of using a cane and carrying a load. Phys Ther, 79, 1163–1173.*

and (2) by slightly changing the lever arm of the abductor muscles through surgery. Changing the center of location of the hip joint, for instance, can decrease the abduction force by more than 40%, thereby decreasing the generated abductor moment by almost 50% (Delp and Maloney, 1993).

Regarding the lever arm length, a short lever arm of the abductor muscle force, as in coxa valga (Fig. 8-4), results in a low ratio and thus an elevated joint reaction force. Moving the greater trochanter laterally during total hip replacement lowers the joint reaction force and increases the lever arm ratio by lengthening the muscle force lever arm (Free and Delp, 1996). Inserting a prosthetic cup deeper into the acetabulum, which decreases the gravitational force lever arm, can increase the ratio as well, thus lowering the joint reaction force. It is difficult, however, to change the lever arm ratio in such a way as to reduce the joint reaction force significantly because the curve formed from plotting the ratios becomes asymptotic when the ratio of the muscle force lever arm to gravitational force lever arm approaches 0.8 (Fig. 8-17).

EFFECT OF EXTERNAL SUPPORT ON HIP JOINT REACTION FORCE

Static analysis of the joint reaction force on the femoral head during walking with a cane demonstrates that a cane should be used on the side opposite the painful or operated hip. Neumann (1998) studied the effects of cane use in 24 subjects with a mean age of 63 years. During walking, the electromyographic activity of the hip abductor muscles was measured. Neumann found that use of a cane on the contralateral side of the affected hip joint, with careful instructions to use with near maximal effort, could reduce the muscle activity by 42% (Fig. 8-18). This calculates to a reduction of approximately one times body weight: from 2.2 times body weight with a cane to 3.4 times body weight without. These studies provide clinicians with important information about how patients with hip problems can moderate hip loading.

Summary

- The hip joint's primary function is to support the weight of the head, arms, and trunk during daily activities.
- The hip joint is a ball-and-socket joint composed of the acetabulum and femoral head.
- Other structures important to the hip joint are the femoral neck, the acetabular labrum, the hip capsule, and the muscles surrounding the hip.
- The thickness and mechanical properties of the cartilage on the femoral head and acetabulum vary based on particular needs and joint forces.
- Hip flexion of at least 120°, abduction of at least 20°, and internal and external rotation of at least 20° are necessary for carrying out daily activities in a normal manner.
- Direct (electrogoniometric measurement and implant devices) and indirect measurements (mathematical models) are used to determine joint forces.
- The magnitude of the joint reaction force acting on the hip varies as the position of the upper body changes in relation to the lower body, though it has been estimated that a force three times body weight acts on the hip joint during a single-leg stance with the pelvis in a neutral position.
- The magnitude of the hip joint reaction force is influenced by the ratio of the abductor muscle force lever arm to the gravitational force lever arm. A low ratio (closer to zero) yields a greater joint reaction force than does a high ratio (one closer to 0.8).

- During a gait cycle, the hip joint reaction force experienced in stance phase is equal to or greater than three to six times body weight, while being approximately equal to body weight during swing phase.
- An increase in gait velocity increases the magnitude of the hip joint reaction force in both swing and stance phase.
- Direct measurements have shown that reaction forces in the hip joint reach levels up to eight times body weight during activities such as running and skiing.
- Gait involves motion in all three planes (sagittal, frontal, and transverse), with rotation of the hip joint occurring throughout the swing and stance phases.
- Range of motion decreases with age due to changes in motor control, loss of motor units, and decreases in fast twitch muscle fibers.
- The most prominent gender differences regarding forces in gait are found in extension and adduction joint moments (higher moments in women), suggesting a narrower step width relative to pelvic width in women.
- The forces acting on an internal fixation device during the activities of daily living vary greatly depending on the nursing care and the therapeutic activities undertaken by the patient.
- The use of a cane on the contralateral side of the affected hip or a brace on the leg can substantially alter and frequently decrease the magnitude of the hip joint reaction force.

Practice Questions

1. Describe the role of the femoral neck in hip joint function.

2. Describe reasons for age-related kinetic and kinematic changes in the hip.

3. In which hand should a cane be held to provide the greatest biomechanical advantage?

4. Describe when hip internal rotation and hip adduction occur in the gait cycle.

5. Which physical activity demonstrates high hip contact forces?

REFERENCES

Andriacchi, T.P., Andersson, G.B.J., Fermier, R.W., Stern, D., Galanie, J.O. (1980). A study of lower-limb mechanics during stair-climbing. *J Bone Joint Surg, 62A*, 749.

Barrack, R.L. (2003). Dislocation after total hip arthroplasty: Implant design and orientation. *J Am Acad Orthop Surg, 11*(2),89–99.

Bergmann, G., Deuretzbacher, G., Heller, M, Graichen, F., Rohlmann, A., Strauss, J., Duda, G.N. (2001). Hip contact forces and gait patterns from routine activities. *J Biomech, 34*(7), 859–871.

Bergmann, G., Graichen, F., Rohlmann, A. (1993). Hip joint loading during walking and running measured in two patients. *J Biomech, 26*(8), 969–990.

Bergmann, G., Graichen, F., Rohlmann, A. (1995). Is staircase walking a risk for the fixation of hip implants? *J Biomech, 28*(5), 535–553.

Boyer, K.A., Beaupre, G.S., Andriacchi, T.P. (2008). Gender differences exist in the hip joint moments of healthy older walkers. *J Biomech, 41*(16), 3360–3365.

Brand, R.A., Pedersen, D.R., Davy, D.T., Kotzar, G.M., Heiple, K.G., Goldberg, V.M. (1994). Comparisons of hip force calculations and measurements in the same patient. *J Arthroplasty, 9*, 45–51.

Callaghan, J.J., Rosenburg, A.G., Rubash, H.E. (2007). *The Adult Hip.* Philadelphia: Lippincott Williams & Wilkins.

Catani, F., Hodge, A., Mann, R.W., Ensini, A., Giannini, S. (1995). The role of muscular co-contraction of the hip during movement. *Chir Organi Mov, 80*(2), 227–236.

Chao, E.Y.S. (2003). Graphic-based musculoskeletal model for biomechanical analyses and animation. *Med Eng Phys, 25*, 201–212.

Crowninshield, R.D., Johnston, R.C., Brand, R.A. (1978). The effects of walking velocity and age on hip kinematics and kinetics. *Clin Orthop Rel Res, 132*, 140–144.

Damm, P., Graichen, F., Rohlmann, A., Bender, A., Bergmann, G. (2010). Total hip joint prosthesis for in vivo measurement of forces and moments. *Med Eng Phys, 32*, 95–100.

Delp, S.L. Maloney, W. (1993). Effects of hip center location on the moment-generating capacity of the muscles. *J Biomech, 26*(5), 485–499.

Draganich, L. F., Andriacchi, T.P., Strongwater, A.M., Galante, J.O. (1980). Electronic measurement of instantaneous foot-floor contact patterns during gait. *J Biomech, 13*, 875.

English, T.A., Kilvington, M. (1979). In vivo records of hip loads using a femoral implant with telemetric output (a preliminary report). *J Biomed Eng, 1*(2), 111.

Ferguson, S.J., Bryant, J.T., Ganz, R., Ito, K. (2003). An in vitro investigation of the acetabular labral seal in hip joint mechanics. *J Biomech. 36*(2):171–1788.

Frankel, V.H. (1960). In *The Femoral Neck: Function, Fracture Mechanisms, Internal Fixation.* Springfield, IL: Charles C Thomas Publisher.

Free, S.A., Delp, S.L. (1996). Trochanteric transfer in total hip replacement: Effects on the moment arms and force-generating capacities of the hip abductors. *J Orthop Res, 14*(2), 245–250.

Heller, M.O., Bergmann, G., Kassi, J.P., Claes, L., Haas, N.P., Duda, G.N. (2005). Determination of muscle loading at the hip joint for use in pre-clinical testing. *J Biomech, 38*(5), 1155–1163.

Hurwitz, D.E., Andriacchi, T.P. (1998). Biomechanics of the hip. In J.J. Callaghan, A.G. Rosenberg, H.E. Rubash (Eds.). *The Adult Hip.* Philadelphia: Lippincott-Raven Publishers, 75–85.

Hurwitz, D.E., Andriacchi, T.P. (1997). Biomechanics of the hip and the knee. In M. Nordin, G.B.J. Andersson, M.H. Pope (Eds.). *Musculoskeletal Disorders in the Workplace. Principles and Practice.* Philadelphia: Mosby–Year Book, 486–496.

Hurwitz, D.E., Foucher, K.C., Andriacchi, T.P. (2003). A new parametric approach for modeling hip forces during gait.*J Biomech, 36*(1), 113–119.

Inman, V.T. (1947). Functional aspects of the abductor muscles of the hip. *J Bone Joint Surg, 29A,* 607.

Johnston, J.D., Noble, P.C., Hurwitz, D.E., Andriacchi, T. (2007). Biomechanics of the hip. In J.J. Callaghan, A.G. Rosenberg, H.E. Rubash (Eds.). *The Adult Hip (2nd ed.).* Philadelphia: Lippincott Williams & Wilkins.

Johnston, R.C., Brand, R.A., Crowninshield, R.D. (1979). Reconstruction of the hip. *J Bone Joint Surg, 61A*(5), 646–652.

Johnston, R.C., Smidt, G.L. (1970). Hip motion measurements for selected activities of daily living. *Clin Orthop, 72,* 205.

Johnston, R.C., Smidt, G.L. (1969). Measurement of hip-joint motion during walking. Evaluation of an electrogoniometric method. *J Bone Joint Surg, 51A,* 1083.

Kelly, B.T., Williams, R.J. III, Philippon, M.J. (2003). Hip arthroscopy: Current indications, treatment options, and management issues. *Am J Sports Med, 31*(6), 1020–1037.

Kotzar, G.M., Davy, D.T., Goldberg, V.M., Heiple, K.G., Berilla, J., Heiple, K.G. Jr, Brown, R.H., Burstein, A.H. (1991). Telemetrized in vivo hip joint force data. A report on two patients after total hip surgery. *J Orthop Res, 9,* 621–633.

Kumagai, M., Shiba, N., Higuchi, F., Nishimura, H., Inoue, A. (1997). Functional evaluation of hip abductor muscles with use of magnetic resonance imaging. *J Orthop Res, 15*(6), 888–893.

Lavigne, M., Kalhor, M., Beck, M., Ganz, R., Leunig, M. (2005). Distribution of vascular foramina around the femoral head and neck junction: Relevance for conservative intracapsular procedures of the hip. *Orthop Clin North Am, 36*(2),171–176.

Lim, L.A., Carmichael, S.W., Cabanela, M.E. (1999). Biomechanics of total hip arthroplasty. *Anat Rec, 257*(3), 110–116.

Maheshwar, A.V., Malik, A., Dorr, L.D. (2007). Impingement of the native hip joint. *J Bone Joint Surg Am, 89*(11):2508–2518.

McMinn, R.H., Huchings, R.H.R. (1988). *Color Atlas of Human Anatomy (2nd ed.).* Chicago: Year Book Medical, 302.

Mulholland, S.J., Wyss, U.P. (2001). Activities of daily living in non-Western cultures: Range of motion requirements for hip and knee joint implants. *Int J Rehabil Res, 24*(3), 191–198.

Murray, M.P. (1967). Gait as a total pattern of movement. *Am J Phys Med, 46,* 290.

Murray, M.P., Kory, R.C., Clarkson, B.H. (1969). Walking patterns in healthy old men. *J Gerontol, 24,* 169–178.

Neumann, D.A. (1998). Hip abductor muscle activity as subjects with hip prostheses walk with different methods of using a cane. *Phys Ther, 78*(5), 490–501.

Ogus, O. (1996). Measurement and relationship of the inclination angle, Alsberg angle and the angle between the anatomical and mechanical axes of the femur in males. *Surg Radiol Anat, 18*(1),29–31.

Paul, J.P. (1966). Biomechanics: The biomechanics of the hip-joint and its clinical relevance. *Proc R Soc Med, 59*(10), 943–948.

Ranawat, A.S., Bryan, T., Kelly, B.T. (2005). Anatomy of the hip: Open and arthroscopic structure and function. *Oper Tech Orthop, 15*(3),160–170.

Röhrle, H., Scholten R., Sigolotto, C., Sollbach, W., Kellner, H. (1984). Joint forces in the human pelvis-leg skeleton during walking. *J Biomech, 17,* 409–424.

Rydell, N.W. (1966). Forces acting on the femoral head prosthesis: A study on strain gauge supplied prostheses in living persons. *Acta Orthop Scand, 37*(Suppl 88), 1–132.

Stansfield, B.W., Nicol, A.C., Paul, J.P., Kelly, I.G., Graichen, F., Bergmann, G. (2003). Direct comparison of calculated hip joint contact forces with those measured using instrumented implants. An evaluation of a three-dimensional mathematical model of the lower limb. *J Biomech, 36*(7), 929–936.

Sutherland, A.G., D'Arcy, S., Smart, D., Ashcroft, G.P. (1999). Abductor muscle weakness and stress around acetabular components of total hip arthroplasty: A finite element analysis. *Int Orthop, 23*(5), 275–278.

Taylor, S J.G., Walker, P.S. (2001). Forces and moments telemetered from two distal femoral replacements during various activities. *J Biomech, 34*(7), 839–848.

van den Bogert, A.J., Read, L., Nigg, B.M. (1999). An analysis of hip joint loading during walking, running and skiing. *Med Sci Sports, 31*(1), 131–142.

Vasavada, A.N., Delp, S.L., Maloney, W.J., Schurman, D.J., Zajac, F.E. (1994). Compensating for changes in muscle length in total hip arthroplasty. Effects on the moment generating capacity of the muscles. *Clin Orthop, 302,* 121–133.

Von Eisenhart-Rothe, R., Eckstein, F., Muller-Gerbl, M., Landgraf, J., Rock, C., Putz, R. (1997). Direct comparison of contact areas, contact stress and subchondral mineralization in human hip joint specimens. *Anat Embryol (Berl), 195*(3), 279–288.

Wang, M.Y., Flanagan, S.P., Song, J.E., Greendale, G.A., Salem, G.J. (2006). Relationships among body weight, joint moments generated during functional activities, and hip bone mass in older adults. *Clin Biomech, 21*(7), 717–725.

Wingstrand, H., Wingstrand, A., Krantz, P. (1990). Intracapsular and atmospheric pressure in the dynamics and stability of the hip. A biomechanical study. *Acta Orthop Scand, 61*(3),231–235.

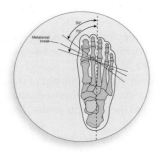

CHAPTER 9

Biomechanics of the Foot and Ankle

Marshall Hagins and Evangelos Pappas

Introduction

The primary task of the foot and ankle is to provide a stable, adaptable, and efficient interface between the body and ground for locomotion. This task requires the foot and ankle to be sufficiently pliable during early stance phase to conform to varying surface terrain, to absorb and translate forces while maintaining superincumbent whole-body stability, and to rapidly achieve sufficient rigidity during the late stance phase to propel the body forward using the rigid lever of the longitudinal arch.

This chapter will first describe motions of the foot and ankle at each region (rearfoot, midfoot, and forefoot) with special emphasis on an understanding of the triplanar motions of pronation and supination. Passive and active control of foot and ankle motion will then be described followed by descriptions of motions and muscular control during the gait cycle. The forces applied to the foot and ankle will be described including clinically relevant information regarding the role of shoe wear. Throughout, the chapter will introduce the clinical application of biomechanics relative to common pathologies and provide clinical case studies for further understanding of these concepts.

Structural Organization of the Foot and Ankle

REARFOOT, MIDFOOT, FOREFOOT

The ankle (talocrural joint) is composed of the articulation of the tibia, fibula, and talus whereas the foot is composed of all bones distal to the ankle joint (28 bones including sesamoids) (Fig. 9-1). The talus is considered a bone of both the ankle and the foot. The foot is most often described as having three functional units: rearfoot, midfoot, and forefoot. The rearfoot comprises the talus and calcaneus, the midfoot comprises the tarsal bones (navicular, three cuneiforms, and cuboid), and the forefoot comprises the metatarsals and phalanges (Fig. 9-1). The subtalar joint is part of the rearfoot, the transverse tarsal joint (talonavicular and calcaneocuboid) and intertarsal joints are part of the midfoot, and the tarsometatarsal joints and all of the more distal joints are part of the forefoot.

THE MEDIAL LONGITUDINAL ARCH

Two models exist to describe the medial longitudinal arch of the foot: the beam model and the truss model (Sarrafian, 1987). The beam model states that the arch is a curved beam made up of interconnecting joints whose structure depends on joint and ligamentous interconnections for stability. Tensile forces are produced on the inferior surface of the beam, and compressive forces are concentrated on the superior surface of the beam (Fig. 9-2). The truss model states that the arch has a triangular structure with two struts connected at the base by a tie rod. The struts are under compression, and the tie rod is under tension (Fig. 9-3). Both models have validity and can be demonstrated clinically.

The structure analogous to the tie rod in the truss model is the plantar fascia. The plantar fascia originates on the medial tuberosity of the calcaneus and spans the transverse tarsal, tarsometatarsal, and metatarsophalangeal joints to insert on the metatarsophalangeal plantar plates and collateral ligaments as well as the sesamoids. Dorsiflexion of the metatarsophalangeal joints places traction on the plantar fascia and causes elevation of the arch through a mechanism known as the "windlass effect" (Hicks, 1954) (Fig. 9-4). During toe-off in the gait cycle, the toes are dorsiflexed passively as the body passes over the foot and the plantar fascia tightens and acts to shorten the distance between the metatarsal heads and the heel, thus elevating the arch. The traction on the plantar fascia also assists in inverting the calcaneus through its attachment on the medial plantar aspect of the calcaneus.

The arch has both passive and active support. Huang et al. (1993) performed an in vitro study of the loaded foot and found that division of the plantar fascia resulted in a 25% decrease in arch stiffness. They found the three most important passive contributors to arch stability in order of importance were the plantar fascia, the long and short plantar ligaments, and the spring (calcaneonavicular) ligament.

Thordarson et al. (1995) performed a dynamic study of arch support by simulating stance phase of gait by applying proportional loads to tendons while the foot was loaded. In this study, the plantar fascia contributed the most to arch stability through toe dorsiflexion while the posterior tibialis contributed the most to active arch support. A clinical study of 14 feet post-plantar fasciotomy at greater than 4 years follow-up showed a decrease in arch height of 4.1 mm, thus supporting the truss model of arch stability (Daly et al., 1992). In a similar cadaveric study, Kitaoka et al. (1997) demonstrated a decrease in arch height and an angular change in the bones of the arch when tension on the posterior tibial tendon was released during simulated stance.

Injury to the plantar fascia (plantar fasciitis) is the most common foot condition. It manifests as pain on the plantar medial aspect of the heel with mechanical loading of the plantar fascia during gait. Typically the pain occurs most with the initial steps after a period of

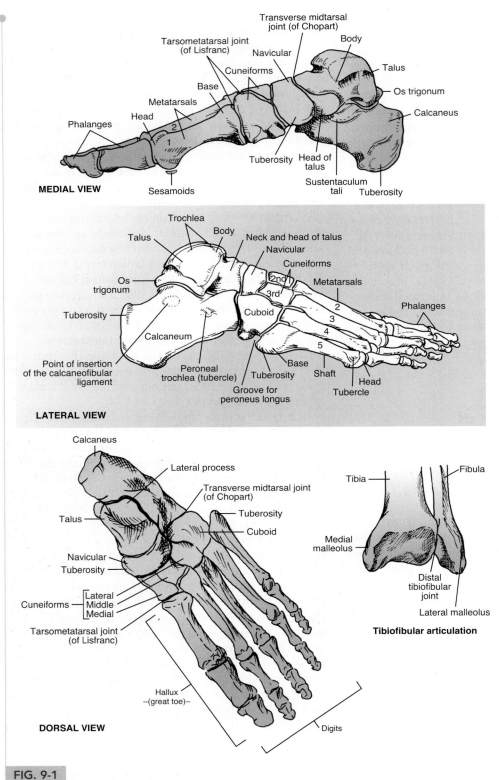

FIG. 9-1

Top. View of the medial aspect of the foot. **Middle.** View of the lateral aspect of the foot.
Bottom left. Superior view of the foot. **Bottom right.** Anterior view of the ankle mortise.

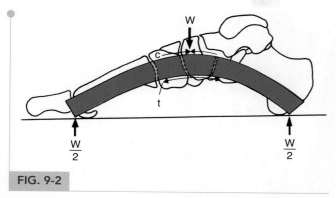

FIG. 9-2

The beam model of the longitudinal arch. The arch is a curved beam consisting of interconnecting joints and supporting plantar ligaments. Tensile forces are concentrated on the inferior beam surface; compressive forces are generated at the superior surface.

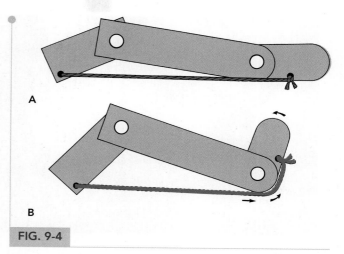

FIG. 9-4

A. Schematic of a truss. The *far left* wooden segment represents the hindfoot, the *middle* wooden segment represents the forefoot, and the *far right* wooden segment is the proximal phalanx. The rope is the plantar fascia. **B.** Dorsiflexion of the proximal phalanx raises the arch through traction on the plantar fascia.

inactivity. This condition is not well understood but is thought to be a direct consequence of excessive repetitive loading of the plantar fascia. Supporting this theory, decreased ankle dorsiflexion and high body mass index are risk factors for development of plantar fasciitis (McPoil et al., 2008).

SOFT TISSUES OF THE FOOT

The soft tissues of the foot are modified to provide traction, cushioning, and protection to the underlying structures. The dorsal skin of the foot is loosely attached as is evident by the sometimes dramatic dorsal foot swelling found during trauma or infection of the foot or ankle. The plantar skin is firmly attached to the underlying bones, joints, and tendon sheaths of the heel and forefoot by specialized extensions of the plantar fascia. This function of the plantar fascia is essential for traction

between the floor and the foot's weight-bearing skeletal structures to occur. During extension of the metatarsophalangeal joints, these plantar fascial ligaments restrict the movement of skin of the forefoot and plantar metatarsal fat pad (Bojsen-Moller and Lamoreux, 1979).

The heel pad is a highly specialized structure designed to absorb shock. The average heel pad area is 23 cm². For the average 70-kg man, the heel loading pressure is 3.3 kg/cm², which increases to 6 kg/cm² with running. At a repetition rate of 1,160 heel impacts per mile, the cumulative effect of running is impressive. These cumulative forces would normally result in tissue necrosis in other parts of the body (Perry, 1983). The heel pad consists of comma-shaped or U-shaped fat-filled columns arrayed vertically. The septae are reinforced internally with elastic transverse and diagonal fibers to produce a spiral honeycomb effect (Fig. 9-5). The multiple small closed cells are arranged to most effectively absorb and dissipate force. With age, septal degeneration and fat atrophy occur, which predispose the calcaneus and foot to injury (Jahss et al., 1992a, 1992b).

Kinematics of the Foot and Ankle

TERMINOLOGY

Due to the projection of the foot anteriorly from the coronal plane of the body, the terminology describing motion of the foot differs in several important ways from standard descriptions of motion in other areas of the body.

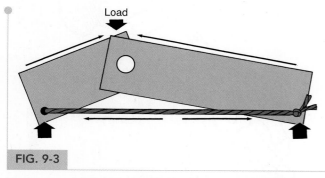

FIG. 9-3

The truss model of the longitudinal arch. The two wooden members, or struts, are connected at the base by a rope, or tie rod. The struts are analogous to bony structures of the foot and the tie rod is analogous to the plantar fascia. The shorter the tie rod, the higher the arch is raised.

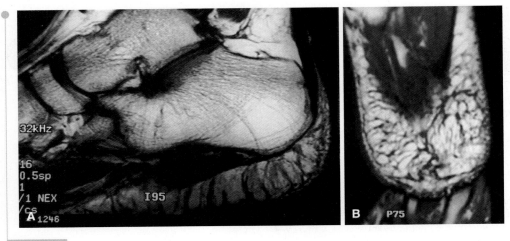

FIG. 9-5

Structure of a normal heel pad as seen on magnetic resonance imaging (MRI). **A.** Lateral view. Note vertically oriented fat-filled columns. **B.** Top view of the heel pad demonstrating the spiral structure of the septae, which separate the fat-filled cells.

First, abduction and adduction occur around a vertical axis rather than an anterior-posterior axis (motion that is called internal and external rotation elsewhere in the body) (Fig. 9-6). Abduction motion of the foot/ankle can be demonstrated by maintaining the foot flat on the floor in a seated position and sliding the surface of the foot laterally (without moving the tibia/fibula). Abduction/adduction motion of the foot/ankle primarily occurs at the subtalar and midtarsal joints and is quite limited. Second, the terms inversion and eversion represent motion in the coronal plane about an anterior/posterior axis (this motion is termed abduction/adduction elsewhere in the body). Inversion/eversion occurs primarily at the subtalar and midtarsal joints and can be demonstrated by moving the plantar surface of the foot to face medially (inversion) or laterally (eversion) (Fig. 9-7). Flexion/extension of the foot is termed dorsiflexion and plantarflexion respectively and occurs around a medial/lateral axis in the sagittal plane. This motion occurs primarily at the talocrural joint.

In addition to the alteration of common terminology just described, there is a second complication associated with an accurate description of ankle and foot motion. Unlike most of the other joints of the body that move around axes that are reasonably congruent with the standard orthogonal axes of the Cartesian coordinate system, the major joints of the ankle/foot (talocrural, subtalar, midtarsal) have axes of motion that are oblique to the standard orthogonal axes. The failure of the foot to conform to a predetermined frame of reference has been the source of much frustration and mystery to students

of kinesiology, particularly in reference to the terms supination and pronation.

Supination and pronation are the appropriate terms to describe motion around the oblique axes of the foot that occur at the talocrural, subtalar, and midtarsal joints. Supination and pronation describe motion

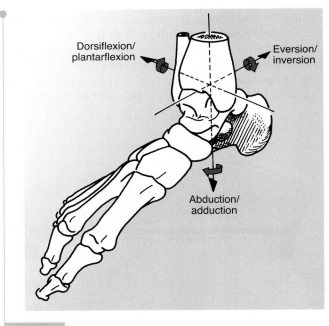

FIG. 9-6

Foot motion occurs around three axes.

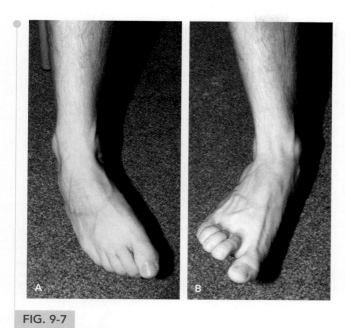

FIG. 9-7

A. During foot supination, the sole faces medially. **B.** During foot pronation, the sole faces laterally.

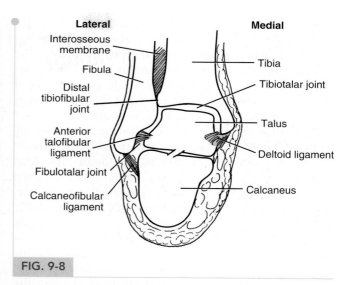

FIG. 9-8

Ankle joint complex composed of the tibiotalar, fibulotalar, and distal tibiofibular joints.

around a *single axis*. The obliquity of the single axis produces a single degree of freedom whose resultant motion can be described as having *components of motion* in three planes.

Supination and pronation are therefore often termed triplanar; meaning that supination has components of plantarflexion, adduction, and inversion and that pronation has components of dorsiflexion, abduction, and eversion. Because supination and pronation are single axis motions, the specific combination of the components of each motion are unvarying. In practical terms this means that supination or pronation are *always* associated with the same three components of motion, and furthermore, that evidence of any one of the three components of motion means that the other two components are also occurring. For example, it is impossible to have inversion of the subtalar joint without also simultaneously having adduction and plantarflexion of the subtalar joint. Non–weight-bearing images of whole foot supination and pronation (occurring primarily at the subtalar and midtarsal joints) are displayed in Figure 9-7.

ANKLE JOINT

The ankle is a simple hinge joint (one degree of freedom) consisting of the talus, medial malleolus, tibial plafond, and lateral malleolus (Fig. 9-8). The axis is set at a slightly oblique angle such that the lateral portion

(lateral malleolus) is posterior and inferior to the medial portion (medial malleolus) (Inman, 1976). The ankle axis can be estimated by palpating the tips of the malleoli (Fig. 9-9). The axis forms an approximately 10° angle with the horizontal medial-lateral axis in the coronal plane and approximately a 6° angle with the horizontal medial-lateral axis in the transverse plane. Technically, as suggested previously, this obliquity to the standard

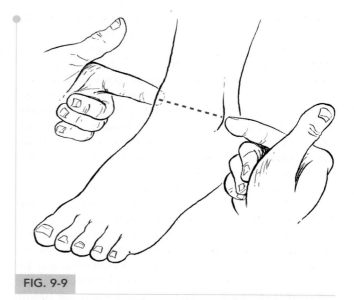

FIG. 9-9

The empirical axis of the ankle joint estimated through palpation of the malleoli. The axis angles downward and posteriorly, moving from medial to lateral.

axes requires the terminology of pronation and supination be applied to the ankle joint. However, the obliquity of the axis is so minor that the vast majority of the motion consists of dorsiflexion and plantarflexion. Consequently, in most clinical situations the components of the other two planes are ignored and the ankle is suggested to function in the sagittal plane alone.

The ankle joint is extremely stable primarily owing to bony congruency and ligamentous support. Bony congruency afforded by the two malleoli and tibial plafond form a "mortise" joint with the dome of the talus. The talus is shaped like a truncated cone, or frustum, with the apex directed medially (Inman, 1976). The talus is 4.2 mm wider anteriorly than posteriorly (Sarrafian, 1993a, 1993b). The anterior increase in dimension is important functionally because during dorsiflexion the anterior portion of the talus is compressed between the tibia and fibula (spreading the mortise slightly) and the ankle joint becomes "close packed" in a position of maximal stability.

Given the direction of the axis and the shape of the mortise, the talus is relatively free to move into dorsiflexion and plantarflexion around a medial-lateral axis but is highly constrained along the vertical and anterior-posterior axes (severely limiting transverse and coronal plane motion at the ankle joint). As will be discussed more fully later, this stability allows torque produced by transverse plane forces of the lower leg to be transmitted across the ankle joint to the subtalar joint, which typically results in subtalar joint pronation or supination.

The lateral ankle ligaments responsible for resistance to inversion and internal rotation are the anterior talofibular ligament, the calcaneofibular ligament, and the posterior talofibular ligament (Fig. 9-10). The superficial and deep deltoid ligaments are responsible for resistance to eversion and external rotation stress. The ligaments responsible for maintaining stability between the distal fibula and tibia are the syndesmotic ligaments. The syndesmotic ligaments consist of the anterior tibiofibular ligament, the posterior tibiofibular ligament, the transverse tibiofibular ligament, and the interosseous ligament (Fig. 9-11).

A wide range of normal motion for the ankle has been reported and depends on whether the motion is measured clinically with a goniometer or whether it is measured radiographically. Goniometric measurements yield a normal motion of 10° to 20° dorsiflexion and 40° to 55° plantarflexion. Lundberg et al. (1989a–1989d) found that the joints of the midfoot contribute 10% to 41% of clinical plantarflexion from neutral to 30° plantarflexion. Therefore, what appears to be clinical ankle plantarflexion is actually occurring distal to the ankle

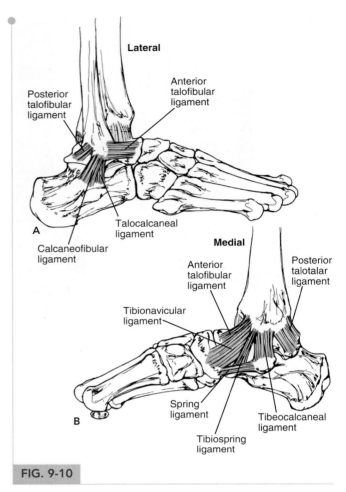

FIG. 9-10

A. Lateral side of the foot and ankle. **B.** Medial side of the foot and ankle. *From Anderson, J. (Ed.). (1978). Grant's Atlas of Anatomy. Baltimore: Lippincott Williams & Wilkins.*

itself. This midfoot motion explains the apparent ability of the foot to dorsiflex and plantarflex following ankle fusion. It also explains the ability of dancers and gymnasts to align the foot with the long axis of the leg while standing on their toes.

Sammarco et al. (1973) performed analyses of instant centers of rotation as well as surface velocities in both normal and diseased ankles. For normal ankles, they found that the ankle axis of rotation does not remain constant during plantarflexion/dorsiflexion but varies slightly while remaining within the talus. Measures of surface motion showed early distraction as dorsiflexion began, followed by posterior talar gliding, until full dorsiflexion was reached and compression of the talus within the tibia/fibula occurred. In arthritic ankles, the direction of displacement of the contact points showed no consistent pattern (Fig. 9-12).

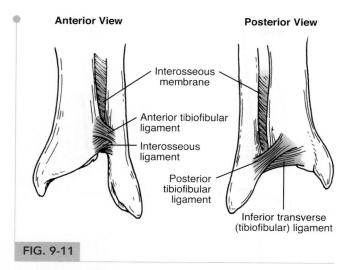

FIG. 9-11

Components of the ankle syndesmosis.

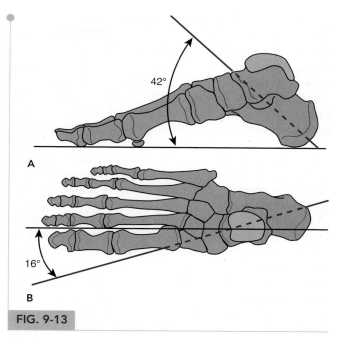

FIG. 9-13

Subtalar joint axis. **A.** Sagittal plane (lateral view). The axis rises up at a 42° angle from the plantar surface. **B.** Transverse plane (top view). The axis is oriented 16° medial to the midline of the foot.

SUBTALAR JOINT

The subtalar joint axis has one degree of freedom and is set at an oblique angle that is oriented upward at an angle 42° from the horizontal and medially 16° from the midline (Fig. 9-13) (Manter, 1941). This almost even split between the anterior-posterior axis and the vertical axis

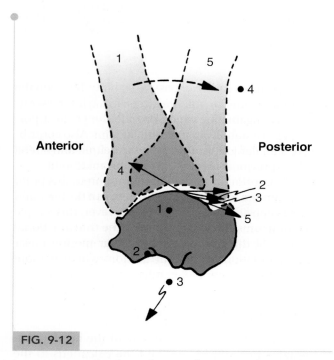

FIG. 9-12

Instant center and surface velocity analysis in an arthritic ankle. The instant centers vary considerably. Joint compression occurs early in motion and distraction occurs in dorsiflexion (velocity 4).

creates almost equal amounts of the component motions of inversion/eversion and abduction/adduction. The small amount of obliquity toward the medial-lateral axis suggests that the subtalar joint has very little motion component of plantarflexion/dorsiflexion. Similar to the talocrural joint in which the small amounts of inversion/eversion and abduction/adduction were not considered clinically meaningful, the plantarflexion/dorsiflexion component of the subtalar joint is considered negligible.

An appreciation of the gross motion of the subtalar joint in a non–weight-bearing position can be obtained by firmly grasping the calcaneus and rocking it side to side in the coronal plane (inversion/eversion) or by rotating it in the transverse plane (abduction/adduction). Although it may appear that these motions can be isolated, as mentioned previously, all components of the single degree of freedom axis always occur simultaneously. Consequently, when performing inversion (supination) of the calcaneus, the component motion of adduction also occurs as these two components are both part of supination. When performing eversion (pronation), the component motion of abduction occurs because these two motions are both part of pronation.

Motions of the subtalar joint (calcaneus relative to the talus) are identical in weight-bearing and non–weight-bearing. However, in non–weight-bearing, the talus is typically stationary and the calcaneus moves on the talus, whereas in weight-bearing the calcaneus is stationary and the talus moves on the calcaneus. The relative motion between the talus and calcaneus remains the same although the active versus stationary role of each bone differs.

As previously described, the superior dome of the talus is firmly stabilized in the mortise of the talocrural joint and is only able to freely dorsiflex and plantarflex. Given that the primary motions of the subtalar joint are inversion/eversion and abduction/adduction, the *combination* of the talocrural and subtalar joints afford freedom of motion in all three planes. The talocrural joint provides primarily for forward progression during locomotion while the subtalar joint provides freedom for the lower leg to rotate in the transverse plane or rock side to side in the coronal plane without requiring the foot to move on the ground. In this manner the foot/ankle provides a stable and fixed platform on the ground with the ability to progress forward, balance, change direction, or operate on uneven surfaces as the talus moves around a fixed calcaneus. In reality, the rocker-bottom shape of the calcaneus allows some coronal plane motion (inversion/eversion) such that true weight-bearing motion is a combination of the talus moving on the calcaneus and the calcaneus moving on the talus.

There is a posterior, middle, and anterior facet connecting the talus to the calcaneus. The posterior facet makes up approximately 70% of the total articular surface of the subtalar joint. The subtalar facets resemble segments of a spiral of Archimedes similar to a right-handed screw in the right foot, so that in non–weight-bearing the calcaneus can be said to translate anteriorly along the subtalar axis as it rotates clockwise during the motion of subtalar supination (Fig. 9-14). Similar to the motion dictated by the threads of a screw, the subtalar joint has a single degree of freedom in which the calcaneus moves in a constrained pattern across all three planes, with inversion/eversion and abduction/adduction the predominant motions. Because subtalar joint motion cuts across all three planes, it is virtually impossible to measure the three components of pronation or supination simultaneously. Clinically, subtalar joint supination is measured by examining the degree of inversion of the calcaneus while subtalar joint pronation is measured by examining the degree of eversion of the calcaneus. Estimates vary, but active measures of inversion/eversion are approximately 22.6°/12.5° degrees, respectively (Grimstone et al., 1993). This suggests that the inversion/eversion ratio is approximately 2:1; however, there is some evidence that this ratio reaches 3:1 when measured passively (Hale et al., 2007).

The posterior and combined anterior and middle facets of the subtalar joint each have their own joint capsule,

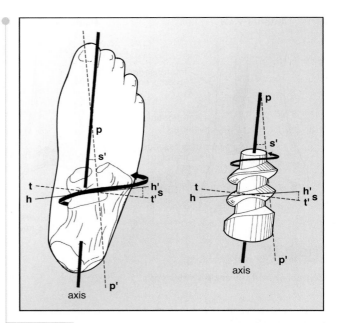

FIG. 9-14

Comparison of the posterior calcaneal facet of the right subtalar joint with a right-handed screw. The *arrow* represents the path of the body following the screw. The horizontal plane in which motion is occurring is *hh'*; *tt'* is a plane perpendicular to the axis of the screw; *s* is the helix angle of the screw, equal to *s'*, which is obtained by dropping a perpendicular (*pp'*) from the axis. As the calcaneus inverts, it rotates clockwise and translates forward along the axis. *Reprinted with permission from Manter, J.T. (1941). Movements of the subtalar and transverse tarsal joints. Anat Rec, 80, 397.*

providing some measure of passive stability. Between the posterior and anterior/middle facets lie the interosseous and cervical ligaments, which provide the greatest portion of passive stability to the subtalar joint. Also contributing to subtalar joint stability are the deltoid and lateral ligaments previously described with the ankle joint.

The description of subtalar joint motion to this point has been limited to the articulation between the talus and calcaneus considered in isolation. However, this simple description is insufficient to describe the true functional movement of the foot because subtalar motion never occurs in isolation to midfoot motion, which is more fully described in later paragraphs.

TRANSVERSE TARSAL JOINT

The transverse tarsal joint consists of the articulation of the talus to the navicular and the calcaneus to the cuboid. Unlike the ankle and subtalar joint, the transverse tarsal joint has two axes resulting in substantial motion in all three planes. Manter (1941) described the

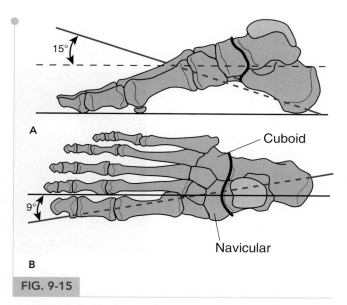

FIG. 9-15

Longitudinal axis of the transverse tarsal joint. Inversion and eversion occur about this axis. **A.** Lateral view. **B.** Top view.

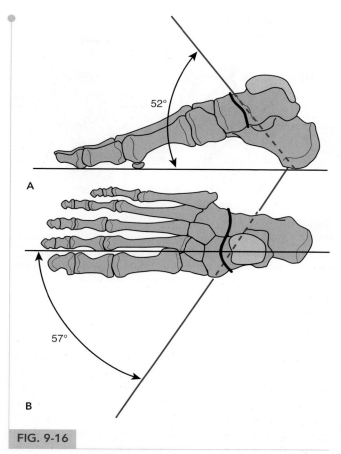

FIG. 9-16

Oblique axis of the transverse tarsal joint. Flexion and extension occur about this axis. **A.** Lateral view. **B.** Top view.

two axes of motion as a longitudinal axis and an oblique axis. The longitudinal axis is oriented 15° upward from the horizontal and 9° medially from the longitudinal axis of the foot. Inversion and eversion occur about the longitudinal axis (Fig. 9-15). The oblique axis is oriented 52° upward from the horizontal and 57° antero-medially. Flexion and extension occur primarily about this axis (Fig. 9-16). In general, the talonavicular joint has much more range of motion than the calcaneo-cuboid joint (Ouzonian and Shereff, 1989). Precise measures of range of motion of the midtarsal joint are difficult to obtain due to the complex nature of the articulation and the small size of the bones. Common estimates suggest that the transverse tarsal joint has similar amounts of pronation and supination as the subtalar joint (Neumann, 2010).

The Relationship between the Transverse Tarsal Joint and the Subtalar Joint

As described fully by Huson (1991) and termed by him the "tarsal mechanism," pronation/supination at the subtalar joint drives motion at the midtarsal joint in a consistent pattern. In this model, the four bones (talus, calcaneus, cuboid, and navicular) form an interlocking chain of motion. Huson describes this as a "constrained" mechanism similar to that which occurs in a gearbox when one cog rotates with its teeth interwoven into the teeth of an adjacent cog. The navicular and cuboid are almost immobile relative to each other (Wolf et al., 2008). Consequently, during subtalar joint supination,

the inversion and adduction forces generated by the cal-caneus and applied to the cuboid are further transmit-ted to the navicular. As the cuboid adducts and inverts underneath the foot it tends to lift and abduct the navic-ular. The navicular force is transmitted to the talus, fur-ther facilitating supination. Similar mechanisms work in reverse for subtalar joint pronation.

Astion et al. (1997) showed the close interplay between the subtalar, talonavicular, and calcaneocuboid joints in a study entailing experimental selective fusion of these joints. Subtalar joint arthrodesis reduced talonavicular motion to 26% of its normal motion and reduced calca-neocuboid motion to 56% of normal. Calcaneocuboid arthrodesis reduced subtalar motion to 92% of normal and talonavicular motion to 67% of normal. Selective fusion of the talonavicular joint had the most profound effect on the remaining joints, reducing their remaining motion to only 2° each. Beaudoin et al. (1991) showed that experimental subtalar fusion resulted in a significant reduction in talona-vicular joint contact and also reduced ankle joint contact.

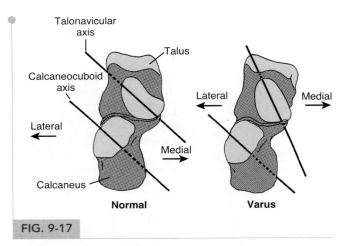

FIG. 9-17

Anteroposterior view of the transverse tarsal joint of the right foot. The anterior articulations of the talar head and calcaneus are shown. The major axes of the talonavicular and calcaneocuboid joints are shown in the neutral position (parallel) and with the heel in varus (convergent).

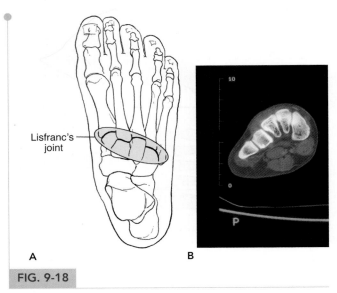

FIG. 9-18

A. Top view of the tarsometatarsal joints, known as the Lisfranc joint. Note the recessed second metatarsal base. **B.** Cross-sectional view of Lisfranc joint seen on computed tomography scan. Note the arch-like structure.

Pronation and supination in the subtalar joint tend to create flexibility or rigidity, respectively, of the transverse tarsal joint. Elftman (1960) showed that the major axes of the calcaneocuboid joint (longitudinal) and talonavicular joint (oblique) are in parallel when the subtalar joint is everted, thus allowing motion of the transverse tarsal joint in a "loose packed" position. As the subtalar joint inverts, the axes of these joints are convergent, thus locking the transverse tarsal joint and providing rigidity to the midfoot (Fig. 9-17) in a "close packed" position. This has substantial functional significance during gait. From the period of midstance to toe-off, the foot becomes a rigid lever through inversion of the subtalar joint and the locking of the transverse tarsal joint, allowing mechanical advantage to the foot for push-off with a rigid lever.

INTERTARSAL AND TARSOMETATARSAL JOINTS

The intertarsal joints include the three cuneonavicular joints, the intercuneiform joints, the cuboideonavicular joint, and the cuneocuboid joint. The intertarsal joints are closely congruent and exhibit minimal gliding motion between one another. Motions between these joints are typically not described. The tarsometatarsal joints, known as Lisfranc joints, are intrinsically stable because of their arch-like configuration best seen in cross-section. The second metatarsal base is recessed into the midfoot, forming a key-like configuration with the intermediate cuneiform (Fig. 9-18). A strong ligament known as the Lisfranc ligament connects the second metatarsal base to the medial cuneiform. The relative rigidity of the second (and to some degree third) metatarsal allows it to serve as the central rigid structure of the longitudinal arch, providing a rigid lever for push-off in late stance. Motion of the first, fourth, and fifth tarsometatarsal joints are much greater with the first tarsometatarsal joint exhibiting approximately 10° of plantar flexion during late stance (Cornwall and McPoil, 2002). Mobility in other planes of motion for these joints is negligible.

Pronator and Supinator Twists of the Forefoot

The flexibility of the most medial and lateral tarsometatarsal joints provide a mechanism for the forefoot to invert and evert independently of the rearfoot. When the first metatarsal plantarflexes 10° at the tarsometatarsal joint through the action of the fibularis longus, it effectively produces a "pronator twist" of the forefoot (with the medial foot moving plantarward and the entire surface of the foot facing more laterally). Similarly, when the ground reaction force of uneven terrain pushes the medial forefoot dorsally, it effectively produces a "supinator twist" of the forefoot. Commonly these "twists" serve two purposes: to better accommodate and adapt to variable terrain and to provide sufficient push-off from the medial border of the foot during the push-off phase of gait.

METATARSOPHALANGEAL JOINT

The five metatarsophalangeal joints are composed of a convex metatarsal head and a concave proximal phalanx. Motions at these joints are primarily dorsiflexion/plantarflexion with lesser amounts of abduction/adduction. Passive range of motion is 65°/40° for dorsiflexion and plantarflexion, respectively (Van Gheluwe et al., 2006), with the exception of greater dorsiflexion for the first metatarsophalangeal joint, which can achieve 85°. During the toe-off phase of normal walking, approximately 60° of dorsiflexion of the first metatarsophalangeal joint is required, although many other tasks require greater amounts of dorsiflexion, such as sitting on the heels or standing on tiptoe. The primary ligaments of the metatarsophalangeal joints are the medial and lateral collateral ligaments. There are also four transverse metatarsal ligaments that bind the metatarsal heads together and provide stability for the entire forefoot.

An analysis of motion of the hallux in the sagittal plane reveals that instant centers of motion often fall within the center of the metatarsal head, with minimal scatter (Fig. 9-19). The arthrokinematics of the first metatarsophalangeal joint are characterized as tangen-

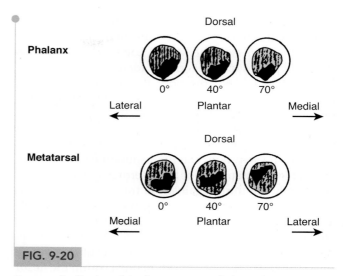

FIG. 9-20

Contact distribution of the first metatarsophalangeal joint in 0° (neutral), 40° of extension, and 70° of extension. **Top.** Joint contact of the proximal phalanx. **Bottom.** Joint contact of the metatarsal head. With increasing extension, joint surface contacts shift dorsally on the metatarsal head.

tial sliding from maximum plantarflexion to moderate dorsiflexion, with some joint compression dorsally at maximum dorsiflexion (Sammarco, 1980; Shereff et al., 1986). Ahn et al. (1997) determined that the metatarsal head surface contact area shifts dorsally with full extension and is associated with joint compression (Fig. 9-20). This explains the characteristic formation of dorsal osteophytes and limited dorsiflexion of the proximal phalanx in cases of hallux rigidus (Fig. 9-21).

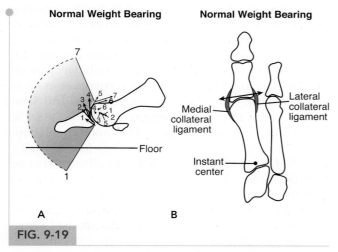

FIG. 9-19

A. Instant center and surface motion analysis of the metatarsophalangeal joint of the hallux in the sagittal plane. Each *arrow* denoting direction of displacement of the contact points corresponds to the similarly numbered instant center. Gliding takes place throughout most of the motion except at the limit of extension, which occurs at toe-off in the gait cycle and with squatting. At full extension, joint compression takes place. The range of motion of the hallux is indicated by the arc. **B.** Instant center analysis of the metatarsophalangeal joint of the hallux in the transverse plane during normal weight-bearing. Gliding (denoted by *arrows*) occurs at the joint surface even though the range of motion is small.

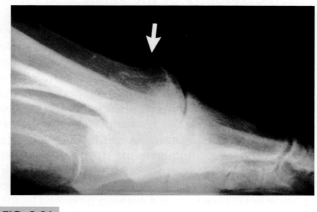

FIG. 9-21

Lateral view of hallux rigidus. Note osteophytes on the dorsal aspect of the metatarsal head, which limit joint extension.

The great toe provides stability to the medial aspect of the foot through the windlass mechanism of the plantar aponeurosis (see The Medial Longitudinal Arch). As the body passes over the foot in toe-off, the metatarsal head is pressed into the floor through the stabilizing action of the fibularis longus. This has been confirmed by force plate analysis in the late stance phase, which shows that pressure under the first metatarsal head increases in this phase of gait (Clark, 1980). Wearing narrow, high-heeled shoes can predispose the individual to mechanical entrapment of the interdigital nerves (more commonly the third) against the transverse intermetatarsal ligaments by compressing the metatarsal heads and creating a painful neuroma (Wu, 1996). Excessive stress on the first metatarsophalangeal joint may result in inflammation or stress fractures and resulting pain on the sesamoid bones that are embedded within the flexor hallucis brevis tendon.

INTERPHALANGEAL JOINT

The hallux has a single interphalangeal joint while the lateral four toes have both a proximal and distal interphalangeal joint. All these joints have a similar concave-on-convex configuration and have a single degree of freedom with a medial-lateral axis producing sagittal plane motions of dorsiflexion and plantarflexion. Collateral ligaments are similar in location and function to those of the metatarsophalangeal joints. Range of motion information is limited regarding these joints, but generally flexion is greater than extension and the proximal joint has slightly greater range of motion than the distal joint.

Passive Stability of the Ankle and Foot

Bony congruency and capsular and ligamentous support provide passive stability to the ankle and foot. Under physiologic loads, the bony congruency of the ankle takes on more importance (Cawley and France, 1991; Stiehl et al., 1993; Stormont et al., 1985). Stormont et al. found that in a loaded state the bony congruency of the ankle provided 30% of rotational stability and 100% of resistance to inversion/eversion. During weight-bearing, the ankle ligaments do not contribute to the stability of the ankle during inversion/eversion, although rotational instability may still occur. Cawley and France showed that the force to cause ankle inversion and eversion increased by 91% and 80%, respectively, with loading. Stiehl et al. (1993) found that loading of the ankle resulted in decreased range of motion (especially plantarflexion), decreased anteroposterior drawer, as well as increased stability

against inversion/version and rotation. In contrast to Stormount et al., Cass and Settles (1994) performed CT scans of loaded cadaveric ankles in an apparatus that did not constrain rotation and demonstrated that an average 20° of talar tilt still occurred in loaded ankles after sectioning both anterior talofibular and calcaneofibular ligaments. They did not feel that the articular surfaces prevented inversion instability during ankle loading. Most studies agree that loading of the ankle results in increased stability as a result of bony congruency, especially in ankle dorsiflexion.

Syndesmotic stability between the tibia and fibula depends on the integrity of both malleoli, the syndesmotic ligaments, and the deltoid ligamentous complex. During ankle dorsiflexion, there is approximately 1 mm of mortise widening and 2° of external rotation of the fibula (Close, 1956). The normal distal fibular migration with loading is 1 mm (Wang et al., 1996). This distal fibular migration serves to deepen the ankle mortise for added bony stability (Scranton et al., 1976). With disruption of the mortise in an external rotation injury, the syndesmotic ligaments and deltoid ligaments are torn, the distal fibula fractures, and the talus displaces laterally. A study of cadaveric ankles by Olgivie-Harris et al. (1994) defined the contribution to resistance of lateral talar displacement by the syndesmotic ligaments to be 35% for the anterior tibiofibular ligament, 40% for the posterior tibiofibular ligament, 22% for the interosseous ligament, and less than 10% for the interosseous membrane.

Subtalar joint and ankle joint inversion are often difficult to separate clinically. The calcaneofibular ligament provides stability to inversion and torsional stresses to both the ankle and subtalar joints. Stephens and Sammarco (1992) provided inversion stress to cadaveric ankles and sequentially sectioned the anterior talofibular and calcaneofibular ligaments. They found that up to 50% of the inversion observed clinically was coming from the subtalar joint. The structures that contribute to stability of the subtalar joint are the calcaneofibular ligament, the cervical ligament, the interosseous ligament, the lateral talocalcaneal ligament, the ligament of Rouviere, and the extensor retinaculum (Harper, 1991).

The lateral ankle ligaments are the most commonly injured and therefore the most frequently studied. The anterior talofibular and calcaneofibular ligaments form a 105° angle with one another (Fig. 9-22). They act synergistically to resist ankle inversion forces. The anterior talofibular ligament is under the greatest tension in plantarflexion, and the calcaneofibular ligament is under the greatest tension in ankle dorsiflexion (Cawley and France, 1991; Inman, 1976; Nigg et al., 1990; Renstrom et al., 1988). The anterior talofibular ligament therefore resists ankle inversion in plantarflexion and

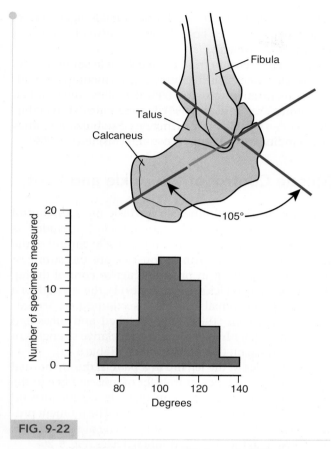

FIG. 9-22

Average angle between the calcaneofibular and talofibular ligaments in the sagittal plane. The average angle is 105° with considerable variation from 70 to 140° among measured subjects.

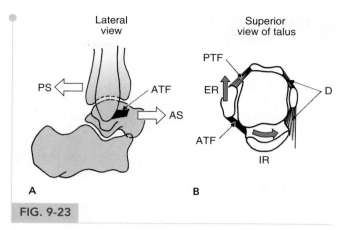

FIG. 9-23

Function of the anterior talofibular ligament (*ATF*). **A.** The ATF limits anterior shift (*AS*) of the talus or posterior shift (*PS*) of the tibia-fibula. **B.** The ATF limits internal rotation (*IR*) of the talus or external rotation (*ER*) of the fibula. *PTF,* posterior talofibular ligament; *D,* deltoid ligament.

the calcaneofibular ligament. These injuries typically occur as a result of landing or falling on a plantarflexed and inverted ankle (Case Study 9-1). However, there is some evidence that not all sprains occur in a plantarflexed position. A recent video analysis of an anterior talofibular ligament injury showed the ankle to be in a position of inversion, adduction, and dorsiflexion (rather than plantarflexion) during occurrence of the sprain (Fong et al., 2009).

the calcaneofibular ligament resists ankle inversion during ankle dorsiflexion. The accessory functions of the anterior talofibular ligament are resistance to anterior talar displacement from the mortise, clinically referred to as anterior drawer, and resistance to internal rotation of the talus within the mortise (Fig. 9-23). The calcaneofibular ligament spans both the lateral ankle joint and lateral subtalar joint, thus contributing to subtalar joint stability (Stephens and Sammarco, 1992). The posterior talofibular ligament is under greatest strain in ankle dorsiflexion and acts to limit posterior talar displacement within the mortise as well as limit talar external rotation (Fig. 9-24) (Sarrafian, 1993a). In vitro testing of unloaded ankles subjected to anterior drawer testing demonstrated that the anterior talofibular ligament was most important in plantarflexion and that the calcaneofibular and posterior talofibular ligaments were most important in ankle dorsiflexion (Bulucu et al., 1991).

Clinically, the most commonly sprained ankle ligament is the anterior talofibular ligament, followed by

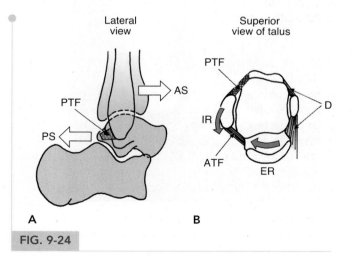

FIG. 9-24

Function of the posterior talofibular ligament (*PTF*). **A.** The PTF limits posterior shift (*PS*) of the talus or anterior shift (*AS*) of the tibia-fibula. **B.** The PTF limits external rotation (*ER*) of the talus or internal rotation (*IR*) of the fibula. *ATF,* anterior talofibular ligament; *D,* deltoid ligament.

Case Study 9-1

Ankle Sprain

A basketball player presents with an injury that resulted from a fall on a plantarflexed and inverted ankle position during a game (Case Study Fig. 9-1).

An abnormally high load in conjunction with the fast loading rate produces the injury. The sprain inversion injury produced by high stress (load per unit of area) in the plantarflexed and inverted direction will most commonly affect the anterotalofibular ligament (failure load ~139N). Sprain of this ligament produces lateral instability in the ankle joint, an abnormal anterior talar displacement from the mortise, and decreasing resistance to internal rotation of the talus within the mortise.

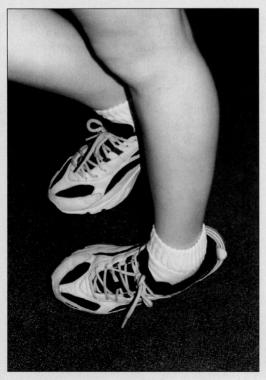

Case Study Figure 9-1

Ataarian et al. (1985) tested the strength of the ankle ligaments by loading cadaver ligaments to failure and found the strength of the various ligaments from weakest to strongest to be anterior talofibular (139 N), posterior talofibular (261 N), calcaneofibular (346 N), and deltoid

(714 N). Therefore, the incidence of ankle ligamentous injury tends to match both the mechanism of injury and ligamentous strength.

Although recovery from a typical ankle sprain occurs within a few weeks, the recovery is often incomplete, with a 73% recurrence rate and 59% disability and residual symptoms (Yeung et al., 1994). Functional ankle instability has been correlated with increased stabilization times after landing from a jump (Ross and Guskiewicz, 2004).

Muscle Control of the Ankle and Foot

Twelve extrinsic and nineteen intrinsic muscles control the foot and ankle (the plantaris muscle is excluded as it generally has no contribution to muscle control of the foot or ankle). The extrinsic muscles are the strongest and most important in providing active control during gait. According to Fick's principle (1911), the strength of a muscle is proportional to its cross-sectional area. Accordingly, Silver et al. (1985) have weighed and measured muscle fiber length to determine the relative strengths of muscles acting on the foot and ankle (Table 9-1).

The muscles of the leg fire in a pattern during normal gait to ensure an efficient transfer of muscle force to the floor and smooth progression of body weight forward along the axis of progression (Fig. 9-25). The moment produced by each muscle tendon unit can be predicted by its relationship to the ankle and subtalar axes (Fig. 9-26).

The soleus and gastrocnemius combine to form the Achilles tendon, which inserts onto the posterior-medial calcaneus and is the strongest plantar flexor of the ankle. Firing of the ankle plantarflexors during midstance acts to slow the forward motion of the tibia over the foot. A mathematical model has predicted peak Achilles tendon forces to be 5.3 to 10 times body weight during running (Burdett, 1982).

TABLE 9-1

Relative Strengths (% of Total Strength) of Muscles Acting on the Foot and Ankle

Plantarflexor Strength Percentage	Dorsiflexor Strength Percentage
Soleus 29.9	Tibialis anterior 5.6
Gastrocnemius 13.7	Extensor digitorum longus 1.7
Flexor hallucis longus 3.6	Extensor hallucis longus 1.2
Flexor digitorum longus 1.8	Peroneus tertius 0.9
Inverters	**Everters**
Tibialis posterior 6.4	Fibularis longus 5.5
	Fibularis brevis 2.6

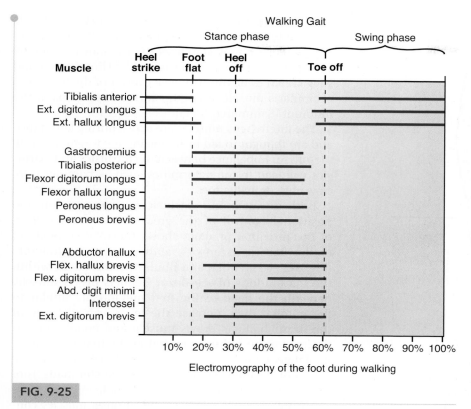

Walking Gait

FIG. 9-25

Electromyography of the musculature of the foot and ankle during one normal gait cycle (heel-strike to heel-strike).

The strongest dorsiflexor of the ankle is the tibialis anterior, which provides eccentric control during stance phase from heel-strike to foot flat to prevent foot slap. During swing phase, the ankle dorsiflexors provide foot clearance from the floor.

The strongest inverter of the foot and ankle is the posterior tibialis muscle. The posterior tibialis is a dynamic supporter of the medial longitudinal arch. It functions to invert the subtalar joint during mid- and late stance, thereby locking the transverse tarsal joint and ensuring rigidity of the foot during toe-off. Loss of this muscle results in acquired pes planus with flattening of the arch, abduction of the forefoot, and eversion of the heel (Fig. 9-27) (Case Study 9-2). Patients with posterior tibialis tendon dysfunction usually are unable to actively invert their heel to form a rigid platform on which to support their weight while attempting a single toe rise.

The primary everters of the foot and ankle are the fibularis longus and brevis. The fibularis longus inserts on the base of the first metatarsal and medial cuneiform and acts to depress (plantarflex) the metatarsal head at the tarsometatarsal joint. Injury or paralysis of this muscle may allow elevation of the first metatarsal

Case Study 9-2

Posterior Tibialis Tendon Dysfunction

A 35-year-old man presents with complaints of lateral ankle pain although he denies any history of trauma. Evaluation of foot alignment during standing reveals a collapsed medial arch, heel valgus, and forefoot abduction. When asked to perform single leg heel raises, the heel does not invert and the patient cannot completely rise on the forefoot. He describes that lately he has noticed pain and fatigue with walking and that he feels that he walks on the inside of his ankle. Inversion of the foot from a plantarflexed position is weak. The patient suffers from posterior tibialis tendon dysfunction, a condition that commonly causes flattened medial arch and lateral ankle pain due to excessive compression (Fig. 9-27; Geideman and Johnson, 2000).

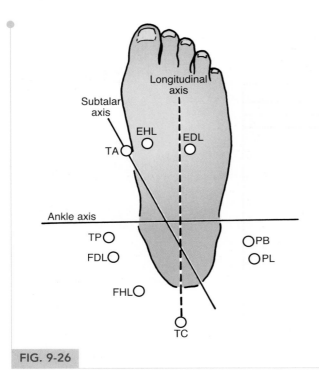

FIG. 9-26

Subtalar and ankle axes in relationship to extrinsic muscles. *EDL,* extensor digitorum longus; *EHL,* extensor hallucis longus; *FDL,* flexor digitorum longus; *FHL,* flexor hallucis longus; *PB,* peroneus brevis; *PL,* peroneus longus; *TA,* tibialis anterior; *TC,* tendon calcaneus; *TP,* tibialis posterior.

head and decrease loads borne by the first metatarsals and can result in the development of a dorsal bunion. The fibularis brevis stabilizes the forefoot laterally by resisting inversion and was found by Hintermann and Nigg (1995) to be the strongest everter of the foot. Loss of fibularis muscle strength can result in varus of the hindfoot (Sammarco, 1995).

The interosseus muscles are active during late stance and are thought to aid in stabilizing the forefoot during toe-off. An imbalance between the intrinsics and extrinsics will lead to toe deformities such as hammer toes, claw toes, or mallet toes.

Both intrinsic and extrinsic muscles mediate the positional control of the great toe. A cross-section of the proximal phalanx shows the relative position of the flexors, extensors, abductors, and adductors (Fig. 9-28). The tibial and fibular sesamoids lie within the toe tendons of the flexor hallucis brevis muscle, beneath the head of the first metatarsal. Similar to the patella, they increase the lever arm distance of the flexor hallucis brevis muscle and enable greater flexion torque to be generated at the first metatarsophalangeal joint by moving the tendon away from the center of the joint. They also act to transfer loads from the ground to the first metatarsal head.

For the toes, the extrinsic and intrinsic muscles contribute to the extensor hood, which controls motion of the metatarsophalangeal and interphalangeal joints (Fig. 9-29).

The lumbricals and interossei are the main intrinsic contributors to the extensor hood. The intrinsics act to flex the metatarsophalangeal joints and extend

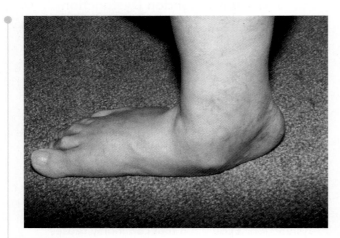

FIG. 9-27

Loss of the medial longitudinal arch in an acquired adult flatfoot secondary to posterior tibial tendon deficiency.

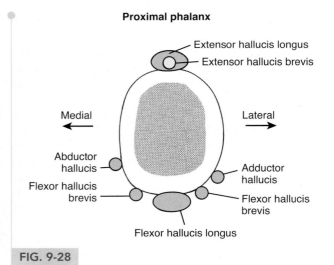

FIG. 9-28

Diagrammatic cross-section of the proximal phalanx of the hallux showing normal positions of the various tendons in relation to the bone.

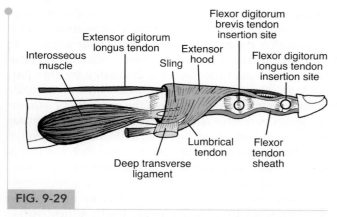

FIG. 9-29

Lateral view of the lesser toe illustrating the extensor hood with contributing muscles and ligaments.

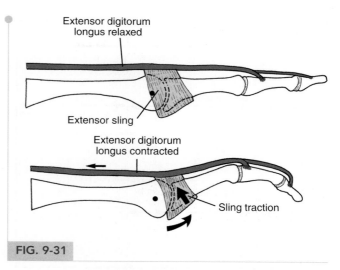

FIG. 9-31

Lateral diagram showing the action of the extensor sling. When the extensor digitorum longus contracts **(bottom)**, the proximal phalanx is lifted into extension through the sagittal bands.

the inter-phalangeal joints (Fig. 9-30). The extrinsic toe extensors extend the metatarsophalangeal joint through the action of the sagittal bands by lifting the proximal phalanges into extension (Fig. 9-31). The flexor digitorum brevis is the primary flexor of the proximal interphalangeal joint. The flexor digitorum longus is the primary flexor of the distal interphalangeal joint.

Foot and Ankle Motion during Gait

The gait cycle consists of a stance phase and a swing phase. The stance phase encompasses 62% of the gait cycle, and the swing phase makes up the remaining 38%. The stance phase is subdivided into heel-strike, foot flat, heel-rise, push-off, and toe-off. Swing phase is divided into acceleration, toe clearance, and deceleration phases (Fig. 9-32). The part of stance phase spent with both feet on the ground is termed double limb support and occurs through the first and last 12% of stance phase (Fig. 9-33).

Normal men have an average gait velocity of 82 m/min and 58 heel-strikes per minute (Waters et al., 1978). Running is defined as that speed at which double stance disappears (float phase) and is typically around speeds exceeding 201 m/min (Fig. 9-34).

During normal walking, the entire lower extremity (including the pelvis, femur, and tibia) rotates internally through the first 15% of stance phase. From heel-strike

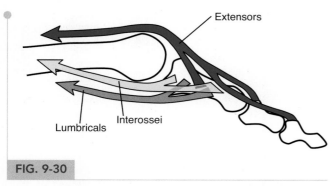

FIG. 9-30

The intrinsic muscles (interossei and lumbricals) act to flex the metatarsophalangeal joint and extend the interphalangeal joints.

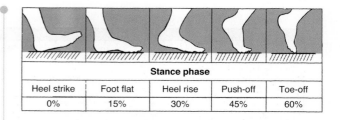

Stance phase				
Heel strike	Foot flat	Heel rise	Push-off	Toe-off
0%	15%	30%	45%	60%

Swing phase			
Acceleration	Toe clearance	Deceleration	Heel strike
70%	85%	100%	

FIG. 9-32

Sixty-two percent of the normal gait cycle is spent in stance phase and 38% is spent in swing phase.

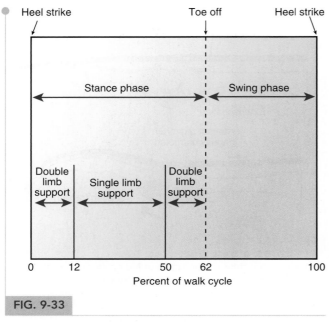

FIG. 9-33

Stance phase consists of two periods of double limb support and one period of single limb support.

through foot flat, the subtalar joint everts, the foot pronates, and the forefoot becomes flexible to absorb shock and adapt to irregularities of the ground floor surface. The subtalar joint everts in part because the point of contact of the heel is lateral to the center of the ankle joint, thus producing a valgus thrust on the subtalar joint. In the middle of stance phase and at push-off, the entire lower extremity begins to reverse and rotate externally as the subtalar joint simultaneously inverts (Fig. 9-35). With inversion of the subtalar joint and supination of the foot, the foot is transformed into a rigid structure capable of propulsion.

The normal pattern of ankle motion has been studied extensively (Lamoreux, 1971; Murray et al., 1964; Stauffer et al., 1977; Wright et al., 1964). At heel-strike, the ankle is in slight plantarflexion. Plantarflexion increases until foot flat, but the motion rapidly reverses to dorsiflexion during midstance as the body passes over the foot. The motion then returns to plantarflexion at toe-off. The ankle again dorsiflexes in the middle of swing phase and changes to slight plantarflexion at heel-strike. Ankle motion during normal walking averages 10.2° dorsiflexion and 14.2° plantarflexion, with a total motion of 25°. Maximum dorsiflexion occurs at 70% stance phase, and maximum plantarflexion occurs at toe-off (Stauffer et al., 1977). Subtalar joint pronation

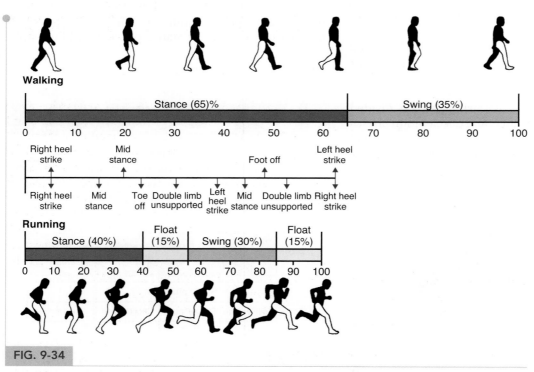

FIG. 9-34

Comparison of walking and running cycles. In the running cycle, stance phase decreases, swing phase increases, double limb support disappears, and a double limb unsupported or float phase develops.

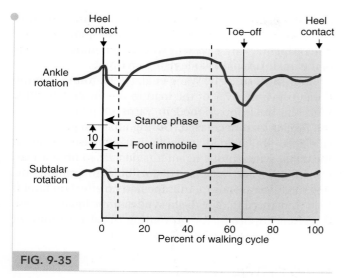

FIG. 9-35

Ankle motion and subtalar rotation during normal walking. Maximal subtalar eversion occurs at foot flat in early stance phase. Maximum subtalar inversion occurs at toe-off.

occurs from heel-strike to foot flat (from 2° of supination to 2° of pronation), reverses direction at approximately 35% of the gait cycle as it moves toward supination, achieving a maximum 6° of supination immediately prior to toe-off (Fujii et al., 2005).

Eversion of the heel is coupled with internal rotation of the tibia whereas inversion is coupled with external rotation of the tibia. However, recent research (Pohl et al., 2007) suggests that these motions are not rigidly connected during walking as the heel everts for a large part of the stance phase while the tibia externally rotates. In contrast, a more direct temporal link of rearfoot eversion and tibia internal rotation exists during running.

The coupling of ankle with hip and knee motion is even less understood. Two biomechanical studies demonstrated that standing on a laterally inclined surface leads to increased pronation, hip adduction, and internal rotation (Pappas and Hagins, 2008), whereas landing on the laterally inclined surface caused increased knee valgus and knee flexion without any significant changes on ankle motion (Hagins et al., 2007).

External rotation of the entire lower extremity occurs during late stance. Contributing to this are forces provided by the swing of the opposite leg and the obliquity of the metatarsal break (Fig. 9-36). The metatarsal break is an oblique axis of 50° to 70° with respect to the long axis of the foot formed by the centers of rotation of the metatarsophalangeal joints. With push-off, the foot and lower extremity externally rotate with respect to the sagittal plane because of this oblique axis.

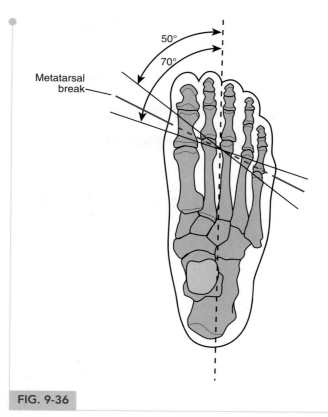

FIG. 9-36

The metatarsal break **(top view),** a generalization of the instant centers of rotation of all five metatarsophalangeal joints, may vary among individuals from 50° to 70° in its orientation to the long axis of the foot.

Muscle Action during Gait

Motions of the foot and ankle during the walking cycle occur as a result of the passive constraints of joints and ligaments and active muscle contraction (Figs. 9-25 and 9-37). At heel-strike, the pretibial musculature fires eccentrically to slow down the descent of the forefoot and prevent foot slap. At midstance, the calf musculature contracts to control the forward movement of the body over the foot. The foot intrinsics contract from midstance to toe-off to aid in rigidity of the forefoot. Toe-off is primarily a passive event. The pretibial musculature again contracts during swing phase to ensure that the foot clears the floor during swing-through.

Kinetics of the Ankle Joint

The reaction forces on the ankle joint during gait are equal to or greater than those of the hip or knee joints. The following static and dynamic analyses give an estimate of the magnitude of the reaction forces acting on the ankle joint during standing, walking, and running.

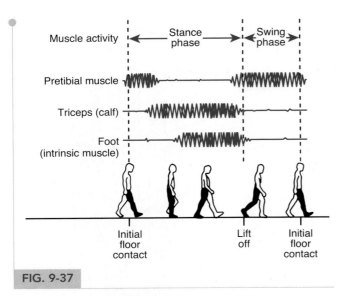

FIG. 9-37

Schematic phasic activity of leg and foot muscles during normal gait.

STATICS

In a static analysis of the forces acting on the ankle joint, the magnitude of the force produced by contraction of the gastrocnemius and the soleus muscles through the Achilles tendon, and consequently the magnitude of the joint reaction force, can be calculated through the use of a free-body diagram. In the following example, the muscle force transmitted through the Achilles tendon and the reaction force on the ankle joint are calculated for a subject standing on tiptoe on one leg. In this example, the foot is considered a free body with three main coplanar forces acting on it: the ground reaction force (W), the muscle force through the Achilles tendon (A), and the joint reaction force on the dome of the talus (J) (Calculation Box 9-1).

The ground reaction force (equaling body weight) is applied under the forefoot and is directed upward vertically. The Achilles force has an unknown magnitude but a known point of application (point of insertion on the calcaneus) and a known direction (along the Achilles tendon). The talar dome joint reactive force has a known point of application on the dome of the talus, but the magnitude and line of direction are unknown. The magnitude of A and J can be derived by designating the forces on a free-body diagram and constructing a triangle of forces. Not surprisingly, these forces are found to be quite large. The joint reactive force is approximately 2.1 times body weight, and the Achilles tendon force reaches approximately 1.2 times body weight. The great force required for rising up on tiptoe explains why the patient with weak gastrocnemius and soleus muscles

has difficulty performing the exercise 10 times in rapid succession. The magnitude of the ankle joint reaction force explains why a patient with degenerative arthritis of the ankle has pain while rising on tiptoe.

An in vitro study by Wang et al. (1996) found that the fibula transmits 17% of the load in the lower extremity. With the ankle positioned in varus or plantarflexion, the fibular load decreased. During ankle valgus or dorsiflexion, fibular load transmission increased. Cutting the distal syndesmotic ligaments decreased fibular load transmission and increased distal fibular migration. Cutting the interosseous membrane had no effect on fibular load transmission. The distal syndesmotic ligaments are therefore important for preventing distal migration of the fibula and maintaining fibular load.

LOAD DISTRIBUTION

The ankle has a relatively large load-bearing surface area of 11 to 13 cm², resulting in lower stresses across this joint than in the knee or hip (Greenwald, 1977). The load distribution on the talus is determined by ankle position and ligamentous integrity. During weight-bearing, 77% to 90% of the load is transmitted through the tibial plafond to the talar dome, with the remainder on the medial and lateral talar facets (Calhoun et al., 1994). As the loaded ankle moves into inversion, the medial talar facet is loaded more. Ankle eversion increases the load on the lateral talar facet. When moving from plantarflexion to dorsiflexion, the centroid of contact area moves from posterior to anterior, and while moving from inversion to eversion the centroid moves from medial to lateral. The total talar contact was greatest and the average high pressure was lowest in ankle dorsiflexion (Calhoun et al., 1994) (Fig. 9-38).

Talar load distribution is also determined by ligamentous forces. Sectioning of the tibiocalcaneal fascicle of the superficial deltoid ligament in a loaded cadaver model resulted in a 43% decrease in talar contact area, a 30% increase in peak pressures, and a 4-mm lateral shift of the centroid (Earll et al., 1996).

DYNAMICS

Dynamic studies of the ankle joint are needed to appreciate the forces that act on the normal ankle during walking and running. Stauffer et al. (1977) found that the main compressive force across the normal ankle during gait is produced by contraction of the gastrocnemius and soleus muscles. The pretibial musculature produces mild compressive forces in early stance of <20% body weight. A compressive force of five times body weight was produced in late stance by contraction

CALCULATION BOX 9-1

The Free-Body Diagram of the Foot

A. On a free-body diagram of the foot, including the talus, the lines of application for W and A are extended until they intersect (intersection point). The line of application for J (*dotted line*) is then determined by connecting its point of application, the tibiotalar contact point, with the intersection point for W and A (Calculation Box Fig. 9-1-1). **B.** A triangle of forces is constructed. Force A is 1.2 times body weight and force J is 2.1 times body weight (Calculation Box Fig. 9-1-2).

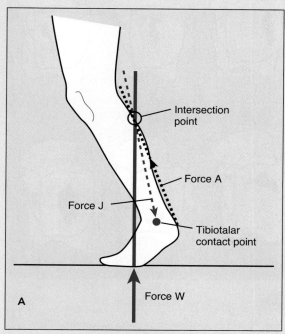

Calculation Box Figure 9-1-1

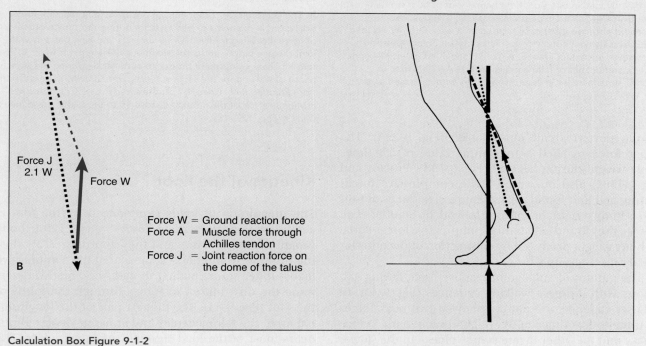

Force J = Joint reaction force on the dome of the talus

Force W = Ground reaction force
Force A = Muscle force through Achilles tendon
Force J = Joint reaction force on the dome of the talus

Calculation Box Figure 9-1-2

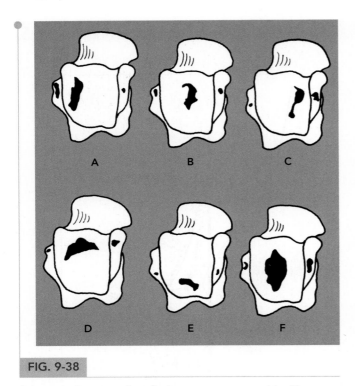

FIG. 9-38

Schematic demonstration of prints on pressure-sensitive film representing high-pressure contact areas on the left talus. **A.** A 490-N load in eversion; note lateral shift of talar contact area. **B.** A 490-N load in neutral version. **C.** A 490-N load in inversion; note medial shift of talar contact area. **D.** A 490-N load in 10° dorsiflexion; note anterior shift of talar contact area and an increase in contact area. **E.** A 490-N load in 30° plantarflexion; note posterior shift of talar contact area. **F.** A 980-N load in neutral; note increase in talar contact area with increased load.

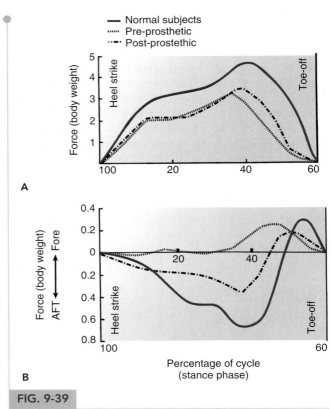

FIG. 9-39

A. The compressive component of the ankle joint reaction force expressed in multiples of body weight during the stance phase of normal walking for five normal subjects and nine patients with joint disease before and after prosthetic ankle replacement. **B.** The fore-aft shear component produced in the ankle during the stance phase of walking for the same subjects. *Reprinted with permission from Stauffer, R.N., Chao, E.Y.S., Brewster, R.L. (1977). Force and motion analysis of the normal, diseased and prosthetic ankle joints. Clin Orthop, 127, 189.*

of the posterior calf musculature (Fig. 9-39A). The shear force reached a maximum value of 0.8 times body weight during heel-off (Fig. 9-39B). Proctor and Paul (1982) also measured ankle compressive forces during gait and found peak compressive forces of four times body weight. In contrast to work by Stauffer et al. (1977), they found substantial compressive forces equal to body weight produced by contraction of the anterior tibial muscle group.

The pattern of ankle joint reactive force during gait differs with different walking cadences (Fig. 9-40). In a faster cadence, the pattern showed two peak forces of three to five times body weight, one in early stance phase and the other in late stance phase. In the slower cadence, only one peak force of approximately five times body weight was reached during the late stance phase (Stauffer et al., 1977). During running, localized ankle forces may be as high as 13 times body weight (Burdett, 1982).

Kinetics of the Foot

The magnitude of loads experienced by the foot is astounding. Peak vertical forces reach 120% body weight during walking, and they approach 275% during running. Manter (1941) measured the compressive loads under static loading in cadaveric feet to determine the distribution of forces through the joints of the foot (Fig. 9-41). The highest part of the longitudinal arch, the talonavicular and cuneonavicular joints, bears most of the load through the tarsal joints. The medial column of the foot, consisting of the talus, navicular, cuneiforms, and first through third metatarsals, bears most of the load. The lateral column, made up of the calcaneocuboid joint and lateral two metatarsals, transmits the lesser load.

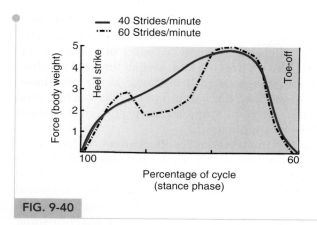

FIG. 9-40

Ankle joint reaction force expressed in multiples of body weight in a normal ankle during the stance phase of gait at two velocities. In the faster cadence, there were two peaks of three to five times body weight, one in early stance and one in late stance phase. In the slower cadence, only one peak force of approximately five times body weight was reached during late stance phase. *Reprinted with permission from Stauffer, R.N., Chao, E.Y.S., Brewster, R.L. (1977). Force and motion analysis of the normal, diseased and prosthetic ankle joints. Clin Orthop, 127, 189.*

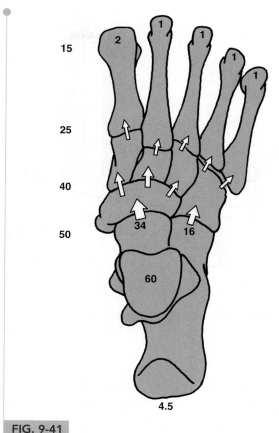

FIG. 9-41

Compressive forces of the foot after a 60-lb load is applied to the talus. Most of the force passes through the talonavicular joint and into the first through third metatarsals.

The distribution of loads under the foot during stance has been the subject of intense investigation for the last half century. Plantar pressure studies by Cavanaugh et al. (1987) of subjects standing barefoot have determined that the distribution of load in the foot is as follows: heel 60%, midfoot 8%, forefoot 28%, and toes 4%. Peak pressures under the heel are 2.6 times greater than forefoot pressures (Fig. 9-42). Forefoot peak pressures occur under the second metatarsal head (Fig. 9-43).

The dynamics of gait exert the primary influence on plantar pressure during walking (Cavanagh et al., 1997). Hutton et al. (1973) studied the progression of the center of pressure across the sole of the foot during gait (Fig. 9-44). During barefoot walking, the center of pressure is initially located in the central heel and accelerates rapidly across the midfoot to reach the forefoot, where the velocity decreases. Peak forefoot pressures are reached at 80% stance phase and are centered under the second metatarsal. At toe-off, the center of pressure is located under the hallux. The metatarsal heads are in contact with the floor at least 50% of stance phase.

The distribution of plantar pressures changes with shoe wear. Shoe wear reduces peak heel pressure by producing a more even distribution of pressure under the heel. With shoes, forefoot load distribution shifts medially with maximum pressure under the first and second

metatarsal heads. The pressures under the toes also increase with shoe wear (Soames, 1985).

The distribution of plantar pressure during running has identified two types of runners characterized by their first point of contact with the ground: rearfoot strikers and midfoot strikers (Fig. 9-45). Rearfoot strikers make initial ground contact with the posterior third of the shoe. The initial contact for the midfoot strikers is in the middle third of the shoe. In both groups, first contact occurs along the lateral border of the foot. Peak pressure does not differ between runner types. The center of pressure is in the distalmost 20% to 40% of the shoe in both contact groups for most of contact time, indicating most time is spent on the forefoot (Cavanagh et al., 1987).

During walking and running, several forces are acting between the foot and the ground: vertical force, fore and aft shear (anteroposterior shear), medial and lateral shear, and rotational torque (Fig. 9-46). The vertical ground reaction force exhibits a double peak following

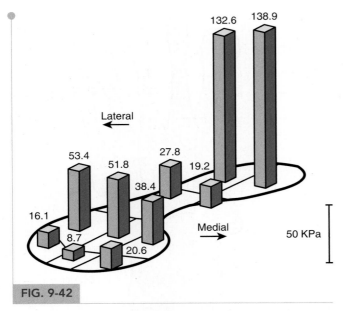

FIG. 9-42

Mean regional peak pressures during standing measured in kilopascals (kPa). The ratio of peak rearfoot to peak forefoot pressures is approximately 2.6:1.

the initial heel-strike spike. The first peak follows heel-strike in early stance, and the second peak occurs in late stance prior to toe-off. The fore and aft shear forces demonstrate initial braking by the foot as the foot places a forward shear force on the ground, followed

by a backward shear on the ground as it pushes off in late stance. Most of the medial-lateral shear is directed laterally because the body's center of gravity is oriented medially over the foot. Medial (internal rotation) torque is generated early in stance as the tibia internally rotates and the foot pronates, followed by lateral (external rotation) torque as the leg externally rotates and the foot supinates.

The second metatarsal suffers stress (fatigue) fractures more commonly than the other metatarsals (Brukner et al., 1996), possibly because it is recessed within the distal tarsal bones and therefore has limited mobility. It is commonly believed among clinicians that athletes with pes cavus are more likely to suffer stress fractures of the tibia due to the limited ability of the rigid foot to absorb impact forces. However, a recent systematic review (Barnes et al., 2008) suggested that athletes with both extremes of foot types (very low and very high arched) are more susceptible to tibial stress fractures.

Effects of Shoe Wear on Foot/Ankle Biomechanics

Western society places great importance on the appearance of footwear, especially among women. Women's footwear is designed to make the foot appear smaller and the leg appear longer by narrowing the toebox and

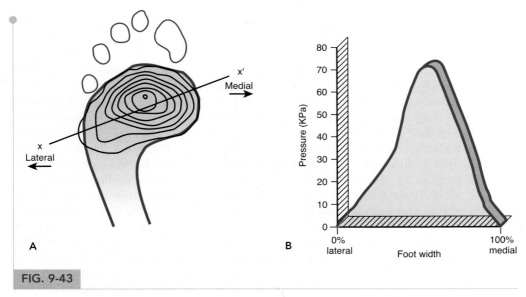

FIG. 9-43

Metatarsal head pressure distribution during standing. **A.** A line (xx') drawn in the contour plot between the approximate locations of the first and fifth metatarsal heads. **B.** The distribution of pressure along the metatarsal head line (xx') indicating maximum pressure under the second metatarsal head.

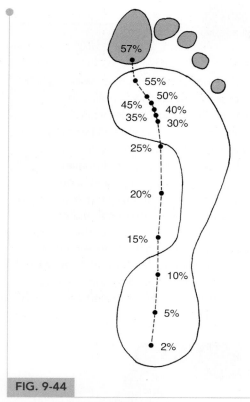

FIG. 9-44

The progression of the center of pressure along the sole of the foot during normal walking is expressed as a *broken line*. Each point on the sole corresponds to a percentage of the gait cycle. Note the rapid progression across the heel and midfoot to reach the forefoot, where most of stance phase is spent. It then progresses rapidly along the plantar aspect of the hallux.

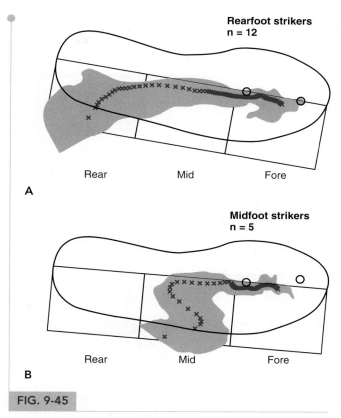

FIG. 9-45

Two types of runners characterized by initial ground contact. **A.** Rearfoot strikers. **B.** Midfoot strikers.

elevating the heel. A narrow toe box compresses the forefoot medially and laterally, thus contributing to the development of hallux valgus, hammer toes, and bunionettes (Case Study 9-3). A study of 356 women by Frey et al. (1993) found that 88% of women with foot pain wore shoes that were on average 1.2 cm narrower than their foot. Women who wore shoes on average 0.5 cm wider than the foot had no symptoms and less deformity. Shoes with elevated heels increase forefoot pressure compared with standing barefoot (Snow et al., 1992). A 1.9-cm heel increased forefoot pressure by 22%, a 5-cm heel increased peak pressure by 57%, and an 8.3-cm heel increased peak pressure by 76%. An elevated heel can cause pain under the metatarsal heads and may also contribute to interdigital neuroma formation. Elevation of the heel also may over time result in Achilles contracture, limited ankle dorsiflexion, and an altered gait. The amount of ankle joint motion in the gait cycle decreases as heel height increases (Murray et al., 1970). Kato and

Watanabe (1981) suggested that the increased popularity of the Western-style shoe in Japan between 1960 and 1980 has resulted in a higher incidence of hallux valgus, which was a rare deformity in that country prior to 1960 when the traditional Japanese clog was more popular.

The athletic shoe industry has evolved into a multibillion dollar industry due to an increase in the number of recreational runners. With it came a plethora of available shoe models that frequently claim to improve performance and reduce injuries by changing the biomechanics of running. Many of these claims revolve around decreasing the rate and excursion of pronation during the stance phase of running and therefore improving alignment of the skeleton. However, epidemiologic and laboratory evidence suggest that the effect of shoe wear on controlling foot motion is questionable (Nigg, 2001). Barefoot running has been suggested as a more natural alternative, and there are examples of world-class runners who run barefoot.

Although there is evidence that barefoot running requires less energy than shod running (Burkett et al., 1985), a laboratory study that used markers attached

Percentage of cycle

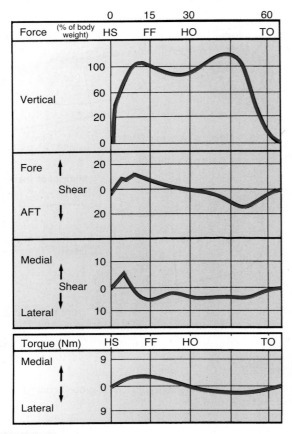

FIG. 9-46

Ground reactive forces acting on the foot during the gait cycle. *HS*, heel-strike; *FF*, foot flat; *HO*, heel-off; *TO*, toe-off. *Reprinted with permission from Mann, R.A. (1982). Biomechanics of running. In AAOS Symposium on the Foot and Leg in Running Sports. St Louis, MO: Mosby, 30–44.*

to the skeleton with intracortical pins suggested that there are practically no differences in eversion between barefoot and shod running (Stacoff et al., 2000), refuting previous research that used shoe and skin markers and had suggested that barefoot running results in decreased total eversion and eversion speed (Stacoff et al., 1991). More recently, rocker bottom shoes have been suggested to result in improved fitness and better walking patterns. Although there is evidence that they decrease plantar pressure, especially of the forefoot, the claims about effects on proximal joints remain largely unproven (Hutchins et al., 2009).

Case Study 9-3

Hallux Valgus

A 50-year-old woman who has been wearing shoes with a narrow toe box for almost 35 years presents with hallux valgus deformity. By compressing the forefoot medially and laterally, these abnormal forces can lead to a hallux valgus deformation. In this way, the proximal phalanx of the first toe shifts laterally and pronates on the first metatarsal head (Case Study Fig. 9-3). This abnormal position of the proximal phalanx decreases its ability to depress the metatarsal head during toe-off. Splints that produce medial forces on the proximal phalanx of the first toe, as well as orthotics, have been used as conservative biomechanical approaches for the treatment of mild and moderate hallux valgus.

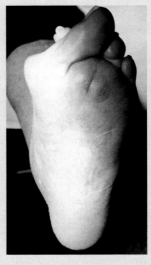

Case Study Figure 9-3 A view of the plantar surface of the foot of a patient with severe hallux valgus. Note calluses underlying the second and third metatarsal heads (transfer lesions), which indicate a transfer of plantar forces away from the first metatarsal head to the lesser metatarsal heads.

Summary

- The foot alternates in form and function between shock-absorbing flexible platform and rigid propulsive lever during different phases of the gait cycle.
- In weight-bearing, ankle dorsiflexion and tibial internal rotation are associated with subtalar eversion

(pronation), and ankle plantarflexion and tibial external rotation are associated with subtalar inversion (supination).

- Subtalar motion occurs around a single axis (i.e., screw-like) with pronation being a triplanar motion consisting of eversion, dorsiflexion, and abduction, and supination being a triplanar motion consisting of inversion, plantarflexion, and adduction.

- Subtalar supination "locks" the transverse joint and causes the foot to become rigid; subtalar eversion "unlocks" the transverse tarsal joint and allows foot flexibility.

- The Lisfranc joint (tarsometatarsal joint) is intrinsically stable and relatively immobile as a result of its arch-like configuration and the key-like structure of the second tarsometatarsal joint.

- The medial longitudinal arch acts like both a beam and a truss. The arch is elevated through the windlass mechanism of the plantar fascia. The posterior tibial tendon provides primary dynamic support to the arch.

- Foot muscle action during standing is relatively silent, but sequential firing of both extrinsic and intrinsic muscles is necessary to produce a normal gait pattern. The anterior tibial musculature fires during early stance to slow foot plantarflexion and prevent foot slap. The posterior calf musculature fires during mid and late stance to control progression of the body over the foot.

- During barefoot standing, the heel bears 60% of the load and the forefoot bears 28%. Forefoot peak pressures occur under the second metatarsal head.

- During walking, the center of pressure moves from the posterolateral heel rapidly across the midfoot to the forefoot with peak pressures under the second or third metatarsal head. At toe-off, the hallux bears the most pressure.

- The heel fat pad is specifically designed to absorb shock during heel-strike. The plantar fascia attaches the skin of the heel and forefoot to the underlying bony and ligamentous structures.

- Ankle joint instant centers of rotation fall within the talus during range of motion. In movement from plantarflexion to dorsiflexion, joint surfaces first distract, then glide and eventually compress at the end of dorsiflexion.

- Ankle joint stability is determined by joint congruency and ligamentous integrity. Ankle stability increases and depends more on bony surface congruency during weight-bearing.

- The anterior talofibular and calcaneofibular ligaments synergistically provide stability against inversion during ankle motion.

- The deltoid ligament prevents ankle eversion, external rotation, and lateral talar shift. It is key in maintaining the integrity of the syndesmosis.

- The fibula bears approximately one-sixth of the force exerted through the lower extremity. The distal syndesmotic ligaments prevent separation of the distal fibula and tibia and help transmit force through the distal fibula on weight-bearing.

- Ankle joint centroid (center of pressure) position changes with ankle flexion-extension and inversion-eversion. Talar surface contact is maximized and joint pressure is minimized in dorsiflexion.

- The forces acting on the ankle can rise to levels exceeding five times body weight during walking and thirteen times body weight during running.

- Narrow shoes and high heels can adversely affect foot mechanics, leading to forefoot deformities, heel pain, and Achilles contracture.

Practice Questions

1. What is the most likely position of the subtalar joint when an athlete is jumping vertically upward right before his foot leaves the ground?

2. What are the motions that can be produced by an eccentric contraction of the soleus?

3. A patient experiences rapid foot plantarflexion immediately after heel-strike during gait that produces an audible sound when the foot comes in contact with the ground (foot slap). What muscle is responsible for this pathology?

REFERENCES

Ahn, T.K., Kitaoka, H.B., Luo, Z.P., et al. (1997). Kinematics and contact characteristics of the first metatarsophalangeal joint. *Foot Ankle Int, 18*, 170.

Astion, D.J., Deland, J.T., Otis, J.C., et al. (1997). Motion of the hindfoot after simulated arthrodesis. *J Bone Joint Surg, 79A*, 241.

Ataarian, D.E., McCrackin, H.J., Devito, D.P., et al. (1985). Biomechanical characteristics of human ankle ligaments. *Foot Ankle, 6*, 54.

Barnes, A., Wheat, J., Milner, C. (2008). Association between foot type and tibial stress injuries: A systematic review. *Br J Sports Med, 42*, 93.

Beaudoin, A.J., Fiore, S.M., Krause, W.R., et al. (1991). Effect of isolated talocalcaneal fusion on contact in the ankle and talonavicular joints. *Foot Ankle, 12*, 19.

Bojsen-Moller, F., Lamoreux, L. (1979). Significance of free dorsiflexion of the toes in walking. *Acta Orthop Scand, 50*, 471.

Brukner P., Bradshaw C., Khan K.M., et al. (1996). Stress fractures: A review of 180 cases. *Clin J Sport Med, 6*, 85.

Bulucu, C., Thomas, K.A., Halvorson, T.L., et al. (1991). Biomechanical evaluation of the anterior drawer test: The contribution of the lateral ankle ligaments. *Foot Ankle, 11*, 389.

Burdett, R.G. (1982). Forces predicted at the ankle during running. *Med Sci Sports Exerc, 14*, 308–316.

Burkett, L.N., Kohrt, W.M., Buchbinder, R. (1985). Effects of shoes and foot orthotics on VO$_2$ and selected frontal plane knee kinematics. *Med Sci Sports Exerc, 17*: 158.

Calhoun, J.H., Eng, M., Li, F., et al. (1994). A comprehensive study of pressure distribution in the ankle joint with inversion and eversion. *Foot Ankle Int, 15*, 125–133.

Cass, J.R., Settles, H. (1994). Ankle in stability: In vitro kinematics in response to axial load. *Foot Ankle Int, 15*(3), 134–140.

Cavanagh, P.R., Morag, E., Boulton, A.J.M., et al. (1997). The relationship of static foot structure to dynamic foot function. *J Biomech, 30*, 243–250.

Cavanagh, P.R., Rodgers, M.M., Iiboshi, A. (1987). Pressure distribution under symptom-free feet during barefoot standing. *Foot Ankle, 7*, 262.

Cawley, P.W., France, E.P. (1991). Biomechanics of the lateral ligaments of the ankle: An evaluation of the effects of axial load and single plane motion on ligament strain patterns. *Foot Ankle, 12*, 92.

Clark, T.E. (1980). *The Pressure Distribution under the Foot during Barefoot Walking [thesis]*. University Park, PA: Penn State University.

Close, J.R. (1956). Some applications of the functional anatomy of the ankle joint. *J Bone Joint Surg, 386*, 761.

Cornwall, M.W., McPoil, T.G. (2002). Motion of the calcaneus, navicular and first metatarsal during the stance phase of walking. *J Am Podiatr Med Assoc, 92*, 67.

Daly, P.J., Kitaoka, H.B., Chao, E.Y.S. (1992). Plantar fasciotomy for intractable plantar fasciitis: Clinical results and biomechanical evaluation. *Foot Ankle, 13*, 188.

Earll, M., Wayne, J., Brodrick, C., et al. (1996). Contribution of the deltoid ligament to ankle joint contact characteristics: A cadaver study. *Foot Ankle Int, 17*, 317.

Elftman, H. (1960). The transverse tarsal joint and its control. *Clin Orthop, 16*, 41.

Fick, R. (1911). Handbuch der Anatomie und Mechanik der Gelenke. Berlin: Jena Fischer.

Frey, C., Thompson, F., Smith, J., et al. (1993). American Orthopaedic Foot and Ankle Society women's shoe survey. *Foot Ankle, 14*, 78.

Fong D., Hong Y., Shima Y., et al. (2009). Biomechanics of supination ankle sprain. *Am J Sports Med, 37*, 822.

Fujii T., Kitaoka H.B., Luo Z.P., et al. (2005). Analysis of ankle-hindfoot stability in multiple planes: An in vitro study. *Foot Ankle Int, 26*, 633.

Geideman, W.M., Johnson, J.E. (2000). Posterior tibial tendon dysfunction, *J Orthop Sports Phys Ther, 3*, 68.

Greenwald, S. (1977). Unpublished data cited in R.N. Stauffer, E.Y.S. Chao, C. Brewster. Force and motion analysis of the normal, diseased, and prosthetic ankle joint. *Clin Orthop, 127*, 189.

Grimstone, S.K., Niggs, B.M., Hanley, D.A., et al. (1993). Differences in ankle joint complex range of motion as a function of age. *Foot Ankle, 14*, 215.

Hagins, M., Pappas, E., Kremenic, I., et al. (2007). The effect of an inclined landing surface on biomechanical variables during a jumping task. *Clin Biomech, 22*, 1030.

Hale, S.A., Hertel, J., Olmsted-Kramer, L.C. (2007). The effect of a 4-week comprehensive rehabilitation program on postural control and lower extremity function in individuals with chronic ankle instability. *J Ortho Sports Phys Ther, 37*, 303.

Harper, M. (1991). The lateral ligamentous support of the subtalar joint. *Foot Ankle, 11*, 354.

Hicks, J.H. (1954). The mechanics of the foot II: The plantar aponeurosis and the arch. *J Anat, 88*, 25.

Hintermann, B., Nigg, B.M. (1995). In vitro kinematics of the axially loaded ankle complex in response to dorsiflexion and plantarflexion. *Foot Ankle Int, 16*, 514.

Huang, C.K., Kitaoka, H.B., An, K.N., et al. (1993). Biomechanical evaluation of longitudinal arch stability. *Foot Ankle, 14*, 353.

Huson, A. (1991). Functional anatomy of the foot. In M.H. Jahss (Ed.). *Disorders of the Foot and Ankle: Medical and Surgical Management*. Philadelphia: WB Saunders, 409.

Hutchins, S., Bowker, P., Geary, N., et al. (2009). The biomechanics and clinical efficacy of footwear adapted with rocker profiles—evidence in literature. *Foot (Edinb), 19*, 165.

Hutton, W.C., Scott, J.R.R., Stokes, I.A.F. (1973). The mechanics of the foot. In L. Klenerman (Ed.). *The Foot and Its Disorders*. Oxford, UK: Blackwell Science, 41.

Inman, V.T. (1976). *The Joints of the Ankle*. Baltimore: Williams & Wilkins.

Jahss, M.H., Kummer, F., Michelson, J.D. (1992a). Investigation into the fat pads of the sole of the foot: Heel pressure studies. *Foot Ankle, 13*, 227.

Jahss, M.H., Michelson, J.D., Desai, P., et al. (1992b). Investigations into the fat pads of the sole of the foot: Anatomy and histology. *Foot Ankle, 13*, 233.

Kato, T., Watanabe, S. (1981). The etiology of hallux valgus in Japan. *Clin Orthop Relat Res, 157*, 78.

Kitaoka, H.B., Luo, Z.P., An, K.N. (1997). Effect of posterior tibial tendon on the arch of the foot during simulated weightbearing: Biomechanical analysis. *Foot Ankle Int, 18*, 43.

Lamoreux, L.W. (1971). Kinematic measurements in the study of human walking. *Bull Prosthet Res, 10*(15), 3.

Lundberg, A., Goldie, I., Kalin, B., et al. (1989a). Kinematics of the ankle/foot complex: Plantarflexion and dorsiflexion. *Foot Ankle, 9*, 194–200.

Lundberg, A., Svensson, O.K., Bylund, C., et al. (1989b). Kinematics of the ankle/foot complex—Part 2: Pronation and supination. *Foot Ankle, 9*, 248.

Lundberg, A., Svenson, O.K., Nemeth, G., et al. (1989c). Kinematics of the ankle/foot complex—Part 3: Influence of leg rotation. *Foot Ankle, 9*, 304.

Lundberg, A., Svenson, O.K., Nemeth, G., et al. (1989d). The axis of rotation of the ankle joint. *J Bone Joint Surg, 71B*, 94–99.

Manter, J.T. (1941). Movements of the subtalar and transverse tarsal joints. *Anat Rec, 80*, 397.

McPoil, T.G., Martin, R.L., Cornwall, M.W., et al. (2008). Heel pain—plantar fasciitis: Clinical practice guidelines linked to the international classification of function, disability, and health from the orthopaedic section of the American Physical Therapy Association, *J Ortho Sports Phys Ther, 38*, A1.

Murray, M.P., Drought, A.B., Kory, R.C. (1964). Walking patterns in normal men. *J Bone Joint Surg, 46A*, 335–360.

Murray, M.P., Kory, R.C., Sepic, S.B. (1970). Walking patterns of normal women. *Arch Phys Med Rehabil, 51*, 637.

Neumann, D.A. (2010). Ankle and Foot. In *Kinesiology of the Musculoskeletal System: Foundations for Rehabilitation (2nd ed.).* St Louis, MO, Mosby Elsevier, 590.

Nigg, B.M. (2001). The role of impact forces and foot pronation: A new paradigm. *Clin J Sports Med, 11*, 2.

Nigg, B.M., Skorvan, G., Frank, C.B., et al. (1990). Elongation and forces of ankle ligaments in a physiological range of motion. *Foot Ankle, 11*, 30.

Olgivie-Harris, D.J., Reed, S.C., Hedman, T.P. (1994). Disruption of the ankle syndesmosis: Biomechanical study of ligamentous restraints. *Arthroscopy, 10*, 558–560.

Ouzonian, T.J., Shereff, M.J. (1989). In vitro determination of midfoot motion. *Foot Ankle, 10*, 140.

Pappas, E., Hagins, M. (2008). The effect of raked stages on standing posture in dancers. *J Dance Med Sci, 12*, 54.

Perry, J. (1983). Anatomy and biomechanics of the hindfoot. *Clin Orthop, 177*, 9.

Pohl, M.B., Messenger, N., Buckley, J.G. (2007). Forefoot, rearfoot and shank coupling: Effects of variations in speed and mode of gait, *Gait Posture, 25*, 295.

Proctor, P., Paul, J.P. (1982). Ankle joint biomechanics. *J Biomech, 15*, 627.

Renstrom, P., Wertz, M., Incavo, S., et al. (1988). Strain on the lateral ligaments of the ankle. *Foot Ankle, 9*, 59.

Ross S.E., Guskiewicz K.M. (2004). Examination of static and dynamic postural stability in individuals with functionally stable and unstable ankles. *Clin J Sport Med, 14*, 332.

Sammarco, G.J. (1980). Biomechanics of the foot. In V.H. Frankel, M. Nordin (Eds.). *Basic Biomechanics of the Musculoskeletal System (2nd ed.).* Philadelphia: Lea & Febiger, 193–219.

Sammarco, G.J. (1995). Peroneus longus tendon tears: Acute and chronic. *Foot Ankle Int, 16*(5), 245–253.

Sammarco, G.J., Burnstein, A.H., Frankel, V.H. (1973). Biomechanics of the ankle: A kinematic study. *Orthop Clin North Am, 4*, 75.

Sarrafian, S.K. (1987). Functional characteristics of the foot and plantar aponeurosis under tibiotalar loading. *Foot Ankle, 8*, 4.

Sarrafian, S.K. (1993a). Functional anatomy of the foot and ankle. In *Anatomy of the Foot and Ankle.* Philadelphia: Lippincott, 474–602.

Sarrafian, S.K. (1993b). Retaining systems and compartments. In *Anatomy of the Foot and Ankle.* Philadelphia: Lippincott, 137–149.

Scranton, P.E., McMaster, J.H., Kelly, E. (1976). Dynamic fibular function: A new concept. *Clin Orthop, 118*, 76–81.

Shereff, M.J., Bejahi, F.J., Kummer, F.J. (1986). Kinematics of the first metatarsophalangeal joint. *J Bone Joint Surg, 68A*, 392.

Silver, R.L., de la Garza, J., Rang, M. (1985). The myth of muscle balance: A study of relative strengths and excursions about the foot and ankle. *J Bone Joint Surg, 67B*, 432.

Snow, R.E., Williams, K.R., Holmes, G.B. Jr. (1992). The effects of wearing high heeled shoes on pedal pressures in women. *Foot Ankle, 13*, 85–92.

Soames, R.W. (1985). Foot pressures during gait. *J Biomech Eng, 7*, 120–126.

Stacoff, A., Kalin, X., Stucci, E. (1991). The effects of shoes on the torsion and rearfoot motion in running. *Med Sci Sports Exer, 23*, 482.

Stacoff, A., Nigg, B.M., Reinschmidt, C., et al. (2000). Tibiocalcaneal kinematics of barefoot versus shod running. *J Biomech, 33*, 1387.

Stauffer, R.N., Chao, E.Y.S., Brewster, R.L. (1977). Force and motion analysis of the normal, diseased and prosthetic ankle joints. *Clin Orthop, 127*, 189.

Stephens, M.M., Sammarco, G.J. (1992). The stabilizing role of the lateral ligament complex around the ankle and subtalar joints. *Foot Ankle, 13*, 130.

Stiehl, J.B., Skrade, D.A., Needleman, R.L., et al. (1993). Effect of axial load and ankle position on ankle stability. *J Orthop Trauma, 7*, 72–77.

Stormont, D.M., Morrey, B.F., An, K.N., et al. (1985). Stability of the loaded ankle. Relation between articular restraint and primary and secondary restraints. *Am J Sports Med, 13*, 295.

Thordarson, D.B., Schmotzer, H., Chon, J., et al. (1995). Dynamic support of the human longitudinal arch. *Clin Orthop, 316*, 165.

Van Gheluwe, B., Dananberg, H.J., Hagman, F., et al. (2006). Effects of the hallux limitus on plantar foot pressure and foot kinematics during walking. *J Am Podiatr Med Assoc, 96*, 428.

Wang, Q.W., Whittle, M., Cunningham, J., et al. (1996). Fibula and its ligaments in load transmission and ankle joint stability. *Clin Orthop, 330*, 261.

Waters, R.L., Hislop, H.J., Perry, J., et al. (1978). Energetics: Application of the study and management of locomotor disabilities. *Orthop Clin North Am, 9*, 351.

Wolf P., Stacoff A., Liu A., et al. (2008). Functional units of the human foot. *Gait Posture, 28*, 434.

Wright, D.G., Desai, S.M., Henderson, W.H. (1964). Action of the subtalar and ankle joint complex during the stance phase of walking. *J Bone Joint Surg, 46A*, 361.

Wu, K.K. (1996). Morton's interdigital neuroma: A clinical review of its etiology, treatment and results. *J Foot Ankle Surg, 35*, 112.

Yeung, M.S., Chan, K.M., So, C.H., et al. (1994). An epidemiological survey on ankle sprain. *Br J Sports Med, 28*, 112.

Biomechanics of the Lumbar Spine

Shira Schecter Weiner, Florian Brunner, and Margareta Nordin

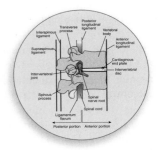

Introduction

The human spine is a complex structure whose principal functions are to protect the spinal cord and transfer loads from the head and trunk to the pelvis, while simultaneously allowing movement and providing stability to the trunk. Each of the 24 vertebrae articulates with the adjacent ones to permit motion in three planes. The spine gains stability from the intervertebral discs and from the surrounding ligaments and muscles; the discs and ligaments provide intrinsic stability, and the muscles provide extrinsic support.

This chapter describes the basic characteristics of the various structures of the spine and the interaction of these structures during normal spine function. Kinematics and kinetics of the spine are also covered. The information in the chapter has been selected to provide an understanding of some fundamental aspects of lumbar spine biomechanics that can be put to practical use.

The Motion Segment: The Functional Unit of the Spine

The functional unit of the spine—the motion segment—consists of two adjacent vertebrae and their intervening soft tissues (Fig. 10-1). The anterior portion of the segment is composed of two superimposed intervertebral bodies, the intervertebral disc, and the longitudinal ligaments (Fig. 10-2). The corresponding vertebral arches, the intervertebral joints formed by the facets, the transverse and spinous processes, and various ligaments make up the posterior portion. The arches and vertebral bodies form the vertebral canal, which protects the spinal cord (Fig. 10-3). The arch consists of two pedicles and the lamina.

THE ANTERIOR PORTION OF THE MOTION SEGMENT

The vertebral bodies are designed to bear mainly compressive loads, and they are progressively larger caudally as the superimposed weight of the upper body increases. The vertebral bodies in the lumbar region are thicker and wider than those in the thoracic and cervical regions; their greater size allows them to sustain the larger loads to which the lumbar spine is subjected.

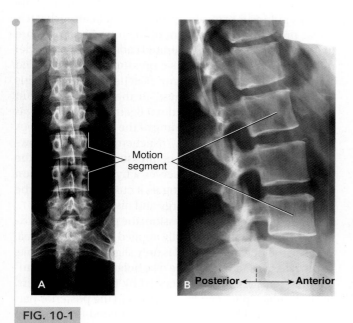

FIG. 10-1

Anteroposterior (**A**) and lateral (**B**) roentgenograms of the lumbar spine. One motion segment, the functional unit of the spine, is indicated.

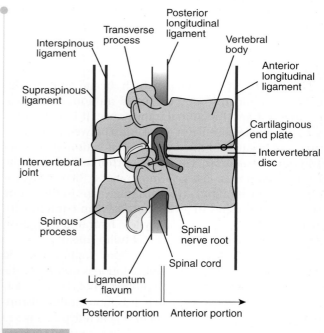

FIG. 10-2

Schematic representation of a motion segment in the lumbar spine (sagittal view). Anterior portion, posterior longitudinal ligament, anterior longitudinal ligament, vertebral body, cartilaginous end plate, intervertebral disc, intervertebral foramen with nerve root. Posterior portion, ligamentum flavum, spinous process, intervertebral joint formed by the superior and inferior facets (the capsular ligament is not shown), supraspinous ligament, interspinous ligament, transverse process (the intertransverse ligament is not shown), arch, vertebral canal (the spinal cord is not depicted).

Anterior

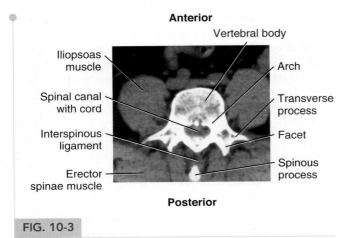

FIG. 10-3

Transverse section of a motion segment at the L4 level viewed by computed tomography. The vertebral body, arch, spinal canal with spinal cord, and transverse processes are clearly seen. The view is taken at a level that depicts only the tip of the spinous process, with the interspinous ligament visible between the spinous process and the facets of the intervertebral joints. Directly anterior to the transverse processes and adjacent to the vertebral body are the iliopsoas muscles. Posterior to the vertebral body, the erector spinae muscles can be seen.

The intervertebral disc, which bears and distributes loads and restrains excessive motion, is of great mechanical and functional importance. It is well suited for its dual role because of its location between the vertebrae and because of the unique composition of its inner and outer structures. The inner portion of the disc, the nucleus pulposus, is a gelatinous mass. Rich in hydrophilic (water-binding) glycosaminoglycans in the young adult, it diminishes in glycosaminoglycan content with age and becomes progressively less hydrated (Ferguson and Steffen, 2003; Urban and McMullin, 1985).

The disc, which has no direct blood supply, relies on diffusion for its nutritional needs. Motion is critical for the diffusion process. Although this process is not clearly understood based on current research, it is thought that disruption of diffusion is the precursor to degenerative changes in the spine (Urban et al., 2004). MRI has been used successfully to study the process of diffusion and has revealed that the end plate controls the process of diffusion (Rajasekaran et al., 2007). Sustained loading has been shown to impair diffusion, with a prolonged recovery time needed for diffusion to return to unloaded conditions (Arun et al., 2009).

The nucleus pulposus lies directly in the center of all discs except those in the lumbar segments, where it has a slightly posterior position. This inner mass is surrounded by a tough outer covering, the annulus fibrosus, composed of fibrocartilage. The crisscross arrangement of the coarse collagen fiber bundles within the fibrocartilage allows the annulus fibrosus to withstand high bending and torsional loads (see Fig. 11-11). Alterations in the disc such as degenerative changes, a normal part of aging, and annular tears will allow for increased intersegmental motion, thereby altering the biomechanical loading of the motion segment (An et al., 2006; Rohlmann et al., 2006). Degenerative disc changes result in increased loading on the facets and changes in the distribution of interdiscal loading. Discs with annular tears display increased rotational moments during loading compared with nondegenerated discs (Haughton et al., 2000; Rohlmann et al., 2006). The end plate, composed of hyaline cartilage, separates the disc from the vertebral body (Fig. 10-2). End-plate fractures, which may occur with degeneration, may also alter the distribution of forces through the motion segment, reducing pressure on the nucleus pulposus while increasing the stress on the posterior annulus fibrosis (Przybyla et al., 2006). The tissues that compose the disc are similar to that of articular cartilage, described in detail in Chapter 3, though recent evidence suggests the unique behavior of the collagen in each of these tissues (Sivan et al., 2006).

During daily activities, the disc is loaded in a complex manner and is usually subjected to a combination of compression, bending, and torsion. Flexion, extension, and lateral flexion of the spine produce mainly tensile and compressive stresses in the disc, whereas rotation produces mainly shear stress.

When a motion segment is transected vertically, the nucleus pulposus of the disc protrudes, indicating that it is under pressure. Measurement of the intradiscal pressure in normal and slightly degenerated cadaver lumbar nuclei pulposi has shown an intrinsic pressure in the unloaded disc of approximately 10 N/cm² (Nachemson, 1960). This intrinsic pressure, or pre-stress, in the disc results from forces exerted by the longitudinal ligaments and the ligamentum flavum. During loading of the spine, the nucleus pulposus acts hydrostatically (Nachemson, 1960), allowing a uniform distribution of pressure throughout the disc; hence, the entire disc serves a hydrostatic function in the motion segment, acting as a cushion between the vertebral bodies to store energy and distribute loads.

In a disc loaded in compression, the pressure is approximately 1.5 times the externally applied load per unit area. Because the nuclear material is only slightly compressible, a compressive load makes the disc bulge laterally; circumferential tensile stress is sustained by the annular fibers. In the lumbar spine, the tensile stress in the posterior part of the annulus fibrosus has been estimated to be four to five times the applied axial compressive load (Fig. 10-4) (Galante, 1967; Nachemson, 1960, 1963). The tensile stress in the annulus fibrosus in the thoracic spine is less than that in the lumbar spine because of differences in disc

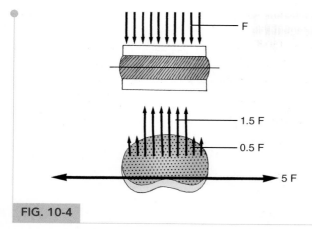

FIG. 10-4

Distribution of stress in a cross-section of a lumbar disc under compressive loading. The compressive stress is highest in the nucleus pulposus, 1.5 times the externally applied load (*F*) per unit area. By contrast, the compressive stress on the annulus fibrosus is only approximately 0.5 times the externally applied load. This part of the disc bears predominantly tensile stress, which is four to five times greater than the externally applied load per unit area. *Adapted with permission from Nachemson, A. (1975). Towards a better understanding of low-back pain: A review of the mechanics of the lumbar disc. Rheumatol Rehabil, 14, 129.*

geometry. The higher ratio of disc diameter to height in the thoracic discs reduces the circumferential stress in these discs (Kulak et al., 1975).

Degeneration of a disc reduces its proteoglycan content and thus its hydrophilic capacity (Fig. 10-5A–C). As the disc becomes less hydrated, its elasticity and its ability to store energy and distribute loads gradually decrease; these changes make the disc(s) more vulnerable to stresses and impact on the loading of other portions of the motion segment (Rohlmann et al., 2006).

THE POSTERIOR PORTION OF THE MOTION SEGMENT

The posterior portion of the motion segment guides its movement. The type of motion possible at any level of the spine is determined by the orientation of the facets of the intervertebral joints to the transverse and frontal planes. This orientation changes throughout the spine.

Except for the facets of the two uppermost cervical vertebrae (Cl and C2), which are parallel to the transverse plane, the facets of the cervical intervertebral joints are oriented at a 45° angle to the transverse plane and are parallel to the frontal plane (Fig. 10-6A). This alignment of the joints of C3 to C7 allows flexion, extension, lateral flexion, and rotation. The facets of the thoracic joints are oriented at a 60° angle to the transverse plane and at a 20° angle to the frontal plane (Fig. 10-6B); this orientation allows

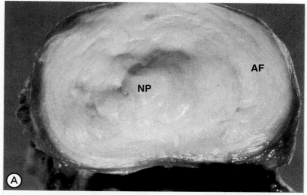

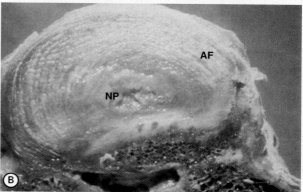

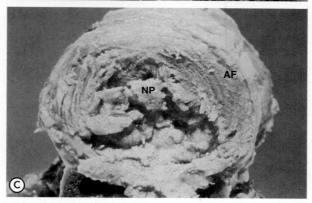

FIG. 10-5

Human intervertebral disc composed of an inner gelatinous mass, the nucleus pulposus (*NP*), and a tough outer covering, the annulus fibrosus (*AF*). **A.** Normal young disc. The gelatinous nucleus pulposus is 80% to 88% water content. Age-related variations in protein-polysaccharides from human nucleus pulposus, annulus fibrosus, and costal cartilage are easy to distinguish from the firmer annulus fibrosus. **B.** Normal mid-age disc. The nucleus pulposus has lost water content, a normal degenerative process. The fibers on the posterior part of the annulus have sustained excessive stress. **C.** Severely degenerated disc. The nucleus pulposus has become dehydrated and has lost its gel-like character. The boundary between the nucleus and the annulus is difficult to distinguish because the degree of hydration is now about the same in both structures.

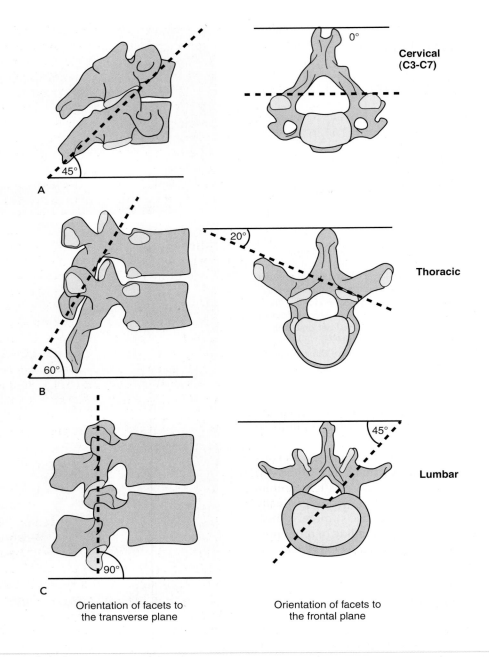

Cervical (C3-C7)

Thoracic

Lumbar

Orientation of facets to
the transverse plane

Orientation of facets to
the frontal plane

FIG. 10-6

Orientation of the facets of the intervertebral joints (approximate values). **A.** In the lower cervical spine, the facets are oriented at a 45° angle to the transverse plane and are parallel to the frontal plane. **B.** The facets of the thoracic spine are oriented at a 60° angle to the transverse plane and at a 20° angle to the frontal plane. **C.** The facets of the lumbar spine are oriented at a 90° angle to the transverse plane and at a 45° angle to the frontal plane. *From White, A.A., Panjabi, M.N. (1978). Clinical Biomechanics of the Spine. Philadelphia: JB Lippincott Co.*

lateral flexion, rotation, and some flexion and extension. In the lumbar region, the facets are oriented at right angles to the transverse plane and at a 45° angle to the frontal plane (Fig. 10-6C) (White and Panjabi, 1978). This alignment allows flexion, extension, and lateral flexion, but almost no rotation. The lumbosacral joints differ from the other lumbar intervertebral joints in that the oblique orientation of the facets allows appreciable rotation (Lumsden and Morris, 1968). The above cited values for facet orientation are only approximations, because considerable variation is found within and among individuals. Researchers have documented that although some variation in facet joint angulation exists, individuals with these variations may be predisposed to more progressive degenerative changes with aging (Boden et al., 1996; Miyazaki et al., 2010).

The facets guide movement of the motion segment and have a load-bearing function, and may have some role in the lateral stability of the motion segment (Okushima et al., 2006). Load sharing between the facets and the disc varies with the position and the health of the spine. The loads on the facets are greatest with axial rotation of the spine (Schmidt et al., 2008; Zhu et al., 2008). With disc degeneration, a greater amount of force is transferred to the facet joints, thereby redistributing the load through the motion segment (Rohlmann et al., 2006). Because the facets are not the primary support structure in extension, if total compromise of these joints occurs, an alternate path of loading is established. This path involves the transfer of axial loads to the annulus and anterior longitudinal ligament as a way of supporting the spine (Haher et al., 1994). High loading of the facets is also present during forward bending, coupled with rotation (El-Bohy and King, 1986). The vertebral arches and intervertebral joints play an important role in resisting shear forces. This function is demonstrated by the fact that patients with deranged arches or defective joints (e.g., from spondylolysis and spondylolisthesis) are at increased risk for forward displacement of the vertebral body (Case Study 10-1) (Adams and Hutton, 1983; Miller et al., 1983). The transverse and spinous processes serve as sites of attachment for the spinal muscles that, when activated, initiate spine motion and provide extrinsic stability.

THE LIGAMENTS OF THE SPINE

The ligamentous structures surrounding the spine contribute to its intrinsic stability (Fig. 10-2). All spine ligaments except the ligamentum flavum have a high collagen content, which limits their extensibility during spine motion. The ligamentum flavum, which connects two adjacent vertebral arches longitudinally, is an exception, having a large percentage of elastin. The elasticity

of this ligament allows it to contract during extension of the spine and to elongate during flexion. Even when the spine is in a neutral position, the ligamentum flavum is under constant tension as a result of its elastic properties. Because it is located at a distance from the center of motion in the disc, it pre-stresses the disc; that is, along with the longitudinal ligaments, it creates an intradiscal pressure and thus helps provide intrinsic support to the spine (Nachemson and Evans, 1968; Rolander, 1966). Research suggests that with degenerative changes such as spondylolisthesis, traction spurs, and disc degeneration, which may lead to instability, altered mechanical stress will increasingly load the ligamentum flavum and cause hypertrophy (Fukuyama et al., 1995).

The amount of strain on the various ligaments differs with the type of motion of the spine. During flexion, the interspinous ligaments are subjected to the greatest strain, followed by the capsular ligaments and the ligamentum flavum. During extension, the anterior longitudinal ligament bears the greatest strain. During lateral flexion, the contralateral transverse ligament sustains the highest strains, followed by the ligament flavum and the capsular ligaments. The capsular ligaments of the facet joints bear the most strain during rotation (Panjabi et al., 1982). Slouching during sitting, a position of relatively low loading, imposes a backward rotation of the pelvis while the trunk is flexed, thereby straining the iliolumbar ligaments (Snijders et al., 2004).

Kinematics

Active motion of the spine as in any joint is produced by the coordinated interaction of nerves and muscles. Agonistic muscles (prime movers) initiate and carry out motion and antagonistic muscles control and modify the motion, while co-contraction of both groups stabilizes the spine. The range of motion differs at various levels of the spine and depends on the orientation of the facets of the intervertebral joints (Fig. 10-6). Motion between two vertebrae is small and does not occur independently; all spine movements involve the combined action of several motion segments. The skeletal structures that influence motion of the trunk are the rib cage, which limits thoracic motion, and the pelvis, which augments trunk movements by tilting.

SEGMENTAL MOTION OF THE SPINE

The vertebrae have six degrees of freedom: rotation about and translation along a transverse, a sagittal, and a longitudinal axis. The motion produced during flexion, extension, lateral flexion, and axial rotation of the spine is a

Case Study 10-1

Spondylolisthesis: Anterior Slippage of One Vertebra in Relation to the Vertebra Below It

A 30-year-old male gymnast complains of severe back pain with radiation to both legs. The pain is associated with periods of strenuous training, and the symptoms decrease with rest or restriction of activity. After a careful examination by a spine specialist and MRI films, a diagnosis was made of spondylolisthesis at the level L5–S1 (Case Study Fig. 10-1-1), with concurrent bilateral pars interarticularis defects of L5 (Case Study Fig. 10-1-2).

Physiologic loads during repeated flexion-extension motion of the lumbar spine caused a fatigue fracture of the pars interarticularis (aspect of the posterior arch of the vertebra that lies between the inferior and superior facets). This bilateral defect leads to an anterior displacement of the vertebra L5 onto S1. As the L5 vertebra begins to slip forward, the center of gravity of the body is displaced anteriorly. To compensate, the lumbar spine above the lesion hyperextends and the upper part of the trunk is displaced backward. Because this is a disease continuum, the abnormal forces placed on the intervertebral disc leads to herniation into the neural foramina, producing moderate stenosis of both L5–S1 nerve roots.

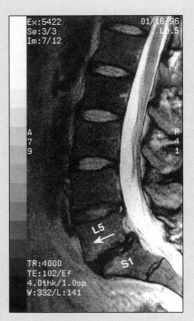

Case Study Figure 10-1-1

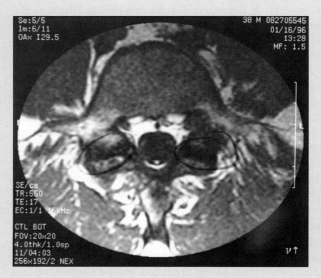

Case Study Figure 10-1-2

complex combined motion resulting from simultaneous rotation and translation.

Range of Motion

Various investigations using autopsy material or radiographic measurements in vivo have shown divergent values for the range of motion of individual motion segments, but there is agreement on the relative amount of motion at different levels of the spine. Representative values from White and Panjabi (1978) are presented in Figure 10-7 to allow a comparison of motion at various levels of the thoracic and lumbar spine. (Representative values for motion in the cervical spine are included for

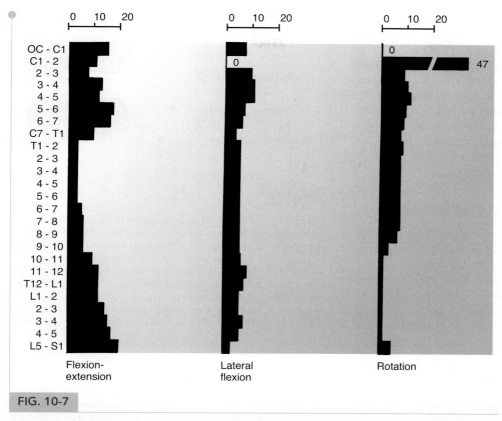

FIG. 10-7

A composite of representative values for type and range of motion at different levels of the spine. *From White, A.A., Panjabi, M.N. (1978). Clinical Biomechanics of the Spine. Philadelphia: JB Lippincott Co.*

comparison.) During all trunk motion the facet joints move most in the primary plane of motion (~4°–6°) with accessory motion to a lesser degree (~2°–3°) in the secondary planes (Kozanek et al., 2009). As degenerative changes of the disc and facet joints occur, these values will fluctuate, increasing or decreasing based on the severity of these changes, with some suggestion of an influence of gender (Fujiwara et al., 2000).

Investigations of the thoracic and lumbar spine show that the range of flexion and extension is approximately 4° in each of the upper thoracic motion segments, approximately 6° in the midthoracic region, and approximately 12° in the two lower thoracic segments. Within the series of the five lumbar vertebrae, the upper lumbar range of motion in flexion and extension is somewhat greater than the lower lumbar region (Kozanek et al., 2009; Li et al., 2009), reaching a maximum of 20° at the lumbosacral level.

Lateral flexion shows the greatest range in each of the lower thoracic segments, reaching 8° to 9°. In the upper thoracic segments, the range is uniformly 6°. Approximately 6° of lateral flexion is also found in each of all lumbar segments, with slightly more motion in the lower lumbar region (Kozanek et al., 2009; Li et al., 2009), except the lumbosacral segment, which demonstrates only 3° of motion.

Rotation is greatest in the upper segments of the thoracic spine, where the range is 9°. The range of rotation progressively decreases caudally, reaching 2° in the lower segments of the lumbar spine. It then increases to 5° in the lumbosacral segment. Newer in vivo studies using other techniques have shown some variations in the number reported previously with a tendency to lower values for each segment (Li et al., 2009).

Surface Joint Motion

Motion between the surfaces of two adjacent vertebrae during flexion-extension or lateral flexion may be analyzed by means of the instant center method of Reuleaux. The procedure is essentially the same as that described for the cervical spine in Chapter 11 (see Figs. 11-18 and 11-19). The path of the instant center during motion impacts on the loading of the various structures of the lumbar spine. The instantaneous center of flexion-extension and lateral flexion in a motion segment of the lumbar spine lies within the disc under normal conditions (Fig. l0-8A) (Cossette et al., 1971; Rolander, 1966). During flexion, the instant center moves caudally, resulting in decreased facet forces, while in extension the reverse is true (Rousseau et al., 2006). With abnormal conditions such as pronounced disc degeneration, the instantaneous center pathway will be altered and move outside of the disc, toward the facet joints (Fig. 10-8B) (Gertzbein et al., 1985; Reichmann et al., 1972; Schmidt et al., 2008).

FUNCTIONAL MOTION OF THE SPINE

Because of its complexity, the movement within a single motion segment is difficult to measure clinically. Approximate values for the normal functional range of motion of the spine can be given. Bible et al. (2010) studied normal motion during activities of daily living with an electrogoniometer. Table 10-1 shows total active range of motion for common activities. Variations among

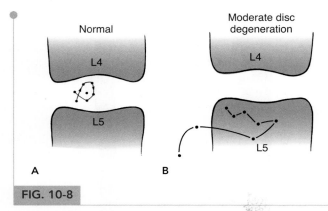

FIG. 10-8

Instant center pathway for a normal cadaver spine **(A)** and a cadaver spine with moderate disc degeneration **(B)**. Instant centers were determined for 3° intervals of motion from maximum extension to maximum flexion. In the normal spine, all instant centers fell within a small area in the disc. In the degenerated spine, the centers were displaced, and hence the surface motion was abnormal. *Reprinted with permission from Gertzbein, S.D., Seligman, J., Holtby, R., et al. (1985). Centrode patterns and segmental instability in degenerative disc disease. Spine, 10, 257.*

TABLE 10-1

Percentage of Total Active ROM for Each Activity of Daily Living

ADL	Flexion/Extension (%)	Lateral Bending (%)	Axial Rotation (%)
	Average Percentage of Full Active ROM		
Stand to sit	37	20	12
Backing car	10	16	18[a]
Reading	4	6	6
Feeding	5	8	9
Socks	22	19	14
Shoes	20	20	16
Sit to stand	39	14	10
Washing hands	12	15	12
Washing hair	9	11	12
Shaving	8	11	9
Make-up	7	11	8
Squatting	52	31	18
Bending	59	29	18
Walking	11	19	19
Up stairs	13	22	20
Down stairs	11	21	18

[a]Only right full active ROM used in rotational percentage for backing up a car. ROM, range of motion; ADL, activities of daily living.
Used with permission from Bible, J.E., Biswas, D.B.A., Miller, C.P., et al. (2010). Normal functional range of motion of the lumbar spine during 15 activities of daily living. *J Spinal Disord Techn, 23,* 106–112.

individuals are large and show a Gaussian distribution in the three planes. The range of motion is strongly age dependent, decreasing by approximately 30% from youth to old age, although with aging, loss of range of motion is noted in flexion and lateral bending while axial rotation motion is maintained with evidence of increased coupled motion (McGill et al., 1999; Trudelle-Jackson et al., 2010). In addition to age, there is some evidence to suggest racial variations in lumbar range of motion (Trudelle-Jackson et al., 2010). Differences have also been noted between the sexes: Men have greater mobility in flexion and extension whereas women are more mobile in lateral flexion (Biering-Sorensen, 1984; Moll and Wright, 1971). Loss of range of motion in the lumbar and/or thoracic spine is compensated for mainly by motion in the cervical spine and hips.

THE MUSCLES

The spinal muscles can be divided into flexors and extensors. The trunk muscles play an important role in the mechanical behavior of the spine, including spine stability and intradiscal pressure (Wilke et al., 1996; Zander et al., 2001). The main flexors are the abdominal muscles (the rectus abdominis muscles, the internal and external oblique muscles, and the transverse abdominal muscle) and the psoas muscles. In general, muscles anterior to the vertebral column act as flexors. The main extensors are the erector spinae muscles, the multifidus muscles, and the intertransversarii muscles attached to the posterior elements. In general, the muscles posterior to the vertebral columns act as extensors (Fig. 10-9). The extensor muscles bridge between each vertebrae and motion segment as well as over several vertebrae and motion segments. When extensor muscles contract symmetrically, extension is produced. When right and left side flexors and extensor muscles contract asymmetrically, lateral bending or twisting of the spine is produced (Andersson and Lavender, 1997).

Muscle coactivation during lifting has been shown to be influenced by stress levels, with variation based on gender (Marras et al., 2000). Increased stress leads to increased muscle activity and results in increased spinal loading,

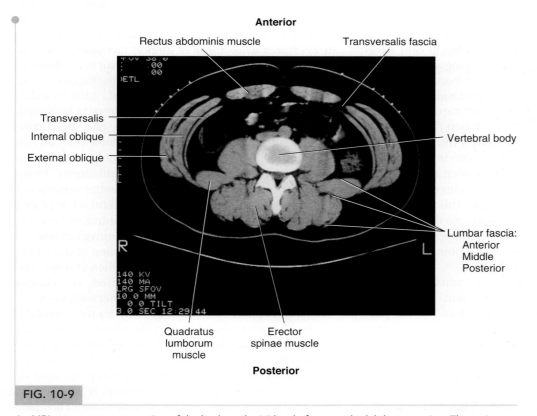

Anterior

Rectus abdominis muscle

Transversalis fascia

Transversalis

Internal oblique

External oblique

Vertebral body

R

L

Lumbar fascia:
Anterior
Middle
Posterior

Quadratus
lumborum
muscle

Erector
spinae muscle

Posterior

FIG. 10-9

An MRI transverse cross-section of the body at the L4 level of a normal adult human spine. The major trunk muscles are shown (R, right; L, left). *Courtesy of Ali Sheihkzadeh, Ph.D., Hospital for Joint Diseases, Mt. Sinai, NYU Health, New York, NY.*

which may provide insight into the correlation between emotional well-being and the onset of low back pain.

Flexion and Extension

During unloaded flexion-extension range of motion, the first 50° to 60° of spine flexion occurs in the lumbar spine, mainly in the lower motion segments (Carlsöö, 1961; Farfan, 1975). Tilting the pelvis forward allows for further flexion. During lifting and lowering a load, this rhythm occurs simultaneously, although a greater separation of these movements is noted during lifting than during lowering (Nelson et al., 1995). Purposeful angulation of the pelvis in lordotic or kyphotic postures will impact on the forces on the lumbar spine, with evidence suggesting a freestyle or slightly flexed pelvic posture to be more advantageous (Arjmand and Shirazi-Adl, 2005). The thoracic spine contributes little to forward flexion of the entire spinal column because of the oblique orientation of the facets (Figs. 10-6 and 10-7), the nearly vertical orientation of the spinous processes, and the limitation of motion imposed by the rib cage.

Flexion is initiated by the abdominal muscles and the vertebral portion of the psoas muscle (Andersson and Lavender, 1997; Basmajian and DeLuca, 1985). The weight of the upper body produces further flexion, which is controlled by the gradually increasing activity of the erector spinae muscles as the forward-bending moment acting on the spine increases. The posterior hip muscles are active in controlling the forward tilting of the pelvis as the spine is flexed (Carlsöö, 1961). When flexion is initiated, the moment arms of the extensor muscles shorten, which adds to the loading on the spine in this position (Jorgensen et al., 2003). It has long been accepted that in full flexion, the erector spinae muscles become inactive once they are fully stretched. In this position, the forward bending moment is counteracted passively by these muscles and by the posterior ligaments, which are initially slack but become taut at this point because the spine has fully elongated (Farfan, 1975).

This silencing of the erector spinae muscles is known as the flexion-relaxation phenomenon (Allen, 1948; Andersson and Lavender, 1997; Floyd and Silver, 1955; Morris et al., 1962). However, Andersson et al. (1996), using wire electrodes inserted in the trunk extensor muscles guided by ultrasound or MRI, showed that in the deep flexed position, the superficial erector spinae muscles relax, sharing the load with the posterior discoligamentous passive structures, while the quadratus lumborum and deep lateral lumbar erector spinae muscles become activated (Fig. 10-10) (Colluca and Hinrichs, 2005). In forced flexion, the superficial extensor muscles become reactivated.

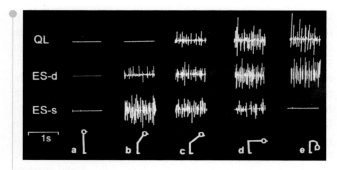

FIG. 10-10

Electromyography of the quadratus lumborum (QL), erector spinae superficial (ES-s), and deep (ES-d) muscles. Wire electrodes were inserted in QL and ES-d; surface electrodes were used for ES-s. Five positions (a–e) of trunk flexion are depicted. In full nonforced trunk flexion (e), the ES-s activity is silent; however, the ES-d and QL are very active to counterbalance the trunk flexion movement. *Courtesy of Eva Andersson, MD, Ph.D, Karolinska Institute, Stockholm, Sweden.*

While prolonged spinal flexion has long been thought to result in back muscle fatigue, it has been shown that this position causes the creep phenomenon in the trunk extensors. With prolonged flexion, muscle activity decreases while range of motion increases by several degrees. This may impair the sensorimotor control mechanism and decrease the ability of the back extensors to protect the spine (Sanchez-Zuriaga et al., 2010). Another study that looked at the effect of fatiguing repetitive trunk extension on the back extensor activity and spine loading concluded that under these conditions there was an insignificant increase in muscular loading of the spine based on recruitment (Sparto and Parnianpour, 1998). The authors further concluded that to better understand the response to fatigue, more emphasis should be placed on investigating the neuromuscular control mechanism and viscoelastic tissue responses to repetitive loading.

From full flexion to upright positioning of the trunk, the pelvis tilts backward and the spine then extends. The sequence of muscular activity is reversed. The gluteus maximus comes into action early together with the hamstrings and initiates extension by posterior rotation of the pelvis. The paraspinal muscles then become activated and increase their activity until the movement is completed (Andersson and Lavender, 1997). Some studies have shown that the concentric exertion performed by the muscles involved in raising the trunk is greater than the eccentric exertion performed by the muscles involved in lowering the trunk (deLooze et al., 1993; Friedebold, 1958; Joseph, 1960). However, this finding has been contradicted in several studies (Marras and Mirka, 1992; Reid and Costigan, 1987). Creswell and Thortensson (1994) support the finding that less electromyographic (EMG) activity is noted during eccentric activity, as in lowering, despite high levels of force generated. The compressive load of the spine caused by the muscle exertion produced by lowering the trunk with a load or resistance can approach the spinal tolerance limits, putting the back at greater risk for injury (Davis et al., 1998).

When the trunk is hyperextended from the upright position, the extensor muscles are active during the initial phase. This initial burst of activity decreases during further extension past erect standing, and the abdominal muscles become active to control and modify the motion (Rohlmann et al., 2006). In forced extension or extension of 15° or more, extensor activity is again required (Floyd and Silver, 1955; Wilke et al., 2003).

Lateral Flexion and Rotation

During lateral flexion of the trunk, motion may predominate in either the thoracic or the lumbar spine. In the thoracic spine, the facet orientation allows for lateral flexion, but the rib cage restricts it (to varying degrees in different people); in the lumbar spine, the wedge-shaped spaces between the intervertebral joint surfaces show variations during this motion (Reichmann, 1971). The spinotransversal and transversospinal systems of the erector spinae muscles and the abdominal muscles are active during lateral flexion; ipsilateral contractions of these muscles initiate the motion and contralateral contractions modify it (Fig. 10-11) (Andersson and Lavender, 1997).

Significant axial rotation occurs at the thoracic and lumbosacral levels but is limited at other levels of the lumbar spine, being restricted by the vertical orientation of the facets (Fig. 10-6C). In the thoracic region, rotation is consistently associated with lateral flexion. During this coupled motion, which is most marked in the upper thoracic region, the vertebral bodies generally rotate toward the concavity of the lateral curve of the spine (White, 1969). Coupling of rotation and lateral flexion also takes place in the lumbar spine, with the vertebral bodies rotating toward the convexity of the curve (Miles and Sullivan, 1961). During axial rotation, the back and abdominal muscles are active on both sides of the spine, as both ipsilateral and contralateral muscles cooperate to produce this movement. High coactivation has been measured for axial rotation (Lavender et al., 1992; Pope et al., 1986).

Pelvic Motion

Functional trunk movements not only involve a combined motion of different parts of the spine but also require the cooperation of the pelvis because pelvic motion is essential for increasing the range of func-

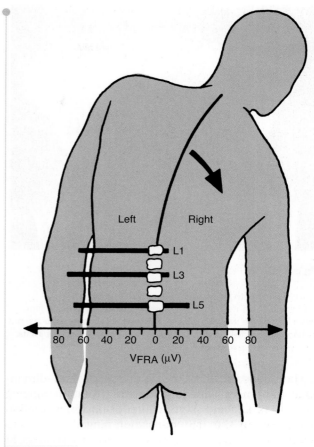

FIG. 10-11

Example of electromyographic activity of the erector spinae muscles collected with surface electrodes during side-bending of the trunk. The figure illustrates trunk bending to the right and muscle activity at the L1, L3, and L5 levels of the lumbar spine. Substantial contralateral muscle activity (left) of the erector spinae muscles is recorded when bending to the right to maintain equilibrium. *Reproduced with permission from Andersson, G.B.J., Ortengren, R., Nachemson, A. (1977). Intradiscal pressure, intra-abdominal pressure and myoelectric back muscle activity related to posture and loading. Clin Orthop, 129, 156.*

tional motion of the trunk. The relationship between pelvic movements and spinal motion is generally analyzed in terms of motion of the lumbosacral joints, the hip joints, or both (Fig. 10-12). Load transfer from the spine to the pelvis occurs through the sacroiliac (SI) joint. Biomechanical analysis of the sacroiliac joints suggests that these joints function mainly as shock absorbers and are important in protecting the intervertebral joints (Wilder et al., 1980). There is evidence of an association between lumbopelvic motion and trunk muscle activity during locomotion at different speeds and modes (running versus walking) (Saunders et al., 2005).

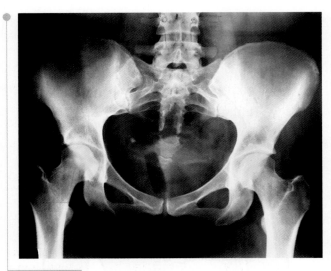

FIG. 10-12

The pelvic ring with its linkage to the spine and the lower extremities. The anteroposterior view of these structures on film gives a hint of the irregular shape of the sacroiliac joint surfaces, but an oblique projection is required for an accurate view of the joints.

When loaded in vitro, the SI joint exhibits three-dimensional movement with joint opening rotation ranging from 0.5° to 1.2° and sacrum anterior-posterior rotation ranging from 0.3° to 0.6°; translation ranged from 0.5 to 0.9 mm (Wang and Dumas, 1998). In vivo analysis of the SI joint using roentgen stereophotogrammetry shows joint rotation mean at 2.5° and translation mean at 0.7 mm with no differences between symptomatic and asymptomatic joints (Bussey et al., 2004; Sturesson et al., 1989).

Muscle forces acting on the SI joint have a stabilizing effect, helping to attenuate the high stress of pelvic loading (Dalstra and Huiskes, 1995; Pel et al., 2008).

Kinetics

Loads on the spine are produced primarily by body weight, muscle activity, pre-stress exerted by the ligaments, and externally applied loads. Simplified calculations of the loads at various levels of the spine can be made with the use of the free-body technique for coplanar forces. Direct information regarding loads on the spine at the level of individual intervertebral discs can be obtained by measuring the pressure within the discs both in vitro and in vivo. Because this method is too complex for general application, a semidirect measuring method is often used. This involves measuring the myoelectric activity of the trunk muscles and correlating this activity with calculated values for muscle contraction forces. The values obtained correlate well with those obtained through intradiscal pressure measurement and can therefore be used to predict the loads on the spine (Andersson and Lavender, 1997; Ortengren et al., 1981; Schultz et al., 1982).

Another method is the use of a mathematical model for force estimation that allows the loads on the lumbar spine and the contraction forces in the trunk muscles to be calculated for various physical activities. The models are useful as predictors of load, for load-sharing analysis under different conditions, to simulate loads, and in spine prosthetic and instrumentation design. The precision of the model depends on the assumption used for the calculations. Two categories of models currently used are the EMG-driven model based on electromyographic trunk muscle recordings and the more traditional biomechanical model based on trunk moments and forces (Chaffin and Andersson, 1991; Lavender et al., 1992; Marras and Granata, 1995; Sheikhzadeh et al, 2008).

STATICS AND DYNAMICS

In the following section, static loads on the lumbar spine are examined for common postures such as standing and sitting and also for lifting, a common activity involving external loads. In the final section, the dynamic loads on the lumbar spine during walking and common strengthening exercises for the back and abdominal muscles are discussed.

Statics

The spine can be considered as a modified elastic rod because of the flexibility of the spinal column, the shock-absorbing behavior of the discs and vertebrae, the stabilizing function of the longitudinal ligaments, and the elasticity of the ligaments flavum. The two curvatures of the spine in the sagittal plane—kyphosis and lordosis—also contribute to the spring-like capacity of the spine and allow the vertebral column to withstand higher loads than if it were straight. A study of the capacity of cadaver thoracolumbar spines devoid of muscles to resist vertical loads showed that the critical load (the point at which buckling occurred) was approximately 20 to 40 N (Gregerson and Lucas, 1967; Lucas and Bresier, 1961). The critical load is much higher in vivo and varies greatly among individuals. The extrinsic support provided by the trunk muscles helps stabilize and modify the loads on the spine in both dynamic and static situations.

LOADING OF THE SPINE DURING STANDING

When a person stands, the postural muscles are constantly active. This activity is minimized when the body segments are well aligned. During standing, the line of gravity of the trunk usually passes ventral to the center

(Cholewicki et al., 1997; Urquhart et al., 2005). However, this activity is readily reduced by the command to stand relaxed (Hodges and Richardson, 1997).

The vertebral portion of the psoas muscles is also involved in producing postural sway (Basmajian and DeLuca, 1985; Nachemson, 1966). The level of activity in these muscles varies considerably among individuals and depends to some extent on the shape of the spine, for example, on the magnitude of habitual kyphosis and lordosis.

The pelvis also plays a role in the muscle activity and resulting loads on the spine during standing (Fig. 10-14). The base of the sacrum is inclined forward and downward. The angle of inclination, or sacral angle, is approximately 30° to the transverse plane during relaxed standing (Fig. 10-14B). Tilting of the pelvis about the transverse axis between the hip joints changes the angle. When the pelvis is tilted backward, the sacral angle decreases and the lumbar lordosis flattens (Fig. 10-14A). This flattening affects the thoracic spine, which extends slightly to adjust the center of gravity of the trunk so that the energy expenditure, in terms of muscle exertion, is minimized. When the pelvis is tilted forward, the sacral angle increases, accentuating the lumbar lordosis and the thoracic kyphosis (Fig. 10-14C). Forward and

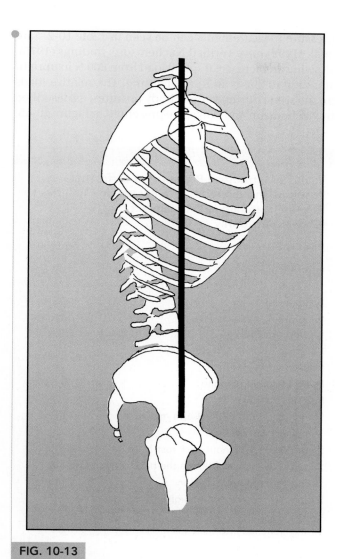

FIG. 10-13

The line of gravity for the trunk (*solid line*) is usually ventral to the transverse axis of motion of the spine, and thus the spine is subjected to a constant forward-bending moment.

of the fourth lumbar vertebral body (Asmussen and Klausen, 1962). Thus, it falls ventral to the transverse axis of motion of the spine and the motion segments are subjected to a forward-bending moment, which must be counterbalanced by ligament forces and erector spinae muscle forces (Fig. 10-13). Any displacement of the line of gravity alters the magnitude and direction of the moment on the spine. For the body to return to equilibrium, the moment must be counteracted by increased muscle activity, which causes intermittent postural sway. In addition to the erector spinae muscles, the abdominal muscles are often intermittently active in maintaining the neutral upright position and stabilizing the trunk

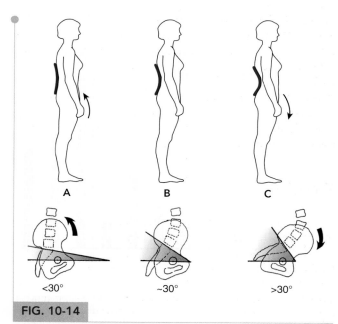

FIG. 10-14

Effect of pelvic tilting on the inclination of the base of the sacrum to the transverse plane (sacral angle) during upright standing. **A.** Tilting the pelvis backward reduces the sacral angle and flattens the lumbar spine. **B.** During relaxed standing, the sacral angle is approximately 30°. **C.** Tilting the pelvis forward increases the sacral angle and accentuates the lumbar lordosis.

backward tilting of the pelvis influences the activity of the postural muscles by affecting the static loads on the spine (Floyd and Silver, 1955; Shirazi-Adl et al., 2002). Ferrara et al. (2005) investigated the effect of unloading the spine in an upright position by using a vest that exerted a distraction force between the pelvis and the ribs. Their findings suggest that intradiscal pressure could be reduced and may have implications for managing low back pain.

COMPARATIVE LOADS ON THE LUMBAR SPINE DURING STANDING, SITTING, AND RECLINING

Body position affects the magnitude of the loads on the spine. As a result of in vivo intradiscal pressure measurement studies conducted by Nachemson (1975), it was shown that these loads are minimal during well-supported reclining, remain low during relaxed upright standing, and rise during sitting. In vivo investigation of intervertebral disc pressure using more sophisticated technology, and based on only one subject, suggested that in relaxed, unsupported sitting, interdiscal pressure is less than in standing (see Figures 10-15 and 10-16; sitting unsupported 0.46 MPa, relaxed standing 0.50 MPa) (Wilke et al., 1999, 2001). Additional pertinent pressure measurements can be seen in Table 10-2. Sato et al. (1999) have verified Nachemson's findings (1975), showing an increase in spinal load from 800 N in upright standing to 996 N in upright sitting. The relative loads on the spine during various body postures, as described by Nachemson and Wilke, are presented in Figure 10-15.

TABLE 10-2

Values of Intradiscal Pressure for Different Positions and Exercises as a Percentage Relative to Relaxed Standing in One Subject (Chosen Arbitrarily as 100%)

Position/Maneuver	Percentage
Lying supine	20
Side-lying	24
Lying prone	22
Lying prone, extended back, supporting elbows	50
Laughing heartily, lying laterally	30
Sneezing, lying laterally	76
Peaks by turning around	140–160
Relaxed standing	100
Standing, performing Valsalva maneuver	184
Standing, bent forward	220
Sitting relaxed, without back rest	92
Sitting actively straightening the back	110
Sitting with maximum flexion	166
Sitting bent forward with thigh supporting the elbows	86
Sitting slouched into the chair	54
Standing up from the chair	220
Walking barefoot	106–130
Walking with tennis shoes	106–130
Jogging with hard street shoes	70–190
Jogging with tennis shoes	70–170
Climbing stairs, one at a time	100–140
Climbing stairs, two at a time	60–240
Walking down stairs, one at a time	76–120
Walking down stairs, two at a time	60–180
Lifting 20 kg, bent over with round back	460
Lifting 20 kg as taught in back school	340
Holding 20 kg close to the body	220
Holding 20 kg, 60 cm away from the chest	360
Pressure increase during the night rest (over a period of 7 hours)	20–48

Adapted with permission from Wilke, H.J., Neef, P., Caimi, M., et al. (1999). New in vivo measurements of pressures in the intervertebral disc in daily life. *Spine, 24*, 755.

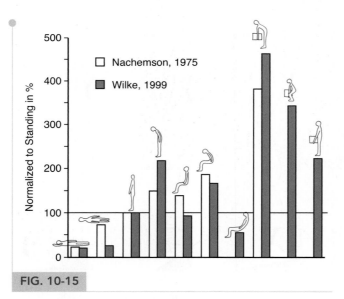

FIG. 10-15

Data from two studies using intradiscal pressure measurements. The relative loads on the third and fourth lumbar discs measured in vivo in various body positions are compared with the load during upright standing, depicted as 100%. *Data from Nachemson, A. (1975). Towards a better understanding of back pain: A review of the mechanics of the lumbar disc. Rheumatol Rehabil, 14, 129; and from Wilke, H.J., Neef, P., Caimi, M., et al. (1999). New in vivo measurements of pressures in the intervertebral disc in daily life. Spine, 24, 755.*

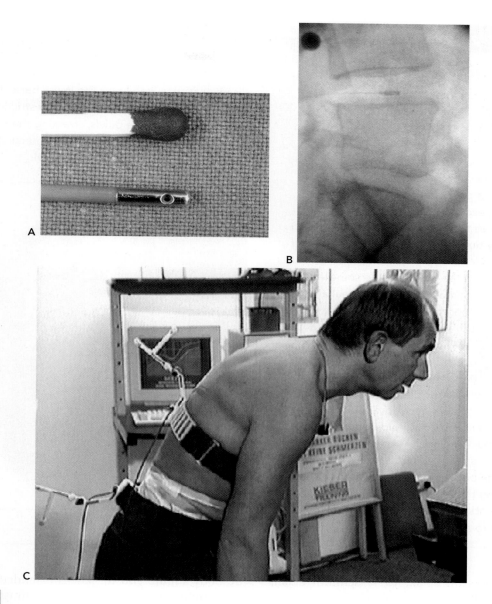

FIG. 10-16

Intradiscal pressure measurements in vivo. **A.** Pressure transducer for the intradiscal pressure measurements in vivo. **B.** Pressure transducer after implantation in L4–L5 for the intradiscal pressure measurements in vivo. **C.** In vivo recording of the intradiscal pressure and motion between the sacrum and thoracolumbar junction was performed with one volunteer. *Used with permission from Wilke, H., Neef, P., Hinz, B., et al. (2001). Intradiscal pressure together with anthropometric data—A data set for the validation of models. Clin Biomech, 16(Suppl 1), S111–S126.*

During relaxed upright standing, the load on the third and fourth lumbar disc is almost twice the weight of the body above the measured level (Nachemson and Elfström, 1970; Nachemson and Morris, 1964; Wilke et al., 1999, 2003). While transitioning from a supine to relaxed standing position, the compressive, tensile, and shear forces vary throughout each disc and at each lumbar level (Wang et al., 2009). Trunk flexion increases the load and the forward-bending moment on the spine. During forward flexion, the annulus bulges ventrally (Klein et al., 1983) and the central portion of the disc moves posteriorly (Krag et al., 1987). More than trunk extension, trunk flexion stresses the posterolateral area of the annulus fibrosus. The addition of twisting motion and accompanying torsional loads further increases the stresses on the disc (Case

Case Study 10-2

Nonspecific Low Back Pain

A 35-year-old male presents with complaints of low back pain with radiation to the posterior aspect of the left thigh, not past the knee. His pain started 3 weeks ago, after working a 12-hour shift, when he twisted his torso while lifting an unusually large yet lightweight box. During the first week of pain, he visited his physician, who prescribed pain medication and recommended that he return to his usual activity as tolerated. Currently, he is still in pain, particularly during sitting or standing for long periods. During a follow-up physician visit, a careful examination showed the patient to be somewhat overweight, with weakness in his abdominal and back muscles and poor flexibility in his hamstrings, psoas, and back muscles. Neurologic tests were normal as was his past and current medical profile, leading to a diagnosis of nonspecific low back pain (Case Study Fig. 10-2).

Combinations of different factors have resulted in this injury. From the biomechanical point of view, although the load lifted was considered light, the vastness of the package and the resultant large lever arm (the distance from the center of gravity of the person to the package) created a larger-than-expected load on the lumbar spine. In addition, weakness in the abdominal and extensor muscles of the spine led to an additional mechanical disadvantage in stabilizing the lower back. Tight psoas and hamstring muscles place restrictions on the mobility of the pelvis, stressing the range of motion in the lumbar region and affecting the normal loads and motions at this level.

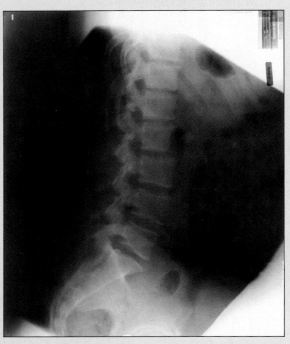

Case Study Figure 10-2

Study 10-2) (Andersson et al., 1977; Schmidt et al., 2007; Shirazi-Adl, 1994; Steffen et al., 1998).

The loads on the lumbar spine are lower during supported sitting than during unsupported sitting. During supported sitting, the weight of the upper body is supported, which reduces the muscle activity, relieving intradiscal pressure (Andersson et al., 1974; Wilke et al., 1999). Backward inclination of the backrest and the use of a lumbar support further reduce the loads. The use of a support in the thoracic region, however, pushes the thoracic spine and the trunk forward and makes the lumbar spine move toward kyphosis to remain in contact with the backrest, thereby increasing the loads on the lumbar spine (Fig. 10-17) (Andersson et al., 1974). Loads on the spine are minimized when an individual assumes a supine position because the loads produced by the body's weight are eliminated (Fig. 10-15). With the body supine and the knees extended, the pull of the vertebral portion of the psoas muscle produces some loads on the lumbar spine. With the hips and knees bent and supported, however, the lumbar lordosis straightens out as the psoas muscle relaxes and the loads decrease (Fig. 10-18).

STATIC LOADS ON THE LUMBAR SPINE DURING LIFTING

The highest loads on the spine are generally produced by external loads, such as lifting a heavy object. Just how much load can be sustained by the spine before damage occurs continues to be investigated. Pioneering studies by Eie (1966) of lumbar vertebral specimens from adult humans showed that the compressive load to vertebrae failure ranged from approximately 5,000 to 8,000 N. On

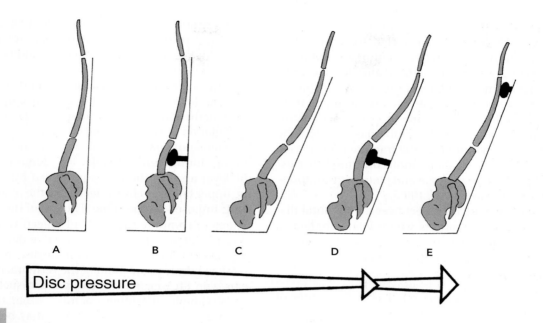

FIG. 10-17

Influence of backrest inclination and back support on loads of the lumbar spine, in terms of pressure in the third lumbar disc, during supported sitting. **A.** Backrest inclination is 90° and disc pressure is at a maximum. **B.** Addition of a lumbar support decreases the disc pressure. **C.** Backward inclination of the backrest is 110°, but with no lumbar support it produces less disc pressure. **D.** Addition of a lumbar support with this degree of backrest inclination further decreases the pressure. **E.** Shifting the support to the thoracic region pushes the upper body forward, moving the lumbar spine toward kyphosis and increasing the disc pressure. *Adapted with permission from Andersson, G.B.J., Ortengren, R., Nachemson, A., et al. (1974). Lumbar disc pressure and myoelectric back muscle activity during sitting. 1. Studies on an experimental chair. Scand J Rehabil Med, 6, 104.*

the whole, values reported subsequently by other authors correspond to those of Eie, although values above 10,000 N and below 5,000 N have been documented (Hutton and Adams, 1982). The application of static bending-shearing moment on lumbar motion segments revealed that bending moment of 620 Nm and shear moment of 156 Nm were tolerated before complete disruption of motion segment occurred. The flexion angle before failure was recorded as 20° with 9 mm of horizontal displacement between the two vertebrae (Osvalder et al., 1990). Both age and degree of disc degeneration influence the range preceding failure. Although the vertebral body strength is relative to the bone mass, with aging the decline in bone strength is more pronounced than is the decline in bone mass (Moskilde, 1993).

Eie (1966) and Ranu (1990) observed that during compressive testing the fracture point was reached in the vertebral body, or end plate, before the intervertebral disc sustained damage. This finding shows that the bone is less capable of resisting compression than is an intact disc. During the testing, a yield point was reached before the vertebra or end plate fractured. When the load was removed at this point, the vertebral body recovered but was more susceptible to damage when reloaded.

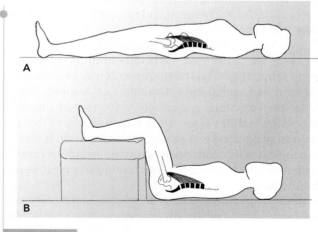

FIG. 10-18

A. When a person assumes a supine position with legs straight, the pull of the vertebral portion of the psoas muscle produces some loads on the lumbar spine. **B.** When the hips and knees are bent and supported, the psoas muscle relaxes and the loads on the lumbar spine decrease.

Evidence exists that the spine may incur microdamage as a result of high loads in vivo (Hansson et al., 1980). In an in vitro study based on dual photon absorptiometry, Hansson et al. observed microfractures in specimens from "normal" human lumbar vertebrae and interpreted this microdamage to be fatigue fractures resulting from stresses and strains on the spine in vivo. Once microdamage, undetectable on radiography, has occurred, the amount of energy required for full fracture is decreased (Lu et al., 2004). In vitro examination confirmed the existence of microdamage near the end plate with compression loading (Hasegawa et al., 1993).

Lifting and carrying an object over a horizontal distance are common situations where loads applied to the vertebral column may be so high as to damage the spine. Several factors influence the loads on the spine during these activities:

1. The position of the object relative to the center of motion in the spine
2. The size, shape, weight, and density of the object
3. The degree of flexion or rotation of the spine
4. The rate of loading

Holding the object close to the body instead of away from it reduces the bending moment on the lumbar spine because the distance from the center of gravity of the object to the center of motion in the spine (the lever arm) is minimized. The shorter the lever arm is for the force produced by the weight of a given object, the lower the magnitude of the bending moment and thus the lower the loads on the lumbar spine (Calculation Box 10-1) (Andersson et al., 1976; Nachemson and Elfström, 1970; Nemeth, 1984; Wilke et al., 1999, 2003).

Even when identical and nonfatiguing repeated lifting tasks are performed, variability in lifting technique of the same subject has been shown in trunk kinematics, kinetics, and spinal load (Granata and Sanford, 1999). When an individual repeatedly performs an identical lift, great variability is recorded, which indicates that the brain may have several motor strategies to do a task. It also indicates the sensitive responsiveness of the muscle system to subtle changes to maintain the performance despite fatigue.

When a person holding an object bends forward, the force produced by the weight of the object plus that produced by the weight of the upper body create a bending moment on the disc, increasing the loads on the spine. This bending moment is greater than that produced when the person stands erect while holding the object (Calculation Box 10-2).

A literature review revealed no significant difference in spinal compression and shear computed forces between stoop or squat lifting (van Dieen et al., 1999). However, it was suggested that loss of balance is more likely during squat lifting, which in turn may add additional stresses on the lumbar spine.

In the following example, the free-body technique for coplanar forces will be used to make a simplified calculation of the static loads on the spine as an object is lifted (Calculation Boxes 10-3A and 10-3B).

Calculations made in this way for one point in time during lifting are valuable for demonstrating how the lever arms of the forces produced by the weight of the upper body and by the weight of the object affect the loads imposed on the spine. The use of the same calculations to compute the loads produced when an 80-kg object is lifted (representing a force of 800 N) yields an approximate load of 10,000 N on the disc, which is likely to exceed the fracture point of the vertebra. Because athletes who lift weights can easily reach such calculated loads without sustaining fractures, other factors, such as intra-abdominal pressure (IAP), may be involved in reducing the loads on the spine in vivo (Davis and Marra, 2000; Hodges et al., 2005; Krajcarski et al., 1999).

Dynamics

Almost all motion in the body increases muscle recruitment and the loads on the spine. This increase is modest during such activities as slow walking or easy twisting but becomes more marked during various exercises and the complexity of dynamic movement and dynamic loading (Nachemson and Elfström, 1970).

WALKING

In a study of normal walking at four speeds, the compressive loads at the L3–L4 motion segment ranged from 0.2 to 2.5 times body weight (Fig. 10-19) (Cappozzo, 1984) and may be explained by activity in the trunk muscles secondary to changes in lumbopelvic motion (Saunders et al., 2005). The loads were maximal around toe-off and increased approximately linearly with walking speed. Muscle action was mainly concentrated in the trunk extensors. Individual walking traits, particularly the amount of forward flexion of the trunk, influenced the loads.

The greater this flexion, the larger the muscle forces and hence the compressive load. Callaghan et al. (1999) corroborated these findings and further showed that walking cadence affects lumbar loading, with increased anterior-posterior shear forces noted as speed increased. Limiting arm swing during walking resulted in increased compressive joint loading and EMG output with decreased lumbar spine motions. In conclusion, because of low tissue loading, walking is a safe and perhaps ideal

CALCULATION BOX 10-1

Influence of the Size of the Object on the Loads on the Lumbar Spine

The size of the object held influences the loads on the lumbar spine. If objects of the same weight, shape, and density but of different sizes are held, the lever arm for the force produced by the weight of the object is longer for the larger object, and thus the bending moment on the lumbar spine is greater (Calculation Box Fig. 10-1-1). In these two situations (Calculation Box Figs. 10-1-1 and 10-1-2), the distance from the center of motion in the disc to the front of the abdomen is 20 cm. In both cases, the object has a uniform density and weighs 20 kg. In the case of Calculation Box Figure 10-1-1, the width of the cubic object is 20 cm; in the case of Calculation Box Figure 10-1-2, the width is 40 cm. Thus, in Case 1, the forward-bending moment acting on the lowest lumbar disc is 60 Nm, as the force of 200 N produced by the weight of the object acts with a lever arm (L_P) of 30 cm (200 N × 0.3 m). In Case 2, the forward-bending moment is 80 Nm, as the lever arm (L_P) is 40 cm (200 N × 0.4 m). [Considering 1 kg ≅ 10 N.]

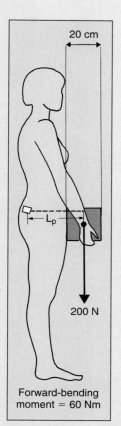
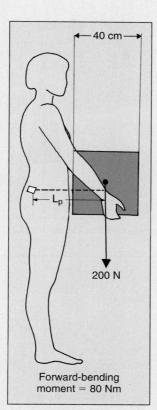

Forward-bending moment = 60 Nm

Calculation Box Figure 10-1-1

Forward-bending moment = 80 Nm

Calculation Box Figure 10-1-2

CALCULATION BOX 10-2

Influence of the Upper Body Position on the Loads at the Lumbar Spine during Lifting

In the two situations shown in Calculation Box Figures 10-2-1 and 10-2-2, an identical object weighing 20 kg is lifted. In Case 1 (upright standing), the lever arm of the force produced by the weight of the object (L_P) is 30 cm, creating a forward-bending moment of 60 Nm (200 N × 0.3 m). The forward-bending moment created by the upper body is 9 Nm; the length of the lever arm (L_W) is estimated to be 2 cm, and the force produced by the weight of the upper body is 450 N. Thus, the total forward-bending moment in Case 1 is equal to 69 Nm (60 Nm + 9 Nm).

In Case 2 (upper body flexed forward), the lever arm of the force produced by the weight of the object (L_P) is increased to 40 cm, creating a forward-bending moment of 80 Nm (200 N × 0.4 m). Furthermore, the force of 450 N produced by the weight of the upper body increases in importance as it acts with a lever arm (L_W) of 25 cm, creating a forward-bending moment of 112.5 Nm (450 N × 0.25 m). Thus, the total forward-bending moment in Case 2 is 192.5 Nm (112.5 Nm + 80 Nm).

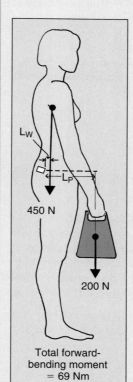
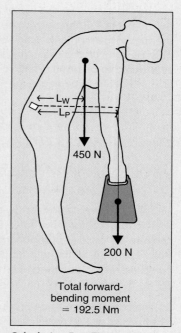

Total forward-bending moment = 69 Nm

Calculation Box Figure 10-2-1

Total forward-bending moment = 192.5 Nm

Calculation Box Figure 10-2-2

CALCULATION BOX 10-3A

Free-Body Diagram Technique for Coplanar Forces. Calculation of the Static Loads on the Spine as an Object Is Lifted

The loads on a lumbar disc will be calculated for one point in time when a person who weighs 70 kg lifts a 20-kg object. The spine is flexed approximately 35°. In this example, the three principal forces acting on the lumbar spine at the lumbosacral level are (1) the force produced by the weight of the upper body (W), calculated to be 450 N (approximately 65% of the force exerted by the total body weight); (2) the force produced by the weight of the object (P), 200 N; and (3) the force produced by contraction of the erector spinae muscles (E), which has a known direction and point of application but an unknown magnitude (Calculation Box Fig. 10-3A-1).

Because these three forces act at a distance from the center of motion in the spine, they create moments in the lumbar spine. Two forward-bending moments (WL_W and PL_P) are the products of (W) and (P) and the perpendiculars from the instant center of rotation to the lines of action of these forces (their lever arms). The lever arm (L_P) for P is 0.4 m and the lever arm (L_W) for W is 0.25 m. A counterbalancing moment (EL_E) is the product of E and its lever arm. The lever arm (L_E) is 0.05 m. The magnitude of E can be found through the use of the equilibrium equation for moments. For the body to be in moment equilibrium, the sum of the moments acting on the lumbar spine must be zero. (In this example, clockwise moments are considered to be positive and counterclockwise moments are considered to be negative.)

Thus,

$$\sum M = 0$$

$$(W \times L_W) + (P \times L_P) - (E \times L_E) = 0$$

$$(450 \text{ N} \times 0.25 \text{ m}) + (200 \text{ N} \times 0.4 \text{ m}) - (E \times 0.05 \text{ m}) = 0$$

$$E \times 0.05 \text{ m} = 112.5 \text{ Nm} + 80 \text{ Nm}$$

Solving this equation for E yields 3,850 N.

The total compressive force exerted on the disc (C) can now be calculated trigonometrically (Calculation Box Fig. 10-3A-2). In the example, C is the sum of the compressive forces acting over the disc, which is inclined 35° to the transverse plane. These forces are as follows:

1. The compressive force produced by the weight of the upper body (W), which acts on the disc inclined 35° ($W \times \cos 35°$)

2. The force produced by the weight of the object (P), which acts on the disc inclined 35° ($P \times \cos 35°$)

3. The force produced by the erector spinae muscles (E), which acts approximately at a right angle to the disc inclination

The total compressive force acting on the disc (C) has a known sense, point of application, and line of action

but an unknown magnitude. The magnitude of C can be found through the use of the equilibrium equation for forces. For the body to be in force equilibrium, the sum of the forces must be equal to zero.

Thus,

$$\sum \text{forces} = 0$$

$$(W \times \cos 35°) + (P \times \cos 35°) + E - C = 0$$

$$(450 \text{ N} \times \cos 35°) + (200 \text{ N} \times \cos 35°) + 3,850 \text{ N} - C = 0$$

$$C = 368.5 \text{ N} + 163.8 \text{ N} + 3,850 \text{ N}$$

Solving the equation for C yields 4,382 N.

The shear component for the reaction force on the disc (S) is found in the same way:

$$(450 \text{ N} \times \sin 35°) + (200 \text{ N} \times \sin 35°) - S = 0$$

$$S = 373 \text{ N}$$

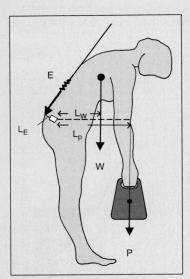

Calculation Box Figure 10-3A-1

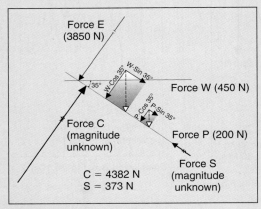

Calculation Box Figure 10-3A-2

CALCULATION BOX 10-3B

Free-Body Diagram Technique for Coplanar Forces. Calculation of the Static Loads on the Spine as an Object Is Lifted

Because C and S form a right angle (Calculation Box Fig. 10-3B-1), the Pythagorean theorem can be used to find the total reaction force on the disc (R):

$$(R) = \sqrt{C^2 + S^2}$$
$$R = 4398 \text{ N}$$

The direction of R is determined by means of a trigonometric function:

$$\sin(\alpha) = C/R$$

$\alpha = \text{archsin} (C/R) = 85°$ where α is the angle between the total vector force on the disc and the disc inclination.

The problem can be graphically solved by constructing a vector diagram based on the known values (Calculation Box Fig. 10-3B-2). A vertical line representing $W + P$ is drawn first; E is added at a right angle to the disc inclination, and R closes the triangle. The direction of R in relation to the disc is determined.

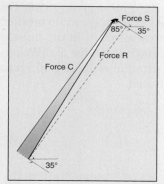

Calculation Box Figure 10-3B-1

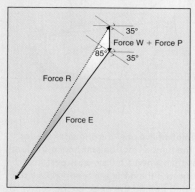

Calculation Box Figure 10-3B-2

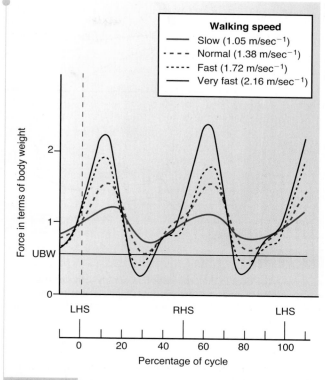

FIG. 10-19

Axial load on the L3–L4 motion segment in terms of body weight for one subject during walking at four speeds. The *horizontal line* (*UBW*) denotes the weight of the upper body, which represents the gravitational component of this load. Loads were predicted using experimental data from photogrammetric measurements along with a biomechanical model of the trunk. *LHS*, left heel strike; *RHS*, right heel strike. *Adapted with permission from Cappozzo, A. (1984). Compressive loads in the lumbar vertebral column during normal level walking. J Orthop Res, 1, 292.*

therapeutic exercise for those with low back pain (Callaghan et al., 1999) while attention to speed of walking can further moderate spinal loads (Cheng et al., 1998).

EXERCISES

During strengthening exercises for the erector spinae and abdominal muscles, the loads on the spine can be high. Although such exercises must be effective for strengthening the trunk muscles concerned, they should be performed in such a way that the loads on the spine are adjusted to suit the condition of the individual and the goals of the strength training program (McGill et al., 2009).

The erector spinae muscles are intensely activated by arching the back in the prone position (Fig. 10-20A) (Pauly, 1966). Loading the spine in extreme positions

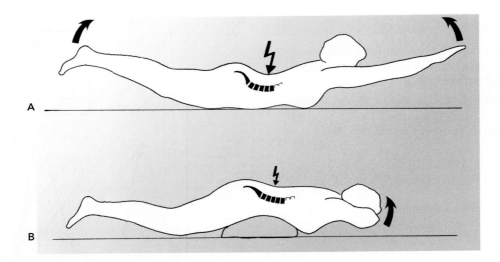

FIG. 10-20

A. Arching the back in the prone position greatly activates the erector spinae muscles but also produces high stresses on the lumbar discs, which are loaded in an extreme position. **B.** Decreasing the arch of the back by placing a pillow under the abdomen allows the discs to better resist stresses because the vertebrae are aligned with each other. Isometric exercise in this position is preferable.

such as this one produces high stresses on spine structures, in particular the spinous process (Adams et al., 1988). Although intradiscal pressure in a prone position with upper body support on the elbows is half that in standing (Wilke et al., 1999), it is recommended for exercise that an initial position that keeps the vertebrae in a more parallel alignment is preferable when strengthening exercises for the erector spinae muscles are performed (Fig. 10-20B). An alternative to strengthening the erector spinae muscles while limiting spine loading is contralateral arm and leg extension, with the spine held isometrically in a neutral position (Callaghan et al., 1998).

The importance of the abdominal muscles in spinal stability and interplay in the production of IAP reinforces the need for strong abdominal flexors (Hodges et al., 2005). Sit-ups are a useful exercise for abdominal muscle strengthening, with many variations practiced and encouraged by health professionals, but certain variations are viewed as harmful to the low back (Cordo et al., 2006; Escamilla et al., 2006). Variations in trunk stability and posture, arm position, and the manner in which these exercises are performed will result in different muscle activation patterns (Cordo et al., 2006; Monfort-Panego et al., 2009). Both bent knee and straight leg sit-ups will produce comparable levels of psoas and abdominal activity, creating compressive spinal loading. Curl-ups, in which the head and shoulders are raised only to the point where the shoulder blades clear the table and lumbar spine motion is minimized (Fig. 10-21)

and often emphasized in rehabilitation programs, are recommended for minimizing compressive lumbar loading (Axler and McGill, 1997; Juker et al., 1998). This modification of the exercise has been shown to be effective in terms of motor unit recruitment in the muscles (Ekholm et al., 1979; Flint, 1965; Partridge and Walters, 1959); all portions of the external oblique and rectus abdominis muscles are activated.

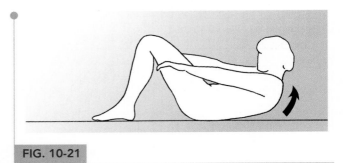

FIG. 10-21

Performing a curl to the point where only the shoulder blades clear the table minimizes the lumbar motion and hence the load on the lumbar spine is less than when a full sit-up is performed. A greater moment is produced if the arms are raised above the head or the hands are clasped behind the neck, as the center of gravity of the upper body then shifts farther away from the center of motion in the spine.

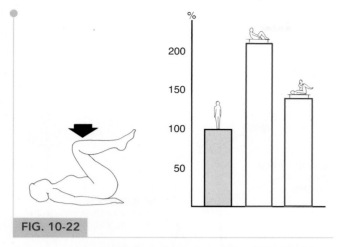

FIG. 10-22

A reverse curl, isometrically performed, provides efficient training of the abdominal muscles and produces moderate stresses on the lumbar discs. The relative loads on the third lumbar disc during a full sit-up and an isometric curl are compared with the load during upright standing, depicted as 100%. *Data from Nachemson, A. (1975). Towards a better understanding of back pain: A review of the mechanics of the lumbar disc. Rheumatol Rehabil, 14, 129.*

To limit the psoas activity, a reverse curl, wherein the knees are brought toward the chest and the buttocks are raised from the table, activates the internal and external oblique muscles and the rectus abdominis muscle (Partridge and Walters, 1959). If the reverse curl is performed isometrically, the disc pressure is lower than that produced during a sit-up, but the exercise is just as effective for strengthening the abdominal muscles (Fig. 10-22). It can be concluded that no single abdominal exercise can optimally train all trunk flexors while minimizing intervertebral joint loading. Instead, a varied program must be prescribed, tailored to the training objectives of the individual (Axler and McGill, 1997; Monfort-Lanego et al., 2009).

When designing a back strengthening exercise program, the most important consideration is the conclusion drawn by the Paris Task Force that has continued to be supported by further investigation (Abenhaim et al., 2000; Chou et al., 2009). The guidelines set forth include recommendations that exercise is beneficial for subacute and chronic low back pain. No particular group or type of exercises has been shown to be most effective.

MECHANICAL STABILITY OF THE LUMBAR SPINE

Mechanical stability for the lumbar spine can be achieved through several means: IAP, co-contraction of the trunk muscles, external support, and surgery. Surgical procedures for lumbar spine stability will not be covered in this section.

Intra-abdominal Pressure

IAP is one mechanism that may contribute to both unloading and stabilization of the lumbar spine. IAP is the pressure created within the abdominal cavity by a coordinated contraction of the diaphragm and the abdominal and pelvic floor muscles. Its unloading mechanism was first proposed by Bartelink in 1957 and Morris et al. in 1961. They suggested that IAP serves as a "pressurized balloon" attempting to separate the diaphragm and pelvic floor (Figs. 10-23A and B). This creates an extensor moment that decreases the compression forces on the lumbar discs. The extensor moment produced by IAP has been calculated in several biomechanical models, with widely varying resulting reductions in extensor moment from 10% to 40% of the extensor load (Anderson et al., 1985; Chaffin, 1969; Eie, 1966; Morris et al., 1961).

Studies using fine-wire EMG of the deeper abdominal muscles found that the transversus abdominis is the primary abdominal muscle responsible for IAP generation (Cresswell, 1993; Cresswell, Blake, Thorstensson, 1994; Cresswell, Grundstrom, Thorstensson, 1992; Hodges et al., 1999). Because the transversus abdominis is horizontally oriented, it creates compression and an increase in IAP without an accompanying flexor moment. An experimental study by Hodges and Gandevia, (2000) gave the first direct evidence for a trunk extensor moment produced by elevated IAP.

It has been demonstrated that the IAP contributes to the mechanical stability of the spine through a coactivation between the antagonistic trunk flexor and extensor muscles, in conjunction with the diaphragm and pelvic floor muscles, leading to increased spinal stiffness (Cholewicki, Juluru, McGill, 1999; Cholewicki, Juluru, Radebold, 1999; Cholewicki, Panjabi, Khachatryan, 1997; Gardner-Morse and Stokes, 1998; Hodges et al., 2005, 2007). As the abdominal musculature contracts, IAP increases and converts the abdomen into a rigid cylinder that greatly increases stability as compared with the multisegmented spinal column (McGill and Norman, 1987; Morris et al., 1961). IAP increases during both static and dynamic conditions such as lifting and lowering, running and jumping, and unexpected trunk perturbations (Cresswell et al., 1992; Cresswell, Oddsson, Thorstensson, 1994,; Cresswell and Thorstensson, 1994; Harman et al., 1988). Current research suggests that the transversus abdominis muscle, together with the diaphragm and pelvic floor muscles, play an important role in stabilizing the spine in preparation for limb movement, regardless of the direction in which movement is anticipated. Transversus abdominis, pelvic floor and diaphragmatic activity appear to occur independently, prior to activity of the primary limb mover or the other abdominal muscles (Hodges et al., 1999, 2007; Hodges and Richardson, 1997).

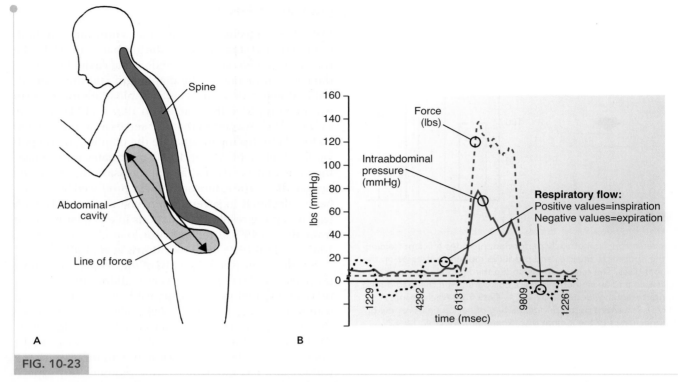

A

B

FIG. 10-23

A. Schematic illustration of the effect of intra-abdominal pressure. An increased pressure will create an extension moment on the lumbar spine. **B.** Intra-abdominal pressure (IAP) (measured by a nasogastric microtip transducer) and respiratory flow (measured through a Pneumotach) during stoop lift of 120 lb (approximately 60 kg). *Solid line,* IAP; *dotted line,* force exerted in pounds; *dashed line,* respiratory flow (negative values delineate expiration and positive values delineate inspiration). Note that the subject inspires before the lift and holds the breath throughout the lift. IAP increases and peaks together with the lifting force, helping to stabilize and unload the lumbar spine. *Courtesy of Markus Pietrek, MD, and Marshall Hagins, PT, MA. Program of Ergonomics and Biomechanics, New York University and Hospital for Joint Diseases, New York, NY.*

Trunk Muscle Co-contraction

To understand the phenomenon of co-contraction during trunk loading, Krajcarski et al. (1999) studied the in vivo muscular response to perturbations at two rates causing a rapid flexion moment. The results of maximum trunk flexion angles and resulting extensor moments were compared. The results showed that with higher levels of loading, muscle co-contraction, spine compression, and trunk stiffness increase. During unexpected loading, a 70% increase in muscle activity has been noted as compared with anticipated loading, which may lead to injury (Marras et al., 1987). Evidence suggests that with expected loading there is increased coactivation of the trunk muscles, which serves to stiffen the spine and better control the flexion moment caused by sudden loading. It is the absence of this preactivation that is believed to result in the increase in muscle activity seen with unexpected loading (Krajcarski et al., 1999). Further investigation into the loading response has revealed that an inverse relationship (i.e., the shorter the warning time, the greater the peak trunk mus-

cle response) exists between peak muscle response and warning time prior to loading (Lavender et al., 1989). While trunk muscle co-contraction of the large trunk muscles is critical to spine stability, increased activity of the deep trunk flexors has been shown to play an important role in lumbopelvic stability. This activity is thought to increase trunk stiffness sufficiently without the need for additional trunk muscle output (McCook et al., 2009).

Loss of spine stability can result from repetitive loading that fatigues the trunk muscles. Muscle endurance is mechanically defined as the point at which fatigue of the muscles is observable, usually through a change in movement pattern. Parnianpour et al. (1988) used an isoinertial triaxial device to study force output and movement patterns when subjects performed a flexion and extension movement of the trunk until exhaustion. The results showed that with fatigue, coupled motion increased in the coronal and transverse planes during the flexion and extension movement. In addition, torque, angular excursion, and angular velocity of the motion decreased. The reduction in the functional

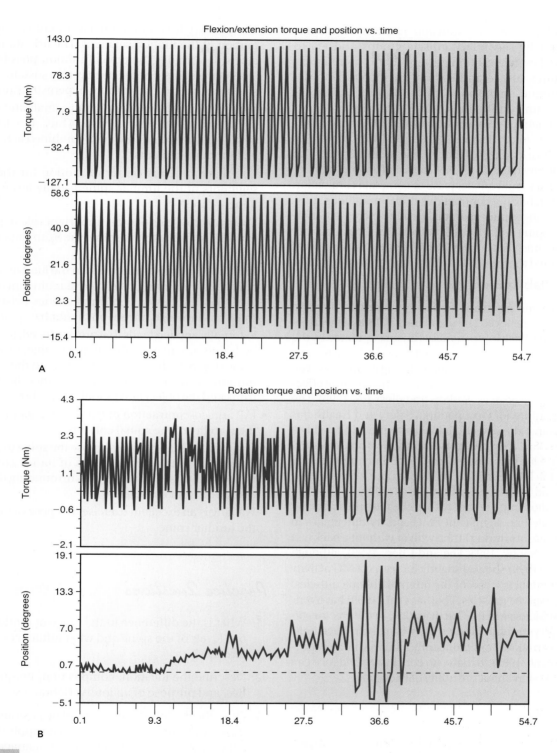

FIG. 10-24

Dynamic (isoinertial) flexion-extension trunk testing until exhaustion for one subject. Torque and position data are depicted for two planes, flexion-extension (**A**) and axial rotation (**B**). Note that flexion-extension torque production is diminishing as is the amount of performed extension of the trunk (**A**). Rotational torque and movement amplitude, increased accessory motion, and torque are shown in **B**. Data for lateral flexion were similar to that for axial rotation and are not shown here. *Adapted with permission from Parnianpour, M., Nordin, M., Kahanovitz, N., et al. (1988). The triaxial coupling of torque generation of trunk muscles during isometric exertions and the effect of fatiguing isoinertial movements on the motor output and movement patterns. Spine, 13:982–92.*

capacity of the flexion-extension muscles was compensated for by secondary muscle groups and led to an increased couple motion pattern that is more injury prone. Figures 10-24A and B show the increase in axial rotation (torque and position) during flexion and extension of the trunk until exhaustion.

In an animal study, Solomonow et al. (1999) induced laxity of the spine in the ligaments, discs, and joint capsule by cyclic repetitive loading of feline in vivo lumbar spines. The cyclic loading resulted in desensitization of the mechanoreceptors with a significant decrease or complete elimination of reflexive stabilizing contractions of the multifidus muscle. These changes may potentially lead to increased instability of the spine and a lack of protective muscular activity even before muscular fatigue is observed. A 10-minute rest period restored the muscular activity to approximately 25%.

External Stabilization

Restriction of motion at any level of the spine may increase motion at another level. The use of back belts as a means of preventing low back injury remains unsubstantiated by the evidence. Originally it was believed to assist in increasing IAP as a way of unloading the spine during lifting; however, inconclusive evidence exists as to the biomechanical effectiveness of these devices (Perkins and Bloswick, 1995). The National Institute for Occupational Safety and Health has advised against the use of back belts to prevent low back injuries (NIOSH, 1994). As well, an orthotic worn to restrict thoracic and lumbar motion may result in compensatory motion at the lumbosacral level (Brown and Norton, 1957; Lumsden and Morris, 1968; Tuong et al., 1998).

Investigation into the effect of back belts on muscle activity has revealed no significant EMG activity differences in the back extensors during lifting with or without a back belt (Ciriello and Snook, 1995; Lee and Chen, 1999), although McGill et al. (1999) showed slightly increased EMG activity in the abdominal (except for the internal oblique muscles) and erector spinae muscles. Thomas et al. (1999) have verified a slight increase in EMG activity (2%) in the erector spinae during symmetric lifting with a back belt. Back belts have not been shown to significantly increase lifting capacity (Reyna et al., 1995). To date, no conclusive evidence can inform a decision about the utility of back belts.

Summary

- The lumbar spine is a highly intricate and complex structure.
- A vertebra-disc-vertebra unit constitutes a motion segment, the functional unit of the spine.
- The intervertebral disc serves a hydrostatic function in the motion segment, storing energy and distributing loads. This function is reduced with disc degeneration.

- The primary function of the facet joint is to guide the motion of the motion segment. The orientation of the facets determines the type of motion possible at any level of the spine. The facets may also sustain compressive loads, particularly during hyperextension.
- Motion between two vertebrae is small and does not occur independently in vivo. Thus, the functional motion of the spine is always a combined action of several motion segments.
- The instantaneous center of motion for the motion segments of the lumbar spine usually lies within the lumbar disc.
- The trunk muscles play an important role in providing extrinsic stability to the spine; the ligaments and discs provide intrinsic stability.
- Body position affects the loads on the lumbar spine. Any deviation from upright relaxed standing increases the load. Forward flexion and simultaneous twisting of the trunk produce high stresses on the lumbar spine and disc.
- Externally applied loads that are produced, for example, by lifting or carrying objects may subject the lumbar spine to very high loads. For the loads on the spine to be minimized during lifting, the distance between the trunk and the object lifted should be as short as possible.
- IAP and co-contraction of trunk musculature increase the stability of the spinal column.
- Trunk muscle fatigue may expose the spine to increased vulnerability as a result of loss of motor control and thereby increased stress on the surrounding ligaments, discs, and joint capsules.
- Walking is an excellent exercise that poses a low load on the lumbar spine.

Practice Questions

1. What is the difference in the range of motion at various levels of the spine and what influences these differences?

2. Describe the location, composition, unique properties, and purpose of an intervertebral disc.

3. Describe the influence of the pelvic position on loading of the lumbar spine. How we can apply these concepts in attempts to prevent and/or manage low back pain?

4. What influences the load on the lumbar spine during lifting and carrying?

5. What are the natural effects of aging on the spine and the disc?

6. Resting, sitting, and standing postures impose various loads on the lumbar spine. What is the practical application of this information?

REFERENCES

Abenhaim, L., Rossignol, M., Valat, J.P., et al. (2000). The role of activity in the therapeutic management of back pain. Report of the International Paris Task Force on Back Pain. *Spine, 25*, 15.

Adams, M.A., Dolan, P., Hutton, W.C. (1988). The lumbar spine in backward bending. *Spine, 13*, 1019–1026.

Adams, M.A., Hutton, W.C. (1983). The mechanical functions of the lumbar apophyseal joints. *Spine, 8*, 327.

Allen, C.E. (1948). Muscle action potentials used in the study of dynamic anatomy. *Br J Phys Med, 11*, 66–73.

An, H.S., Masuda, K., Inoue, N. (2006). Intervertebral disc degeneration: Biological and biomechanical factors. *J Orthop Sci, 11*, 541–552.

Anderson, C.K., Chaffin, D.B., Herrin, G.D., et al. (1985). A biomechanical model of the lumbosacral joint during lifting activities. *J Biomech, 18*, 571–584.

Andersson, B.J., Ortengren, R., Nachemson, A., et al. (1974). Lumbar disc pressure and myoelectric back muscle activity during sitting. I. Studies on an experimental chair. *Scand J Rehab Med, 16*, 104–114.

Andersson, E.A., Oddsson, L.I.E., Grundström, H., et al. (1996). EMG activities of the quadratus lumborum and erector spinae muscles during flexion-relaxation and other motor tasks. *Clini Biomech, 11*, 392–400.

Andersson, G.B., Ortengren, R., Nachemson, A. (1977). Intradiscal pressure, intra-abdominal pressure and myoelectric back muscle activity related to posture and loading. *Clin Orthop Rel Res, 129*, 156–164.

Andersson, G.B.J., Lavender, S.A. (1997). Evaluation of muscle function. In J. W. Frymoyer (Ed.), *The Adult Spine. Principles and Practice (2nd Ed.)*. New York: Lippincott-Raven, pp. 341–380.

Andersson, G.B.J., Ortengren, R., Nachemson, A. (1976). Quantitative studies of back loads in lifting. *Spine, 1*, 178.

Arjmand, N., Shirazi-Adl, A. (2005). Biomechanics of changes in lumbar posture in static lifting. *Spine, 30*, 2637–2648.

Arun, R., Freeman, B.J., Scammell, B.E., et al. (2009). 2009 ISSLS Prize Winner: What influence does sustained mechanical load have on diffusion in the human intervertebral disc? An in vivo study using serial postcontrast magnetic resonance imaging. *Spine, 34*, 2324–2337.

Asmussen, E., Klausen, K. (1962). Form and function of the erect human spine. *Clin Orthop, 25*, 55–63.

Axler, C.T., McGill, S.M. (1997). Low back loads over a variety of abdominal exercises: Searching for the safest abdominal challenge. *Med Sci Sports Exerc, 29*, 804–811.

Basmajian, J.V., DeLuca, C.J. (1985). *Muscles Alive*. Baltimore: Williams & Wilkins.

Bible, J.E., Biswas, D.B.A., Miller, C.P., et al. (2010). Normal functional range of motion of the lumbar spine during 15 activities of daily living. *J Spinal Disord Tech, 23*, 106–112.

Biering-Sorensen, F. (1984). Physical measurements as risk indicators for low-back trouble over a one-year period. *Spine, 9*, 106–119.

Boden, S.D., Riew, K.D., Yamaguchi, K., et al. (1996). Orientation of the lumbar facet joints: Association with degenerative disc disease. *J Bone Joint Surg, 78A*, 403–411.

Brown, T., Norton, P.L. (1957). The immobilizing efficiency of back braces; their effect on the posture and motion of the lumbosacral spine. *J Bone Joint Surg, 39A*, 111–139.

Bussey, M.D., Yanai, T., Milburn, P. (2004). A non-invasive technique for assessing innominate bone motion. *Clin Biomech, 19*, 85–90.

Callaghan, J.P., Gunning, J.L., McGill, S.M. (1998). The relationship between lumbar spine load and muscle activity during extensor exercises. *Phys Ther, 78*, 8–18.

Callaghan, J.P., Patla, A.E., McGill, S.M. (1999). Low back three-dimensional joint forces, kinematics, and kinetics during walking. *Clin Biomech, 14*, 203–216.

Cappozzo, A. (1984). Compressive loads in the lumbar vertebral column during normal level walking. *J Ortho Res, 1*, 292–301.

Carlsöö, S. (1961). The static muscle load in different work positions: An electromyographic study. *Ergonomics, 4*, 193–211.

Chaffin, D.B. (1969). A computerized biomechanical model: Development of and use in studying gross body actions. *J Biomech, 2*, 429–441.

Chaffin, D.B., Andersson, G.B.J. (1991). *Occupational Biomechanics (2nd ed.)*. New York: John Wiley & Sons, Inc., pp. 171–263.

Cheng, C.K., Chen, H.H., Chen, C.S., et al. (1998). Influences of walking speed change on the lumbosacral joint force distribution. *Biomed Mater Eng, 8*, 155–165.

Cholewicki, J., Juluru, K., McGill, S.M. (1999). Intra-abdominal pressure mechanism for stabilizing the lumbar spine. *J Biomech, 32*, 13–17.

Cholewicki, J., Juluru, K., Radebold, A., et al. (1999). Lumbar spine stability can be augmented with an abdominal belt and/or increased intra-abdominal pressure. *Eur Spine J, 8*, 388–395.

Cholewicki, J., Panjabi, M.M., Khachatryan, A. (1997). Stabilizing function of trunk flexor-extensor muscles around a neutral spine posture. *Spine, 22*, 2207–2212.

Chou, R., Atlas, S.J., Stanos, S.P., et al. (2009). Nonsurgical interventional therapies for low back pain: A review of the evidence for an American Pain Society clinical practice guideline. *Spine, 34*, 1078–1093.

Ciriello, V.M., Snook, S.H. (1995). The effect of back belts on lumbar muscle fatigue. *Spine, 20*, 1271–1278.

Colloca, C.J., Hinrichs, R.N. (2005). The biomechanical and clinical significance of the lumbar erector spinae flexion-relaxation phenomenon: A review of literature. *J Manipulative Physiol Ther, 28*, 623–631.

Cordo, P.J., Hodges, P.W., Smith, T.C., et al. (2006). Scaling and non-scaling of muscle activity, kinematics, and dynamics in sit-ups with different degrees of difficulty. *J Electromyogr Kinesiol, 16*, 506–521.

Cossette, J.W., Farfan, H.F., Robertson, G.H., et al. (1971). The instantaneous center of rotation of the third lumbar intervertebral joint. *J Biomech, 4*, 149–153.

Cresswell, A.G. (1993). Responses of intra-abdominal pressure and abdominal muscle activity during dynamic trunk loading in man.[Erratum appears in *Eur J Appl Physiol* 1993;67(1), 97]. *Eur J Appl Physiol Occup Physiol, 66*, 315–320.

Cresswell, A.G., Blake, P.L., Thorstensson, A. (1994). The effect of an abdominal muscle training program on intra-abdominal pressure. *Scand J Rehab Med, 26*, 79–86.

Cresswell, A.G., Grundstrom, H., Thorstensson, A. (1992). Observations on intra-abdominal pressure and patterns of abdominal intra-muscular activity in man. *Acta Physiol Scand, 144*, 409–418.

Cresswell, A.G., Oddsson, L., Thorstensson, A. (1994). The influence of sudden perturbations on trunk muscle activity and intra-abdominal pressure while standing. *Exp Brain Res, 98*, 336–341.

Cresswell, A.G., Thorstensson, A. (1994). Changes in intra-abdominal pressure, trunk muscle activation and force during isokinetic lifting and lowering. *Eur J Appl Physiol Occup Physiol, 68*, 315–321.

Dalstra, M., Huiskes, R. (1995). Load transfer across the pelvic bone. *J Biomechanics, 28*, 715–724.

Davis, K.G., Marras, W.S. (2000). The effects of motion on trunk biomechanics. *Clin Biomech, 15*, 703–717.

Davis, K.G., Marras, W.S., Waters, T.R. (1998). Evaluation of spinal loading during lowering and lifting. *Clin Biomech, 13*, 141–152.

de Looze, M.P., Toussaint, H.M., van Dieen, J.H., et al. (1993). Joint moments and muscle activity in the lower extremities and lower back in lifting and lowering tasks. *J Biomech, 26*, 1067–1076.

Eie, N. (1966). Load capacity of the low back. *J Oslo City Hosp, 16*, 73–98.

Ekholm, J., Arborelius, U., Fahlcrantz, A., et al. (1979). Activation of abdominal muscles during some physiotherapeutic exercises. *Scand J Rehab Med, 11*, 75–84.

El-Bohy, A.A., King, A.I. (1986). Intervertebral disc and facet contact pressure in axial torsion. In S.A. Lantz & A.I. King (Eds.), *1986 Advances in Engineering*. New York: American Society of Mechanical Engineers, pp. 26–27.

Escamilla, R.F., Babb, E., DeWitt, R., et al. (2006). Electromyographic analysis of traditional and nontraditional abdominal exercises: Implications for rehabilitation and training. *Phys Ther, 86*, 656–671.

Farfan, H.F. (1975). Muscular mechanism of the lumbar spine and the position of power and efficiency. *Orthop Clin North Am, 6*, 135–144.

Ferguson, S.J., Steffen, T. (2003). Biomechanics of the aging spine. *Eur Spine J, 12*, S97–S103.

Ferrara, L., Triano, J.J., Sohn, M.J., et al. (2005). A biomechanical assessment of disc pressures in the lumbosacral spine in response to external unloading forces. *Spine, 5*, 548–553.

Flint, M.M. (1965). Abdominal muscle involvement during the performance of various forms of sit-up exercise. An electromyographic study. *Am J Phys Med, 44*, 224–234.

Floyd, W.F., Silver, P.H.S. (1955). The function of the erectors spinae muscles in certain movements and postures in man. *J Physiol, 129*(1), 184–203.

Friedebold, G. (1958). Die AktiviUt normaler Rilckenstreckmuskulatur im Elektromyogramm unter verschiedenen Haltungsbedingungen; Eine Studie zur Skelettmuskelmechanik. *Z Orthop, 90*(1), 1–18.

Fujiwara, A,. Lim, T.H., An, H.S., et al. (2000). The effect of disc degeneration and facet joint osteoarthritis on the segmental flexibility of the lumbar spine. *Spine, 25*, 3036–3044.

Fukuyama, S., Nakamura, T., Ikeda, T., et al. (1995). The effect of mechanical stress on hypertrophy of the lumbar ligamentum flavum. *J Spinal Disord, 8*, 126–130.

Galante, J.O. (1967). Tensile properties of the human lumbar annulus fibrosus. *Acta Orthop Scand, 100*, 1–91.

Gardner-Morse, M.G., Stokes, I. A. (1998). The effects of abdominal muscle coactivation on lumbar spine stability. *Spine, 23*, 86–91.

Gertzbein, S.D., Seligman, J., Holtby, R., et al. (1985). Centrode patterns and segmental instability in degenerative disc disease. *Spine, 10*, 257–261.

Granata, K.P., Sanford, A.H. (1999). Lumbar-pelvic coordination is influenced by lifting task parameters. *Spine, 25*, 1413–1418.

Gregersen, G., Lucas, D. (1967). An in vivo study of the axial rotation of the human thoracolumbar spine. *J Bone Joint Surg, 49A*, 247–262.

Haher, T.R., O'Brien, M., Dryer, J.W., et al. (1994). The role of the lumbar facet joints in spinal stability. Identification of alternative paths of loading. *Spine, 19*, 2667–2670.

Hansson, T., Roos, B., Nachemson, A. (1980). The bone mineral content and ultimate compressive strength of lumbar vertebrae. *Spine, 5*, 46–55.

Harman, E.A., Frykman, P.N., Clagett, E.R., et al. (1988). Intra-abdominal and intra-thoracic pressures during lifting and jumping. *Med Sci Sports Exerc, 20*, 195–201.

Hasegawa, K., Takahashi, H.E., Koga ,Y., et al. (1993). Mechanical properties of osteopenic vertebral bodies monitored by acoustic emission.[Erratum appears in Bone 1993 Nov–Dec;14(6), 891]. *Bone, 14*, 737–743.

Haughton, V.M., Schmidt, T.A., Keele, K., et al. (2000). Flexibility of lumbar spinal motion segments correlated to type of tears in the annulus fibrosus. *J Neurosurg, 92*, 81–86.

Hodges, P., Cresswell, A., Thorstensson, A. (1999). Preparatory trunk motion accompanies rapid upper limb movement. *Exp Brain Res, 124*, 69–79.

Hodges, P.W., Eriksson, A.E., Shirley, D., et al. (2005). Intra-abdominal pressure increases stiffness of the lumbar spine. *J Biomechanics, 38*, 1873–1880.

Hodges, P.W., Gandevia, S.C. (2000). Activation of the human diaphragm during a repetitive postural task. *J Physiol, 1*, 165–175.

Hodges, P.W., Richardson, C.A. (1997). Feedforward contraction of transversus abdominis is not influenced by the direction of arm movement. *Exp Brain Res, 114*, 362–370.

Hodges, P.W., Sapsford, R., Pengel, L.H. (2007). Postural and respiratory functions of the pelvic floor muscles. *Neurourol Urodyn, 26*, 362–371.

Hutton, W.C., Adams, M.A. (1982). Can the lumbar spine be crushed in heavy lifting? *Spine, 7*, 586–590.

Jorgensen, M.J., Marras, W.S., Gupta, P., et al. (2003). Effect of torso flexion on the lumbar torso extensor muscle sagittal plane moment arms. *Spine, 3*, 363–369.

Joseph, J. (1960). *Man's Posture: Electromyographic Studies.* Springfield: Charles C. Thomas.

Juker, D., McGill, S., Kropf, P., et al. (1998). Quantitative intramuscular myoelectric activity of lumbar portions of psoas and the abdominal wall during a wide variety of tasks. *Med Sci Sports Exerc, 30*, 301–310.

Klein, J. A., Hickey, D.S., Huskins, D.W. (1983). Radial bulging of the annulus fibrosus and the function and failure of the intervertebral disc. *J Biomech, 16*, 211–217.

Kozanek, M., Wang, S., Passias, P.G., et al. (2009). Range of motion and orientation of the lumbar facet joints in vivo. *Spine, 34*(19), E689–E696.

Krag, M.H., Seroussi, R.E., Wilder, D.G., et al. (1987). Internal displacement distribution from in vitro loading of human thoracic and lumbar spinal motion segments: Experimental results and theoretical predictions. *Spine, 12*, 1001–1007.

Krajcarski, S.R., Potvin, J.R., Chiang, J. (1999). The in vivo dynamic response of the spine to perturbations causing rapid flexion: Effects of pre-load and step input magnitude. *Clin Biomech, 14*, 54–62.

Kulak, R.F., Schultz, A.B., Belytschko, T., et al. (1975). Biomechanical characteristics of vertebral motion segments and intervertebral discs. *Orthop Clin North Am, 6*, 121–133.

Lavender, S.A., Mirka, G.A., Schoenmarklin, R.W., et al. (1989). The effects of preview and task symmetry on trunk muscle response to sudden loading. *Human Factors, 31*, 101–115.

Lavender, S.A., Tsuang, Y.H., Andersson, G.B. (1992). Trunk muscle co-contraction while resisting applied moments in a twisted posture. *Ergonomics, 36*(10), 1145–1157.

Lee, Y.H., Chen, C.Y. (1999). Lumbar vertebral angles and back muscle loading with belts. *Ind Health, 37*, 390–397.

Li, G., Wang, S., Passias, P., et al. (2009). Segmental in vivo vertebral motion during functional human lumbar spine activities. *Eur Spine J, 18*, 1013–1021.

Lu, W.W., Luk, K.D., Cheung, K.C., et al. (2004). Microfracture and changes in energy absorption to fracture of young vertebral cancellous bone following physiological fatigue loading. *Spine, 29*, 1196–1201.

Lucas, D.B., Bresier, B. (1961). *Stability of the Ligamentous Spine.* Biomechanics Laboratory, University of California, San Francisco and Berkeley. Technical Report 40. San Francisco: The Laboratory.

Lumsden, R.M., Morris, J.M. (1968). An in vivo study of axial rotation and immoblization at the lumbosacral joint. *J Bone Joint Surg, 50A*, 1591–602.

Marras, W.S., Davis, K.G., Heaney, C.A., et al. (2000). The influence of psychosocial stress, gender, and personality on mechanical loading of the lumbar spine. *Spine, 25*, 3045–3054.

Marras, W.S., Granata, K.P. (1995). A biomechanical assessment and model of axial twisting in the thoracolumbar spine. *Spine, 20*, 1440–1451.

Marras, W.S., Mirka, G.A. (1992). A comprehensive evaluation of trunk response to asymmetric trunk motion. *Spine, 17*(3), 318–326.

Marras, W.S., Rangarajulu, S.L., Lavender, S.A. (1987). Trunk loading and expectation. *Ergonomics, 30*, 551–562.

McCook, D.T., Vicenzino, B., Hodges, P.W. (2009). Activity of deep abdominal muscles increases during submaximal flexion and extension efforts but antagonist co-contraction remains unchanged. *J Electromyogr Kinesiol, 19*, 754–762.

McGill, S.M., Karpowicz, A., Fenwick, C.M., et al. (2009). Exercises for the torso performed in a standing posture: Spine and hip motion and motor patterns and spine load. *J Strength Cond Res, 23*, 455–464.

McGill, S.M., Norman, R.W. (1987). Reassessment of the role of intra-abdominal pressure in spinal compression. *Ergonomics, 30*, 1565–1588.

McGill, S. M., Yingling, V.R., Peach, J. P. (1999). Three-dimensional kinematics and trunk muscle myoelectric activity in the elderly spine—A database compared to young people. *Clin Biomech, 14*, 389–395.

Miles, M., Sullivan, W.E. (1961). Lateral bending at the lumbar and lumbosacral joints. *Anat Rec, 139*(3), 387–398.

Miller, J.A., Haderspeck, K.A., Schultz, A.B. (1983). Posterior element loads in lumbar motion segments. *Spine, 8*, 331–337.

Miyazaki, M., Morishita, Y., Takita, C., et al. (2010). Analysis of the relationship between facet joint angle orientation and lumbar spine canal diameter with respect to the kinematics of the lumbar spinal unit. *J Spinal Disord Tech* [Epub.].

Moll, J.M., Wright, V. (1971). Normal range of spinal mobility. An objective clinical study. *Ann Rheum Dis, 30*, 381–386.

Monfort-Panego, M., Vera-Garcia, F.J., Sanchez-Zuriaga, D., et al. (2009). Electromyographic studies in abdominal exercises: A literature synthesis. *J Manipulative Physiol Ther, 32*, 232–244.

Morris, J.M., Benner, G., Lucas, D.B. (1962). An electromyographic study of the intrinsic muscles of the back in man. *J Anat Lond, 96*, 509–520.

Morris, J.M., Lucas, D.B., Bresier, B. (1961). Role of the trunk in stability of the spine. *J Bone Joint Surg, 43A*, 327–351.

Mosekilde, L. (1993). Vertebral structure and strength in vivo and in vitro. *Calcif Tissue Int, 53*, Suppl 1, S121–S125.

Nachemson, A. (1966). Electromyographic studies on the vertebral portion of the psoas muscle; with special reference to its stabilizing function of the lumbar spine. *Acta Orthopaedica Scand, 37*, 177–190.

Nachemson, A. (1963). The influence of spinal movements on the lumbar intradiscal pressure and on the tensile stresses in the annulus fibrosus. *Acta Orthop Scand, 33*(1–4), 1–104.

Nachemson, A. (1960). Lumbar intradiscal pressure. Experimental studies on post-mortem material. *Acta Orthopaedica Scand Suppl, 43*, 1–104.

Nachemson, A. (1975). Towards a better understanding of lowback pain: A review of the mechanics of the lumbar disc. *Rheumatol Rehabil, 14*, 129–143.

Nachemson, A., Elfström, G. (1970). *Intravital Dynamic Pressure Measurements in Lumbar Discs: A Study of Common Movements, Maneuvers and Exercises.* Stockholm: Almquist & Wiksell, 1–40.

Nachemson, A.L., Evans, J.H. (1968). Some mechanical properties of the third human lumbar interlaminar ligament (ligamentum flavum). *J Biomechanics, 1*, 211–220.

Nachemson, A., Morris, J.M. (1964). In vivo measurements of intradiscal pressure. Discometry, a method for the determination of pressure in the lower lumbar discs. *J Bone & Joint Surg 4A*, 1077–1092.

Nelson, J.M., Walmsley, R.P., Stevenson, J.M. (1995). Relative lumbar and pelvic motion during loaded spinal flexion/extension. *Spine, 20*, 199–204.

Nemeth, G. (1984). On hip and lumbar biomechanics. A study of joint load and muscular activity. *Scand J Rehab Med Suppl, 10*, 1–35.

NIOSH. (1994). *Workplace Use of Back Belts: Review and Recommendations.* http://www.cdc.gov/niosh/94-122.html. [NIOSH Publication 1994-122, serial online.]

Okushima, Y., Yamazaki, N., Matsumoto, M., et al. (2006). Lateral translation of the lumbar spine: In vitro biomechanical study. *J Appl Biomech, 22*, 83–92.

Ortengren, R., Andersson, G.B., Nachemson, A.L. (1981). Studies of relationships between lumbar disc pressure, myoelectric back muscle activity, and intra-abdominal (intragastric) pressure. *Spine, 6*, 98–103.

Osvalder, A.L., Neumann, P., Lovsund, P., et al. (1990). Ultimate strength of the lumbar spine in flexion–An in vitro study. *J Biomechanics, 23*, 453–460.

Panjabi, M.M., Goel, V.K., Takata, K. (1982). Physiologic strains in the lumbar spinal ligaments. An in vitro biomechanical study. 1981 Volvo Award in Biomechanics. *Spine, 7*, 192–203.

Parnianpour, M., Nordin, M., Kahanovitz, N., et al. (1988). The triaxial coupling of torque generation of trunk muscles during isometric exertions and the effect of fatiguing isoinertial movements on the motor output and movement patterns. 1988 Volvo Award in Biomechanics. *Spine, 13*, 982–992.

Partridge, M.J., Walters, C.E. (1959). Participation of the abdominal muscles in various movements of the trunk in man: An electromyographic study. *Phys Ther Rev, 39*, 791–800.

Pauly, J.E. (1966). An electromyographic analysis of certain movements and exercises. I. Some deep muscles of the back. *Anat Rec, 155*, 223–234.

Pel, J.J., Spoor, C.W., Pool-Goudzwaard, A.L., et al. (2008). Biomechanical analysis of reducing sacroiliac joint shear load by optimization of pelvic muscle and ligament forces. *Ann Biomed Eng, 36*, 415–424.

Perkins, M.S., Bloswick, D.S. (1995). The use of back belts to increase intraabdominal pressure as a means of preventing low back injuries: A survey of the literature. *Int J Occup Environ Health,1*(4), 326–335.

Pope, M.H., Andersson, G.B., Broman, H., et al. (1986). Electromyographic studies of the lumbar trunk musculature during the development of axial torques. *J Orthop Res, 4*, 288–297.

Przybyla, A., Pollintine, P., Bedzinski, R., et al. (2006). Outer annulus tears have less effect than endplate fracture on stress distributions inside intervertebral discs: Relevance to disc degeneration. *Clin Biomech, 21*, 1013–1019.

Rajasekaran, S., Naresh-Babu, J., Murugan, S. (2007). Review of postcontrast MRI studies on diffusion of human lumbar discs. *J Magn Reson Imaging, 25*, 410–418.

Ranu, H.S. (1990). Measurement of pressures in the nucleus and within the annulus of the human spinal disc: Due to extreme loading.[Erratum appears in Proc Inst Mech Eng 1991;205(1):following 53]. *Proc Inst Mech Eng H, 204*, 141–146.

Reichmann, S. (1971). Motion of the lumbar articular processes in flexion-extension and lateral flexions of the spine. *Acta Morphol Neerl Scand, 8*, 261–272.

Reichmann, S., Berglund, E., Lundgren, K. (1972). Das Bewegungszentrum in der LendenwirbelsAule bei Flexion und Extension. *Z Anat Entwicklungsgesch, 138*(3), 283–287.

Reid, J.G., Costigan, P.A. (1987). Trunk muscle balance and muscular force. *Spine, 12*, 783–786.

Reyna, J.R. Jr., Leggett, S.H., Kenney, K., et al. (1995). The effect of lumbar belts on isolated lumbar muscle. Strength and dynamic capacity. *Spine, 20*, 68–73.

Rohlmann, A., Zander, T., Schmidt, H., et al. (2006). Analysis of the influence of disc degeneration on the mechanical behaviour of a lumbar motion segment using the finite element method. *J Biomech, 39*, 2484–2490.

Rolander, S.D. (1966). Motion of the lumbar spine with special reference to the stabilizing effect of posterior fusion. An experimental study on autopsy specimens. *Acta Orthop Scand, 90*, 1–144.

Rousseau, M.A., Bradford, D.S., Hadi, T.M., et al. (2006). The instant axis of rotation influences facet forces at L5/S1 during flexion/extension and lateral bending. *Eur Spine J, 15*, 299–307.

Sanchez-Zuriaga, D.P., Adams, M.A., Dolan, P.P. (2010). Is activation of the back muscles impaired by creep or muscle fatigue? *Spine, 35*, 517–525.

Sato, K., Kikuchi, S., Yonezawa, T. (1999). In vivo intradiscal pressure measurement in healthy individuals and in patients with ongoing back problems. *Spine, 24*, 2468–2474.

Saunders, S.W., Schache, A., Rath, D., et al. (2005). Changes in three dimensional lumbo-pelvic kinematics and trunk muscle activity with speed and mode of locomotion. *Clin Biomech, 20*, 784–793.

Schmidt, H., Heuer, F., Wilke, H.J. (2008). Interaction between finite helical axes and facet joint forces under combined loading. *Spine, 33*, 2741–2748.

Schmidt, H., Kettler, A., Heuer, F., et al. (2007). Intradiscal pressure, shear strain, and fiber strain in the intervertebral disc under combined loading. *Spine, 32*, 748–755.

Schultz, A., Andersson, G., Ortengren, R., et al. (1982). Loads on the lumbar spine. Validation of a biomechanical analysis by measurements of intradiscal pressures and myoelectric signals. *J Bone Joint Surg, 64A*, 713–720.

Sheikhzadeh, A., Parnianpour, M., Nordin, M. (2008). Capability and recruitment patterns of trunk during isometric uniaxial and biaxial upright exertion. *Clin Biomech, 23*, 527–535.

Shirazi-Adl, A. (1994). Biomechanics of the lumbar spine in sagittal/lateral moments. *Spine, 19*, 2407–2414.

Shirazi-Adl, A., Sadouk, S., Parnianpour, M., et al. (2002). Muscle force evaluation and the role of posture in human lumbar spine under compression. *Eur Spine J, 11*, 519–526.

Sivan, S., Merkher, Y., Wachtel, E., et al. (2006). Correlation of swelling pressure and intrafibrillar water in young and aged human intervertebral discs. *J Orthop Res, 24*, 1292–1298.

Snijders, C.J., Hermans, P.F., Niesing, R., et al. (2004). The influence of slouching and lumbar support on iliolumbar ligaments, intervertebral discs and sacroiliac joints. *Clin Biomech, 19*, 323–329.

Solomonow, M., Zhou, B.H., Baratta, R.V., et al. (1999). Biomechanics of increased exposure to lumbar injury caused by cyclic loading: Part 1. Loss of reflexive muscular stabilization. *Spine, 24*, 2426–2434.

Sparto, P.J., Parnianpour, M. (1998). Estimation of trunk muscle forces and spinal loads during fatiguing repetitive trunk exertions. *Spine, 23*, 2563–2573.

Steffen, T., Baramki, H.G., Rubin, R. (1998). Lumbar intradiscal pressure measured in the anterior and posterolateral annular regions during asymmetrical loading. *Clin Biomech, 13*(7), 495–505.

Sturesson, B., Selvik, G., Uden, A. (1989). Movements of the sacroiliac joints. A roentgen stereophotogrammetric analysis. *Spine, 14*, 162–165.

Thomas, J.S., Lavender, S.A., Corcos, D.M., et al. (1999). Effect of lifting belts on trunk muscle activation during a suddenly applied load. *Hum Factors, 41*, 670–676.

Trudelle-Jackson, E., Fleisher, L.A., Borman, N., et al. (2010). Lumbar spine flexion and extension extremes of motion in women of different age and racial groups: The WIN Study. *Spine*, Epub.

Tuong, N.H., Dansereau, J., Maurais, G., et al. (1998). Three-dimensional evaluation of lumbar orthosis effects on spinal behavior. *J Rehabil Res Dev, 35*, 34–42.

Urban, J.P., McMillan, J.F. (1985). Swelling pressure of the intervertebral disc: Influence of proteoglycan and collagen contents. *Biorheology, 22*(2), 145–157.

Urban, J.P., Smith, S., Fairbank, J.C. (2004). Nutrition of the intervertebral disc. *Spine, 29*, 2700–2709.

Urquhart, D.M., Hodges, P.W., Allen, T.J., et al. (2005). Abdominal muscle recruitment during a range of voluntary exercises. *Man Ther, 10*, 144–153.

van Dieen, J.H., Hoozemans, M.J., Toussaint, H.M. (1999). Stoop or squat: A review of biomechanical studies on lifting technique. *Clin Biomech, 14*, 685–696.

Wang, M., Dumas, G.A. (1998). Mechanical behavior of the female sacroiliac joint and influence of the anterior and posterior sacroiliac ligaments under sagittal loads. *Clin Biomech, 13*(4-5), 293–299.

Wang, S., Xia, Q., Passias, P., et al. (2009). Measurement of geometric deformation of lumbar intervertebral discs under in-vivo weightbearing condition. *J Biomech, 42*, 705–711.

White, A. (1969). Analysis of the mechanics of the thoracic spine in man. An experimental study of autopsy specimens. *Acta Orthop Scand Suppl, 127*, 1–105.

White, A.A., Panjabi, M.N. (1978). *Clinical Biomechanics of the Spine*. Philadelphia: J.B. Lippincott.

Wilder, D.G., Pope, M.H., Frymoyer, J.W. (1980). The functional topography of the sacroiliac joint. *Spine, 5*, 575–579.

Wilke, H., Neef, P., Hinz, B., et al. (2001). Intradiscal pressure together with anthropometric data—A data set for the validation of models. *Clin Biomech, 16*, Suppl 1, S111–S126.

Wilke, H.J., Neef, P., Caimi, M., et al. (1999). New in vivo measurements of pressures in the intervertebral disc in daily life. *Spine, 24*, 755–762.

Wilke, H.J., Rohlmann, A., Neller, S., et al. (2003). ISSLS prize winner: A novel approach to determine trunk muscle forces during flexion and extension: A comparison of data from an in vitro experiment and in vivo measurements. [Erratum appears in Spine, 2004 Aug 15;29(16), 1844]. *Spine, 28*, 2585–2593.

Wilke, H.J., Wolf, S., Claes, L.E., et al. (1996). Influence of varying muscle forces on lumbar intradiscal pressure: An in vitro study. *J Biomech, 29*, 549–555.

Zander, T., Rohlmann, A., Calisse, J., et al. (2001). Estimation of muscle forces in the lumbar spine during upper-body inclination. *Clin Biomech, 16*, Suppl 1, S73–S80.

Zhu, Q.A., Park, Y.B., Sjovold, S.G., et al. (2008). Can extra-articular strains be used to measure facet contact forces in the lumbar spine? An in-vitro biomechanical study. *Proc Inst Mech Eng H, 222*, 171–184.

Biomechanics of the Cervical Spine

Ronald Moskovich

Introduction

Knowledge of spinal biomechanics advanced exponentially during the second half of the twentieth century, facilitated by a more sophisticated view of the spine and the development of complex models by which function, injury, and disease could be better understood. A two-column model of the spine was described by Sir Frank Holdsworth (1963), and later, a three-column model by Denis (1983) further refined the principles of spinal stability. The computer age has produced powerful methods for modern biomechanical modeling, providing surgeons with the ability to assess the stability of a construct prior to implantation. Due to the modern day proliferation of biomechanical studies, the application of cervical biomechanical knowledge spans many industries and supports improved medical diagnoses and treatment that is more effective. Future technological and electronic advances will continue to build on basic biomechanical principles, many of which are outlined in this chapter.

Component Anatomy and Biomechanics

ANATOMY

The intricate and elegant design of the cervical spine uniquely contributes to the structure of the human body and profoundly enhances its function. The cervical spine has a critical structural role in that it supports the skull, acts as a shock absorber for the brain, and protects the brainstem, spinal cord, and various neurovascular structures as they transit the neck and when they enter and exit the skull. Part of its biomechanical function is to facilitate the transfer of weight and bending moments of the head. The vertebral column also provides a multitude of muscle and ligamentous attachments for complex movement and stability. The neuromuscular control afforded by the muscle attachments, when combined with the numerous articulations of the cervical spine, allows for a wide range of physiologic motion that maximizes the range of motion of the head and neck and serves to integrate the head with the rest of the body and the environment.

The spine consists of 33 vertebrae divided into five regions: cervical (7) (Fig. 11-1), thoracic (12), lumbar (5), sacral (5 fused segments), and coccygeal (approximately 4). The two most cranial vertebrae, C1 (atlas) and C2 (axis), are atypical and have a unique structural role in the articulation between the head and the cervical spine. The atlanto-occipital joint, between C1 and the

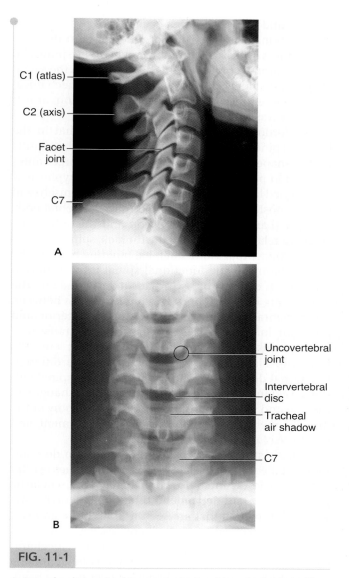

FIG. 11-1

A. Lateral radiograph of the cervical spine. Note the lordosis. The facet joints are aligned obliquely only to the frontal or coronal plane; hence their excellent visualization in the lateral view. **B.** Anteroposterior view of the cervical spine.

occipital bone of the skull, is also a functional part of the cervical spine. There are five typical cervical vertebrae, C3 to C7, that are similar in structure and function.

The spine has four curves when viewed in the sagittal plane. The cervical and lumbar regions are convex anteriorly (lordotic), while the thoracic and sacral regions are convex posteriorly (kyphotic). The lordotic curves develop after birth as the infant's spine straightens out, which facilitates development of a bipedal posture. Although there is a harmonious progression of these curves from one to another, which may help distribute stresses and strains, injuries occur more commonly at the junctional areas because of differences in the relative stiffness of each anatomic segment of the spine. The spinal anatomy changes from the occipital to the sacral region, reflecting varying kinematic and biomechanical functional requirements: Form follows function, or is it the other way around?

The lordosis in the cervical spine, like that in the lumbar spine, is maintained predominantly by slightly wedge-shaped intervertebral discs that are larger anteriorly than posteriorly. In contrast, thoracic kyphosis is maintained largely by the vertebral bodies themselves in that the posterior portion of the thoracic vertebral body is larger than the anterior portion and this inherently creates a relative kyphosis of the thoracic spine.

The conceptual biomechanical building block of the spinal column is the functional spinal unit or motion segment. It consists of two adjacent vertebrae and the intervening intervertebral disc and ligaments between the vertebrae. These ligaments are the anterior and posterior longitudinal ligaments; the intertransverse, interspinous, and supraspinous ligaments; and the facet capsular ligaments. As a result of the different functional demands of various parts of the spinal column, segmental variation is expressed by changes in the size and shape of the vertebrae, the anatomy of the discoligamentous structures, and the alignment and structure of the facet joints.

Biologic structures behave differently than do common engineering materials. Collagenous tissues exhibit both viscoelastic and anisotropic behavior. Viscoelastic properties are rate-dependent (time-dependent) behaviors under loading that are seen in both bone and soft tissues; mechanical strength increases with increased rates of loading. Anisotropy is the alteration in mechanical properties that is seen when bone is loaded along different axes. Anisotropic behavior occurs as a result of the dissimilar longitudinal and transverse crystalline microstructure of bone.

Osseous Structures

The occiput-C1-C2 complex constitutes the upper cervical spine and is responsible for approximately 40% of cervical flexion and 60% of cervical rotation. The occipital condyles articulate with the congruent, concave lateral masses of the atlas, imparting intrinsic stability to the joint. The atlas, or C1, is a bony ring consisting of an anterior and posterior arch that is attached to the two lateral masses of the atlas (Fig. 11-2). The superior surfaces of the lateral masses, which face cranially and inward, form an articulation with the caudally and outward-facing occipital condyles of the skull

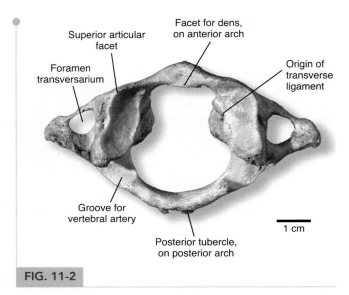

FIG. 11-2

Bony architecture of the atlas. *Bar* = 1 cm.

(Fig. 11-3). The articular surfaces can be considered part of a sphere, with the center of rotation located above the articular surfaces (Kapandji, 1974). The primary motion permitted by this articulation is flexion and extension, accounting for much of the sagittal range of motion of the cervical spine.

Extension of the occipitocervical joint is limited by the bony anatomy; flexion is limited primarily by posterior ligamentous structures (Figs. 11-4 and 11-5). The tectorial membrane does not limit cervical flexion per se, but rather, helps to ensure that the odontoid process does not encroach excessively into the cervical canal (Tubbs et al., 2007). The anterior tubercle on the arch of C1 serves as an attachment for the longus colli muscle, a flexor of the neck. The posterior arch of the atlas is a modified lamina that is grooved on its superior surface for the passage of the vertebral arteries as they enter into the foramen magnum after piercing the posterior atlantooccipital membrane.

The C1–2 articulation is primarily responsible for rotation in the cervical spine. As there is no intervertebral disc between C1 and C2, and the facet joints are incongruent condylar articulations, stability at this level is predicated on intact osseoligamentous structures. The body of C2 projects superiorly to form the odontoid process, or dens (Fig. 11-6). The projection of the body of C2 and the dens has a characteristic oblong appearance on lateral cervical radiographs and is often a helpful anatomic landmark. The dens articulates with and is restrained within a socket formed by the transverse ligament of C1 and the anterior arch of C1. The transverse ligaments run from the anterior arch of C1, behind

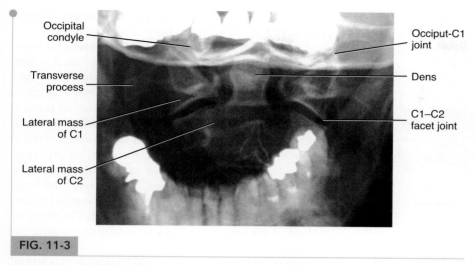

Occipital condyle

Transverse process

Lateral mass of C1

Lateral mass of C2

Occiput-C1 joint

Dens

C1–C2 facet joint

FIG. 11-3

The open-mouth radiograph demonstrates the occiput-C1 and the atlantoaxial articulations. Note the symmetric spacing between the lateral masses of C1 and the dens. Asymmetry or widening of these spaces may occur after rotatory disturbances or fractures of the C1 ring.

the dens, and prevent anterior translation of C1 on C2. The other ligaments at the C1–2 articulation are the alar, apical, and accessory alar ligaments. The alar ligaments, which are symmetrically placed on both sides of the dens, attach the dens to the occiput to prevent excessive rotation. The left alar ligament prevents right rotation and vice versa. To some extent, the alar ligaments also act to limit motion during side bending (Dvorak and

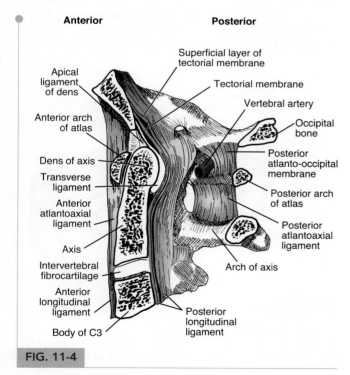

Anterior   Posterior

Apical ligament of dens

Anterior arch of atlas

Dens of axis

Transverse ligament

Anterior atlantoaxial ligament

Axis

Intervertebral fibrocartilage

Anterior longitudinal ligament

Body of C3

Superficial layer of tectorial membrane

Tectorial membrane

Vertebral artery

Occipital bone

Posterior atlanto-occipital membrane

Posterior arch of atlas

Posterior atlantoaxial ligament

Arch of axis

Posterior longitudinal ligament

FIG. 11-4

Median sagittal section through the occipital bone and the first three cervical vertebrae, showing the articulations and surrounding ligaments.

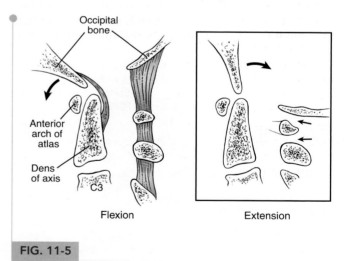

Occipital bone

Anterior arch of atlas

Dens of axis

C3

Flexion   Extension

FIG. 11-5

Tracings of lateral flexion and extension radiographs showing the occiput, C1, C2, and C3. The substantial relative motion between the occiput, C1, and C2 can be seen. *Large arrows* indicate the direction of motion. *Small arrows* indicate that approximation of the posterior elements limit occipitocervical extension. In contradistinction, maximum flexion is controlled by tautness of the ligaments. *Reprinted with permission from Moskovich, R., Jones, D.A. (1999). Upper cervical spine instrumentation. Spine: State of the Art Reviews, 13(2), 233–253.*

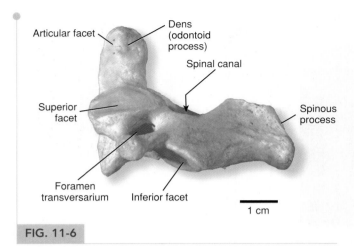

FIG. 11-6

The axis vertebra, or C2. The superior facet articulation permits multiplanar motion, while the inferior facet is aligned to articulate with a more typical cervical facet, which is more constrained. There is a smooth surface on the front of the dens for articulation with the anterior ring of C1. *Bar* = 1 cm.

Panjabi, 1987). The apical ligament also connects the dens to the occiput.

Unlike the two most cranial vertebrae, the anatomy of the third through the sixth cervical vertebrae is similar (Fig. 11-7). These four cervical vertebrae consist of a body, two pedicles, two lateral masses, two laminae, and a spinous process. The seventh cervical vertebra is slightly different in that it has a transitional form. It is called the vertebra prominens and has a larger spinous process that is not bifid like those of C3 to C6.

The anterior components of a subaxial cervical motion segment are the vertebral bodies and the disc. The cervical vertebral body is oval and is wider mediolaterally than anteroposteriorly. The transverse processes of the cervical spine are unique in that they all contain a transverse foramen for the passage of the vertebral artery. The transverse processes of the subaxial cervical vertebrae have two projections, the anterior and posterior tubercles, which serve as attachment points for anterior and posterior muscles, respectively. The large anterior tubercle of C6, referred to as the carotid tubercle, can be an important surgical landmark. The superior surface of the transverse process provides a groove for the exiting nerve root.

Each pedicle connects the vertebral body to a lateral mass, that portion of bone containing the superior and inferior facets. The facet joints regulate the movements of the spine and play a critical role in spinal stability. Those of the cervical spine are oriented at approximately 45° to the coronal plane and are located in the sagittal plane (Figs. 11-8 and 11-9). This orientation allows greater flexion than lateral bending or rotation in the cervical spine. The facet joints resist most of the shear forces and approximately 16% of the compressive forces acting on the spine (Adams and Hutton, 1980). The laminae also arise from the lateral masses. The lateral masses have important surgical implications in the

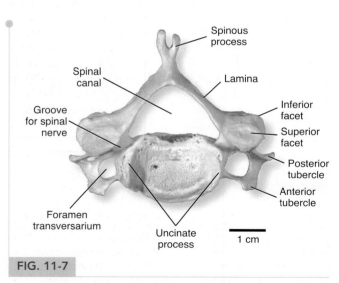

FIG. 11-7

Superior view of a typical cervical vertebra, representative of C3 to C6 (C7, the vertebra prominens, differs slightly in that it has a prominent nonbifid spinous process).

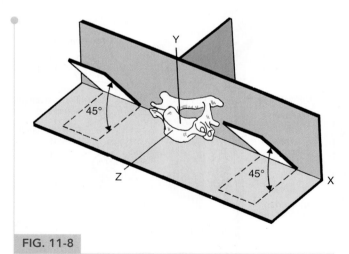

FIG. 11-8

Orientation of the facets of a typical cervical vertebra in three planes. The facets are oriented at a 45° angle to the transverse plane and the frontal plane and are at right angles to the sagittal plane. Y indicates the craniocaudal axis, Z the anteroposterior axis, and X the mediolateral axis. *Adapted from White, A.A., III, Panjabi, M.M. (1990). Clinical Biomechanics of the Spine. Philadelphia: JB Lippincott Co.*

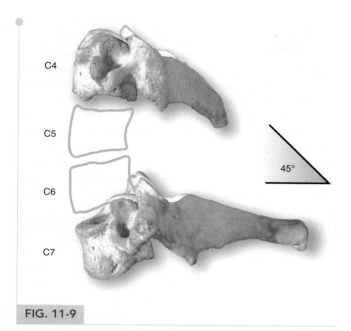

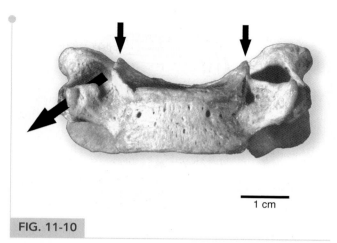

FIG. 11-10

Anterior view of a sixth cervical vertebra. The *short arrows* indicate the uncinate processes, and the *long arrow* indicates the pathway of the sixth cervical nerve root. The facet joints are located posteriorly. *Bar* = 1 cm.

FIG. 11-9

Lateral photograph of a fourth and seventh cervical vertebra. The facet joint alignment is fairly close to 45° from the transverse plane. Note also the difference in size of the spinous processes, which are a reflection of the size and importance of the muscle attachments.

subaxial cervical spine, because they contain a relatively large amount of bone and are easily accessible for the placement of screws, as opposed to the pedicles, which are difficult to cannulate safely in the neck.

The superior surfaces of the cervical vertebrae are saddle-shaped because of the uncinate processes, which are bony protuberances that arise from the lateral margins of the superior end plates (Fig. 11-10). The uncovertebral joints (joints of Luschka) develop during spinal maturation but are not universally demonstrable and are thought to occur only in obligatory or facultative bipedal animals. Head rotation is used to look around in bipeds, whereas side bending is used by quadrupeds, suggesting a role of these joints in facilitating or limiting rotation.

Intervertebral Discs

The intervertebral discs are highly specialized structures that contribute up to one-third of the height of the vertebral column and form specialized joints between the cartilaginous end plates of the adjacent vertebral bodies. Activities such as running and jumping apply short-duration, high-amplitude loads to the intervertebral discs, whereas normal physical activity and upright stance result in the application of long-duration, low-magnitude loads. Discs are able to withstand greater-than-normal

loads when compressive forces are rapidly applied, based on the biomechanical principles of viscoelasticity. This property protects the disc from catastrophic failure until extremely high loads are applied.

The nucleus pulposus is centrally located within the disc and consists of almost 90% water in young individuals. The water content is highest at birth and decreases to approximately 70% as the disc degenerates with age. The rest of the nucleus pulposus consists of proteoglycan and collagen, which is exclusively type II collagen. Type II collagen fibrils are thought to be able to absorb compressive forces better than type I collagen fibrils.

Proteoglycans consist of a protein core attached to polysaccharide (glycosaminoglycan) chains. The polysaccharides are either keratin sulfate or chondroitin sulfate. The core protein, with its attached polysaccharides, is aggregated to hyaluronic acid through a link protein. The proteoglycans in the intervertebral discs are similar to those in articular cartilage, except that the proteoglycans present in the intervertebral discs have shorter polysaccharide chains as well as shorter core proteins. The nucleus pulposus contains more proteoglycan than does the annulus fibrosus. With increasing age and disc degeneration, the total proteoglycan content decreases.

The annulus fibrosus is the outer portion of the disc. Its water content is slightly less than that of the nucleus, being only approximately 78% water in younger individuals. With age, the water content falls to approximately 70%, like that of the nucleus pulposus in older persons. The annulus consists of collagen that is arranged in approximately 90 concentric lamellar bands. The collagen fibers in these sheets run at approximately 30° to the disc or 120°

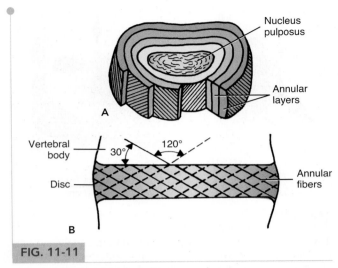

FIG. 11-11

Schematic drawings of an intervertebral disc that show the crisscross arrangement of its fibers. **A.** Concentric layers of the annulus fibrosus are depicted as cut away to show the alternating orientation of the collagen fibers. **B.** The layers of the annular fibers are oriented at a 30° angle to the vertebral body and at 120° angles to each other. *Adapted from White, A.A., III, Panjabi, M.M. (1990). Clinical Biomechanics of the Spine. Philadelphia: JB Lippincott Co.*

to each other in the adjacent bands. This unique orientation confers strength to the annulus while permitting some flexibility (Fig. 11-11). The composition of the collagen in the annulus is approximately 60% type II collagen and approximately 40% type I collagen. As the disc ages, the collagen undergoes irreducible cross-linking and the relative amount of type I collagen increases, replacing type II collagen in the disc.

MECHANICAL PROPERTIES

Vertebrae

The mechanical properties of bone and soft tissues differ, with strength, stiffness, and the relation of stress to strain the primary, measureable attributes underlying their mechanical and functional differences. Stress-strain curves are used to determine the relative loading behavior of bone. Stress is the load per unit area of a perpendicularly applied load. Strain is the change in length per unit of original length, usually expressed as a percentage.

Cortical bone is stiffer than cancellous bone and can withstand greater stresses before failure. When the strain in vivo exceeds 2% of the original length, cortical bone fractures; cancellous bone can withstand somewhat

greater strains before fracturing. The greater ability to withstand strain is because of the structure of cancellous bone: Its porosity varies from 30% to 90% compared with cortical bone, which has values of 5% to 30% (Carter and Hayes, 1977). Vertebral compression strength increases from the upper cervical to the lower lumbar levels.

The mineral content of vertebrae decreases with increasing age at a relatively constant rate (Hansson and Roos, 1986; Hansson et al., 1980) and impacts bone fragility. With the development of osteoporosis, often occurring in late middle age and in the elderly, abnormal bone loss ensues, placing people at risk for facture from otherwise minor trauma and falls or "silent" vertebral compression. A 25% decrease in osseous tissue results in a greater than 50% decrease in the strength of the vertebrae (Bell, 1967). Because the cortical shell of a vertebra is responsible for only approximately 10% of its strength during compression, good-quality cancellous bone is critically important (McBroom et al., 1985).

When bone is loaded in vivo, contraction of the muscles attached to the bone can alter the stress distribution in the bone. Bending moments are applied to the vertebral bodies during motions. During flexion, tensile stresses are applied to the posterior cortex and compression to the anterior cortex of the vertebral body. To perform lifting tasks, typically flexion-extension motions, the back muscles are required to develop considerable forces (Schultz et al., 1982). Stresses in a typical cervical vertebra change from tensile to compressive in a region approximately 0.5 to 1 cm anterior to the posterior longitudinal ligament (Pintar et al., 1995). Because bone is weaker and fails earlier in tension than in compression, posterior paraspinal muscle contraction can decrease the tensile stress on bone by producing a compressive stress that reduces or neutralizes the posterior cortical tensile stresses. This allows the vertebrae to sustain higher loads than would otherwise be possible. However, bone will often fail prior to damage occurring to the intervertebral disc under compressive loading. Finite element modeling of the cervical spine indicates that the increase in end-plate stresses may be the initiating factor for failure of this component under compressive loads (Yoganandan et al., 1996).

Intervertebral Discs

Intervertebral discs exhibit viscoelastic properties (creep and relaxation) and hysteresis (Kazarian, 1975). The term hysteresis is derived from Ὑστέρησις an ancient Greek word meaning deficiency or lagging behind. All viscoelastic structures exhibit hysteresis, a phenomenon in which there is a loss of energy when a structure is subject to repetitive loading and unloading cycles. Creep occurs more slowly in healthy discs than in degenerated or herniated discs, suggesting that degenerated discs are less viscoelastic in nature (Kazarian, 1972). The hysteresis loops are also smaller in older discs.

In the lumbar spine, discs lose and regain approximately 20% of their water every day (Botsford et al., 1994), with most of the loss taking place during the first hour of the morning (Dolan and Adams, 2001). After stress profiling cervical discs, specimens were subjected to sustained creep compressive loading for two hours to expel water from the disc (Skrzypiec et al., 2007). The nucleus pressures were reduced by 17%, 22%, and 37% in the neutral, flexed, and extended postures, respectively. Creep similarly reduced maximum stresses in the posterior annulus by 13%, 20%, and 29%, respectively. Maximum stresses in the annulus (relative to nucleus pressure) tended to increase following creep, and the effects of posture were exaggerated. Creep loading reduces the height of intervertebral discs and transfers loading to the facet joints, and, presumably, to the uncovertebral joints in the cervical spine.

Ligaments

Clinical stability of the spine depends primarily on the soft tissue components, especially in the cervical spine. The spinal ligaments are functional mainly in distraction along the line of their fibers. Ligament strength and limited extensibility help maintain stability, especially around the craniocervical junction. One study found that alar ligaments have an in vitro strength of 200 N, and the transverse ligaments have an in vitro strength of 350 N (Dvorak, Schneider, et al., 1988). Serial ligament transection studies suggest that ligaments that lie close to the intervertebral centers of rotation are stronger and play a critical role in stabilizing the spinal column and protecting the neural tissues from injury (Panjabi et al., 1975).

In another study, the tensile strength of cervical spinal ligaments was measured on a Materials Testing Solution (MTS) system at a distraction speed of 1 cm/s (Myklebust et al., 1988). The tests produced sigmoidal-shaped force-deformation curves that indicated viscoelastic behavior of the spinal ligaments. The apical ligament, which connects the apex of the odontoid of C2 to the occiput, failed at 125 to 423 N, and the alar ligaments, which connect the superior-lateral aspect of the odontoid to the occiput, failed at 231 to 445 N. The anterior longitudinal ligament (ALL) was strongest at the high cervical and the lower thoracic and lumbar regions. The tectorial membrane exhibited an average failure load of 76 N. The anterior and posterior atlanto-occipital membranes demonstrated strengths consistent with their equivalent structures at lower levels, with mean values of 233 N and 83 N, respectively. The vertical cruciate ligament showed an absolute mean strength of 436 N. The strength of the ligaments is related to both the anatomic demands and the flexibility required, which is a classic example of form following function.

The ligaments all have high collagen content except for the ligamentum flavum, which is exceptional in having a large percentage of elastin. The ligamentum flavum is under tension even when the spine is in a neutral position or somewhat extended, and thus, pre-stresses the disc to some degree, which provides some intrinsic support to the spine (Nachemson and Evans, 1968; Rolander, 1966). Myklebust et al. (1988) noted that many spinal ligaments were quite extensible; this, however, being more so in the cervical and lumbar spine than the thoracic. The elastic properties also assist in limiting the inward buckling of these ligaments during extension, which could potentially compress the neural elements.

Traumatic injuries may occur at higher speeds than the common sprains and strains, and injury patterns may differ due to the viscoelastic properties of ligaments. Ivancic et al. (2007) studied cervical bone-ligament-bone specimens (age range: 71–92 years) that were elongated to complete rupture at an average (SD) peak rate of 723 (106) mm/s. They concluded that high-speed elongation may cause cervical ligaments to fail at a higher peak force and smaller peak elongation, and they may be stiffer and absorb less energy, as compared with a slow elongation rate. Panjabi, Crisco, et al. (1998) compared slow and fast elongation rates of alar and transverse ligaments of 11 fresh human cadavers ranging in age from 37 to 53 years (mean, 49 years). The strain and energy absorbed decreased to less than one-tenth, while the stiffness increased greater than tenfold for both the alar and transverse ligaments, as the extension rate increased. There was a similar incidence of failure by ligament tears or bone avulsion at high loading rates, suggesting that bone and ligament probably have comparable strengths at these loading rates. The force-elongation curves for bone-ligament-bone specimens clearly illustrate the viscoelastic properties of the tissues (Fig. 11-12). The strain and energy absorbed by the transverse ligament preparations were lower than for the alar ligaments. Differences may be related to the presence of elastin fibers in the transverse ligaments and their alignment, compared to the almost exclusive parallel-aligned collagen content of the alar ligament (Dvorak, Schneider, et al., 1988).

Muscle

Muscular strength and control is imperative to maintain head and neck balance. In the cervical spine, muscle strength also has a role in reducing stresses

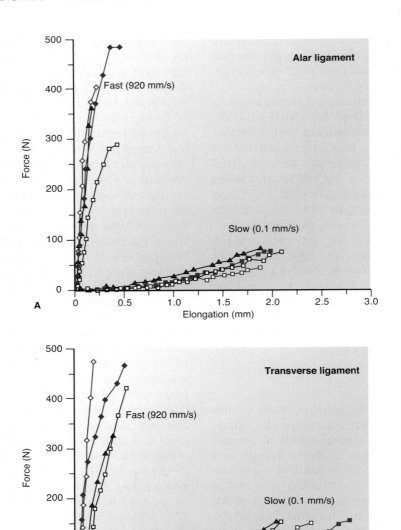

FIG. 11-12

A. Dens-alar ligament-occiput complex force-elongation curves at slow (0.1 mm/s) and fast (920 mm/s) extension rates. **B.** Atlas-transverse ligament-atlas complex force-elongation curves at slow (0.1 mm/s) and fast (920 mm/s) extension rates. *Reprinted with permission from Panjabi, M.M., Crisco, J.J., III, Lydon, C., et al. (1998). The mechanical properties of human alar and transverse ligaments at slow and fast extension rates. Clin Biomech 13, 112–120.*

on bones. During various motions, bending moments are applied to the vertebral bodies. In flexion, tensile stresses are applied to the posterior cortex and compression to the anterior cortex of the vertebral body. Substantial loads on the cervical spine have been calculated during neck flexion, particularly in the lower cervical motion segments.

Harms-Ringdahl (1986) calculated the bending moments generated around the axes of motion of the atlanto-occipital joint and the C7–T1 motion segment in seven subjects with the neck in five positions: full flexion, slight flexion, neutral, head upright with the chin tucked in, and full extension. The load on the junction between the occipital bone and C1 was lowest during extreme extension (ranging from an extension moment of 0.4 Nm to a flexion moment of 0.3 Nm). It was highest during extreme flexion (0.9–1.8 Nm), but this was only a slight increase over that produced when the neck was in the neutral position. The load on the C7–T1 motion segment was low with the neck in the neutral position but became even lower when the head was held upright with the chin tucked in (ranging from an extension moment of 0.8 Nm to a flexion moment of 0.9 Nm). The load increased somewhat during extreme extension (ranging from 1.1–2.4 Nm) and substantially during slight flexion (reaching 3.0–6.2 Nm). The greatest loads were produced during extreme extension, with moments ranging from 3.7 to 6.5 Nm.

In the same study, surface electrode electromyography was used to record activity over the erector spinae muscles of the cervical spine, with the neck in the same five positions described previously. Interestingly, the values obtained showed very low levels of muscle activity for all positions, even during extreme flexion, in which the flexion moment on the C7–T1 motion segment increased more than threefold over the neutral position. The fact that the electromyographic levels over the neck extensors were low in this and other studies (Fountain et al., 1966; Takebe et al., 1974) suggests that the flexing moment is balanced by passive connective tissue structures, such as the joint capsules and ligaments. This phenomenon is seen in many other joints in which passive support is provided by the ligaments.

The values for the moments computed by Harms-Ringdahl (1986), however, are approximately 10% of the maximal values measured by Moroney and Schultz (1985) in 14 male subjects, who resisted maximal and submaximal loads against the head while in an upright sitting position. The mean maximal voluntary moments were 10 Nm during axial rotation of the cervical spine, 12 to 14 Nm during flexion and lateral bending, and 30 Nm during extension. Calculations of the maximum (compressive) reaction forces on the C4–5 motion segment ranged from 500 to 700 N during flexion, rotation, and lateral bending and rose to 1,100 N during extension. Anteroposterior and lateral shear forces reached 260 N and 110 N, respectively. Calculated moments and forces generally correlated well with mean-measured myoelectric activities at eight sites around the perimeter of the neck at the C4 level.

Muscles play a critical role in basic postural homeostasis, as can be observed in both historical and present-day clinical settings. In unique observational studies in the 1950s of severely affected poliomyelitis patients, improvements in respiratory assistance for those with respiratory paralysis led to higher survival rates and a large number of patients who sustained

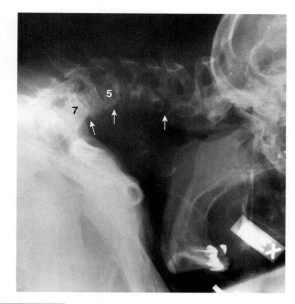

FIG. 11-13

Lateral cervical radiograph of a 68-year-old woman, who presented with severe torticollis. She denied any history of injury and did not have evidence of a structural vertebral abnormality, infection, tumor, or inflammatory disease. Pseudo-subluxations (*arrows*) are evident subaxially as a consequence of the marked kyphosis. Her neck was twisted only because she had a severe cervical flexion deformity, and she was unable to see forward except by turning her head to one side. She was neurologically intact. It was possible to extend her neck to a relatively neutral position using gentle traction. Following posterior fusion from C2 to C7, she returned to a normal independent life. Muscle biopsy was consistent with senile myopathy. *From Moskovich R. Cervical instability (rheumatoid, dwarfism, degenerative, others). In K.H. Bridwell, R.L. DeWald (Eds.). The Textbook of Spinal Surgery (2nd ed.). Philadelphia: Lippincott–Raven Publishers, 969–1009.*

complete paralysis of the cervical musculature. Patients with completely flail cervical spines were unable to support their heads unless adequate support was provided and actually remained in bed despite the good function of their extremities (Perry and Nickel, 1959). Similarly, severe cervical kyphosis occasionally is seen in elderly patients who do not have an obvious structural etiology when investigated radiologically. Some of these patients were found to have marked cervical extensor muscle weakness that has been attributed to senile cervical myopathy (Fig. 11-13) (Simmons and Bradley, 1988).

Neural Elements

Biomechanics of the neural elements has not been as well studied as the biomechanics of the osteoligamentous

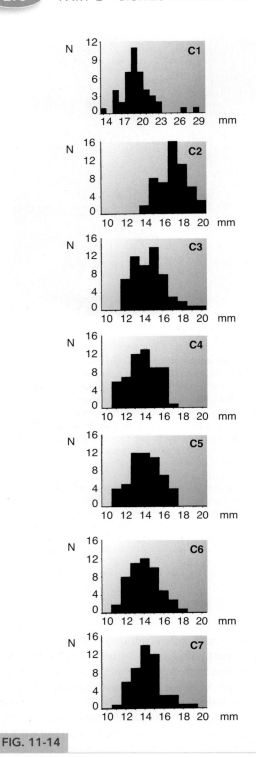

FIG. 11-14

Histograms of the C1 to C7 sagittal canal diameters. Apart from the C1 plot, the same scale is used on all of the horizontal axes so that the distributions of the diameters can be compared. *Reprinted with permission from Moskovich, R., Shott, S., Zhang, Z.H. (1996). Does the cervical canal to body ratio predict spinal stenosis? Bull Hosp Jt Dis, 55, 61–71.*

vertebral column, but our knowledge base is growing. To date, certain basic parameters have been established. The cervical spine undergoes significant changes in length during flexion and extension (Breig et al., 1966; Reill, 1960). Thus, while there is some longitudinal elasticity to the spinal cord, it tolerates axial translation poorly. It is the translatory forces that typically result in neurologic injury. A compressive tolerance between 2.75 and 3.44 kN is estimated for the adult cervical spine before significant neurologic injury occurs (Myers and Winkelstein, 1995).

Spinal cord injuries can also result from extreme or sudden flexion-extension movements, especially in the face of a shallow spinal canal. Head flexion alone has been shown to result in significant increases in the intramedullary spinal cord pressure in canines (Kitahara et al., 1995). Neurologic injuries may result from anteroposterior compression of the spinal cord and are more common if the spinal canal is stenotic. Flexion motions can result in injuries when the spinal cord makes contact with cervical osteophytes, and extension motions may result in a pincer-like compression of the cord between (anterior) osteophytes and (posterior) invaginated ligamentum flavum. Anterior or central spinal cord injuries may ensue.

Although a diagnosis of spinal stenosis may be made based on the absolute size of the spinal canal, imaging of the neuraxis itself may be of greater value. Contrast-enhanced computerized tomography, myelography, and magnetic resonance imaging (MRI) can demonstrate actual impingement or distortion of the spinal cord. Studies performed in flexion and extension may enhance the value of the information by demonstrating the contribution of any dynamic soft tissue component to the impingement. The exact size of the bony cervical spinal canal and the vertebral body was measured in 368 cadaveric adult male vertebrae (Moskovich et al., 1996). This study used well-validated parametric statistical methods to determine that the mean sagittal diameter of the spinal canal for C3 to C7 was close to 14 mm (14.07 ± 1.63 mm; $N = 272$) (Figs. 11-14 and 11-15). The mean ratio of the sagittal canal diameter to the vertebral body diameter (canal to body [c-b] ratio) was 86.68 ± 13.70. Thirty-one percent of subaxial vertebrae would be diagnosed as having spinal stenosis if a c-b ratio of less than 80% were considered abnormal. Another study also found a high false-positive error rate for the c-b ratio, with 49% of 80 asymptomatic football players having a c-b ratio of less than 80% at one or more cervical levels (Herzog et al., 1991, 1991). Yet another group evaluated the reliability of the c-b ratio using plain lateral radiographs and CT scans (Blackley et al., 1999). Results confirmed that a poor correlation exists between the true diameter of

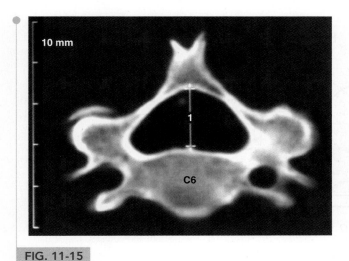

FIG. 11-15

Axial computerized tomographic scan of a sixth cervical vertebra, which was not part of the study detailed in the text. The anteroposterior diameter of the spinal canal measured 13.96 mm in this specimen.

the canal and the c-b ratio. The variability in anatomic morphology means that the use of ratios from anatomic measurements within the cervical spine is not reliable in determining the true diameter of the cervical canal.

Spinal cord injury without radiographic abnormalities (SCIWORA) has also been described, especially in children (Dickman et al., 1991; Osenback and Menezes, 1989; Pang and Pollack, 1989). The etiology of such injuries is unknown, but one mechanism may be longitudinal traction. The unusually elastic biomechanics of the pediatric bony spine allows deformation of the musculoskeletal structures beyond physiologic extremes, permitting direct cord trauma followed by spontaneous reduction of the bony spine (Kriss and Kriss, 1996). The isolated spinal cord resists tension poorly; axial tensile forces to failure for three adult spinal cord specimens were reported to be 278 N ± 90 (Yoganandan et al., 1996). Lower forces may result in direct neural injury or vascular disruption.

Kinematics

Kinematics is study of the motion of rigid bodies without taking into consideration other relevant forces. Kinematics of the spine describes the physiologic and pathologic motions that occur in the various spinal units. The traditional unit of study in kinematics is the motion segment, or the functional spinal unit. As described earlier, each motion segment consists of two adjacent vertebrae and their intervening soft tissues (Fig. 11-16).

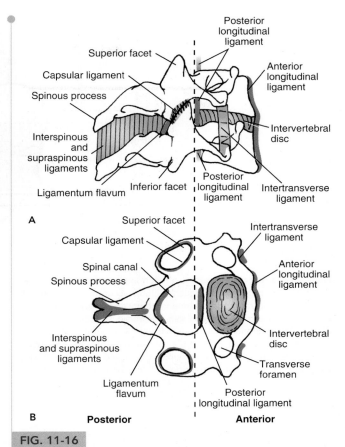

FIG. 11-16

Schematic representations of a cervical motion segment composed of two typical cervical vertebrae (C4 and C5), the intervertebral disc, and surrounding ligaments. The broken line divides the motion segment into anterior and posterior components. **A.** Lateral view. **B.** Superior view. *Adapted with permission from White, A.A., III, Johnson, R.M., Panjabi, M.M., et al. (1975). Biomechanical analysis of clinical stability in the cervical spine. Clin Orthop Relat Res, 109, 85–96.*

Basic biomechanical testing involves the application of forces to a vertebral body and the subsequent measurement of the movements that occur (Fig. 11-17). Movements can be either rotational or translational. A degree of freedom is defined as a motion in which a rigid body can either translate back and forth along a straight line or rotate around a particular axis. Thus, each vertebral body may either translate or rotate in each of three orthogonal planes, for a total of six degrees of freedom (Fig. 11-18) (Panjabi et al., 1981). When either rotation on or translation of a body along one axis is consistently associated with a simultaneous rotation on or translation along another axis, the motions are coupled. Coupled motions are normally expressed as displacements in the

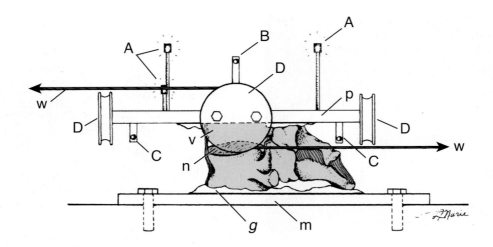

FIG. 11-17

Diagram of a test rig to evaluate a functional spinal unit using video photogrammetry. This technique facilitates accurate measurements of motion without the measurements themselves having an effect on the displacements of the mobile vertebrae. **A.** Light-emitting diodes (*LEDs*). **B** and **C.** Guide bars for application of tensile and compressive forces. **D.** Pulley for application of torques. Weights are attached to guide wires (*w*), which go around pulleys to produce torque on the upper vertebral body (*v*). Intervertebral disc (*n*); acrylic cement (*g*) that attaches lower aluminum plate (*m*) to test rig, which is rigidly bolted to the support frame. The upper plate (*p*) and upper vertebral body (*v*) are the moveable elements to which the loads and torques are applied. LEDs are rigidly attached to the upper plate, and their movement is recorded by two video cameras. *Reprinted with permission from Raynor, R.B., Moskovich, R., Zidel, P., et al. (1987). Alteration in primary and coupled neck motions after facetectomy. Neurosurgery, 21, 681–687.*

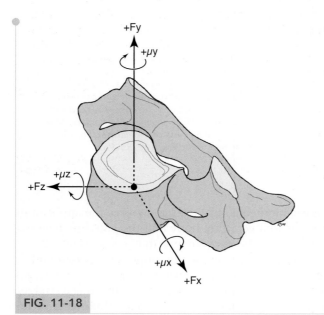

FIG. 11-18

A vertebral body showing the three primary Cartesian axes, x, y, and z. Along each axis a positive force, *+F*, is denoted by the direction of the *arrow*. The *curved arrows* indicate the direction of a positive torque, *+μ*. *Reprinted with permission from Raynor, R.B., Moskovich, R., Zidel, P., et al. (1987). Alteration in primary and coupled neck motions after facetectomy. Neurosurgery, 21, 681–687 (Modified from Raynor et al., 1987).*

X, Y, or Z directions and rotations about the three orthogonal axes. Testing whole spines requires more complex analysis, and has yielded interesting results; analysis of spinal motion segments, however, remains important for basic understanding of spinal biomechanics.

RANGE OF MOTION

Measurements of cervical range of motion are based on radiographic studies or postmortem investigations. Inclinometers and various optoelectronic and electromagnetic devices used clinically for noninvasive evaluation of cervical spine motion are not as accurate; in particular, coupled motion is poorly quantified (Roozmon et al., 1993). The established range of active axial rotation to one side at C1-2 is 27° to 49° (mean = 39°); passive rotation is 29° to 46° (mean = 41°) (Dvorak et al., 1987; Dvorak, Froehlich, et al., 1988; Penning and Wilmink, 1987). These measurements account for approximately 50% of the total cervical rotation.

Another stereoradiographic study of neck motion in men found a mean of 105° axial rotation between the occiput and the C7 vertebra. Seventy percent of the total axial rotation occurred between the occiput and the C2 vertebra. Each motion segment between the C2 and C7 vertebrae averaged from 4° to 8° rotation (Mimura et al., 1989).

Less well appreciated is the fact that a considerable amount of flexion and extension occurs at the C1–2 articulation; 5° to 20° of flexion and extension occur with means of 12° (active) to 15° (passive) (Dvorak, Schneider, et al., 1988).

Approximately 90° of axial rotation takes place in the subaxial cervical spine (C3–C7), about 45° to each side of neutral. Even greater lateral flexion is possible: approximately 49° to each side of neutral, giving a total of about 98°. The range of flexion and extension is approximately 64°, approximately 24° of extension and 40° of flexion. An experimental study on autopsy specimens concluded the motion in each plane is fairly evenly distributed throughout the motion segments (Lysell, 1969). The mean total range of anteroposterior translation in subaxial spinal motion segments is 3.5 ± 0.3 mm, divided unequally: 1.9 mm for anterior shear and 1.6 mm for posterior shear. Lateral shear loading results in a mean total range of lateral motion of 3.0 mm ± 0.3 mm, divided equally between right and left; tension results in 1.1 mm of distraction and compression, 0.7 mm of loss of vertical height (Panjabi et al., 1986). The figures quoted account for radiographic magnification; the actual motion being less. Measurement technique also affects the differences reported. A summary of studies of angular and translational segmental motion is presented in Table 11-1. The values were calculated from the various authors' data, and readers are referred to the original papers for further supportive data (Bhalla and Simmons, 1969; Dvorak, Froehlich, et al., 1988; Frobin et al., 2002; Penning, 1978; Reitman et al., 2004).

Dynamic in vivo imaging studies reveal that segmental motion varies depending on the anatomic level and existing pathology. Kinetic MRIs of cervical spines of symp-tomatic patients in axially loaded, upright neutral (0°), flexion (40°), and extension (–20°) positions revealed, however, that in normal cervical spines, most of the total angular mobility was attributed to C4–5 and C5–6. Mobility was significantly reduced in spondylotic segments in patients with severe disc degeneration (Miyazaki et al., 2008). A three-dimensional MRI study of cervical rotation (in 15° increments) noted statistically significant hypomobility in the C5–6 and C6–7 segments, rather than midcervical (Nagamoto et al., 2009). Coupled motions were maintained, including the spondylotic levels.

The great flexibility of the cervical spine allows the head to be positioned in many ways, permitting one, with equal ease, to gaze at an airplane overhead, glance over one's shoulder, or look for an object under a table. An analysis of the combined motion of the cervical spine using an electrogoniometer produced a remarkably large range of motion: $122° \pm 18°$ of flexion and extension, $144° \pm 20°$ of axial rotation, and $88° \pm 16°$ of lateral flexion (Feipel et al., 1999). All primary motions were reduced with age. Gender had no influence on cervical motion range.

The active range of cervical spine motion required to perform daily functional tasks was studied in healthy adults (Bennett et al., 2000). A cervical spine range-of-motion device was fastened to the subject's head with a Velcro strap, and a magnet was placed on the patient's shoulders to calibrate the instrument for measurement of cervical motion. Of the 13 daily functional tasks performed, tying shoes (flexion-extension, 66.7°), backing up a car (rotation, 67.6°), washing hair in the shower (flexion-extension 42.9°), and crossing the street

TABLE 11-1

Comparison of Reported Mean Values and Standard Deviations of the Intervertebral Angulations for Cervical Flexion and Extension at Each Spinal Level

Intervertebral Angulation	n	C2–3 Angulation		C3–4 Angulation		C4–5 Angulation		C5–6 Angulation		C6–7 Angulation	
		Degrees	s.d.	Degrees	s.d.	Degrees	s.d.	Degrees	s.d.	Degrees	s.d
Bhalla and Simmons (1969)	22	9	0.9	15	1.7	23	1.4	19	1.5	18	1.1
Penning (1978)	20	12	—	18	—	20	—	20	—	15	—
Dvorak, Froehlich, et al. (1988)	28	10	2.5	15	3	19	3.5	20	3.5	19	3.5
Frobin et al. (2002)	*	8.2	3.3	14.2	4.4	16.3	5.2	16.6	6.3	10.9	6.5
Reitman et al. (2004)	140	9.9	3.7	15.2	3.2	16.9	3.8	15.8	4.2	13.5	5.3

*Varies by level: C2–3: $n = 91$; C3–4: $n = 126$; C4–5: $n = 128$; C5–6: $n = 119$; C6–7: $n = 33$

(rotation of head left, 31.7°, and rotation of head right, 54.3°) required the greatest active range of motion of the cervical spine. Of interest, several tasks were not found to produce the degrees of motion expected, and these included reading a newspaper (flexion-extension, 19.9°), writing at a table (flexion-extension, 26.2°), and reaching for objects overhead (flexion-extension, 4.3°). Side-bending was not found to be a significant movement in completion of the tasks but was coupled with rotation in one of the tasks (looking left and right to cross a street).

SURFACE JOINT MOTION

The motion between the joint surfaces of two adjacent vertebrae may be analyzed by means of the instant center technique of Reuleaux. The method may be used to analyze surface motion of the cervical spine during flexion-extension and lateral flexion.

In a normal cervical spine, the instant center of flexion-extension is located in the anterior part of the lower vertebra in each motion segment. Instant center analysis indicates that tangential motion (gliding) takes place between the facet joints as the cervical spine is flexed and extended (Fig. 11-19). A consequence of these motions is that the size of the intervertebral foramina increases with flexion and decreases with extension (Fielding, 1957). These alterations have been quantified in a cadaver study that found there were statistically significant reductions of 10% and 13% in foraminal diameter, at 20° and 30° of extension, respectively. Conversely, in flexion there were statistically significant increases of 8% and 10% at 20° and 30° of flexion, respectively (Yoo et al., 1992). One practical application of these data relates to cervical collars used for the relief of neck pain. Conventional foam collars tend to place patients in slight extension, which may aggravate the symptoms. Turning a foam collar around, with the Velcro and narrow part anterior, puts the neck in slight flexion, which may increase the size of the intervertebral foramina and thereby relieve some of the pressure on an inflamed nerve root.

The instant center of motion of the cervical spine may be displaced as a result of pathologic processes such as disc degeneration or ligament impairment. In such cases, instant center analysis may reveal distraction and jamming (compression) of the facet joint surfaces during flexion-extension instead of gliding (Fig. 11-20).

COUPLED MOTION OF THE CERVICAL SPINE

Atlantoaxial Segment

The coupling characteristics of the atlantoaxial spinal motion segments are particularly important because this area of the neck is extremely mobile. The dens is constrained within the osteoligamentous ring of the atlas, causing the C1–2 lateral masses to articulate similarly to the condyles of

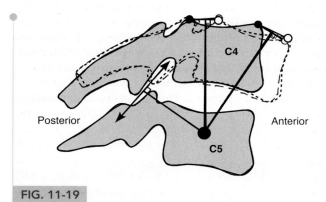

FIG. 11-19

Analysis of the surface motion of the facet joints of the C4–5 motion segment during flexion-extension. The schematic drawing represents superimposed roentgenograms of the motion segment in the neutral position and in slight flexion. The upper vertebra (C4) is considered to be the moving body, and the subjacent vertebra (C5) is the base vertebra. Two points have been identified and marked on the moving body in the neutral position (*solid outline* of C4), and the same two points have been marked on the second roentgenogram with the motion segment slightly flexed (*dashed outline* of C4). Lines connecting the two sets of points have been drawn, and their perpendicular bisectors have been added. The intersection of the perpendicular bisectors identifies the instant center of motion (*large solid dot*) for the degree of flexion under study. The perpendicular bisector (*arrowed line*) of a line drawn from the center of motion to the contact point of the facet joint surfaces indicates tangential motion, or gliding.

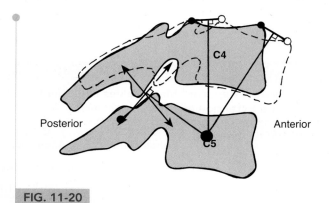

FIG. 11-20

Schematic drawing representative of a roentgenogram of the C4–5 motion segment of a patient injured in a rear-end auto collision. The instant center of flexion-extension at this level (represented by the *large solid dot*) has been displaced from the anterior to the posterior part of C5 as a result of the injury process, which impaired the ligaments (compare with Fig. 11-18). The analysis of surface motion shows compression and distraction of the facet joints with flexion and extension.

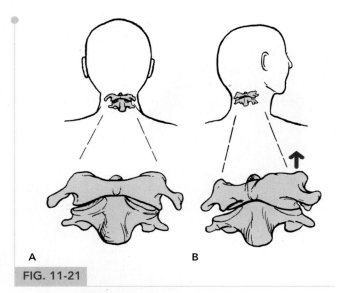

A   **B**

FIG. 11-21

Coupling of rotation and axial translation is depicted schematically. **A.** C1 and C2 are in the neutral position. **B.** C1 rises fractionally on C2 (*arrow*) as the head is rotated away from the midline. *Adapted with permission from Fielding, J.W. (1957). Cineroentgenography of the normal cervical spine. J Bone Joint Surg Am, 39A, 1280–1288.*

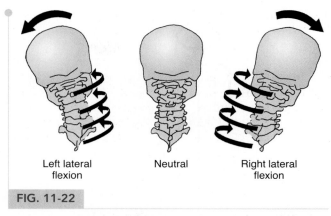

Left lateral   Neutral   Right lateral
flexion   flexion

FIG. 11-22

Coupled motion during lateral bending is depicted schematically. When the head and neck are flexed to the left, the spinous processes shift to the right, indicating rotation. The converse is also illustrated. *Adapted from White, A.A III, Panjabi, M.M. (1990). Clinical Biomechanics of the Spine (2nd ed.). Philadelphia: JB Lippincott Co.*

the knee, with some sliding and rolling during flexion and extension. The instant centers of both rotation and flexion-extension lie in the center of the dens itself. Rotation at C1–2 is coupled with both vertical translation along the Y axis (Fig. 11-21) and a degree of anteroposterior displacement (Werne, 1957). This implies that the C1–2 joint is most stable in the neutral position, and, if rotated, attempts should be made to return it to the reduced position when performing an arthrodesis.

Subaxial Spine

The coupling patterns in the lower cervical spine are such that in lateral bending to the left, the spinous processes move to the right, and in lateral bending to the right, they move to the left (Figs. 11-22 and 11-23) (Lysell, 1969; Moroney et al., 1988). At C2, there are 2° of coupled axial rotation for every 3° of lateral bending, which gives a ratio of 2:3 or 0.67. At C7, there is 1° of coupled axial rotation for every 7.5° of lateral bending, which gives a ratio of 2:15 or 0.13 (White and Panjabi, 1990). Results of finite element modeling indicate that the facet joints and uncovertebral joints are the major contributors to coupled motion in the lower cervical spine and that the uncinate processes effectively reduce motion coupling and primary cervical motion (in the same direction as load application), especially in response to axial rotation

and lateral bending loads. Uncovertebral joints appear to increase primary cervical motion, showing an effect on cervical motion opposite to that of the uncinate processes (Clausen et al., 1997). Coupling of flexion-extension with transverse translation may be visualized radiographically (Fig. 11-24). During flexion, the vertebral body normally shifts forward; the facets glide up and over one another with pseudosubluxation. Changes in normal coupling patterns occur following pathologic changes or surgical intervention.

Tensile load application to the cervical spine occurs in therapeutic traction, but more commonly so during trauma. Deployment of passive vehicular restraint systems, such as airbags, may induce tensile forces in the neck. Isolated intervertebral discs fail at 569 N ± 54, and intact human cadaver cervical spines fail at 3,373 N ± 464 (Yoganandan et al., 1996). Active muscular contraction, however, is likely to raise these figures considerably.

Abnormal Kinematics

Abnormal kinematics generally refers to excessive motion within functional spinal units; however, abnormal kinematics may also refer to atypical patterns of motion, such as abnormal coupling or paradoxical motion. Paradoxical motion is seen when the overall pattern of motion of one aspect of the spine is in one direction and the local pattern is in the opposite. For instance, paradoxical flexion is seen when flexion occurs at a single functional spinal unit although the spine as a whole is extended. These types of abnormal motions describe a pattern of movement known as instability.

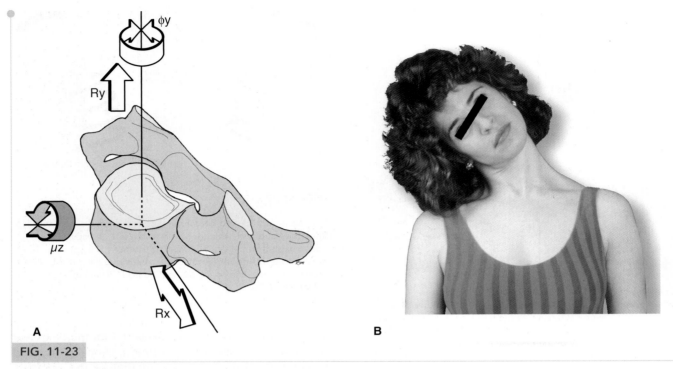

FIG. 11-23

A. This diagram illustrates some of the other coupled motions, which occur in response to a torque (*μz*) about the z-axis (lateral bending). Lateral translation (*Rx*) and vertical motion (*Ry*) occur, as well as horizontal rotation (*Φy*), which results in motion of the spinous processes to the right or the left. **B.** The subject is bending to her right, demonstrating the large range of normal cervical motion possible (approximately 50°).

SPINAL STABILITY

The concept of spinal stability is an intriguing and sometimes confusing notion. Medical practitioners are frequently asked to look at a series of radiographs and make a determination as to whether the spine is stable. Stability is determined by many factors. There are different anatomic considerations in different regions of the spine. Certainly, ligamentous anatomy plays a large part in the stability of the spine, but the muscular and bony elements of the spine also play important roles. Exactly what is stability, how is it determined, and what happens if it is not present? The term spinal stability has acquired different meanings, depending on the setting in which it has been used. White and Panjabi (1990) describe the term clinically as the ability of the spine under physiologic loads to maintain its pattern of displacement, so that there is no initial or additional neurologic deficit, no major deformity, and no incapacitating pain. Instability can be analyzed by considering kinematic instability and structural or component instability. Kinematic instability focuses either on the quantity of motion (too much or too little) or the quality of motion present (alterations in the normal pattern), or

both. Component instability addresses the clinical biomechanical role of the various anatomic components of the functional spinal unit. In this type of instability, loss or alteration of various anatomic portions determines the presence of instability (Box 11-1).

Sir Frank Holdsworth's account (1963) of a simple two-column concept of spinal stability provided a constructive basis for describing and analyzing the basic biomechanics of the spine. The synarthroses between the vertebral bodies rely for their stability on the strong annulus fibrosus. The diarthrodial apophyseal joints are stabilized by the capsule, by the interspinous and supraspinous ligaments, and by the ligamenta flava. This group of ligaments is called the posterior ligament complex; The spine largely depends on this ligamentous complex for stability.

Subsequently, Denis (1983) described a classification system for thoracolumbar fractures that can also be applied to a biomechanical analysis of spinal stability. In this description, the spinal elements are divided into three regions that form three spinal columns:

1. The anterior column consists of the anterior longitudinal ligament, the anterior annulus fibrosus, and the anterior half of the vertebral body.

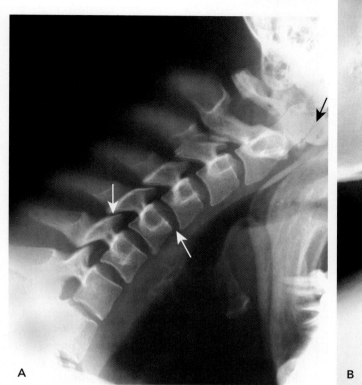
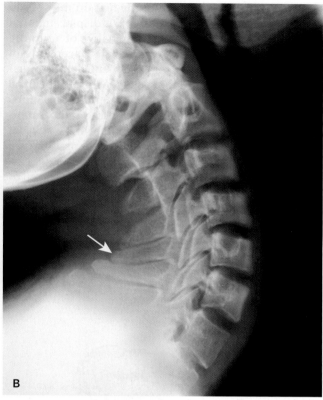

A

B

FIG. 11-24

Coupling of flexion-extension with transverse translation of the cervical spine is visible radiographically. **A.** During flexion, the vertebral body shifts forward (*small white arrow*); the facets glide up and over one another with moderate subluxation at full flexion (*large white arrow*). Up to 2.5 mm of transverse translation may normally occur at the C1–2 articulation during flexion-extension; no translation is evident in this example (*black arrow*). **B.** During extension, the reverse occurs; and the spinous processes limit motion, as they touch at full extension (*arrow*). The size of the intervertebral foramina increases with flexion and decreases with extension.

BOX 11-1 Conceptual Types of Instability

Kinematic instability
  Motion increased
  Instantaneous axes of rotation altered
  Coupling characteristics changed
  Paradoxical motion present
Component instability
  Trauma
  Tumor
  Surgery
  Degenerative changes
  Developmental changes
Combined instability
  Kinematic
  Component

2. The middle column consists of the posterior longitudinal ligament, the posterior half of the vertebral body, and the posterior annulus fibrosus.

3. The posterior column consists of the pedicles, facet joints, laminae, and spinous processes, as well as the interspinous and supraspinous ligaments. Their functional roles are not mutually exclusive, but the anterior and middle columns form the primary weight-bearing zone of the spine, with the posterior column providing the guiding and stabilizing elements.

Occipitoatlantoaxial Complex

The transverse ligament of the atlas completes the socket into which the dens is inserted. The ligament allows the dens to rotate but limits its anterior translation. The ligament is nonelastic and will not permit more than 2 to

3 mm of anterior subluxation of the first on the second vertebra (Fielding et al., 1974). Anterior displacement of C1 on C2 of 3 to 5 mm is usually indicative of a rupture of the transverse ligament, whereas displacements of 5 to 10 mm suggest accessory ligament damage; displacement greater than 10 mm occurs with rupture of all the ligaments (Fielding et al., 1976). Anterior translations or displacements of C1 on C2 are assessed radiographically by measuring the distance from the anterior ring of the atlas to the back of the dens (atlantodental interval or ADI) (Fig. 11-25) (Case Study 11-1). Posterior subluxation of the atlas can occur only if the dens is fractured or if there is an os odontoideum or hypoplastic dens.

Some diseases can weaken or destroy the transverse ligament. Most notably, synovitis in rheumatoid arthritis can create a pannus, which helps to destroy

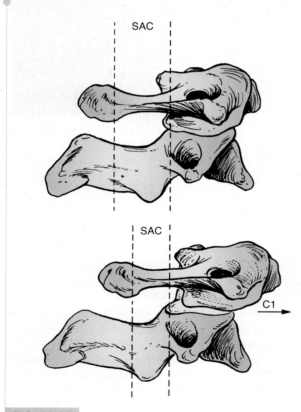

FIG. 11-25

The atlantodental interval (*ADI*) is inversely related to the space available for the spinal cord (*SAC*), which is denoted by the *dotted lines.* Anterior atlantoaxial subluxation causes a reduction in the SAC. Normal measurements for the ADI are less than 3 mm in adults or 4 mm in children. *Reprinted with permission from Moskovich, R. (1994). Atlanto-axial instability. Spine: State of the Art Reviews, 8, 531–549.*

Case Study 11-1

Atlantoaxial Instability without Fracture

A 30-year-old woman had a traumatic injury as a result of a forced flexion movement in a car accident. Continuous and severe neck pain occurred after the accident. She visited the emergency room, and after a careful examination and radiographic evaluation, an anterior displacement of C1 on C2 was discovered (Case Study Fig. 11-1).

A severe anterior dislocation of the atlas on the axis was confirmed after measuring the atlantodental interval (6 mm). For this case, no fracture of the atlas or the axis was detected, and, thus, a deficiency of the transverse ligament may be assumed. The patient returned to normal activities after undergoing a posterior C1–2 arthrodesis.

Clinical instability of the spine depends mainly on the soft tissue components. The cervical spine is a very mobile area, especially at the atlantoaxial level. Cervical subluxations and dislocations resulting from injuries of the osteoligamentous complex affect spinal stability and mobility. In addition, subluxation may narrow the spinal canal and cause neurologic impairment. Because pure ligamentous atlantoaxial injuries are unlikely to heal and stabilize, surgical treatment should be considered.

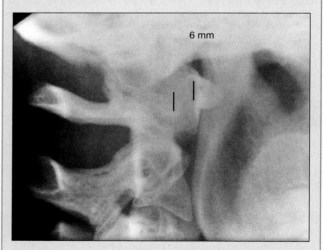

Case Study Figure 11-1 Lateral radiograph demonstrating an increased atlantodental interval of 6 mm after trauma.

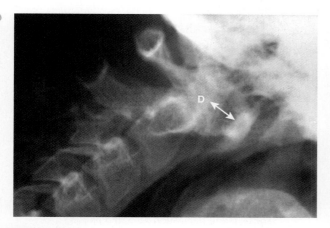

FIG. 11-26

Flexion lateral radiograph of a patient with rheumatoid arthritis. The dens (*D*) has been eroded and the transverse ligament is incompetent, resulting in atlantoaxial subluxation. The markedly widened atlanto-dens interval is indicated by the *arrowed line*.

the atlantoaxial articulation as well as the transverse ligament (Figs. 11-26 and 11-27). Patients with Down syndrome are also susceptible to weakened transverse ligaments and must be carefully assessed clinically and radiographically before being allowed to participate in sporting events such as the Special Olympics.

Steel's rule of thirds (1968) is a guide to the amount of atlantoaxial displacement that can occur before spinal cord compression ensues. The internal anteroposterior diameter of the atlas is approximately 3 cm; of that, the dens occupies approximately 1 cm and the spinal cord approximately 1 cm, leaving 1 cm of space for soft tissue and for normal movement to occur (Fig. 11-28).

Subaxial Cervical Spine

Bailey (1963) stated that the musculature of the spine and the intervertebral discs were the most significant anatomic structures providing cervical stability. Holdsworth (1963) emphasized the importance of the supraspinous and interspinous ligaments as well as the nuchal ligament. The nuchal ligament is thought to play a major role in proprioception and correct functioning of the erector spinae muscles. Experimental section of the ligaments in sequence from either anterior to posterior or posterior to anterior suggests that if a functional spinal unit has all of its anterior elements plus one additional structure or all of its posterior elements plus one anterior structure intact, it will probably remain stable under normal physiologic loads. To provide for some clinical margin of safety, any motion segment should

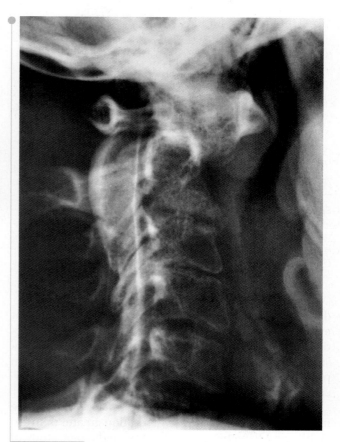

FIG. 11-27

Postmyelogram cervical radiograph of a patient with long-standing rheumatoid arthritis and fixed atlantoaxial subluxation. The space available for the cord (*SAC*) has been reduced to 6.5 mm, and compression of the proximal spinal cord is evident by examining the subdural space, which is outlined in white by the injected contrast medium. The patient developed cervical myelopathy, which necessitated decompression by transoral resection of the dens and posterior stabilization.

be considered unstable in which all of the anterior elements or all the posterior elements are destroyed or are unable to function (Panjabi, White, and Johnson 1975; White et al., 1975). The clinical stability of various injuries must be assessed individually. The importance of clinical evaluation cannot be underestimated, because significant spinal cord damage may occur after trauma, even in the absence of fractures or ligamentous injuries (Gosch et al., 1972; Schneider et al., 1954). Valuable guidelines for the determination of clinical instability in the lower cervical spine have been provided in the form of a scoring system checklist (Box 11-2).

Using this scale, the measurement of translation takes into account variations in magnifications and is based on a tube-to-film distance of 183 cm. The 11° rotation is

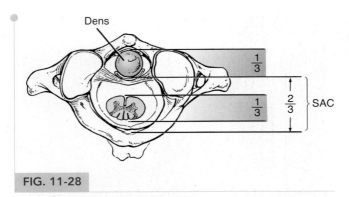

FIG. 11-28

Space available for the cord (*SAC*) is approximately two-thirds of the anteroposterior diameter of the spinal canal. One-third is taken up by the dens, one-third by the cord itself, and one-third is free space. *Reprinted with permission from Moskovich, R. (1994). Atlanto-axial instability. Spine: State of the Art Reviews, 8, 531–549.*

---

defined as 11° greater than the amount of rotation that exists at the motion segment above or below the functional spinal unit in question. The 3.5-mm value represents the radiographic measurement of the maximum permissible translation when the radiographic magnification is taken into account (Panjabi et al., 1986).

A criterion for identifying abnormal cervical spine motion has been intervertebral rotation greater than 20° (White and Panjabi, 1990). In a large radiologic study of healthy volunteers, Reitman et al. (2004) found that 72 of the 644 levels analyzed (11%) demonstrated greater than 20° of intervertebral rotation. Analysis of coupled motions revealed relatively little variation in the ratio of the anterior displacements to intervertebral rotations (0.17 mm/degree, SD = 0.04).

Shear between vertebrae was also calculated. The authors concluded the guideline of 2.7 mm of shear (equivalent to 3.5 mm when evaluated with radiographic magnification) suggested by previous investigators may be too low for all but the C6–7 level (White et al., 1975). Based on Reitman et al. (2004), acceptable normal values for shear at C2–3, C3–4, and C5–6 would approximate 3.5 mm (or 4.5 mm in a standard 30% magnified film). At C4–5, there is even greater motion, and acceptable values could be considered as high as 4.2 mm (or 5.5 mm in a standard 30% magnified film). The treating physician must use good clinical judgment and be cognizant of the increased values for acceptable intervertebral translation when applying the checklist quoted in Box 11-2.

In case of questionable injury, when flexion and extension maneuvers should not be performed, a stretch test can be done to assess cervical integrity. A lateral cervical spine radiograph is taken at a standardized tube distance of 180 cm, and incremental

---

BOX 11-2 Checklist for the Diagnosis of Clinical Instability in the Middle and Lower Cervical Spine

Element	Point Value[a]
Anterior elements destroyed or unable to function	2
Posterior elements destroyed or unable to function	2
Positive stretch test	2
Radiographic criteria	4
Flexion and extension radiographs	
Sagittal plane translation >3.5 mm or 20% (2 pts)	
Sagittal plane rotation >20° (2 pts)	
or	
Resting radiographs	
Sagittal plane displacement >3.5 mm or 20% (2 pts)	
Relative sagittal plane angulation >11° (2 pts)	
Developmentally narrow spinal canal	1
Abnormal disc narrowing	1
Spinal cord damage	2
Nerve root damage	1
Dangerous loading anticipated	1

[a]Total of 5 or more = clinically unstable.
Note that acceptable sagittal plane displacement may be up to 4.5 mm at C2–3, C3–4, and C5–6 on plain radiographs, and 3.5 mm at C6–7 (Reitman et al., 2004).
Modified from White, A.A., III and Panjabi, M.M. (1990). *Clinical Biomechanics of the Spine (2nd ed.)*. Philadelphia: JB Lippincott Co, 314.

5-kg weights are applied as traction to the skull using cranial tongs (Fig. 11-29). Radiographs are taken after each additional weight has been applied. An abnormal stretch test is defined as differences of greater than 1.7 mm interspace separation or greater than 7.5° change in angle between the pre-stretch condition and the application of one-third body weight.

Applied Biomechanics

A thorough understanding of biomechanical principles is an important aspect of the treating physician's knowledge base because the normal structure and function of the spinal column is frequently altered during surgery. Whether treatment is a decompressive cervical laminectomy, a posterior foraminotomy with partial facetectomy,

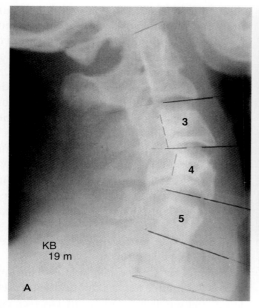
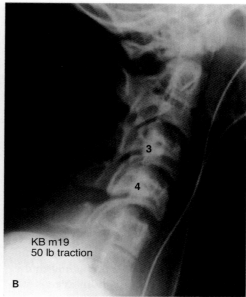

FIG. 11-29

Stretch test. **A.** Lateral radiograph of the cervical spine of a 19-year-old man who was admitted with multiple injuries and neurologic deficit compatible with an anterior cord syndrome. The radiograph shows increased angulation at C3–4 and a block vertebra at C5–6. Flexion-extension radiographs were contraindicated for fear of exacerbating his neurologic injury. **B.** The stretch test was performed to ascertain whether significant instability existed. No abnormal distraction occurred at the interspace in question. His other injuries and fractures were treated routinely, and he was given a soft collar to wear for 6 weeks. He made a slow but steady recovery, with almost complete resolution of his upper extremity deficit and no evidence of instability 1 year after the accident. *Reprinted with permission from Moskovich, R. (1997). Cervical instability (rheumatoid, dwarfism, degenerative, others). In K.H. Bridwell, R.L. DeWald (Eds.). The Textbook of Spinal Surgery (2nd ed.). Philadelphia: Lippincott–Raven Publishers, 969–1009.*

or an anterior cervical fusion, all of these interventions have certain ramifications with which one must be familiar. This knowledge not only benefits patient care but also is valuable in planning and executing treatment.

DECOMPRESSION

Cervical laminectomy often is performed to decompress the spinal cord. The compression may be caused by a stenotic process and can result in neurologic symptoms such as radiculopathy or myelopathy. Other posterior decompressive procedures, such as partial or full facetectomies for visualization or decompression of nerve root pathology, also are commonly performed. Development of postlaminectomy kyphosis is well known in children and may develop in 17% to 25% of adults (Herkowitz, 1988). Postlaminectomy spinal deformity may occur in up to 50% of children who undergo laminectomies for spinal cord tumors (Lonstein, 1977). Simulated finite element analysis on cervical spines indicates that the primary cause of postlaminectomy deformity is resection of one or more spinous processes,

as well as the posterior ligamentous structures, such as the ligamentum flavum or interspinous or supraspinous ligaments. The removal of these structures causes the tensile forces normally present in the cervical spine to become unbalanced and places extra stress on the facet joints. Results indicate that either a kyphotic or a hyperlordotic cervical deformity may ensue, depending on the center of balance of the head (Saito et al., 1991).

Although reports exist of patients undergoing multiple-level cervical laminectomy with no evidence of clinical instability or deformity on long-term follow-up (Jenkins, 1973), most biomechanical studies indicate some degree of instability when the posterior elements are resected. Multilevel cervical laminectomy induces significant increases in total column flexibility associated with increased segmental flexural sagittal rotations. In a cadaveric laminectomy model, the mean stiffness of the intact cervical column was significantly greater than the mean stiffness for the laminectomized specimen, and there were consistently greater rotations as compared with the intact specimen (3.6° vs 8.0°) at every cervical spine level (Cusick et al., 1995).

The loss of facet joints alone causes a significant decrease in coupled motions that result from lateral bending. A moment about the anteroposterior axis results in a significant reduction in lateral displacement, a decrease in vertical displacement, and a decrease in rotation about the vertical axis. Partial facetectomy (<50%) did not, however, significantly alter flexion and extension movements (Raynor et al., 1987). Another anatomic study demonstrated that progressive laminectomy with resection of more than 25% of the facet joints resulted in significantly increased cervical flexion-extension, axial torsion, and lateral bending motion when compared with the intact spine (Nowinski et al., 1993).

Several studies using three-dimensional finite element models showed that facetectomy has a greater effect on annulus stress than on intervertebral joint stiffness. Based on these models, it was concluded that a significant increase in annulus stresses and segmental mobility may occur when bilateral facet resection exceeds 50% (Kumaresan et al., 1997; Voo et al., 1997). Decompression by cervical laminoplasty, in which the facet joints are not sacrificed and the laminae are reconstructed, results in maintenance of flexion-extension and lateral bending stability, with a marginal increase in axial torsion. Iatrogenic injury is less likely to result if the capsules of the remaining facet joints and anterior elements remain intact.

Cervical subluxations and dislocations resulting from injury may narrow the spinal canal and cause neurologic impairment. In some cases, adequate reduction and realignment of the vertebrae, followed by stabilization, may decompress the neural elements without resecting bone (Fig. 11-30).

ARTHRODESIS

Spinal arthrodesis is indicated in many disease processes, such as spinal instability, neoplasm, and post-traumatic and degenerative conditions of the spine. The goal of arthrodesis is to achieve a solid bony union between two or more vertebrae. In many cases, internal fixation is used to achieve initial stabilization as well as to correct deformity.

An important principle regarding vertebral arthrodesis is that the stability established by internal fixation is a prelude to the biologic process of fusion. The ideal biologic environment for arthrodesis is influenced by several factors. Mechanical protection of the graft in the intervertebral space may increase the fusion rate and maintain structural alignment. Internal fixation by no means supplants the need for the surgeon to perform a thorough and careful preparation of the vertebrae and use optimal grafting techniques. With few exceptions, internal fixation that is not ultimately supported and

protected by a solid fusion will fatigue and fail after a finite number of cycles. The "race" is to achieve a solid fusion before fatigue failure of the fixation occurs.

The choice of a surgical approach to the cervical spine, as well as whether an anterior, posterior, or combined arthrodesis should be performed, depends on the particular pathology. When performing a fusion, it is important for the surgeon to understand the biomechanical properties of different types of fusion constructs. By limiting local motion, cervical arthrodesis affects adjacent motion segments. The effect of fusion may be mitigated by the fact that the motion segment itself may already be ankylosed or extremely stiff as a result of degeneration or pathology. Theoretically, there is an increase in motion at nearby nonfused levels. Subsequent degeneration may occur at other motion segments (Cherubino et al., 1990; Hunter et al., 1980).

Fuller et al. (1998) evaluated the distribution of motion across the mobile cervical motion segments after a simulated segmental arthrodesis in cadaver cervical spines. The authors simulated one-, two-, and three-level fusions in human cervical spines. They then moved the cervical spines through a nondestructive 30° sagittal range of motion and compared this range of motion with that of nonfused cervical spines. The findings of this study were interesting in that sagittal plane rotation was not increased disproportionately at the cervical motion segments immediately adjacent to a segmental arthrodesis. Although the authors acknowledged certain limitations of the study, they proposed that a cervical fusion causes a fairly uniform increase in motion across all remaining open cervical motion segments; therefore, an increased potential for degenerative change may exist at all cervical levels.

Another study was performed describing the incidence, prevalence, and radiographic progression of symptomatic adjacent level disease after cervical arthrodesis (Hilibrand et al., 1999). Adjacent level disease was defined as the development of a new radiculopathy or myelopathy that was referable to a motion segment adjacent to the site of a prior anterior cervical arthrodesis. The findings revealed that symptomatic adjacent level disease occurred at a relatively constant incidence of 2.9% per year. A survivorship analysis revealed that approximately 26% of patients who had an anterior cervical arthrodesis would have new disease at an adjacent level within 10 years of the operation. The study also demonstrated that more than two-thirds of the patients who developed adjacent-level cervical disease experienced failure of nonoperative treatment and needed an additional procedure performed.

Total disc arthroplasty is intended to preserve physiologic motion of the spine and reduce the stress transfer patterns to other functional spinal units induced by

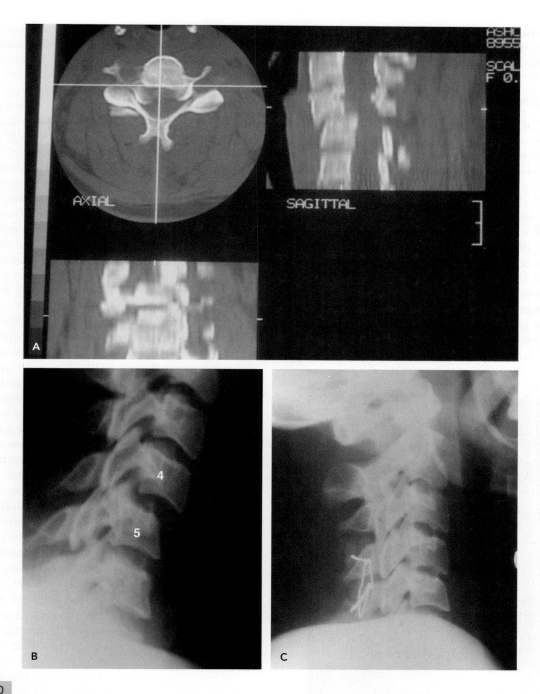

FIG. 11-30

Unilateral facet dislocation in a 24-year-old woman who was involved in a motor vehicle accident. The degree of vertebral subluxation is less than half the AP diameter of the vertebral body. Her spinal cord was compressed, and she presented with an incomplete spinal cord injury. **A.** Computerized tomographic scans and reformatted images demonstrated a canal compromise at C5–6. An attempt to realign the spine was made using longitudinal traction of up to approximately one-third of her body weight. **B.** The lateral radiograph with traction applied shows persistent malalignment. An open reduction was performed using a posterior exposure of the vertebra. Once the spine was realigned, the posterior tension band was recreated using interspinous wiring, and autogenous bone graft was inserted. **C.** The postoperative radiograph demonstrates restoration of normal vertebral relationships. The fixation provided good stability and enabled the patient to mobilize early. A good clinical outcome was achieved. *Reprinted with permission from Moskovich, R. (1997). Cervical instability (rheumatoid, dwarfism, degenerative, others). In K.H. Bridwell, R.L. DeWald (Eds.). The Textbook of Spinal Surgery (2nd ed.). Philadelphia: Lippincott–Raven Publishers, 969–1009.*

arthrodesis (Bartels et al., 2008; Galbusera et al., 2008). Long-term studies will be vital to inform us whether it is possible to alter the natural history of the aging spine.

Cervical Spine Fixation

Arthrodesis of the cervical spine may be indicated various reasons, most commonly for trauma and degenerative diseases. Much research has been performed to analyze the biomechanical advantages of anterior approaches, posterior approaches, or a combined procedure. With the advent of newer technologies, internal fixation systems have become available that can satisfactorily stabilize the cervical spine from either approach.

Anterior discectomy or vertebrectomy at one or more levels is usually followed by an anterior cervical arthrodesis. The excised disc or bone must be replaced with a structural graft or prosthesis to restore anterior column support. Osseous replacement may be in the form of autogenous or allogenic bone, commonly from the iliac crest if autogenous or from the fibula or iliac

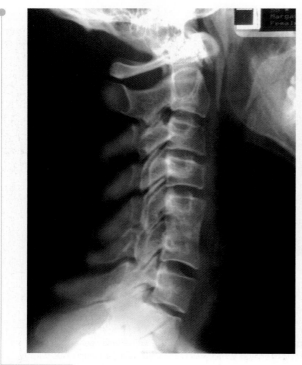

FIG. 11-31

Lateral radiograph of the cervical spine of a 35-year-old woman who had an anterior cervical discectomy at C5–6 and interbody arthrodesis using autogenous iliac crest tricortical bone graft. Note the maintenance of lordosis at the fused segment and the integration and remodeling of the graft.

crest if allogenic (Fig. 11-31). Tricortical iliac crest bone graft, formerly the gold standard, is associated with high donor site morbidity. The more cortical nature of fibula grafts may result in delayed incorporation compared with iliac crest grafts. The immediate postoperative strength of any of these bone grafts under axial compression on an MTS reveals that they will adequately support the loads required in the cervical spine. The physical properties of human bank bone seem to be preferable to autologous bone grafts, especially in older patients (Wittenberg et al., 1990). The biologic incorporation of allograft appears to be satisfactory, and its use obviates the need to harvest autologous bone from the iliac crest, sparing patients additional surgical trauma and potential complications. The volume-related stiffness of several titanium, carbon fiber, and PEEK (polyetheretherketone) cages was higher than that of iliac bone graft when tested biomechanically in a mature goat cervical spine model (Gu et al., 2007).

Calcium phosphate (bone) ceramics form a strong bond with host bone because of a zone of apatite microcrystals deposited perpendicular to the hydroxylapatite ceramic surface (Jarcho, 1981). Synthetic hydroxylapatite blocks used for interbody cervical arthrodesis in goats produced similar fusion rates and biomechanical stiffness when compared with autogenous bone (Pintar et al., 1994). It is interesting to note that the goat holds its head erect and thus loads the cervical spine similarly to human bipeds. Goat models for cervical spine biologic/biomechanical testing are therefore popular.

A canine thoracic anterior arthrodesis model yielded contrary results when tested biomechanically: Autogenous iliac crest grafts were stiffer in all motions than were ceramic graft substitutes (Emery et al., 1996). Naturally grown coral is also a useful bone graft substitute once it has been processed. It is available commercially in two porosities (200-mm and 500-mm pore size). The grafts of lower porosity had a compressive strength comparable with bicortical iliac crest grafts although they were much more brittle. They therefore can be considered appropriate for clinical use with respect to their compressive strength (Table 11-2).

An increasing variety of prosthetic interbody cages are becoming available. Additional anterior (Figs. 11-32 and 11-33) or posterior internal fixation using plates and screws for added support may be used. Two of the factors that influence the tendency of an intervertebral implant to subside are the shape of the implant, especially its contact area at the implant–end plate interface, together with preparation of the end plates (Wilke et al., 2000). The addition of an anterior cervical plate to a cylindrical interbody cage significantly improves segmental stability and subsidence in a porcine cervical model (Hakalo et al., 2008).

TABLE 11-2

Compressive Strength of Various Interbody Graft Materials

Graft Type	Mean ± Standard Deviation	
Fibular strut	5,070 ± 3,250 N	Fibular strut significantly stronger than crest or rib grafts: $P < 0.05$
Anterior iliac crest	1,150 ± 487 N	
Posterior iliac crest	667 ± 311 N	
Rib	452 ± 192 N	
Hydroxylapatite 200-mm pore size	1,420 ± 480 N	Pore size of 200 mm significantly stronger than 500-mm pore size: $P < 0.05$
Hydroxylapatite 500-mm pore size	338 ± 78 N	

Adapted with permission from Wittenberg, R.H., Moeller, J., White A.A. III, (1990). Compressive strength of autogenous and allogenous bone grafts for thoracolumbar and cervical spine fusion. *Spine, 15*(10), 1073–1077.

Posterior arthrodeses of the cervical spine are commonly performed after trauma but may also be used to treat degenerative, inflammatory, or neoplastic conditions. Unlike anterior fixation devices, which are mainly used in the subaxial cervical spine, posterior fixation devices can extend up to the occiput (Fig. 11-34). Posterior wire fixation is often used as a posterior tension band (Sutterlin et al., 1988). Screw and plate or rod fixation techniques are commonly performed to provide stable segmental fixation (Bambakidis et al., 2008; Frush et al., 2009).

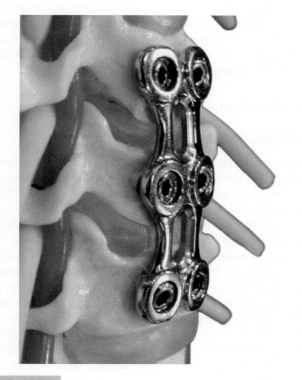

FIG. 11-32

An example of a titanium anterior cervical plate applied over two motion segments on a model of the cervical spine. The screws do not penetrate the posterior cortex and are locked to the plate to prevent backing out.

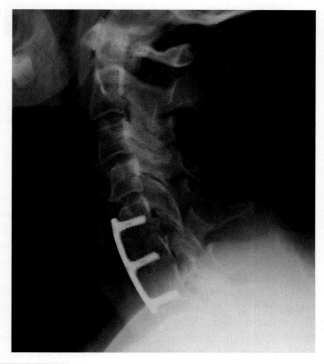

FIG. 11-33

Lateral cervical radiograph of a 62-year-old man eight years after undergoing an anterior cervical arthrodesis at C5–6 and C6–7 with allograft bone and a titanium anterior cervical plate. The proximal intervertebral disc spaces and cervical alignment are maintained.

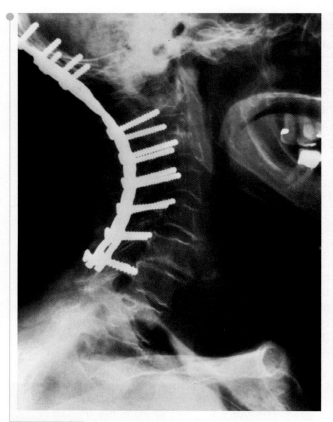

FIG. 11-34

Posterior occipitocervical fixation using a titanium plate and screw fixation. The patient is an elderly woman who had metastatic melanoma to the C2 vertebra. The body and dens were fractured resulting in occipitocervical settling and severe, uncontrollable pain. A halo-vest was applied but did not provide adequate support or pain relief. Posterior segmental fixation permitted immediate postoperative independent mobilization and rapid pain relief.

Techniques for atlantoaxial fixation include wire fixation and interlaminar clamp fixation (Moskovich and Crockard, 1992). Screw fixation techniques are technically more challenging but confer increased translational and axial rotational stability. Both transarticular screw fixation and C1 lateral mass to C2 pedicle screw fixation provide very stable atlantoaxial fixation with resultant high bony union rates (Figs. 11-35 and 11-36). The security of the fixation facilitates monosegmental fusion and reduces or eliminates the need for external support postoperatively. Atlantoaxial fixation in cadaver C1–2 functional spinal units using sublaminar wire with one median graft (Gallie) produced significantly more rotation in flexion, extension, axial rotation, and lateral bending than did wire fixation with two bilateral grafts (Brooks), bilateral posterior interlaminar clamps, and transarticular screw fixation (Magerl) (Grob et al., 1992).

The Magerl transarticular screw fixation tended to permit the least amount of rotation, as might be expected.

Various anterior cervical plating systems are currently available. Investigators demonstrated that in a single-level procedure, an anterior cervical plate serves as a load-sharing device rather than as a load-shielding device, enabling graft consolidation as observed in other clinical studies (Rapoff et al., 1999). Bone union may be expected to occur at a lower rate if the bones are shielded from compressive forces; however, clinical experience with anterior cervical plates has generally demonstrated similar or improved fusion rates compared with stand-alone bone grafts.

Biomechanical stability of seven different cervical reconstruction methods was assessed using 24 calf cervical spine segments (Kotani et al., 1994). The findings do not support the use of exclusive anterior methods in either posterior or three-column instability. Clinical use of anterior plate fixation without posterior fixation for three-column cervical injuries, however, has resulted in satisfactory clinical outcomes in other studies, and modern cervical plates and load-sharing techniques have reduced the need for circumferential fusion except for complex cases (Ripa et al., 1991). These results underscore the need for good clinical evaluation and investigations that take into account the fourth-dimension—time. Fusion is a biologic process that occurs over time and supersedes the importance of in vitro or computer-simulated biomechanical studies.

Biomechanics of Cervical Trauma

AIRBAG INJURIES

Motor vehicle accidents continue to be the leading cause of injury-related deaths in the United States. In 1984, the National Highway Traffic Safety Administration (NHTSA) required that automatic occupant protection devices (airbags or automatic seat belts) be placed in all automobiles in the 1987 to 1990 model years. In 1993, the passenger-side airbag was introduced. Studies generally concluded that front seat occupants are adequately protected against frontal impact if belts are worn in an airbag-equipped vehicle (King and Yang, 1995). Not long after airbag devices became available, airbag injuries involving front seat passengers began to be described, and many child deaths and serious injuries had been attributed to passenger-side airbags (Marshall et al., 1998). Passenger-side airbags pose a lethal threat to children riding in the front seat of an automobile (Giguere et al., 1998; McCaffrey et al., 1999; Mohamed and Banerjee, 1998). The collisions were frequently low-speed accidents in which the driver sustained no or only minor injuries. The pattern of injury in the rear-facing infant

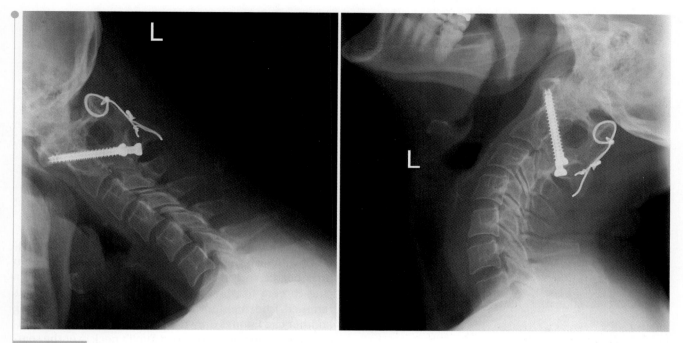

FIG. 11-35

A 55-year-old woman underwent an atlantoaxial fusion; nine years postoperatively she remained able to continue all normal activities. She had rheumatoid arthritis with C1–2 anterior subluxation, which was treated operatively using transarticular screw fixation and posterior arthrodesis. The flexion and extension radiographs demonstrate that the anatomy of the adjacent levels appears normal as does the cervical flexibility and stability.

car seat placed in the front was often massive skull injury and cerebral hematoma as a result of the proximity of their heads to the airbag, whereas in forward-facing child car seats children sustained many more cervical injuries. Two of the older children had autopsy findings of atlanto-occipital dislocation, and one sustained a "near decapitation" injury, demonstrating the vulnerability of the pediatric cervical spine to the explosive forces of an expanding airbag that hyperextends the child's fragile neck. NHTSA (Box 11-3) guidelines emphasize that children of any age should be properly secured in the back seat (NHTSA, 1996).

A mathematical simulation was performed to study the potential of head and neck injury to an unbelted driver restrained by an airbag (Yang et al., 1992). It was found that when the standard 20°-angle steering wheel was used, neck joint torques decreased by 22%. The resultant head acceleration increased 41% from the baseline study when a vertical steering wheel was used. If the vertical dimension of the airbag was reduced by 10%, neck joint torques increased by 14%, while head acceleration showed a slight decrease of 9%. Although ideal dimensions and inflation rates for airbags remain elusive, their use has resulted in a significant overall reduction in head and neck injuries.

WHIPLASH SYNDROME

Whiplash syndrome is a complex set of symptoms that may present after an acceleration hyperextension injury. These injuries typically occur when a car is struck from behind, but may also be caused by side or head-on collisions (Barnsley et al., 1994). The acceleration of the car seat pushes the torso of the occupant forward, with the result that the unsupported head falls backward, resulting in an extension strain to the neck. A secondary flexion injury may occur if the vehicle just struck then strikes another vehicle in front and, just as suddenly, decelerates again, throwing the occupant forward once more. Crowe coined the term "whiplash" in 1928 in a lecture on neck injuries caused by rear-end automobile collisions in the United States. He later reported that he regretted using the term (Breck and Van Norman, 1971) because it describes only the manner in which the head was moved suddenly to produce a sprain in the neck and does not describe a specific injury pattern. Whiplash syndrome is the more correct term but whiplash injury remains an accepted, familiar term used in the literature and by physicians and patients. Although

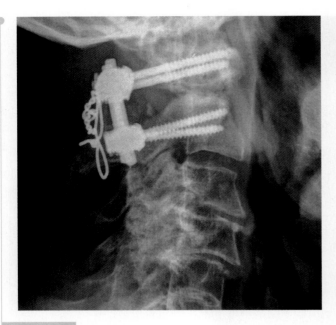

FIG. 11-36

An active 95-year-old man sustained a fracture at the base of the dens after he tripped and fell. The fracture was still mobile after 3 months of immobilization, necessitating operative treatment. He had C1 lateral mass and C2 pedicle screw fixation with autologous bone graft. Full reduction of the displaced fracture with immediate stabilization was achieved. Early mobilization and return to normal function was possible.

whiplash syndrome is a common traumatic event, the pathology is poorly understood. Often the severity of whiplash trauma does not correlate with the seriousness of the clinical problem, which can include neck and shoulder pain, dizziness, headache, and blurring of vision (Brault et al., 1998; Ettlin et al., 1992; Panjabi, Cholewicki, Nibu, Grauer, et al., 1998; Sturzenegger et al., 1994). Apart from a frequently observed loss of physiologic lordosis, a radiographic examination of the cervical spine is often normal. Even newer technology such as MRI is not always able to reveal a soft tissue injury. MRI examination of the cerebrum and cervical spinal column performed two days after a whiplash neck sprain injury in 40 patients did not detect any pathology connected to the injury, nor was the MRI able to predict symptom development or outcome (Borchgrevink et al., 1997). Injuries that have been documented include interspinous ligament tears, spinous process fractures, disc rupture, ligamentum flavum rupture, facet joint disruption, and stretching of the anterior muscles. The diagnosis and management of whiplash injuries are often confounded by concomitant psychosocial and medicolegal issues as well (Wallis et al., 1998).

One of the early (unpublished) biomechanical studies of whiplash was done by the late Dr. Irving Tuell, an orthopaedic surgeon in Seattle, Washington. He used a ciné camera to photograph himself driving as he was rammed from behind by his surgical resident driving another car. Frames drawn from that movie clearly demonstrate the hyperextension of Tuell's neck over the seatback of his car (Fig. 11-37). One can also see the effect of inertial forces on the mandible, as his jaw snaps open while the acceleration forces his head backward. This mechanism may explain the temporomandibular joint injuries that are a common accompaniment of cervical whiplash injuries.

Using clinical and MRI examinations, it was concluded that the "limit of harmlessness" for stresses arising from rear-end impacts with regard to velocity changes lies between 10 and 15 km/h (Castro et al., 1997). After being rammed from behind, the mean acceleration of the target vehicles was from 2.1 to 3.6 g. Maximal extension was reached when the head contacted the headrest; the angle between the head and upper body varied from 10° to 47° (mean, 20°). In the absence of a headrest, the maximal recorded extension was 80°.

The electromechanical delay (EMD), the time between the onset of muscle activity and an external manifestation of the resulting muscle force, is about 10 ± 15 ms. The long EMD times reported in some of the muscle literature are unreliable because they are influenced by unknown factors of the apparatus on which they were recorded (Corcos et al., 1992). Accepting that the data may be obscured by artifact, reaction times for sternocleidomastoid muscle are reported at about 75 to 90 ms (Brault et al., 2000).

Brault et al. (2000) enhanced Dr. Tuell's study by exposing 42 subjects (21 men and 21 women, 20–40 years of age) to collisions of 4 and 8 km/h speed change while measuring the kinematic response of the head and torso and performing electromyography of the sternocleidomastoid

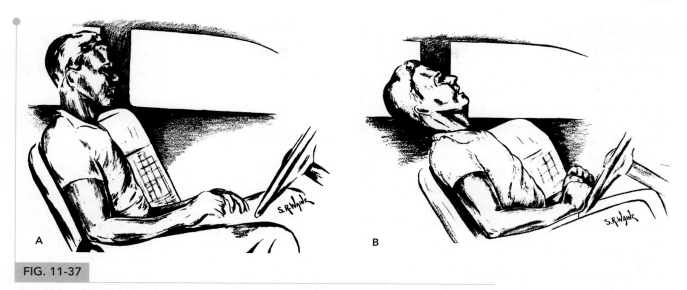

FIG. 11-37

Hand-drawn cells traced from two frames of a ciné movie of Dr. Irving Tuell driving his car. Before **(A)** and after **(B)** being rammed from behind by another car driven by his resident.

and cervical paraspinal muscles. Data indicated that cervical muscles contract rapidly in response to impact, and the potential exists for muscle injury due to lengthening contractions; i.e., muscles fire fast enough to have an influence on injury pattern (Table 11-3; Case Study 11-2).

In other studies, a reproducible whiplash trauma model, using whole cervical spine specimens mounted on a bench-top sled, was used to simulate rear-end collision, with increasing horizontal accelerations applied to the sled (Panjabi, Cholewicki, Nibu, Babat, et al., 1998; Panjabi, Cholewicki, Nibu, Grauer, et al., 1998). Both sled and head kinematics can be measured using potentiometers and accelerometers. Using this whiplash model, an S-shaped curve was described in which the lower cervical spine hyperextended and the upper cervical spine flexed (Grauer et al., 1997). The investigators felt that the injury was incurred during the hyperextension phase in the lower cervical spine.

Correct positioning of adjustable headrests behind the skull, not behind the neck, is important. If the risk of injury is assumed to be proportional to neck extension, a low headrest position carries a relative injury risk of 3.4 in rear-end crashes, compared with 1.0 for the favorable condition (Viano and Gargan, 1996). If all adjustable headrests were placed in the up position, the relative risk would be lowered to 2.4, a 28.3% reduction in whiplash injury risk. An initial head restraint gap of, at most, 5 to 6 cm is needed to reduce neck loads and motions to within the physiologic range (Sendur et al., 2005; Stemper et al., 2006). Active headrests help to automatically counter forces during an accident. They are an improvement on fixed headrests but may not be fully activated at the time of peak spinal motions if the gap between the head and the headrest is greater than 8.0 cm, thus reducing the potential protective effect (Ivancic et al., 2009).

TABLE 11-3			
Muscle Onset Times (ms) from Bumper Contact in Rear-end Collisions as a Function of Gender, Muscle Group, and Speed Changes			
Gender	*Muscle group*	*4 km/h*	*8 km/h*
Female	Sternocleidomastoid	87 (10)	79 (9)
	Paraspinal	94 (12)	82 (7)
Male	Sternocleidomastoid	95 (8)	83 (8)
	Paraspinal	99 (8)	85 (11)

Note: Differences between all three factors were significant at $P < 0.05$.
Standard deviations in parentheses.

Case Study 11-2

Whiplash Syndrome

A 32-year-old man was injured in a car accident when his vehicle was struck from behind. As there was no headrest, his unsupported head fell backward, resulting in an extension strain to the neck.

The patient presented with severe neck and shoulder pain accompanied by sleep disturbance. The injured right sternocleidomastoid muscle was approximately twice as large as the left sternocleidomastoid muscle. After muscle testing against resistance, the pain increased (Case Study Fig. 11-2)

To decelerate the posterior rotating head, a moment and a force must be developed by active and passive stabilizers, joint surfaces, and the intervertebral disc. The moment and force give rise to tension, compression, and shear stresses and strains in various parts of the neck, causing damage. During the collision, the muscle tension increases faster than the velocity of muscle stretch due to lengthening contraction. A tension inappropriate for the length-tension curve occurs and partial rupture of the sternocleidomastoid muscle is produced.

Rest, icing of the muscle, and an extended rehabilitation program were indicated.

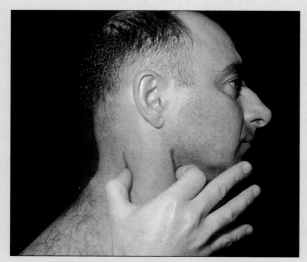

Case Study Figure 11-2

Reprinted with permission from Frankel, V. H. (1972). Whiplash injuries to the neck. In C. Hirsh, Y. Zotterman (Eds.) Cervical Pain. New York: Pergamon Press, 97–111.

Significantly more complex acceleration injuries ascribed to high G-force activities occur in pilots of F-16 fighter planes. A 1-year prevalence of neck injury of 56.6% and a career prevalence of neck injury of 85.4% has been cited (Albano and Stanford, 1998).

Conclusion

Maintenance of neurologic homeostasis and protection of the spinal cord, nerves, and vessels, along with support and protection of the skull, are the ultimate tasks of the cervical spine. An appreciation of the biomechanical principles presented should afford a greater understanding to physicians and allied health professionals involved in the treatment of cervical spine pathology. With increasing technological advances in our society, we remain vulnerable not only to common types of cervical trauma but also to idiosyncratic methods of injury related to these new technologies. Our participation in activities involving high speed and high-risk behavior place us in increasing jeopardy from acceleration and deceleration injury. Cervical spine injuries caused by sports, accidents, and the unintended consequences of devices such as automobile airbags are being critically evaluated. It is vital to pursue rational treatments for these disorders based on sound biomechanical principles.

Summary

- A functional spinal unit or motion segment consists of two adjacent vertebrae and the intervening intervertebral disc and ligaments between the vertebrae.

- Coupled spinal motions are the consistent simultaneous rotation on or translation along a different axis from that of the primary motion.

- Intervertebral discs exhibit viscoelastic properties (creep and relaxation) and hysteresis.

- Discs are able to withstand greater-than-normal loads when compressive forces are rapidly applied, which protects the disc from catastrophic failure until extremely high loads are applied.

- Vertebral body compression strength increases from upper cervical to lower lumbar levels.

- The mean sagittal diameter of the male adult spinal canal for C3–7 is close to 14 mm; the spinal cord diameter is about 10 mm.

- The ligamentum flavum is under tension, even when the spine is in a neutral position or somewhat extended, pre-stressing the disc and providing some intrinsic support to the spine.

- Muscles play a critical role in basic postural homeostasis. Patients with paralyzed cervical muscles are unable to support their heads.

- The spinal cord has some longitudinal elasticity but tolerates axial translation poorly. It is the translatory forces that typically result in neurologic injury.

- Instant center analysis indicates that tangential motion (gliding) takes place between the facet joints as the cervical spine is flexed and extended. The size of the intervertebral foramina increases with flexion and decreases with extension.

- Kinematic instability refers to the quantity of motion (too much or too little) or the quality of motion present (alterations in the normal pattern), or both. Component instability addresses the clinical biomechanical role of the various anatomic structures of the functional spinal unit.

- Any motion segment should be considered unstable in which all of the anterior elements or all the posterior elements are destroyed or unable to function.

- A significant increase in annulus stresses and segmental mobility may occur when bilateral facet resection exceeds 50%.

- Appropriate use of internal fixation helps to increase the fusion rate and maintain structural alignment.

- Front seat occupants are adequately protected against frontal impact if seatbelts are worn in an airbag-equipped vehicle. Passenger-side air bags pose a lethal threat to children riding in the front seat of an automobile.

- Whiplash syndrome is a complex set of symptoms that may present after an acceleration hyperextension injury.

Practice Questions

1. What are the essential differences between the anisotropic and viscoelastic properties of biologic structures?

2. What are the functional demands on the human cervical spinal column (name and describe) and what are the anatomic modifications inherent to the atlantoaxial complex that facilitate them?

3. What biochemical changes occur in the nucleus of the aging intervertebral disc?

4. What is the effect of muscle dysfunction on cervical postural homeostasis and what maintenance techniques would you suggest to optimize healthy function?

5. Where is the location of the instant center of flexion-extension within the motion segments of the subaxial spine?

6. What structures compose each of the so-called three spinal columns?

7. Where should infants and children be seated in a motor vehicle and how are they positioned for ultimate protection? Describe what protection this positioning affords.

8. Along which Cartesian axis is the spinal cord more tolerant to strain?

REFERENCES

Adams, M.A., Hutton, W.C. (1980). The effect of posture on the role of the apophysial joints in resisting intervertebral compressive forces. *J Bone Joint Surg Br, 62B*, 358–362.

Albano, J.J., Stanford, J.B. (1998). Prevention of minor neck injuries in F-16 pilots. *Aviat Space Environ Med, 69*, 1193–1199.

Bailey, R.W. (1963). Observations of cervical intervertebral disc lesions in fractures and dislocations. *J Bone Joint Surg Am, 45A*, 461.

Bambakidis, N.C., Feiz-Erfan, I., Horn, E.M. et al. (2008). Biomechanical comparison of occipitoatlantal screw fixation techniques. *J Neurosurg Spine, 8*, 143–152.

Barnsley, L., Lord, S., Bogduk, N. (1994)., Whiplash injury. *Pain, 58*, 283–307.

Bartels, R.H., Donk, R.D., Pavlov, P., et al. (2008). Comparison of biomechanical properties of cervical artificial disc prosthesis: A review. *Clin Neurol Neurosurg, 110*, 963–967.

Bell, G.H. (1967). Variation in the strength of vertebrae with age and their relation to osteoporosis. *Calcif Tissue Res, 1*, 75.

Bennett, S.E., Schenk, R.J., Simmons, E.D. (2002). Active range of motion utilized in the cervical spine to perform daily functional tasks. *J Spinal Disord Tech, 15*(4), 307–311.

Bhalla, S.K., Simmons, E.H. (1969). Normal ranges of intervertebral-joint motion of the cervical spine. *Can J Surg, 12*, 181–187.

Blackley, H.R., Plank, L.D., Robertson, P.A. (1999). Determining the sagittal dimensions of the canal of the cervical spine. The reliability of ratios of anatomical measurements. *J Bone Joint Surg Br, 81B*, 110–112.

Borchgrevink, G., Smevik, O., Haave, I., et al. (1997). MRI of cerebrum and cervical columna within two days after whiplash neck sprain injury. *Injury, 28*, 331–335.

Botsford, D.J., Esses, S.I., Ogilvie-Harris, D.J. (1994). In vivo diurnal variation in intervertebral disc volume and morphology. *Spine, 19*, 935–940.

Brault, J.R., Siegmund, G.P., Wheeler, J.B. (2000). Cervical muscle response during whiplash: Evidence of a lengthening muscle contraction. *Clin Biomech, 15*, 426–435.

Brault, J.R., Wheeler, J.B., Siegmund, G.P., et al. (1998). Clinical response of human subjects to rear-end automobile collisions [published erratum appears in Arch Phys Med Rehabil 1998 Jun;79(6):723]. *Arch Phys Med Rehabil, 79*, 72–80.

Breck, L. W., Van Norman, R. W. (1971). Medicolegal aspects of cervical spine sprains. *Clin Orthop, 74*, 124–8:124–128.

Breig, A., Turnbell, I., Hassler, O. (1966). Effect of mechanical stress on the spinal cord in cervical spondylosis. *J Neurosurg, 25*, 45–56.

Carter, D.R., Hayes, W.C. (1977). The compressive behavior of bone as a two-phase porous structure. *J Bone Joint Surg Am, 59A*, 954–962.

Castro, W.H., Schilgen, M., Meyer, S., et al. (1997). Do "whiplash injuries" occur in low-speed rear impacts? *Eur Spine J, 6*, 366–375.

Cherubino, P., Benazzo, F., Borromeo, U., et al. (1990). Degenerative arthritis of the adjacent spinal joints following anterior cervical spinal fusion: Clinicoradiologic and statistical correlations. *Ital J Orthop Traumatol, 16*, 533–543.

Clausen, J.D., Goel, V.K., Traynelis, V.C., et al. (1997). Uncinate processes and Luschka joints influence the biomechanics of the cervical spine: Quantification using a finite element model of the C5-C6 segment. *J Orthop Res, 15*, 342–347.

Corcos, D.M., Gottlieb, G. L., Latash, M. L., et al. (1992). Electromechanical delay: An experimental artifact. *J Electromyogr Kinesiol, 2*, 59–68.

Cusick, J. F., Pintar, F. A., Yoganandan, N. (1995). Biomechanical alterations induced by multilevel cervical laminectomy. *Spine, 20*, 2392–2398.

Denis, F. (1983) The three column spine and its significance in the classification of acute thoracolumbar spinal injuries. *Spine, 8*, 817–831.

Dickman, C.A., Zabramski, J. M., Hadley, M. N., et al. (1991). Pediatric spinal cord injury without radiographic abnormalities: Report of 26 cases and review of the literature. *J Spinal Disord, 4*, 296–305.

Dolan, P., Adams, M. A. (2001). Recent advances in lumbar spinal mechanics and their significance for modelling. *Clin Biomech 16 Suppl, 1*:S8-S16.

Dvorak, J., Froehlich, D., Penning, L., et al. (1988). Functional radiographic diagnosis of the cervical spine: Flexion/extension. *Spine, 13*, 748–755.

Dvorak, J., Hayek, J., Zehnder, R. (1987). CT-functional diagnostics of the rotatory instability of the upper cervical spine. Part II. An evaluation of healthy adults and patients with suspected instability. *Spine, 12*(8), 726–731.

Dvorak, J., Panjabi, M. M. (1987). Functional anatomy of the alar ligaments. *Spine, 12*, 183–189.

Dvorak, J., Schneider, E., Saldinger, P., et al. (1988). Biomechanics of the craniocervical region: The alar and transverse ligaments. *J Orthop Res, 6*, 452–461.

Emery, S. E., Fuller, D. A., Stevenson, S. (1996). Ceramic anterior spinal fusion. Biologic and biomechanical comparison in a canine model. *Spine, 21*, 2713–2719.

Ettlin, T. M., Kischka, U., Reichmann, S., et al. (1992). Cerebral symptoms after whiplash injury of the neck: A prospective clinical and neuropsychological study of whiplash injury. *J Neurol Neurosurg Psychiatry, 55*, 943–948.

Feipel, V., Rondelet, B., Le Pallec, J., et al. (1999). Normal global motion of the cervical spine: An electrogoniometric study. *Clin Biomech, 14*, 462–470.

Fielding, J. W. (1957). Cineroentgenography of the normal cervical spine. *J Bone Joint Surg Am, 39A*, 1280–1288.

Fielding, J. W., Cochran, G.B., Lawsing, J. F. III, et al. (1974). Tears of the transverse ligament of the atlas: A clinical and biomechanical study. *J Bone Joint Surg Am, 56A*, 1683–1691.

Fielding, J. W., Hawkins, R. J., Ratzan, S. A. (1976). Spine fusion for atlanto-axial instability. *J Bone Joint Surg Am, 58A*, 400–407.

Fountain, F. P, Minear, W. L., Allison, R. D. (1966). Function of longus colli and longissimus cervicis muscles in man. *Arch Phys Med, 47*, 665–669.

Frobin, W., Leivseth, G., Biggemann, M., et al. (2002). Sagittal plane segmental motion of the cervical spine. A new precision measurement protocol and normal motion data of healthy adults. *Clin Biomech, 17*, 21–31.

Frush, T. J., Fisher, T. J., Ensminger, S. C., et al. (2009). Biomechanical evaluation of parasagittal occipital plating: Screw load sharing analysis. *Spine, 34*, 877–884.

Fuller, D. A., Kirkpatrick, J. S., Emery, S. E., et al. (1998). A kinematic study of the cervical spine before and after segmental arthrodesis. *Spine, 23*, 1649–1656.

Galbusera, F., Bellini, C. M., Brayda-Bruno, M., et al. (2008). Biomechanical studies on cervical total disc arthroplasty: A literature review. *Clin Biomech, 23*, 1095–1104.

Giguere, J. F., St-Vil, D., Turmel, A., et al. (1998). Airbags and children: A spectrum of C-spine injuries. *J Pediatr Surg, 33*, 811–816.

Gosch, H. H., Gooding, E., Schneider, R. C. (1972). An experimental study of cervical spine and cord injuries. *J Trauma, 12*, 570–575.

Grauer, J. N., Panjabi, M. M., Cholewicki, J., et al. (1997). Whiplash produces an S-shaped curvature of the neck with hyperextension at lower levels. *Spine, 22*, 2489–2494.

Grob, D., Crisco, J. J. III, Panjabi, M. M., et al. (1992). Biomechanical evaluation of four different posterior atlantoaxial fixation techniques. *Spine, 17*(5), 480–490.

Gu, Y. T., Jia, L. S., Chen, T. Y. (2007). Biomechanical study of a hat type cervical intervertebral fusion cage. *Int Orthop, 31*, 101–105.

Hakalo, J., Pezowicz, C., Wronski, J., et al. (2008). Comparative biomechanical study of cervical spine stabilisation by cage alone, cage with plate, or plate-cage: A porcine model. *J Orthop Surg (Hong Kong), 16*, 9–13.

Hansson, T., Roos, B. (1986). Age changes in the bone mineral of the lumbar spine in normal women. *Calcif Tissue Int, 38*, 249–251.

Hansson, T., Roos, B., Nachemson, A. (1980). The bone mineral content and ultimate compressive strength of lumbar vertebrae. *Spine, 5*, 46–55.

Harms-Ringdahl, K. (1986). An assessment of shoulder exercise and load-elicited pain in the cervical spine. Biomechanical analysis of load - EMG - methodological studies of pain provoked by extreme position. Karolinska Institute, University of Stockholm.

Herkowitz, H. N. (1988). A comparison of anterior cervical fusion, cervical laminectomy, and cervical laminoplasty for the surgical management of multiple level spondylotic radiculopathy. *Spine, 13*, 774–780.

Herzog, R. J., Wiens, J. J., Dillingham, M. F., et al. (1991). Normal cervical spine morphometry and cervical spinal stenosis in asymptomatic professional football players. Plain film radiography, multiplanar computed tomography, and magnetic resonance imaging. *Spine, 16*, S178–S186.

Hilibrand, A. S., Carlson, G. D., Palumbo, M. A., et al. (1999). Radiculopathy and myelopathy at segments adjacent to the site of a previous anterior cervical arthrodesis. *J Bone Joint Surg Am, 81*, 519–528.

Holdsworth, F.W. (1963). Fractures, dislocations, and fracture-dislocations of the spine. *J Bone Joint Surg Br, 45B*, 6–20.

Hunter, L., Braunstein, E., Bailey, R. (1980). Radiographic changes following anterior fusion. *Spine, 5*, 399–401.

Ivancic, P. C., Coe, M. P., Ndu, A. B., et al. (2007). Dynamic mechanical properties of intact human cervical spine ligaments. *Spine J, 7*, 659–665.

Ivancic, P. C., Sha, D., Panjabi, M. M. (2009). Whiplash injury prevention with active head restraint. *Clin Biomech, 24*, 699–707.

Jarcho, M. (1981). Calcium phosphate ceramics as hard tissue prosthetics. *Clin Orthop, 157*, 259–278.

Jenkins, D. H. (1973). Extensive cervical laminectomy. Long-term results. *Br J Surg, 60*, 852–854.

Kapandji, I.A. (1974). *The Physiology of the Joints. The trunk and the vertebral column.* Edinburgh: Churchill Livingstone.

Kazarian, L. E. (1975). Creep characteristics of the human spinal column. *Orthop Clin North Am, 6*, 3.

Kazarian, L. (1972). Dynamic response characteristics of the human intervertebral column: An experimental study of autopsy specimens. *Acta Orthop Scand, 146*, 146.

King, A. I., Yang, K. H. (1995). Research in biomechanics of occupant protection. *J Trauma, 38*, 570–576.

Kitahara, Y., Iida, H., Tachibana, S. (1995). Effect of spinal cord stretching due to head flexion on intramedullary pressure. *Neurol Med Chir, 35*, 285–288.

Kotani, Y., Cunningham, B. W., Abumi, K., et al. (1994). Biomechanical analysis of cervical stabilization systems. An assessment of transpedicular screw fixation in the cervical spine. *Spine, 19*, 2529–2539.

Kriss, V. M., Kriss, T. C. (1996). SCIWORA (spinal cord injury without radiographic abnormality) in infants and children. *Clin Pediatr, 35*, 119–124.

Kumaresan, S., Yoganandan, N., Pintar, F. A. (1997). Finite element analysis of anterior cervical spine interbody fusion. *Biomed Mater Eng, 7*, 221–230.

Lonstein, J. E. (1977). Post-laminectomy kyphosis. *Clin Orthop, 128*, 93–100.

Lysell, E. (1969). Motion in the cervical spine. An experimental study on autopsy specimens. *Acta Orthop Scand Supplementum, 123*, 1–61.

Marshall, K. W., Koch, B. L., Egelhoff, J. C. (1998). Air bag-related deaths and serious injuries in children: Injury patterns and imaging findings. *Am J Neuroradiol, 19*, 1599–607.

McBroom, R. J., Hayes, W. C., Edwards, W. T., et al. (1985). Prediction of vertebral body compressive fracture using quantitative computed tomography. *J Bone Joint Surg Am, 67A*, 1206–1214.

McCaffrey, M., German, A., Lalonde, F., et al. (1999). Air bags and children: A potentially lethal combination. *J Pediatr Orthop, 19*, 60–64.

Mimura, M., Moriya, H., Watanabe, T., et al. (1989). Three-dimensional motion analysis of the cervical spine with special reference to the axial rotation. *Spine, 14*, 1135–1139.

Miyazaki, K., Tada, K., Matsuda, Y., et al. (1989). Posterior extensive simultaneous multisegment decompression with posterolateral fusion for cervical myelopathy with cervical instability and kyphotic and/or S-shaped deformities. *Spine, 14*, 1160–1170.

Miyazaki, M., Hong, S. W., Yoon, S. H., et al. (2008). Kinematic analysis of the relationship between the grade of disc degeneration and motion unit of the cervical spine. *Spine, 33*, 187–193.

Mohamed, A. A., Banerjee, A. (1998). Patterns of injury associated with automobile airbag use. *Postgrad Med J, 74*, 455–458.

Moroney, S. P., Schultz, A. B. (1985). Analysis and measurement of loads on the neck. *Trans Orth Res Soc, 10*, 329.

Moroney, S. P., Schultz, A. B., Miller, J., et al. (1988). Load displacement properties of lower cervical spine motion segments. *J Biomech, 21*, 769–779.

Moskovich, R., Crockard, H. A. (1992). Atlantoaxial arthrodesis using interlaminar clamps. An improved technique. *Spine, 17*, 261–267.

Moskovich, R., Shott, S., Zhang, Z. H. (1996). Does the cervical canal to body ratio predict spinal stenosis? *Bull Hosp Jt Dis, 55*, 61–71.

Myers, B. S., Winkelstein, B. A. (1995). Epidemiology, classification, mechanism, and tolerance of human cervical spine injuries. *Crit Rev Biomed Eng, 23*, 307–409.

Myklebust, J. B., Pintar, F., Yoganandan, N., et al. (1988). Tensile strength of spinal ligaments. *Spine, 13*, 526–531.

Nachemson, A., Evans, J. H. (1968). Some mechanical properties of the third human lumbar interlamina ligament (ligamentum flavum). *J Biomech, 1*, 201.

Nagamoto, Y., Ishii, T., Sakaura, H., et al. (2009). In vivo 3-dimensional kinematics of the cervical spine. *09 Dec 3; Cervical Spine Research Society.*

National Highway Traffic Safety Administration. (1996). Air bag alert. *Ann Emerg Med, 28*, 242.

Nowinski, G. P., Visarius, H., Nolte, L. P., et al. (1993). A biomechanical comparison of cervical laminaplasty and cervical laminectomy with progressive facetectomy. *Spine, 18*, 1995–2004.

Osenbach, R. K., Menezes, A. H. (1989). Spinal cord injury without radiographic abnormality in children. *Pediatr Neurosci, 15*, 168–174.

Pang, D., Pollack, I. F. (1989). Spinal cord injury without radiographic abnormality in children–The SCIWORA syndrome. *J Trauma, 29*, 654–664.

Panjabi, M. M., Cholewicki, J., Nibu, K., Babat, L.B., Dvorak, J. (1998). Simulation of whiplash trauma using whole cervical spine specimens. *Spine, 23*, 17–24.

Panjabi, M. M., Cholewicki, J., Nibu, K., Grauer, J., Vahldiek, M. (1998). Capsular ligament stretches during in vitro whiplash simulations. *J Spinal Disord, 11*, 227–232.

Panjabi, M. M., Crisco, J. J. III, Lydon, C., et al. (1998). The mechanical properties of human alar and transverse ligaments at slow and fast extension rates. *Clin Biomech, 13*, 112–120.

Panjabi, M. M., Kraig, M. H., Goel, V. K. (1981). A technique for measurement and description of three-dimensional six degree-of-freedom motion of a body joint with an application to the human spine. *J Biomechanics, 14*, 447–460.

Panjabi, M. M., Summers, D. J., Pelker, R. R., et al. (1986). Three-dimensional load-displacement curves due to forces on the cervical spine. *J Orthop Res, 4*, 152–161.

Panjabi, M. M., White, A. A. III, Johnson, R. M. (1975). Cervical spine mechanics as a function of transection of components. *J Biomechanics, 8*, 327–336.

Penning, L. (1978). Normal movements of the cervical spine. *Am J Roentgenol, 130*, 317–326.

Penning, L., Wilmink, J. T. (1987). Rotation of the cervical spine. A study in normal people. *Spine, 12*, 732–738.

Perry, J., Nickel, V. L. (1959). Total cervical-spine fusion for neck paralysis. *J Bone Joint Surg Am, 41A*, 37–60.

Pintar, F. A., Maiman, D. J., Hollowell, J. P., et al. (1994). Fusion rate and biomechanical stiffness of hydroxylapatite versus autogenous bone grafts for anterior discectomy. An in vivo animal study. *Spine, 19*, 2524–2528.

Pintar, F.A., Yoganandan, N., Pesigan, M., et al. (1995). Cervical vertebral strain measurements under axial and eccentric loading. *J Biomech Eng, 117*, 474–478.

Rapoff, A. J., O'Brien, T. J., Ghanayem, A. J., et al. (1999). Anterior cervical graft and plate load sharing. *J Spinal Disord, 12*, 45–49.

Raynor, R. B., Moskovich, R., Zidel, P., et al. (1987). Alteration in primary and coupled neck motions after facetectomy. *Neurosurg, 21*, 681–687.

Reill, J. (1960). Effects of flexion-extension movement of the head and spine upon the spinal cord and nerve roots. *J Neurol Neurosurg Psychiatry, 23*, 214–221.

Reitman, C. A., Mauro, K. M., Nguyen, L., et al. (2004). Intervertebral motion between flexion and extension in asymptomatic individuals. *Spine, 29*, 2832–2843.

Ripa, D. R., Kowall, M. G., Meyer, P. R. Jr., et al. (1991). Series of ninety-two traumatic cervical spine injuries stabilized with anterior ASIF plate fusion technique. *Spine, 16*(3), 46–55.

Rolander, S. D. (1966). Motion of the lumbar spine with special reference to the stabilizing effect of posterior fusion. An experimental study on autopsy specimens. *Acta Orthop Scand, 90*, 1–144.

Roozmon, P., Gracovetsky, S. A., Gouw, G. J., et al. (1993). Examining motion in the cervical spine. I: Imaging systems and measurement techniques. *J Biomed Eng, 15*, 5–12.

Saito, T., Yamamuro, T., Shikata, J., et al. (1991). Analysis and prevention of spinal column deformity following cervical laminectomy. I. Pathogenetic analysis of postlaminectomy deformities. *Spine, 16*, 494–502.

Schneider, R. C., Cherry, G., Pantek, H. (1954). The syndrome of acute central cervical spinal cord injury with special reference to the mechanism involved in hyperextension injuries of the cervical spine. *J Neurosurg, 11*, 546–577.

Schultz, A., Anderson, G., Örtengren, R., et al. (1982). Loads on the lumbar spine: Validation of a biomechanical analysis by measurements of intradiscal pressure and myoelectric signals. *J Bone Joint Surg Am, 64A*, 713–720.

Sendur, P., Thibodeau, R., Burge, J., et al. (2005). Parametric analysis of vehicle design influence on the four phases of whiplash motion. *Traffic Inj Prev, 6*, 258–266.

Simmons, E. H., Bradley, D. D. (1988). Neuro-myopathic flexion deformities of the cervical spine. *Spine, 13*, 756–762.

Skrzypiec, D. M., Pollintine, P., Przybyla, A., et al. (2007). The internal mechanical properties of cervical intervertebral discs as revealed by stress profilometry. *Eur Spine J, 16*, 1701–1709.

Steel, H. H. (1968). Anatomical and mechanical considerations of atlanto-axial articulation. In Proceedings of the American Orthopaedic Association [abstract] Steel, H. H. *J Bone Joint Surg Am, 50A*, 1481–1482.

Stemper, B. D., Yoganandan, N., Pintar, F. A. (2006). Effect of head restraint backset on head-neck kinematics in whiplash. *Accid Anal Prev, 38*, 317–323.

Sturzenegger, M., DiStefano, G., Radanov, B. P., et al. (1994). Presenting symptoms and signs after whiplash injury: The influence of accident mechanisms. *Neurology, 44*, 688–693.

Sutterlin, C. Ed., McAfee, P. C., Warden, K. E., et al. (1988). A biomechanical evaluation of cervical spinal stabilization methods in a bovine model. Static and cyclical loading. *Spine, 13*, 795–802.

Takebe, K., Vitti, M., Basmajian, J. V. (1974). The functions of semispinalis capitis and splenius capitis muscles: An electromyographic study. *Anat Rec, 179*, 477–480.

Tubbs, R. S., Kelly, D. R., Humphrey, E. R., et al. (2007). The tectorial membrane: Anatomical, biomechanical, and histological analysis. *Clin Anat, 20*, 382–386.

Viano, D. C., Gargan, M. F. (1996). Headrest position during normal driving: Implication to neck injury risk in rear crashes. *Accid Anal Prev, 28*, 665–674.

Voo, L. M., Kumaresan, S., Yoganandan, N., et al. (1997). Finite element analysis of cervical facetectomy. *Spine, 22,* 964–969.

Wallis, B. J., Lord, S. M., Barnsley, L., et al. (1998). The psychological profiles of patients with whiplash-associated headache. *Cephalalgia, 18,* 101–105.

Werne, S. (1957). Studies in spontaneous atlas dislocation. *Acta Orthop Scand Suppl, 23,* 1–105.

White, A. A. III, Johnson, R. M., Panjabi, M. M., Southwick, W.O. (1975). Biomechanical analysis of clinical stability in the cervical spine. *Clin Orthop Relat Res, 109,* 85–96.

White, A. A. III, Panjabi, M. M. (1990). *Clinical biomechanics of the spine.* (2 ed.). Philadelphia: J.B. Lippincott Co., 314.

Wilke, H.J., Kettler, A., Goetz, C., et al. (2000). Subsidence resulting from simulated postoperative neck movements: An in vitro investigation with a new cervical fusion cage. *Spine, 25,* 2762–2770.

Wittenberg, R.H., Moeller, J., White, A.A. III. (1990). Compressive strength of autogenous and allogenous bone grafts for thoracolumbar and cervical spine fusion. *Spine, 15*(10), 1073–1077.

Yang, K.H., Latouf, B.K., King, A.I. (1992). Computer simulation of occupant neck response to airbag deployment in frontal impacts. *J Biomech Eng, 114,* 327–331.

Yoganandan, N., Pintar, F.A., Maiman, D.J., et al. (1996). Human head-neck biomechanics under axial tension. *Med Eng Phys, 18,* 289–294.

Yoo, J.U., Zou, D., Edwards, W.T., et al. (1992). Effect of cervical spine motion on the neuroforaminal dimensions of human cervical spine. *Spine, 17,* 1131–1136.

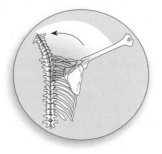

CHAPTER 12

Biomechanics of the Shoulder

Charles J. Jordan, Laith M. Jazrawi, and Joseph D. Zuckerman

Introduction

The shoulder girdle is the link between the upper extremity and the trunk. It acts in conjunction with the elbow to position the hand in space and is the most dynamic and mobile joint in the body (Fig. 12-1). It consists of the glenohumeral, acromioclavicular, sternoclavicular, and scapulothoracic articulations and the muscular structures that act on them. The absence of bony constraints allows a wide range of motion at the expense of stability, which is provided for entirely by soft tissues.

The biomechanics of the shoulder are complex, and a complete discussion requires an analysis of the four aforementioned articulations in addition to the surrounding muscles, ligaments, and cartilage. This chapter describes the anatomy of the shoulder complex and shows how its structure allows for efficient biomechanical function.

Kinematics and Anatomy

To produce the intricate motions necessary for normal hand positioning in space, the four joints with their associated components act together in a way that produces mobility greater than that afforded by any one individual articulation. The ability of the shoulder complex to position the upper extremity is further augmented by movement of the spine. A discussion follows of the types and ranges of motion for the shoulder complex as a whole, with subsequent sections discussing the manner in which motion is achieved at each of the articulations.

RANGE OF MOTION OF THE SHOULDER COMPLEX

Shoulder range of motion is traditionally measured in terms of flexion and extension (elevation of the humerus anteriorly or posteriorly away from the side of the thorax in the sagittal plane), abduction (elevation in the coronal plane), and internal-external rotation (rotation about the long axis of the humerus) (Fig. 12-2). Although during functional activities these individual motions are rarely seen in isolation, we can better understand the complex motions of the shoulder by analyzing the separate components needed to achieve any one position.

Although forward elevation of 180° is theoretically possible, the average value in men is 167° and in women it is 171°. Extension or posterior elevation averages 60° (Boone and Azen, 1979). These values are limited by tension on the joint capsule. Abduction in the coronal plane is limited by bony impingement of the greater tuberosity on the acromion. Forward elevation in the plane of the scapula, therefore, is considered to be more functional

because in this plane the inferior portion of the capsule is most lax and the musculature of the shoulder is optimally aligned for elevation of the arm (Fig. 12-3). Although shoulder range of motion normally decreases as part of the aging process, physical activity can counteract this process (Murray et al., 1985).

STERNOCLAVICULAR JOINT

The sternoclavicular joint consists of the medial end of the clavicle and the most superolateral aspect of the manubrium, linking the upper extremity directly to the thorax. In addition, a small facet is present inferiorly on the base of the clavicle that articulates with the first rib. This is a true synovial joint that has a saddle-like shape and contains a fibrocartilaginous articular disc or meniscus that divides it into two compartments.

Although the joint itself has little intrinsic or bony stability, the articular disc in conjunction with anterior, posterior, costoclavicular, and interclavicular ligaments maintains joint congruity (Fig. 12-4). The costoclavicular ligament, which runs between the undersurface of the

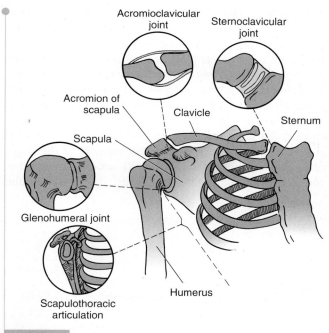

FIG. 12-1

Schematic depiction of the bony structures of the shoulder and their four articulations. The circular insets show front views of the three synovial joints—sternoclavicular, acromioclavicular, and glenohumeral—and a lateral view of the scapulothoracic joint, a bone-muscle-bone articulation. *Adapted from De Palma, A.F. (1983). Biomechanics of the shoulder. In Surgery of the Shoulder (3rd ed.). Philadelphia: JB Lippincott Co, 65–85.*

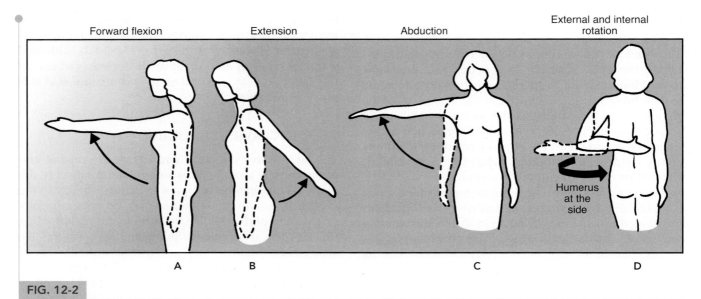

FIG. 12-2

A. Forward flexion. The humerus is in the sagittal plane.
B. Extension. The humerus is in the sagittal plane. **C.** Abduction.
The humerus is in the frontal plane. **D.** Rotation around the

long axis of the humerus. External and internal rotation with the
humerus at the side. Internal rotation is shown with the arm behind
the back, which is a functionally important form of this motion.

medial end of the clavicle and the first rib, primarily resists
superior displacement, while its anterior portion resists
posterior displacement of the clavicle and is thought to be
the major constraint in limiting sternoclavicular motion.
The anterior and posterior sternoclavicular ligaments are
the primary restraint to anterior and posterior translation,
respectively. They secondarily resist superior displacement. The interclavicular ligament connects the supero-
medial aspect of the clavicles and assists in restraining
the joint superiorly. The posterior portion of the interclavicular ligament also assists with anterior restraint of
the sternoclavicular joint. Specifically, the interclavicular

ligament tightens with arm depression and is lax when
the arm is elevated (Morrey and An, 1990). The disc prevents medial displacement of the clavicle, which may
occur when carrying objects at the side, as well as inferior displacement via articular contact. Although these
structures act as important stabilizers, they still allow for

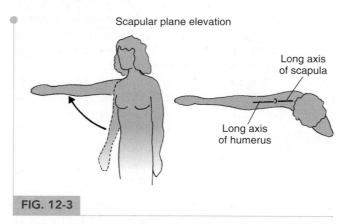

FIG. 12-3

Elevation in the scapular plane, which is midway between forward
flexion and abduction. The humerus is in the plane of the scapula.

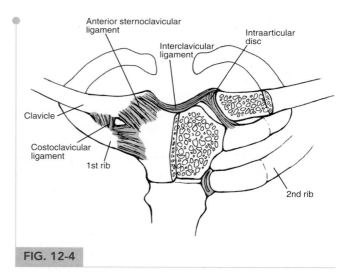

FIG. 12-4

Demonstration of the anatomy of the sternoclavicular joint.
*Adapted from Oatis, C.A. (2008). Kinesiology: The Mechanics and
Pathomechanics of Human Movement. Baltimore, MD: Lippincott
Williams & Wilkins, 128.*

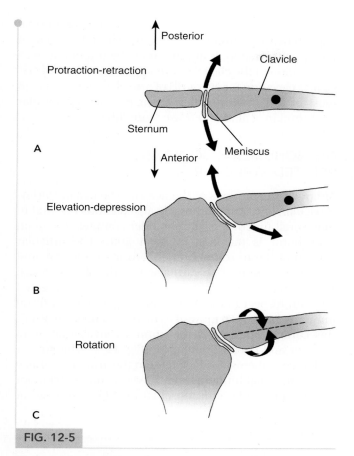

FIG. 12-5

Motion at the sternoclavicular joint. **A.** Top view showing the clavicular protraction and retraction (anteroposterior gliding) in the transverse plane around a longitudinal axis (*solid dot*) through the costoclavicular ligament, not shown. Motion takes place between the sternum and the meniscus. **B.** Anterior view showing clavicular elevation and depression (superoinferior gliding) in the frontal plane around a sagittal axis (*solid dot*) through the costoclavicular ligament, not shown. Motion occurs between the clavicle and the meniscus. **C.** Anterior view depicting the clavicular rotation around the longitudinal axis of the clavicle.

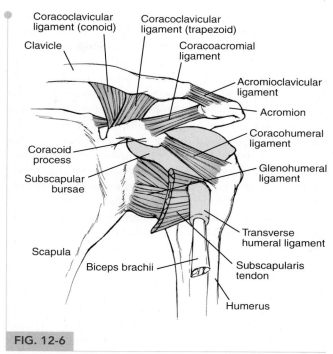

FIG. 12-6

The coracoclavicular ligament complex consists of the larger and heavier trapezoid ligament, which is oriented laterally, and the smaller conoid ligament, situated more medially. *Adapted from Hamill, J., Knutzen, K.M. (2008). Biomechanical Basis of Human Movement. Baltimore, MD: Lippincott Williams & Wilkins, 141.*

significant motion including up to 50° of axial rotation and 35° of both superior-inferior elevation and anterior-posterior translation (Fig. 12-5).

ACROMIOCLAVICULAR JOINT

The acromioclavicular joint (Fig. 12-6) lies between the lateral end of the clavicle and the acromion of the scapula (the lateral and anterior extension of the scapular spine) and is subject to the high loads transmitted from the chest musculature to the upper extremity. It too is a synovial joint, but has a planar configuration. A wedge-shaped articular disc, whose function is poorly understood, is found within the joint originating from the superior aspect. Both sides of the articular surface are covered with fibrocartilage, and the joint itself slopes inferomedially, causing the lateral end of the clavicle to slightly override the acromion.

A weak fibrous capsule encloses the joint and is reinforced superiorly by the acromioclavicular ligament. The acromioclavicular ligament acts primarily to restrain both axial rotation and posterior translation of the clavicle. Most of the joint's vertical stability is provided by the coracoclavicular ligaments that suspend the scapula from the clavicle (Fukuda et al., 1986). The coracoclavicular ligaments consist of the posteromedially directed conoid and the anterolaterally directed trapezoid ligaments, which are distinct structures that serve different biomechanical functions. The smaller conoid ligament acts to limit superior-inferior displacement of the clavicle. The quadrilaterally shaped trapezoid is the larger and stronger of the two ligaments and is found lateral to the conoid; it resists axial compression or motion about a horizontal axis. The coracoacromial ligament lies on the lateral side of the acromioclavicular joint and runs from the most lateral aspect of the coracoid to the medial aspect of the acromion.

Although clavicular rotation does occur with arm elevation, Rockwood (1975) found little relative motion between the clavicle and acromion. This has been attributed to synchronous clavicular and scapular rotation with little resultant relative motion at the acromioclavicular joint; most scapulothoracic motion occurs via the sternoclavicular joint. However, a recent study used three-dimensional magnetic resonance images (3D-MRI) to determine the motion of the AC joint during shoulder abduction. This test showed that although there appears to be no more than 3.5 mm of joint translation, the AC joint may rotate up to 35° at maximal shoulder abduction (Sahara et al., 2006). Thus, the thought that rigid fixation or fusion of the acromioclavicular joint produces little loss of overall shoulder function may now be called into question.

CLAVICLE

The clavicle lies between the two aforementioned articulations, acting as a strut connecting the thorax to the upper extremity. It is an S-shaped, double-curved bone: The medial two-thirds of the body is convex anteriorly while the lateral end is concave. It protects the underlying brachial plexus and vascular structures and serves as an attachment site for many of the muscles that act on the shoulder. The clavicle also provides for the normal appearance and contour of the upper chest. Elevation of the upper extremity is accompanied by rotation as well as elevation of the clavicle, with approximately 4° of clavicular elevation for every 10° of arm elevation, and with most of this motion occurring at the sternoclavicular joint (Inman et al., 1944) (Case Study 12-1).

GLENOHUMERAL JOINT AND RELATED STRUCTURES

Although the motions at the acromioclavicular, sternoclavicular, and scapulothoracic articulations are vital to the overall function of the shoulder complex, the main contributor is the glenohumeral joint. The articular surface of the proximal humerus forms a 120° arc and is covered with hyaline cartilage as is the glenoid fossa. The humeral head is retroverted or posteriorly directed 30° with respect to the intercondylar plane of the distal humerus and has an upward or medial inclination of 45°; this configuration gives the humerus an overall more anterior and lateral orientation (Fig. 12-7). The greater and lesser tuberosities lie lateral to the articular surface of the proximal humerus that serves as the attachment site for the rotator cuff musculature. The long head of

---

Case Study 12-1

Separated Shoulder

A 26 year-old soccer player presents after tripping over the ball and landing directly onto his right shoulder. He complains of pain over the right acromioclavicular joint, with significant swelling and a noticeable prominence. Radiographs reveal a dislocation of the acromioclavicular joint, with the clavicular end of the joint displaced superiorly approximately 100% (Case Study Fig. 12-1). This injury is commonly referred to as an "AC separation" or a "separated shoulder." This is a very common injury, constituting almost half of all athletic shoulder injuries seen by doctors. Although there are various degrees of injury, this patient demonstrates a prominence over the joint and radiographically has a dislocated AC joint with 100% translation. These findings suggest that all of the static stabilizers of the joint have been disrupted, including the acromioclavicular and the coracoclavicular (conoid and trapezoid) ligaments. Many of these injuries can be treated nonoperatively in a sling, but in some cases studies have shown that, depending on the severity of injury and the patient's activity level, surgical fixation or CC ligament reconstruction may be indicated to prevent long-term pain and loss of function (Simovitch, 2009).

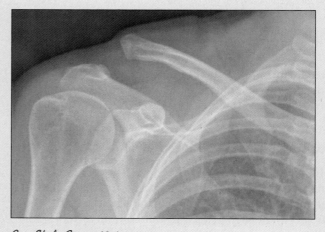

Case Study Figure 12-1

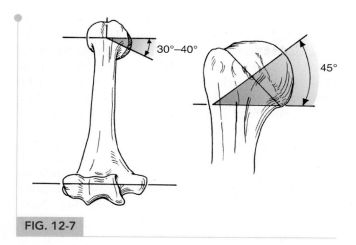

FIG. 12-7

The two-dimensional orientation of the articular surface of the humerus with respect to the bicondylar axis. *Reprinted with permission from Rockwood, C., Matsen, F. (1990). The Shoulder. Philadelphia: WB Saunders, 219.*

the biceps tendon traverses the bicipital groove (which lies between the tuberosities) beneath the transverse humeral ligament (Fig. 12-8).

The proximal humerus articulates with the glenoid fossa, which itself is retroverted 7° and superiorly inclined 5° relative to the plane of the scapula (Fig. 12-9A and B). This slight superior inclination provides a significant degree of geometric stability, helping to resist inferior subluxation or dislocation (Itoi et al., 1992). This position of the glenoid is dictated by the inclination and retroversion of the scapular neck, and its importance is demonstrated by several studies. In a recent biomechanical study, Chadwick et al. (2004) demonstrated that in cases of scapular neck malunion (changes in its normal alignment secondary to fracture), the resultant shortening and altered positioning of the stabilizing muscles around the shoulder is associated with significantly altered mechanics and loss of strength and function compared to normal controls.

The glenoid fossa is shallow and is able to contain only approximately one-third of the diameter of humeral head. The bony architecture is augmented by the cartilaginous surface, which is thicker peripherally than it is centrally, acting to slightly but significantly increase the depth of the glenoid as a whole. Although the congruity between the articular surfaces of the proximal humerus and glenoid was previously thought to be somewhat imprecise, stereophotogrammetric studies show this articulation to be precise, with the deviation from sphericity of the convex humeral articular surface and concave glenoid articular surface being less than 1% (Soslowsky et al., 1992). Less than 1.5 mm of translation

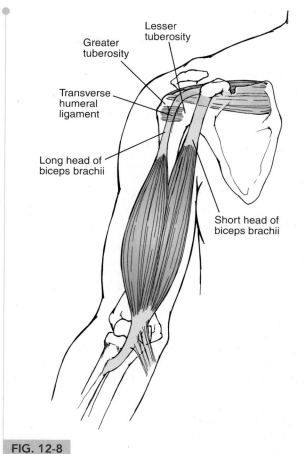

FIG. 12-8

The long head of the biceps goes in the bicipital groove between the greater and lesser tuberosities. The transverse humeral ligament helps stabilize the biceps tendon in the groove. *Adapted from Oatis, C.A. (2008). Kinesiology: The Mechanics and Pathomechanics of Human Movement. Baltimore, MD: Lippincott Williams & Wilkins, 222.*

of the humeral head on the glenoid surface has been demonstrated in normal subjects during a 30° arc of motion (Poppen and Walker, 1976); thus, motion at the glenohumeral joint is almost purely rotational. Given the paucity of bony constraint, stability is instead provided by the capsular, ligamentous, and muscular structures that surround the glenohumeral joint.

Glenoid Labrum

The glenoid labrum (Fig. 12-10) is a fibrocartilaginous rim that acts to deepen the glenoid, providing 50% of the overall depth of the glenohumeral joint (Warner, 1993). It has a triangular configuration when viewed in cross section and has firm attachments inferiorly to the underlying bone, having looser and more variable

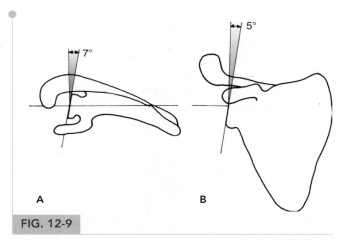

FIG. 12-9

A. The glenoid is retroverted 7° with respect to the plane perpendicular to the scapular plane. **B.** The glenoid faces superiorly approximately 5°. *Reprinted with permission from Simon, S.R. (Ed). (1994). Orthopaedic Basic Science. Rosemont, IL: AAOS, 526–527.*

attachments in its superior and anterosuperior portions. The superior portion of the glenoid labrum is confluent with the tendon of the long head of the biceps and, along with the adjacent supraglenoid tubercle, serves as its site of insertion (Moore, 1999).

Measurements of the force needed to dislocate the humeral head under constant compressive pressure have shown that with an intact labrum, the humeral head resists tangential forces of approximately 60% of the compressive load; resection of the labrum reduces the effectiveness of compression-stabilization by 20% (Lippitt et al., 1993). Detachment of the superior labrum with anterior-posterior extension (i.e., SLAP [superior labral anterior-posterior] lesion) can occur from traction (repetitive overhead activities or a sudden pull on the arm) or compression (a fall onto an outstretched arm). This lesion can be a cause of severe pain and shoulder instability (Itoi et al., 1996) as a result of significant increases in glenohumeral translation when compared with the intact shoulder (Pagnani et al., 1995) (Fig. 12-11).

Joint Capsule

The glenohumeral joint capsule has a significant degree of inherent laxity with a surface area that is twice that of the humeral head (Warner, 1993). This redundancy allows for a wide range of motion. Medially, the capsule attaches both directly onto (anteroinferiorly) and beyond the glenoid labrum and laterally it reaches to the anatomic neck of the humerus. Superiorly, it is attached at the base of the coracoid, enveloping the long head of the biceps tendon and making it an intra-articular structure (Fig. 12-11).

The capsule also has a stabilizing role, tightening with various arm positions. In adduction, the capsule is taut superiorly and lax inferiorly; with abduction of the upper extremity, this relationship is reversed and the inferior

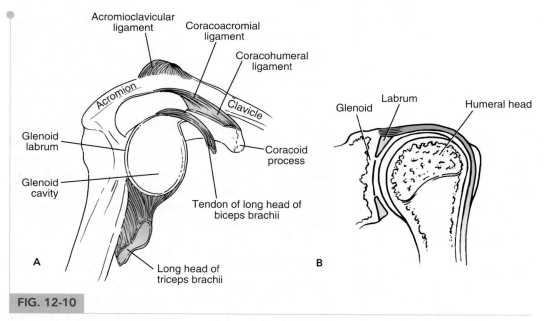

FIG. 12-10

A. The glenoid labrum is attached to the underlying bony glenoid and is confluent at its area with the long head of the biceps tendon. **B.** The labrum has a triangular configuration when viewed in cross section and serves to effectively deepen the glenoid, increasing the stability of the glenohumeral joint.

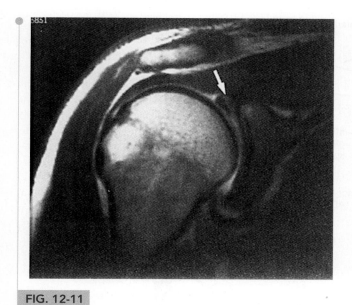

FIG. 12-11

Oblique coronal image showing a tear involving the insertion of the long head of the biceps tendon and the superior labrum (*arrow*).

capsule tightens. As the arm is externally rotated, the anterior capsule tightens while internal rotation induces tightening posteriorly. The posterior capsule in particular has been shown to be crucial in maintaining glenohumeral stability, acting as a secondary restraint to anterior dislocation (particularly in positions of abduction) as well as acting as a primary posterior stabilizing structure (Itoi et al., 1996).

Glenohumeral and Coracohumeral Ligaments

The three glenohumeral ligaments (superior, middle, and inferior) are discrete extensions of the anterior glenohumeral joint capsule and are critical to shoulder stability and function (Fig. 12-12). The superior glenohumeral ligament originates from the anterosuperior labrum, just anterior to the long head of the biceps, and inserts onto the lesser tuberosity. It is present in most shoulders but only well developed in 50%. The superior glenohumeral ligament acts as the main restraint to inferior translation with the arm in the resting or adducted position (Warner et al., 1992).

The coracohumeral ligament originates from the lateral side of the base of the coracoid to insert on the anatomic neck of the humerus (Fig. 12-6) (Cooper et al., 1993). This structure lies anterior to the superior glenohumeral ligament and reinforces the superior aspect of the joint capsule. These ligaments span the rotator interval between the subscapularis and supraspinatus. Some research indicates that these structures have a secondary role in preventing inferior translation of the shoulder while in the adducted, neutrally rotated position. The functional significance of the coracohumeral ligament, however, seems to be related to the overall development of the glenohumeral ligaments in a given individual, having a larger role in those with a less-developed superior glenohumeral ligament (Warner et al., 1992).

The middle glenohumeral ligament originates inferior to the superior glenohumeral ligament (at the 1 o'clock to 3 o'clock position, right shoulder) and inserts further laterally on the lesser tuberosity. Great variability in the anatomy of this structure has been demonstrated, and it is absent in as many as 30% of shoulders (Curl and Warren, 1996). It may originate from the anterosuperior portion of the labrum, the supraglenoid tubercle, or the scapular neck. Morphologic variants have been described, including a cord-like variant (clearly distinct from the anterior band of the inferior glenohumeral ligament) and a sheet-like variant (blending with the anterior band of the inferior glenohumeral ligament). Functionally, the middle glenohumeral ligament acts as a secondary restraint to inferior translations of the glenohumeral joint with the arm in the abducted and externally rotated position (Warner et al., 1992). It also serves as a restraint to anterior translation, having its maximal effect with the arm abducted 45°.

The inferior glenohumeral ligament originates from the inferior aspect of the labrum and inserts on the anatomic neck of the humerus. It has been shown to have three distinct components (O'Brien et al., 1990): an anterior band originating from 2 to 4 o'clock (right shoulder), a posterior component originating from 7 to 9 o'clock (right shoulder), and an axillary pouch (Fig. 12-12). The inferior glenohumeral ligament has the greatest functional significance, acting as the primary anterior stabilizer of the shoulder with the arm in 90° of abduction. As the arm is abducted and externally rotated, the anterior band of the inferior glenohumeral ligament tightens, resisting anterior translation. With internal rotation of the abducted arm, the posterior band becomes taut and posterior translation is resisted. The inferior glenohumeral ligament complex also serves to resist inferior translation of the glenohumeral joint with the arm in the abducted position. Variability as to the size and attachment sites of the glenohumeral ligaments has been demonstrated (Warner et al., 1992); however, the clinical significance of this has yet to be fully elucidated although it has been suggested that absence of the middle glenohumeral ligament may predispose to instability (Steinbeck et al., 1998). For additional information, see Box 12-1.

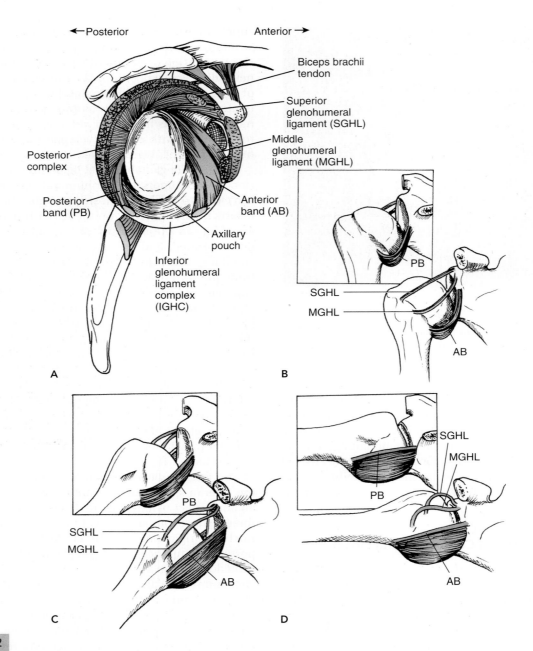

FIG. 12-12

A. Schematic drawing of the shoulder capsule illustrating the location and extent of the inferior glenohumeral ligament complex (IGHLC). *Reprinted with permission from O'Brien, S.J., Neves, M.C., Amoczky, S.P., et al. (1990). The anatomy and histology of the inferior glenohumeral ligament complex of the shoulder. Am J Sports Med, 18, 579–584.* **B.** The superior glenohumeral ligament is the primary restraint to inferior translation in the adducted shoulder at neutral rotation. In this position, the middle glenohumeral ligament and the anterior and posterior bands of the inferior glenohumeral ligament complex remain lax. **C.** The anterior band is the primary restraint resisting inferior translation of

the shoulder at 45° abduction and neutral rotation. In this position, the superior glenohumeral ligament, the middle glenohumeral ligament, and posterior band are lax. **D.** At 90° abduction, the anterior and posterior bands of the inferior glenohumeral ligament cradle the humeral head to prevent inferior translation. The posterior band is more significant in external rotation, whereas the anterior band plays a greater role in internal rotation. *Reprinted with permission from Warner, J.P., Deng, X.H., Warren, R.F., et al. (1992). Static capsuloligamentous restraints to superior-inferior translations of the glenohumeral joint. Am J Sports Med, 20, 675–678.*

Experimental Techniques: Ligament Cutting Studies

Ligament cutting studies have been instrumental in furthering our knowledge regarding the contribution of a given anatomic structure to overall glenohumeral stability (Curl and Warren, 1996). In this technique, cadaveric specimens are biomechanically tested before and after selectively cutting sequential structures. A force is then applied in a given arm position, and the translation that occurs is measured. From this information, the relative contribution that a given structure provides to overall stability can be determined. When a particular pattern of shoulder instability is then identified, the physician can infer which anatomic structures may be deficient or disrupted so that an appropriate treatment plan can be implemented.

Additional Constraints to Glenohumeral Stability

Synovial fluid acts via cohesion and adhesion to further stabilize the glenohumeral joint. Synovial fluid adheres to the articular cartilage overlying the glenoid and proximal humerus, causing the two surfaces to slide along one another. The synovial fluid provides a cohesive force between these two, making it difficult to pull them apart (Simon, 1994). Under normal conditions, the intra-articular pressure within the glenohumeral joint is negative, acting to pull the overlying capsule and glenohumeral ligaments toward the center of the joint. If the integrity of the glenohumeral joint capsule is compromised (e.g., venting the capsule) or if a significant effusion exists (normally the glenohumeral joint contains less than 1 mL of fluid), significant increases in translation are observed (Kumar and Balasubramaniam, 1985). Specifically, venting the capsule reduces the force needed for anterior humeral head translation by 55%, for posterior translation by 43%, and for inferior translation by 57% (Gibb et al., 1991) (see Case Study 12-2 later).

SCAPULOTHORACIC ARTICULATION

The scapula is a flat, triangular bone that lies on the posterolateral aspect of the thorax between the second and seventh ribs. It is angled 30° anterior to the coronal plane of the thorax and is rotated slightly toward the midline at its superior end and tilted anteriorly with respect to the sagittal plane (Fig. 12-13) (Laumann, 1987). The spine of the scapula gives rise laterally to the acromion process that articulates with the distal clavicle at the acromioclavicular joint. The coracoclavicular ligaments and muscular attachments help to support the scapula and stabilize it against the thorax (Fig. 12-6). This allows for a wide range of scapular motion, including protraction, retraction, elevation, depression, and rotation (Case Study 12-2).

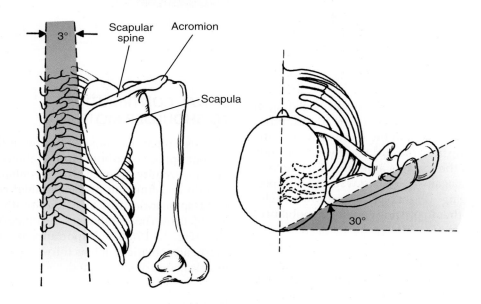

FIG. 12-13

Scapular orientation on the chest wall. Left, 30° anterior. Right, 3° upward. *Reprinted with permission from Warner, J. J. P. (1993). The gross anatomy of the joint surfaces, ligaments,* labrum, and capsule. In: Matsen, F.A., Fu, F.H., Hawkins, R.J. (Eds.). The Shoulder: A Balance of Mobility and Stability. Rosemont, IL: AAOS.

Case Study 12-2

Shoulder Instability

While skiing, a 21-year-old man fell onto his right upper extremity, causing forceful abduction and external rotation. He noted acute pain in the arm and was unable to move it. Physical examination revealed loss of the normal contour of the shoulder and painful range of motion. Radiographs showed an anterior dislocation with posterior superior humeral head impaction fracture. The patient underwent closed reduction. Postreduction radiographs confirmed reduction of the humeral head with a small bony avulsion fracture of the anterior-inferior glenoid rim, which represents a detachment of the anterior labrum in the area of the superior band of the glenohumeral ligament insertion (Case Study Fig. 12-2). Structural alteration of bony geometry, ligaments, and labrum resulted in shoulder instability. The detachment of the anterior labrum and the superior band of the ligament insertion affected primarily the resistance to anterior translation of the humeral head, resulting in anterior dislocation. In addition, a concomitant capsule lesion affected the intra-articular negative pressure necessary to pull the humeral head inward. After conservative management, the patient did well for 6 months

until he sustained a recurrent dislocation. The patient subsequently opted for operative treatment that included repair of the fracture and capsule and a complete period of rehabilitation with emphasis on joint stability and proprioception.

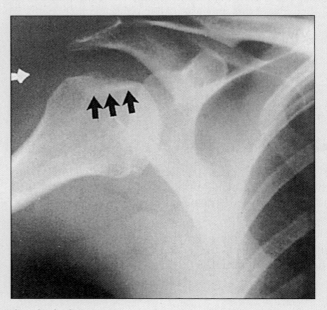

Case Study Figure 12-2

The scapulothoracic articulation involves gliding of the scapula on the posterior aspect of the thorax. Interposed between the scapula and the thoracic wall lie the subscapularis (arising from the costal surface of the body of the scapula) and the serratus anterior, which help to stabilize the scapula against the chest wall and thus prevent "scapular winging" (Fig. 12-14). These two muscles glide along one another to provide enhanced mobility of the shoulder complex.

Elevation of the arm involves motion at both the glenohumeral and scapulothoracic articulations. Although the contribution from each varies according to arm position and the specific task being performed, the average ratio of glenohumeral to scapulothoracic motion is 2:1 (Tibone et al., 1994). Elevation of the arm also induces complex rotatory motion of the scapula, with anterior rotation during the first 90° followed by posterior rotation with a total arc of approximately 15° (Morrey and An, 1990).

SPINAL CONTRIBUTION TO SHOULDER MOTION

Although often overlooked, motion of the thoracic and lumbar spine contributes to the ability to position the upper extremity in space, thereby enhancing the overall motion and function of the shoulder complex. Flexion of the spine away from an extremity attempting to reach an object overhead enhances the range of motion attainable (Fig. 12-15). The importance of spinal motion in overhead activities such as throwing and racquet sports has also been demonstrated.

Kinetics

Numerous muscles act on the various components of the shoulder complex to provide both mobility and dynamic stability. Dynamic stabilization occurs via several possible

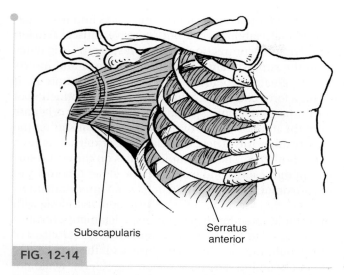

FIG. 12-14

Anterior view of the scapulothoracic articulation, a bone-muscle-bone articulation between the scapula and thorax. During scapular motion, the subscapularis muscle, which attaches broadly to the costal surface of the scapula, glides on the serratus anterior muscle, which originates on the first eight ribs and inserts into the costal surface of the scapula along the length of its vertebral border.

mechanisms (Moffey and An, 1990) including passive muscle tension, or via a barrier effect of the contracted muscle, compressive forces brought about by muscular contraction, joint motion that induces tightening of the passive

or ligamentous constraints, or via a redirection of the joint force toward the center of the glenoid.

To understand muscle function and force transmission, one must consider a given muscle's orientation, size, and activity. Given the multiple articulations present in the shoulder complex, any given muscle may span multiple joints. Depending on the position of the upper extremity, the relationship of a muscle with regard to any joint may change, altering its effect on that joint and the resultant forces or motions produced.

MUSCULAR ANATOMY

The shoulder musculature can be thought of in layers. The outermost layer consists of the deltoid and pectoralis major (Fig. 12-16). The deltoid forms the normal, rounded contour of the shoulder and is triangular in shape, with anterior, middle, and posterior heads. Each portion of the deltoid is activated differently for specific activities. The deltoid originates from the lateral third of the clavicle, acromion, and scapular spine and inserts on the anterolateral aspect of the humerus. The anterior head acts as a strong flexor and internal rotator of the humerus, the middle head as an abductor, and the posterior head as an extensor and external rotator. The pectoralis major lies

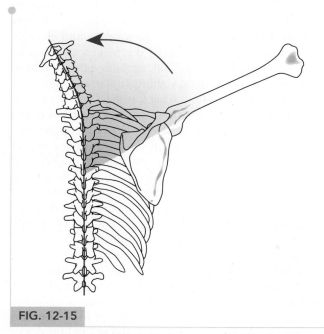

FIG. 12-15

Lateral bending of the spine enhances the ability to position the upper extremity.

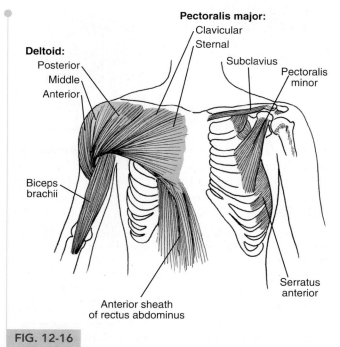

FIG. 12-16

Anterior view showing the superficial muscles (left shoulder) and the deep muscles beneath the deltoid and pectoralis muscles (right shoulder).

over the anterior chest wall and has two heads, a clavicular head originating from the side of the clavicle and a sternocostal head originating from the sternum, manubrium, and the upper costal cartilages. The two heads converge at the sternoclavicular joint. The pectoralis major inserts at the lateral lip of the intertubercular groove of the humerus and acts to adduct and internally rotate the humerus. Secondarily, the clavicular head acts as a flexor or forward elevator of the humerus while the sternocostal head extends the humerus. The pectoralis minor lies deep to the pectoralis major, functioning as an important scapular stabilizer. The pennate subclavius muscle lies inferior to the clavicle and may assist in clavicular motions. It has a tendinous origin from the anteromedial aspect of the first rib and inserts on the undersurface of the medial clavicle (Morrey and An, 1990).

Beneath this outer layer lies the rotator cuff musculature: the supraspinatus, infraspinatus, subscapularis, and teres minor (Fig. 12-17). These four muscles act to abduct and rotate the humerus and act as important glenohumeral stabilizers via both passive muscle tension and dynamic contraction. The supraspinatus originates from the supraspinatus fossa of the scapula and inserts on the greater tuberosity of the proximal humerus. It forms a force couple with the deltoid during abduction of the humerus. The infraspinatus and teres minor originate from the inferior aspect of the scapula and insert on the greater tuberosity. These muscles act as external rotators of the humerus. The subscapularis lies on the costal surface of the scapula and inserts on the lesser tuberosity of the proximal humerus. It functions as an important internal rotator of the humerus. The subscapularis, along with the middle and inferior glenohumeral ligaments, has also been shown to act as an important anterior stabilizer of the glenohumeral joint, particularly with the arm held at 45° of abduction. The teres major (Fig. 12-18), while not part of the rotator cuff, also originates from the scapula, but at its inferior angle coursing inferior to the teres minor and then passing anteriorly to insert on the humerus at the medial lip of the intertubercular groove. It functions to assist with arm adduction and internal rotation.

The biceps muscle is also integral to the motion of the shoulder complex. It is composed of two heads: a short head that originates from the tip of the coracoid process of the scapula and a long head that originates from the superior glenoid labrum and supraglenoid tubercle (Fig. 12-8). The tendon of the long head of the biceps lies within the glenohumeral joint and descends between the greater and lesser tuberosities, joining the short head to insert on the bicipital tuberosity of the radius. The biceps functions to flex and supinate the forearm and elevates the humerus. The long head of the biceps also acts as a humeral head depressor and, as such, plays a role in maintaining glenohumeral stability (Itoi et al., 1994).

Several muscles lie on the back and act directly on the scapula (Fig. 12-18). The outermost layer consists of the trapezius that covers the posterior neck and uppermost portion of the trunk, inserting on the superior aspect of the lateral one third of the clavicle, acromion, and scapular spine. The trapezius serves to elevate, retract, and rotate the scapula. The latissimus dorsi covers the inferior portion of the back, inserting on the floor of the intertubercular groove of the humerus. It acts to extend, adduct, and internally rotate the humerus.

Below these muscles lie the levator scapulae superiorly, which elevate and inferiorly rotate the scapula, and the rhomboid muscles below, which retract and rotate the scapula. Both of these muscle groups act to assist the serratus anterior (which lies laterally on the chest and intercostal muscles inserting onto the medial border of the anterior surface of the scapula) in keeping the scapula fixed to the trunk.

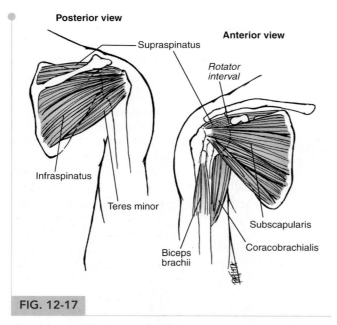

Posterior view

Supraspinatus

Anterior view

Rotator interval

Infraspinatus

Teres minor

Biceps brachii

Subscapularis

Coracobrachialis

FIG. 12-17

Anterior view. "Rotator interval" is a term introduced in 1970 to indicate the space between the supraspinatus and subscapularis tendons. The coracohumeral ligament lies superficially along its anterior edge, where it is readily available for release as indicated. The long head of the biceps lies deep along its posterior edge and serves as a guide to this interval during surgery. **Posterior view.** The two external rotators of the humerus, the infraspinatus and teres minor muscles, which are also the posterior wall of the rotator cuff. Note the median raphe of the infraspinatus, which is often mistaken at surgery for the border between the infraspinatus and the teres minor. *Adapted from Oatis, C.A. (2008). Kinesiology: The Mechanics and Pathomechanics of Human Movement. Baltimore, MD: Lippincott Williams & Wilkins, 168.*

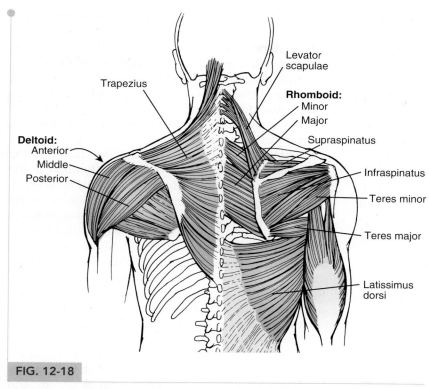

FIG. 12-18

Posterior view showing the superficial muscles (left shoulder) and underlying muscles (right shoulder).

INTEGRATED MUSCULAR ACTIVITY OF THE SHOULDER COMPLEX

Electromyography allows for the quantification of muscular activity during dynamic conditions. This permits insight into the level of muscular activity but does not directly indicate forces generated. A complete understanding of the latter requires knowledge of the moment arm (measured as the distance between the instantaneous center of rotation of the joint and the distance of muscular pull) and the physiologic cross-section of the involved muscle (measured as the muscular volume divided by its length). In the shoulder complex, each motion is associated with movement at multiple articulations and the constantly changing relationships of the muscular origins and insertions.

Given the paucity of osseous stability at the glenohumeral joint, force generated by one muscle (the primary agonist) requires the activation of an antagonistic muscle so that a dislocating force does not result (Simon, 1994). The antagonist usually accomplishes this via an eccentric contraction whereby the muscle is lengthened while actively contracting or via the production of a neutralizing force of equal magnitude but in the opposite direction. The relationship between two such muscles is also referred to as a force couple (Fig. 12-19). Around the glenohumeral joint, there is a force couple in the coronal plane (between the deltoid and the inferior portion of the rotator cuff) and in the transverse plane between the subscapularis muscle anteriorly and the posterior rotator cuff musculature (the infraspinatus and teres minor).

Relative motion is produced by an imbalance between the agonist and antagonist that produces torque. The degree of torque and the resultant angular velocity produced is determined by the relative activation of two such muscles or muscle groups. Resultant muscular forces are determined via an understanding of the cross-sectional area of the activated muscles involved and their orientation at the time of activation.

Forward Elevation

The most basic motion of the shoulder complex involves elevation of the arm in the scapular plane. This motion has been studied in depth via both electromyography and stereophotogrammetry. The muscles of the shoulder girdle have subsequently been grouped according to relative importance with regard to this motion. The first

Anterior view

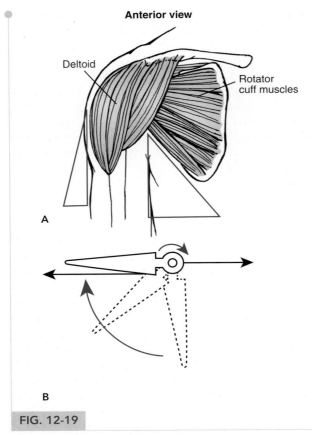

Deltoid

Rotator
cuff muscles

A

B

FIG. 12-19

The deltoid and the oblique rotator cuff muscles (infraspinatus, subscapularis, and teres minor) combine to produce elevation of the upper extremity by means of a force couple (two forces equal in magnitude but opposite in direction). With the arm at the side **(A)**, the directional force of the deltoid is upward and outward with respect to the humerus while the force of the oblique rotator cuff muscles is downward and inward. These two rotational forces can be resolved into their respective vertical and horizontal components. The horizontal force of the deltoid acting below the center of rotation of the glenohumeral joint is opposite in direction to the horizontal force of the oblique rotators, which is applied above the center of rotation. These forces acting in opposite directions on either side of the center of rotation produce a powerful force couple, as illustrated by the arm signal **(B)**. The vertical forces offset each other, thereby stabilizing the humeral head on the glenoid and allowing elevation to take place.

grouping includes the deltoid (specifically the anterior and middle heads), trapezius (inferior portion), supraspinatus, and serratus anterior (Simon, 1994). The second grouping consists of the middle portion of the trapezius, the infraspinatus, and the long head of the biceps. A third group consists of the posterior head of the deltoid, the clavicular head of the pectoralis major, and the superior portion of the trapezius. The fourth and final group includes the sternal head of the pectoralis major, the latissimus dorsi, and the long head of the triceps.

The interrelationship between the muscular forces involved in shoulder elevation was first studied by Inman, who found that the deltoid and supraspinatus work synergistically while the remainder of the rotator cuff musculature provides a humeral depressing force to counter subluxation of the humeral head (Inman et al., 1944). Thus, the vertically oriented pull of the deltoid is offset by a net inferior force created by the infraspinatus, subscapularis, and teres minor.

Electromyographic studies have shown that both the supraspinatus and deltoid are active throughout the range of arm elevation. The supraspinatus, however, is felt to have a larger role in initiating abduction. As the arm is progressively elevated from the side, the moment arm of the deltoid improves, resulting in a larger force in relation to the supraspinatus (Fig. 12-20). The percentage of shearing or vertical force created by the deltoid likewise decreases with increasing amounts of abduction. The angle of pull of the supraspinatus is more constant at approximately 75°, acting not only to elevate or abduct the arm but also to compress the humeral head within the glenoid. This is supported by several studies that have reported that the humeral head may translate superiorly 1 to 3 mm during abduction from 0° to 30°, but then demonstrate a net inferior translation from 30° to 60°, and then remain nearly constant throughout the remainder of the arc of motion (Graichen et al., 2000). The remaining rotator cuff muscles also contribute to humeral head depression, pulling almost directly inferiorly at approximately 45°, resulting in forces that equally compress and depress the humeral head to maintain glenohumeral stability (Fig. 12-21). As a result, it has been shown that during active scapular plane abduction, the humeral head

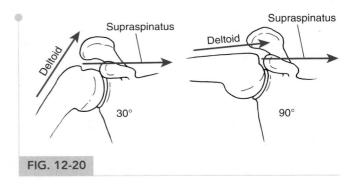

Supraspinatus

Deltoid

Supraspinatus

Deltoid

30°

90°

FIG. 12-20

As the arm is abducted to 90°, the direction of pull of the deltoid approximates that of the supraspinatus. Therefore, patients with a large tear of the rotator cuff often can actively maintain the arm abducted to 90° but may not be capable of actively abducting to 90°. *Reprinted with permission from Simon, S.R. (Ed). (1994). Orthopaedic Basic Science. Rosemont, IL: AAOS, 527.*

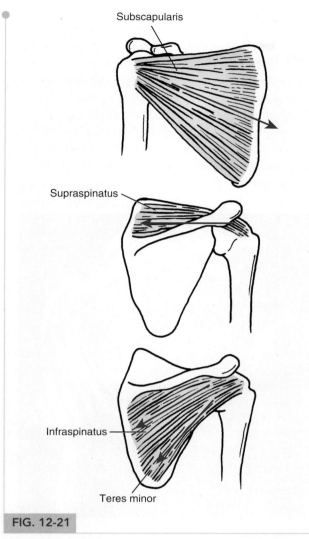

FIG. 12-21

The angle of pull of the subscapularis **(top)** is approximately 45°. The angle of pull of the infraspinatus **(bottom)** is also approximately 45°, and the teres minor **(bottom)** is approximately 55°. These vectors result in nearly equal glenohumeral joint compression and humeral head depression. The supraspinatus **(center)** is essentially horizontal in its orientation, resulting in compression of the glenohumeral joint.

remains almost completely centered within the glenoid throughout a full arc of motion.

Selective anesthetic block of the axillary nerve (and resulting deltoid paralysis) demonstrates that forward elevation is possible albeit significantly weakened. Likewise, a suprascapular nerve block and the resultant supraspinatus paralysis induced have a similar effect. However, a block of both nerves results in a loss of arm elevation (Colachis and Strohm, 1971; Howell et al., 1986) (see Case Study 12-3).

When pure abduction is compared with pure forward elevation, the same basic relationships are seen with the rotator cuff acting to stabilize the glenohumeral joint while the deltoid provides the necessary torque. Forward flexion results in activation of the anterior and middle deltoid (73% and 62% activity, respectively) with stability provided mainly by the supraspinatus, infraspinatus, and latissimus dorsi, the latter being particularly active (25% activation) with forward flexion beyond 90°. Pure abduction requires similar muscular activity; however, the subscapularis shows increased activation as it acts as the prime stabilizer via eccentric contraction.

External Rotation

The primary external rotator of the humerus is the infraspinatus, with significant contributions made by the posterior head of the deltoid and the teres minor. With any given amount of shoulder abduction, electromyography reveals the prime external rotator to be the infraspinatus. The subscapularis is similarly active but serves an antagonistic role as the main stabilizer preventing anterior displacement of the humeral head with external rotation. As the amount of shoulder abduction is increased, the posterior deltoid increases in efficiency as an accessory external rotator of the humerus secondary to improvement of its moment arm (Case Study 12-3).

Internal Rotation

Internal rotation of the shoulder is accomplished by the subscapularis, sternal head of the pectoralis major, latissimus dorsi, and teres major. The subscapularis is active during all phases of internal rotation, with decreased relative activity seen with extremes of abduction. In the same way, activity of the sternal head of the pectoralis major and the latissimus dorsi decreases with abduction. However, the posterior and middle heads of the deltoid compensate with increased eccentric activity during internal rotation while the arm is abducted.

Extension

Extension of the upper extremity is accomplished by the posterior and middle heads of the deltoid. The supraspinatus and subscapularis are also continually active throughout arm extension, resisting forces via eccentric activity that would tend to cause anterior dislocation.

Scapulothoracic Motion

Motion at the scapulothoracic articulation allows maintenance of deltoid tension, allowing it to maintain optimal power regardless of arm position. With forward

Case Study 12-3

Subacromial Impingement Syndrome and Rotator Cuff Tear

A 63-year-old right-hand-dominant woman presents with right shoulder pain of 6 months' duration. The patient noted increasing pain at night and with overhead activities of daily living such as putting on a shirt and brushing her hair. The pain is unresponsive to pain killers. Physical examination revealed diffuse tenderness over the shoulder and deltoid with painful forward elevation above 60° and internal rotation limited to her ipsilateral greater trochanter. She had a positive Neer sign (painful forward elevation between 60° and 120°) and a positive Hawkins sign (pain with passive abduction and internal rotation). A subacromial impingement test was positive with near complete relief of pain and restoration of forward elevation to 150° but persistent weakness on resisted shoulder abduction. A diagnosis of subacromial impingement syndrome is made. The patient was initially managed with conservative treatment. At 6 weeks follow-up, the patient had persistent pain, and an MRI of her shoulder revealed a full thickness tear of the supraspinatus tendon. The patient opted for operative management (Case Study Fig.12-3).

Biomechanically, rotator cuff tears have been likened to a suspension bridge, whereby the free margin of the tear corresponds to the cable and the residual attachments correspond to the supports at each end of the cable's span. This configuration allows the muscle that has been torn at its insertion to still exert its effects by way of the "spans" of the bridge. Thus, patients may have a "functional" rotator cuff tear in which they are still able to perform overhead activities. If the tearing force present at the two supports is great enough, however, worsening of the tear will occur.

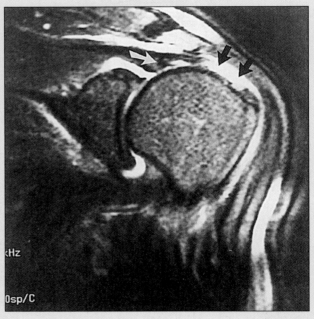

Case Study Figure 12-3

---

elevation at the arm, the scapula rotates, increasing stability at the glenohumeral joint and decreasing the tendency for impingement of the rotator cuff beneath the acromion (Fig. 12-22). A rotational force couple (two equal, noncollinear, parallel but oppositely direct forces) between the upper trapezius, levator scapulae, and upper serratus anterior with concomitant contraction of the lower trapezius and serratus anterior leads to scapular rotation that is necessary for full forward elevation (Fig. 12-23) (Simon, 1994).

LOADS AT THE GLENOHUMERAL JOINT

The glenohumeral joint is considered to be a major load-bearing joint. Although calculations of the exact forces acting on it are challenging given the many involved muscular structures and possible positions attainable, several simplifying assumptions allow an estimation of the magnitude of these forces. A free-body diagram of a person holding the upper extremity held at 90° of abduction can be used as an example; it is assumed that only the deltoid muscle is active and that it acts at a distance of 3 cm from the center of rotation of the humeral head. Three forces are then considered: the deltoid muscle force (D), the weight of the arm (equivalent to 0.05 body weight [BW] acting at the center of gravity of the limb, 3 cm), and the joint-reactive force at the glenohumeral joint (J). The joint reactive force and the force of the deltoid, being nearly parallel, are considered to be a force couple and are of equal but opposite magnitude. The force on the glenohumeral joint when holding the arm at 90° of abduction can be estimated to be one-half of body

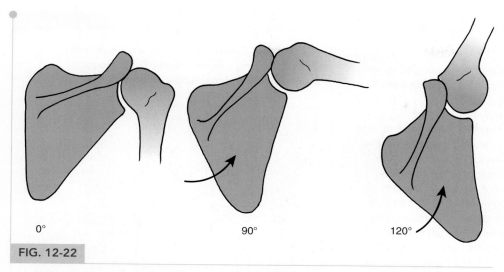

FIG. 12-22

Forward elevation or abduction 0–120° of the arm requires synchronous rotation of the scapula. *Reprinted with permission from Simon, S.R. (Ed). (1994). Orthopaedic Basic Science. Rosemont, IL: AAOS.*

weight (see Calculation Box 12-1, Case A). If a weight (*W*) of 2 kg is added (equivalent to 0.025 body weight of an 80-kg male) to the hand of the outstretched extremity held at 90° of abduction, a similar calculation can be made (see Calculation Box 12-1, Case B).

Experimentally, loads at the glenohumeral joint and the forces necessary for arm elevation have been determined. These forces have been found to be greatest at 90° of elevation, with the deltoid force equivalent to 8.2 times the weight of the arm and the joint reactive force

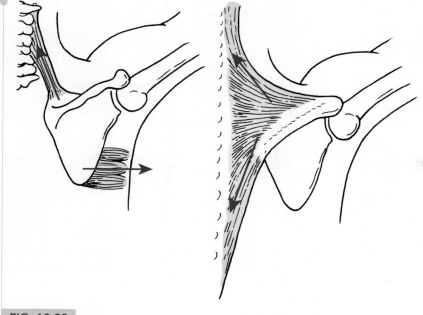

FIG. 12-23

Rotation of the scapula is produced by the synergistic contractions of the lower portion of the serratus anterior and the lower trapezius, with the upper trapezius, levator scapulae, and upper serratus anterior. *Reprinted with permission from Simon, S.R. (Ed). (1994). Orthopaedic Basic Science. Rosemont, IL: AAOS.*

CALCULATION BOX 12-1

Calculation of Reaction Forces

Estimates of the reaction force on the glenohumeral joint are obtained with the use of simplifying assumptions (Poppen and Walker, 1978).

Case A. In this example, the arm is in 90° of abduction, and it is assumed that only the deltoid muscle is active. The force produced through the tendon of the deltoid muscle (*D*) acts at a distance of 3 cm from the center of rotation of the joint (indicated by the *hollow circle*). The force produced by the weight of the arm is estimated to be 0.05 times body weight (*BW*) and acts at a distance of 30 cm from the center of rotation. The reaction force on the glenohumeral joint (*J*) may be calculated with the use of the equilibrium equation that states that for a body to be in moment equilibrium, the sum of the moments must equal zero. In this example, the moments acting clockwise are considered to be positive and counterclockwise moments are considered to be negative.

$$\sum M = 0$$

$$(30\,cm \times .05\,BW) + (60\,cm \times .025\,BW) - (D \times 3\,cm) = 0$$

$$D = \frac{(30\,cm \times .05\,BW)}{3\,cm}$$

$$D = 0.5\,BW$$

D is approximately one-half body weight. Because *D* and *J* are almost parallel but opposite, they form a force couple and are of equal magnitude; thus, the joint reaction force is also approximately one-half body weight.

Case B. Similar calculations can be made to determine the value for *D* when a weight equal to 0.025 times body weight is held in the hand with the arm in 90° of abduction.

$$\sum M = 0$$

$$(30\,cm \times 0.5\,BW) + (60\,cm \times .025\,BW) - (D \times 3\,cm) = 0$$

$$D = \frac{(30\,cm \times .05\,BW) + (60\,cm \times .025\,BW)}{3\,cm}$$

$$D = 1\,BW$$

Once again, *D* and *J* are essentially equal and opposite, forming a force couple. Thus, the joint reaction force is approximately equal to body weight.

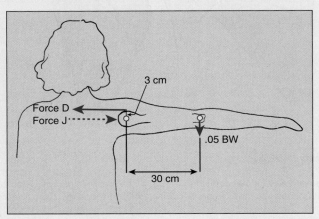

Calculation Box Figure 12-1-1

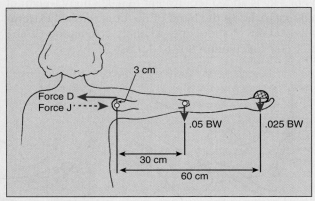

Calculation Box Figure 12-1-2

equivalent to 10.2 times the weight of the arm (Inman et al., 1944). More recent investigations of these same forces using the assumption that the muscular force is proportional to its area multiplied by its activity as determined by electromyography have yielded similar values, with a maximal joint reactive force of 89% of body weight seen at 90° of elevation in the scapular plane (Poppen and Walker, 1978).

THE BIOMECHANICS OF PITCHING

Pitching has been divided into five stages: wind up, early cocking, late cocking, acceleration, and follow-through (Tibone et al., 1994). The deltoid has been found to be responsible for elevation and abduction of the humerus in the early phases, followed by increased activation of the rotator cuff musculature in the late cocking phase acting both to rotate the humerus and prevent anterior subluxation of the glenohumeral joint (Barnes and Tullos, 1978). Specifically, the supraspinatus acts in the late cocking phase to draw the humeral head toward the glenoid, the infraspinatus and teres minor pull the humeral head posteriorly, and the subscapularis both prevents excessive external rotation of the humerus and contracts eccentrically to relieve stress on the anterior shoulder (Tibone et al., 1994). The importance of scapular (and thus glenoid) stabilization has also been recognized, and the serratus anterior has been shown to fire actively in the late cocking phase; this provides a stable platform for humeral motion. Thus, coordinated, sequential activation of the shoulder musculature is needed to prevent anterior subluxation of the glenohumeral joint and the overuse tendinitis that can ensue.

Summary

- The shoulder consists of the glenohumeral, acromioclavicular, sternoclavicular, and scapulothoracic articulations and the muscular structures that act on them to produce the most mobile joint in the body.

- The sternoclavicular joint, which connects the medial end of the clavicle to the manubrium, links the upper extremity to the thorax. The articular disc within the joint and the ligaments that surround it provide stability while allowing for significant rotation of the clavicle. Little relative motion is seen between the clavicle and acromion at the acromioclavicular joint.

- The glenohumeral joint, while known to be a precise articulation, is inherently unstable because the glenoid fossa is shallow and is able to contain only approximately one third of the diameter of the humeral head. Stability is instead provided by the capsular, muscular, and ligamentous structures that surround it.

- The three glenohumeral ligaments (superior, middle, and inferior) are discrete extensions of the anterior glenohumeral joint capsule and are critical to shoulder stability and function.

- The inferior glenohumeral ligament has the most functional significance (particularly the anterior band), acting as the primary anterior stabilizer of the shoulder when the arm is abducted 90°.

- The integrity of the shoulder capsule and the negative intra-articular force it maintains also plays an integral part in maintaining shoulder stability.

- Elevation of the arm involves motion at both the glenohumeral joint and the scapulothoracic articulation.

- Movements of the spine assist the shoulder in positioning the upper extremity in space.

- The muscles around the shoulder contribute to stability via a barrier effect by producing compressive forces on the glenohumeral joint and by eccentric contraction.

- The glenohumeral joint is a major load-bearing joint with forces equivalent to one-half body weight produced when holding the arm in an outstretched position.

Practice Questions

1. A young, healthy active male sustains an anterior shoulder dislocation and over time develops anterior instability and recurrent dislocations. What are the structures most likely damaged by the injury that now contribute to the instability?

2. A 65-year-old woman presents to her physician after falling. She complains of severe right shoulder pain and inability to lift her arm overhead, with special difficulty with initiating abduction and forward elevation. She has noted shoulder pain and weakness with activity for several years, which is now acutely worsened. What is her most likely injury?

3. A 20 year-old woman presents to her physician after falling onto her shoulder while skiing. She is diagnosed with an acromioclavicular separation, with greater than 100% superior displacement of the clavicle. What structures must be damaged to allow this degree of displacement?

REFERENCES

Barnes, D.A., Tullos, H.S. (1978). An analysis of 100 symptomatic baseball players. *Am J Sports Med, 6*, 62–67.

Boone, D.C., Azen, S.P. (1979). Normal range of motion of joints in male subjects. *J Bone Joint Surg, 61*, 756–759.

Chadwick, E.K., van Noort, A., van Der Helm, F.C. (2004). Biomechanical analysis of scapular neck malunion: A simulation study. *Clin Biomech, 19*(9), 906–912.

Colachis, S.C. Jr., Strohm, B.R. (1971). Effect of suprascapular and axillary nerve blocks on muscle force in upper extremity. *Arch Phys Med Rehabil, 52*, 22–29.

Cooper, D.E., O'Brien, S.J., Amoczky, S.P., et al. (1993). The structure and function of the coracohumeral ligament: An anatomic and microscopic study. *J Shoulder Elbow Surg, 2*, 70–77.

Curl, L.A., Warren, R.F. (1996). Glenohumeral joint stability: Selective cutting studies on the static capsular restraints. *Clin Orthop, 330*, 54–65.

DeLee, J., Drez, D. (1994). *Orthopaedic Sports Medicine. Principles and Practice.* Philadelphia: WB Saunders, 464.

De Palma, A.F. (1983). Biomechanics of the shoulder. In *Surgery of the Shoulder (3rd ed.).* Philadelphia: JB Lippincott Co, 65–85.

Fukuda, K., Craig, E.V., An, K., et al. (1986). Biomechanical study of the ligamentous system of the acromioclavicular joint. *J Bone Joint Surg, 68*, 434–440.

Gibb, T.D., Sidles, J.A., Harryman, D.T. 2nd, et al. (1991). The effect of capsular venting on glenohumeral laxity. *Clin Orthop, 268*, 120–127.

Graichen, H., Stammberger, T., Bonel, H., et al. (2000). Glenohumeral translation during active and passive elevation of the shoulder: A 3-D open MRI study. *J Biomech, 33*(5), 609–613.

Hollinshead, W.H. (1969). *Anatomy for Surgeons (vol 3).* New York: Harper & Row.

Howell, S.M., Imobersteg, A.M., Seger, D.H., et al. (1986). Clarification of the role of the supraspinatus muscle in shoulder function. *J Bone Joint Surg, 68A*, 398–404.

Inman, V.T., Saunders, J.B., Abbott, L.C. (1944). Observations on the function of the shoulder joint. *J Bone Joint Surg, 26A*, 1–30.

Itoi, E., Hsu, H.C., An, K.N. (1996). Biomechanical investigation of the glenohumeral joint. *J Shoulder Elbow Surg, 5*, 407–424.

Itoi, E., Motzkin, N.E., Morrey, B.F., et al. (1994). Stabilizing function of the long head of the biceps in the hanging arm position. *J Shoulder Elbow Surg, 3*, 135–142.

Itoi, E., Motzkin, N.E., Morrey, B.F., et al. (1992). Scapular inclination and inferior stability of the shoulder. *J Shoulder Elbow Surg, 1*, 131–139.

Kumar, V.P., Balasubramaniam, P. (1985). The role of atmospheric pressure in stabilizing the shoulder: An experimental study. *J Bone Joint Surg, 67B*, 719–721.

Laumann, U. (1987). Kinesiology of the shoulder joint. In R. Kolbel, et al. (Eds.), *Shoulder Replacement.* Berlin, Germany: Springer-Verlag, 1987.

Lippitt, S.B., Vanderhooft, J.E., Harris, S.L., et al. (1993). Glenohumeral stability from concavity-compression: A quantitative analysis. *J Shoulder Elbow Surg, 2*, 27–35.

Lucas, D.B. (1973). Biomechanics of the shoulder joint. *Arch Surg, 107*, 425.

Matsen, F., Fu, F., Hawkins, R. (Eds.) (1992). *The Shoulder: A Balance of Mobility and Stability.* Rosemont, IL: AAOS. [Vail, Colorado: Workshop Supported by the American Academy of Orthopaedic Surgeons, the National Institute of Arthritis and Musculoskeletal Skin Diseases, the American Shoulder and Elbow Surgeons, the Orthopaedic Research and Education Foundation.]

Moffey, B.F., An, K.N. (1990). Biomechanics of the shoulder. In C.A. Rockwood, F.A. Matsen III (Eds.). *The Shoulder.* Philadelphia: WB Saunders, 208–245.

Moore, K.L. (1999). *Clinically Oriented Anatomy (4th ed.).* Philadelphia: Lippincott Williams & Wilkins.

Morrey, B.F., An, K.N. (1990). Biomechanics of the shoulder. In C.A. Rockwood, F.A. Matsen III (Eds.). *The Shoulder.* Philadelphia: WB Saunders.

Murray, M.P., Gore, D.R., Gardner, G.M., et al. (1985). Shoulder motion and muscle strength of normal men and women in two age groups. *Clin Orthop, 192*, 195.

Neer, C. (1990). *Shoulder Reconstruction.* Philadelphia: WB Saunders, 29.

O'Brien, S.J., Neves, M.C., Amoczky, S.P., et al. (1990). The anatomy and histology of the inferior glenohumeral ligament complex of the shoulder. *Am J Sports Med, 18*, 579–584.

Pagnani, M.J., Deng, X.H., Warren, R.F., et al. (1995). Effect of lesions of the superior portion of the glenoid labrum on glenohumeral translations. *J Bone Joint Surg, 77A*, 103–110.

Poppen, N.K., Walker, P.S. (1978). Forces at the glenohumeral joint in abduction. *Clin Orthop, 135*, 165–170.

Poppen, N.K., Walker, P.S. (1976). Normal and abnormal motion of the shoulder. *J Bone Joint Surg, 58A*, 195–201.

Rockwood, C.A., Jr. (1975). Dislocations about the shoulder. In C.A. Rockwood Jr., D.P. Green (Eds.). *Fractures (vol 1, 1st ed.).* Philadelphia: JB Lippincott Co, 624–815.

Rockwood, C., Matsen, F. (1990). *The Shoulder.* Philadelphia: WB Saunders, 219.

Sahara, W., Sugamoto, K., Murai, M., et al. (2006). 3D Kinematic analysis of the acromioclavicular joint during arm abduction using vertically open MRI. *J Orthop Res, 24*(9), 1823–1831.

Simon, S.R. (Ed). (1994). *Orthopaedic Basic Science.* Rosemont, IL: American Association of Orthopaedic Surgeons, 527.

Simovitch, R., Sanders, B., Ozbaydar, M., et al. (2009). Acromioclavicular joint injuries: Diagnosis and management. *J Amer Acad Orthop Surg, 17*(4), 207–219.

Soslowsky, L.J., Flatow, E.L., Bigliani, L.U., et al. (1992). Articular geometry of the glenohumeral joint. *Clin Orthop, 285*, 181–190.

Steinbeck, J., Liljenqvist, U., Jerosh, J. (1998). The anatomy of the glenohumeral ligamentous complex and its contribution to anterior shoulder stability. *J Shoulder Elbow Surg, 7*(2), 122–126.

Tibone, J., Patek, R., Jobe, F.W., et al. (1994). The shoulder: Functional anatomy, biomechanics and kinesiology. In J.C. DeLee, D. Drez (Eds.). *Orthopaedic Sports Medicine*. Philadelphia: WB Saunders.

Warner, J.P. (1993). The gross anatomy of the joint surfaces, ligaments, labrum and capsule. In F.A. Matsen, F.H. Fu, R.J. Hawkins (Eds.). *The Shoulder: A Balance of Mobility and Function*. Rosemont, IL: American Association of Orthopaedic Surgeons, 7–27.

Warner, J.P., Deng, X.H., Warren, R.F., et al. (1992). Static capsuloligamentous restraints to superior-inferior translations of the glenohumeral joint. *Am J Sports Med, 20*, 675–678.

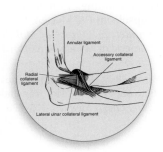

CHAPTER 13

Biomechanics of the Elbow

Laith M. Jazrawi, Joseph D. Zuckerman,
Brett H. Young, and Michael S. Day

Introduction

The elbow is a complex joint that functions as a fulcrum for the forearm lever system that is responsible for positioning the hand in space. A detailed understanding of the biomechanics of elbow function is essential for the clinician to treat effectively pathologic conditions affecting the elbow joint.

Anatomy

The elbow joint complex allows two types of motion: flexion-extension and pronation-supination. The humeroulnar and humeroradial articulations allow elbow flexion and extension and are classified as ginglymoid or hinged joints. The proximal radioulnar articulation allows forearm pronation and supination and is classified as a trochoid joint. The elbow joint complex, when considered in its entirety, is therefore a trochleoginglymoid joint. The trochlea and capitellum of the distal humerus are internally rotated 3° to 8° (Fig. 13-lC) and in 94° to 98° of valgus with respect to the longitudinal axis of the humerus (Fig. 13-lA). The distal humerus is anteriorly angulated

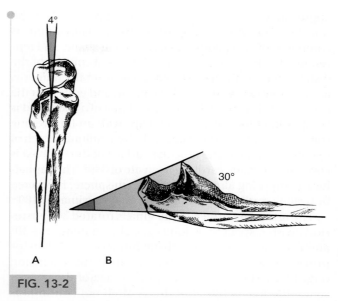

FIG. 13-2

Angular orientation of the proximal ulna in the AP (**A**) and lateral (**B**) planes.

30° along the long axis of the humerus (Fig. 13-1B). The articular surface of the ulna is oriented in approximately 4° to 7° of valgus angulation with respect to the longitudinal axis of its shaft (Fig. 13-2A).

The distal humerus is divided into medial and lateral columns that terminate distally with the trochlea connecting the two columns (Fig. 13-3). The medial column diverges from the humeral shaft at a 45° angle and ends

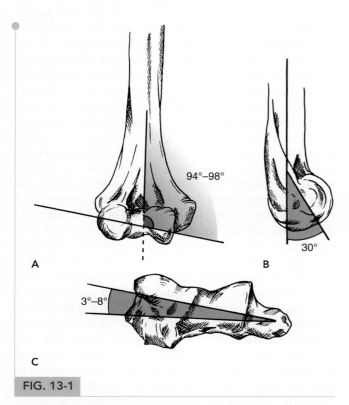

FIG. 13-1

Angular orientation of the distal humerus in the anteroposterior (**A**), lateral (**B**), and axial (**C**) projections.

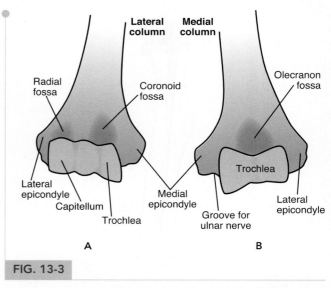

FIG. 13-3

Anterior (**A**) and posterior (**B**) projections of the distal humerus highlighting the medial and lateral columns.

approximately 1 cm proximal to the distal end of the trochlea. The distal one-third of the medial column is composed of cancellous bone, is ovoid in shape, and represents the medial epicondyle. The lateral column of the distal humerus diverges at a 20° angle from the humerus at the same level as the medial column and ends with the capitellum. The trochlea takes the shape of a spool and is composed of medial and lateral lips with an intervening sulcus. This sulcus articulates with the semilunar notch of the proximal ulna. The articular surface of the trochlea is covered by hyaline cartilage in an arc of 330°. The capitellum, composing almost a perfect hemisphere, is covered by hyaline cartilage forming an arc of approximately 180°.

The articular surface of the ulna is rotated 30° posteriorly with respect to its long axis. This matches the 30° anterior angulation of the distal humerus, which helps provide stability to the elbow joint in full extension (Fig. 13-2). The arc of articular cartilage of the greater sigmoid notch is 180°, but this is often not continuous in its midportion. In more than 90% of individuals, this area is composed of fatty, fibrous tissue (Walker, 1977). As Morrey noted (1986), this anatomic feature explains the propensity for fractures to occur in this area, since this portion of the greater sigmoid notch is not supported by stronger subchondral bone.

The radial neck is angulated 15° from the long axis in the anterior-posterior plane away from the bicipital tuberosity (Fig. 13-4). Four-fifths of the radial head is covered by hyaline cartilage. The anterolateral one-fifth lacks articular cartilage and strong subchondral bone, explaining the increased propensity for fractures to occur in this region.

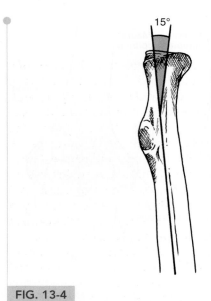

15°

FIG. 13-4

Angulation of radial head/neck in relation to radial shaft.

Kinematics

Elbow flexion and extension take place at the humeroulnar and humeroradial articulation. The normal range of flexion-extension is from 0° to 146° with a functional range of 30° to 130°. The normal range of forearm pronation-supination averages from 71° of pronation to 81° of supination (Morrey et al., 1981). As the elbow is flexed, the maximum angle of supination increases, while the maximum angle of pronation decreases (Shaaban et al., 2008). Most activities are accomplished within the functional range of 50° pronation to 50° supination. Clinically, patients can tolerate flexion contractures of up to 30°, which is consistent with the functional range values described previously. Flexion contractures greater than 30° are associated with complaints of significant loss of motion. There is a considerable and rapid loss of the ability to reach in space with flexion contractures greater than 30° (Fig. 13-5) (An and Morrey, 1991).

The axis of rotation for flexion-extension has been shown by several investigators to be at the center of the trochlea, supporting the concept that elbow flexion can be represented as a uniaxial hinge. Ewald and Ishizaki, on the other hand, discovered a changing axis of rotation with elbow flexion (Ewald, 1975; Ishizuki, 1979). London demonstrated that the axis of rotation passes through the center of concentric arcs outlined by the bottom of the trochlear sulcus and the periphery of the capitellum (London, 1981). He also noted that the surface joint motion during flexion-extension was primarily of the gliding type and that with the extremes of flexion-extension (the final 5°–10° of both flexion and extension), the axis of rotation changed and the gliding/sliding joint motion changed to a rolling-type motion. The rolling occurs at the extremes of flexion and extension as the coronoid process comes into contact with the floor of the humeral coronoid fossa and the olecranon contacts the floor of the olecranon fossa. In addition, internal axial rotation of the ulna has been shown to occur during early flexion and external axial rotation during terminal flexion, demonstrating that the elbow cannot be truly represented as a simple hinge joint. In conclusion, there is evidence to suggest that the elbow has a changing center of rotation during flexion-extension and functions as a loose rather than a "pure" hinge joint (Duck et al., 2003).

Despite variation in findings among investigators, Morrey, Tanaka, and An (1991) have stated that the deviation of the center of joint rotation is minimal and the reported variation is probably due to limitations in experimental design. Therefore, the ulnohumeral joint could be assumed to move as a uniaxial articulation except at the extremes of flexion-extension. The axis of rotation of flexion-extension occurs about a tight locus

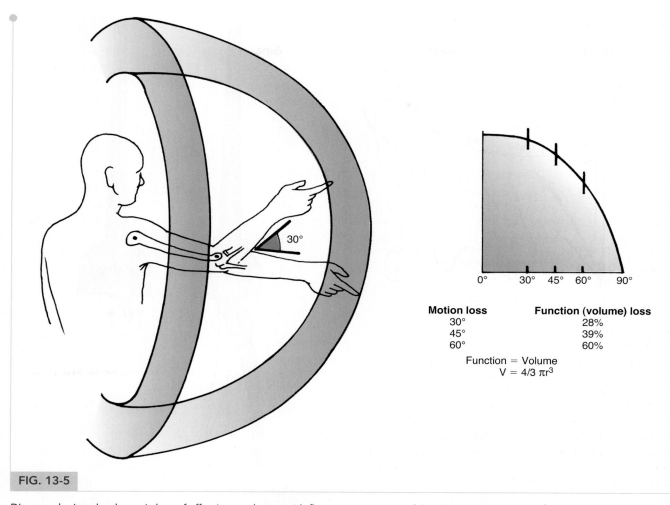

Motion loss
30°
45°
60°

Function (volume) loss
28%
39%
60%

Function = Volume
V = 4/3 πr³

FIG. 13-5

Diagram depicts the dramatic loss of effective reach area with flexion contractures of the elbow greater than 30°.

of points measuring 2 to 3 mm in its broadest dimension and is in the center of the trochlea and capitellum on the lateral view. It is approximated by a line passing through the center of the lateral epicondyle and trochlea and then through the anteroinferior aspect of the medial epicondyle (Fig. 13-6) (Morrey and Chao, 1976). These factors should be taken into account during joint replacement procedures of the elbow as well as when placing hinged external fixators across the elbow joint (Figgie et al., 1986).

Pronation and supination take place primarily at the humeroradial and proximal radioulnar joints with the forearm rotating about a longitudinal axis passing through the center of the capitellum and radial head and the distal ulnar articular surface. This axis is oblique in relation to the anatomic axis of the radius and ulna. During pronation-supination, the radial head rotates within the annular ligament and the distal radius rotates

around the distal ulna in an arc outlining the shape of a cone. Carret et al. (1976) studied the instant centers of rotation at the proximal and distal radioulnar joints with the forearm in varying degrees of pronation and supination. They found that the proximal instant center of rotation varied with differences in the curvature of the radial head among individuals. Chao and Morrey (1978) investigated the effect of pronation and supination on the position of the ulna and found no significant axial rotation or valgus deviation of this bone during forearm rotation when the elbow was fully extended. O'Driscoll et al. (1991) demonstrated that internal axial rotation of the ulna occurs with pronation while external axial rotation occurs with supination. Kapandji (1982) has suggested that both the distal radius and ulna rotate about the axis of pronation-supination with the ulnar arc of rotation being significantly smaller than the radial arc of rotation. Galik et al. (2007) showed that the pronation-supination

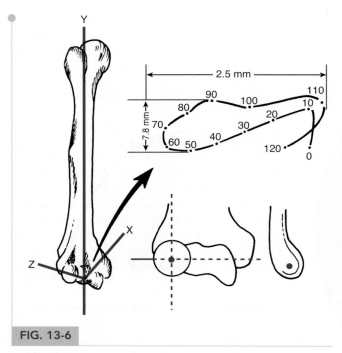

FIG. 13-6

Dimensions of the locus of the instant center of rotation. As depicted, the axis of rotation runs through the center of the trochlea and capitellum.

axis of the forearm is nearly constant and is not affected by annular ligament transection. Ray et al. (1951) demonstrated some varus-valgus movement of the distal ulna with rotation on an axis extending from the radial head through the index finger (Fig. 13-7).

Palmer et al. (1982) have demonstrated proximal radial migration with forearm pronation. This has been supported by observations at elbow arthroscopy and in vitro biomechanical studies such those by Fu et al. (2009). They showed proximal translation of the radius to be minimized with the arm in supination. They also demonstrated that proximal translation decreased as the elbow was flexed from 0° to 90°. In addition, due to the ovoid shape of the radial head, its axis is displaced laterally in pronation by 2 mm to allow space for the medial rotation of the radial tuberosity (Kapandji, 1982).

Carrying Angle

The valgus position of the elbow in full extension is commonly referred to as the carrying angle. The carrying angle is defined as the angle between the anatomic axis of the ulna and the humerus measured in the anteroposterior (AP) plane in extension or simply the orientation of the ulna with respect to the humerus, or vice versa, in

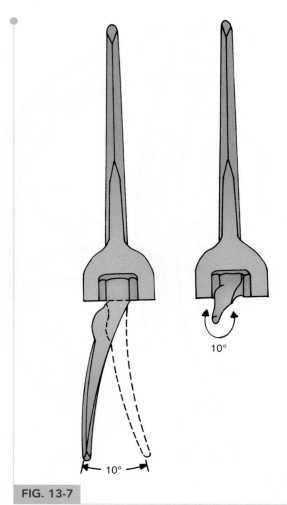

FIG. 13-7

Diagram of semiconstrained total elbow replacement allowing a variable amount of toggle in the varus/valgus and axial planes. The design takes into consideration the fact that the motion of the elbow cannot be purely represented as a simple hinge.

full extension (Fig. 13-8). The angle is less in children as compared to adults and greater in females as compared to males, averaging 10° and 13° of valgus, respectively, with a wide distribution in both (Atkinson and Elftman, 1945; Mall, 1905). Steindler (1955) reported a gradual increase in the carrying angle with age but found no statistical difference between men and women in this rate of increase or the carrying angle. Chang et al. (2008) found increased carrying angle to be an independent risk factor for nontraumatic ulnar neuropathy at the elbow.

There is controversy regarding the change in the carrying angle as the elbow is flexed. An et al. (1984) have noted this controversy arises from the various reference systems used to determine the carrying angle. They noted that when the carrying angle is defined as either

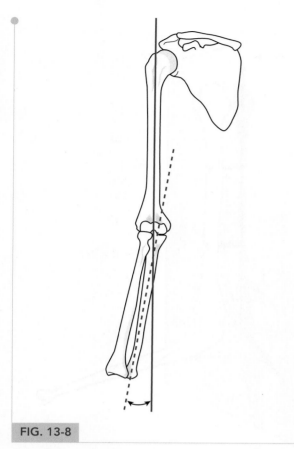

FIG. 13-8

The carrying angle of the elbow, formed by the interception of the long axes of the humerus and the ulna with the elbow fully extended and the forearm supinated. Valgus angulation normally ranges from 10° to 15°.

the angle formed between the long axis of the humerus and ulna on a plane containing the humerus or vice versa, the carrying angle changes minimally with flexion. If the carrying angle is defined as the abduction-adduction angle of the ulna relative to the humerus using Eulerian angles to describe arm motion, the carrying angle decreases with joint flexion changing to varus in extreme flexion (Fig. 13-9).

Elbow Stability

Valgus forces at the elbow are resisted primarily by the anterior band of the medial collateral ligament (MCL). The MCL complex consists of an anterior bundle, posterior bundle, and the transverse ligament (Fig. 13-10). Neutral forearm rotation results in greater valgus laxity than pronation or supination (Safran et al., 2005). Other authors have shown increased laxity in

pronation compared to supination (Pomianowski et al., 2001). The anterior bundle of the MCL tightens in extension whereas the posterior bundle tightens in flexion. This occurs because the medial collateral ligament complex does not originate at the center of the axis of elbow rotation (Fig. 13-11). The anterior band of the MCL originates from the inferior surface of the medial epicondyle of the distal humerus and inserts along the medial edge of the olecranon. With an intact anterior band, the radial head does not offer significant additional resistance to valgus stress. However, with a transected or disrupted anterior band, the radial head becomes the primary restraint to valgus stress, emphasizing its function as a secondary stabilizer in elbows with an intact MCL (Palmer et al., 1982). Despite studies by Morrey (Morrey et al., 1988, 1991) demonstrating the secondary valgus stabilizing effect of the radial head, several investigators have noted increased valgus laxity after radial head excision (Coleman et al., 1987; Gerard et al., 1984; Johnston, 1962; Morrey et al., 1979). However, this does not seem to be clinically disabling (Hotchkiss, 1997). The stability can be restored to near native levels with radial head arthroplasty, only if the collateral ligaments are intact (Beingessner et al., 2004).

The throwing motion illustrates the role of the MCL in a common functional activity. Baseball pitchers are frequently at risk for MCL injury due to the repetitive valgus stress placed on their elbows by the nature of the throwing motion. Recent investigations suggest that increased valgus torque at the elbow is associated with late trunk rotation, reduced shoulder external rotation, and increased elbow flexion (Aguinaldo and Chambers, 2009).

Selective ligament transection studies have shown that in extension, resistance to valgus stress is shared equally by the MCL, capsule, and joint articulation. In flexion, the primary resistor to valgus stress is the MCL (Morrey and An, 1983). In extension, the elbow articulation provides most of the resistance to varus stress followed by the anterior capsule. Hull et al. (2005) showed that resistance to varus displacement decreased after removal of more than 50% of the coronoid, particularly at lower elbow flexion angles. In a "terrible triad" model (elbow dislocation with injury to the lateral collateral ligament [LCL] complex, radial head, and coronoid), the same group demonstrated that LCL repair and radial head replacement did not overcome the varus instability created by a loss of 75% of the coronoid (Fern et al., 2009).

In flexion, the elbow articulation remains the primary restraint to varus stress followed by the anterior capsule and LCL, respectively, with the LCL contributing only 9% (Table 13-1). Elbow extension is limited primarily by the anterior capsule and anterior bundle of MCL. Excision of the olecranon fossa fat pad has been shown to provide

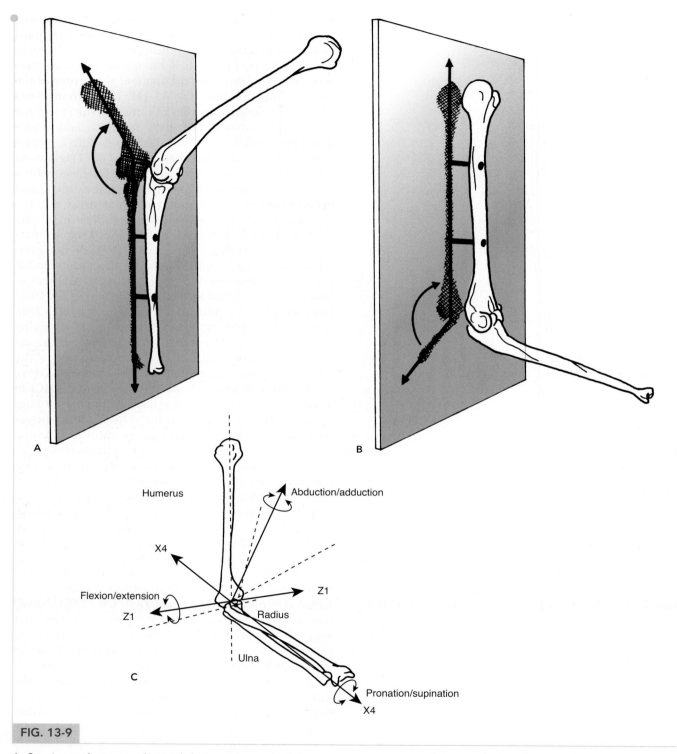

FIG. 13-9

A. Carrying angle measured as angle between long axis of ulna and long axis of projection of humerus on plane containing the ulna. **B.** Carrying angle measured as the angle formed between the long axis of the humerus and long axis of the projection of the ulna on plane containing the humerus. **C.** Eulerian angle measurement of ulnar motion in reference to humerus. Abduction/adduction rotates about the axis orthogonal to both the Z and X4 axes; flexion/extension rotates about the Z1 axis; forearm axial rotation takes place about the X4 axis.

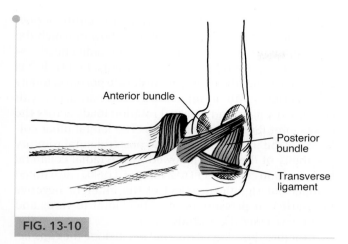

FIG. 13-10

Medial collateral ligament complex containing anterior and posterior bundles as well as a transverse component.

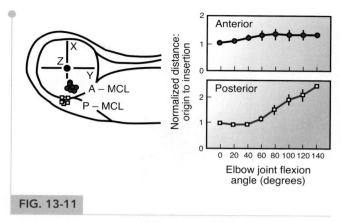

FIG. 13-11

Origin of the anterior and posterior bundles of the MCL. Since the MCL does not originate on the axis of elbow rotation, there are changes in its length as a function of elbow flexion. The anterior bundle, which is closer to the axis of rotation, is the most isometric.

5° of additional extension (Walker, 1977). Furthermore, Morrey et al. (1991) have demonstrated an almost linear decrease in ulnohumeral joint stability with serial removal of 25% to 100% of the olecranon. More recent studies have focused on the treatment of valgus extension overload and subsequent osteophyte formation. Kamineni et al. (2004) showed that the strain on the anterior bundle of the MCL increased with olecranon resection beyond 3 mm, with a marked increase occurring at 9 mm of resection (Kamineni et al., 2003). In contrast, Levin et al. (2004) found that 12 mm of olecranon resection did not significantly increase MCL strain, although methodological differences among these studies make direct comparison difficult.

In addition to the static ligamentous stabilizers of the elbow, the flexor-pronator muscles on the medial side of the elbow have been shown to contribute to dynamic valgus stability (Park and Ahmad, 2004; Lin et al., 2007;

Hsu et al., 2008; Udall et al., 2009). Several investigators have found the flexor carpi ulnaris (FCU) to be the most significant stabilizer, by creating a significant varus moment that unloads and protects the MCL (Park and Ahmad, 2004; Hsu et al., 2008; Lin et al., 2007). In contrast, Udall et al. (2009) showed that flexor digitorum superficialis (FDS) reduced the valgus angle more than other flexor-pronator muscles. Seiber et al. (2009) reported that medial elbow musculature affects elbow stability to a greater degree when the forearm is in supination, although the MCL contributes more than twice as much to valgus stability.

The LCL complex consists of the radial collateral ligament that originates from the lateral epicondyle and inserts on the annular ligament; the lateral ulnar collateral ligament, which originates from the lateral epicondyle and passes superficial to the annular ligament,

TABLE 13-1

Percent Contribution of Restraining Force during Displacement (Rotational or Distractional)

Position	Stabilizing Element	Distraction	Varus	Valgus
Extension	MCL	12	—	31
	LCL	10	14	—
	Capsule	70	32	38
	Articulation	—	55	31
Flexion	MCL	78	—	54
	LCL	10	9	—
	Capsule	8	13	10
	Articulation	—	75	33

MCL, medial collateral ligament complex; LCL, lateral collateral ligament complex.

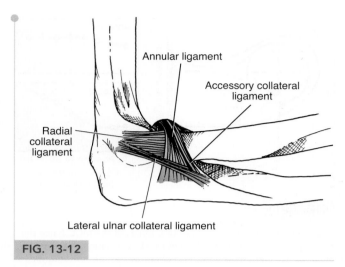

FIG. 13-12

The lateral collateral ligament complex.

inserting on the supinator crest of the ulna; and the accessory lateral collateral ligament (Fig. 13-12). The origin of the LCL complex lies at the center of the axis of elbow rotation, explaining its consistent length throughout the flexion-extension arc (Fig. 13-13). Although Morrey and An (1983) have demonstrated only a minimal contribution of the LCL complex to varus stability, others have shown the LCL complex to be an important stabilizer of the humeroulnar joint with forced varus and external rotation (Daria et al., 1990; Dunning et al., 2001; Durig et al., 1979; Josefsson et al., 1987; O'Driscoll et al., 1990b; Olsen et al., 1996; Osborne and Cotterhill, 1966). When either the radial collateral or lateral ulnar collateral ligament is transected, there is no increase in varus laxity (Dunning et al., 2001). The lateral elbow musculature also contributes to varus stability, particularly when the forearm is in pronation (Seiber et al., 2009).

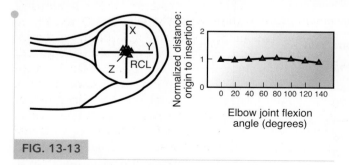

FIG. 13-13

Origin of the lateral collateral ligament complex at the elbow axis of rotation. The ligament remains isometric throughout the flexion/extension range of the elbow.

O'Driscoll et al. (1990b) described the entity of posterolateral rotatory instability of the elbow in which the ulna supinates on the humerus and the radial head dislocates in a posterolateral direction (Fig. 13-14). It has been shown that the elbow can dislocate posterolaterally or posteriorly with an intact MCL. This can occur with combined valgus and external rotation loads across the elbow joint (Sojbjerg et al., 1989). The lateral ulnar collateral is the primary restraint to posterolateral rotatory instability of the elbow followed by the radial collateral ligament and capsule. O'Driscoll et al. (1990a) also noted a small but significant effect of the inherent negative intra-articular pressure of the elbow joint to varus and rotation stresses (Case Study 13-1).

Structures limiting passive flexion include the capsule, triceps, coronoid process, and the radial head. Structures limiting elbow extension include the olecranon process and the anterior band of the MCL. Passive resistance to pronation-supination is provided in large part by the antagonist muscle group on stretch rather than ligamentous structures (Braune and Flugel, 1842). Others have shown that the quadrate ligament provides restraint to forearm rotation (Spinner and Kaplan, 1970).

Longitudinal stability of the forearm is provided by both the interosseous membrane and the triangular fibrocartilage. Lee et al. (1992) demonstrated marked proximal migration of the radius only after 85% of the interosseous membrane was sectioned. Hotchkiss et al. (1989) demonstrated increased stiffness of the interosseous membrane with forearm supination and noted that the triangular fibrocartilage complex was responsible for 8% of longitudinal forearm stiffness while the central band of the interosseous membrane provided 71%. DeFrate et al. (2001) showed that interosseous membrane transfers more force from the radius to the ulna in supination than in pronation or neutral rotation, regardless of flexion angle. Reardon et al. (1991) demonstrated in cadavers that with removal of the radial head alone, proximal radial migration was 0.4 mm. When combined with interosseous membrane transection alone, proximal radial migration increased to 4.4 mm. Radial head resection when combined with triangular fibrocartilage complex (TFCC) transection caused 2.2 mm of proximal radial migration. The combination of radial head resection, interosseous membrane transection, and TFCC transection led to the greatest increase in proximal radius migration of 16.8 mm.

The coronoid process also plays a role in longitudinal stability and has been shown to prevent posterior displacement of the ulna. When more than 50% of the coronoid process is resected in vitro, elbows displace more readily in the face of an axial load, especially at elbow flexion angles of 60° or more (Closkey et al., 2000).

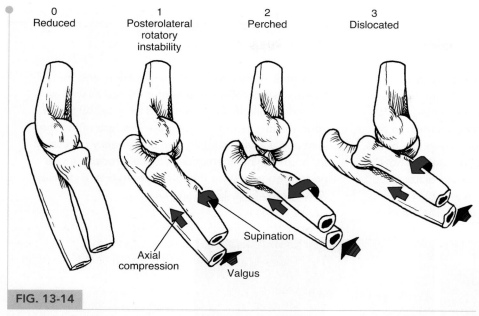

0
Reduced

1
Posterolateral
rotatory
instability

2
Perched

3
Dislocated

Supination

Axial
compression

Valgus

FIG. 13-14

Clinical stages of posterolateral rotatory instability of the elbow.

Kinetics

The primary flexor of the elbow is the brachialis, which arises from the anterior aspect of the humerus and inserts on the anterior aspect of the proximal ulna (Fig. 13-15). The biceps arises via a long head tendon from the supraglenoid tubercle and a short head tendon from the coracoid process of the scapula and inserts on the bicipital tuberosity of the radius. It is active in flexion when the forearm is supinated or in the neutral position. The brachioradialis, which originates from the lateral two thirds of the distal humerus and inserts on the distal aspect of the radius near the radial styloid, is active during rapid flexion movements of the elbow and when weight is lifted during a slow flexion movement (Basmajian and Latif, 1957). The brachialis, biceps, brachioradialis, and extensor carpi radialis are the major flexors of the elbow, the brachialis possessing the greatest work capacity (An et al., 1981), though the biceps may be preferentially recruited during fast exercise protocols (Kulig et al., 2001).

The primary extensor of the elbow, the triceps, is composed of three separate heads. The long head originates from the infraglenoid tubercle, and the medial and lateral heads originate from the posterior aspect of the humerus (Fig. 13-15). The three heads coalesce to form one tendon that inserts onto the olecranon process of the ulna. The medial head is the primary extensor, and the lateral and long heads act in reserve (Basmajian, 1969). The anconeus muscle, which arises from the posterolateral aspect of the distal humerus and inserts onto

the posterolateral aspect of the proximal ulna, is also active in extension. This muscle is active in initiating and maintaining extension. While the triceps, anconeus, and flexor carpi ulnaris are active in extension, the triceps has the largest work capacity of all the elbow extensors (An et al., 1981).

Muscles involved in supination of the forearm include the supinator, biceps, and the lateral epicondylar extensors of the wrist and fingers. The primary muscle involved in supination is the biceps brachii. The biceps generates four times more torque with the forearm in the pronated position than in the supinated position (Haugstvedt et al., 2001). The supinator arises from the lateral epicondyle of the humerus and the proximal lateral aspect of the ulna and inserts into the anterior aspect of the supinated proximal radius.

Muscles involved in pronation include the pronator quadratus (PQ) and pronator teres (PT). PQ and PT are active throughout the whole rotation, being most efficient around the neutral position of the forearm (Haugstvedt et al., 2001). The pronator quadratus originates from the volar aspect of the distal ulna and inserts onto the distal and lateral aspect of the supinated radius. The pronator teres is more proximally located, arising from the medial epicondyle of the humerus and inserting onto the lateral aspect of the midshaft of the supinated radius. The pronator quadratus is the primary pronator of the forearm regardless of its position. The pronator teres is a secondary pronator when rapid pronation is required or during resisted pronation (Basmajian, 1969).

Case Study 13-1

Elbow Fracture Dislocation

A 16-year-old male gymnast falls onto an outstretched arm, producing abnormal loads in the elbow complex. The axial loading during the fall onto the outstretched position caused a fracture on the radial head, altering the articular congruity of the radiocapitellum joint and the stability of the elbow.

Instability of the joint occurs in posterolateral dislocation (Case Study Figs. 13-1A and B). The ulna supinates onto the humerus, the radial head dislocates in the posterolateral direction, and the lateral ulnar collateral ligament is injured as well as the radial collateral ligament and capsule. All these abnormalities lead to an increase in stress within the joint and a loss of stability and congruency necessary for normal joint kinematics. Surgical procedure was performed to restore the joint congruity and stability (Case Study Figs. 13-1C and D).

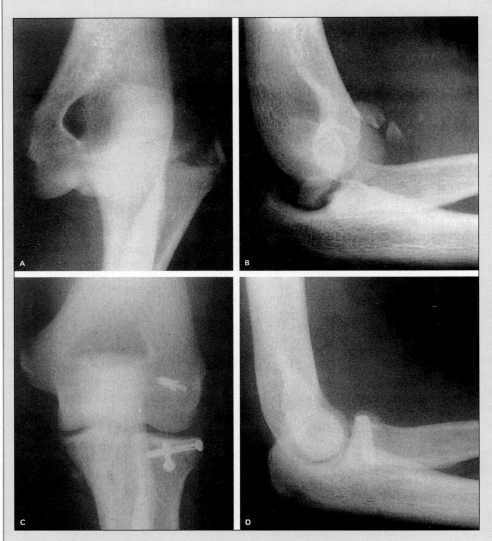

Case Study Figure 13-1 **A.** Anteroposterior radiograph that confirms posterolateral elbow dislocation. **B.** Lateral radiography that shows fracture on the radial head and capitellum. **C** and **D.** Posterior and lateral view. Postoperative radiograph. Congruence of the joint has been achieved.

MUSCLES OF RIGHT UPPER EXTREMITY

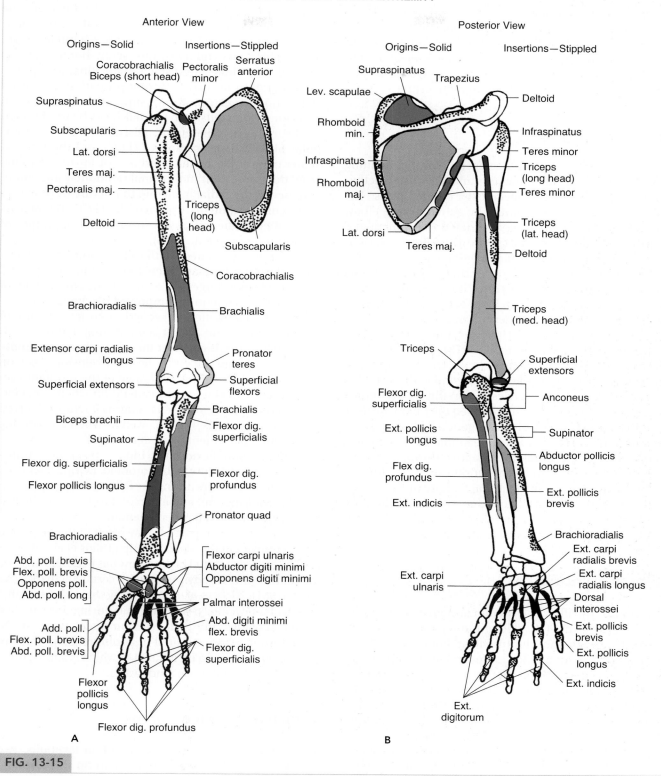

FIG. 13-15

Origin and insertions of muscles of the upper extremity. **A.** Anterior view. **B.** Posterior view.

In a study examining elbow strength in normal individuals, supination strength was shown to be 20% to 30% greater than pronation strength (Askew et al., 1986). Consistent with muscle cross-sectional area and moment arms, flexion strength was 30% greater than extension strength. Last, males were consistently 40% stronger than females in elbow strength testing.

Electromyography

Electromyography (EMG) has been helpful in defining the contributions of elbow musculature during activities of daily living and specifically defined tasks. The biceps brachii is only minimally active during elbow flexion when the forearm is pronated (Basmajian and Latif, 1957; Funk et al., 1987; Maton and Bouisset, 1977; Stevens et al., 1973). Brachialis activity, however, is not affected by forearm rotation during flexion (Funk et al., 1987; Stevens et al., 1973). The brachioradialis is also active during flexion. This activity is enhanced when the forearm is in a neutral or pronated position (Basmajian and Travill, 1961; DeSousa et al., 1961; Funk et al., 1987; Stevens et al., 1973). EMG data demonstrate that the medial head of the triceps and anconeus are active during elbow extension with the lateral and long heads of the triceps serving as secondary extensors. Morrey (1993) concluded the following from the EMG data: (1) The biceps is generally less active in full pronation of the forearm, secondary to its role as a supinator; (2) the brachialis is active throughout flexion and is believed to be the workhorse of flexion; (3) there is an increase in electrical activity of the triceps with increased elbow flexion, due to the stretch reflex; and (4) the anconeus is active in all positions and is considered to be a dynamic joint stabilizer.

Elbow Joint Forces

Halls and Travill (1964) demonstrated that in intact cadaver forearms, 43% of longitudinal forces are transmitted through the ulnotrochlear joint and 57% are transmitted through the radiocapitellar joint. Ewald et al. (1977) determined that the elbow joint compressive force was eight times the weight held by an outstretched hand. An and Morrey (1991) determined that during strenuous weightlifting, the resultant force at the ulnohumeral joint ranges from one to three times body weight. The coronoid process bears 60% of the total compressive stress when the elbow joint is extended (Chantelot et al., 2008). Force transmission through the radial head is greatest between 0 and 30°

of flexion and is greater in pronation than supination (Morrey et al., 1988). In extension, the force on the radial head decreases from 23% (of total load) in neutral rotation to 6% in full supination (Chantelot et al., 2008). This is secondary to the "screw-home" mechanism of the radius with respect to the ulna, with proximal migration occurring during pronation and distal translation occurring during supination. As has been mentioned previously, the radial head bears the load at the radiocapitellar joint. Disruption of the triangular fibrocartilage complex (TFCC) and the interosseous membrane in the presence of an intact radial head does not result in proximal radioulnar migration. Absence of the radial head due to fracture or resection and a concomitant disruption of the TFCC and interosseous membrane will result in proximal migration of the radius (Sowa et al., 1995).

The force generated at the elbow joint is greatest when flexion is initiated. Increased flexion strength and decreased elbow forces are seen with the elbow at 90° of flexion. This is due to the improved mechanical advantage of the elbow flexors secondary to lengthening of the flexion moment arm. It is interesting that the resultant force vector direction at the elbow changes by more than 180° though the entire flexion extension (Pearson et al., 1963). Clinically, this change in the resultant vector should be taken into consideration when considering internal fixation of distal humerus fractures (Morrey, 1994; Pearson et al., 1963) as well as when considering total joint replacement (Goel et al., 1989).

During elbow flexion, the ulna is posteriorly translated as contact occurs at the coronoid. During the forced extension that occurs during the follow-through phase of the throwing motion, impaction of the olecranon against the olecranon fossa has been demonstrated in the overhead athlete. This impaction may result in the formation of osteophytes at the olecranon tip (Tullos et al., 1972).

The force generated in the elbow has been shown to be up to three times body weight with certain activities (An et al., 1981). Nicol et al. (1977), using three-dimensional biomechanical analysis, found that during dressing and eating activities, the joint reaction forces were 300 N. Rising from a chair resulted in a joint reaction force of 1,700 N and pulling a table 1,900 N, which is almost three times body weight (Case Study 13-2).

Articular Surface Forces

Contact areas of the elbow occur at four locations: Two are located at the olecranon and two on the coronoid (Fig. 13-16) (Stormont et al., 1985). The humeroulnar contact area increases from elbow extension to flexion.

Lateral Epicondylitis (Tennis Elbow)

A 57-year-old woman, an avid tennis player, developed a gradual onset of pain in the right elbow, which was exacerbated by playing tennis.

A high strain rate resulting from continuous flexion/extension of the elbow in combination with pronation/supination of the forearm generated repetitive microtraumas. This overuse injury exceeded the reparative process in tendons that insert in the lateral epicondyle, developing a lateral epicondylitis. The patient was initially treated nonoperatively with physical therapy and a tennis elbow wrist band for 6 months with no resolution of symptoms. She ultimately required surgery (Case Study Fig. 13-2).

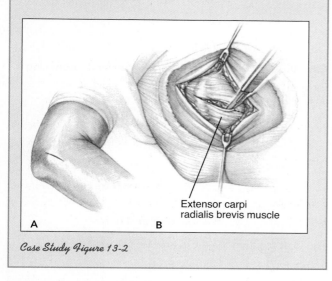

Extensor carpi radialis brevis muscle

A   B

Case Study Figure 13-2

In addition, the radial head also increases its contact area with the capitellum from extension to flexion. During valgus/varus loads to the elbow, Morrey et al. (1988) demonstrated the varus/valgus pivot point to be located at the midpoint of the lateral aspect of the trochlea.

Calculation of Joint Reaction Forces at the Elbow

Because several muscles participate in producing flexion and extension of the elbow, a few simplifying assumptions must be made for joint reaction forces to be estimated in certain static and dynamic situations. In the following static example, the simplified free-body technique for

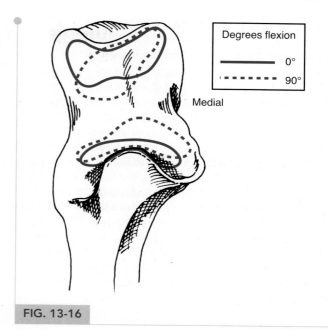

Degrees flexion

——— 0°
- - - - 90°

Medial

FIG. 13-16

Contact areas in the sigmoid fossa during elbow flexion demonstrating that the contact areas move toward the center of the sigmoid fossa during elbow flexion.

coplanar forces is used to calculate the joint reaction force at the elbow during flexion with and without an object in the hand (see Calculation Box 13-1). The elbow is flexed 90°; it is assumed that the predominant elbow flexors are the brachialis and the biceps and that the force produced through the tendons of these muscles (*M*) acts perpendicular to the longitudinal axis of the forearm. The distance between the center of rotation of the elbow joint and point of insertion of the tendons of these muscles (the lever arm of *M*) is approximately 5 cm. The mass of the forearm (2 kg) produces a gravitational force (*W*) equal to 20 N. The lever arm of *W*, the distance from the center of rotation of the elbow to the midpoint of the forearm, is 13 cm. The force produced by any weight held in the hand (*P*) acts at a distance of 30 cm from the center of rotation of the elbow joint.

The muscle force required to keep the elbow in the flexed position (*M*) is calculated with the equilibrium equation for moments. The equilibrium equation for forces is then used to calculate the joint reaction force on the trochlear fossa (*J*). When no object is held in the hand, the muscle force is calculated to be 52 N and the joint reaction force, 32 N. By contrast, when a 1-kg weight is held in the hand, producing a gravitational force (*P*) of 10 N at a distance of 30 cm from the center of elbow rotation, the required muscle forces (*M*) increase to 112 N and the joint reaction force more than doubles, reaching 82 N. Thus, small loads applied to the hand dramatically increase the elbow joint reaction force.

CALCULATION BOX 13-1

Joint Reaction Force: Elbow Flexion

The reaction force on the elbow joint during elbow flexion with and without an object in the hand can be calculated by means of the simplified free-body technique for coplanar forces and the equilibrium equations that state that the sum of the moments and the sum of the forces acting on the elbow joint must be zero. The primary elbow flexors are assumed to be the biceps and the brachialis muscles. The force produced through the tendons of these muscles (M) acts at a distance of 5 cm from the center of rotation of the joint (indicated by the *hollow circle*). The force produced by the weight of the forearm (W), taken to be 20 N, acts at a distance of 13 cm from the center of rotation. The force produced by any weight held in the hand (P) acts at a distance of 30 cm from the center of rotation (Calculation Box Fig. 13–1-1).

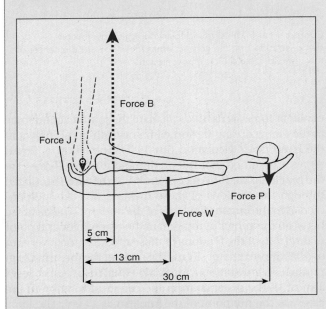

Calculation Box Figure 13-1-1

Case A. No object is held in the hand. M is calculated with the equilibrium equation for moments. Clockwise moments are considered to be positive, whereas counterclockwise moments are considered to be negative.

$$\sum M = 0.$$

$$(13\ cm \times W) + (30\ cm \times P) - (5\ cm \times B) = 0$$

If $W = 20$ N and $P = 0$.

$$B = \frac{13\ cm \times 20\ N}{5\ cm}$$

B is calculated to be 52 N.

J, the reaction force on the trochlear fossa of the ulna, can now be calculated by means of the equilibrium equation for forces. Gravitational forces are negative; forces in the opposite direction are positive.

$$\sum F = 0.$$

$$B - J - W - P = 0$$

$$J = 52\ N - 20\ N - 0\ N$$

J is found to be 32 N.

Case B. An object of 1 kg is held in the hand, producing a force of 10 N (P).

$$\sum M = 0.$$

If $W = 20$ N and $P = 10$ N.

$$(13\ cm \times 20\ N) + (30\ cm \times 10\ N) - (5\ cm \times B) = 0$$

$$B = \frac{260\ N\ cm + 300\ N\ cm}{5\ cm}$$

B is found to be 112 N.

The joint reaction force can now be calculated.

$$\sum F = 0.$$

$$B - W - P - J = 0$$

$$J = B - W - P$$

$$J = 112\ N - 20\ N - 10\ N$$

J is found to be 82 N. Thus, in this example, a 1-kg object held in the hand with the elbow flexed 90° increases the joint reaction force by 50 N.

CALCULATION BOX 13-2

Joint Reaction Force: Elbow Extension

The joint reaction force during elbow extension can be calculated by means of the same method:

$$\sum M = 0.$$

$$(13\ cm \times W) - (3\ cm \times T) = 0$$

If $W = 20\ N$

$$T = \frac{13\ cm \times 20\ N}{3\ cm}$$

T is found to be 87 N.

$$\sum F = 0.$$

$$J - T - W = 0$$

$$J = T + W$$

$$J = 87\ N + 20\ N$$

J is found to be 107 N. Thus in this example the joint reaction force during elbow extension is 75 N greater that during elbow flexion (Calculation Box Fig. 13-2-1).

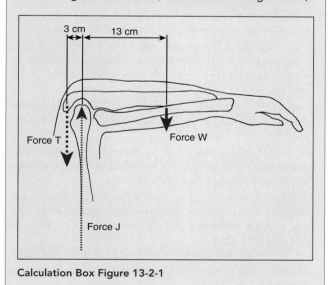

Calculation Box Figure 13-2-1

An estimation of the joint reaction force can also be made for the elbow during extension. In the case study, the elbow is held in 90° of flexion with the forearm positioned over the head and parallel to the ground (Calculation Box 13-2). In this position, action of the elbow extensors is required to offset the gravitational force on the forearm. It is assumed that the triceps is the predominant extensor and that the force through the tendon of this muscle acts perpendicular to the longitudinal axis of the forearm. Therefore, the three main coplanar forces acting on the elbow include the force produced by the weight of the arm (*W*), the tensile force exerted through the tendon of the triceps muscle (*M*), and the joint reaction force on the trochlear fossa of the ulna (*J*). The distance between the center of rotation of the elbow and the point of insertion of the tendon of the triceps muscle (the lever arm of *M*) is approximately 3 cm.

M and *J* are calculated with the equilibrium equations. The joint reaction force for the elbow in extension is 107 N, compared with 32 N in flexion. This more-than-threefold increase can be explained by the fact that the lever arm for the elbow extensor force is shorter than that for the flexor force—3 cm as opposed to 5 cm. Thus, a greater muscle force (87 N as opposed to 52 N) is required for the forearm to be maintained in the extended position, and as a result the joint reaction force is greater.

Summary

- The elbow joint complex consists of three articulations: the humeroulnar, humeroradial, and proximal radioulnar. It allows two types of motion: flexion-extension and pronation-supination.

- The functional range of elbow motion is 30° to 130° of flexion-extension and 50° to 50° of pronation-supination with most activities of daily living accomplished within this range. There is a significant and rapid loss of the ability to reach in space with flexion contractures of the elbow greater than 30°.

- The axis of rotation for flexion-extension occurs about a tight locus of points measuring 2 to 3 mm in its broadest dimension and is located in the center of the trochlea and capitellum in the lateral view.

- The elbow has a changing center of rotation during flexion extension and cannot be truly represented as a simple hinge joint.

- The carrying angle of the elbow is defined as the angle between the anatomic axis of the ulna and humerus in the AP plane and in full elbow extension. It averages between 10° and 15° of valgus.

- The primary stabilizer to valgus stress at the elbow is the anterior band of the medial collateral ligament with the radial head acting as a secondary stabilizer. The primary restraint to varus stress is the elbow articulation. The lateral ulnar collateral ligament is the main stabilizer to posterolateral rotatory instability of the elbow.

- The primary flexor of the elbow is the brachialis whereas the primary extender is the triceps. The anconeus is active in initiating and maintaining flexion and is considered to act as a dynamic joint stabilizer. The main source of supination is the biceps brachii. The pronator quadratus is the primary pronator of the forearm regardless of position of the forearm or degree of elbow flexion.
- Force generated in the elbow has been shown to be up to three times body weight when performing activities of daily living.

Practice Questions

1. A patient suffers a minimally displaced radial head fracture that does not require surgery, but the surgeon wants to immobilize the elbow. What elbow position would minimize the force seen across the radial head and minimize the chance of fracture displacement?

2. A professional baseball pitcher complains of pain along the inside of his elbow. What force is likely the cause and what structure is likely damaged?

3. A weightlifter wants to focus a workout on the brachialis muscle and not the biceps. How should he go about isolating this muscle?

REFERENCES

Aguinaldo, A.L., Chambers, H. (2009). Correlation of throwing mechanics with elbow valgus load in adult baseball pitchers. *Am J Sports Med, 37*(10), 2043–2048.

An, K.N., Hui, F.C., Morrey, B.F., Linscheid, R.L., Chao, E.Y. (1981). Muscles across the elbow joint: A biomechanical analysis. *J Biomech, 14*, 659–669.

An, K.N., Morrey, B.F. (1991). Biomechanics. In B.F. Morrey (Ed.). *Joint Replacement Arthroplasty.* New York: Churchill Livingstone, 257–273.

An, K.N., Morrey, B.F., Chao, E.Y.S. (1984). Carrying angle of the human elbow. *Joint J Orthop Res, 1*, 369–378.

Askew, L.J., An, K.N., Morrey, B.F., Chao, E.Y. et al. (1987). Isometric strength in normal individuals. *Clin Orthop, 222*, 261–266.

Atkinson, W.B., Elftman, H. (1945). The carrying angle of the human arm as a secondary sex character. *Anat Record, 91*, 49–52.

Basmajian, J.V. (1969). Recent advances in the functional anatomy of the upper limb. *Am J Phys Med, 48*, 165.

Basmajian, J.V., Latif, S. (1957). Integrated actions and functions of the chief flexors of the elbow. *J Bone Joint Surg, 39A*, 1106.

Basmajian, J.V., Travill, A.A. (1961). Electromyography of the pronator muscles in the forearm. *Anat Rec, 139*, 45.

Beingessner, D.M., Dunning, C.E., Gordon, K.D., Johnson, J.A., King, G.J. (2004). The effect of radial head excision and arthroplasty on elbow kinematics and stability. *J Bone Joint Surg Am, 86-A*(8), 1730–1739.

Braune, W., Flugel, A. (1842). Uber pronation and supination des menschlichen voderarms und der hand. *Arch Anat Physiol Anat Rbt.*

Carret, J.P., Fischer, L.P., Gonon, G.P., Dimnet, J. (1976). Etude cinematique de la prosupination au niveau des articulations radiocubitales (radio ulnaris). *Bull Assoc Anat (Nancy), 60*, 279–295.

Chang, C.W., Wang, Y.C., Chu, C.H. (2008). Increased carrying angle is a risk factor for nontraumatic ulnar neuropathy at the elbow. *Clin Orthop Relat Res, 466*(9), 2190–2195.

Chantelot, C., Wavreille, G., Dos Remedios, C., Landejerit, B., Fontaine, C., Hildebrand, H. (2008). Intra-articular compressive stress of the elbow joint in extension: An experimental study using Fuji films. *Surg Radiol Anat, 30*(2), 103–111.

Chao, E.Y., Morrey, B.F. (1978). Three dimensional rotation of the elbow. *J Biomech, 11*, 57–73.

Closkey, R.F., Goode, J.R., Kirschenbaum, D., Cody, R.P. (2000). The role of the coronoid process in elbow stability. A biomechanical analysis of axial loading. *J Bone Joint Surg Am, 82A*(12), 1749–1753.

Coleman, D.A., Blair, W.F., Shurr, D. (1987). Resection of the radial head for fracture of the radial head: A long-term follow-up of seventeen cases. *J Bone Joint Surg, 69A*, 385–392.

Daria, A., Gil, E., Delgado, E., Alonso-Llames, M. (1990). Recurrent dislocation of the elbow. *Int Orthop, 14*, 41–45.

DeFrate, L.E., Li, G., Zavontz, S.J., Herndon, J.H. (2001). A minimally invasive method for the determination of force in the interosseous ligament. *Clin Biomech (Bristol, Avon), 16*(10), 895–900.

DeSousa, O.M., DeMoraes, J.L., DeMoraes, V.F.L. (1961). Electromyographic study of the brachioradialis muscle. *Anat Rec, 139*, 125.

Duck, T.R., Dunning, C.E., King, G.J., Johnson, J.A. (2003). Variability and repeatability of the flexion axis at the ulnohumeral joint. *J Orthop Res, 21*(3), 399–404.

Dunning, C.E., Zarzour, Z.D., Patterson, S.D., Johnson, J.A., King, G.J. (2001). Ligamentous stabilizers against posterolateral rotatory instability of the elbow. *J Bone Joint Surg Am, 83A*(12), 1823–1828.

Durig, M., Muller, W., Rüedi, T.P., Gauer, E.F. (1979). The operative treatment of elbow dislocation in the adult. *J Bone Joint Surg, 61A*, 239–244.

Ewald, F.C. (1975). Total elbow replacement. *Orthop Clin North Am, 6*, 685–696.

Ewald, F.C., Thomas, W.H., Sledge, C.B., et al. (1977). Non-constrained metal to plastic total elbow arthroplasty in rheumatoid arthritis. In *Joint Replacement in the Upper Limb.* London, UK: Institution of Mechanical Engineers, 77–81.

Fern, S.E., Owen, J.R., Ordyna, N.J., Wayne, J.S., Boardman, N.D. 3rd. (2009). Complex varus elbow instability: A terrible triad model. *J Shoulder Elbow Surg, 18*(2), 269–274.

Figgie, H.E., III, Inglis, A.E., Mow, V.C. (1986). A critical analysis of alignment factors affecting functional outcome in total elbow arthroplasty. *J Arthroplasty, 1*, 169.

Fu, E., Li, G., Souer, J.S., Lozano-Calderon, S., Herndon, J.H., Jupiter, J.B., Chen, N.C. (2009). Elbow position affects distal radioulnar joint kinematics. *J Hand Surg Am, 34*(7), 1261–1268.

Funk, D.A., An, K.N., Morrey, B.F., Daube, J.R. (1987). Electromyographic analysis of muscles across the elbow joint. *J Orthop Res, 5*(4), 529.

Galik, K., Baratz, M.E., Butler, A.L., Dougherty, J., Cohen, M.S., Miller, M.C. (2007). The effect of the annular ligament on kinematics of the radial head. *J Hand Surg Am, 32*(8), 1218–1224.

Gerard, Y., Schernburg, F., Nerot, C. (1984). Anatomical, pathological and therapeutic investigation of fractures of the radial head in adults [abstract]. *J Bone Joint Surg, 64B*, 141.

Goel, V.K., Lee, I.K., Blair, W.F. (1989). Stress distribution in the ulna following a hinged elbow arthroplasty. *J Arthroplasty, 4*, 163.

Halls, A.A., Travill, A. (1964). Transmission of pressures across the elbow joint. *Anat Rec, 150*, 243–248.

Haugstvedt, J.R., Berger, R.A., Berglund, L.J. (2001). A mechanical study of the moment-forces of the supinators and pronators of the forearm. *Acta Orthop Scand, 72*(6), 629–634.

Hotchkiss, R.N. (1997). Displaced fractures of the radial head: Internal fixation or excision? *J Am Acad Orthop Surg, 5*, 1–10.

Hotchkiss, R.N., An, K.N., Sowa, D.T., Basta, S., Weiland, A.J. (1989). An anatomic and mechanical study of the interosseous membrane of the forearm: Pathomechanics of proximal migration of the radius. *J Hand Surg, 14A*, 256–261.

Hsu, J.E., Peng, Q., Schafer, D.A., Koh, J.L., Nuber, G.W., Zhang, L.Q. (2008). In vivo three-dimensional mechanical actions of individual. *J Appl Biomech, 24*(4), 325–332.

Hull, J.R., Owen, J.R., Fern, S.E., Wayne, J.S., Boardman, N.D. 3rd. (2005). Role of the coronoid process in varus osteoarticular stability of the elbow. *J Shoulder Elbow Surg, 14*(4), 441–446.

Ishizuki, M. (1979). Functional anatomy of the elbow joint and three-dimensional quantitative motion analysis of the elbow joint. *Nippon Seikeigeka Gakkai Zasshi, 53*, 989–996.

Johnston, G.W. (1962). A follow-up of one hundred cases of fracture of the head of the radius with a review of the literature. *Ulster Med J, 31*, 51–56.

Josefsson, P.O., Johnell, O., Wenderberg, B. (1987). Ligamentous injuries in dislocations of the elbow joint. *Clin Orthop, 221*, 221–225.

Kamineni, S., ElAttrache, N.S., O'Driscoll, S.W., Ahmad, C.S., Hirohara, H., Neale, P.G., An, K.N., Morrey, B.F. (2004). Medial collateral ligament strain with partial posteromedial olecranon resection. A biomechanical study. *J Bone Joint Surg Am, 86A*(11), 2424–2430.

Kamineni, S., Hirahara, H., Powmianowski, S., Neale, P.G., O'Driscoll, S.W., ElAttrache, N., An, K.N., Morrey, B.F. (2003). Partial posteromedial olecranon resection: A kinematic study. *J Bone Joint Surg Am, 85-A*(6), 1005–1011.

Kapandji, I.A. (1982). *The Physiology of Joints (vol 1)*. Edinburgh, UK: Churchill Livingstone.

Kulig, K., Powers, C.M., Shellock, F.G., Terk, M. (2001). The effects of eccentric velocity on activation of elbow flexors: Evaluation by magnetic resonance imaging. *Med Sci Sports Exerc, 33*(2), 196–200.

Lee, D.H., Greene, K.S., Bidez, M.W., et al. (1992). Role of the forearm interosseous membrane. Paper presented at: 47th Annual Meeting of the American Society for Surgery of the Hand; Phoenix, AZ, 42.

Levin, J.S., Zheng, N., Dugas, J., Cain, E.L., Andrews, J.R. (2004). Posterior olecranon resection and ulnar collateral ligament strain. *J Shoulder Elbow Surg, 13*(1), 66–71.

Lin, F., Kohli, N., Perlmutter, S., Lim, D., Nuber, G.W., Makhsous, M. (2007). Muscle contribution to elbow joint valgus stability. *J Shoulder Elbow Surg, 16*(6), 795–802.

London, J.T. (1981). Kinematics of the elbow. *J Bone Joint Surg, 63A*, 529–535.

Mall, F.P. (1905). On the angle of the elbow. *Am J Anat, 4*, 391–404.

Maton, B., Bouisset, S. (1977). The distribution of activity among the muscles of a single group during isometric contraction. *Eur J Appl Physiol, 37*, 101.

Morrey, B.F. (1986). Applied anatomy and biomechanics of the elbow joint. *AAOS Instructional Course Lectures (vol 35)*. St Louis, MO: Mosby, 59–68.

Morrey, B.F. (1994). Biomechanics of the elbow and forearm. In J.C. Delee, D. Drez (Eds.). *Orthopaedic Sports Medicine*. Philadelphia: WB Saunders, Chapter 17.

Morrey, B.F. (1993). *The Elbow and Its Disorders* (2nd ed.). Philadelphia: WB Saunders.

Morrey, B.F., An, K.N. (1983). Articular and ligamentous contributions to stability of the elbow joint. *Am J Sports Med, 11*, 315–319.

Morrey, B.F., An, K.N., Stormont, T.J. (1988). Force transmission through the radial head. *J Bone Joint Surg, 70A*, 250–256.

Morrey, B.F., Askew, L.J., An, K.N., Chao, E.Y. (1981). A biomechanical study of functional elbow motion. *J Bone Joint Surg, 63A*, 872.

Morrey, B.F., Chao, E.Y. (1976). Passive motion of the elbow joint. *J Bone Joint Surg, 58A*, 501.

Morrey, B.F., Chao, E.Y., Hui, F.C. (1979). Biomechanical study of the elbow following excision of the radial head. *J Bone Joint Surg, 61A*, 63–68.

Morrey, B.F., Tanaka, S., An, K.N. (1991). Valgus stability of the elbow. *Clin Orthop, 265*, 187.

Nicol, A.C., Berme, N., Paul, J.P. (1977). A biomechanical analysis of elbow joint function. In *Joint Replacement in the Upper Limb*. London, UK: Institution of Mechanical Engineers, 45–51.

O'Driscoll, S.W., Bell, D.F., Morrey, B.F. (1991). Posterolateral rotatory instability of the elbow. *J Bone Joint Surg, 73A*(3), 440.

O'Driscoll, S.W., Morrey, B.F., An, K.N. (1990b). Intrarticular pressure and capacity of the elbow. *Arthroscopy, 6*(2), 100.

O'Driscoll, S.W., Morrey, B.F., An, K.N. (1990a). The pathoanatomy and kinematics of posterolateral instability (pivot-shift) of the elbow. *Orthop Trans, 14*, 306.

Olsen, B.S., Søjbjerg, J.O., Dalstra, M., Sneppen, O. (1996). Kinematics of the lateral ligamentous constraints of the elbow joint. *J Shoulder Elbow Surg, 5*(5), 333–341.

Osborne, G., Cotterill, P. (1966). Recurrent dislocation of the elbow. *J Bone Joint Surg Br, 48B,* 340–346.

Palmer, A.K., Glisson, R.R., Werner, F.W. (1982). Ulnar variance determination. *J Hand Surg, 7,* 376.

Park, M.C., Ahmad, C.S. (2004). Dynamic contributions of the flexor-pronator mass to elbow valgus stability. *J Bone Joint Surg Am, 86-A*(10), 2268–2274.

Pearson, J.R., McGinley, D.R., Butzel, L.M. (1963). A dynamic analysis of the upper extremity: Planar motions. *Human Factors, 5,* 59.

Pomianowski, S., O'Driscoll, S.W., Neale, P.G., Park, M.J., Morrey, B.F., An, K.N. (2001). The effect of forearm rotation on laxity and stability of the elbow. *Clin Biomech (Bristol, Avon), 16*(5), 401–407.

Ray, R.D., Johnson, R.J., Jameson, R.M. (1951). Rotation of the forearm. An experimental study of pronation and supination. *J Bone Joint Surg, 33A,* 993–996.

Reardon, J.P., Lafferty, M., Kamaric, E., et al. (1991). Structures influencing axial stability to the forearm: The role of the radial head, interosseous membrane, and distal radio-ulnar joint. *Orthop Trans, 15,* 436–437.

Safran, M.R., McGarry, M.H., Shin, S., Han, S., Lee, T.Q.(2005). Effects of elbow flexion and forearm rotation on valgus laxity of the elbow. *J Bone Joint Surg Am, 87*(9), 2065–2074.

Seiber, K., Gupta, R., McGarry, M.H., Safran, M.R., Lee, T.Q. (2009). The role of the elbow musculature, forearm rotation, and elbow flexion in elbow stability: An in vitro study. *J Shoulder Elbow Surg, 18*(2), 260–268.

Shaaban, H., Pereira, C., Williams, R., Lees, V.C. (2008). The effect of elbow position on the range of supination and pronation of the forearm. *J Hand Surg Eur, 33*(1), 3–8.

Sojbjerg, J.O., Helmig, P., Jaersgaard-Andersen, P. (1989). Dislocation of the elbow: An experimental study of the ligamentous injuries. *Orthopedics, 12,* 461–463.

Sowa, D.T., Hotchkiss, R.N., Weiland, A.J. (1995). Symptomatic proximal translation of the radius following radial head resection. *Clin Orthop, 317,* 106–113.

Spinner, M., Kaplan, E.B. (1970). The quadrate ligament of the elbow: Its relationship to the stability of the proximal radioulnar joint. *Acta Orthop Scand, 41,* 632.

Steindler, A. (1955). *Kinesiology of the Human Body Under Normal and Pathological Conditions.* Springfield, IL: Charles C Thomas Publisher.

Stevens, A., Stijns, H., Reybrouck,T., Bonte, G., Michels, A., Rosselle, N., Roelandts, P., Krauss, E., Verheyen, G. (1973). A poly-electromyographical study of the arm muscles at gradual isometric loading. *Electromyogr Clin Neurophysiol, 13,* 46S.

Stormont, T.J., An, K.N., Morrey, B.F., et al. (1985). Elbow joint contact study: A comparison of techniques. *J Biomech, 18*(5), 329.

Tullos, H.S., Erwin, W., Woods, G.W., Wukasch, D.C., Cooley, D.A., King, J.W. (1972). Unusual lesions of the pitching arm. *Clin Orthop, 88,* 169.

Udall, J.H., Fitzpatrick, M.J., McGarry, M.H., Leba, T.B., Lee, T.Q. (2009). Effects of flexor-pronator muscle loading on valgus stability of the elbow with an intact, stretched, and resected medial ulnar collateral ligament. *J Shoulder Elbow Surg, 18*(5), 773–778.

Walker, P.S. (1977). *Human Joints and Their Artificial Replacement.* Springfield, IL: Charles C Thomas, Publisher.

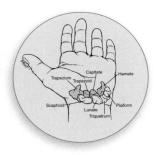

Biomechanics of the Wrist and Hand

Ann E. Barr and Jane Bear-Lehman

Introduction

The wrist, or carpus, is the collection of bones and soft tissue structures that connects the hand to the forearm. This joint complex is capable of a substantial arc of motion that augments hand and finger function, yet it possesses a considerable degree of stability. The wrist functions kinematically by allowing for changes in the location and orientation of the hand relative to the forearm and kinetically by transmitting loads from the hand to the forearm and vice versa.

Although the function of all joints of the upper extremity is to position the hand to allow it to perform daily life tasks, the wrist appears to be the key to hand function. Stability of the wrist is essential for proper functioning of the digital flexor and extensor muscles, and wrist position affects the ability of the fingers to flex and extend maximally and to grasp effectively during prehension.

The hand is a highly complex and multifaceted mobile effector organ that allows us to grasp and manipulate objects within the hand and to serve or provide the body with support (Jones and Lederman, 2006; Neumann, 2010; Wilson, 1998). It is valued and judged for its performance and appearance in delicate prehensile tasks to powerful grasp patterns. It is remarkably mobile and adaptable as it conforms to the shape of objects to be grasped or studied, emphasizes or gestures an idea to be expressed, or shows an act of love or affection (Tubiana, 1984). The hand helps us explore and understand small objects through within-the-hand manipulations between the fingers and the thumb. In other words, the within-the-hand manipulation skills of the thumb and the fingers (supported by interplay between the muscular and sensory system) facilitates our understanding of and appreciation for small objects, e.g., the ability to clasp a necklace together, ready a contact lens, and so on. Additionally, we rely on the hand to support, stabilize, or brace an object to free the other hand for specific task engagement, or our hand can provide the needed support as we rise up from a chair.

The hand is the final link in the mechanical chain of levers that begins at the shoulder. The mobility and stability of the shoulder, the elbow, and the wrist, all operating in different planes, allow the hand to move within a large volume of space and to reach all parts of the body with relative ease. The unique arrangement and mobility of the 19 bones and 14 joints of the hand provide the structural foundation for the hand's extraordinary functional adaptability.

Anatomy of the Wrist and Hand

WRIST ARTICULATIONS

The wrist joint complex consists of the multiple articulations of the eight carpal bones with the distal radius, the structures within the ulnocarpal space, the metacarpals, and each other (Fig. 14-1). The soft tissue structures

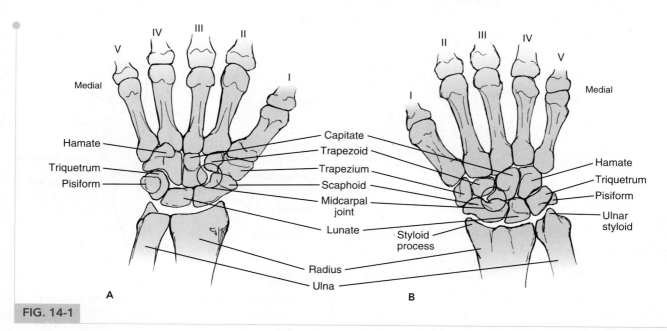

FIG. 14-1

Schematic drawings of the wrist joint complex showing the eight carpal bones and their articulations with the distal radius, the metacarpal bones of the hand, and each other. Palmar view **(A)** and dorsal view **(B)** of the right hand. The *arrows* indicate the line of the midcarpal joint.

surrounding the carpal bones include the tendons that cross the carpus or attach to it and the ligamentous structures that connect the carpal bones to each other and to the bony elements of the hand and forearm.

The eight carpal bones are divided into the proximal and distal rows. The bones of the distal row from radial to ulnar are the trapezium, trapezoid, capitate, and hamate. The distal carpal row forms a relatively immobile transverse unit that articulates with the metacarpals to form the carpometacarpal joints. All four bones in the distal row fit tightly against each other and are held together by stout interosseous ligaments. The more mobile proximal row consists of the scaphoid, lunate, and triquetrum. This row articulates with the distal radius and soft tissue triangular fibrocartilage to form the radiocarpal joint. The proximal component of the radiocarpal joint is the concave surface of the distal radius and the triangular fibrocartilage (also known as the articular disc). The distal components are the convex surfaces of the scaphoid, lunate, and, during extreme ulnar deviation, triquetrum (Neumann, 2010). The scaphoid spans both rows anatomically and functionally and articulates exclusively with the radius. The lunate articulates in part with the ulnar

triangular fibrocartilage. The eighth carpal bone, the pisiform, is a sesamoid bone that mechanically enhances the wrist's most powerful motor, the flexor carpi ulnaris, and forms its own small joint with the triquetrum. Between the proximal and distal rows of carpal bones is the midcarpal joint, and between adjacent bones of these rows are the intercarpal joints (Fig. 14-1). The palmar surface of the carpus as a whole is concave, constituting the floor and walls of the carpal tunnel (Fig. 14-2).

The distal radius, lunate, and triquetrum articulate with the distal ulna through a ligamentous and cartilaginous structure, the ulnocarpal or triangular fibrocartilage complex (TFCC). The components of this complex are illustrated in Figure 14-3, and its functional role will be discussed in detail along with ligamentous function.

HAND ARTICULATIONS

The finger and thumb are the elementary components of the hand (Fig. 14-4). Because each digital unit extends into the middle of the hand, the term digit ray is used to indicate the entire chain; each finger is composed of one metacarpal and three phalanges and the thumb is composed of one metacarpal and two phalanges. The digital rays are numbered from the radial to the ulnar side: I (thumb), II (index finger), III (middle finger), IV

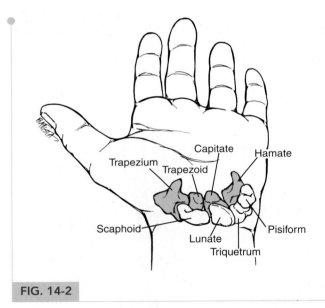

FIG. 14-2

Longitudinal view of the left hand from proximal to distal showing the palmar surface of the bones. This concave surface constitutes the floor and walls of the carpal tunnel, through which the median nerve and flexor tendons pass. The carpal tunnel is bordered laterally by the prominent tubercle of the trapezium and medially by the hook of the hamate. The motor branch of the ulnar nerve (not shown) winds around the base of the hook before entering the deep palmar compartment. *Adapted from Oatis, C.A. (2008). Kinesiology: The Mechanics and Pathomechanics of Human Movement. Baltimore, MD: Lippincott Williams & Wilkins, 259.*

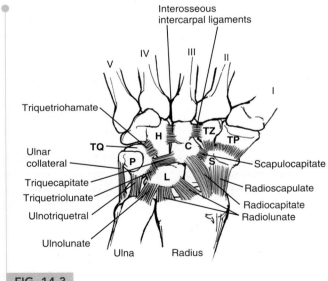

FIG. 14-3

Longitudinal section (frontal plane) of the right wrist and hand viewed from the palmar side. The components of the triangular fibrocartilage complex are visible between the distal ulna and the lunate and triquetrum. *S,* scaphoid; *L,* lunate; *TQ,* triquetrum (the pisiform is not shown); *H,* hamate; *C,* capitate; *TZ,* trapezoid, *TP,* trapezium, *P,* prestyloid recess.

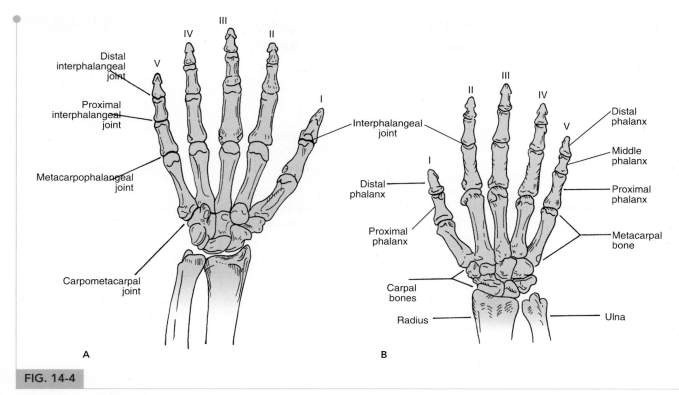

FIG. 14-4

Schematic drawing of the skeleton of the hand. The finger rays are numbered from the radial (medial) to the ulnar (lateral) side.

A. Palmar (anterior) view of the right hand. The joints are labeled.
B. Dorsal (posterior) view of the right hand. The bones are labeled.

(ring finger), and V (little finger). Each digital ray articulates proximally with a particular carpal bone to form the carpometacarpal (CMC) joint. The next joint in each ray—the metacarpophalangeal (MCP) joint—links the metacarpal bone to the proximal phalanx. Between the phalanges of the fingers, a proximal (PIP) and a distal (DIP) interphalangeal joint are found; the thumb has only one interphalangeal (IP) joint. The thenar eminence at the palmar side of the first metacarpal is formed by the intrinsic muscles of the thumb. Its ulnar counterpart, the hypothenar eminence, is created by muscles of the little finger and an overlying fat pad.

ARCHES OF THE HAND

The bones of the hand are arranged in three arches (Fig. 14-5), two transverse and one longitudinal (Neumann, 2010; Tubiana, 1984). The proximal transverse arch, with the capitate as its keystone, lies at the level of the distal carpus and is relatively fixed. The distal transverse arch, with the head of the third metacarpal as its keystone, passes through all of the metacarpal heads and is more mobile. The two transverse arches are connected by the rigid portion of the longitudinal arch, which is composed

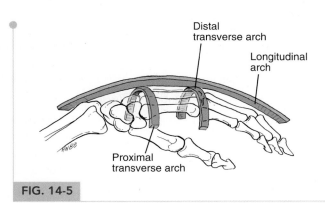

FIG. 14-5

The three skeletal arches of the hand (mediolateral view). The relatively fixed proximal transverse arch passes through the distal carpus at the level of the distal carpal row. The more mobile distal transverse arch passes through the metacarpal heads. The longitudinal arch is composed of the four finger rays and the proximal carpus. *Adapted with permission from Strickland, J.W. (1987). Anatomy and kinesiology of the hand. In E.E. Fess, C.A. Philips (Eds.). Hand Splinting: Principles and Methods (2nd ed.). St Louis, MO: Mosby, 3–41.*

TABLE 14-1

Muscles of the Wrist and Hand

Muscles of the Wrist	
Muscle	**Action**
Flexors	
Flexor carpi ulnaris	Flexion of wrist; ulnar deviation of hand
Flexor carpi radialis	Flexion of wrist; radial deviation of hand
Palmaris longus	Tension of the palmar fascia
Extensors	
Extensor carpi radialis longus and brevis	Extension of wrist; radial deviation of hand
Extensor carpi ulnaris and brevis	Extension of wrist; ulnar deviation of hand
Pronators-supinators	
Pronator teres	Forearm pronation
Pronator quadratus	Forearm pronation
Supinator	Forearm supination
Brachioradialis	Pronation or supination, depending on position of forearm
Muscles of the Hand	
Muscle	**Action**
Extrinsic Muscles	
Flexors	
Flexor digitorum superficialis	Flexion of PIP and MCP joints
Flexor digitorum profundus	Flexion of DIP, PIP, and MCP joints
Flexor pollicis longus	Flexion of IP and MCP joints of thumb
Extensors	
Extensor pollicis longus	Extension of IP and MCP joints of thumb; secondary adduction of the thumb
Extensor pollicis brevis	Extension of MCP joint of thumb
Abductor pollicis longus	Abduction of thumb
Extensor indicis proprius	Extension of index finger
Extensor digitorum communis	Extension of fingers
Extensor digiti quinti proprius	Extension of V finger
Intrinsic Muscles	
Interossei (all)	Extension of PIP and DIP joints and flexion of MCP joints
Dorsal interossei	Spread of index and ring fingers away from long finger
Palmar interossei	Adduction of index, ring, and little fingers toward long finger
Lumbricals	Extension of PIP and DID joints and flexion of MCP 2–5 finger
Thenar Muscles	
Abductor pollicis brevis	Abduction of thumb
Flexor pollicis brevis	Flexion and rotation of thumb
Opponens pollicis	Rotation of first metacarpal toward palm
Hypothenar Muscles	
Abductor digiti quinti	Abduction of little finger (extension of PIP and DIP joints)
Flexor digiti quinti brevis	Flexion of proximal phalanx of little finger and forward rotation of fifth metacarpal
Adductor pollicis	Adduction of thumb

Modified from Strickland, J.W. (1987). Anatomy and kinesiology of the hand. In E.E. Fess, C.A. Philips (Eds.). *Hand Splinting: Principles and Methods (2nd ed.)*. St Louis, MO: Mosby, 3–41.

of the four digital rays and the proximal carpus. The second and third metacarpal bones form the central pillar of this arch. The longitudinal arch is completed by the individual digital rays, and the mobility of the thumb and fourth finger and fifth finger rays around the second and third fingers allows the palm to flatten or cup itself to accommodate objects of various sizes and shapes (Fess et al., 2005).

Although the extrinsic flexor and extensor muscles are largely responsible for changing the shape of the working hand, the intrinsic muscles of the hand are primarily responsible for maintaining the configuration of the three arches (refer to Table 14-1 for a listing of the muscles of the wrist and the hand as well as the corresponding muscle actions). A collapse in the arch system resulting from bone injury, rheumatic disease, or paralysis of the intrinsic muscles can contribute to severe disability and deformity.

NERVE AND BLOOD SUPPLY OF THE WRIST AND HAND

The covering of the hand is important because of its physical qualities, sensory properties, and microcirculation (Moore and Dalley, 2006). The skin on the dorsum or the back of the hand differs and is distinct from the skin that covers the palmar surface. Dorsal skin is mobile, often regarded as very fine, and highly flexible, allowing for a wide array of articular movements. In contrast, palmar skin is thick, glabrous, and inelastic. Palmar skin plays a significant role in hand perceptibility or the perception of touch, safety of the upper limb through sensory protection, and in providing support for the limb during weight-bearing.

The wrist and the hand are innervated for motor and sensory function by three peripheral nerves, the radial, median, and ulnar, that descend from the brachial plexus (Fig. 14-6). Motor or sensory impairment correlates to the location of the injury or impingement along a particular nerve tract. The radial nerve primarily innervates those muscles that facilitate extension of the wrist and the digits, namely the long wrist extensors. Impairment to the radial nerve can cause wrist drop and produce wrist instability that impedes hand grasp. In terms of sensory function, the radial nerve supplies the skin along the radial sphere of the forearm and the hand, and sensory impairment in radial nerve denervation minimally impedes hand function. The median nerve primarily innervates the long wrist and hand extrinsic flexors. Thus, high-level median nerve impairment affects the radial flexor muscles of the hand more greatly than it does the flexor capacity along the ulnar side. The median nerve is most critical to fine motor hand function because of the motor and sensory supply that it provides. The median nerve is often regarded as the eyes of the hand because it is responsible for the innervation of the first three digits of the hand on the palmar surface. Without adequate sensation in these digits, fine motor skill is compromised or lost (Case Study 14-1).

The ulnar nerve is regarded as the hand's power source for hand grasp. It innervates the muscles along the ulnar sphere, including the ulnar hand flexors and most of the hand intrinsics, particularly those responsible for digital adduction and abduction. The ulnar nerve is known for its ability to protect the upper limb as it innervates the skin surface along the ulnar border. Most resting patterns for the upper limb or hand use are performed with the upper limb positioned so that the ulnar borders of the forearm, wrist, and hand are in direct contact with the environment or giving support to the body by direct and sustained weight-bearing or contact onto a surface.

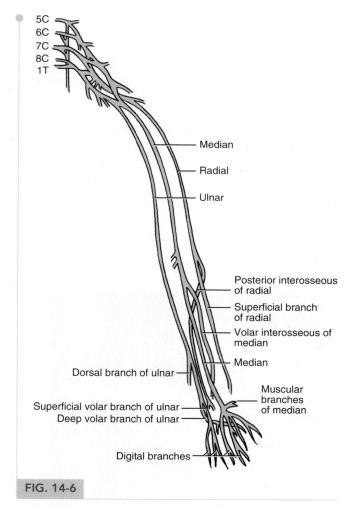

FIG. 14-6

Radial, median, and ulnar peripheral nerves descending from the brachial plexus.

Case Study 14-1

Carpal Tunnel Syndrome in a Clerical Worker

A 26-year-old female administrative assistant presented with complaints of right hand and wrist discomfort of gradual onset and intermittent progressive severity over the past two years. Her symptoms are worsened by prolonged hours of typing on her computer keyboard and use of the mouse. She presented no symptoms in her left hand. A detailed examination shows numbness and tingling of the volar aspect of the distal right forearm and wrist, swelling of the wrist, grip weakness, and positive Phalen and Tinel signs. The clinical evaluation suggests compression of the median nerve within the carpal tunnel producing associated motor and sensory changes. A nerve conduction velocity test later confirmed a diagnosis of right carpal tunnel syndrome (CTS). The patient was placed on light duty with restrictions on typing tasks and was scheduled for CTS release surgery (Case Study Fig. 14-1).

This case illustrates the potential influence of work tasks involving highly repetitive hand and finger movements for prolonged periods of time on the delicate structures of the hand and wrist. The increased intracompartmental pressure within the carpal tunnel associated with ergonomic risk factors such as repetitive loading with sustained and awkward wrist positions are major factors contributing to median nerve compression in this case.

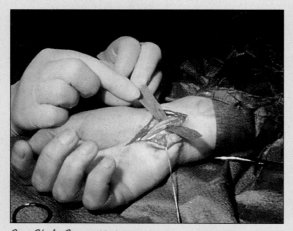

Case Study Figure 14-1

Irritation or compression of the ulnar nerve along its path from the elbow to the wrist may impair hand function (Case Study 14-2).

The hand as an organ of touch has a myriad of sensory receptors within the skin that are uniquely positioned to provide the central nervous system with information about finger motion, about objects held or manipulated within the hand, or about the environmental context such as temperature (Jones and Lederman, 2006). Microscopic study of the palmar, nonglabrous skin shows that it possesses highly specialized pilary ridges with many types of sensory receptors and nerve fiber properties (Cauna, 1954). The sensory receptors change from free nerve ending to encapsulated receptors or mechanoreceptors. There are more receptors than nerve fibers, and each fiber is connected to several receptors (Mountcastle, 1968). Furthermore, sensory information is transmitted over quickly or slowly adapting nerve fiber properties. It is also known that sensation does not have the same value within the hand; certain zones or regions have more receptivity to a stimulus than do others. For example, the sensory acuity is considered to be of a more specialized quality in the specific anatomic regions required for very fine motor prehension: the ulnar half of the digital pulp of the thumb, the radial half of the digital pulp of both the index and middle fingers, and the ulnar border of the little finger. It is essential to be aware of these specialized regions and their critical role in terms of the restoration of hand function following injury (Tubiana, 1984; Wilson, 1998).

Blood is dually supplied to the wrist and the hand by the ulnar and radial arteries, which join or communicate together after each has individually entered into the hand. The skin of the hand is supplied by both a deep and a superficial plexus. The general pattern of the blood supply to the wrist and the hand does not differ from that which is found in other parts of the body. What differs in terms of cutaneous circulation relates to the hand's distal location from the heart and to its constant exposure to thermal and postural variations (Moore and Dalley, 2006). Similar to the highly complex and varied sensory receptors noted within the hand, particularly within the palmar skin, the hand hosts a complex and dense capillary system (Cauna, 1954). This dense system allows for more variation in capillary pressure than in other parts of the body. Capillary pressure depends on a number of factors such as arteriolar tone, venous return, the position of the wrist and the hand, and temperature (Cauna, 1954; Moore and Dalley, 2006; Tubiana, 1984). Hand injury or disease that alters or threatens the cycle of vasodilatation-vasoconstriction can cause progressive wrist and hand edema that leads to stiffness or causalgia.

Case Study 14-2

Work-related Cubital Tunnel Syndrome

A 58-year-old college professor presented with frequent numbness and tingling in the lateral aspect of her left hand with muscle spasms in her left thumb and index finger during activities involving prolonged simultaneous wrist extension and elbow flexion, such as typing or holding the telephone receiver. She reported difficulty sleeping due to hand discomfort. She uses her computer four to eight hours per day at work, and plays computer games and engages in real-time text messaging with her daughter while at home. She also knits and does fine needlework as a hobby. She is right-hand dominant.

The patient examination included tests for peripheral nerve integrity, joint integrity and mobility, muscle performance, presence of pain, sensory integrity, and an ergonomic review of body mechanics and hand use patterns. The presence of thumb and index finger muscular spasms prompted a specific clinical examination of median and ulnar neural status and testing for focal hand dystonia.

Results of the physical examination were negative for median nerve compression/irritation. Positive findings for ulnar nerve compression irritation included tenderness to palpation over the left medial epicondyle with reproduction of numbness and tingling in the lateral aspect of the left hand, a positive Tinel sign over the left cubital tunnel, a palpable and cord-like ulnar nerve in the ulnar groove, and a positive ulnar nerve tension test with reproduction of pain, numbness, and tingling in the left hand. During her physical examination, the patient was not able to reproduce her thumb and index finger spasm; however, her demonstration of the digit positions during the spasms showed left thumb adduction and index finger MCP flexion ("intrinsic plus" position), indicating affected muscles innervated by the ulnar nerve (i.e., adductor pollicis and first dorsal and/or palmar interossei). This dystonic pattern has been reported in musicians with focal hand dystonia emanating from ulnar neuropathy (Charness et al., 1996). A nerve conduction velocity test of the ulnar nerve confirmed slowing of motor nerve conduction by >10 m/s across the elbow. All other test sites across the forearm and wrist for the median and ulnar nerves were normal.

The patient does not suffer from any strength deficits at present. Her plan of care focused on resolution of her presenting sensory symptoms through a self-management program that included a night splint for the elbow and wrist to prevent extreme elbow flexion during sleep (Case Study Fig. 14-2), nerve gliding and posture exercises, and modifications to her physical workstation setup to decrease aggravation of the healing ulnar nerve from contact with the work surface or an improper keyboard height.

This case illustrates the complex signs and symptoms that must be differentiated in a clinical examination of work-related musculoskeletal disorders and the effects of proximal nerve irritation/compression on distal wrist and hand function. The treatment plan in this case was focused on prevention and management to avoid future more serious impairments that may require more invasive treatments, such as surgery.

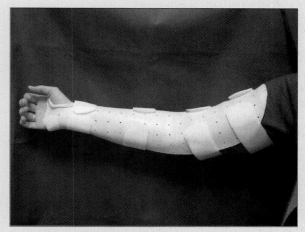

Case Study Figure 14-2

Control of the Wrist and Hand

Active control of the wrist and hand is achieved through coordinated action of both extrinsic musculature, originating from the forearm and humeral segments, and intrinsic musculature, originating from the carpal and hand segments. This muscular control fulfills needs for both mobility and stability during functional wrist and hand activities. The muscles of the wrist and hand are summarized in Table 14-1. No muscles are intrinsic to the carpus; therefore, passive mechanisms derived from bony morphology, ligamentous function, and tendinous expansions play major roles in controlling carpal and digital movements during hand activities. In this sense, the carpus acts as a bridge for muscle action and load transmission between the hand and forearm segments.

Several anatomic features contribute to the stability and control of the various articulations of the hand. The coordinated actions of the extrinsic and intrinsic muscles of the hand permit control of the digit rays; a dorsal tendinous complex known as the extensor assembly contributes to the control and stability of the joints, and a well-developed flexor tendon sheath pulley system facilitates smooth and stable flexion of these joints. The bony and ligamentous asymmetry of the MCP joints lends the hand its functional versatility. The IP joints gain stability from the shape of their articular contours and from special ligamentous restraints.

PASSIVE CONTROL MECHANISMS

Bony Mechanisms

The mobile proximal row of the carpus between the forearm and the distal carpal row forms an intercalated segment that is subject to "zigzag" collapse under compressive load (Neumann, 2010). Intricate ligamentous constraints and the precise opposition of multifaceted articular surfaces counteract these tendencies and afford stability.

In the sagittal plane of the wrist, both the scaphoid and the lunate are wedge-shaped with the palmar aspect of both bones being wider than the dorsal aspect (Kauer, 1980). Because compression tends to squeeze a wedge to its narrowest portion, both the lunate and the scaphoid would tend to be displaced palmward and rotate into extension with compression caused by contraction of the long flexors and extensors.

As both the scaphoid and lunate tend to be forced into extension, stabilization forces must be directed primarily toward flexion. It is here that the contribution of the scaphoid spanning both distal and proximal carpal rows can be appreciated. The natural tendency of the scaphoid to extend is stabilized at the midcarpal level;

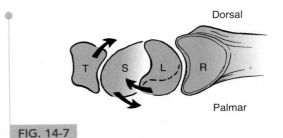

FIG. 14-7

Schematic drawing of the trapezoid (*T*), scaphoid (*S*), lunate (*L*), and radius (*R*) in a sagittal view. The tendency of the wedge-shaped lunate (palmar pole larger than dorsal pole) to rotate into extension is counteracted by the scaphoid, which provides a palmar-flexing force induced by the trapezium and trapezoid. *Adapted with permission from Taleisnik, J. (1985). The Wrist. New York: Churchill Livingstone.*

the trapezium and trapezoid articulate with the dorsal aspect of the scaphoid, pushing its distal pole down into flexion. Hence, the scaphoid counteracts the extension tendency of the lunate, lending some stability to the biarticular carpal complex (Fig. 14-7).

This arrangement has an advantage over a symmetric intercalated segment because instability is focused in only one direction and can be countered by a single force applied in the opposite direction or flexion (Kauer, 1980; Kauer and Landsmeer, 1981). This mechanism is consistent with the use of the finger and wrist flexors during hand function.

Ligamentous Mechanisms

WRIST LIGAMENTS

As in other joints, the function of the wrist ligaments is to maintain intracarpal alignment, both static and dynamic, and transmit loads originating in proximal or distal segments (Neumann, 2010). The palmar ligaments (Fig. 14-8A) are thick and strong, whereas the dorsal ligaments (Fig. 14-8B) are much thinner and fewer in number (Berger, 2002; Neumann, 2010).

The highly developed, complex ligamentous system of the wrist can be divided into extrinsic and intrinsic components (Table 14-2). The extrinsic ligaments run from radius to carpus and from carpus to metacarpals. The intrinsic ligaments originate and insert on the carpus.

The palmar extrinsic ligaments include the radial collateral ligament, the palmar radiocarpal ligaments, and components of the triangular fibrocartilage complex (TFCC). The radial collateral ligament is actually more palmar than lateral and is viewed as the most lateral of all palmar radiocarpal fascicles rather than as a collateral ligament per se, because the function of a true

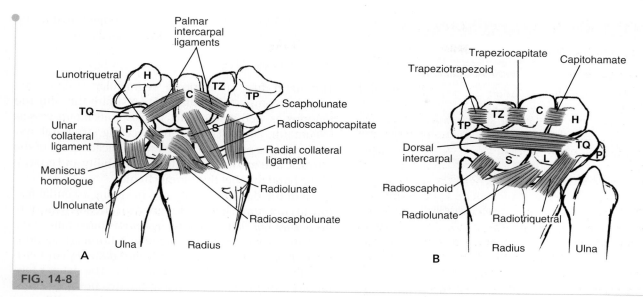

FIG. 14-8

The ligaments of the wrist. **A.** The palmar wrist ligaments (right hand). Extrinsic ligaments: radioscaphocapitate ligament, radial collateral ligament, radiolunate ligament, radioscapholunate ligament, ulnolunate ligament, meniscus homologue (radiotriquetral ligament), ulnar collateral ligament; the superficial palmar radiocarpal ligament and the triangular fibrocartilage are not shown. Intrinsic ligaments: scapholunate ligament, lunotriquetral ligament, and palmar intercarpal (deltoid, or V) ligaments. The short palmar intrinsics are not shown. **B.** The dorsal wrist ligaments of the right hand. Extrinsic ligaments: radiotriquetral, radiolunate, and radioscaphoid fascicles of the dorsal radiocarpal ligament. Intrinsic ligaments: dorsal intercarpal, trapeziotrapezoid, trapeziocapitate, and capitohamate fascicles of the short intrinsic ligaments. The scaphotrapezium ligament is not shown. *S*, scaphoid; *L*, lunate; *TQ*, triquetrum; *H*, hamate; *C*, capitate; *TZ*, trapezoid, *TP*, trapezium, *P*, prestyloid recess.

collateral ligament is not functionally advantageous in the wrist (Neumann, 2010).

The palmar radiocarpal ligaments are arranged in superficial and deep layers. In the superficial layer, most fibers assume a V shape, providing restraint and support. The deep ligaments are three strong fascicles named according to their points of origin and insertion: the radioscaphocapitate (or radiocapitate) ligament, which supports the waist of the scaphoid; the radiolunate ligament, which supports the lunate; and the radioscapholunate ligament, which connects the scapholunate articulation with the palmar portion of the distal radius. This ligament checks scaphoid flexion and extension.

The dorsal extrinsic ligaments include the three bands of the dorsal radiocarpal ligament. Originating from the rim of the radius, these three fascicles insert firmly into the lunate (radiolunate ligament), triquetrum (radiotriquetral ligament), and scaphoid (radioscaphoid ligament), respectively.

The intrinsic ligaments can be grouped into three categories (short, long, and intermediate) according to their length and the relative intercarpal movement they allow. Overall, the palmar intrinsic ligaments are thicker and stronger than the dorsal ones.

The three short intrinsic ligaments—palmar, dorsal, and interosseous—are stout, unyielding fibers that bind the adjacent carpal bones tightly. These strong ligaments are responsible for maintaining the four bones of the distal carpal row as an integrated kinematic unit (Berger, 2002). Three intermediate intrinsic ligaments are located between the lunate and triquetrum, the scaphoid and lunate, and the scaphoid and trapezium.

Of the two long intrinsic ligaments—dorsal intercarpal and palmar intercarpal—the palmar is the more important. Also called the deltoid, or V, ligament, it stabilizes the capitate because it attaches to its neck and fans out proximally to insert into the scaphoid and triquetrum. The dorsal intercarpal ligament originates from the triquetrum and courses laterally and obliquely to insert on the scaphoid and trapezium (Berger, 2002; Neumann, 2010).

TRIANGULAR FIBROCARTILAGE COMPLEX

The components of the TFCC are the meniscus homologue, the triangular fibrocartilage (articular disc), the palmar ulnocarpal ligament (consisting of the ulnolunate and ulnotriquetral ligaments), the ulnar collateral ligament, and the poorly distinguishable dorsal and palmar

TABLE 14-2

Ligaments of the Wrist

Extrinsic Ligaments	Intrinsic Ligaments
Proximal (radiocarpal)	Short
Radio collateral	Palmar
Palmar radiocarpal	Dorsal
Superficial	Intermediate
Deep	Lunotriquetral
Radioscaphocapitate (radiocapitate)	Scapholunate
Radiolunate	Scaphotrapezium
Radioscapholunate	Long
Ulnocarpal complex	Palmar intercarpal (V, deltoid)
Meniscus homologue (radiotriquetral)	Dorsal intercarpal
Triangular fibrocartilage (articular disc)	—
Ulnar collateral ligament	—
Ulnolunate ligament	—
Dorsal radiocarpal	—
Distal (carpometacarpal)	—

Modified from Taleisnik, J. (1985). *The Wrist.* New York: Churchill Livingstone.

radioulnar ligaments (Fig. 14-3). The meniscus homologue and the triangular fibrocartilage have a common origin from the dorsoulnar corner (sigmoid notch) of the radius. From there, the meniscus courses toward the palm and around the ulnar border of the wrist to insert firmly into the triquetrum, while the triangular fibrocartilage extends horizontally to insert into the base of the ulnar styloid process. Between the meniscus homologue and the triangular fibrocartilage there is often a triangular area, the prestyloid recess, which is filled with synovium. Dorsally, the TFCC has a weak attachment to the carpus except where some of its fibers join the tendon sheath of the flexor carpi ulnaris dorsolaterally. The ulnolunate ligament connects the palmar border of the triangular fibrocartilage with the lunate. The ulnar collateral ligament arises from the ulnar styloid process and extends distally to the base of the fifth metacarpal bone.

The functions of the TFCC are to stabilize the distal radioulnar joint, reinforce the ulnar aspect of the wrist, form the ulnar aspect of the proximal articulating surface of the radiocarpal joint, and transmit approximately 16% of the compressive forces when the wrist is in the neutral position (Berger, 2002; Haugstvedt et al., 2006; Kleinman, 2007; Moritomo et al., 2008; Neumann, 2010). Such internal compressive forces can reach as

much as four times the weight of an object held in the hand (Calculation Box 14-1) (Karnezis, 2005).

HAND LIGAMENTS

The hand has an intricate retinacular system that encloses, compartmentalizes, and restrains the joint and tendons as well as the skin, nerves, and blood vessels (Neumann, 2010). This interconnecting structural system encircles each digit to create balanced forces of the intrinsic and extrinsic musculature, stability and control of the hand for effective biomechanics.

All of the digital articulations have one essential feature in common: They are designed to function in flexion. Each joint has firm collateral ligaments bilaterally and a thick anterior capsule reinforced by a fibrocartilaginous structure known as the palmar (volar) plate. By comparison, the dorsal capsule is thin and lax. The palmar tendinous apparatus, composed of the two flexor tendons, is much stronger than the dorsal extensor assembly, and even the skin is thicker on the palmar side.

Digital Flexor Tendon Sheath Pulley System

Most tendons in the hand are restrained to some extent by tendon sheaths and retinacula that keep them close to the bone so that they maintain a relatively constant moment arm, rather than "bowstringing" across the joints. The pulley system of the flexor tendon sheath in the finger is the most highly developed of these restraints.

As they extend from their muscles, the digital flexor tendons pass through the carpal tunnel, along with the tendon of the flexor pollicis longus and the median nerve, before fanning out toward their respective digits. The flexor superficialis tendon inserts on the middle phalanx, and the flexor profundus inserts on the distal phalanx. In each digit, these two tendons, surrounded by their synovial sheaths, are held against the phalanges by a fibrous sheath. At strategic locations along the sheath are five dense annular pulleys (designated as Al, A2, A3, A4, and A5) and three thinner cruciform pulleys (Cl, C2, and C3) (Fig. 14-9). These pulleys allow for a smooth curve so that no sharp or angular bends exist in the course of the tendon. Local points of high pressure, stress raisers, between tendon and sheath are therefore minimized.

At the point where the A3 pulley traverses the PIP joint, the tension in the tendon generated by joint flexion either pulls the pulley away from its attachment to the bone or pulls the bone away from the joint. This is no problem in a normal, stable joint, but when the joint becomes unstable, as in a patient with rheumatoid arthritis, there could be a danger of severe PIP subluxation.

To appreciate the magnitude of these subluxating forces and how they increase with increased flexion,

CALCULATION BOX 14-1

Wrist Loads during Weight Lifting

Magnitude of the radiocarpal joint reaction force with flexion position above horizontal at the elbow joint (Calculation Box Figs. 14-1-1 and 14-1-2).

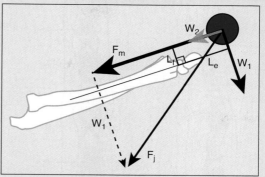

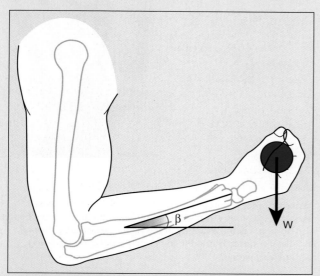

Calculation Box Figure 14-1-1 Lateral view. The elbow is flexed to 115° from the anatomic position (i.e., the angle β is 15° above horizontal), and a weight, W, is held in the palm of the hand.

Calculation Box Figure 14-1-2 Free-body diagram showing the forces for the handheld weight (W_1 and W_2), the wrist and digital flexors (F_m), and the radiocarpal joint reaction force (F_j). Also shown are the moment arms for the wrist extension force (L_e) and the wrist flexion force (L_f), as well as the triangular solution to resolve the vector components of F_j. The wrist flexors will have to balance the moment caused by W_1 only, since W_2 acts in the same direction as F_m (i.e., to flex the wrist). Therefore,

$$F_m\,L_f = W_1\,L_e$$

Substituting $W \cos \beta$ for W_1 and rearranging the equation gives

$$F_m = (W \cos \beta)(L_e/L_f)$$

F_j is equal to the vector sum of all the forces. Using the Pythagorean theorem, as depicted in the right triangle of the forces,

$$F_j = \sqrt{(F_m + W_2)^2 + W_1^2} = \sqrt{(W\cos\beta)(L_e/L_f) + (W\sin\beta)^2 + (W\cos\beta)^2}$$

Radiographic measurements have shown that $L_e/L_f = 4$.

Therefore, for $\beta = 15°$, $F_j = 4.235W$.

Adapted from Karnezis, I.A, (2005). Clin Biomech, 20, 270.

---

consider two separate flexed positions of a PIP joint: 60° and then 90°. At 60°, the two limbs of the flexor tendon form an angle of 120° (Calculation Box 14-2). At that point, the tension in the restraining pulley must equal the tension in the tendon for the system to be in equilibrium. At 90° of flexion, however, the pulley must sustain 40% more tension than the tendon (Calculation Box Fig. 14-2-2) (Brand, 1985; Brand and Hollister, 1992).

Digital Collateral Ligaments

The common essential feature of the articulations of the digits is that they function in the direction of flexion and have two firm collateral ligaments and a thick reinforced anterior capsule. The anterior fibrocartilage is known as the palmar or volar plate (Neumann, 2010; Tubiana, 1984). There are significant differences between the interphalangeal and metacarpophalangeal articulations of the digits as well as significant differences between the same level for each digit (Hakstian and Tubiana, 1967; Kucynski, 1968; Landsmeer, 1955; Smith and Kaplan, 1967; Tubiana, 1984).

A unique feature of the MCP joint is its asymmetry, which is apparent both in the bony configuration of the metacarpal head (Fig. 14-10) and in the location of the radial and ulnar collateral ligament attachments to it (Landsmeer, 1955). The collateral ligaments of the MCP joint extend obliquely forward from the proximal attachment at the dorsolateral aspect of the metacarpal head to their insertion on the palmolateral aspect of the base of the proximal phalanx. The bilateral

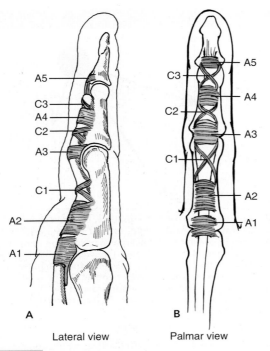

A Lateral view   **B** Palmar view

FIG. 14-9

Schematic drawings of the components of the digital flexor tendon sheath. The five strong annular pulleys (A1, A2, A3, A4, A5) are important in ensuring efficient digital motion by apposing the tendons to the phalanges. The three thin, pliable cruciate pulleys (C1, C2, C3) allow flexibility of the sheath while maintaining its integrity. **A.** Mediolateral view. *Adapted from Doyle J.R. Hand. In: Doyle J.R, Botte M.J., eds. (2003). Surgical anatomy of the hand and upper extremity. Philadelphia: Lippincott Williams & Wilkins, 2003:522–666, with permission.* **B.** Palmar view of the sheath without its tendons. *Adapted from Oatis, C.A. (2008). Kinesiology: The Mechanics and Pathomechanics of Human Movement. Baltimore, MD: Lippincott Williams & Wilkins, 344.*

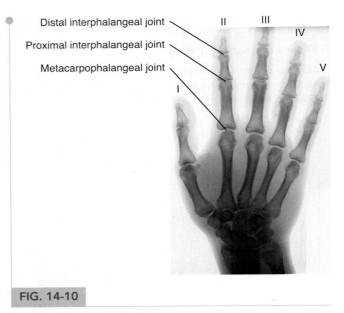

FIG. 14-10

Posteroanterior roentgenogram of the right hand and wrist revealing asymmetry in the configuration of the metacarpal heads. Also apparent is the disparity in the diameters of the proximal and distal surfaces of the PIP joints, the distal surfaces being considerably wider.

asymmetry in the site of the attachment of these ligaments manifests itself particularly in the asymmetric range of abduction-adduction in these joints. The asymmetric bilateral arrangement of the interossei also contributes to the overall asymmetry of the MCP joints (Fig. 14-11).

Quantification of the length changes in the collateral ligaments during MCP joint motion was looked at by Minami et al. (1984), who used biplanar roentgenographic techniques to analyze the lengths of the dorsal, middle, and palmar thirds of the radial and ulnar collateral ligaments of the index finger at various degrees of joint flexion. When the MCP joint was flexed from 0° to 80°, the dorsal portion of the ligaments lengthened 3 to 4 mm, the middle portion elongated slightly, and the palmar portion shortened 1 to 2 mm. When the MCP joint moved into hyperexten-

sion, the dorsal portion of the ligaments shortened 2 to 3 mm, the middle third shortened slightly, and the palmar third lengthened slightly. Thus, the dorsal portions of both collateral ligaments appear to provide the principal restraining force when the MCP joint is flexed, whereas the palmar portions provide a restraining force during MCP extension. Therapists will often splint the traumatized or postsurgical MCP joint into 60° to 70° of flexion to place a relative stretch on the collateral ligaments to prevent extension contracture when immobilization is required.

The collateral ligaments are found to be slack when the MCP joints are held positioned in extension and are taut when the MCPs are positioned in flexion. By placing the MCPs into full flexion, the cam configuration of the metacarpal head tightens the collateral ligaments and the lateral mobility or "play" observed when the MCPs are held into extension is limited (Neumann, 2010). Therefore, the fingers cannot be spread or abducted unless the hand is open, or flattened (Moore and Dalley, 2006).

The transverse intermetacarpal ligament, which connects the palmar plates, gives additional stability to the MCP region (Fig. 14-12). The extensor tendons are linked to this transverse structure by the transverse laminae, which hold them in position on the dorsal side of the MCP joint.

CALCULATION BOX 14-2

Flexor Tendon Sheath Pulley System at the PIP Joint

Magnitude of subluxating forces and increment with flexion position at the PIP joint. (Calculation Box Figs. 14-2-1 and 14-2-2).

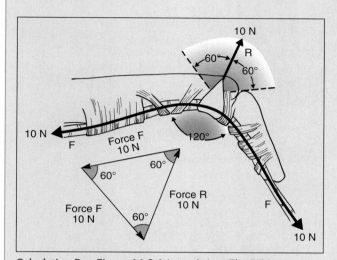

Calculation Box Figure 14-2-1 Lateral view. The PIP joint is flexed 60°. With the system in equilibrium, the resultant force (*R*) in the pulley system is equal to the vector sum of the two components of the tensile force (*F*) in the flexor tendon (i.e., 10 N). These three forces are presented graphically in an equilateral triangle of forces.

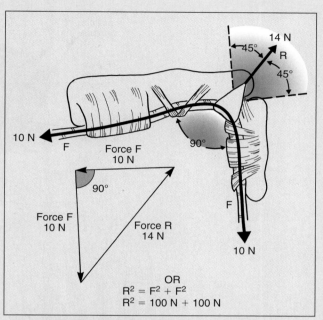

$$R^2 = F^2 + F^2$$
$$R^2 = 100 \text{ N} + 100 \text{ N}$$

Calculation Box Figure 14-2-2 Lateral view. The PIP joint is flexed 90°. A triangle of forces shows that the resultant force R in the pulley system equals 14 N. Therefore, *R* equals 1.4 F. The value for *R* is also found by use of the Pythagorean theorem, which states that in a right triangle, the square of the hypotenuse equals the sum of the squares of the sides. *Adapted with permission from Brand, P.W. (1985). Clinical Mechanics of the Hand. St Louis, MO: Mosby, 30–60.*

Volar Plate

In addition to the role of the collateral and accessory collateral ligament, attention is brought to the function of the palmar or volar plate (Fig. 14-13). The accessory collateral ligaments are just palmar to the radial and ulnar collateral ligaments, which originate from the metacarpal and insert into the thick palmar fibrocartilaginous plate. This plate on the volar surface of the MCP is firmly attached to the base of the proximal phalanx, and it is loosely attached to the volar surface of the neck of the metacarpal. This anatomic alignment allows for the volar plate to slide proximally like a moving visor during MCP flexion (Neumann, 2010; Strickland, 1987). The volar plates are connected by the transverse intermetacarpal ligaments that then connect each plate to its neighbor. They serve to reinforce the joint capsule anteriorly, prevent impingement of the flexor tendons during MCP flexion, and limit hyperextension of the MCP joint.

Tendinous Mechanisms

DIGITAL EXTENSOR ASSEMBLY SYSTEM

The long extensor tendons are flat structures that emerge from their synovial sheaths at the dorsal side of the carpus and run over the MCP joint; they are held in this position by the sagittal bands. At the dorsum of the proximal phalanx, these extensor tendons and parts of the interossei interweave so as to form a tendinous complex, the extensor assembly (also known as the extensor mechanism), which extends over both IP joints (Fig. 14-14).

Trifurcation of the long extensor tendon and fanning of interosseous fibers result in the formation of one medial and two lateral bands. The medial band (or central slip) runs dorsally over the trochlea of the proximal phalanx and inserts into the base of the middle phalanx. The two lateral bands course alongside the shoulders of the PIP joint. These bands pursue their way distally and

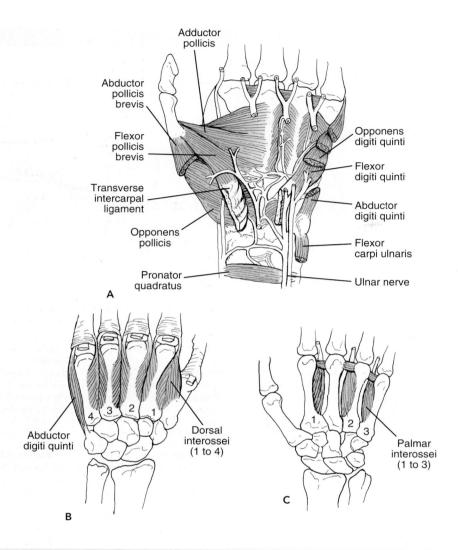

FIG. 14-11

The intrinsic muscles of the hand. **A.** Palmar view of the left hand. **B.** Dorsal view of the left hand showing the four dorsal interossei and the abductor digiti quinti. These muscles abduct the fingers (i.e., move them away from the midline of the hand). **C.** Palmar view of the left hand showing the three palmar interossei. These muscles adduct the second, fourth, and fifth fingers, flex the MCP joint, and extend the PIP joint. *Adapted with permission from Strickland, J.W. (1987). Anatomy and kinesiology of the hand. In E E. Fess, C.A. Philips (Eds.). Hand Splinting. Principles and Methods (2nd ed.). St Louis, MO: Mosby, 3–41; and Caillet, R. (1982). Hand Pain and Impairment (3rd ed.). Philadelphia: FA Davis.*

merge over the dorsum of the middle phalanx, forming the terminal tendon, which inserts into the dorsal tubercle of the distal phalanx. This terminal tendon is linked to the proximal phalanx by means of the oblique retinacular ligaments. These ligaments originate from the proximal phalanx and run laterally around the PIP joint, just palmar to the center of motion of this joint in the extended position, to join the terminal tendon.

Illustrating the action of the extensor assembly in coupling PIP and DIP joint motion, Landsmeer (1949) described the "release of the distal phalanx" (Fig. 14-15).

If a finger is flexed at the PIP joint only, the whole trifurcated extensor assembly is pulled distally, following the central slip. This slip alone is taut because the distal pull occurs at the middle phalanx; the lateral bands remain slack but are allowed to shift distally over the same distance. Only part of the slack of the lateral bands is required for flexion of the PIP joint because these bands run closer to the center of motion of this joint than does the central slip. Therefore, some of the slack will remain, allowing passive or active flexion of the distal phalanx but no active extension. The "released" distal phalanx is

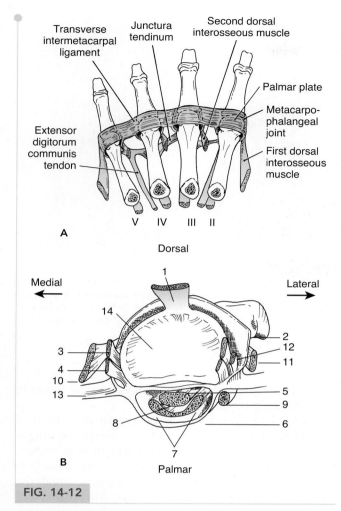

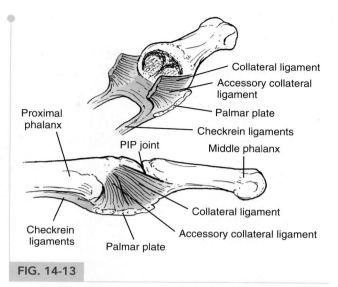

FIG. 14-13

Oblique **(top)** and mediolateral **(bottom)** views of the PIP joint. This joint gains stability from a strong, three-sided ligamentous support system produced by the collateral ligament, the accessory collateral ligament, and the palmar fibrocartilaginous plate (volar plate), which is anchored to the proximal phalanx by proximal and lateral extensions known as the checkrein ligaments. *Adapted with permission from Strickland, J.W. (1987). Anatomy and kinesiology of the hand. In E.E. Fess, C.A. Philips (Eds.). Hand Splinting. Principles and Methods (2nd ed.). St Louis, MO: Mosby, 3–41.*

FIG. 14-12

A. Fibrous structures of the proximal transverse (*MCP*) arch (palmar view of the right hand). *Adapted with permission from Tubiana, R. (1984). Architecture and functions of the hand. In R. Tubiana, J.-M. Thomine, E. Mackin (Eds.). Examination of the Hand and Upper Limb. Philadelphia: WB Saunders, 1–97.* **B.** Capsuloligamentous structures of the MCP joint (transverse view of the proximal phalangeal joint surface, middle finger, left hand). *1,* extensor digitorum communis tendon; *2,* sagittal band; *3,* collateral ligament; *4,* accessory collateral ligament; *5,* volar plate; *6,* flexor tendon sheath; *7,* flexor digitorum superficialis tendon; *8,* flexor digitorum profundus tendon; *9,* lumbrical muscle; *10,* dorsal interosseous muscle; *11,* dorsal interosseous muscle; *12,* insertion of dorsal interosseous muscle into base of phalanx; *13,* transverse intermetacarpal ligament; *14,* articular surface of proximal phalanx. *Adapted from Zancolli, E. (1979). Structural and Dynamic Bases of Hand Surgery (2nd ed.). Philadelphia: JB Lippincott Co, 3–63.*

the functional basis for the coupled flexion and extension of the DIP and PIP joints.

Conversely, if the DIP is actively flexed, the entire extensor assembly is displaced distally. This relaxes the central slip and simultaneously increases the tension in the oblique retinacular ligaments, a tension that creates a flexion force at the PIP joint. Because the central slip is already unloaded, flexion of this joint is then unavoidable. The release of the distal phalanx is fundamental for pulp-to-pulp pinch. It also allows, through intermittent contraction of the flexor profundus, a change from pulp-to-pulp to tip-to-tip pinch, a mechanism used in precision handling such as needlework and active tactile exploration.

Sarrafian et al. (1970) used strain gauges to measure the tension in different parts of the extensor mechanism during finger flexion and further elaborated on this phenomenon. They found an increase in the central slip tension beyond 60° of PIP flexion; at 90° of flexion there was total relaxation of the lateral bands.

ACTIVE CONTROL MECHANISMS

Muscular Mechanisms of the Wrist

The wrist joint complex is surrounded at its periphery by the 10 wrist tendons, whose muscles and their actions are listed in Table 14-1. The three flexors and three extensors are the motors of the wrist, controlling radial and ulnar deviation as well as wrist flexion and extension. Four additional muscles control pronation and supination of

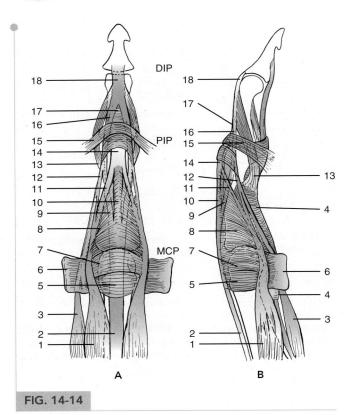

FIG. 14-14

Schematic drawing of the digital extensor assembly. *MCP,* metacarpophalangeal joint; *PIP,* proximal interphalangeal joint; *DIP,* distal interphalangeal joint. *1,* interosseous muscle; *2,* extensor digitorum communis tendon; *3,* lumbrical muscle; *4,* flexor tendon fibrous sheath; *5,* sagittal band; *6,* intermetacarpal ligament; *7,* transverse fibers of interosseous hood; *8,* oblique fibers of interosseous hood; *9,* lateral band of long extensor tendon; *10,* medial band of long extensor tendon; *11,* central band of interosseous tendon; *12,* lateral band of interosseous tendon; *13,* oblique retinacular ligament; *14,* medial band of long extensor tendon in central slip; *15,* transverse retinacular ligament; *16,* lateral band of long extensor tendon; *17,* triangular ligament; *18,* terminal tendon. **A.** Dorsal view. Just proximal to the PIP joint, the long extensor tendon (extensor digitorum communis tendon) within the central slip trifurcates into one medial and two lateral bands. The medial band inserts into the base of the middle phalanx. The lateral bands converge over the dorsum of the middle phalanx to form the terminal tendon, which inserts on the distal phalanx. **B.** Sagittal view. The oblique retinacular ligaments, which originate from the proximal phalanx, course laterally around the PIP joint just palmar to the center of rotation of flexion-extension, then join the terminal tendon. *Adapted with permission from Tubiana, R. (1984). Architecture and functions of the hand. In R. Tubiana, J.-M. Thomine, E. Mackin (Eds.). Examination of the Hand and Upper Limb. Philadelphia: WB Saunders, 1–97.*

the forearm. Eight of the muscles originate from the forearm, and two, the brachialis and extensor carpi radialis longus, originate above the elbow. Except for the flexor carpi ulnaris tendon, which attaches to the pisiform, all

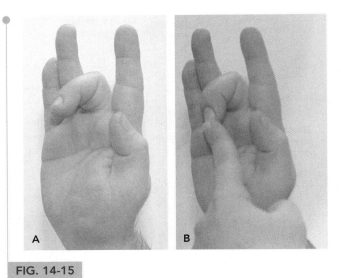

FIG. 14-15

Release of the distal phalanx. **A.** All fingers are extended, and the PIP joint of the middle finger is flexed. The DIP joint of this finger is totally out of control. **B.** The DIP joint is very loose and can be flexed or extended only passively.

of the wrist muscle tendons traverse the carpal bones to insert on the metacarpals.

Each wrist tendon has a substantial amplitude of excursion. The extensor carpi radialis brevis and longus each have an excursion of 14 mm and 19 mm, respectively, through the 90° arc of motion from 45° wrist flexion to 45° of wrist extension (Tang, 1999). Through the same arc of motion, the extensor carpi ulnaris excursion is approximately 7 mm, the flexor carpi radialis excursion is approximately 22 mm, and the flexor carpi ulnaris excursion is approximately 25 mm. Impairment of the excursion of any of these tendons owing to adhesions after trauma or surgery can seriously limit wrist motion.

The arrangement of digital and wrist extensor and flexor systems around the wrist axis makes for antagonist groupings of motor forces that afford positional stability during finger movements and grip. For example, contraction of the flexor digitorum profundus and superficialis results in a wrist flexion as well as a finger flexion movement. Therefore, the wrist extensors (extensor carpi radialis brevis and longus and extensor carpi ulnaris) contract to stabilize the wrist during gripping (Neumann, 2010).

The contributions of the extensor carpi ulnaris, extensor pollicis brevis, and abductor pollicis longus were assessed electromyographically during wrist flexion (Kauer, 1979, 1980). In addition to demonstrating their expected muscle actions, these muscles were found to function as a dynamic "adjustable collateral system" that acts as a true collateral support, the extensor carpi ulnaris for the ulnar side of the wrist and the extensor pollicis brevis and abductor pollicis longus for the radial side. In this way, active control mechanisms serve to fulfill the void left by the lack of collateral ligamentous restrictions while still affording considerable positional variability for functional hand activities (Case Study 14-3).

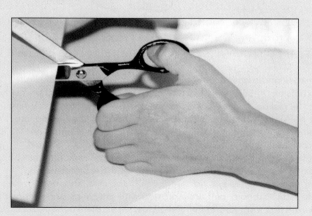
Muscular Mechanisms of the Hand

The digital rays are controlled by the extrinsic and intrinsic muscles (Table 14-1). The larger and stronger movements of the hand and its fingers for grasping or pointing are produced by the extrinsic muscles (Moore and Dalley, 2006). The muscle bellies of the extrinsic muscles are primarily located near the elbow and in the forearm, and their tendons pass into the hand and into the digital rays. The intrinsic muscle bellies and tendons are entirely confined within the hand (Fig. 14-11). The intrinsic muscles position the digits for more powerful motions controlled by the extrinsic muscles and provide more delicate motor tasks such as handwriting and musical instrument playing (Moore and Dalley, 2006). Although the contribution of the intrinsic or extrinsic system is distinctly different, the coordinated functioning of the two muscle systems is essential for the satisfactory performance of the hand in a wide range of prehensile and nonprehensile tasks.

The values most often cited for the strengths of the extrinsic muscles of the hand were reported by Von Lanz and Wachsmuth (1970). Their values (Table 14-3) show that the strength of the finger flexors is more than twice that of the extensors.

Kinematics

The multiplicity of wrist articulations and the complexity of carpal motion make it difficult to calculate the instant center of motion for the primary axes of flexion-extension

TABLE 14-3

Strength Values of the Extrinsic Muscles of the Hand

Muscle	Strength (Nm)
Flexor pollicis longus	12
Extensor pollicis longus	1
Abductor pollicis longus	
As a wrist flexor	1
As a wrist abductor	4
Extensor pollicis brevis	1
Flexor digitorum superficialis	48
Flexor digitorum profundus communis	45
Extensor digitorum communis	17
Extensor indicis proprius	5

Data from Von Lanz, T., Wachsmuth, W. (1970). Functional anatomy. In J.H. Boyes (Ed.). *Bunnell's Surgery of the Hand (5th Ed.)*. Philadelphia: JB Lippincott Co.

or radial-ulnar deviation. Various studies have placed the instant center of rotation in the head of the capitate, with the flexion-extension axis oriented from the radial to the ulnar styloid process and the radial-ulnar deviation axis oriented orthogonal to the flexion-extension axis. This kinematic model is undoubtedly an oversimplification of the complex carpal motions during wrist movement, but it appears to adequately describe functional wrist movement (Brumbaugh et al., 1982, Moritomo et al., 2006; Youm and Yoon, 1979).

The hand is an extremely mobile organ that can coordinate many movements in relation to each of its components. The blending of hand and wrist movements enables the hand to mold itself to the shape of an object being palpated or grasped. The great mobility of the hand is the result of the articular contours, the position of the bones in relation to one another, and the actions of an intricate system of muscles.

WRIST RANGE OF MOTION

The articulations of the wrist joint complex allow motion in two planes: flexion-extension (palmar flexion and dorsiflexion) in the sagittal plane and radial-ulnar deviation (abduction-adduction) in the frontal plane. Combinations of these motions are also possible.

Although small amounts of axial rotation are possible and may exist in some individual wrists, from a practical standpoint such rotation does not occur through the carpal complex (Youm et al., 1978). Axial rotation of the hand, expressed as pronation and supination, results instead from motion arising at the proximal and distal radioulnar and the radiohumeral joints (Volz et al., 1980).

In recent years, in vivo, three-dimensional imaging techniques, such as 3D computed tomography and 3D computer imaging, have become available to study individual carpal motion during wrist movements in cardinal and functional planes of motion. Among the most well-studied of the functional planes is the so-called dart thrower's motion, which describes the path of motion from a position of radial deviation and extension to that of ulnar deviation and flexion (Crisco et al., 2005).

Flexion and Extension

The normal wrist range of motion is 65° to 80° of flexion and 55° to 75° of extension, but it can vary widely among individuals. Owing to a slight palmar tilt of the distal radial plates, flexion exceeds extension by an average of 10°.

Investigators have found various values for the contribution of the proximal and distal carpal rows to the total arc of flexion and extension. In a cadaveric study, Kaufmann

et al. (2006) found that, in flexion, 75% of wrist motion occurred at the radioscaphoid joint, and 50% occurred at the radiolunate joint, suggesting that the proximal carpal bones do not move as a fixed unit during wrist flexion. In extension, this finding was even more pronounced, with 92% of the wrist motion occurring at the radioscaphoid joint whereas only 52% occurred at the radiolunate joint. Furthermore, the scaphoid and capitate tended to move together, whereas midcarpal motion between the lunate and capitate was evident (Figure 14-16).

Radial and Ulnar Deviation

The total arc of radial-ulnar deviation is approximately 50° to 60°, 15° to 20° radialward and 35° to 40° ulnarward (Neumann, 2010). The distal carpal row follows the finger

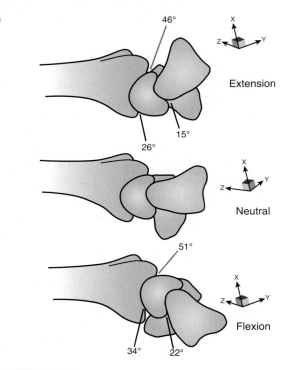

FIG. 14-16

Images of radiocarpal and midcarpal joints during wrist flexion-extension. In 50° of wrist extension (**A**) the motion occurs primarily at the radiocarpal joint, with 92% of the overall motion occurring at the radioscaphoid joint and 52% occurring at the radiolunate joint relative to neutral (**B**). In 68° of wrist flexion (**C**), motion occurs at both the radiocarpal and midcarpal joints: 75% and 50% of overall motion at the radioscaphoid and radiolunate joints, respectively. Note that motions for the radiolunate and capitolunate joints are indicated with *black lines,* and motions for the radioscaphoid joint are indicated with *gray lines. Adapted with permission from Kaufmann, R A., Pfaeffle, J., Blankenhorn, B.D., et al. (2006). Kinematics of the midcarpal and radiocarpal joint in flexion and extension: An in vitro study. J Hand Surg, 31A, 1142.*

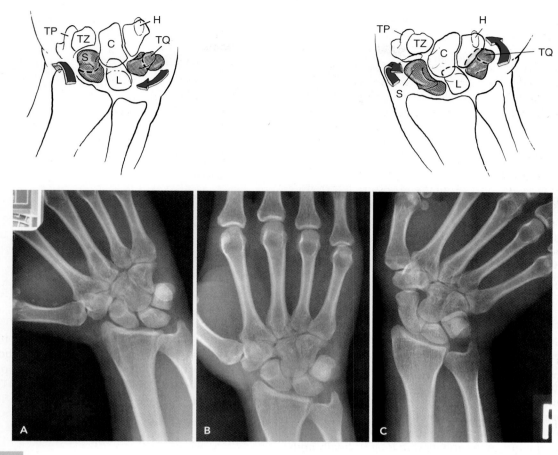

FIG. 14-17

Roentgenograms of the right wrist and hand (dorsal view) showing the position of the carpal bones in radial deviation **(A)**, in the neutral position **(B)**, and in ulnar deviation **(C)**. *Arrows* in the schematic drawings above roentgenograms **A** and **C** indicate general movement of the bones of the proximal row with wrist motion. In radial deviation, the bones of the proximal row are flexed toward the palm. The scaphoid appears foreshortened, the lunate appears triangular, and the triquetrum is proximal in relation to the hamate. In ulnar deviation, the bones of the proximal row are extended. The scaphoid appears elongated, the shape of the lunate appears trapezoidal, and the triquetrum is distal in relation to the hamate. *TP*, trapezium; *TZ*, trapezoid; *C*, capitate; *H*, hamate; *TQ*, triquetrum; *L*, lunate; *S*, scaphoid. *Roentgenogram courtesy of Alex Norman, M.D.; drawings adapted with permission from Taleisnik, J. (1985). The Wrist. New York: Churchill Livingstone.*

rays during both radial and ulnar deviation, whereas the proximal carpal row glides in the direction opposite to hand movement with greater excursion during ulnar deviation. Radial and ulnar deviation occurs primarily at the midcarpal joint. For example, Kaufman et al., (2005) observed 22° of capitate radial deviation motion but only 4° of scaphoid and 3° of lunate radial deviation using a 3D computed tomography technique in cadaveric specimens.

During radial deviation, the scaphoid undergoes flexion (palmward rotation of its distal pole) as a result of its encroachment on the radial styloid process (Fig. 14-17A). This scaphoid motion is transmitted across the proximal row through the scapholunate ligament. Thus, in radial deviation the scaphoid flexes and so does the proximal carpal row. This conjunct movement of the scaphoid and proximal carpal row is reversed toward extension during ulnar deviation (Fig. 14-17C). During ulnar deviation, the triquetrum is displaced palmward by the proximal migration of the hamate. The motion of the triquetrum in turn causes the lunate to extend.

A double-V system formed by the palmar intercarpal ligament and the radiolunate and ulnolunate ligaments renders support during radial-ulnar deviation (Fig. 14-18). The apex of the proximal V is at the lunate and that of the distal V at the capitate. In ulnar deviation, the medial arm of the proximal V, the ulnolunate ligament, becomes

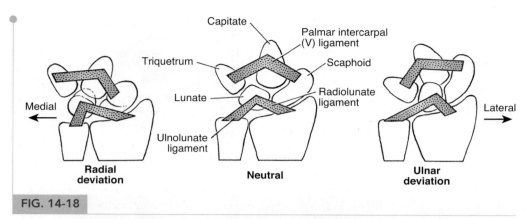

FIG. 14-18

Diagrammatic representation of the changes in alignment of the double-V system formed by the ulnolunate and radiolunate ligaments and the palmar intercarpal (V, or deltoid) ligament with the wrist in radial deviation, the neutral position, and ulnar deviation (palmar view of right hand). *Adapted with permission from Taleisnik, J. (1985). The Wrist. New York: Churchill Livingstone.*

somewhat transverse and inhibits radial displacement of the lunate, while the lateral arm, the radiolunate ligament, orients longitudinally and limits lunate extension. The V configuration is now an L. The distal V also becomes an L but in the opposite direction. The lateral intrinsic ligamentous fibers connecting the scaphoid and capitate become somewhat transverse to check the central ulnar translation of the capitate during this motion. The medial fibers from triquetrum to capitate shift longitudinally and control capitate flexion. In radial deviation, the opposite configurations apply (Neumann, 2010).

Dart Thrower's Motion

The arc of motion from radial extension to ulnar flexion is known as the dart thrower's motion (DTM), and it has evolutionary as well as functional significance. The DTM movement arc is used in tasks with varying levels of fine motor control and force demands, from throwing a ball to swinging an axe. Recent in vivo studies have begun to compare and contrast carpal motions during DTM and the cardinal plane movements of radial-ulnar deviation and flexion-extension (e.g., Crisco et al., 2005; Moritomo et al., 2006).

During DTM, the axes of motion for the radiocarpal and midcarpal joints converge, thereby allowing for synergistic motion through a relatively wide arc (Moritomo et al., 2006). Furthermore, the motions of the scaphoid and lunate bones are minimized throughout the arc of the DTM compared with either flexion-extension or radial-ulnar deviation (Crisco et al., 2005). Therefore, during DTM, the proximal carpal row forms a stable base on which movements requiring either or both strength

and agility can be performed. Crisco et al. (2005) suggest that rehabilitation efforts following common injuries or surgeries involving the scaphoid and lunate bones may lead to improved function more quickly while protecting these carpal bones if movements along the DTM movement path are initiated early.

Forearm Pronation and Supination

The motions of forearm pronation and supination, although not part of wrist motion proper, play an intricate role in hand and wrist function. Average range of motion of pronation-supination is 160° (75° of pronation and 85° of supination). The axis of pronation-supination lies oblique to both the radius and the ulna, passing through the center of the humeral capitulum and the midpoint of the head of the ulna (Neumann, 2010). During pronation and supination, the radial head rotates within the fibro-osseous ring formed by the annular ligament and the radial notch of the ulna and spins with respect to the lateral humeral condyle. Distally, the radius glides with respect to the ulna at the distal radioulnar joint. The dorsal and palmar capsular ligaments of the distal radioulnar joint limit motion at the extremes of pronation and supination, respectively (Neumann, 2010).

DIGITAL RANGE OF MOTION

The varying shapes of the CMC, MCP, and IP joints of the thumb and the fingers are responsible for the differences in degrees of freedom at these joints. The unique orientation of the thumb, the large web space, and the special configuration of the thumb CMC joint affords this digit great mobility and versatility.

Fingers

The second and third metacarpals are linked to the trapezoid and capitate and to each other by tight-fitting joints that are basically immobile (Fig. 14-19). As a result, these metacarpal and carpal bones constitute the "immobile unit" of the hand. The articulations of the fourth and fifth metacarpals with the hamate permit a modest amount of motion: 10° to 15° of flexion-extension at the fourth CMC joint and 20° to 30° at the fifth. Limited palmar displacement, or descent, of these metacarpals

may take place. This motion allows cupping of the hand and is essential for gripping.

The MCP joints of the four fingers are unicondylar diarthrodial joints (Figs. 14-10 and 14-19), allowing motion in three planes: flexion-extension (sagittal plane), abduction-adduction (frontal plane), and slight pronation-supination (transverse plane), which is coupled with abduction-adduction (Hagert, 1981).

The range of MCP flexion from the zero position is approximately 90° (Fig. 14-20A), but this value differs among the fingers. The fifth finger demonstrates the most flexion (approximately 95°) and the second (index) finger, approximately 70° (Batmanabane and Malathi, 1985). Extension beyond the zero position varies considerably and depends on joint laxity.

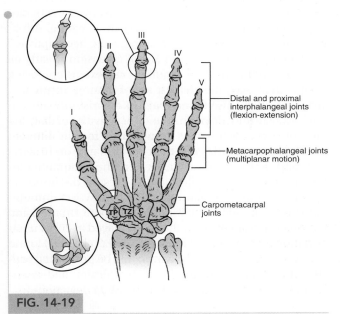

FIG. 14-19

Schematic representations of the joints of the finger rays (dorsal view of the right hand). The CMC joint between the first metacarpal and trapezium (*TP*) is composed of two saddle-shaped surfaces, the convexity of one fitting tightly into the concavity of the other (**inset** shows enlargement). This arrangement allows for movement of the thumb in a wide arc of motion. The tight-fitting joints that link the second and third metacarpals with the trapezoid (*TZ*) and capitate (*C*), respectively, and with each other are relatively immobile, rendering these four bones the "immobile unit" of the hand. The joints between the fourth and fifth metacarpals and the hamate (*H*) permit a modest amount of flexion and extension. The unicondylar configuration of the MCP joints of the four fingers allows motion in three planes and combinations thereof. By contrast, the tongue-and-groove articular contours of the bicondylar hinge joints between the phalanges limit motion to one plane (flexion-extension) and contribute to the stability of these joints in resisting shear and rotary forces (**inset** shows enlargement of a typical IP joint in an oblique view).

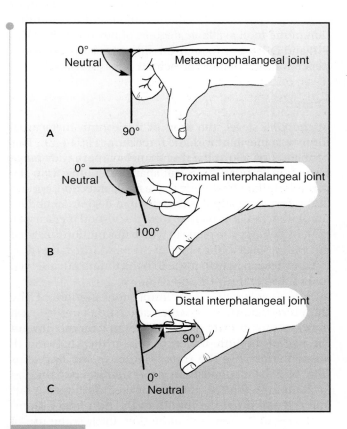

FIG. 14-20

Flexion of the three joints of the finger, beginning with the neutral position in which the extended fingers are in the plane of the dorsal hand and wrist. **A.** Flexion of the MCP joint, averaging 70° to 90°. **B.** Flexion at the PIP joint, averaging 100° or more. **C.** Flexion at the DIP joint, averaging 90°. *Adapted with permission from Joint Motion Method of Measuring and Recording. Rosemont, IL, American Academy of Orthopaedic Surgeons, 1965.*

The PIP and distal joints of the four digits are bicondylar hinge joints as a result of the tongue-and-groove fit of their articular surfaces (Figs. 14-10 and 14-19). These surfaces are closely congruent throughout the range of flexion-extension, which is the only motion possible in these joints. Flexion is measured from the zero position with the finger in the plane of the hand. The largest range of flexion, 110° or more, occurs in the PIP joint (Fig. 14-20B). Flexion of approximately 90° takes place in the DIP joint (Fig. 14-20C). Extension beyond the zero position, termed hyperextension, is a regular feature of the DIP and PIP joints, although it depends largely on ligamentous laxity, especially in the PIP joint.

The range of motion of the MCP, PIP, and DIP joints is often reported individually for each of the three joints. In addition to this, composite measurements scores are often reported. These summation scores for either active or passive movement—TAM (total active movement) or TPM (total passive movement)—represent the summation of the total available degrees of flexion at the MCP, PIP, and DIP joints for a given digit minus the extension deficit for each of the represented three joints.

Thumb

At the CMC level, the base of the thumb metacarpal forms a saddle joint with the trapezium (Fig. 14-19). This configuration allows the thumb metacarpal a wide range of motion through a conical space extending from the plane of the hand palmarly to the radial direction. Motion of the first metacarpal is described in degrees of abduction, either radial or palmar, from the second metacarpal, thereby defining the plane in which this motion is carried out with respect to the plane of the hand. The terms flexion and extension with respect to the thumb are reserved for motions of the MCP and IP joints.

Functionally, the most important motion of the thumb is opposition, in which abduction coupled with rotation at the CMC joint moves the thumb toward the pad of the little finger; flexion at the MCP and IP joints then brings the thumb closer to the fingertips (Fig. 14-21). Full opposition is then noted when the pad of the thumb touches the pad of the fifth finger. A lateral orientation of the thumb to the fifth finger represents use of flexion and adduction, for it is the use of the opponens muscles that defines thumb opposition, which is considered complete only when pad-to-pad contact is made.

The MCP joint of the thumb resembles those of the fingers. The range of flexion from the zero position varies considerably among individuals, from as little as 30° to as much as 90°; extension from the zero position is approximately 15° (Batmanabane and Malathi, 1985). The IP joint of the thumb, the most distal joint, resembles and performs similarly to the analogous distal joints in the fingers.

FUNCTIONAL WRIST MOTION

Because the joints proximal to the wrist may provide compensatory motion, even a considerable loss of wrist motion may not interfere significantly with activities of daily living. An electrogoniometric study of the range of wrist flexion-extension required for accomplishing 14 activities showed that an arc of 45° (10° of flexion to 35° of extension) was sufficient for performing most of them (Brumfield and Champoux, 1984). Seven activities of personal care that require placing the hand at various locations on the body were accomplished within a range of 10° of flexion to 15° of extension, and most were performed with the wrist slightly flexed. Other necessary activities requiring an arc of wrist motion, such as eating, drinking, using a telephone, and reading, were accomplished by motion of 5° of flexion to 35° of extension. Nearly all of these continuous tasks required only extension. Rising from a chair used the greatest arc of motion, nearly 63°. Volz et al. (1980) also found that loss of wrist mobility did not seriously impede performance of activities of daily living. Volunteers with wrists immobilized in four different positions were asked to rate their performance on 10 activities, and performance averages were then computed for each position of immobilization. The results disclosed the least compromise of hand function with wrists immobilized in 15° of extension (88% of normal performance) and the greatest disability with the wrists placed in 20° of ulnar deviation (71% of normal function). Ryu et al. (1991) found that most activities of daily living could be performed with 70% of total wrist motion (40° each of flexion and extension, and 40° of combined radial and ulnar deviation).

Interaction of Wrist and Hand Motion

Wrist motion is essential for augmenting the fine motor control of the fingers and hand. Positioning the wrist in the direction opposite that of the fingers alters the functional length of the digital tendons so that maximal finger movement can be attained. Wrist extension is synergistic to finger flexion and increases the length of the finger flexor muscles, allowing increased flexion with stretch (Fig. 14-22A) (Tubiana, 1984). Conversely, some flexion of the wrist puts tension on the long extensors, causing the fingers to open automatically and aiding full finger extension (Fig. 14-22B).

The synergistic movements of the wrist extensors and the more powerful digital flexors are facilitated by the architecture of the wrist. The digital flexor tendons cross

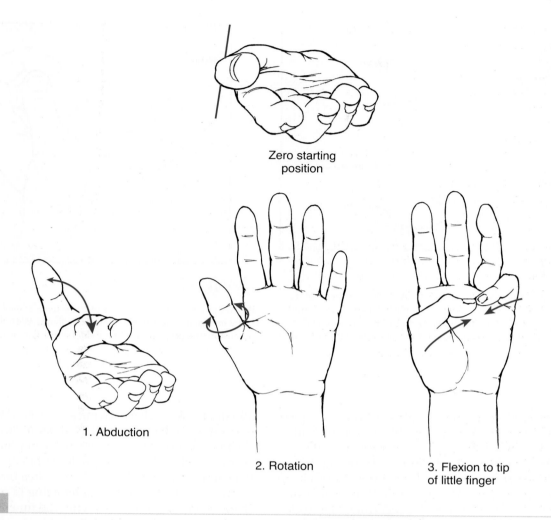

Zero starting
position

1. Abduction

2. Rotation

3. Flexion to tip
of little finger

FIG. 14-21

Opposition of the thumb, which begins with the extended thumb in line with the index finger, is the combined motions of abduction and rotation of the carpometacarpal (*CMC*) joint. Flexion in the metacarpophalangeal (*MCP*) and interphalangeal (*IP*) joints then brings the tip of the thumb closer to the fifth finger. Palmar displacement, or descent, of the fourth and fifth metacarpals and flexion in the MCP and IP joints of the fifth finger result in tip-to-tip contact between the thumb and fifth finger. *Adapted from Oatis, C.A. (2008). Kinesiology: The Mechanics and Pathomechanics of Human Movement. Baltimore, MD: Lippincott Williams & Wilkins, 279.*

the wrist within the depths of the carpal arch and are held close to the axis of wrist flexion-extension, affecting wrist position minimally. By contrast, the extrinsic wrist flexors and extensors are positioned widely about the periphery to provide maximal moment arms for positioning the wrist.

As the wrist changes its position and the functional lengths of the digital flexor tendons are altered, the resultant forces in the fingers vary, affecting the ability to grip. Volz et al. (1980) evaluated electromyographically the relationship of grip strength and wrist position. Grip strengths of 67, 134, 201, and 268 N were analyzed with the wrist in five positions: 40° and 20° of flexion, neutral, and 20° and 40° of extension. They found that grip strength was greatest at approximately 20° of wrist extension and was least at 40° of wrist flexion. With the wrist in 40° of extension and in the neutral position, grip strength was slightly less than the maximal values. Studies by Hazelton et al. (1975) of the influence of wrist position on the force produced at the middle and distal phalanges revealed that the greatest force was generated with the wrist in ulnar deviation, the next greatest in extension, and the least in palmar flexion. O'Driscoll et al. (1992) have shown that the optimal, self-selected wrist position that maximizes grip strength is 35° of extension and 7° of ulnar deviation. Li (2002) extended these findings to show that wrist position also had an effect on individual

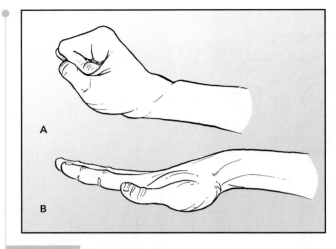

FIG. 14-22

Role of wrist position in finger function. **A.** Slight extension of the wrist allows the flexor muscles to attain maximal functional length, permitting full flexion. **B.** Slight flexion of the wrist places tension on the digital extensor tendons, automatically opening the hand and aiding full finger extension.

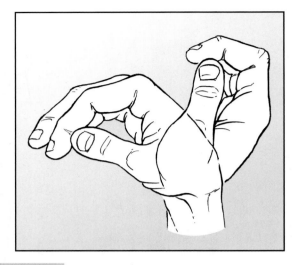

FIG. 14-23

When the wrist is flexed, the tip of the thumb is level with the distal interphalangeal joint of the index finger. With the wrist in extension, the pulps of the thumb and index finger come passively into contact.

finger force as well as total force production with peak finger forces recorded at 20° of wrist extension and 5° of ulnar deviation. Taken together, the results of these studies suggest that for grip to be effective and have maximal force, the wrist must be stable and must be in slight extension and ulnar deviation. This conclusion is consistent with the findings concerning load transmission through the ulnar TFCC structures.

The position of the wrist also changes the position of the thumb and fingers, thus affecting the ability to grip. When the wrist is flexed with the hand relaxed, the pulp of the thumb reaches only the level of the DIP of the index finger; with the wrist extended, the pulps of the thumb and index finger are passively in contact, creating an optimal situation for gripping or pinching (Fig. 14-23).

Patterns of Prehensile Hand Function

The hand provides us with an intricate and complex capacity to support, manipulate, and engage in prehensile functions. One hand can stabilize an object to provide the opportunity for our other hand to perform dexterous actions on the object or we call on our hands to hold our tired head up or provide the needed pillar support as we transition from sitting to standing. Our hands engage in an array of nonprehensile actions such as pointing, repetitive tasks such as scratching or typing, continuous

and fluid tasks such as playing a clarinet or handwriting (Jones and Lederman, 2006; Neumann, 2010). Prehensile movements of the hand are those in which an object is seized and held partly or wholly within the compass of the hand. Such movements are used in a broad range of purposive activities involving handling of objects of all shapes and sizes. Efficient prehensile function depends on a multitude of factors, the most important of which are the following:

1. Mobility of the first CMC joint and, to a lesser extent, of the fourth and fifth MCP joints
2. Relative rigidity of the second and third CMC joints
3. Stability of the longitudinal finger and thumb arches
4. Balanced synergism and antagonism between the long extrinsic muscles and the hand intrinsic muscles
5. Adequate sensory input from all areas of the hand
6. The precise relationships among the length, mobility, and position of each digital ray

Many attempts have been made to classify different patterns of prehensile hand function. Napier (1956) identified two distinct patterns of prehensile movement in the normal hand: power grip and precision grip. He emphasized that the fundamental requisite to prehension—stability—can be met by either posture.

Power grip, or power grasp, is a forceful act performed with the finger flexed at all three joints so that

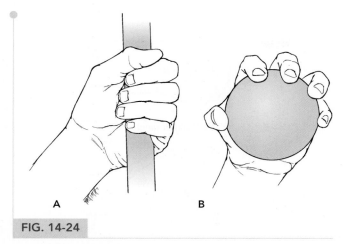

FIG. 14-24

The two fundamental patterns of prehensile hand function. **A.** A typical power grip. The adducted thumb forms a clamp with the partly flexed fingers and the palm. The palmar descent of metacarpals IV and V and additional flexion in their respective MCP joints enable these fingers to hold the object firmly against the palm. Counterpressure is applied by the thumb, which lies approximately in the plane of the palm. The wrist is deviated ulnarward and dorsiflexed slightly to increase the tension in the flexor tendons. Grip of an object along the oblique palmar axis (palmar groove), as shown here, involves a larger area of contact, and thus more control, than does grip along the transverse palmar axis. **B.** A typical precision maneuver. The object is pinched between the flexor aspects of the fingers and the thumb. The fingers are semiflexed and the thumb is abducted and opposed. The wrist is dorsiflexed.

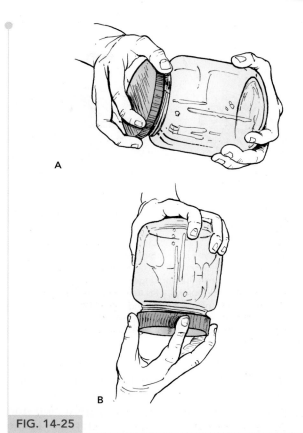

FIG. 14-25

The two fundamental patterns of hand function are used in unscrewing the lid of a tightly closed jar. **A.** As the motion is begun, the right hand assumes a power grip posture. **B.** As the lid loosens, the hand assumes a precision posture to perform the final stages of unscrewing.

the object is held between the finger and the palm, with the thumb positioned on the palmar side of the object to force it securely into the palm (Fig. 14-24A). It is usually performed with the wrist deviated ulnarly and dorsiflexed slightly to augment the tension in the flexor tendons.

Precision grip involves the manipulation of small objects between the thumb and the flexor aspects of the fingers in a finely controlled manner (Fig. 14-24B). The wrist position varies so as to increase the manipulative range. The fingers are generally in a semiflexed position, and the thumb is palmarly abducted and opposed. Certain prehensile activities involve both power and precision grips (Fig. 14-25).

As a refinement of Napier's classification, Landsmeer (1962) suggested that the precision grip be termed "precision handling" because it involves no forceful gripping of the object and is a dynamic process without a static phase. In both power grip and precision handling, full opposition of the thumb to the ring and little fingers is obtained via palmar displacement of the metacarpals of these digits.

A variant of precision handling is the often-used "dynamic tripod" (Capener, 1956), wherein the thumb, index finger, and middle finger have a dynamic action, working in close synergy for precision handling of the object, while the ring and little fingers are used largely for support and static control (Fig. 14-26). A further refinement is pinching a small object between the thumb and index finger. Such maneuvers are commonly classified as tip pinch, palmar pinch, lateral (or key) pinch, and pulp (or ulnar) pinch, depending on the parts of the phalanges brought to bear on the object being handled (Fig. 14-27).

Another important distinction between power grip and precision handling is the fundamentally different position of the thumb in each posture. In the power grip, the thumb is adducted; in precision handling, it is palmarly abducted (Fig. 14-24). The relationship of the hand to the forearm also differs strikingly. In the

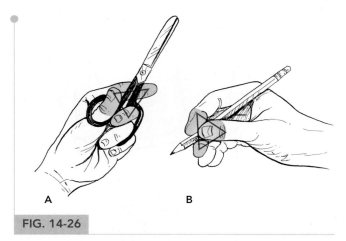

FIG. 14-26

The "dynamic tripod," a type of precision handling where the thumb, index finger, and middle finger work in close synergy for precision handling of the object while the ring finger and the little finger provide support and static control. This functional configuration is demonstrated in the use of scissors (**A**) and a pencil (**B**).

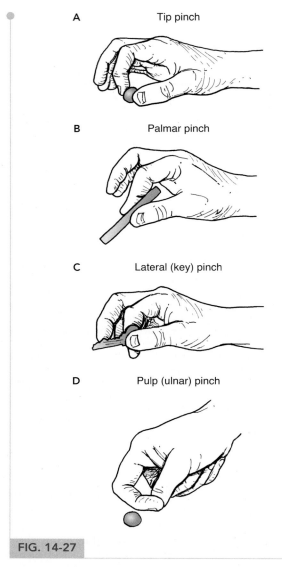

FIG. 14-27

Examples of precision handling in which small objects are pinched between the thumb and index finger. These grips are classified according to the parts of the phalanges brought to bear on the object handled. **A.** Tip-to-tip pinch. **B.** Palmar pinch. **C.** Lateral (key) pinch. **D.** Pulp (ulnar) pinch.

power grip (Fig. 14-24A), the hand is usually deviated ulnarly and the wrist is held in an approximately neutral position so that the long axis of the thumb coincides with that of the forearm. In this way, pronation and supination can be transmitted from the forearm to the object. In precision handling (Fig. 14-24B), the hand is generally held midway between wrist radial and ulnar deviation, and the wrist is markedly reflected in the posture of the thumb. When the demand for precision is minimal or absent, the thumb is wrapped over the dorsum of the middle phalanges of the digits and acts purely as a reinforcing mechanism. When an element of precision is required in what is predominately a power grip, such as the fencing grip (Fig. 14-28A), the thumb is adducted and aligned with the long axis of the cylinder so that, by means of small adjustments of posture, it can control the direction in which the force is being applied. At the other extreme of the power grip range is the coal-hammer grip (Fig. 14-28B), the crudest form of prehensile function, where the thumb is wholly occupied in reinforcing the clamping action of the digits. An example of this extreme in an empty hand is the bunched fist (Fig. 14-28C).

Rotating the thumb into an opposing position is a requirement of almost every hand function, whether it requires a strong grip or a delicate precision pinch. In some instances, however, the thumb may not be involved at all, as in the hook grip in which the fingers are flexed so that their pads lie parallel and slightly away from the palm, together forming a hook. This posture requires relatively little muscle activity to maintain and

is used when precision requirements are minimal and when power must be exerted continuously for long periods. Functionally, the hook grip pattern has limited potential and is not used very often. An example of its use is to carry an attaché case or a suitcase by its handle. In contrast, the individual whose hand intrinsic muscles are paralyzed or severely weakened relies on the hook grasp for all functional task completion. The hook grasp is the only grasp pattern available when the hand intrinsics are not working.

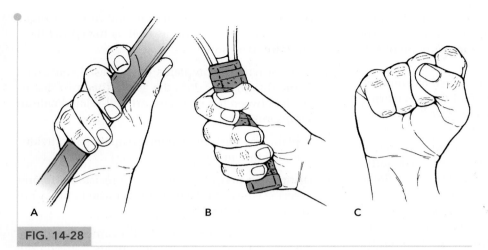

FIG. 14-28

The fencing grip (**A**) is a power grip in which the element of precision plays a large part. Instead of being wrapped over the dorsum of the digits, the thumb is aligned with the long axis of the cylinder so that it can control the direction in which the force is applied. In so doing, the thumb loses its effect as a powerful buttress on the radial side of the hand. Hence, some of the power grip is sacrificed in the interest of precision. The coal-hammer grip (**B**) and the bunched fist (**C**) are examples of a strong power grip with no element of precision. The ulnar deviation characteristic of a power grip is apparent in both cases.

Summary

- The wrist is a complicated joint complex consisting of the multiple articulations of the eight carpal bones with the distal radius, the structures of the TFCC, the metacarpals, and each other. The carpal bones are conventionally divided into a proximal and a distal row.

- Motions at the wrist include flexion-extension and radial-ulnar deviation. Stability during radial-ulnar deviation is provided by a double-V system formed by the palmar intrinsic ligament and the radiolunate and ulnolunate ligaments.

- The dart thrower's motion, the arc from radial extension to ulnar flexion, is a functional movement arc throughout which the proximal carpal row forms a stable base for powerful grip and fine motor control tasks.

- The proximal and distal carpal rows form a bimuscular, biarticular chain that is subject to collapse under compression. Stability is provided by precise opposition of the articular surfaces and intricate intrinsic and extrinsic ligament constraints.

- The extensor carpi ulnaris, extensor pollicis brevis, and abductor pollicis longus act as a dynamic collateral system to provide wrist stability during functional hand movements.

- Wrist position affects the ability of the fingers to flex and extend maximally and to grasp effectively.

- The TFCC plays a significant role in cushioning compressive loads across the wrist joint.

- The flexor carpi ulnaris is the most powerful wrist motor and tends to place the wrist in a position of flexion and ulnar deviation.

- The finger rays of the hand are arranged in three arches: one longitudinal and two transverse. Derangement or collapse of the arch system as a result of bone injury, rheumatic disease, or paralysis of the intrinsic muscles of the hand can contribute to severe disability and deformity.

- The hand is the principal instrument of touch. The combination of sensibility and motor function gives the hand its great importance as an organ of information and accomplishment.

- The trapezoid, capitate, and second and third metacarpals, with their tight-fitting articulations, form the immobile unit of the hand. In their articulation with the hamate, the fourth and fifth metacarpals are permitted a modest amount of palmar displacement, a motion essential for gripping.

- The most important motion of the thumb is opposition, in which abduction coupled with rotation at the CMC joint moves the thumb toward the tip of the little finger.

- The finger rays are controlled by the coordinated action of the extrinsic and intrinsic muscle systems. The operation of each ray is not completely independent of its neighbor's.

- The components of the digital extensor assembly, especially the oblique retinacular ligaments, account for the release of the distal phalanx and the coupling of PIP and DIP joint motion.
- A unique feature of the MCP joints is their asymmetry, reflected in the bony configuration of the metacarpal heads, in the attachments of the collateral ligaments, and in the arrangement of the interossei.
- The MCP joints are stabilized primarily by the radial and ulnar collateral ligaments and also by the transverse intermetacarpal ligament, which links the palmar plates to each other.
- The digital flexor tendon sheath pulley system is essential for maintaining a relatively constant moment arm for the finger flexors and for minimizing stress raisers between tendon and sheath. The second and fourth annular pulleys play a particularly important role in this respect.
- The flexor superficialis tendon has a greater overall excursion than does the flexor profundus. The excursion of the flexors is larger than that of the extensors, and the excursion of the extrinsic muscle tendons is generally greater than that of the intrinsic tendons.
- Extra excursion required at any one joint owing to disruption of the pulley system results in inadequate excursion and subsequent weakness in the more distal joints.
- The strength of the finger flexors is more than twice that of the extensors.
- Efficient prehensile function depends on the mobility of the thumb CMC joint and the fourth and fifth MCP joints, relative rigidity of the second and third CMC joints, balanced synergism-antagonism between the extrinsic and intrinsic muscles, and adequate sensory input. The relative lengths of the metacarpals and phalanges and the finger rays as a whole are also important.
- The position of the thumb and the relationship between hand and forearm are the most important differences between power grip and precision handling.

Practice Questions

1. Describe what is meant by the frequently used concept that the hand is a sensory organ with extraordinary functional adaptability that has a large reaching capacity.

2. Distinguish the motor and sensory contributions of the three main neural roots (median, ulnar, and radial) into the hand and delineate what the hand could be capable of if neural integrity was interrupted for each of the three neural roots for high- and for low-level injury as well as mixed median-ulnar interruption.

3. When a patient complains of increased sensitivity in the thumb, how does the astute therapist distinguish between neural versus arterial-related painful symptoms?

4. Explain how the human hand is capable of individual digital function.

5. Distinguish between power and precision grip and describe why both are needed and whether they are unique and distinct.

6. Describe the role of the intrinsic ligaments and carpal bone shapes in providing stability between the forearm and the distal carpal role under compressive loading from extrinsic muscle contraction.

7. What are the components and the function of the triangular fibrocartilage complex?

8. Describe the motion of the carpal bones during dart thrower's motion and discuss the functional importance of this motion.

9. What are the wrist ranges of motion most consistent with good hand function?

10. How does wrist position affect grip force, and what are the underlying muscle mechanics that determine these effects?

REFERENCES

American Academy of Orthopaedic Surgeons (1965). *Joint Motion. Method of Measuring and Recording*. Chicago: AAOS. [Reprinted by the British Orthopaedic Association, 1966.]

Batmanabane, M., Malathi, S. (1985). Movements at the carpometacarpal and metacarpophalangeal joints of the hand and their effect on the dimensions of the articular ends of the metacarpal bones. *Anat Rec, 213*, 102.

Berger, R.A. (2002). Anatomy and kinesiology of the wrist. In J.M. Hunter, E.J. Macklin, A.D. Callahan (Eds.). *Rehabilitation of the Hand and Upper Extremity (5th ed.)*. Philadelphia: Mosby, 77– 87.

Brand, P.W. (1985). *Clinical Mechanics of the Hand*. St. Louis, MO: Mosby, 30–60.

Brand, P.W., Hollister, A. (1992) *Clinical Mechanics of the Hand (2nd ed.)*. St Louis, MO: Mosby.

Brumbaugh, R.B., Crowninshield, R.D., Blair, W.F., et al. (1982). An in vivo study of normal wrist kinematics. *J Biochem Eng, 104*, 176.

Brumfield, R.H., Champoux, J.A. (1984). A biomechanical study of normal functional wrist motion. *Clin Orthop, 187*, 23.

Caillet, R. (1982). *Hand Pain and Impairment* (*3rd ed.*). Philadelphia: FA Davis.

Capener, N. (1956). The hand in surgery. *J Bone Joint Surg, 38B*, 128.

Cauna, N. (1954). Nature and functions of the papillary ridges of the digital skin. *Anat Rec, 119*, 449.

Charness, M.E., Ross, M.H., Shefner, J.M. (1996). Ulnar neuropathy and dystonic flexion of the fourth and fifth digits: Clinical correlation in musicians. *Muscle Nerve, 19*, 431.

Crisco, J.J., Coburn, J.S., Moore, D.C., et al. (2005). In vivo radiocarpal kinematics and the dart thrower's motion. *J Bone Joint Surg, 87A*, 2729.

Doyle, J.R., Blythe, W. (1975). The finger flexor tendon sheath and pulleys: Anatomy and reconstruction. In *AAOS Symposium on Tendon Surgery in the Hand*. St Louis, MO: Mosby.

Fess, E.E., Gettle, K.S., Philips, C.A., et al. (2005). *Hand and Upper Extremity Splinting: Principles & Methods*. St Louis, MO: Elsevier Mosby.

Hagert, C.-G. (1981). Anatomical aspects on the design of metacarpophalangeal implants. *Reconstr Surg Traumatol, 18*, 92.

Hakstian, R.W., Tubiana, R. (1967). Ulnar deviation of the fingers. The role of joint structure and function. *J Bone Joint Surg, 49A*, 299–316.

Haugstvedt, J.-R., Berger, R.A., Nakamura, T., et al. (2006). Relative contributions of the ulnar attachments of the triangular fibrocartilage complex to the dynamic stability of the distal radioulnar joint. *J Hand Surg, 31A*, 445.

Hazelton, F.T., Smidt, G.L., Flatt, A.E., et al. (1975). The influence of wrist position on the force produced by the finger flexors. *J Biomech, 8*, 301.

Jones, L.A., Lederman, S.J. (2006). *Human Hand Function*. New York: Oxford.

Karnezis, I.A. (2005). Correlation between wrist loads and the distal radius volar tilt angle. *Clin Biomech, 20*, 270.

Kauer, J.M.G. (1979). The collateral ligament function in the wrist joint. *Acta Morphol Neer Scand, 17*, 252.

Kauer, J.M.G. (1980). Functional anatomy of the wrist. *Clin Orthop, 149*, 9.

Kauer, J.M.G., Landsmeer, J.M.F. (1981). Functional anatomy of the wrist. In R. Tubiana (Ed.). *The Hand (vol 1)*. Philadelphia: WB Saunders.

Kaufmann, R., Pfaeffle, J. Blankenhorn, B., et al. (2005). Kinematics of the midcarpal and radiocarpal joints in radioulnar deviation: An in vitro study. *J Hand Surg, 30A*, 937.

Kaufmann, R., Pfaeffle, J., Blankenhorn, B., et al. (2006). Kinematics of the midcarpal and radiocarpal joints in flexion and extension: An in vitro study. *J Hand Surg, 31A*, 1142.

Kleinman, W.B. (2007). Stability of the distal radioulnar joint: Biomechanics, pathophysiology, physical diagnosis, and restoration of function. What we have learned in 25 years. *J Hand Surg, 32A*, 1086.

Kucynski, K. (1968). The upper limb. In R. Passmore, J.S. Robson (Eds.). *A Companion to Medical Studies (vol I)*. Oxford: Blackwell Scientific Publications.

Landsmeer, J.M.F. (1955). Anatomical and functional investigations on the articulation of the human fingers. *Acta Anat Suppl, 25*(24), 1–69.

Landsmeer, J.M.F. (1949). The anatomy of the dorsal aponeurosis of the human finger and its functional significance. *Anat Rec, 104*, 31.

Landsmeer, J.M.F. (1962). Power grip and precision handling. *Ann Rheum Dis, 21*, 164.

Li, Z.M. (2002). The influence of wrist position on individual finger forces during forceful grip. *J Hand Surg, 27A*, 886.

Minami, A., An, K.-N., Cooney, W.P. 3rd, et al. (1984). Ligamentous structures of the metacarpophalangeal joint: A quantitative anatomic study. *J Orthop Res, 1*, 361.

Moore, K.L., Dalley, A.F. (2006). *Clinically Oriented Anatomy (5th ed.)*. Philadelphia: Lippincott Williams & Wilkins.

Moritomo, H., Murase, T., Arimitsu, S., et al. (2008). Change in the length of the ulnocarpal ligaments during radiocarpal motion: Possible impact on triangular fibrocartilage complex foveal tears. *J Hand Surg, 33A*, 1278.

Moritomo, H., Murase, T., Goto, A., et al. (2006). In vivo three-dimensional kinematics of the midcarpal joint of the wrist. *J Bone Joint Surg, 88A*, 611.

Mountcastle, V.B. (1968). *Medical Physiology (vol II, 12th ed.)*. St Louis, MO: Mosby, 1345–1371.

Napier, J.R. (1956). The prehensile movements of the human hand. *J Bone Joint Surg, 38B*, 902–913.

Neumann, D.A. (2010). *Kinesiology of the Musculoskeletal System: Foundations for Rehabilitation (2nd ed.)*. St Louis, MO: Mosby Elsevier.

O'Driscoll, S.W., Horii, E., Ness, R., et al. (1992). The relationship between wrist position, grasp size, and grip strength. *J Hand Surg, 17A*, 169.

Palmer, A.K., Werner, F.W. (1981). The triangular fibrocartilage complex of the wrist—Anatomy and function. *J Hand Surg, 6*, 153.

Ryu, J., Cooney, W.P., Askew, L.J., et al. (1991). Functional ranges of motion of the wrist. *J Hand Surg, 16A*, 409.

Sarrafian, S.K., Kazarian, L.E., Topouzian, L.K., et al. (1970). Strain variation in the components of the extensor apparatus of the finger during flexion and extension: A biomechanical study. *J Bone Joint Surg, 52A*, 80.

Smith, R.J., Kaplan, E.B. (1967). Rheumatoid deformities at the metacarpo-phalangeal joints of the fingers: A correlative study of anatomy and physiology. *J Bone Joint Surg, 49A*, 31.

Strickland, J.W. (1987). Anatomy and kinesiology of the hand. In E.E. Fess, C.A. Philips (Eds.). *Hand Splinting: Principles and Methods (2nd ed.)*. St Louis, MO: Mosby, 3–41.

Tang, J.B., Ryu, J., Omokawa, S., et al. (1999). Biomechanical evaluation of wrist motor tendons after fractures of the distal radius. *J Hand Surg, 24A*, 121.

Taleisnik, J. (1985). *The Wrist*. New York: Churchill Livingstone.

Tubiana, R. (1984). Architecture and functions of the hand. In R. Tubiana, J.-M. Thomine, E. Mackin (Eds.). *Examination of the Hand and Upper Limb*. Philadelphia: WB Saunders, 1–97.

Volz, R.G., Lieb, M., Benjamin, J. (1980). Biomechanics of the wrist. *Clin Orthop, 149*, 112.

Von Lanz, T., Wachsmuth, W. (1970). Functional anatomy. In J.H. Boyes (Ed.). *Bunnell's Surgery of the Hand (5th Ed.)*. Philadelphia: JB Lippincott Co.

Wilson, F.R. (1998). *The Hand: How Its Use Shapes the Brain, Language, and Human Culture*. New York: Vintage.

Youm, Y., McMurtry, R.Y., Flatt, A.E., et al. (1978). Kinematics of the wrist: I. An experimental study of radial-ulnar deviation and flexion-extension. *J Bone Joint Surg, 60A*(4), 423–431.

Youm, Y., Yoon, Y.S. (1979). Analytical development in investigation of wrist kinematics. *J Biomech, 12*, 613.

Zancolli, E. (1979). *Structural and Dynamic Bases of Hand Surgery (2nd ed.)*. Philadelphia: JB Lippincott, 3–63.

PART 3

Applied Biomechanics

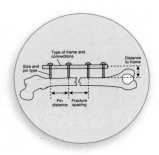

C H A P T E R   **15**

Biomechanics of Fracture Fixation

Frederick J. Kummer

Introduction

Fracture Stability and Healing
Fracture Healing
Surgical Factors

Fixation Devices and Methods

Summary

Practice Questions

Suggested Reading

References

Introduction

The study of fracture fixation biomechanics can be divided into two main areas: (1) criteria for achieving fracture stability and promotion of bone healing and (2) the characterization of techniques and devices intended to mechanically stabilize a fracture. An understanding of the biomechanical principles involved in these areas will aid the engineer in implant design and assist the surgeon in selecting the most effective technique and device to obtain successful results in patients.

Fracture Stability and Healing

The clinical goal of effective fracture treatment is rapid healing, without significant deformity or limb shortening, to restore the patient to a prefracture level of function. In the elderly, rapid mobilization is essential to prevent the deleterious consequences of bed stay. The first goal of treatment is fracture stabilization. This is determined by the location and type of fracture, the muscle and body forces acting on it, and the various passive soft tissue constraints such as ligaments and fascia. Some simple fractures are inherently stable with low loading and thus require minimal treatment, such as a sling for a clavicle or a cast, while others, such as a midshaft comminuted femur fracture, require major surgical intervention and insertion of an internal fixation device for adequate fixation. Although an osteotomy (a surgically created fracture for the correction of a deformity) allows close approximation of fracture ends, typical fractures are often fragmented and usually lack inherent stability. Interdigitation of the bone ends can facilitate stability, such as when a tapered bone end is inserted into the medullary cavity (creating a displacement deformity, however).

Traditional methods for the treatment of fractures are externally applied and include traction, cast, and braces. External forces or constraints applied to the injured limb act to stabilize the fracture (by limiting muscle or soft tissue forces leading to deformity) and maintain alignment of the limb. However, in many cases, as a result of the nature of the fracture or the patient's condition, an internal or external fixation device attached directly to the bone is required to achieve adequate fracture stabilization. The designs and use of these fixation devices rely on an understanding of bone healing and the loads and forces to which the device is subjected. The relationship of biomechanical forces borne by the device to those borne by the bone (load bearing, load sharing) influences fracture healing and device survival.

FRACTURE HEALING

Controversy currently exists about whether completely rigid fixation is the optimal condition for bone healing. Micromotion has been shown to aid healing. Healing results even in cases of gross motion such as that seen in rib fractures. Rigid fixation may lead to delayed healing, bone atrophy, and a lack of external stimuli necessary for the healing process.

Although gross motion between two or more bone fragments usually leads to nonunion and fibrocartilage tissue formation, there is a low level of displacement (micromotion) that appears advantageous to healing by providing a mechanical signal that stimulates the biologic repair processes. The amount of local strain in the healing region (change in length divided by the original length) is thought to determine the nature of the tissues formed (e.g., fibrocartilage or bone). The optimal frequency, waveform, and total number of cycles of this signal currently are being investigated. Several methods to promote healing by externally stimulating a fracture with ultrasound or electromagnetic fields are in clinical use (see Case Study 15-1).

Recent studies have examined the use of biologic agents such as growth factors to promote fracture healing (Simpson et al., 2006).These can be injected directly into the fracture or used in biodegradable coatings on fixation devices. The specific factor (or factors), amount, and timing of delivery are the main questions. Concern also exists regarding the process of stress shielding that occurs when the fixation device carries all or most of the mechanical load and thus, by Wolff's law, promotes localized osseous resorption as a result of the resultant unloading of the bone around the device. This is often referred to as load bearing versus load sharing. However, much of the initial osteopenia seen beneath fracture plates is thought to be caused by vascular disruption during their application (Perren, 2002).

Bone healing in the presence of a gap with minimal movement passes through several stages of repair with a concomitant increase in mechanical strength as mineralization increases: hematoma and inflammation, callus formation, replacement by woven bone, and finally remodeling into lamellar or trabecular bone. Callus can form both periosteally and endosteally and enlarges the diameter of the bone at the fracture site. Although callus is less strong and stiff than mature bone, this increased diameter can increase stiffness in bending and torsion at the fracture site as a result of the increased moments of inertia. Direct bone apposition caused by compression with rigid fixation, in which the initial repair stages seen in a gap

Case Study 15-1

Ultrasound Treatment for Fracture Healing

A 40-year-old woman involved in a motor vehicle collision in December sustained a left tibiofibular fracture treated with external fixation. In January, low-intensity pulsed ultrasound (US) was initiated to promote fracture healing (Case Study Fig. 15-1A). In March, three months postfracture and two months after the initiation of pulsed US application, early healing is detected (*arrow,* Case Study Fig. 15-1B). By May, five months after injury and four months following the initiation of US, the bone healing is successful (Case Study Fig. 15-1C).

Pulsed low-intensity ultrasound has been successfully used for fracture repair (Frankel, 1998). Ultrasound is an acoustic radiation at frequencies above the limit of human hearing. Its acoustic radiation, in the form of pressure waves, provides micromechanical stress and force to the bone and surrounding tissue. This mechanical stimulation plays a major role in bone healing because bone reacts to the amount and direction of force and remodels to adapt to the applied stress and its direction (Wolff, 1986).

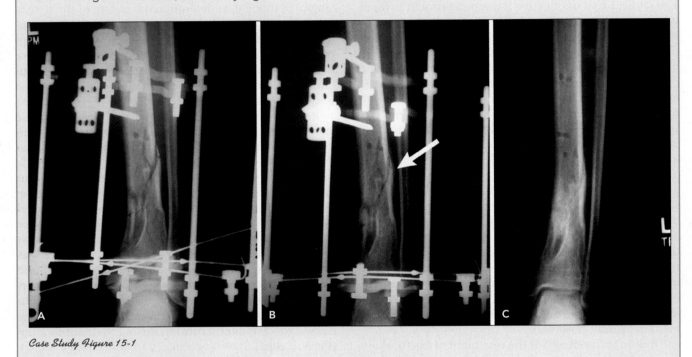

Case Study Figure 15-1

are eliminated or minimized, heals by a remodeling process and can take longer because vascularity must be re-established.

The other important factor for healing is adequate blood supply, which necessitates the surgeon preserving the vascular supply of the bone (e.g., periosteum) and providing conditions for early revascularization by careful operative technique (e.g., soft tissue preservation). Numerous studies have demonstrated a direct relationship between the quantity and quality of microvascular structures in the healing region and the rate of formation and resultant mechanical properties of the new bone.

SURGICAL FACTORS

Various factors determine the optimal fixation method for a specific fracture application. A principle factor is the mechanical loading, specifically the types (tension, bending, and/or torsion) and the magnitude of forces to which the fixation will be subjected and whether these forces will be cyclic, requiring additional strength of fixation to account for possible device fatigue (see Case Study 15-2). Another important factor is the bone quality, which determines the strength available to support the fixation device. Other factors relate to surgical and anatomic considerations,

Case Study 15-2

Fixation Plate Failure

An internal contemporary fixation plate was inserted into the arm of a 25-year-old man who sustained a fracture of the radius. The plate was fractured as a result of fatigue 20 years later. Repeated loading and unloading of a material will cause it to fail, even if the loads are less than the ultimate stress (Simon, 1994). Each loading cycle produces a minute amount of microdamage that accumulates with repetitive loads until the material fails. Mechanical considerations as to the magnitude and repetition of the loads to which the fixation will be subjected should be considered, along with the fatigue life of the material. This is recorded on a curve of stress versus number of cycles. Thus, higher stresses produce failure in fewer cycles (loading to the ultimate stress produces failure in one cycle), whereas lower stresses are tolerated for an extended period (Case Study Fig. 15-2).

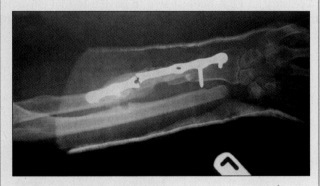

Case Study Figure 15-2

for example, the exposure (possible scarring and vascular compromise), whether the device will fit adequately within the soft tissues, and if neurovascular structures are at risk. The nature of the original injury and the amount of soft tissue damage also determine treatment techniques.

Evaluation of fixation strength can be accomplished by laboratory testing of actual implants in cadaver bone (sometimes animal); composite bones (urethane foams and fiberglass/epoxy) are being increasingly used for testing. One difficulty of such testing is adequately simulating in the test model the complex in vivo, cyclic forces on the device. Another difficulty is in simulating the biologic

repair processes that would act to stabilize the fixation over time. Computer modeling such as finite element analysis can be used as an initial method to evaluate fixation methods and device designs but requires quantified parameters of bone modulus and strength, which may be lacking, for an exact solution. Cadaver studies also can determine the neurovascular anatomic structures at risk.

Clinical trials are the other major method used to evaluate the efficacy of a particular fixation method. However, care must be taken to adopt the appropriate techniques to quantify data and design the trial (number of patients, adequate follow-up) so that the many variables can be properly analyzed and proper statistical significance determined.

Fixation Devices and Methods

Wires, staples, pins, plates, and screws are the common implant devices used to achieve fracture fixation. These are usually made of stainless steel (316L), sometimes of titanium alloy (Ti-6A1–4V), or occasionally of cobalt-chromium alloy. Each metal has advantages and disadvantages such as strength, modulus (stiffness), corrosion resistance, and ease of imaging (MRI, CAT). Sometimes there is a "race" between the healing of the bone and the fracture, usually by fatigue, of the device. There is current interest in the clinical application of biodegradable polymers such as polylactic acid. Polymers are more flexible than metals and would lead to greater load bearing by the healing fracture; biodegradable materials do not have to be removed in a secondary operation and their mechanical properties gradually decrease with time, thus avoiding stress shielding. However, their mechanical strength is much less than that of metal and some of the degradation products have shown untoward biologic responses. Research continues into the use of various glues, cements, and adhesives for fracture fixation, some of which are also biodegradable.

Wire fixation (solid or cable) used as cerclage or a bone suture is a common application; in both cases multiple wires are required to provide stable, three-dimensional fixation. This requires achieving equal tension during tightening because loosening at one or more sites can provide a locus for motion and possible nonunion or cause malpositioning. Problems with wire fracture fixation include the necessity and surgical complexity of making a bone hole and passing the wire, breakage during tightening or later as a result of fatigue (cyclic loading), and cut-through of the bone. For cerclage applications, there is concern about compromise of the periosteal blood supply and the resulting increased healing time required for revascularization.

Some recent developments are wire tensioning/twisting instruments and the use of crimping systems to avoid the problems with twisting or knot tying. There are also new oriented polymers (Spectra) that do not stretch to the same degree as traditional suture materials. Sutures can be used with suture anchor systems to attach soft tissue and eliminate the difficulty of looping a suture through bone and suture abrasion against the bone.

Staples alone usually do not provide sufficient mechanical stability for permanent fixation, and their use often requires predrilling holes for the staple legs. Pneumatic-driven staples can be used to rapidly tack fragments prior to a more rigid fixation but need careful control of the insertion driving force to prevent untoward damage to the bone. Some staple designs have been developed that can effect compression during insertion, such as staple fabrication from nitinol (an alloy that changes shape when heated to body temperature).

Kirschner wires (K-wires) are normally used to hold fragments of bone prior to rigid fixation and for percutaneous pinning of small bone fractures, but, in general, they lack sufficient mechanical stability for their use as a primary fixation in weight-bearing bones. At least two wires should be used for each bone fragment, and they should not be inserted in a parallel manner to prevent "pistoning" of the bone fragment along the wires (Fig. 15-1). Threaded pins provide additional stability because they minimize sliding of the bone fragments, but their removal is more difficult. Occasionally, pinning is used in combination with sutures looped and tightened around the pin ends or through loops in the pins. This "tension band" technique provides significantly increased mechanical stability of the fixation.

The two basic types of screws are cortical and cancellous and are distinguished by their thread design; Cancellous screws have a greater distance between adjacent threads (pitch) and the ratio of outer thread diameter to body diameter (Fig. 15-2). The major intrinsic factors that influence screw-holding power are outer thread diameter, thread configuration, and thread length; extrinsic factors are bone quality, type, and screw insertion orientation and driving torque (Fig. 15-3). The inherent holding power of a screw is a function of the outer thread diameter times the length its threads are in contact with the bone. When used to hold two bone fragments together, screws commonly are used in a lag modality in which the proximal portion of the screw remains free within one fragment (either by using a screw design having no proximal threads or by enlargement of the hole in the proximal fragment, which should require the use of a washer under the screw head for adequate support). Insertion torque determines the force with which bone fragments are held together, which, in turn, creates the friction that prevents their motion. Control of torque is important to prevent stripping of the bone or screw head torsional failure. Pretapping of the screws is usually not necessary and has been shown to have minimal effect on their holding ability; many screws are self-tapping owing to a modification to the design of the leading threads.

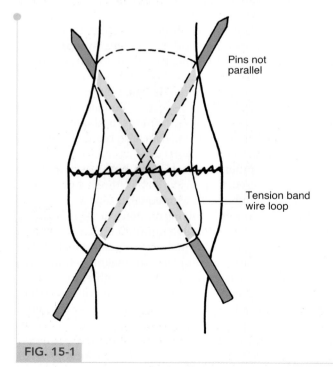

Pins not parallel

Tension band wire loop

FIG. 15-1

Tension band wiring of two K-wires; tightening the wire loop applies compression to the fixation. K-wires are inserted in a skewed configuration for stability.

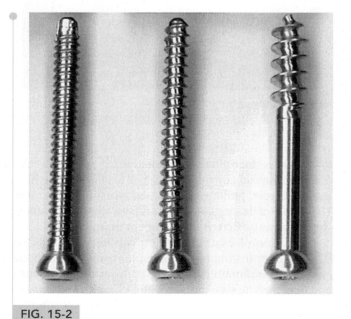

FIG. 15-2

Types of bone screws. **Left** to **Right:** Cortical, cancellous, and cancellous lag.

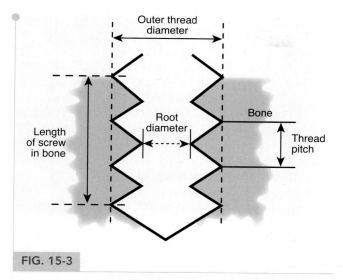

FIG. 15-3

Screw parameters. For screw pull-out, the bone must shear along the outer diameter (*dotted line*).

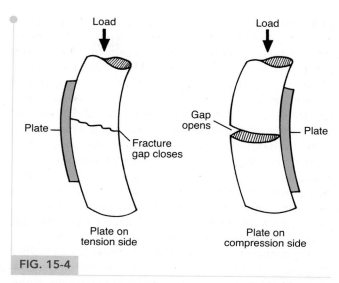

FIG. 15-4

Effect of plate placement. A plate located on the compression side causes the fracture to gap when loaded.

Because of anatomic constraints, surgical exposure, or the orientation of the ends of the bone fragments, screws cannot always be inserted perpendicular to the fracture. In this case, the holding power of the screw is decreased and a shear component of the holding force is created, which can act to destabilize fracture alignment. Usually two or more screws are required for stability, although one screw has been suggested for some applications if sufficient interfragment approximation can be achieved to create adequate friction between the bone surfaces. The quality of bone also determines screw-holding ability; cortical bone is approximately ten times stronger than cancellous bone. The thickness of the cortex and the degree of osteopenia (bone density) are thus critical for fixation strength and influence the number of screws required for adequate stability. Using screws in a bicortical manner appreciably increases the strength of fixation.

Anatomic constraints limit the number or size of screws that can be applied in a given region. As a result, screws are often combined with plates to achieve adequate stability and increased strength of fixation that is enhanced by friction between the plate and bone. The optimal site for a single plate application is on the side of the bone subjected to tension; usually two plates are applied to achieve better fixation stability as load directions vary with activities (Fig. 15-4). Plate designs vary with applications and intended location such as an expanded end for condylar fixation. Owing to anatomic constraints such as soft tissue thickness, occasionally, thinner plates are used (such as for forearm fracture stabilization), but these plates possess sufficient stiffness (function of the width of the plate times its thickness cubed) to prevent undesired fracture motion as a result of bending loads (Fig. 15-5).

Screws should be inserted into a plate with a torque driver and the tightness of all screws rechecked; if this is not done, one screw could bear most of the load and possibly fail. Some plates use a specially designed screw hole slot, countersunk to accommodate the screw head, whose center is offset with respect to the screw head to obtain interfragmentary compression as the screw is

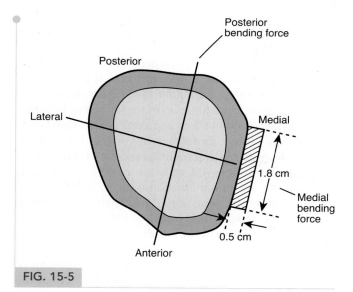

FIG. 15-5

Effect of loading direction on plate stiffness. The rigidity of the plate is EI, where E is the modulus of the plate material and I is the moment of inertia of the plate. $I = bh^3/12$ (I_1, posterior bending; I_2, medial bending; $I_1 = 0.5 \times 1.8^3/12 = 0.243$; $I_2 = 1.8 \times 0.5^3/12 = 0.01875$), where b is the base dimension and h is its height. Thus, the plate is 13 times more rigid in posterior bending than in medial bending.

tightened. An alternative method to achieve compression is to pre-bend the plate before application so that when the attachment screws are tightened, the bone fragments are approximated as the plate straightens. Some new plate designs use threaded holes to engage the screw so that bicortical screw insertion is not as essential for maximum fixation stability. These plates do not have to contact the bone, and thus the entire load across the fracture can be borne by the screws (which can be a site of fixation failure).

Plates can also be used to span gaps created by severe fractures or tumor surgery and are frequently used with bone grafts for this application. Unless the graft is exactly sized, the plate will bear the entire load across the defect. The bending moment on the plate-screw fixation linearly increases with defect size, and thus the plate requires adequate stabilization, particularly at the more highly loaded ends where at least three screws into bone are needed. Long, multiple-holed plates enable the selection of the best osseous sites for screw purchase and should permit anchoring of the graft by at least two additional screws.

The major surgical considerations for the use of plates are the requirements of a large exposure for their insertion and the possibility of compromising periosteal blood supply by the exposure or plate insertion (some plate designs have inferior feet or ridges to minimize this possibility). There is also interest in polymeric plates that would be more flexible to achieve a greater degree of micromotion at the fracture, which could be advantageous to bone healing and minimize stress shielding.

Hip fracture devices can be internal or external in their application; the most common external device is a side plate affixed to the femur supporting an internal lag screw inserted through the femoral neck and into the head across the fracture. An important factor is the ability of the device to slide to consolidate the fracture during healing. This is usually the function of the lag screw; however, some designs also have a plate that accommodates sliding. Internal devices for fixation are usually intramedullary (IM) nails with one or more lag screws. In comparison with external devices, IM nails are subject to less loading forces because their location is closer to the neutral bending axis of the bone (Fig. 15-6). Their size is critical because their bending and torsional stiffness are proportional to the diameter to the fourth power. This is why one large nail provides more rigid fixation than do multiple smaller rods. Size, amount of curvature, and amount of reaming are also important because the stability of the fixation relies on load transfer to the bone; often, distal and proximal screws inserted through the bone and nail are used to increase torsional stability (Fig. 15-7). Bending of the nail as a result of insertion in a curved medullary cavity can make insertion of the distal screws difficult.

External fixation devices are also used for fracture stabilization; multiple transcutaneous pins are inserted into

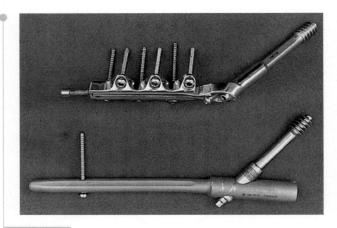

FIG. 15-6

Typical intramedullary and extramedullary devices. **Top.** Medoff sliding plate. **Bottom.** Intramedullary hip screw.

the bone and stabilized with an external bar(s) or ring(s). Factors that influence the mechanical stability and rigidity of these constructs are the number, diameter, orientation, and length of these pins and their relation with respect to the fracture. These factors, however, are subject to surgical considerations such as neurovascular structures for frame

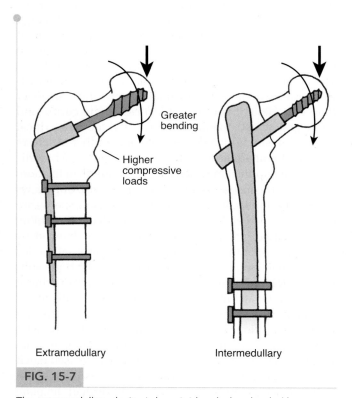

FIG. 15-7

The extramedullary device is less rigid and when loaded has greater deflection, creating higher medial stresses in the femur.

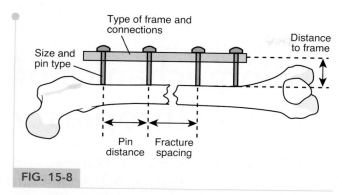

FIG. 15-8

Typical external fixator showing the variables that influence fixation stability.

application. Large, short pins located close to the fracture provide the most rigid fixation (Fig. 15-8).

Spinal implants used for deformity correction or fracture fixation consist of various combinations of rods, wires, plates, and screws. The junctions between these components are often the site of failure, such as fatigue or fretting as a result of cyclic loading. A specific problem is that the size and the location of appropriate sites for device attachment to the spine are limited. This is important because these devices are subject to appreciable forces during flexion and extension of the neck, and torso and fixation failure can occur with serious consequences.

Summary

- Although new techniques are being studied that promote fracture healing, some type of rigid or semirigid (micromotion) device is necessary for mechanical stability and correct bone orientation. Gross motion, instability, and inadequate blood supply may lead to fracture nonunion.
  - Factors determining optimal fixation for specific fracture applications are the following:
  - Mechanical considerations, such as the types and magnitude of forces to which the fixation will be subjected and their duration
  - Bone quality (strength)
  - Surgical, anatomic, and clinical considerations
  - The type and extent of the bone fracture and the amount of soft tissue damage

Practice Questions

1. An implant manufacturer has made a titanium bone plate 25% thicker than a stainless steel plate to account for titanium's lower modulus. Are the plates similar in stiffness?

2. Why does one large nail provides more rigid fixation than do multiple smaller rods?

3. What can occur if the sliding screw (see Fig. 15-7) is locked by an internal set screw in the intramedullary nail or plate?

SUGGESTED READING

Bong, M.R., Kummer, F.J., Koval, K.J., et al. (2007). Intramedullary nailing of the lower extremity: Biomechanics and biology. *J Am Acad Orthop Surg, 15*(2), 97–106.

Burstein, A.H., Wright, T.M. (1994). *Fundamentals of Orthopaedic Biomechanics.* Baltimore: Williams & Wilkins.

Dycheyne, P., Hastings, G.W. (Eds.). (1984). *Functional Behavior of Orthopaedic Biomaterials.* Boca Raton, FL: CRC Press.

Egol, K.A., Kubiak, E.N., Fulkerson, E., et al. (2004). Biomechanics of locked plates and screws. *J Orthop Trauma, 18*(8), 488–493.

Gozna, E.R., Harrington, I.J., Evans, D.C. (1982). *Biomechanics of Musculoskeletal Injury.* Baltimore: Williams & Wilkins.

Mow, V.C., Hayes, W.C. (Eds.). (1991). *Basic Orthopaedic Biomechanics.* New York: Raven Press.

Özkaya, N., Nordin, M. (1998). *Fundamentals of Biomechanics* (2nd ed.). New York: Springer-Verlag.

Radin, E.L. (Ed.). (1992). *Practical Biomechanics for the Orthopedic Surgeon.* New York: Churchill Livingstone.

Tencer, A.F., Johnson, K.D. (1994). *Biomechanics in Orthopedic Trauma.* London, UK: M. Dunitz.

White, A.A., Panjabi, M.M. (Eds.). (1990). *Clinical Biomechanics of the Spine.* Philadelphia: JB Lippincott Co.

REFERENCES

Frankel, V.H. (1998). Results of prescription of pulse ultrasound therapy in fracture management. In Z. Szabo, J. Lewis, G. Fantini, R. Savalgi (Eds.). *Surgical Technology International VII. International Developments in Surgery and Surgical Research.* San Francisco: Universal Medical Press, Inc. Surgical Technology International.

Perren, S.M. (2002). Evolution of the internal fixation of long bone fractures. The scientific basis of biological internal fixation: Choosing a new balance between stability and biology. *J Bone Joint Surg Br, 84*(8), 1093–1110.

Simon, S.R. (1994). *Orthopedic Basic Science.* Rosemont, IL: American Academy of Orthopaedic Surgeons.

Simpson, A.H., Mills, L., Noble, B. (2006). The role of growth factors and related agents in accelerating fracture healing. *J Bone Joint Surg Br, 88*(6), 701–705.

Wolff, J. (1986). *Das Gesetz der Transformation der Knochen [The law of bone remodeling].* P. Maquet, R. Foulong (Trans.). Berlin, Germany: Springer-Verlag [Original work published in 1892].

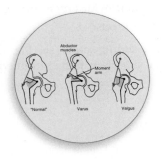

Biomechanics of Arthroplasty

Kharma C. Foucher, Markus A. Wimmer, and
Gunnar B.J. Andersson

Introduction

The success of modern total hip and knee arthroplasty can be attributed in large part to advanced biomechanical knowledge generated over nearly half a century. Because of developments in implant designs, materials, fixation techniques, and surgical approaches, most of the tens of thousands of patients who receive a total hip replacement (THR) or total knee replacement (TKR) each year can enjoy relief of pain and restoration of basic function with a low likelihood of mechanical failure. This success, however, has led an expansion of the pool of patients considered appropriate surgical candidates over the years. For example, many contemporary arthroplasty candidates are younger and more active than those seen in previous years.

Our range of biomechanical information must continually be updated and expanded to meet the needs of all patients, promoting healthy function and preventing complications. Mechanical problems associated with total joint replacements include issues related to wear of the bearing surface, mechanical failure of the implant, and loosening of the implant from the bone. A common feature of these clinical problems is that they are all related directly or indirectly to the magnitude or pattern of forces acting on the artificial implants. Forces acting at the hip or knee joint are dependent on the external forces acting on the limb and the internal forces primarily generated by muscle contraction. Joint forces have been measured with implanted transducers or estimated using inverse dynamics and analytic methods. This chapter will focus on a discussion of these joint forces after total hip and knee arthroplasty and several important factors that influence joint loading.

Hip

FORCES AT THE HIP JOINT

Implant failure after total hip arthroplasty (THA) is most often caused by aseptic loosening (Malchau et al., 2002). Aseptic (meaning not secondary to infection) loosening is closely related to the associated complications of polyethylene wear and periprosthetic bone loss. All three may have a partially mechanical etiology. Implant loosening has been related to the magnitude and direction of the forces on the femoral stem. Polyethylene wear is related to the forces transmitted along the path that the head of the femur traces across the acetabulum during walking and other activities (Davey et al., 2005). Bone loss may also be caused by stress shielding, the process by which more force is transmitted to the stiffer implant than to

the surrounding bone, leading to bone resorption. Forces at the femoral head, forces directed along the femoral shaft by the pull of muscle, the direction of the applied forces, and the number of cycles of force application are all important factors for the biomechanical environment of the artificial hip (Case Study 16-1). We will first discuss those factors that can be measured directly.

Direct Measurement of Hip Forces

Hip forces can be measured directly from implants instrumented with strain gauges or other force transducers. Rydell published the first such work in 1966. Since then in vivo hip forces have also been measured by several others (Bergmann et al., 1993, 1995, 2001; Davy et al., 1988; English and Kilvington, 1979; Kotzar et al., 1991, 1995) during various activities (Table 16-1). During level walking, peak forces of 1.8 to 4.6 times body weight have been measured. Forces can be much higher during other activities—Bergmann et al. (2004) recorded forces of 7.2 times body weight (BW) during an instance of stumbling in one subject. During level walking, the force at the hip joint generally reaches an initial peak in early stance and a second peak in late stance (Fig. 16-1). These peaks are usually similar in magnitude; however, different muscles are active during these two phases of the gait cycle. Therefore, depending on the application, it can be important to distinguish between the two force peaks.

Hip forces are determined by forces produced by internal structures—primarily muscles but including the capsule and ligaments—and external forces (i.e., intersegmental forces from the ground reaction force during walking). Although external forces can be measured during gait analysis, there is no way to directly measure the internal forces acting on the hip. For some clinical or research applications, a more detailed understanding of the hip force environment is desirable. Analytic models, therefore, remain an important tool to predict both implant forces and hip muscle forces despite the availability of fairly detailed in vivo force data. Numeric models can be applied to larger numbers of subjects, subjects with other hip pathologies aside from total hip replacement, or subjects with no hip pathology at all. The next section will discuss analytic hip force modeling.

Analytic Hip Force Modeling

The basic concept behind most muscle force modeling strategies at the hip, as well as other joints, is that the internal forces and moments (which cannot be directly measured) must be equal and opposite to the external forces (which can be measured). In other words, the joint must be in mechanical equilibrium at all times. Therefore

Cemented Total Hip Replacement

A 65-year-old woman presented with chronic pain caused by osteoarthrosis in the hip. Severe cartilage degeneration and reactive periarticular bony changes leading to loss of congruency had been affecting her hip function and daily activities such as gait. After a careful documentation of the patient's history, physical examination, and radiographic information, a cemented total hip replacement is performed (Case Study Fig. 16-1).

The picture shows the stage of the total hip arthroplasty after surgery. The head-neck angle is in valgus. This valgus position will decrease bending moments in the stem prosthesis but will increase the joint reaction force as a result of reduction in the mechanical advantage of the abductor muscles (lever arm shortens [*c*]). The figure is marked with the line of action of the abductor muscles (*A*), the line of gravity (*white line*), the line of the lever arm (*b*) from the center of rotation to the line of gravity, and the line of action of the reaction force at the estimated point of contact of the femoral head and the acetabular unit (*J*). In addition, the line of gravity (*white line*) is depicted with the line of action of gravity force (*W*).

The picture was made during a single-leg position, in which the line of gravity moves toward the hip, decreasing the abductor moment arm relative to the body weight. Thus, the patient adapts her gait and posture to decrease the demand on the abductor muscles by leaning over her leg and thereby limping a little.

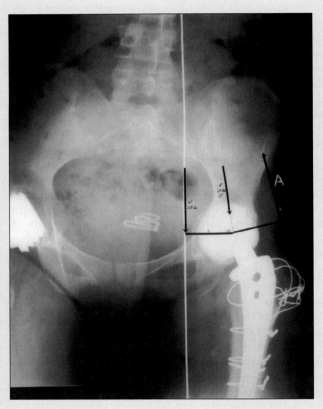

Case Study Figure 16-1

most analytic models establish a system of equilibrium equations where one side represents the external forces and moments (knowns) and the other side represents the internal forces and moments (unknowns). The problem is that there are many more muscles crossing the hip joint than there are equations—the moment balance problem is statically indeterminate and has an infinite number of solutions. To solve a statically indeterminate problem, one must either somehow reduce the number of unknowns (reduction methods) or apply other mathematical constraints (optimization methods) (Table 16-2).

Reduction methods involve combining individual muscles into a few muscle groups based on function and anatomy, or simply reducing the number of muscles in the model. This method is a very intuitive approach for problems at the knee or elbow, which are or can be approximated as hinge joints (Morrison, 1970; Schipplein and Andriacchi, 1991). The hip is a spheroidal (ball-and-socket) joint that allows motion in all three directions. Also, the muscles that cross the hip often have lines of action that allow significant force production in more than one direction (e.g., the tensor fascia lata, which is nearly as strong a hip abductor as it is a hip flexor). Because of this complexity, the statically indeterminate problem at the hip is more commonly solved using optimization methods.

TABLE 16-1

Hip Contact Forces Measured in Vivo in Patients with Instrumented Implants

Activity	Typical Peak Force (Body Weights)	Number of Patients	Time since Surgery (Months)	Reference
Hip Force				
Walking normal to fast speeds	2.7–3.6	2	1–2	Kotzar et al., 1991
Stair climbing	2.6			
Walking slow speed (crutches)	2.6	1	1	Davy et al., 1988
Ascending stairs	2.6			
Walking	2.7–4.3	2	8–33	Bergmann et al., 1993, 1995
Ascending stairs	3.4–5.5			
Descending stairs	3.9–5.1			
Walking	1.8–3.3	2	6	Rydell, 1966
Walking slow speed	2.7	1	15	English and Kilvington, 1979
Walking normal to fast speeds	3.5 (traditional) 3.2 (AL-minimally invasive) 2.9 (PL-minimally invasive)	5 per surgery type	2.3–12	Glaser et al., 2008
Walking fast speed	3.6	2	36	Stansfield et al., 2003
Walking slow speed	2.0	2	36	Stansfield et al., 2003
Walking slow speed unassisted	2.5	1	2	Brand et al., 1994
Walking normal speed unassisted	3.2	1	2	Brand et al., 1994
Walking	2.38	4	11–31	Bergmann et al., 2001
Ascending stairs	2.51	3	11–31	Bergmann et al., 2001
Descending stairs	2.6	3	11–31	Bergmann et al., 2001
Walking slow speed	2.42	1	11–31	Bergmann et al., 2001

AL, anterolateral; PL, posterolateral.

Optimization methods require making a reasonable guess about how the central nervous system (CNS) distributes the muscle forces and converting this assumption into one or more mathematical criteria to select a single solution from the infinite number possible. The optimization criterion is usually the mathematical formulation of a physical parameter. Some models have used criteria that limit allowable muscle and/or ligament forces (Heller et al., 2001, 2005; Rhrle et al., 1984; Seireg and Arvikar, 1975), limit muscle stress (Crowninshield and Brand, 1981; Pedersen et al., 1997), or both (Stansfield et al., 2003), or maximize endurance (Pedersen et al.,1997).

Since it is not possible to know what criteria are used by the CNS to "assign" muscle forces for a given person at a given time, it is not possible to really know whether the specific optimization criteria selected are physically appropriate. An alternative parametric approach has been proposed to exploit the natural redundancy in the neuromusculoskeletal system

(Hurwitz et al., 2003). This model predicts a solution space, rather than a single force solution, that contains all the physiologically possible combinations of muscle activity that could result in the external moments measured. To solve the equilibrium equations, muscles are combined into three groups based on function, but these equations are established and then solved thousands of times as all physiologically reasonable potential groupings are attempted and the relative contribution of individual muscles within groups are scaled up or down. The solution space is constrained to disallow solutions where a muscle would not be required to push instead of pull. Additional constraints can be applied as desired. This solution method tries to maintain the best features of both solution approaches by keeping the simplicity and computational speed of reduction methods and allowing any desired mathematical constraints without discarding physically possible solutions. Both the parametric and optimization approaches generally predict forces that are

Activities: walking, jogging, stairs, stumbling

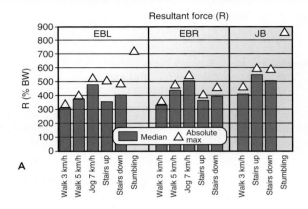

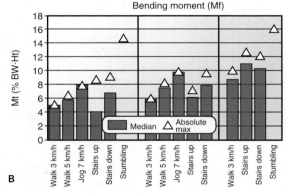

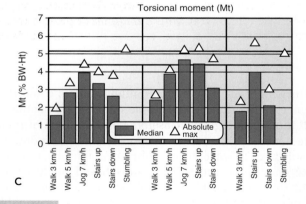

FIG. 16-1

Joint loading during different activities for two patients, EB (*EBL,* left hip; *EBR,* right hip) and JB. The median load values (*column* height) and absolute maxima (*triangle*) are given. The resultant force is in the top graph **(A),** the bending moment in the frontal plane is in the middle graph **(B),** and the torsional moment in the transverse plane is in the bottom graph **(C).** The *gray bar* indicates the fixation strength of cementless implants (14). *Reprinted with permission from Bergmann G., Graichen, F., Rohlmann, A. (1995). Is staircase walking a risk for the fixation of hip implants? J Biomech, 28(5), 535–553.*

comparable to in vivo measurements, so the choice of the best approach depends on the application.

For completeness, it should be noted that several models have approached the indeterminate problem using neither reduction nor optimization. These models solve the indeterminate problem by either modeling individual muscle recruitment based on the external moments by using dynamic solution techniques (for example, Anderson and Pandy, 2001). Dynamic solution methods also usually predict contact forces that are similar to reported in vivo measurements. Their main disadvantage is their relative computational expense. Although they may be useful in studying unusual activities, they are not practical for modeling simple activities, like walking, in large numbers of subjects.

FACTORS THAT INFLUENCE THE HIP LOADING ENVIRONMENT

The magnitude, location, and frequency of hip loading all contribute to the lifespan of an artificial hip implant. Therefore it is important to understand what factors, in turn, determine these aspects of loading. Some of these can be influenced or selected by the surgeon (e.g., the geometry of the reconstructed hip joint). Others are primarily under the patients' control (e.g., activity level and type). Continued improvements in our understanding of the factors that determine hip loading and implant longevity can lead to not only better implant designs and surgical techniques, but also better rehabilitation regimens and patient recommendations. In this section, we focus on the role of gait patterns and joint geometry and briefly discuss activity.

Gait Biomechanics

Many researchers have found that patients with total hip replacements often have various functional biomechanical deficits including problems with balance or slower walking speeds (Majewski et al., 2005; Sicard-Rosenbaum et al., 2002). THR patients may also have reduced ground reaction forces or patterns of muscle activity, joint forces, or motion compared to either the unoperated hip or to normal subjects (Foucher et al., 2007, 2008; Long et al., 1993; Shih et al., 1994). In fact, the patterns of biomechanical deficits seen after surgery are very similar to those seen in patients with osteoarthritis (Fig. 16-2). This suggests both that some gait abnormalities seen after surgery arise because of the disease that lead to the hip replacement rather than the THR itself and that THR does not fully cure abnormal hip biomechanics, even when it relieves pain and basic disability. In most cases, the functional problems that are detected

TABLE 16-2

Analytic Methods of Estimating Peak Hip Contact Force

Activity	Magnitude (BW)	Method	Reference
Walking	5.5	Optimization	Seireg and Arvikar, 1975
Walking	4.8	Reduction	Paul and McGrouther, 1976
Stair ascending	7.2		
Stair descending	7.1		
Walking slow speed with cane	2.2	Optimization	Brand and Crowninshield, 1980
Walking slow speed without cane	3.4		
Walking	5.0	Optimization	Crowninshield et al., 1978
Stair climbing	7.4		
Chair rising	3.3		
Walking	4.6	Optimization	Collins, 1994
Walking	4.0	Optimization	Anderson and Pandy, 2001
Walking fast speed	4.3	Optimization	Stansfield et al., 2003
Walking slow speed	3.2	Optimization	Stansfield et al., 2003
Chair rising	2.0	Optimization	Stansfield et al., 2003
Walking	5.26	Optimization	Heller et al., 2005
Stair climbing	8.8	Optimization	Heller et al., 2005

through gait analysis are not noticeable by the patient; however, there is still debate over whether there may be any implications for the longevity of the implant or even the other healthy joints.

In a basic sense, it is desirable to avoid excessive forces on the implant (leaving aside the definition of "excessive"). It has been shown that external moments during gait are correlated with implant forces (Foucher et al., 2009). Therefore, the fact that many external moments are less than normal after THR surgery may be beneficial if looked at from the implant's perspective—these gait adaptations may serve to reduce the loads on

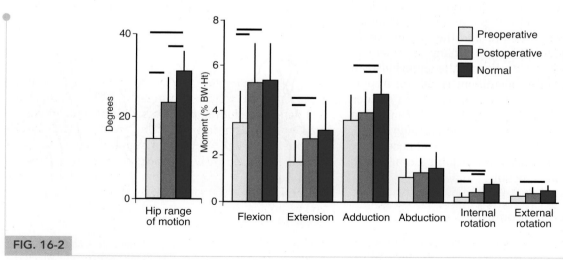

FIG. 16-2

Postoperative gait retained similarities to preoperative gait and was different from normal. Range of motion in degrees and peak external moments in % body weight × height are shown for preoperative THR patients; the same patients one year after surgery; and age-, height-, weight-, and gender-matched normal subjects. *Bars* indicate means with standard deviations for each group. *Lines above* indicate significant differences between groups. *Adapted from Foucher, K.C., Hurwitz, D.E., Wimmer, M.A. (2007). Preoperative gait adaptations persist one year after surgery in clinically well-functioning total hip replacement patients. J Biomech, 40(15), 3432–3437.*

the implant and may be protective. In addition, walking with a decreased hip range of motion may minimize the out-of-plane force components (anterior-posterior) and thus the torques on the implant stems. Torsional loads may be especially dangerous to the stability of implant fixations. Thus, decreased sagittal plane motion during walking often reported in patients with total hip replacements (Foucher et al., 2007; Murray et al., 1972; Stauffer et al., 1974) may be beneficial for implant stability by reducing the rotational moments about the implant stem.

Despite the potential benefits of "abnormal" gait patterns, they may not be without risks or disadvantages. It has been shown that subjects with one total hip replacement are up to 400 times more likely to get their contralateral knee or hip replaced than subjects who have not had a THR, even when the initial disease process is considered to be unilateral (Lazansky, 1967; Shakoor et al., 2002; Umeda et al., 2009). The cause of this may be because people with hip replacements put a disproportionate amount of load on the limb contralateral to their operated hip (McCrory et al., 2001; Shakoor et al., 2003; White and Lifeso, 2005). For example, in a study of patients with total hip replacements, the external knee adduction moment was significantly higher on the contralateral knee than on the knee on the same side as the operated hip (Shakoor et al., 2003). This is important because the knee adduction moment is an established biomechanical marker of both the severity and progression of knee osteoarthritis.

The bony geometry of the limb on the side of the operated hip may also play a role (Umeda et al., 2009). For example, hip offset, the perpendicular distance between the head of the femur and the femoral shaft, can be changed with THR surgery. Changing offset can alter the alignment of the entire leg (the angle between the hip, knee, and ankle); alignment is also an important factor that contributes to the development of knee osteoarthritis (OA) (Hurwitz et al., 2002). Since the two limbs are physically (and thus biomechanically) linked, changes in the biomechanics of one leg will necessarily affect the other. Other implications of joint geometry will be discussed in the next section.

Reconstructed Joint Geometry

The geometry between the pelvis, proximal femur, and the rest of the leg can all be affected by total hip arthroplasty. Changes in these relationships can have a profound effect on the line of force transmission between joints and the moment-generating capacity of the muscles that cross the hip. Joint geometry considerations make up a key component of preoperative surgical planning. Surgeons can define or select the size of the femoral head, the length of the neck of the implant, and angle between the neck and the shaft. The offset (the perpendicular distance between the femoral head and the shaft), the abductor muscle moment arm, and the location of the hip center relative to the pelvis can be indirectly affected. The acetabular cup can be placed in a position of more or less tilt or anteversion. All of these changes can affect the amount of force or the location of force transmission.

Alterations in hip center location have a large effect on the moment-generating capacity of the muscles and the resultant hip force (Doehring et al., 1996; Lenaerts et al., 2008) (Fig. 16-3). Predicted joint forces are minimized when the joint center is moved medially, inferiorly, and anteriorly (Fig. 16-4). This position maximizes the moment-generating capacity of the abductors and brings the hip center closer to the line of action of the foot floor reaction force, thus decreasing the external moment that needs to be balanced by muscle forces (Doehring et al., 1996; Johnston et al., 1979). In general, the analytic and experimental results on the effect of joint geometry on hip joint forces are consistent with clinical patient studies. Clinical studies have associated inferior functional outcomes with superior placement of the joint center (Box and Noble, 1993) and have associated decreases in abductor strength and loss of passive hip flexion motion with superior placements of the joint centers unless compensated with an increased neck length. Higher femoral loosening rates have been

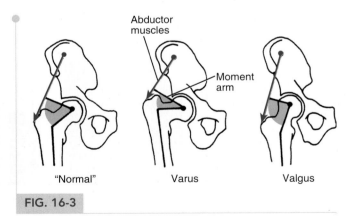

FIG. 16-3

The abductor mechanism changes with head-neck angle or neck length. A valgus neck angle decreases the moment arm, whereas a varus neck angle or an increased neck length increases the moment arm. *Reprinted with permission from Hurwitz, D.E., Andriacchi, T.P. (1998). Biomechanics of the hip. In J. Callaghan, A. Rosenberg, H. Rubash (Eds.). The Adult Hip. New York: Raven Press, 75–86.*

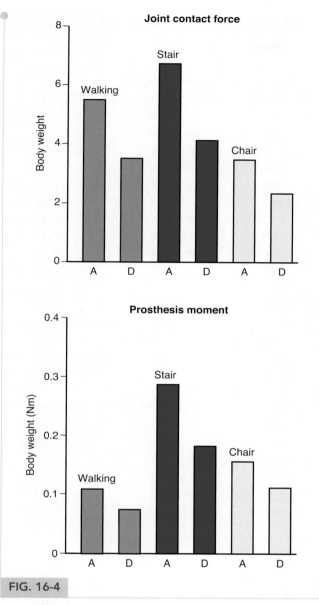

FIG. 16-4

Peak contact forces at the hip and peak moments about the neck-stem junction of the prosthesis during level walking, stair climbing, and rising from a chair under the following conditions: *A,* normal location of the hip center; *D,* hip center 20 mm medial, 20 mm inferior, and 10 mm anterior. Peak forces were based on an optimization approach that minimized muscle stresses. *Reprinted with permission from Johnston, R.C., Brand, R.A., Crowninshield, R.D. (1979). Reconstruction of the hip. J Bone Joint Surg, 61A(5), 646.*

associated with joint centers placed superior and lateral as opposed to those placed in an anatomic position (Yoder et al., 1988).

Increasing the neck length or advancing the greater trochanter can partially compensate for these losses in

muscle moment-generating capacities (Delp et al., 1994). Other considerations may, however, limit a surgeon's options. For example, the pathologic conditions that can lead to total hip replacement (i.e., osteoarthritis or hip dysplasia) can affect the range of potential locations of the hip center. Although a certain joint position may be recommended by biomechanical analyses, it may not be practical to achieve these positions in practice.

It may someday be possible to use other biomechanical strategies to optimize hip forces when a total hip replacement patient ends up with less than optimal joint geometry. It has been shown that gait has an independent effect on hip forces even after considering joint geometry (Foucher et al., 2009). In one study, the external moments during gait were generally more predictive of hip forces during walking than were radiographic measurements (Table 16-3). There is evidence that gait patterns can be changed before or after surgery with appropriate rehabilitation interventions (Schroter et al., 1999; White and Lifeso, 2005). If potentially dangerous hip forces are a concern in a patient after THR surgery, gait retraining may offer an avenue to improve the hip loading environment, whereas joint geometry cannot be changed without further surgery.

Activity Level

In addition to the magnitude of hip forces, the number of loading cycles is a critical determinant of implant wear and longevity. Younger (and thus presumably more active) people have joined the pool of potential THA candidates because improvements in surgical techniques, implant designs, and materials have increased the expected lifetime of each total hip implant. Today many older people are also quite active. Total hip implants must be able to withstand the demands of contemporary patients. Patients in all age groups expect THA to allow them to maintain an active lifestyle. Number of steps walked per day, or year, is often considered to be a surrogate marker of activity and an analogue of number of loading cycles for in vitro testing.

Wear simulation studies often assume that patients with THRs take one million steps per year. However, several recent studies have suggested that this figure may dramatically underestimate patient activity in some cases. Most recent studies have measured steps from 5,078 up to 12,288 per day, depending on time from surgery and evaluation methodology (Table 16-4). Some patients who seek out minimally invasive surgical approaches (MIS) may be more active still. In a recently evaluated group of subjects only three weeks after MIS THR, subjects walked from as few as 146 to as many as 16,392 steps per day (6,005 ± 4,175), which extrapolates

TABLE 16-3

Relationships between Gait Parameters, Hip Loading, and Joint Geometry after Total Hip Replacement

	Mean ± SD Coefficient of Variation	First Contact Force Peak	Second Contact Force Peak	Peak Implant Twisting Moment
Flexion moment	77.2 ± 30.2	$R^2 = 0.61$	$R^2 = 0.26$	$R^2 = 0.10$
	39%	$p < 0.001$	$p = 0.006$	$p = 0.125$
Extension moment	39.5 ± 16.1	$R^2 = 0.08$	$R^2 = 0.64$	$R^2 = 0.02$
	41%	$p = 0.158$	$p < 0.001$	$p = 0.494$
Adduction moment	57.0 ± 14.8	$R^2 = 0.50$	$R^2 = 0.67$	$R^2 = 0.07$
	26%	$p < 0.001$	$p < 0.001$	$p = 0.189$
Abduction moment	18.1 ± 9.3	$R^2 = 0.13$	$R^2 = 0.01$	$R^2 = 0.06$
	51%	$p = 0.063$	$p = 0.557$	$p = 0.244$
Internal rotation moment	6.5 ± 3.2	$R^2 = 0.08$	$R^2 = 0.30$	$R^2 = 0.10$
	48%	$p = 0.155$	$p = 0.003$	$p = 0.115$
External rotation moment	6.1 ± 3.9	$R^2 = 0.24$	$R^2 < 0.01$	$R^2 = 0.52$
	63%	$p = 0.010$	$p = 0.684$	$p < 0.001$
Vertical joint center position	16.5 ± 6.1	$R^2 = 0.08$	$R^2 = 0.06$	$R^2 = 0.33$
	37%	$p = 0.143$	$p = 0.232$	$p = 0.002$
Horizontal joint center position	33.1 ± 4.0	$R^2 < 0.01$	$R^2 < 0.01$	$R^2 = 0.06$
	12%	$p = 0.909$	$p = 0.888$	$p = 0.241$
Abductor moment arm	50.9 ± 9.7	$R^2 = 0.27$	$R^2 = 0.02$	$R^2 = 0.12$
	19%	$p = 0.005$	$p = 0.510$	$p = 0.083$
Offset	43.9 ± 5.3	$R^2 < 0.01$	$R^2 < 0.01$	$R^2 < 0.01$
	12%	$p = 0.748$	$p = 0.714$	$p = 0.948$
Height	1.7 ± 0.1	$R^2 = 0.42$	$R^2 = 0.07$	$R^2 = 0.14$
	5%	$p < 0.001$	$p = 0.188$	$p = 0.060$
Weight	850 ± 155	$R^2 = 0.30$	$R^2 = 0.17$	$R^2 = 0.04$
	18%	$p = 0.003$	$p = 0.033$	$p = 0.326$

Mean, standard deviation (SD), and coefficient of variation for each external moment (Nm), radiographic measurement (mm), height (m) and weight (N) with correlations between these parameters and hip loading. Bold text indicates statistical significance. Used with permission from Foucher, K.C. et al. (2009). *J Orthop Res*, 27(12), 1576–1582.

to between 53,290 and 5,983,080 steps per year. Also, this patient cohort was not statistically significantly less active than controls (Foucher et al., 2010).

Although some patients are certainly much more active than previously thought, it is important to note that activity level is quite variable within and between study populations. Many studies report a large difference in step count between the most active and least active subjects. It has been suggested that the differences in wear volume seen in retrieved implants can be explained by varying activity levels among subjects (Schmalzried et al., 1998). It is important to note that activities other than simple walking may also be important to consider, although most wear studies take only walking into account.

WHY STUDY HIP FORCES? COMPLICATIONS INFLUENCED BY HIP BIOMECHANICS

In this section, to reiterate the importance of understanding hip biomechanics, we will discuss the major causes of total hip arthroplasty complications that are influenced by postoperative hip biomechanics: periprosthetic bone loss and implant wear. Developments in materials and implant design have reduced the likelihood of bone loss and wear; however, it remains important to understand the underlying concepts. Periprosthetic bone loss and implant wear remain the leading causes of failure and ultimate revision of total hip replacements. Bone loss is also an emerging issue in hip

TABLE 16-4

Studies Measuring Activities for Patients after Surgery

Activity	Average Steps/Day (Standard Deviation)	Number of Patients Tested	Time since Surgery	Duration of Measurement Period	Measurement Method	Reference
Normal daily activities	5,078 (3,156)	100	>6 months	Unknown	Pedometer	Zahiri et al., 1998
Normal daily activities	5,194 (not reported)	111	>6 months	7 days	Pedometer	Schmalzried et al., 1998
Normal daily activities	6,878 (3,736)	33	>2 years	4 days	Pedometer	Silva et al., 2002
Normal daily activities	10,438 (4,388)	33	>2 years	4 days	SAM	Silva et al., 2002
Walking	12,288 (not reported for population)	105	3.4 years (avg.)	5–14 days	SAM	Kinkel et al., 2009

Zahiri et al. analyzed a group of patients who had either knee or hip replacements. SAM, Step Activity Monitor.

resurfacing (Huo et al., 2009)—an increasingly popular and somewhat more conservative surgical reconstruction for the arthritic hip. Finally, although more wear-resistant bearing surfaces have been introduced over the past decade, long-term data are only beginning to emerge. Early reports suggest that similar issues may be at play, albeit on a smaller scale.

Periprosthetic bone loss associated with uncemented femoral stems has been well documented (Bryan et al., 1996; Engh et al., 1992; Krger et al., 1998), and concerns have been raised with regard to the long-term clinical implications of this phenomenon. Osteolysis, stress shielding, and generalized limb unloading all may play a role in periprosthetic bone loss. Wear particles from polyethylene and other implant materials are found in the joint fluid and adjacent tissues. These wear particles lead to a foreign-body reaction with increased macrophage activity and intercellular secretion of mediators that stimulate osteoclasts and result in periprosthetic bone loss (Jasty, 1993).

Stress shielding results from a decrease in the stress distribution in the femoral bone as a result of the presence of the implant stem, which has a greater or equivalent mechanical stiffness as compared with the femur. Changes in the loading environment result in bone remodeling. Once bone ingrowth occurs in the cementless prosthesis, load transfer can occur through these areas of bony attachment. However, bone remodeling does not result in the restoration of normal cortical strain levels (Engh et al., 1992). The fit of the prosthesis within the femoral canal (Jasty et al., 1994), as well as the

material properties (stiffness) of the stem (Cheal et al., 1992; Weinans et al., 1992), affects the amount of stress shielding (Fig. 16-5).

Bone loss also can result from limb disuse. Postoperative subjects with total hip replacements continue to walk asymmetrically, with decreased forces on the operated side as compared with the contralateral side (Bryan et al., 1996; Long et al., 1993). Tibial bone loss is unaffected by

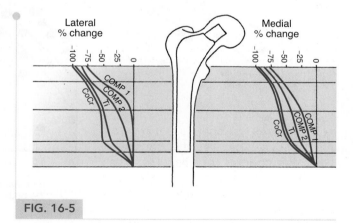

FIG. 16-5

Percent change in the strain energy density within the medial and lateral regions of the medial and lateral cortex from a three-dimensional finite element model. The greatest reduction in strain energy density was found for the femur with the stiffest implant (CoCr). The femur with the most flexible implant (Composite 1) had the smallest reduction in strain energy density (*Comp1* and *Comp2*, composite materials; *CoCr*, cobalt chromium; *Ti*, titanium) (Personal communication, R.N. Natarajan).

implant characteristics and most likely results from generalized limb disuse associated with asymmetries in joint loading conditions. A 16% decrease in proximal tibial bone mineral content has been demonstrated in subjects with long-term total hip replacements, and this decrease has been related to an asymmetry in the peak vertical intersegmental knee force during gait (Bryan et al., 1996). Preoperative gait mechanics of patients with hip osteoarthritis have been shown to be correlated with the bone mineral density of the proximal femur (Fig. 16-6) (Hurwitz et al., 1998). The greater the bone loss preoperatively, the less stiff the femur and the more likely that stress shielding and associated bone resorption will occur postoperatively. Autopsy studies have also shown that the lower the bone mineral density or content of the contralateral femur, the greater the reduction in periprosthetic bone on the affected side, which further implies that preoperative bone mineral density influences the extent of postoperative bone loss.

In addition to the bone mineral density of the host bone and the loading environment, the material properties of the implant itself have a large effect on the likelihood of bone loss, wear, and consequent implant failure. Since the early 2000s, most total hip acetabular components have been made with highly cross-linked polyethylene. Cross-linking is usually accomplished by radiating the polyethylene in various environments (McKellop et al., 1999) and results in greater wear resistance in most studies. Results from the first decade of service are so far very positive. It is possible that other biomechanical issues may overtake wear as the biggest problem in hip arthroplasty. For example, some studies suggest that highly cross-linked polyethylene may be more brittle and may therefore be more susceptible to crack formation and other failure modes (Furmanski et al., 2009). Other bearing surfaces, including ceramics and metal on metal, are also in use and are subjects of active study. A deeper investigation of these

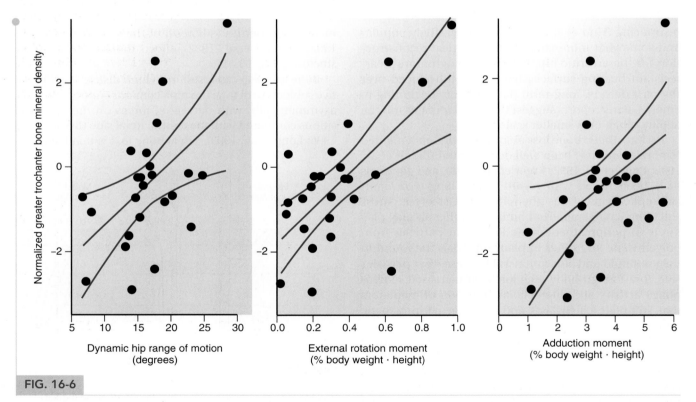

FIG. 16-6

The hip range of motion and external adduction and external rotation moments were significant predictors of the normalized bone mineral density of the greater trochanter in the patients with hip osteoarthritis. The abductors are the primary structures responsible for balancing the adduction moment. Because the abductors insert on the greater trochanter, a reduced adduction moment may reflect reduced forces in this region and may result in bone loss. Similarly, the anterior fibers of the gluteus medius and minimus are recognized as primary internal rotators. Thus, the decreased external rotation moment in early stance may also be reflective of decreased abductor muscle forces. *Modified with permission from Hurwitz, D.E., Foucher, K.C., Sumner, D.R., et al. (1998). Hip motion and moments during gait relate directly to proximal femoral bone mineral density in patients with osteoarthritis. J Biomech, 31(10), 919–925.*

materials is recommended for the interested reader. Further study of other clinical issues in total hip arthroplasty that relate to hip biomechanics is also recommended. These include the effects of surgical approach, dislocation, implant design, and rehabilitation.

Knee

MOTION AND FORCES AT THE KNEE JOINT

Several principal considerations for THR are also applicable to TKR. For example, the articulating materials are typically the same (cobalt-chromium on polyethylene) and clinical failure scenarios are very similar. Therefore, in this section, we will supplement previous information and focus on the differences. Obviously, one of the biggest differences between hip and knee are the joint kinematics. Whereas the hip joint in THR is modeled as a technical ball-and-socket joint, the anatomy of the knee joint does not have such a straightforward analogue in technical applications. Its anatomy is complex, although it has a preferred rotation direction, namely flexion-extension (FE) in the sagittal plane. For this reason, it has been often simply referred to as a hinge joint. However, anteroposterior (AP) translation in the sagittal plane, internal-external (IE) rotation in the transverse plane, and varus-valgus motions in the frontal plane are also very important to its overall function and cannot be neglected.

There are many TKR designs, but to be clinically successful, all of these designs need to allow various kinematic patterns that are executed during different activities of daily living. Neglect of the three-dimensional freedom of the natural knee joint generates constraints that in turn stress the interface between prosthetic device and bone and typically cause early aseptic loosening. Too little constraint, on the other hand, leaves the TKR mechanically unstable, without proper support for the patient. It may also wear out quickly due to excessive motion at the articulation. Knowledge of the magnitude and cyclic nature of motions and forces is therefore crucial in the design of TKR.

Knee Kinematics

The relative motion at the knee joint can be described by three translations and three rotations, which constitute six degrees of freedom at the joint (Kapandji, 1970). Typically, flexion and extension are highly reproducible intraindividually during human locomotion in healthy subjects as well as in TKR patients (Ngai and Wimmer, 2009). During level walking, the knee is almost fully extended at heel-strike. Following heel-strike, the knee begins to bend, reaching a maximum of about 15° to

20° knee flexion during midstance. At this point, the direction of the angular progression reverses and the knee fully extends again (~45% of gait cycle). The joint reverses direction once more to start the pre-swing phase and continues flexion into toe-off, which occurs at approximately 63% of the gait cycle (compare with Fig. 17-4A top). Often, TKR patients show gait adaptations with movement patterns that deviate from normal. Deficits are observed at heel-strike, where full extension may not be reached. Also, midstance extension is typically not as pronounced as seen during normal gait (Fig. 17-5).

Flexion of the knee progresses as a combination of rolling, sliding, and spin of the femoral condyles over the tibial plateau. Experiments demonstrating this mechanism were performed as early as 1836 by the Weber brothers. They evaluated the relative motion between the femoral condyles and the tibial surface by placing markers on the corresponding points of contact on both surfaces (Weber and Weber, 1836). Nearly a hundred years later it was demonstrated that the ratio of rolling to sliding varies during flexion and extension (Strasser, 1908). As depicted in Figure 16-7, rolling predominates early in flexion (0°–20°), whereas sliding becomes dominant at flexion angles beyond 30° (Draganich et al., 1987). This mechanism is called femoral rollback. One of the models to explain the mechanism is the crossed-four-bar

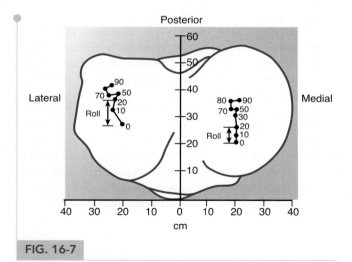

FIG. 16-7

The tibial-femoral contact moves posteriorly with knee flexion. The contact on the lateral side moves posteriorly much more during flexion (0°–20°) than does the medial side because the lateral femoral condyle is rolling on a larger radius than is the medial femoral condyle. Beyond 20°, sliding motion begins on both condyles. *Reprinted with permission from Andriacchi, T.P., Stanwyck, T.S., Galante, J.O. (1986). Knee biomechanics and total knee replacement. J Arthroplasty, 1(3), 211–219.*

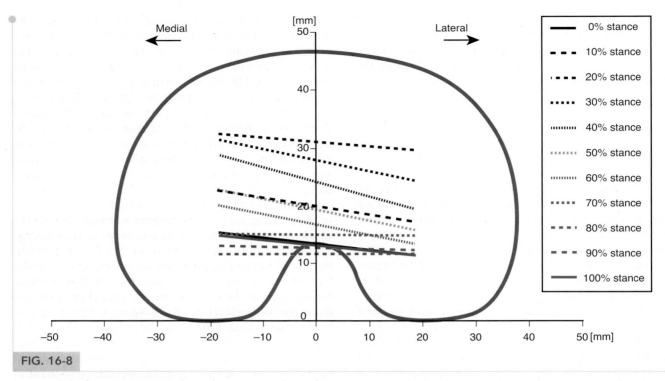

FIG. 16-8

Contact path of a representative subject during stance phase of gait. Medial and lateral contact points are connected via a line to animate AP translation and superimposed IE rotation. Note that contact at heel-strike occurs posteriorly (0% stance) with anteriorly directed sliding and that the pivot point is on the lateral side during most of the stance phase.

linkage, considered by Müller (1983) and O'Connor and Zavatsky (1990). In this model, the insertions of both cruciate ligaments are rigidly attached to the femur and the tibia. They are represented by two crossed bars, which are not linked but held fixed firmly at their anatomic points of insertion (Fig. 16-8).

In vivo investigations have demonstrated that substantial angular and linear motions occur about all six degrees of freedom of the joint during activities of daily living. In general, the motion pattern is activity specific. In a study by Lafortune et al. (1992), the motion pattern during walking was characterized using bone pins that were inserted into the tibia and femur. Obviously this invasive technique was limited to very few subjects. Therefore, numerous studies have used fluoroscopy to track the knee motions of normal and TKR subjects during deep knee bending, step up task, lunging, kneeling, and treadmill walking. However, due to the limited field of view of the x-ray camera system, it has been difficult to investigate functional activities of daily living with such technologies.

More recently, level walking has been studied using the point cluster technique. Here, reflective markers are attached to the skin as point clouds allowing the analysis of rigid body motion with minimal soft tissue noise (Andriacchi et al., 1998). Similar to Lafortune's study, it was found that in a fixed tibial reference frame, the femur translates anteriorly following heel-strike and changes direction at approximately 15% to 20% gait to translate posteriorly. At toe-off, the femur changes direction again to anterior translation entering swing phase and concludes the gait cycle translating posteriorly. Hence, there are three AP translation direction changes during one gait cycle. Notably, these directions do not follow the femoral rollback theory, suggesting that the knee is largely driven by internal and external forces. Ligaments and anatomic surface features provide the constraints, which define the envelope of knee motion during various activities.

TKR subjects with less constraining prostheses demonstrate similar motion patterns; however, they display an increased range of AP translation in stance phase (Ngai et al., 2009). Figure 16-8 shows the contact path of a representative subject. The connecting line of medial and lateral contact points throughout stance are shown in 5% time increments. During another secondary motion, internal-external (IE) rotation of the tibia, there is 2° to 4° of rotation during stance, which has been shown to be quite variable from subject to subject for TKR patients (Ngai et al., 2009) (Fig 16-7). From the figure, it can be seen that the predominant center of rotation is on the lateral side of the knee (Fig. 16-8).

These results are consistent with those of Koo and Andriacchi (2008), who found that during walking, the natural knee shows mostly lateral pivoting. There is more IE rotation of the knee during the swing phase than stance phase of gait. Ngai (2010) reported 13.9° ± 1.2° for TKR patients. It is interesting that the femur exhibits internal rotation with respect to the tibia entering swing and externally rotates at the end of swing. These rotation patterns contradict the "screw-home" mechanism, which specifies the internal tibial rotation with increasing knee flexion and external tibial rotation on knee extension. In cadaver studies, it has been found that the knee moves a greater distance on the lateral plateau as compared with the medial plateau. As a result, during rollback, the femur rotates externally during knee flexion and does the reverse during knee extension, which has been termed the "screw-home" mechanism (Shaw and Murray, 1974). In summary, these data demonstrate that motions of the natural knee and most prosthetic replacements are activity dependent and cannot be generalized based on knee flexion/extension.

Knee Kinetics

Typically, at heel-strike, there is an external flexion-extension moment tending to extend the knee joint (compare with Fig. 17-4A bottom). As the knee moves into midstance, the external moment reverses direction, demanding the action of the extensor muscles. The external moment reverses direction again during late midstance, activating the flexor muscles. Finally, at toe-off, the extensors have to become active once more.

When the knee is near full extension, the patellar ligament is anteriorly angled (Draganich et al., 1987). Due to this orientation of the ligament, the quadriceps muscles pull the tibia forward against the resistance of the anterior cruciate ligament (ACL). Patients who no longer have an ACL (true for most TKR patients) often adapt by reducing the quadriceps force (Andriacchi, 1990). The reasons for such an adaptation are not completely understood and could be of a protective or pathologic nature. Its influence on the contact kinematics has just been recently revealed: Lower peak flexion moments during the stance phase of gait were correlated with higher AP translation during stance (Ngai, 2010). This new finding implies that a well-functioning quadriceps helps to reduce secondary sliding between the tibia and femur, which is particularly important for TKR longevity (and reduced implant wear).

During walking, the associated adduction moment produces an asymmetric load distribution in the frontal joint plane and forces the knee into varus (Fig. 16-9). This can be related to the vector of the ground reaction force, which typically passes the knee medially. The collateral ligaments help to balance the loads between the medial and lateral condyles of the knee. Proper tensioning of the lateral collateral ligament is therefore critical to allow proper function and prevent lift-off of the lateral condyle in TKR (Schipplein and Andriacchi, 1991). Also the risk of leaving the knee in residual varus is evident: Residual varus results in increased adducting moments and, thus, higher medial compartment loads leading to subsequent failure of the implant (Andriacchi et al., 1986) (Fig. 16-10).

The rotation in the transverse plane moment points primarily into the external direction during stance. Although low in magnitude, this moment is difficult to balance by muscle structures of the knee joint alone and requires the assistance of cruciate and collateral ligaments. In the absence of both cruciates (as is typical for TKR), the patients may adapt their gait and lower the moment.

Forces at the Knee

Numerous analytic and numeric models have been introduced for both the natural and the prosthetic human knee. The basic concepts behind such models using external joint kinetics have already been described (see Analytic Hip Force Modeling). All these models assumed frictionless surfaces, an acceptable approach for THR and the natural knee. For TKR, however, shear forces play an important role in the failure mechanisms of the joint. Shear forces can damage the polyethylene articulation and may stress the underlying bone bed. Shear forces may arise due to friction at the artificial articulation, which can be 100-fold higher compared with cartilage (Fisher et al., 1994; Unsworth, 1993). In addition, the tibiofemoral contact path alters the mechanical efficiency of the muscles crossing the knee joint.

Using a mathematical approach (Wimmer, 1999; Wimmer and Andriacchi, 1997; Wimmer et al., 1998), it was shown that there is a substantial influence of gait kinematics and kinetics on the generated shear contact forces. Briefly, the model was used to calculate the compressive (normal) and the tractive (shear) forces at the knee from kinematic and kinetic measurements taken during the stance phase of gait of patients following total knee arthroplasty. Being based on a previously developed approach (Schipplein and Andriacchi, 1991) and gait kinetics common to patients following TKR, the model considered secondary kinematics (pure rolling or sliding in the anteroposterior direction) between femur and tibia and introduced friction. It is important to understand that tangential shear forces can be generated during both

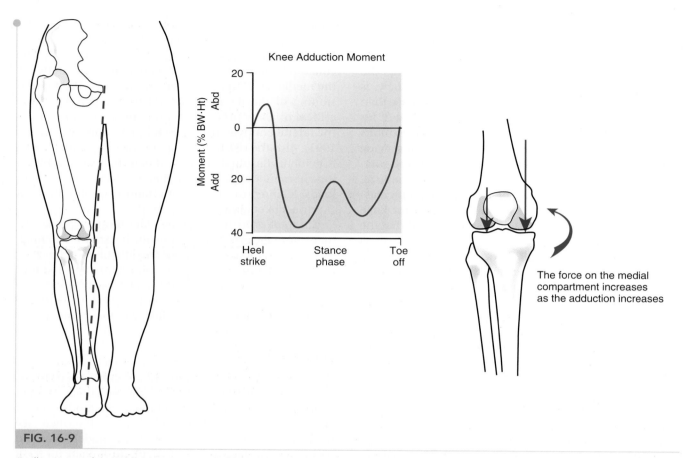

FIG. 16-9

An illustration of the adduction moment during walking and the resultant greater load across the medial compartment of the knee compared with the lateral compartment. *Reprinted with permission from Andriacchi, T.P. (1993). Functional analysis of pre- and post-knee surgery: Total knee arthroplasty and ACL reconstruction. J Biomech Eng, 115, 575–581.*

rolling and sliding (Johnson, 1987). The calculated contact forces for a patient with normal kinetic patterns are plotted in Figure 16-11. Tractive forces close to 0.5 BW at the contact peak during midstance and 0.3 BW during the peak just before toe-off were reached. Table 16.2 provides a summary of internal compressive and shear forces as published in the literature. It should be noted that many published values are above the measured internal forces as discussed further later, and the research findings should be revisited once more as direct measurements become available.

As described previously for the hip joint, our laboratory developed a parametric approach to calculate contact forces. Such approach has been recently applied to the knee joint, as well. Again, the indeterminate problem is solved by categorizing muscles into functional groups at each instance of stance. Unlike previous reduction methods, muscles in each group are not required to have the same force or line of action. Within each functional group, the relative activation level of muscles is allowed

to vary so that many combinations of possible muscle forces are tried when balancing the equilibrium equations. The parametric variation results in a solution space of contact forces calculated for muscle activation levels throughout their physiologic range. Details of this model can be found elsewhere (Lundberg, Foucher, and Wimmer, 2009). We recently had the opportunity to compare the model output against force readings of an instrumented total knee. Input kinematics and kinetics to the parametric model were measured during gait analysis simultaneously with the telemetric force data (Lundberg, Foucher, Andriacchi, and Wimmer, 2009). The parametric model compared very well to instrumented total knee output (Figure 16-12).

There have been recent activities to directly measure knee forces with the help of implants instrumented with strain gauges. D'Lima et al. (2006) found axial contact forces ranging from 2.2 to 2.8 times BW. Higher forces are induced by more vigorous activities (Mündermann et al., 2008). Axial forces increased to 3.5 to 8 times BW

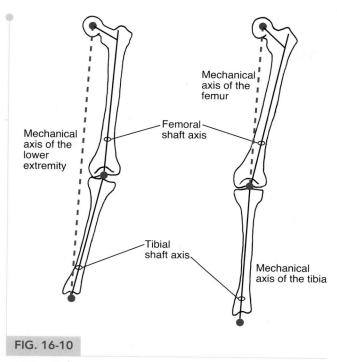

FIG. 16-10

Left. In a lower extremity with varus deformity, the mechanical axis passes medial to the knee. **Right.** When alignment is normal, the mechanical axis of the femur is in line with the mechanical axis of the tibia (tibial shaft axis). The line represented by the mechanical axes of the femur and the tibia is coincident with the mechanical axis of the lower extremity in this situation. *Reprinted with permission from Krachow, K.A. (1995). Surgical principles in total knee arthroplasty: Alignment, deformity, approaches and bone cuts. In J.J. Callagan, D.A. Dennis, W.G. Paprosky, et al. (Eds.). Orthopaedic Knowledge Update: Hip and Knee Reconstruction. Rosemont, IL: AAOS, 269–276.*

for stair ascent and downhill walking. These first-generation instrumented knee prostheses could measure only axial contact forces. Bergmann and others (Heinlein et al., 2009) developed a device that is capable of measuring loads and moments for all six degrees of freedom. For level walking, axial force measurements were similar to D'Lima's data. Shear forces were approximately 0.3 times BW. During stair maneuvers, all force readings increased by about 30%.

FACTORS THAT INFLUENCE KNEE MOTION AND LOAD

Gait

As evident from the previous discussion, the kinetic pattern of the knee will affect contact load. Hence, the gait style matters and will affect implant longevity. Also,

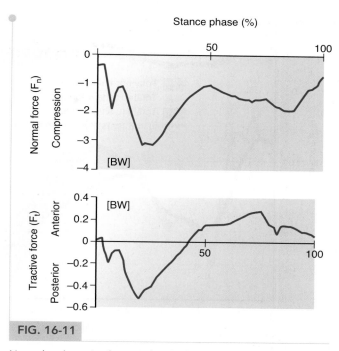

FIG. 16-11

Normal and tractive force at the tibial plateau during stance phase of gait of a representative subject with normal gait. Note the biphasic shape of the traction force with a direction change around midstance.

knee kinematics during gait is—at least in part—driven by knee kinetics (Ngai and Wimmer, 2009). Hilding et al. (1996) reported that patients who walked with higher peak flexion moments demonstrated increased tibial component migration, which put them at risk for aseptic loosening. However, such relationships are complex and need to be evaluated in context with prosthetic design and patient cohort.

Activity

Knee kinematics and kinetics of TKR are driven by the specific activity. Biomechanical analyses of activities of daily living are therefore essential. Examples are level walking, stair climbing and descent, sitting and rising from a chair, and transitional stop-start movements, such as those that would be performed in a kitchen. Orozco and Wimmer (2010) recently pointed to their importance regarding implant wear. Although the frequency of other activities than walking is low, they have a considerable portion (~20%) of the overall cumulative load. In addition, since wear is a function of load and motion, activities like chair and stair maneuvers could be more detrimental to the polyethylene liner than walking (see Case Study 16-2).

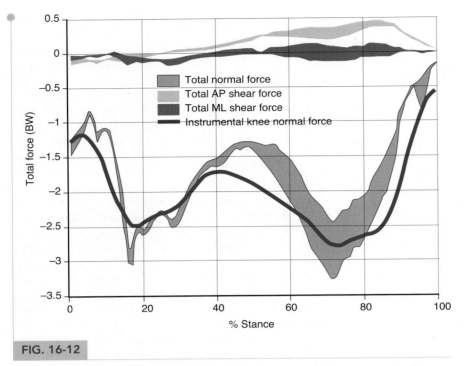

FIG. 16-12

Comparison of the results from the parametric model (solid solution spaces) to the force output from an instrumented TKR for the same input data.

Case Study 16-2

Knee Arthroplasty

A 70-year-old man presents with suffering from disabling left knee pain as a result of a severe genu varus and progressive left knee joint degeneration. The abnormal load resulting from the genu varus deformity creates a load imbalance characterized by a decrease in contact area at the lateral left tibia plateau and an increase in the contact stresses at the medial tibial plateau resulting in progressive wear in the medial tibiofemoral compartment (Case Study Fig. 16-2).

The patient has an impairment and is unable to walk more than half a mile. He wishes to maintain his active lifestyle. A first trial of conservative treatment was unsuccessful. A careful examination ensures the presence of a functional extensor mechanism. Moreover, complete imaging studies confirm the severe articular surface degeneration. The decision of knee arthroscopy was made to avoid a more severe joint degeneration and improve the patient's lifestyle.

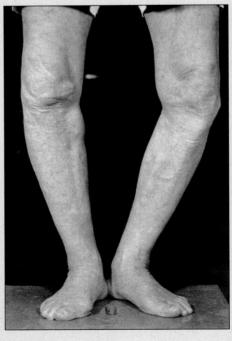

Case Study Figure 16-2

Surgical Factors

Malalignment and improper soft tissue balance are known to cause elevated and eccentric loads between femur and tibia (Dorr et al., 1985). This increases the contact stresses of polyethylene tremendously, often resulting in accelerated wear. Malrotation of the femoral and tibial components have been shown to affect loads not only at the tibiofemoral but also at the patellofemoral joint (Wasielewski et al., 1994). Already small amounts (1°–4°) of internal femoral rotation were associated with maltracking due to incorrect or excessive loading, whereas large amounts (7°–17°) were associated with patellar dislocation or patellar component failure (Berger et al., 1998).

Prosthetic Design

The surgeon has the possibility to choose from various designs, some of them more and less conforming. Although desirable with regard to material stress, conformity is most frequently attached to varying degrees of constraint by virtue of the motion requirements. The amount of constraint in the prosthesis can markedly affect the contact mechanics of the joint. Banks and Hodge (2004) found significant differences in the contact kinematics between cruciate-retaining (less conforming) and posterior-stabilized (more conforming) designs. It has been suggested that limited anteroposterior movement of the femur on the tibia influences function during daily activities (Andriacchi and Galante, 1988). While it is generally accepted that stresses resulting from compressive forces inside the bulk polyethylene are reduced with higher conformity (Bartel et al., 1995), limited freedom of motion can increase tangential (constraint) forces on the tibial plateau, thus causing more harm than good.

WHY DO IMPLANTS FAIL? WEAR IS A PRIMARY REASON

Polyethylene wear of the tibial insert and the resulting mechanical and biologic complications remains a problem and constitutes a recognized cause of failure limiting long-term success. Although the precise mechanisms by which wear particles induce periprosthetic osteolysis are still not fully elucidated and remain an active research area, the particle characteristics, such as composition, size, shape, and overall number (especially for those in the most biologically active submicrometer-size range) are known to play a role in the cell and tissue response (Jacobs et al, 2006).

In a study by Sharkey et al. (2002), polyethylene wear accounted for 25% of the reasons for TKR failure. Several factors can influence the wear of ultrahigh molecular weight polyethylene (UHMWPE) inserts. These can be grouped into material factors (e.g., type of resin, method of consolidation, shelf life, method of sterilization, manufacturing process, or level of oxidation), surgical technique factors (e.g., implant alignment, soft tissue balance), and patient factors (e.g., patient activity, gait pattern) and have been summarized elsewhere (Wimmer et al., 1998). In this section we will reiterate knee biomechanics and its influence on wear.

As discussed previously, the knee joint undergoes rolling/sliding motions during flexion/extension maneuvers. Therefore, the contact location between the femur and the tibia is always in motion and stresses fluctuate at the bearing surface. The fluctuating stresses that result from complex knee kinematics are responsible for surface fatigue-driven wear mechanisms such as pitting and delamination (Figure 16-13). Material properties of the articulation play a major role to this effect and oxidative embrittlement of polyethylene is an unwanted process. Several groups have therefore started to manipulate polyethylene with cross-linking or the addition of vitamin E to stabilize its properties over time (Brach Del Prever et al., 2009).

Knee kinematics and contact forces can similarly affect the amount and type of adhesive and abrasive wear processes observed. Adhesive and abrasive wear mechanisms generate submicron-sized particles (Figure 16-14) that can migrate from the joint space into the periprosthetic tissue leading to systemic inflammatory reactions. This systemic response causes osteolysis and consequent loosening of the prosthetic device

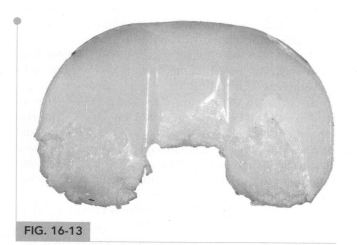

FIG. 16-13

Tibial polyethylene plateau with delamination damage—probably due to misalignment.

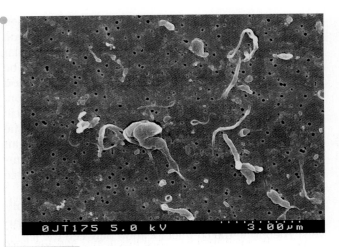

FIG. 16-14

Recovered polyethylene wear particles from total knee replacement. Due to their small size, these particles are phagocytosed by macrophages, which may cause inflammatory reactions.

(Jacobs et al., 2006). It is generally agreed that the amount of wear increases with sliding distance. In addition, the shape of the motion pathway matters for polyethylene wear. An increase in cross-shear, the amount of movement perpendicular to the main motion direction, results in increased wear (Bragdon et al., 1996).

The relationship between wear and motion direction is nonlinear and has been recently theoretically described (Schwenke et al., 2008). Because the molecular chains in the polyethylene surface layer align along the direction of shear, more energy is required to rupture the molecules in their longitudinal direction than perpendicularly. Hence mathematically, such a behavior could be described with a double sinusoidal function (based on the circular vector theory), where zero cross-shear occurs when the motion is along the main polyethylene fibril orientation, and maximum cross-shear occurs in the motion direction perpendicular to the fibril orientation. In summary, kinematic and kinetic inputs have been shown to affect the wear seen in actual patients and in joint simulators. For example, McEwen et al. (2005) showed that increased displacements and rotations during TKR wear testing caused a significant increase in the wear rate.

In summary, contact load and motion pathway affect implant wear and are influenced by surgical technique, implant design, and patient-specific factors. There is a growing body of literature that explores these relationships and should be further monitored.

Summary

- The effect of forces on the stability of a total joint replacement depends not only on its magnitude but also on its orientation and point of application.
- Magnitude, orientation, and point of application of the loads at joints influence the stresses, bending moments, and rotational moments of the implant and are critical for the implant stability, wear, and longevity.
- Understanding the dynamic loads during daily activities provides critical information for addressing clinical problems such as mechanical loosening of implants, amount of wear, bone resorption, and the choice of rehabilitation and surgical protocols.
- The evolution of total joint replacement has been aided by information generated from biomechanical studies.

Practice Questions

1. Is gait normal after total hip or knee arthroplasty? Should normal gait biomechanics be the goal after total hip or knee arthroplasty?

2. List and briefly describe three factors that cause implant loosening after hip or knee arthroplasty.

3. Describe three strategies for approaching the statically indeterminate equilibrium problem faced when trying to model hip or knee joint forces. What are some pros and cons of each approach?

4. How does patient activity level affect hip and knee implant wear?

5. Wear and its consequences are the major cause of implant failure. List two strategies for reducing wear through manipulating material properties.

REFERENCES

Anderson, F.C., Pandy, M.G. (2001). Dynamic optimization of human walking. *J Biomech Eng, 123*(5), 381–390.

Andriacchi, T.P. (1990). Dynamics of pathological motion: Applied to the anterior cruciate deficient knee. *J Biomech, 23,* 99.

Andriacchi, T.P., Alexander, E.J., Toney, M.K., et al. (1998). A point cluster method for in vivo motion analysis: Applied to a study of knee kinematics. *J Biomech Eng, 120*(6), 743–749.

Andriacchi, T.P., Galante, J.O. (1988). Retention of the posterior cruciate in total knee arthroplasty. *J Arthroplasty, 3* Suppl, S13–19.

Andriacchi, T.P., Stanwick, T.S., Galante, J.O. (1986). Knee biomechanics and total knee replacement. *J Arthroplasty, 1,* 211.

Banks, S.A., Hodge, W.A. (2004). Implant design affects knee arthroplasty kinematics during stair-stepping. *Clin Orthop, 426,* 187–193.

Bartel, D.L., Rawlinson, J.J., Burstein, A.H., et al. (1995). Stresses in polyethylene components of contemporary total knee replacements. *Clin Orthop, 317*, 76–82.

Berger, R.A., Crossett, L.S., Jacobs, J.J., et al. (1998). Malrotation causing patellofemoral complications after total knee arthroplasty. *Clin Orthop, 356*, 144–153.

Bergmann, G., Deuretzbacher, G., Heller, M., et al. (2001). Hip contact forces and gait patterns from routine activities. *J Biomech, 34*(7), 859–871.

Bergmann, G., Graichen, F., Rohlmann, A. (2004) Hip joint contact forces during stumbling. *Langenbecks Arch Surg, 389*(1), 53–59.

Bergmann, G., Graichen, F., Rohlmann, A. (1993). Hip joint loading during walking and running, measured in two patients. *J Biomech, 26*(8), 969–990.

Bergmann, G., Graichen, F., Rohlmann, A. (1995). Is staircase walking a risk for the fixation of hip implants? *J Biomech, 28*(5), 535–553.

Box, G., Noble, P.C. (1993). The position of the joint center and the functional outcome of total hip replacement [abstract]. In *Transactions of the 39th Annual Meeting, Orthopaedic Research Society*; February 15–18, San Francisco, CA.

Brach Del Prever, E.M., Bistolfi, A., Bracco, P., et al. (2009). UHMWPE for arthroplasty: Past or future? *J Orthop Traumatol, 10*, 1–8.

Bragdon, C.R., O'Connor, D.O., Lowenstein, J.D., et al. (1996). The importance of multidirectional motion on the wear of polyethylene. *Proc Inst Mech Eng H, 210*(3), 157–165.

Brand, R.A., Crowninshield, R.D. (1980). The effect of cane use on hip contact force. *CORR, 147*, 181–184.

Brand, R.A., Pedersen, D.R., Davy, D.T. (1994). Comparison of hip force calculations and measurements in the same patient. *J Arthroplasty, 9*(1), 45–51.

Bryan, J.M., Sumner, D.R., Hurwitz, D.E., et al. (1996). Altered load history affects periprosthetic bone loss following cementless total hip arthroplasty. *J Ortho Res, 14*(5), 762–768.

Cheal, E.J., Spector, M., Hayes, W.C. (1992). Role of loads and prosthesis material properties on the mechanics of the proximal femur after total hip arthroplasty. *J Orthop Res, 10*(3), 405–422.

Cheng, C.K., Huang, C.H., Liau, J.J., et al. (2003). The influence of surgical malalignment on the contact pressures of fixed and mobile bearing knee prostheses—a biomechanical study. *Clin Biomech, 18*, 231–236.

Crowninshield, R.D., Brand, R.A. (1981). A physiologically based criterion of muscle force prediction in locomotion. *J Biomech, 14*(11), 793–801.

Crowninshield, R.D., Johnston, R.C., Andrew, J.G., et al. (1978). A biomechanical investigation of the human hip. *J Biomechanics, 11*(1–2), 75–85.

Davey, S., Orr, J., Buchanan, F., et al. (2005). The effect of patient gait on the material properties of UHMWPE in hip replacements. *Biomaterials, 26*(24), 4993–5001.

Davy, D.T., Kotzar, G.M., Brown, R.H., et al. (1988). Telemetric force measurements across the hip after total arthroplasty. *J Bone Joint Surg, 70A*(1), 45–50.

Delp, S.L., Komattu, A.V., Wixson, R.L. (1994). Superior displacement of the hip in total joint replacement: Effects of prosthetic neck length, neck-stem angle, and anteversion angle on the moment-generating capacity of the muscles. *J Orthop Res, 12*(6), 860–870.

D'Lima, D.D., Patil, S., Steklov, N., et al. (2006). Tibial forces measured in vivo after total knee arthroplasty. *J Arthroplasty, 21*, 255–262.

Doehring, T.C., Rubash, H.E., Shelley, F. J., et al. (1996). Effect of superior and superolateral relocations of the hip center on hip joint forces. An experimental and analytical analysis. *J Arthroplasty, 11*(6), 693–703.

Dorr, L.D., Conaty, J.P., Schreiber, R., et al. (1985). Technical factors that influence mechanical loosening of total knee arthroplasty. In L.D. Dorr (Ed.): *The Knee.* Baltimore: University Park Press, 121–135.

Draganich, L.F., Andriacchi, T.P., Andersson, G.B.J. (1987). Interaction between intrinsic knee mechanics and the knee extensor mechanism. *J Orthop Res, 5*, 539.

Engh, C.A., McGovern, T.F., Bobyn, J.D., et al. (1992). A quantitative evaluation of periprosthetic bone-remodeling after cementless total hip arthroplasty. *J Bone Joint Surg, 74A*(7), 1009–1020.

English, T.A., Kilvington, M. (1979). In vivo records of hip loads using a femoral implant with telemetric output (a preliminary report). *J Biomed Eng, 1*(2), 111–115.

Fisher, J., Dowson, D., Hamdzah, H., et al. (1994). The effect of sliding velocity on the friction and wear of UHMWPE for use in total artificial joints. *Wear, 175*, 219–225.

Foucher, K.C., Wimmer, M.A., Moisio, K.C., et al. (2010). Full activity recovery but incomplete biomechanical recovery after minimally invasive total hip arthroplasty. In *Transactions of the 56th Annual Meeting of the Orthopaedic Research Society*, March 6–9, New Orleans, LA, p. 1924.

Foucher, K.C., Hurwitz, D.E., Wimmer, M.A. (2007). Preoperative gait adaptations persist one year after surgery in clinically well-functioning total hip replacement patients. *J Biomech, 40*(15), 3432–3437.

Foucher, K.C., Hurwitz, D.E., Wimmer, M.A. (2008). Do gait adaptations during stair climbing result in changes in implant forces in subjects with total hip replacements compared to normal subjects? *Clin Biomech (Bristol, Avon), 23*(6), 754–761.

Foucher, K.C., Hurwitz, D.E., Wimmer, M.A. (2009). Relative importance of gait vs. joint positioning on hip contact forces after total hip replacement. *J Ortho Res, 27*(12), 1576–1582.

Furmanski, J., Anderson, M., Bal, S., et al. (2009). Clinical fracture of cross-linked UHMWPE acetabular liners. *Biomaterials, 30*(29), 5572–5582.

Glaser, D., Dennis, D.A., Komistek, R.D., et al. (2008). In vivo comparison of hip mechanics for minimally invasive versus traditional total hip arthroplasty. *Clin Biomech (Bristol, Avon). 23*(2):127–34. Epub 2007 Nov 26.

Heinlein, B., Kutzner, I., Graichen, F., et al. (2009). ESB Clinical Biomechanics Award 2008: Complete data set of total knee replacement loading floor level walking and stair climbing measured in vivo with a follow-up of 6–10 months. *Clin Biomech, 24*, 315–326.

Heller, M.O., Bergmann, G., Deuretzbacher, G., et al. (2001). Musculo-skeletal loading conditions at the hip during walking and stair climbing. *J Biomech, 34*(7), 883–893.

Heller, M.O., Bergmann, G., Kassi, J.P., et al. (2005). Determination of muscle loading at the hip joint for use in pre-clinical testing. *J Biomech, 38*(5), 1155–1163.

Hilding, M.B., Lanshammer, H., Ryd, L. (1996). Knee joint loading and tibial component loosening. *J Bone Joint Surg, 78B*, 66.

Huo, M., Parvizi, J., Bal, B.S., et al. (2009). What's new in total hip arthroplasty. *J Bone Joint Surg, 91A*(10), 2522–2534.

Hurwitz, D.E., Foucher, K.C., Andriacchi, T.P. (2003). A new parametric approach for modeling hip forces during gait. *J Biomech, 36*(1), 113–119.

Hurwitz, D.E., Foucher, K.C., Sumner, D.R., et al. (1998). Hip motion and moments during gait relate directly to proximal femoral bone mineral density in patients with hip osteoarthritis. *J Biomech, 31*(10), 919–925.

Hurwitz, D.E., Ryals, A.B., Case, J.P., et al. (2002). The knee adduction moment during gait in subjects with knee osteoarthritis is more closely correlated with static alignment than radiographic disease severity, toe out angle and pain. *J Ortho Res, 20*(1), 101–107.

Jacobs, J.J., Hallab, N.J., Urban, R.M., et al. (2006). Wear particles. *J Bone Joint Surg, 88-A*(Suppl 2), 99–102.

Jacobs, J.J., Roebuck, K.A., Archibek, M., et al. (2001). Osteolysis: Basic science. *Clin Orthop, 393*, 71–77.

Jasty, M. (1993). Clinical reviews: Particulate debris and failure of total hip replacements. *J Appl Biomater, 4*(3), 273–276.

Jasty, M., O'Connor, D.O., Henshaw, R.M., et al. (1994). Fit of the uncemented femoral component and the use of cement influence the strain transfer the femoral cortex. *J Ortho Res, 12*(5), 648–656.

Johnson, K.L. (1987). *Contact Mechanics* (*2nd ed.*). Cambridge, UK: Cambridge University Press.

Johnston, R.C., Brand, R.A., Crowninshield, R.D. (1979). Reconstruction of the hip. A mathematical approach to determine optimum geometric relationships. *J Bone Joint Sur, 61A*(5), 639–652.

Kapandji, I.A. (1970). The knee. In I.A. Kapandji (Ed.): *The Physiology of the Joints*. Paris, France: Editions Maloine, 72–135.

Kinkel, S., Wollmerstedt, N., Kleinhans, J.A., et al. (2009). Patient activity after total hip arthroplasty declines with advancing age. *Clin Orthop Relat Res, 467*(8):2053–8. Epub 2009 Feb 27.

Koo, S., Andriacchi, T.P. (2008). The knee joint center of rotation is predominantly on the lateral side during normal walking. *J Biomech, 41*(6), 1269–1273.

Kotzar, G.M., Davy, D.T., Berilla, J., et al. (1995). Torsional loads in the early postoperative period following total hip replacement. *J Ortho Res, 13*(6), 945–955.

Kotzar, G.M., Davy, D.T., Goldberg, V.M., et al. (1991). Telemeterized in vivo hip joint force data: A report on two patients after total hip surgery. *J Ortho Res, 9*(5), 621–633.

Krger, H., Venesmaa, P., Jurvelin, J., et al. (1998). Bone density at the proximal femur after total hip arthroplasty. *Clin Orthop Relat Res*, (352), 66–74.

Lafortune, M.A., Cavanagh, P.R., Sommer, H.J., et al. (1992). Three-dimensional kinematics of the human knee during walking. *J Biomech, 25*, 347.

Lazansky, M.G. (1967). A method for grading hips. *J Bone Joint Surg, 49B*(4), 644–651.

Lenaerts, G., De Groote, F., Demeulenaere, B., et al. (2008). Subject-specific hip geometry affects predicted hip joint contact forces during gait. *J Biomech, 41*(6), 1243–1252.

Long, W.T., Dorr, L.D., Healy, B., et al. (1993). Functional recovery of noncemented total hip arthroplasty. *Clin Orthop Relat Res*, (288), 73–77.

Lundberg, H.J., Foucher, K.C., Andriacchi, T.P., et al. (2009). Comparison of numerically modeled knee joint contact forces to instrumented total knee prosthesis forces. In *Proceedings of the ASME 2009 Summer Bioengineering Conference* (SBC2009–206791); June 17–21; Lake Tahoe, CA.

Lundberg, H.J., Foucher, K.C., Wimmer, M.A. (2009). A parametric approach to numerical modeling of TKR contact forces. *J Biomech, 42*, 541–545.

Majewski, M., Bischoff-Ferrari, H.A., Gruneberg, C., et al. (2005). Improvements in balance after total hip replacement. *J Bone Joint Surg, 87B*(10), 1337–1343.

Malchau, H., Herberts, P., Eisler, T., et al. (2002). The Swedish Total Hip Replacement Register. *J Bone Joint Surg, 84-A*(Suppl 2), 2–20.

McCrory, J.L., White, S.C., Lifeso, R.M. (2001). Vertical ground reaction forces: Objective measures of gait following hip arthroplasty. *Gait Posture, 14*(2), 104–109.

McEwen, H.M., Barnett, P.I., Bell, C.J., et al. (2005). The influence of design, materials and kinematics on the in vitro wear of total knee replacements. *J Biomech, 38*(2), 357–365.

McKellop, H., Shen, F.W., Lu, B., et al. (1999). Development of an extremely wear-resistant ultra high molecular weight polyethylene for total hip replacements. *J Ortho Res, 17*(2), 157–167.

Mündermann, A., Dyrby, C., D'Lima, D., et al. (2008). In vivo knee loading characteristics during activities of daily living as measured by an instrumented total knee replacement. *J Ortho Res, 26*(9), 1167–1172.

Morrison, J.B. (1970). The mechanics of the knee joint in relation to normal walking. *J Biomech, 3*(1), 51–61.

Müller, W. (1983). *The Knee: Form, Function, and Ligament Reconstruction*. Berlin, Germany: Springer-Verlag.

Murray, M.P., Brewer, B.J., Zuege, R.C. (1972). Kinesiologic measurements of functional performance before and after McKee-Farrar total hip replacement. A study of thirty patients with rheumatoid arthritis, osteoarthritis, or avascular necrosis of the femoral head. *J Bone Joint Surg, 54A*(2), 237–256.

Naudie, D.D., Ammeen, D.J., Engh, G.A., et al. (2007). Wear and osteolysis around total knee arthroplasty. *J Am Acad Orthop Surg, 15*(1), 53–64.

Ngai, V. (2010). *Assessment of In Vivo Gait Patterns on Wear of Total Knee Replacements* [Ph.D. Thesis]. Chicago, IL: University of Illinois.

Ngai, V., Wimmer, M.A. (2009). Are TKR Knee Kinematics Influenced by Gait Kinetics? Paper Presented at: 55th Annual

Meeting of the Orthopaedic Research Society; February 22–25; Las Vegas, NV.

Ngai, V., Schwenke, T., Wimmer, M.A. (2009). In-vivo kinematics of knee prostheses patients during level walking compared with the ISO force-controlled simulator standard. *Proc Inst Mech Eng H, 223*(7), 889–896.

O'Connor, J.J., Zavatsky, A. (1990). Kinematics and mechanics of the cruciate ligaments of the knee. In V.C. Mow, A. Ratelitte, S. L.-Y. Woo (Eds.). *Biomechanics of Diathrodial Joints (vol II)*. New York: Springer, 197–241.

Orozco, D.A., Wimmer, M.A. (2010). Cumulative loading of TKR during activities of daily living: The contribution of chair and stair maneuvers [abstract]. Paper presented at: 56th Annual Meeting of the Orthopaedic Research Society; March 6–9; New Orleans, LA.

Paul, J.P., McGrouther, D.A. (1975). Forces transmitted at the hip and knee joint of normal and disabled persons during a range of activities. *Acta Orthop Belg, 41 Suppl 1*(1):78–88.

Pedersen, D.R., Brand, R.A., Davy, D.T. (1997). Pelvic muscle and acetabular contact forces during gait. *J Biomech, 30*(9), 959–965.

Rhrle, H., Scholten, R., Sigolotto, C., et al. (1984). Joint forces in the human pelvis-leg skeleton during walking. *J Biomech, 17*(6), 409–424.

Rydell, N.W. (1966). Forces acting on the femoral head-prosthesis. A study on strain gauge supplied prostheses in living persons. *Acta Orthop Scand, 37*(Suppl 88), 1–132.

Schipplein, O.D., Andriacchi, T.P. (1991). Interaction between active and passive knee stabilizers during level walking. *J Orthop Res, 9*, 113.

Schmalzried, T.P., Szuszczewicz, E.S., Northfield, M.R., et al. (1998). Quantitative assessment of walking activity after total hip or knee replacement. *J Bone Joint Surg, 80A*(1), 54–59.

Schroter, J., Guth, V., Overbeck, M., et al. (1999). The 'Entlastungsgang'. A hip unloading gait as a new conservative therapy for hip pain in the adult. *Gait Posture, 9*(3), 151–157.

Schwenke, T., Wimmer, M.A., Uth, T., et al. (2008). Experimental determination of cross-shear dependency in polyethylene wear. In *Transactions of the 54th Annual Meeting of the Orthopaedic Research Society;* March 2–5; San Francisco, CA, p. 1890.

Seireg, A., Arvikar, R.J. (1975). The prediction of muscular load sharing and joint forces in the lower extremities during walking. *J Biomech, 8*(2), 89–102.

Shakoor, N., Block, J.A., Shott, S., et al. (2002). Nonrandom evolution of end-stage osteoarthritis of the lower limbs. *Arthritis Rheum, 46*(12), 3185–3189.

Shakoor, N., Hurwitz, D.E., Block, J.A., et al. (2003). Asymmetric knee loading in advanced unilateral hip osteoarthritis. *Arthritis Rheum, 48*(6), 1556–1561.

Sharkey, P.F., Hozack, W.J., Rothman, R.H., et al. (2002). Insall Award Paper: Why are total knee arthroplasties failing today? *Clin Orthop Relat Res, (404)*, 7.

Shaw, J.A., Murray, D.G. (1974). The longitudinal axis of the knee and the role of the cruciate ligaments in controlling transverse rotation. *J Bone Joint Surg, 56A*, 1603.

Shih, C.H., Du, Y.K., Lin, Y.H., et al. (1994). Muscular recovery around the hip joint after total hip arthroplasty. *Clin Orthop Relat Res, (302)*, 115–120.

Sicard-Rosenbaum, L., Light, K.E., Behrman, A.L. (2002). Gait, lower extremity strength, and self-assessed mobility after hip arthroplasty. *J Gerontol. A Biol Sci Med Sci, 57*(1), M47–51.

Silva, M., Shepherd, E.F., Jackson, W.O., et al. (2002). Average patient walking activity approaches 2 million cycles per year: Pedometers under-record walking activity. *J Arthroplasty, 17*(6):693–7.

Stansfield, B.W., Nicol, A.C., Paul, J.P., et al. (2003). Direct comparison of calculated hip joint contact forces with those measured using instrumented implants. An evaluation of a three-dimensional mathematical model of the lower limb. *J Biomech, 36*(7), 929–936.

Stauffer, R.N., Smidt, G.L., Wadsworth, J.B. (1974). Clinical and biomechanical analysis of gait following Charnley total hip replacement. *Clin Orthop Relat Res, (99)*, 70–77.

Strasser, H. (1908). *Lehrbuch der Muskel— und Gelenkmechanik.* Berlin, Germany: Springer-Verlag.

Umeda, N., Miki, H., Nishii, T., et al. (2009). Progression of osteoarthritis of the knee after unilateral total hip arthroplasty: Minimum 10-year follow-up study. *Arch Orthop Trauma Surg, 129*(2), 149–154.

Unsworth, A. (1993). Lubrication of human joints. In V. Wright, E.L. Radin (Eds.). *Mechanics of Human Joints: Physiology, Pathophysiology, and Treatment.* New York: Marcel Dekker Inc, 137–162.

Wasielewski, R.C., Galante, J.O., Leighty, R.M., et al. (1994). Wear patterns on retrieved polyethylene tibia inserts and their relationship to technical considerations during total knee arthroplasty. *Clin Orthop, 299*, 31.

Weber, W., Weber, E. (1836). *Mechanik der menschlichen Gehwerkzeuge*, Göttingen.

Weinans, H., Huiskes, R., Grootenboer, H.J. (1992). Effects of material properties of femoral hip components on bone remodeling. *J Orthop Res, 10*(6), 845–853.

White, S.C., Lifeso, R.M. (2005). Altering asymmetric limb loading after hip arthroplasty using real-time dynamic feedback when walking. *Arch Phys Med Rehab, 86*(10), 1958–1963.

Wimmer, M.A. (1999). *Wear of the Polyethylene Component Created by Rolling Motion of the Artificial Knee.* Shaker Verlag GmbH, Germany.

Wimmer, M.A., Andriacchi, T.P. (1997). Tractive forces during rolling motion of the knee: Implications for wear in total knee replacement. *J Biomech, 30*, 131–137.

Wimmer, M.A., Andriacchi, T.P., Natarajan, R.N., et al. (1998). A striated pattern of wear in ultrahigh-molecular-weight polyethylene components of Miller-Galante total knee arthroplasty. *J Arthroplasty, 13*(1), 8–16.

Yoder, S.A., Brand, R.A., Pedersen, D.R., et al. (1988). Total hip acetabular component position affects component loosening rates. *Clin Orthop Relat Res, (228)*, 79–87.

Zahiri, C.A., Schmalzried, T.P., Szuszczewicz, E.S., et al. (1998). Assessing activity in joint replacement patients. *J Arthroplasty, 13*(8):890–895.

CHAPTER 17

Biomechanics of Gait

Sherry I. Backus, Allison M. Brown, and Ann E. Barr

Introduction

Bipedal locomotion, or gait, is a functional task requiring complex interactions and coordination among most of the major joints of the body, particularly of the lower extremity. This fundamental task has been the subject of study by scientists for several centuries, with respect to description of both typical body movements and of pathologic conditions and therapeutic interventions. Gait analysis and gait training in one form or another is a staple of physical therapy and rehabilitation medicine practice. As technological advances become both more sophisticated and affordable, detailed biomechanical analyses of gait increasingly can be performed in a clinical setting. This means that the biomechanics of gait needs to be more broadly understood by both clinicians and researchers. In the pages that follow, the anatomic characteristics of the major joints of the lower limb and trunk will be summarized and their behavior during level walking in healthy adults will be described. More detailed anatomy of the relevant joints and tissues can be found in other chapters of this book.

Anatomic Considerations

HIP

During gait, motion about the coxofemoral, or hip, joint is triaxial: Flexion-extension occurs about a mediolateral axis; adduction-abduction occurs about an anteroposterior axis; and internal-external rotation occurs about a longitudinal axis. Although flexion-extension movements are of the highest amplitude, motions in the other two planes are substantial and consistent both within and between individuals. In addition, impairments in any or all of the three movement planes can cause problematic deviations of the typical gait pattern at the hip and other joints.

KNEE

In the case of the knee, three degrees of freedom of angular rotation are also possible during gait. The primary motion is knee flexion-extension about a mediolateral axis. Knee internal-external rotation about a longitudinal axis and adduction-abduction (varus-valgus) about an anteroposterior axis may also occur, but with less consistency and amplitude among healthy individuals owing to soft tissue and bony constraints to these motions.

ANKLE AND FOOT

Ankle motion is restricted by the morphologic constraints of the talocrural joint, which permits only plantarflexion (extension) and dorsiflexion (flexion). Although frequently modeled in gait analysis as a rigid segment, the foot is required to act as both a semirigid structure (as a spring during weight transfer and a lever arm during push-off) and a rigid structure that permits adequate stability to support body weight.

The movements of the ankle, subtalar, tarsal, metatarsal, and phalangeal joints contribute to the smooth progression of the body's center of mass through space. There are constant adjustments in these joints in response to the characteristics of the supporting terrain and to the actions of the muscles that cross them, which provides a smooth interaction between the body and the wide variety of supporting surfaces encountered when walking. The loss of normal motion or muscular function at these joints has a direct effect not only on the foot and ankle but also on the remainder of the joints of the lower extremity.

UPPER BODY

The pelvis and thorax may be considered separately or, as in many studies in the literature, as a rigid unit comprising the head, arms, and trunk (pelvis plus thorax), or HAT, segment. The upper limbs and head have not received as much attention as the trunk and lower limbs in the literature. Studies that do exist indicate that shoulder motions occur primarily as flexion-extension and internal-external rotation at the glenohumeral joints. These motions are generally passive and occur as result of lower body movement (Pontzer et al., 2009) rather than active, purposeful motion. Reciprocal arm swing also limits the angular momentum of the trunk. Elbow flexion-extension and forearm pronation-supination occur. Cervical spine motion is primarily in flexion-extension and rotation to stabilize visual gaze or facilitate the vestibulo-ocular reflex as the body is propelled through the environment.

Methods of Gait Analysis

The information presented in this chapter is summarized from the scientific and clinical literature in which various laboratory methods have been used to measure gait characteristics, including stride analysis, angular kinematic analysis, force plate and foot pressure analysis, and electromyographic (EMG) analysis. In stride analysis, the temporal sequence of stance and swing are quantified using either simple tools, such as a stopwatch and ink and paper, or electromechanical instruments, such as pressure-sensitive switches imbedded in shoe inserts or applied to the bottom of the foot. Stride analysis data are used to calculate basic time-distance variables, which will be described in detail.

Angular kinematic analysis uses electrogoniometry, accelerometry, and optoelectronic techniques. Electrogoniometers are available in uniaxial and multiaxial configurations and are attached directly to the body segments on either side of the joint or joints of interest for the direct measurement of angular displacement. Accelerometers are attached to the body segments of interest for the direct measurement of segmental acceleration from which segmental velocities and displacements are then derived. Optoelectronic techniques involve the use of video or digital cameras to capture images of an individual walking. Such systems usually include the use of reference markers, which are attached to the subject, to estimate the location of joint axes and to assist in digitization. Such camera systems require careful calibration to locate anatomic markers and are often permanently installed in a "gait laboratory." Force plate and foot pressure analysis techniques involve the recording of information at the foot-floor interface during the stance phase of gait. Force plates measure the resultant ground reaction force beneath the foot and the location of its point of application in the plane of the supporting surface. Pressure plates or insoles measure the load distribution beneath the foot during stance. Force plates are often combined with angular kinematic methods for the calculation of kinetic variables, such as joint moments.

EMG is used to record muscle activation during walking. Both surface and intramuscular sensing techniques are used in gait analysis. EMG is typically combined with stride or angular kinematic analysis to provide information about phasic muscle activation patterns. EMG helps to explain the motor performance underlying the kinematic and kinetic characteristics of gait.

Gait Cycle

Bipedal locomotion is a cyclic activity consisting of two phases for each limb, stance and swing. Gait is relatively symmetric with regard to angular motions of the major joints, muscle activation patterns, and load bearing of the lower extremities (Crenshaw and Richards, 2006; Sadeghi et al., 2000) and, as a result, is efficient in translating the body's center of mass in the overall direction of locomotion. A full gait cycle or a stride is defined by the occurrence of a sequential stance phase and swing phase by one limb (Figs. 17-1 and 17-2). The limits of a stride can be demarcated by the occurrence of a specific gait event (e.g., initial contact) on one limb to the next occurrence of that same event on the ipsilateral limb. Typically, heel contact or initial contact is used as the event that defines the limits of a stride.

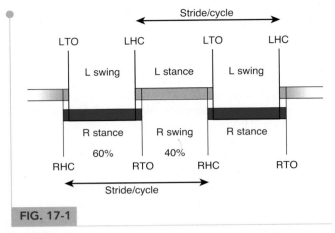

FIG. 17-1

Schematic diagram of the temporal sequence of the gait cycle or stride showing complete right (*shaded bars*) and left strides. *HC*, heel/initial contact; *TO*, toe-off; *R*, right; *L*, left. The areas of overlap between HC and TO represent periods of double limb support, which coincide with the occurrence of pre-swing on the trailing limb and loading response on the leading limb. In the case of the right stride, initial double limb support (lasting ~10% of the stride) occurs from RHC to LTO, and terminal double limb support (lasting ~10% of the stride) occurs from LHC to RTO. *Reprinted with permission from Barr, A.E. (1998). Gait analysis. In J. Spivak, J. Zuckerman (Eds.). Orthopaedics—A Comprehensive Study Guide. New York: McGraw-Hill.*

Stance phase occupies 60% of the stride and consists of two periods of double limb support (initial and terminal), when the contralateral foot is in contact with the ground, and an intermediate period of single limb support, when the contralateral limb is engaged in swing phase. Stance can be divided into six events and periods. Initial contact or heel contact is defined as the instant the foot makes contact with the floor. Loading response is an interval during which the sole of the foot comes into contact with the floor and the weight of the body is accepted onto the supporting limb. The loading response period coincides with the end of initial double limb support at approximately 10% to 12% of the stride. Midstance is the period during which the tibia rotates over the stationary foot in the direction of locomotion. The beginning of midstance coincides with single limb support and lasts from approximately 10% to 30% of the stride. Terminal stance is the period during which the weight of the body is transferred from the hind and midfoot regions onto the forefoot. It occurs from 30% to 50% of the stride and coincides with the beginning of terminal double limb support. Pre-swing occurs simultaneously with terminal double limb support and lasts from approximately 50% to 60% of the stride. During pre-swing, weight is transferred onto the contralateral limb in preparation for swing phase. The end of pre-swing

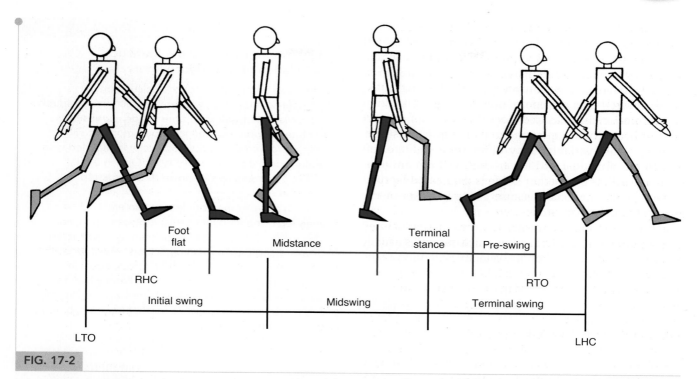

Foot flat **Midstance** **Terminal stance** **Pre-swing**

RHC RTO

Initial swing Midswing Terminal swing

LTO LHC

FIG. 17-2

Schematic diagram of the spatial sequence of the gait cycle or stride showing stance phase on the right and swing phase on the left. *HC*, heel/initial contact; *TO,* toe-off; *R,* right; *L,* left. Stance phase is demarcated by two events, HC and TO, and broken into four periods: loading response (foot flat) (from ~0%–10% of the stride), midstance (from ~10%–30% of the stride), terminal stance (from ~30%–50% of the stride), and pre-swing (from ~50%–60% of the stride). Swing phase is demarcated by two events, TO and HC, and broken into three periods: initial swing (from ~60%–70% of the stride), midswing (from ~70%–85% of the stride), and terminal swing (from ~85%–100% of the stride). *Reprinted with permission from Barr, A.E. (1998). Gait analysis. In J. Spivak, J. Zuckerman (Eds.). Orthopaedics—A Comprehensive Study Guide. New York: McGraw-Hill.*

corresponds to toe-off at which moment the foot breaks contact with the floor, thereby demarcating the beginning of swing phase.

Swing phase occupies 40% of the gait cycle and is decomposed into three periods. Initial swing lasts from approximately 60% to 73% of the stride (approximately one third of swing phase), from toe-off until the swinging foot is opposite the stance foot. Midswing ends when the tibia of the swinging limb is vertically oriented and lasts from 73% to 87% of the stride. Terminal swing lasts from 87% to 100% of the stride and ends at the moment of initial contact.

Time-Distance Variables

Time-distance variables are derived from the temporal and spatial occurrence of the stance and swing phases. Normal values are provided in Table 17-1.

Stride time refers to the time it takes to perform a single stride. Stride length refers to the distance covered by

a stride in the direction of locomotion. Step is defined as the occurrence of an event on one foot until the next occurrence of that same event on the opposite foot. It is most commonly delineated by sequential contralateral

TABLE 17-1

Ranges of Normal Values for Time-Distance Parameters of Adult Gait at Free Walking Velocity

Stride or cycle time	1.0–1.2 m/sec[a]
Stride or cycle length	1.2–1.9 m[b]
Step length	0.56–1.1 m[a]
Step width	7.7–9.6 cm[a]
Cadence	90–140 steps/minute[b]
Velocity	0.9–1.8 m/sec[b]

[a]Values adapted from multiple sources as summarized in Craik, R.L., Oatis, C.A. (1995). *Gait Analysis: Theory and Application.* St Louis, MO: Mosby.
[b]Values adapted from Whittle, M.W. (1991). *Gait Analysis: An Introduction.* Oxford, UK: Butterworth-Heinemann.
Reprinted with permission from Barr, A.E. (1998). Gait analysis. In J. Spivak, J. Zuckerman (Eds.). *Orthopedics: A Comprehensive Study Guide.* New York: McGraw-Hill.

initial contact. Laterality is determined by the swinging limb; for example, right step is delineated by left initial contact to the subsequent right initial contact. Step length refers to the distance covered by a step in the direction of locomotion. Step width refers to the distance covered by a step perpendicular to the direction of locomotion as measured from the points of contact on the heels. Two sequential steps comprise a stride. Although step variables may differ from right to left within an individual, stride variables will remain constant regardless of whether stride is delineated by right or left initial contacts, because stride consists of the sums of right and left steps.

Cadence is a measure of step frequency that is defined as the number of steps taken per unit time and is usually expressed in steps per minute. Velocity is defined as the distance covered in the direction of locomotion per unit time and is usually expressed in meters per second.

Angular Kinematics

This discussion will focus on joint angular displacements about the motion axes of the major lower limb and axial segments during level walking. Figures 17-3 through 17-7 show examples of angular displacements at these motion segments over the course of the stride in a healthy adult population (Calculation Box 17-1).

HIP

At initial contact, the hip is flexed approximately 30° (Fig. 17-3A, top). Throughout stance phase, the hip extends until it reaches approximately 10° of extension by terminal stance. During pre-swing and throughout most of the swing phase, the hip flexes to a peak of approximately 35°, and then begins to extend just prior to the next initial contact as the lower limb is extended for placement of the foot on the ground.

The hip is neutral with respect to adduction-abduction at initial contact (Fig. 17-3B, top). By the end of initial double limb support or early midstance, the hip achieves its maximum adduction position of approximately 5°. Throughout the remainder of stance, the hip abducts to approximately 10° at toe-off, then steadily adducts throughout swing in preparation for the next initial contact.

Hip rotational motions are more variable across individuals during gait (Fig. 17-3C, top). At initial contact, the hip is externally rotated approximately 5° and remains so throughout loading response and early midstance. It begins to internally rotate to within 2° of neutral rotation by the middle of terminal stance, then

---

CALCULATION BOX 17-1

Bootstrap Method for the Statistical Calculation of Confidence Intervals

The bootstrap method for the calculation of confidence intervals is an iterative technique whereby a population of gait data time history curves (e.g., joint angular displacement or joint moment with respect to percent of the gait cycle) is sampled with replacement (Lenhoff et al., 1999; Olshen et al., 1989). This sampling is known as a bootstrap iteration.

Each curve in the population of interest is first analyzed using a Fourier series representation, and a mean curve for the entire population is constructed by averaging the Fourier coefficients. Then, for each bootstrap iteration, a sample of curves, equal in number to the population of curves from which the sample is taken, is randomly selected with replacement and a multiplier is calculated for that iteration, b, using the following formula:

$$M[b] = \max_{(0 \le t_j \le t_{max})}$$

$$\alpha \bar{F}(t_j)_{af} - \bar{F}(t_j)_{bs} \beta / (\sigma[((t_j))]_{bs}$$

where $\bar{F}(t_j)_{af}$ is the mean of all curves at point t_j of the gait cycle, $\bar{F}(t_j)_{bs}$ is the mean of the bootstrap sample at the same point of the gait cycle, and $\sigma[(t]_j)_{bs}$ is the standard deviation of the bootstrap sample at the same point of the gait cycle (Lenhoff et al., 1999).

After the final bootstrap iteration, the multipliers, $M[b]$, are sorted by magnitude and an M value is selected corresponding to the desired confidence limit at each point in time. For example, if a 90% confidence interval is desired, the value for M is selected such that it is larger than 90% of the remaining M values at a given point in the gait cycle. The standard deviation of the population mean is then multiplied by the appropriate M value at each point in time of the gait cycle to obtain the confidence interval envelope.

The stability of the confidence intervals obtained by the bootstrap method increases as the population size increases and as the number of bootstrap iterations increases. The curves depicted in Figures 17-3 through 17-5 were analyzed using the bootstrap method. For more specific computational details, see Lenhoff et al. (1999).

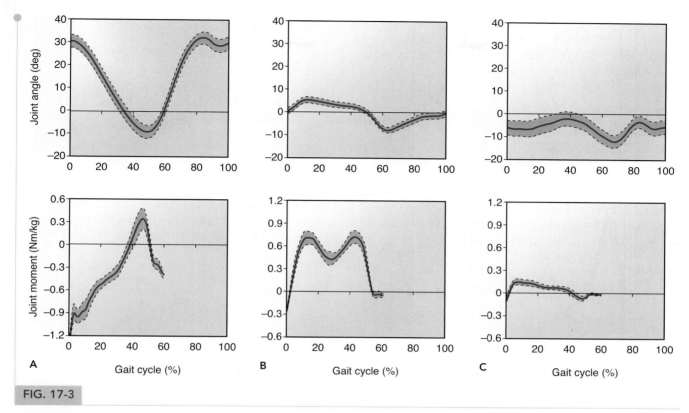

FIG. 17-3

Angular displacements and moments of the hip during level walking at freely chosen velocity among normal subjects (25 males, 4 females; 15 to 35 years of age). *Solid lines* indicate mean values (ordinate) over the course of a single stride (abscissa). *Dashed lines* indicate 90% bootstrap confidence intervals. **A top.** Hip flexion (<0) and extension (>0) position. **A bottom.** Hip extensor (<0) and flexor (>0) moment. **B top.** Hip abduction (<0) and adduction (>0) position. **B bottom.** Hip adductor (<0) and abductor (>0) moment. **C top.** Hip external rotation (<0) and internal rotation (>0) position. **C bottom.** Hip internal rotator (<0) and external rotator (>0) moment.

reverses direction and externally rotates, as the heel begins to rise, to its peak of 15° of external rotation during initial swing. As the limb swings past the opposite stance leg during midswing, its hip internally rotates to within 3° of neutral, then it oscillates between 3° and 5° of external rotation during terminal swing. Except for perhaps a brief period during the middle of terminal swing, the hip never achieves an internally rotated position during gait.

KNEE

At initial contact, the knee is almost fully extended, then it gradually flexes to its support phase peak flexion of approximately 20° during the early portion of midstance (Fig. 17-4A, top). During the latter portion of midstance, it again extends almost fully, and then flexes to approximately 40° during pre-swing. Immediately following toe-off, the knee continues to flex to its peak flexion of 60° to

70° at midswing, then extends again in preparation for the next initial contact.

In the adduction-abduction plane of motion, the knee is quite stable during stance phase because of the presence of bony and ligamentous constraints in the relatively extended knee position. Individual skeletal alignment plays a major role in adduction-abduction movements at the knee. In the normal sample presented in Figure 17-4B (top), which is predominantly males (25 males/29 subjects), the knee remains in a slightly adducted (varus) position throughout stance but fluctuates only within 2° to 3° of neutral. During pre-swing and initial swing, as weight is shifted onto the opposite limb, the knee may abduct (move into valgus) as much as 10°, but it then regains its adducted position by terminal swing.

Internal and external rotation about the knee during gait, as in the case of adduction-abduction, is determined primarily by bony and ligamentous mechanisms and is

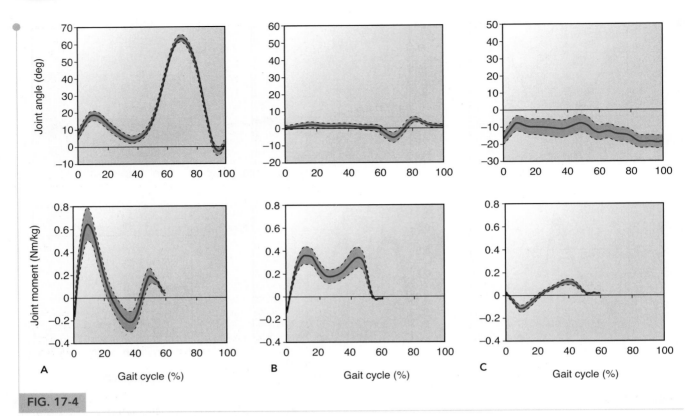

FIG. 17-4

Angular displacements and moments of the knee during level walking at freely chosen velocity among normal subjects (25 males, 4 females; 15 to 35 years of age). *Solid lines* indicate mean values (ordinate) over the course of a single stride (abscissa). *Dashed lines* indicate 90% bootstrap confidence intervals. **A top.** Knee extension (<0) and flexion (>0) position. **A bottom.** Flexor (<0) and extensor (>0) moment. **B top.** Knee abduction (<0) and adduction (>0) position. **B bottom.** Knee adductor (<0) and abductor (>0) moment. **C top.** Knee external rotation (<0) and internal rotation (>0) position. **C bottom.** Knee internal rotator (<0) and external rotator (>0) moment.

variable across individuals. In addition, the placement of reference markers during optoelectronic gait analysis may introduce specific offsets into angular calculations. For example, in the data presented in Figure 17-4, there was an external rotation offset as a result of placement of the ankle markers on the medial and lateral malleoli. Such technical differences may result in slight discrepancies between the absolute value of joint angular position at the knee as reported by different laboratories, although the relative displacement range and overall patterns of motion should be similar.

In the normal sample of predominantly males (25 males/29 subjects) depicted in Figure 17-4C (top), the knee is maintained in an externally rotated position throughout stance and fluctuates between 10° and 20°. Rotational motions about the knee are strongly coupled with flexion-extension motions. A comparison of Figure 17-4A–C (top) illustrates that during periods when the knee is flexing, it also internally rotates; whereas during periods when the knee is extending, it

also externally rotates. This coupling is related to the bony morphology of the femoral condyles and tibial plateaus as well as the displacements induced in this articulation by, especially, the anterior and posterior cruciate ligaments.

ANKLE AND FOOT

Talocrural Joint

At initial contact, the ankle joint is neutral or slightly plantarflexed 3° to 5° (Fig. 17-5, top). From initial contact to loading response, the ankle plantarflexes (i.e., extends) to a maximum of 7° as the foot is lowered to the supporting surface. Throughout midstance, the ankle dorsiflexes (i.e., flexes) to a maximum of 15° as the lower leg rotates anteriorly and medially over the supporting foot. During terminal stance and pre-swing, the ankle plantarflexes to approximately 15° as body weight is transferred onto the contralateral limb. Immediately following toe-off, the

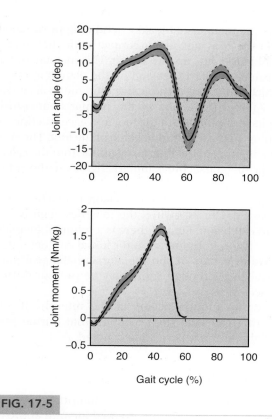

FIG. 17-5

Angular displacements and moments of the ankle during level walking at freely chosen velocity among normal subjects (25 males, 4 females; 15 to 35 years of age). *Solid lines* indicate mean values (ordinate) over the course of a single stride (abscissa). *Dashed lines* indicate 90% bootstrap confidence intervals. **Top.** ankle dorsiflexion (flexion) (<0) and plantarflexion (extension) (>0) position. **Bottom.** Ankle plantarflexor (extensor) (<0) and dorsiflexor (flexor) (>0) moment.

ankle rapidly dorsiflexes to the neutral position to attain toe clearance and then may plantarflex slightly during terminal swing in preparation for initial contact.

Subtalar Joint

The subtalar joint rotates in both stance and swing (Fig. 17-6, bottom), but it is the motion during stance that influences the weight-bearing alignment of the entire lower extremity. Like the ankle joint, the arc of motion at the subtalar joint is small compared with the knee and the hip, but it is the motion present at this joint that permits the foot to adapt to various surfaces. The subtalar joint functions as a mitered hinge during gait to transmit internal and external rotation from the tibia to rotations (eversion and inversion) about the foot. The subtalar joint also transmits inversion and eversion from the foot to external and internal rotation about the tibia.

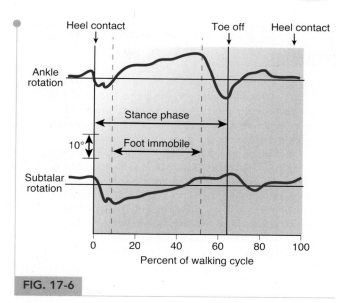

FIG. 17-6

Ankle and subtalar rotations during normal walking in a single subject. *Reprinted with permission from Wright, D.G., Desai, S.M., Henderson, W.H. (1964). Action of the subtalar and ankle-joint complex during the stance phase of walking. J Bone Joint Surg, 46A(2), 361–382.*

During loading response, the subtalar joint begins everting until peak eversion is reached by early midstance (Fig. 17-6, bottom). Peak eversion averages 4° to 6°. This rapid eversion is followed by gradual inversion, with peak inversion achieved by pre-swing. The foot drifts back to neutral during swing followed by minimal inversion during the last 20% of the stride.

Subtalar eversion is one of the mechanisms for shock absorption as body weight is transferred onto the supporting foot during loading response and early midstance. Subtalar eversion is a normal passive response to initial contact with the heel. Because the body of the calcaneus is lateral to the longitudinal axis of the tibia at initial contact, as load is applied to the talus, eversion occurs at the subtalar joint. Eversion of the subtalar joint unlocks the midtarsal joint to produce a relatively flexible forefoot.

When the body's center of mass is translated more laterally as stance progresses, the calcaneal support of the talus is decreased and the calcaneus inverts. This is coupled with internal tibial rotation resulting from the shape of the ankle joint. Subtalar inversion helps to bring about stability of the foot during single limb stance (Box 17-1).

Midtarsal Joint

Motion about the transverse axis of the midtarsal joint affects the longitudinal arch of the foot. Following forefoot contact during loading response, the longitudinal

BOX 17-1 Joint Compensation: In-toeing and Out-toeing

There is an important interrelationship between the motion at the ankle joint and the subtalar joint during gait that permits compensation between the joints. If this compensatory mechanism fails, there is increased stress in these joints and possibly an increased incidence of secondary degenerative arthritis. For example, the degree of in-toeing (internal rotation at the ankle joint) and out-toeing (external rotation at the ankle) affects the amount of motion required at the subtalar joint. In the case of an individual with excessive out-toeing, the range of motion required at the ankle joint is decreased, and at the subtalar joint, the motion required is increased. This occurs because the greatest motion will always occur about the axis that is closest to perpendicular to the plane of progression. With out-toeing, the ankle joint axis is even less perpendicular to the plane of progression than normal. The subtalar joint axis becomes oriented more perpendicular to the plane of progression and subsequently undergoes a larger angular excursion. The reverse occurs with increased in-toeing; the ankle joint axis becomes more perpendicular to the plane of progression and the range of motion required at the ankle joint is increased. At the subtalar joint, the motion required is then decreased. One of the compensations seen clinically for the loss of ankle range of motion is increased out-toeing, so that the motion required for walking can occur at the subtalar joint.

arch flattens during single limb support. The restoration of the arch occurs with heel rise.

Midtarsal extension is another of the mechanisms for shock absorption as body weight is lowered onto the stance limb during loading response and early midstance. This motion, which accompanies forefoot contact at the onset of midstance, occurs after subtalar eversion.

Finally, the interaction between the subtalar joint and the midtarsal joint is such that if motion at the subtalar joint is limited, then motion at the midtarsal joint will be limited. Similarly, when motion at the talonavicular joint is prevented, almost no motion is permitted at the subtalar joint.

Forefoot and Interphalangeal Joints

At initial contact, the toes are off the ground with the metatarsophalangeal joints in 25° of extension. The toes then flex to neutral after forefoot contact at the end of loading response. A neutral position is maintained throughout midstance. During terminal stance, as the heel rises, the metatarsophalangeal joints (collectively known as the metatarsal break) extend to approximately 21° while the toes remain in contact with the ground and the hind foot lifts up into the air. This metatarsophalangeal extension places tension on the plantar aponeurosis, which in turn exerts a passive hind foot (calcaneal) inversion force. Tightening of the plantar aponeurosis also results in supination of the foot and accentuation, or heightening, of the longitudinal arch of the foot. The subsequent stiffening of the intertarsal joints from the calcaneus to the metatarsal break imparts rigidity to the entire foot and facilitates push-off.

A maximum of 58° of toe extension is reached during pre-swing. During swing, the toes flex slightly but remain in extension. Finally, there is a minimal increase in toe extension in preparation for initial contact. Little or no flexion occurs at the metatarsophalangeal joint during walking, although some may be present during athletic activities.

Little or no motion occurs at the interphalangeal joints during gait with the exception that during pre-swing, slight flexion is occasionally noted.

TRUNK AND PELVIS

At initial contact, the pelvis is tilted anteriorly approximately 7° (Fig. 17-7A, bottom), rotated forward approximately 5° (Fig. 17-7C, bottom), and is level from right to left. During the loading response, the pelvis tilts upward on the stance limb side to a maximum of 5°, and then returns to neutral at the next initial contact of the swinging limb (Fig. 17-7B, bottom). During stance phase, the pelvis rotates backward on the stance limb side and tilts anteriorly (bottom of Fig. 17-7C and B, respectively). The total excursion for anteroposterior tilt is approximately 5°; for lateral tilting, approximately 10°; and for forward and backward rotation, approximately 10°.

Trunk motion during gait is opposite in direction, or out of phase, to the motions of the pelvis (Fig. 17-7A–C, top). For example, at initial contact, the trunk is rotated backward approximately 3° while the pelvis is rotated forward approximately 5°. The amplitudes of the angular displacements of the trunk segment as reflected in the movement of the shoulder girdle are only slightly attenuated in comparison with the pelvic movements, as can be easily appreciated by comparing the top (trunk) to bottom (pelvis) plots in Figure 17-7.

CENTER OF MASS

The body's center of mass remains located within the pelvis anterior to the sacrum throughout the gait cycle. It undergoes sinusoidal displacements in all three planes with peak-to-peak excursions of approximately 3 cm in

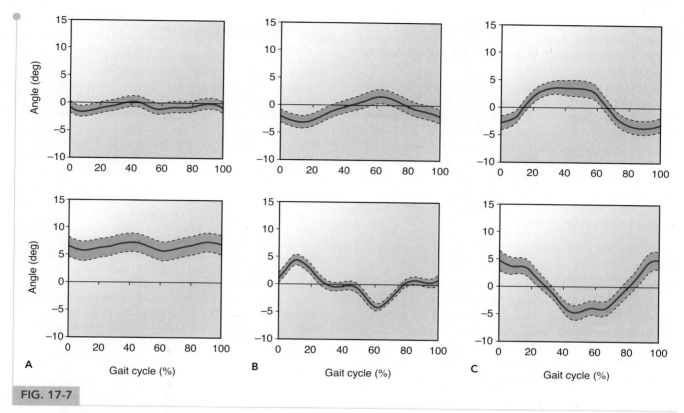

FIG. 17-7

Angular rotation of the trunk and pelvis during level walking at freely chosen velocity among normal subjects (25 males, 4 females; 15 to 35 years of age). *Solid lines* indicate mean values (ordinate) over the course of a single stride (abscissa). *Dashed lines* indicate 90% bootstrap confidence intervals. **A top.** Trunk posterior (<0) and anterior (>0) lean. **A bottom.** Pelvis posterior (<0) and anterior (>0) tilt. **B top.** Trunk downward (< 0) and upward (>0) tilt with respect to the stance limb. **B bottom.** Pelvis downward (<0) and upward (>0) tilt with respect to the stance limb. **C top.** Trunk backward (<0) and forward (>0) rotation with respect to the stance limb. **C bottom.** Pelvis backward (<0) and forward (>0) rotation with respect to the stance limb.

the vertical direction, 4 cm in the lateral direction, and 2 cm in the anteroposterior direction.

Segmental Kinetics

In gait analysis, the human body is modeled as a mechanical system of anatomic segments linked together by the joints. Kinetic computations in gait analysis make use of angular kinematic and force data.

JOINT MOMENTS

A moment is defined as the vector cross-product of a force vector and the perpendicular distance of the joint center from the line of action of that force vector. Moments are frequently expressed in newton-meters per kilogram (Nm/kg) of body weight in gait analysis (i.e., normalized to body weight). The effect of moments is to cause a tendency for joint rotation. Moments can be defined either as external or internal. In this discussion, the term moment will refer to the internal moment generated about the joint in question. A knee extensor moment, for example, refers to the internal moment of force that tends to rotate the knee joint in the direction of extension and occurs when the line of action of the tibiofemoral reaction force vector passes posteriorly to the axis of knee flexion-extension (i.e., when the external moment tends to cause knee flexion). External moments are those moments that act on the segment such as gravity, inertia and the ground reaction force. Activation of the knee extensors is required to counterbalance the tendency for knee flexion caused by the external flexion moment. Internal moments are assumed to be generated by the muscles, soft tissues, and joint contact forces acting on the joint and are inferred from inverse dynamics calculations of external moments. As such, the internal moment is an expression of the net effect of internal active and passive structures and is strictly accurate in the case where a muscle group is contracting unopposed by antagonist activation.

At certain periods during normal gait and for longer periods during gait in many pathologic conditions, agonist-antagonist coactivation may be present. In such cases, reported values for net internal moments will underestimate the actual muscular forces occurring. However, this terminology is prevalent in the literature and useful for the calculation of other kinetic variables. Plots of the internal moments occurring about the hip, knee, and ankle joints during level walking in healthy adults are depicted in Figures 17-3 through 17-5.

Hip

At initial contact, there is an extensor moment about the hip that fluctuates at first, then stabilizes at approximately 5 Nm/kg (Fig. 17-3A, bottom). This extensor moment, associated with concentric gluteal activity, persists through early midstance. The hip moment then reverses to a flexor moment in the latter third of midstance and for the remainder of stance there is a hip flexor moment that peaks at approximately 1 Nm/kg near the end of terminal stance.

Although the hip moment about the anteroposterior axis is adductor at initial contact (Fig. 17-3B, bottom), it rapidly reverses to an abductor moment of approximately 0.7 Nm/kg during loading response. As the opposite limb swings near the midline of the body during midstance, the stance limb hip abductor moment decreases to approximately 0.4 Nm/kg, but it once again increases to 0.7 Nm/kg during terminal stance (Box 17-2).

By the end of loading response, the peak hip external rotator moment of approximately 0.18 Nm/kg is achieved (Fig. 17-3C, bottom). The external rotator moment gradually decreases until the middle of terminal stance. Throughout the remainder of terminal stance and pre-swing, a slight hip internal rotator moment occurs.

Knee

At initial contact, there is a small knee flexor moment (Fig. 17-4A, bottom). During early midstance, an extensor moment peak of approximately 0.6 Nm/kg occurs. During terminal stance, a second low-amplitude extensor moment of 0.2 Nm/kg occurs (Case Study 17-1).

As was the case with adduction-abduction about the knee, knee adduction-abduction moments are controlled primarily through bone and soft tissue constraints. Therefore, the terminology for these moments at the knee refers to passive restraints, not to muscular control.

BOX 17-2 Gait Deviations

Loss of hip abductor strength or pain of the coxofemoral joint as a result of arthritic degeneration results in profound gait deviations. One possible pathologic gait pattern is the Trendelenburg gait, which results from the failure of the hip abductors to produce a sufficient abductor moment during loading response and terminal stance. This pattern is easily observed as a lateral drop of the pelvis on the side opposite the weakness during stance on the weak side. Another way of describing this pattern is excessive adduction of the weak hip during stance phase.

Another pathologic gait pattern seen with abductor weakness or coxofemoral pain is the lateral lurch. In this pattern, the trunk is displaced toward the affected stance limb during loading response, where it remains throughout terminal stance. This is observed as excessive lateral displacement of the trunk toward the affected side. The result of this gait deviation is to reduce the required hip abductor moment by displacing the body's center of mass closer to the hip adduction-abduction rotation axis.

Both of these gait deviations effectively reduce compression across the coxofemoral joint by reducing contraction force of the hip abductors, thereby alleviating joint pain. The Trendelenburg pattern is a simple mechanical result of hip abductor weakness. The lateral lurch is a compensation for hip abductor weakness.

An abductor moment persists about the knee throughout stance with two peaks of approximately 0.4 Nm/kg during loading response and terminal stance (Fig. 17-4B, bottom). During midstance, the knee abductor moment decreases to approximately 0.2 Nm/kg. In individuals who achieve abduction (valgus) positions of the knee, the moment profile may be shifted toward adductor moments, and midstance adductor moments may occur.

A knee internal rotator moment peak of 0.18 Nm/kg occurs at the transition between loading response and midstance. The knee rotation moment then reverses direction during the latter portion of midstance, reaching an external rotator moment peak of approximately 0.15 Nm/kg during terminal stance.

Ankle

Immediately after initial contact, there is a slight dorsiflexor (i.e., flexor) moment of approximately 0.2 Nm/kg about the ankle that rapidly reverses to a plantarflexor (i.e., extensor) moment for the remainder of stance (Fig. 17-5, bottom). The plantarflexor moment peak is approximately 1.6 Nm/kg at 45% of the stride, or the latter portion of terminal stance.

Case Study 17-1

Gait Adaptations in an Individual with Anterior Cruciate Ligament Deficiency

Some individuals with anterior cruciate ligament (ACL) deficiency demonstrate a "quadriceps avoidance" gait pattern associated with reduction of the stance phase knee extensor moment by as much as 140% (Andriacchi and Birac, 1993; Wexler et al., 1998). The angular motion and moment data plotted in Figure Case Study 17-1 show just such a clinical example.

The subject of this analysis was a 60-year-old man who had sustained a partial tear of the right ACL approximately 10 years prior to the gait analysis. The injury was not surgically repaired. The subject had minor complaints of functional deficits, primarily on descending stairs.

The flexion-extension motion plot (average of three trials) of the affected right knee shows a flattening and a reduction of the support phase peak knee flexion. The knee is maintained in 10° to 15° of flexion throughout midstance. Associated with this motion adaptation is a marked decrease of the knee extensor moment during the early stance phase (see corresponding right knee moment plot).

This adaptation is hypothesized to prevent unrestrained anterior translation of the tibia by the patellar tendon through reduction in quadriceps activation. EMG analysis of the quadriceps muscles in studies of ACL deficiency is consistent with this hypothesis. This effective behavioral mechanism to reduce mechanical instability of the knee seems to be subconscious on the part of injured subjects. However, it may be possible to train individuals to use this adaptation in a conservative management program for ACL injury (Andriacchi and Birac, 1993). Although this is an example where a quadriceps avoidance gait pattern was present, the pattern is not consistently seen in all individuals with ACL deficiency (Knoll et al., 2004; Lindstrom et al., 2009; Roberts et al., 1999).

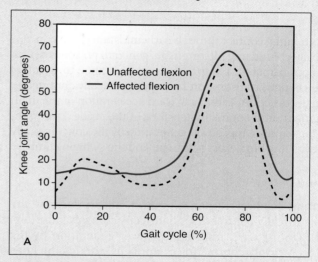

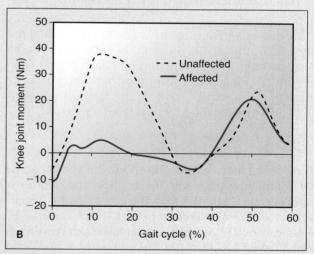

Case Study Figure 17-1 Effect of anterior cruciate ligament injury on knee motion and moments about the flexion-extension axis. Data were obtained during three trials of walking at freely chosen velocity and averaged. **A.** The difference between the ACL-deficient knee (affected) and unaffected knee flexion angles. **B.** The corresponding knee moments with values greater than zero representing knee extensor moments and values less than zero representing knee flexor moments.

JOINT POWER

Joint power is defined as the product of joint angular velocity and the corresponding internal moment at a given point in time and is expressed in watts per kilogram of body weight (W/kg). It fluctuates continuously throughout the gait cycle and can be either negative or positive in value. Joint power indicates the generation or absorption of mechanical energy by muscle groups and other soft tissues. If both muscle activation patterns and joint powers are known, then the type of muscle contraction, eccentric or concentric, can be inferred with power generation most commonly associated with concentric muscle contraction and power absorption associated with eccentric contraction.

Hip

From initial contact through early midstance, the concentrically contracting hip extensors generate power to a peak of approximately 1 watt/kg. From midstance to terminal stance, power absorption by the eccentrically contracting hip flexors controls the backward acceleration of the thigh segment until approximately 50% of the stride. From pre-swing to midswing, power generation by the concentrically contracting hip flexors acts to pull off the swinging limb.

Knee

During loading response, power absorption by the eccentrically contracting quadriceps controls knee flexion. During early midstance, power generation by the concentrically contracting quadriceps extends the knee while the contralateral limb is engaged in swing. During pre-swing, power absorption by the eccentrically contracting quadriceps controls knee flexion while the stance limb unloads in preparation for swing and the transfer of body weight onto the contralateral limb. During terminal swing, power absorption by the eccentrically contracting hamstrings controls the forward acceleration of the swinging thigh, leg, and foot segments.

Ankle

During midstance, power absorption by the eccentrically contracting plantarflexors controls the tibia as it rotates over the stationary foot. During pre-swing, a high-magnitude power generation peak of 2 to 3 watts/kg by the concentrically contracting plantarflexors represents approximately two thirds of the total energy generated during walking and is believed to contribute significantly to propulsion in gait.

WORK AND ENERGY TRANSFER

Work is defined as the integral of power with respect to time and is expressed in Joules/kg of body weight. Work is an estimate of the flow of mechanical energy from one body segment to another and is used to determine overall mechanical energy efficiency during gait. When work is positive in value, the internal moment and joint angular velocity are acting in the same direction, a concentric muscle contraction is indicated, and mechanical energy is being generated. When work is negative in value, the internal moment and joint angular velocity are acting in opposite directions, an eccentric contraction is indicated, and

mechanical energy is being absorbed. During periods of energy generation, the muscle is working on the limbs to produce movement. During periods of energy absorption, the limbs are working on the muscles, which must then contract to resist the tendency for muscle elongation.

Muscular Control

Muscle activation patterns are also cyclic during gait (Figs. 17-8–17-10). Muscle contraction type varies between the eccentric control of joint angular accelerations, such as in hamstrings activation during terminal swing, and the concentric initiation of movement, such as in tibialis anterior activation in pre-swing. In normal individuals, agonist-antagonist coactivation is of relatively short duration and occurs during periods of kinematic transition (e.g., terminal swing to initial contact). The presence of prolonged or out-of-phase agonist-antagonist co-activation during gait in individuals with pathology may indicate skeletal instability as well as motor control deficiencies.

HIP

During early stance phase, the hip extensors act concentrically while the hip abductors stabilize the lateral aspect of the coxofemoral joint (Fig. 17-8). The gluteus maximus shows increasing activation intensity from initial contact to the middle of loading response that tapers off by the end of loading response. The gluteus medius (and probably the gluteus minimus) increases activation intensity through loading response and tapers off by the end of midstance (Gottschalk et al., 1989; Wootten et al., 1990). The posterior fibers of the tensor fascia lata are moderately activated at the onset of loading response while the anterior fibers become activated later and persist into terminal stance (Gottschalk et al., 1989).

During pre-swing and initial to midswing, the hip flexors act to advance the limb, particularly when walking velocity is changing. The adductor longus is activated earliest in terminal stance and persists the longest to early midswing (Perry, 1992; Winter and Yack, 1987). The rectus femoris is the second hip flexor activated during pre-swing and remains activated a short time into early initial swing. The iliacus, sartorius, and gracilis have short periods of activation predominantly during initial swing (Perry, 1992).

The hip adductors are activated during transitions between stance and swing, as are the hamstring muscle group (Winter and Yack, 1987; Wootten et al., 1990). This activation pattern can be interpreted as the dynamic control of the swinging limb that is tending to flex and abduct at the hip. The function of the muscles during such periods is to control the acceleration of the rotating joints to ensure the precise placement of the foot on the

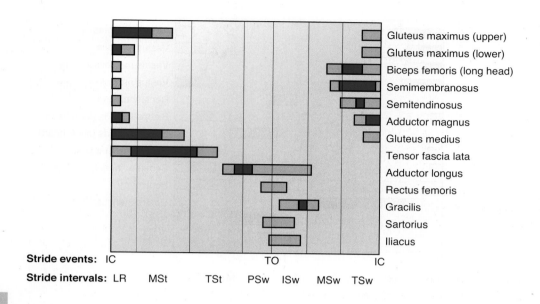

| Stride events: | IC | | | | TO | | | IC |
| Stride intervals: | LR | MSt | TSt | PSw | ISw | MSw | TSw |

FIG. 17-8

Phasic pattern of electromyographic (EMG) activity of the muscles of the hip during level walking in healthy adults. *Gray regions* represent activation less than 20% of maximum voluntary contraction. *Black regions* represent activation greater than 20% of maximum voluntary contraction. *IC*, initial contact; *TO*, toe-off; *LR*, loading response; *MSt*, midstance; *TSt*, terminal stance; *PSw*, pre-swing; *ISw*, initial swing; *MSw*, midswing; *TSw*, terminal swing. *Adapted with permission from Perry, J. (1992). Gait Analysis: Normal and Pathological Function. Thorofare, NJ: SLACK Inc.*

support surface in anticipation of the upcoming stance phase. This explains the hamstrings and adductor magnus activity during terminal swing.

KNEE

During stance phase, the quadriceps muscle group (vasti) is relied on to control the tendency for knee flexion collapse with weight acceptance and single limb support (Fig. 17-9). This muscle group is activated during terminal swing and then acts eccentrically during weight acceptance as the knee rotates from the fully extended position at initial contact to its peak support phase flexion of approximately 20° during loading response (Winter and Yack, 1987; Wootten et al., 1990). Thereafter, the quadriceps act concentrically to extend the knee through early midstance as the body's center of mass is raised vertically over the supporting limb and the anterior orientation of the ground reaction force vector precludes the need for further muscular control of knee flexion.

Most of the hamstrings muscles are activated in late midswing or terminal swing (Wootten et al., 1990). Their function at the knee is probably to control the angular acceleration into knee extension. This is consistent with their presumed action at the hip, or the control of hip flexion in preparation for the upcoming stance phase. The short head of the biceps femoris is activated earlier than are the other

hamstrings muscles in early midswing and probably assists in flexing the knee for foot clearance (Perry, 1992).

The gracilis (Perry, 1992) and sartorius (Perry, 1992; Winter and Yack, 1987) muscles also may contribute to swing phase knee flexion when they are activated during late pre-swing, initial swing, and early midswing. However, these muscles may very well be acting as primary hip flexors during this period.

ANKLE AND FOOT

Talocrural Joint

From EMG studies on the muscles that cross the ankle, the dorsiflexor muscles are shown to be firing concentrically during swing to allow for foot clearance and eccentrically during loading response to control the placement of the foot by ankle plantarflexion (Fig. 17-10). The plantarflexors are consistently firing eccentrically during stance to control the advancement of the tibia over the foot, to stabilize the knee, and concentrically to assist push-off (Sutherland, 1966; Sutherland et al., 1980).

The onset of muscle activity in the dorsiflexors begins just prior to foot-off during pre-swing. Winter and Yack (1987) reported that these muscles remain active throughout swing and loading response with peak electrical activity seen in the first 15% of the gait cycle during weight acceptance, when they must assist in controlling

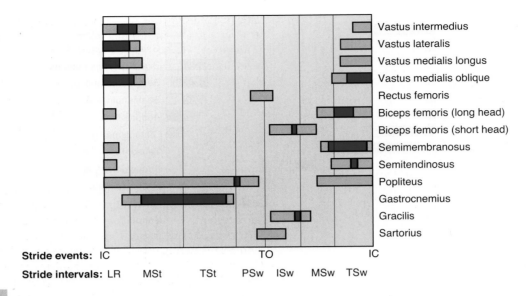

FIG. 17-9

Phasic pattern of electromyographic (EMG) activity of the muscles of the knee during level walking in healthy adults. *Gray regions* represent activation less than 20% of maximum voluntary contraction. *Black regions* represent activation greater than 20% of maximum voluntary contraction. *IC,* initial contact; *TO,* toe-off; *LR,* loading response; *MSt,* midstance; *TSt,* terminal stance; *PSw,* pre-swing; *ISw,* initial swing; *MSw,* midswing; *TSw,* terminal swing. *Adapted with permission from Perry, J. (1992). Gait Analysis: Normal and Pathological Function. Thorofare, NJ: SLACK Inc.*

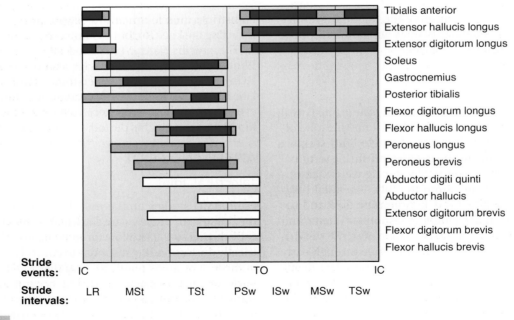

FIG. 17-10

Phasic pattern of electromyographic (EMG) activity of the muscles of the ankle and foot during level walking in healthy adults. *Gray regions* represent activation less than 20% of maximum voluntary contraction. *Black regions* represent activation greater than 20% of maximum voluntary contraction. *White bars* for the intrinsic muscles of the foot indicate phasic data only for which relative intensity as a percent of maximum voluntary contraction is not shown. *IC,* initial contact; *TO,* toe-off; *LR,* loading response; *MSt,* midstance; *TSt,* terminal stance; *PSw,* pre-swing; *ISw,* initial swing; *MSw,* midswing; *TSw,* terminal swing. *Adapted with permission from Perry, J. (1992). Gait Analysis: Normal and Pathological Function. Thorofare, NJ: SLACK Inc.*

The peak electrical activity of the dorsiflexors corresponds to a high demand on the pretibial muscles as body weight is transferred onto the supporting foot. These muscles fire eccentrically to decelerate the rate of ankle plantarflexion. If there is inappropriate timing or insufficient force of contraction in the pretibial muscles, a drop-foot or foot-slap gait pattern may be present. In addition, restrained ankle plantarflexion provides some shock absorption during the loading response. During swing, the tibialis anterior and the toe extensors function to dorsiflex the foot for toe clearance. Loss of normal function in the pretibial muscle during swing frequently results in increased knee and hip flexion, or a steppage gait pattern.

the fall of the center of mass (Fig. 17-10). They are virtually silent during mid- and terminal stance (Box 17-3).

The soleus and the medial head of the gastrocnemius begin activation at approximately 10% of the gait cycle, as single limb support begins (Fig. 17-10). They continue firing throughout the stance phase until pre-swing, when single limb support ends and the opposite foot makes contact with the ground. The lateral head of the gastrocnemius may not begin activation until midstance. During midstance, the plantarflexors eccentrically contract to restrain forward motion of the tibia. During terminal stance, as the heel begins to rise, the gastrocnemius continues to contract to begin active ankle plantarflexion. During this phase, they provide a stable tibia over which the femur may advance. Peak electrical activity is seen at 50% of the gait cycle (Winter and Yack, 1987).

Although both the soleus and gastrocnemius muscles share a common insertion, the role of the soleus is somewhat different from that of the gastrocnemius because of the soleus' origin on the tibia. The soleus, as a one-joint muscle, provides a direct link between the tibia and the calcaneus and is thought to be the dominant decelerating plantarflexion force. The gastrocnemius, as a two-joint muscle, plays a direct role in knee flexion during midstance (Fig. 17-9) (Sutherland, 1966; Sutherland et al., 1980). The remaining five posterior muscles are smaller in size and as perimalleolar muscles lie closer to the ankle joint (Fig. 17-10). These five muscles are the tibialis posterior, flexor hallucis longus, flexor digitorum longus, peroneus longus, and peroneus brevis. These muscles play a greater role at the subtalar joint and the foot than at the ankle but still create a plantarflexion force at the ankle joint.

The tibialis posterior begins firing at initial contact and remains active through single limb stance until opposite initial contact (Murley et al., 2009; Semple et al., 2009).

The flexor digitorum longus begins firing next at opposite toe-off and also continues until opposite initial contact. The flexor hallucis longus is active from 25% of the gait cycle into pre-swing. Peroneus brevis and longus activity begins early in stance and continues into terminal stance or pre-swing. Note that the activity in these muscles is subject to considerable variation across individuals.

The posterior calf muscles function as a group and cease functioning by 50% of the gait cycle when opposite initial contact has occurred. The continuation of plantarflexion past this point probably serves to balance the body because the opposite foot has already accepted the body's weight. In a small group of healthy adults, Sutherland et al. (1980) used a nerve block to the tibial nerve to further describe the role of the ankle plantarflexors, particularly the gastrocnemius and soleus during gait. They concluded that these muscles did not serve as a propulsion mechanism during pre-swing. Instead, they concluded that these muscles should be thought of as maintaining forward progression, step length, and gait symmetry. If the plantarflexors do not function normally, an increase in ankle dorsiflexion is seen with a shortened step by the swinging limb. In addition, the swinging limb strikes the ground prematurely as a result of the lack of restraint of the tibial movement of the stance limb.

In summary, during the first arc of plantarflexion following initial contact, the dorsiflexors are firing eccentrically to decelerate the rate of plantarflexion and footfall onto the ground. During the first arc of dorsiflexion, the plantarflexors are firing eccentrically to control the rate of dorsiflexion and tibial progression over the stationary foot. During the second arc of plantarflexion, just prior to weight transfer onto the opposite limb, the plantarflexors are firing to maintain walking velocity and step length. Finally, during the last arc of motion, dorsiflexion during swing, the dorsiflexors are firing concentrically to allow foot clearance.

SUBTALAR JOINT

As the foot makes contact with the floor, subtalar eversion occurs as a shock-absorbing mechanism. The inverters fire to decelerate this eversion (Fig. 17-10). The tibialis anterior acts to restrain the subtalar joint during loading response. With its greatest activity seen during loading response, the tibialis anterior is quiet by midstance.

Although the activation patterns of the tibialis posterior have been reported with various patterns of activity by different investigators, there is agreement that it is a stance phase muscle. It becomes active during loading response and remains active throughout stance until early pre-swing. Perry (1992) proposed that the activation during loading response provides early subtalar control. In addition, the variability of activity in this muscle may be indicative of its function as a reserve force to supplement insufficient varus control by the ankle muscles.

Soleus activity is seen during midstance with progressively increasing activity in terminal stance. Despite its major function as an ankle plantarflexor, this muscle also has considerable inversion leverage, especially as a result of its large cross-sectional area. By pre-swing there is a rapid decline in activity, and the muscle remains quiet during swing. The long toe flexors are the last inverters to be activated. The flexor digitorum longus and the flexor hallucis longus begin activation during midstance and cease firing during pre-swing.

The muscles that are responsible for eversion at the subtalar joint are the extensor digitorum longus, peroneus tertius, peroneus longus, and peroneus brevis. The first two lie anterior to the subtalar joint axis, while the last two lie posterior to the subtalar joint axis. The extensor digitorum longus is active during loading response and quiet with the onset of midstance. Little information is available about the firing of the peroneus tertius, but Perry (1992) reports similar timing as the extensor digitorum longus.

The peroneus longus and brevis initiate activity during forefoot loading and demonstrate their peak activity during terminal stance. Both the timing and intensity of the EMG signals in the peroneus brevis and longus are closely coordinated. Activity in these muscles ceases by the middle of pre-swing. The peroneus longus has peak electrical activity at 50% of the gait cycle during push-off.

Midtarsal Joint

The midtarsal joint is supported primarily by the tibialis posterior. Because the activity of the long toe flexors and the lateral plantar intrinsic muscles begins before the toes flex, these muscles may well contribute to the support of the midtarsal joint.

Forefoot and Interphalangeal Joints

The flexor digitorum longus and the flexor hallucis longus begin activation during midstance and cease firing during pre-swing. These muscles stabilize the metatarsophalangeal joints and add toe support to supplement forefoot support. The intrinsic muscles of the forefoot and interphalangeal joints include the abductor hallucis, adductor hallucis, flexor digitorum brevis, flexor hallucis brevis, and abductor digiti quinti. These muscles become active at approximately 20% to 30% of the walking cycle and cease when the foot leaves the ground. These muscles aid in the stabilization of the longitudinal arch and of the toes at the metatarsophalangeal joint (Mann and Inman, 1964; Perry, 1992).

Although the kinematics, kinetics, and muscular control of the major joints have been presented separately, they are functionally interrelated during gait. The musculoskeletal system must undergo highly integrated, precisely coordinated actions in both timing and amplitude for efficient locomotion to occur. This requires not only an intact musculoskeletal system, or physical plant, but also a functioning nervous system, or controller. The nervous system must be able to instantaneously assess pertinent aspects of the external and internal environments to act or respond appropriately to various functional contexts. Motion limitations or other pathology of any participating joint will have a subsequent effect on all other participating joints. It is the complex integration of anatomy, biomechanics, and muscular control that permits normal walking.

Summary

- Motion about the hip joint during gait occurs in all three planes: flexion-extension, adduction-abduction, and internal-external rotation. Flexion-extension is the primary motion occurring about the knee joint during gait, although varus-valgus and rotational motions are present to a lesser extent.
- Because of its bony morphology, the talocrural joint undergoes only plantarflexion and dorsiflexion during gait. Subtalar, midtarsal, and phalangeal motion further assist in providing adaptation to the support surface as well as rigidity for propulsion.
- The upper body, including the pelvis and trunk, undergoes sinusoidal displacements in all three cardinal planes. The trunk and pelvis rotate in opposite directions while the head usually remains stable. The swinging of the arms involves shoulder flexion-extension and rotation, elbow flexion-extension, and forearm pronation-supination.
- The gait cycle, or stride, is defined as the occurrence of an event on one lower limb until the next occurrence of the same event on the same lower limb. It is most typically demarcated by sequential ipsilateral initial contact. Stance phase occupies 60% of the stride and is divided into six events or periods: initial contact, loading response, midstance, terminal stance, pre-swing, and toe-off. Swing phase occupies 40% of the stride and is divided into periods: initial swing, midswing, and terminal swing.
- Internal joint moments indicate the net moment of force generated by muscles, bones, and passive soft tissues that counteract the tendency for joint rotation caused by gravity.

- Joint power is the product of joint angular velocity and the corresponding internal moment at a given point in time. It indicates the generation or absorption of mechanical energy by muscle groups and other soft tissues.
- Coactivation of agonist-antagonist muscle groups usually occurs during periods of kinematic transition when a joint may be reversing the direction of rotation.
- Motion limitations or disorders of motor control affecting any of the lower limb segments will potentially alter the patterns of movement and motor control at all other joints during gait.

Practice Questions

1. When describing an individual's gait, variables such as step length and step time may differ from right to left, whereas variables such as stride length and stride time may not. Please explain why this is the case.

2. Describe the location and motion of the body's center of mass throughout the gait cycle.

3. In gait analysis, angular kinematics and ground force values are used for the calculation of which variables?

4. During loading response, there is power absorption occurring at the knee. Please describe which muscle group is contributing to this power absorption and what motion (flexion vs. extension) is being prevented.

REFERENCES

Andriacchi, T.P., Birac, D. (1993). Functional testing in the anterior cruciate ligament-deficient knee. *Clin Orthop Rel Res*, (288), 40–47.

Barr, A.E. (1998). Gait analysis. In J.M. Spivak, J.D. Zuckerman (Eds.). *Orthopaedics: A Comprehensive Study Guide (1st ed.)*. New York: McGraw-Hill.

Craik, R., Oatis, C.A. (1994). *Gait analysis: Theory and application (1st ed.)*. St Louis, MO: Mosby.

Crenshaw, S.J., Richards, J.G. (2006). A method for analyzing joint symmetry and normalcy, with an application to analyzing gait. *Gait Posture*, 24, 515–521.

Gage, J.R. (1991). *Gait analysis in cerebral palsy*. London, UK: Mac Keith Press.

Gottschalk, F., Kourosh, S., Leveau, B. (1989). The functional anatomy of tensor fasciae latae and gluteus medius and minimus. *J Anat, 166*, 179–189.

Knoll, Z., Kocsis, L., Kiss, R.M. (2004). Gait patterns before and after anterior cruciate ligament reconstruction. *Knee Surg Sports Traumatol Arthrosc, 12*, 7–14.

Lenhoff, M.W., Santner, T.J., Otis, J.C., et al. (1999). Bootstrap prediction and confidence bands: A superior statistical method for analysis of gait data. *Gait Posture, 9*, 10–17.

Lindstrom, M., Fellander-Tsai, L., Wredmark, T., et al. (2009). Adaptations of gait and muscle activation in chronic ACL deficiency. *Knee Surg Sports Traumatol Arthrosc, 18*(1), 106–114.

Mann, R.A., Inman, V.T. (1964). Phasic activity of intrinsic muscles of the foot. *J Bone Joint Surg, 46A*, 469–481.

Murley, G.S., Buldt, A.K., Trump, P.J., et al. (2009). Tibialis posterior EMG activity during barefoot walking in people with neutral foot posture. *J Electromyogr Kinesiol, 19*, e69–e77.

Olshen, R.A., Biden, E.N., Wyatt, M.P., et al. (1989). Gait analysis and the bootstrap. *Ann Stat, 17*, 1417–1440.

Perry, J. (1992). *Gait Analysis: Normal and Pathological Function (1st ed.)*. Thorofare, NJ: SLACK Inc.

Pontzer, H., Holloway, J.H., 4th, Raichlen, D.A., et al. (2009). Control and function of arm swing in human walking and running. *J Exp Biol, 212*, 523–534.

Roberts, C.S., Rash, G.S., Honaker, J.T., et al. (1999). A deficient anterior cruciate ligament does not lead to quadriceps avoidance gait. *Gait Posture, 10*, 189–199.

Sadeghi, H., Allard, P., Prince, F., et al. (2000). Symmetry and limb dominance in able-bodied gait: A review. *Gait Posture, 12*, 34–45.

Semple, R., Murley, G.S., Woodburn, J., et al. (2009). Tibialis posterior in health and disease: A review of structure and function with specific reference to electromyographic studies. *J Foot Ankle Res, 2*, 24.

Sutherland, D.H. (1966). An electromyographic study of the plantar flexors of the ankle in normal walking on the level. *J Bone Joint Surg, 48A*, 66–71.

Sutherland, D.H., Cooper, L., Daniel, D. (1980). The role of the ankle plantar flexors in normal walking. *J Bone Joint Surg, 62A*, 354–363.

Wexler, G., Hurwitz, D.E., Bush-Joseph, C.A., et al. (1998). Functional gait adaptations in patients with anterior cruciate ligament deficiency over time. *Clin Orthop Rel Res*, (348), 166–175.

Whittle, M.W. (2007). *An Introduction to Gait Analysis (4th ed.)*. Oxford, UK: Butterworth-Heinemann.

Winter, D.A., Yack, H.J. (1987). EMG profiles during normal human walking: Stride-to-stride and inter-subject variability. *Electroencephalogr Clin Neurophysiol, 67*, 402–411.

Wootten, M.E., Kadaba, M.P., Cochran, G.V. (1990). Dynamic electromyography. II. Normal patterns during gait. *J Orthop Res, 8*, 259–265.

Wright, D.G., Desai, S.M., Henderson, W.H. (1964). Action of the subtalar and ankle-joint complex during the stance phase of walking. *J Bone Joint Surg, 46A*(2), 361–382.

Index

Note: Page number followed by f and t indicates figure and table respectively.